second edition

PRINCIPLES OF
ELECTRICAL
ENGINEERING

Vincent Del Toro

Professor of Electrical Engineering
School of Engineering
The City College of the City University of New York

Prentice-Hall, Inc., Englewood Cliffs, N.J.

To Franca

PRENTICE-HALL INTERNATIONAL, INC., *London*
PRENTICE-HALL OF AUSTRALIA, PTY. LTD., *Sydney*
PRENTICE-HALL OF CANADA, LTD., *Toronto*
PRENTICE-HALL OF INDIA PRIVATE LIMITED, *New Delhi*
PRENTICE-HALL OF JAPAN, INC., *Tokyo*

contents

preface
to the second edition

There are two major revisions appearing in this edition: (1) the inclusion of a substantial amount of new material and (2) the reorganization of the original material.

Experience with the first edition disclosed several areas in which reorganization of the subject matter could easily enhance the teaching as well as the studying aspects of the book. The material on electronics has been arranged in a way that now allows the options of either focusing exclusively on semiconductor electronics or else including as much of vacuum-tube electronics as is deemed appropriate. If only semiconductor electronics is to be taught, this can be readily achieved without any loss of continuity by assigning Chaps. 7, 9, and 11 in that order. Chapter 7 is concerned with an exposition of the input-output characteristics of semiconductor devices. Chapter 9 treats the analysis of electronic circuits in which the semiconductor device appears as a circuit element. This is then followed in Chapter 11 by a treatment of the more important special applications of electronic circuits. Because of the far greater importance of semiconductor electronic circuitry in today's technology, attention in Chapter 11 is confined solely to applications involving semiconductor devices.

The treatment of vacuum-tube electronics proceeds in a manner similar to that employed for the semiconductor devices. One Chapter (Chapter 8) is devoted to an explanation of the theory that underlies the input–output characteristics exhibited by these devices, while Chapter 10 concerns itself with their behavior in typical electronic circuits. This material is included for completeness and for the convenience of those who desire to teach vacuum-tube electronics for whatever reasons.

Those fundamental laws relating to electronics, which in the first edition appeared in the first chapter, have now been placed more appropriately in Chapters 7 and 8. Continuity certainly will be better served by this change. Moreover, consistent with this theme, the chapter on Laplace transforms was moved to a place just preceding those chapters where it is first needed.

The material on electromechanical energy conversion has been moved ahead so that it now becomes Part III rather than Part IV. Furthermore, in the interest of facilitating the assignment and handling of the subject matter, it is now divided into eight chapters instead of three. In curricula where relatively little time can be allotted to the study of electromechanical energy conversion, Chapters 12–14 can be used to provide an understanding of the essential elements of the subject matter. In situations where ample time is available, Chapter 14 can be followed by Chapters 15–18 where attention is directed to a description of the operation, characteristics, performance and applications of the various types of electrical machines. In addition to these modifications, some of the heaviness associated with the derivation of the basic torque equations encountered early in the treatment of energy conversion has been lifted by placing the details of the derivations in the appendix.

New material appears in each of the four parts of the book. However, the major portion is to be found in the section devoted to Electronics. Research in this area has continued at an impressive rate and the results have led to the introduction and acceptance by the industry of new and important devices. Accordingly, in this revised edition, considerable attention is given to field effect transistors (FET's), metal oxide semiconductor field effect transistors (MOSFET's) and integrated circuits (IC's). New material also appears in the chapter on electronic circuitry (Chap. 9). There the hybrid equivalent circuit of the transistor is replaced by the simpler *h*-parameter version. Moreover, throughout the treatment, emphasis is put on the derivation of generalized expressions for current gain and voltage gain for all devices and amplifiers. Linear modelling also receives a good deal of attention.

The section on Circuit Theory has been expanded to include an exposition of duality as well as circuit response to pulse and impulse driving forces.

Another significant change concerns the inclusion of a chapter on direct energy conversion. Thus in Chapter 19 attention is focused on the generation of electrical energy from non-electromechanical sources. This leads to a discussion of the solar cell, the fuel cell, the thermoelectric cell and magnetohydrodynamic energy conversion.

Finally, in Part IV, which now is devoted to control systems and computers, there appears an improved method of magnitude scaling that should go a long way towards eliminating the mystery of scaling that seems to surround the study of the analog computer.

The text material contained herein can be used in a variety of ways depending on the goals to be achieved and the amount of time available. Moreover, because the material extends over a wide range, a considerable amount of flexibility is possible. Listed below are some suggestions that might serve as guide lines in the use of the text material:

1. One-quarter course on Electric Circuit Theory
   Chaps. 1–6
2. One-quarter course on Electronics
   Chaps. 7–11
3. One-quarter course in Electromechanical Energy Conversion
   Chaps. 12–19
4. One-quarter course on Control Systems
   Chaps. 20–23
5. One-semester course in Electrical Engineering (no dynamics)
   Chaps. 1–3
   Chap. 6 (select topics as time permits)
   Chap. 7
   Chap. 9 (select topics as time permits)
   Chaps. 12–14
6. Two-semester course on Electrical Engineering
   1st Semester (Course on Circuits & Electronics)
   Chaps. 1–7
   Chap. 9
   Chap. 11
   2nd Semester (Course on Machines & Systems)
   Chaps. 12–21
   Chap. 23
7. One-semester course on Circuits and Energy Conversion
   Chaps. 1–4
   Chap. 5 (secs. 5–1 to 5–4)
   Chap. 6
   Chaps. 12–19
8. One-semester course on Circuits & Systems
   Chaps. 1–5
   Chap. 6 (secs. 6–1 to 6–10)
   Chaps. 20–23

The pace in options 1 to 4 ought to be moderate thus allowing more time for the solution and discussion of additional problems. Option 5 presents a fuller schedule but the pace should prove to be manageable. Options 6, 7, & 8 are much more demanding. Unless the students bring an exceptionally strong background to these courses, it will likely be necessary to eliminate material in several chapters. This will be found particularly necessary if ample class discussion of assigned problems is desired.

preface
to the first edition

The introduction of new devices and the application of basic principles to new situations have been occurring at as rapid a rate in electrical engineering as in the mathematical and physical sciences. However, the amount of time allotted in the curriculum for studying the subject matter is usually no greater than it was in the past and in some instances even less. If students leaving engineering colleges today are to be essentially up-to-date, it becomes necessary every so often to make available text material that adequately reflects the new emphasis on devices and principles. *Principles of Electrical Engineering* was written to achieve this goal.

Twenty years ago the vacuum tube was the cornerstone in the complex structure of electronics. Today, the semiconductor (transistor) is the electronic device of primary importance. Perhaps a few years from now the transistor will be supplanted by an even more revolutionary unit. However, the basic laws governing the behavior of equipment in associated circuitry continue to be valid. For example, Kirchhoff's laws are applicable in a lumped parameter circuit whether the circuit includes a vacuum tube or a transistor. The difference between the two devices is effectively described by the variance in their external characteristics. Thus the manner in which current varies with voltage at the terminals of a semiconductor involves a different principle—diffusion current—than that which applies for the vacuum tube. But the relationship of the vacuum tube or the semiconductor to the other elements of a circuit is still governed by the same unchanging laws of circuit theory.

Emphasis on the fundamental laws of electrical engineering is a theme that is established in the first chapter and referred to often in the development of the subject matter of this book. It is the objective of Chapter 1 to stress the fact that the whole science of electrical engineering is based upon a few experimentally established fundamental laws. Once these principles are grasped and understood, a base is built from which the study of the various areas of electrical engineering follows with considerable ease. Thus electric circuit theory is seen to develop naturally from five basic laws: Coulomb's law, which leads to the concept of capacitance; Ohm's law, which leads to the concept of resistance; Faraday's law, which leads to the concept of inductance; and finally Kirchhoff's voltage and current laws, which provide a systematic formulation of the principles involved in the first three laws. In a similar fashion, the subject matter of electro-mechanical energy conversion is derived from two basic laws: Faraday's law and Ampere's law. The situation in electronics is the same, although a two-part division is necessary. In vacuum-tube electronics, it is the Richardson-Dushman equation and the Langmuir–Child law that are fundamental in describing the properties and characteristics of the vacuum devices. In semiconductor electronics, on the other hand, it is the Boltzmann relationship that is important.

A second major emphasis in this book concerns the use of the Laplace transform as the mathematical language. The Laplace transform has many advantages to offer, and it requires no more than an understanding of integration as a background. Through the Laplace transform, a consistent, systematic, and readily interpretable language is made available for analytical and design purposes. Primarily, the Laplace transform is used in this book to solve linear differential equations, and its application in this connection often means that the transformed solution of a problem may be written in a single step. This method stands in sharp contrast to the classical method of solving such equations. Moreover, the systematic formulation afforded by the Laplace transform permits treating forcing functions and external disturbances in a unified and uniform fashion, thus placing stress on the close kinship the Laplace transform bears to the handling of initial conditions. As a mathematical tool, the Laplace transform provides a generality in analysis that is difficult to match by other procedures. Thus it is an easy matter in the study of electric circuits, for example, to treat the more general problem of the transient response (Chapter 5) before the sinusoidal steady-state response (Chapter 6). This approach also results in other noteworthy dividends. The Laplace-transform formulation leads naturally to the important concepts of transfer functions and transform impedance as operators that not only characterize the system or circuit but also lead directly to the required response functions.

As one gains experience in the use of the Laplace transform, the analytical formulations and the interpretation of the resulting mathematical solutions come easily and naturally irrespective of the area of electrical engineering being studied.

Thus the meaning and the interpretation of a transformed solution is the same whether it be concerned with circuits, amplifiers, control systems, or computers.

Chapter 2 furnishes a treatment of the Laplace transform that is sufficiently detailed for the purposes of this book. As the direct Laplace transforms of many useful functions are developed in the chapter, they are tabulated for future use. No attempt is made to evaluate the inverse Laplace transform, because this requires a knowledge of integration in the complex plane, which is usually beyond the scope of undergraduate courses in mathematics for engineers. Instead, the inverse operation is performed through the use of tables. There should be no objection to such a procedure, since we often resort to the use of appropriate tables in evaluating logarithms or certain difficult integrals.

Finally, it is worthwhile to note that the use of the Laplace transform does not involve the loss of any physical insight to a problem. Initial conditions are handled in a direct and systematic fashion irrespective of the order of the system. Then, when the transformed solution is obtained, one may apply the initial- and final-value theorems to get information about the variable as well as its derivatives.

This book is written for a one-year course in electrical engineering for nonelectrical majors. However, it can also be used for electrical majors, particularly when the amount of time allotted to this material is restricted as it is in an engineering science curriculum. A satisfactory syllabus for electrical majors is possible for two reasons. One, no compromise was made in the rigor or depth of the treatment of the subject matter. Prepublication experience with this material has indicated that the nonelectrical majors are in favor of such an approach. Two, sections specifically for electrical majors have been included in most of the chapters. The beginning of each of these sections is indicated by a double slash style mark in line with the left margin. A single slash mark precedes the subtitles, and a double slash mark in line with the right margin identifies the close of the section.

The book is divided into four major parts: Electric Circuit Theory; Electronics; Control Systems and Computers; and Electromechanical Energy Conversion. This selection was made to meet the needs of the nonelectrical majors in terms of today's requirements. The treatment of electronics gives due recognition to the importance of semiconductor devices by placing more emphasis on them than on vacuum tubes. This is the first extensive treatment of the subject matter in a book of this kind. A second area of major departure occurs in Part III where there is presented the first stand-alone treatment of control systems. More and more, nonelectrical majors, such as mechanical, aeronautical, and chemical engineers, are in need of firmer and more extensive background in this area.

Part I is devoted to an exposition of the theory of electric circuits. This is begun in Chapter 3 by a discussion of circuit parameters as they derive from the fundamental laws. An outstanding feature here is that each parameter is treated from three viewpoints: circuit, energy, and geometry. In this way a complete rather than a partial picture results. How often does it happen that a student knows how to identify inductance in terms of current and voltage (circuit viewpoint), but has no idea about what to do to change its value (geometrical viewpoint)? Chapter 4 concerns itself with the elementary network theorems. In the interest of placing maximum emphasis on the theorems, networks consisting solely of resistors are used, thereby freeing the treatment of the clutter of complex numbers. The subject matter is developed in a logical fashion. The response of a network is found first by the more cumbersome methods of network reduction and superposition. Improvements in the solution process, through the use of mesh currents and node-pair voltages, are then revealed. When the situation warrants, further simplifications are effected by the use of Thevenin's and Norton's theorems. The chapter is amply illustrated with examples. Chapter 5 deals with the forced and transient responses of circuits to standard forcing functions. An important feature of this chapter is the solution of a second-order system in terms of figures of merit that make the solution universally applicable to all linear second-order systems irrespective of composition. Chapter 6, designed to serve both electrical and nonelectrical majors well, presents an extensive treatment of the important topic of the sinusoidal steady-state response of circuits.

Part II is devoted to an exposition of electronics. Chapter 7 describes the external characteristics and control capabilities of diodes and triodes of both the semiconductor and vacuum-tube types. Once the theoretical explanations are given, the devices are then expressed in terms of appropriate circuit parameters so that the powerful tools of circuit analysis may be conveniently applied in the circuits where they appear. Chapter 8 covers the chief topics necessary to an understanding of electronic circuits. It is concerned with the behavior and performance calculations of circuits that use diodes and triodes for control and amplifying purposes. The topic is handled with the systems concept in mind, the amplifier being discussed not for its own sake, but rather as a part of an overall system. Accordingly the two-stage, then the three-stage, and finally the four-stage amplifiers are analyzed, with the latter arrangement representing an intercommunication system. In Chapter 9 attention is directed to some special topics and electronic circuit applications, including transistor logic circuits.

Part III deals with control systems and computers. The underlying principles of automatic control are carefully explained and illustrated in Chapter 10 with examples from various fields of engineering. This is followed in Chapter 11 with a general treatment of the dynamic behavior of control systems. Here the student is exposed to systems analysis by dealing directly with the differential equation approach as well as with the transfer function approach. The advantages and the implementation of such procedures as error-rate, output-rate, and integral-error control are also described. The combination of Chapters 10 and 11 should provide a sufficiently strong background in control system theory where the time in the curriculum is limited. Chapter 12 is recommended for a more extensive treatment of the subject matter. Here the technique of describing the dynamic behavior of a system by the sinusoidal steady-state frequency response is described. Part III ends with a discussion of the manner in which the analog computer can be used to solve differential equations, linear or nonlinear.

Electromechanical energy conversion is the topic of Part IV, which begins with a chapter on magnetic theory and circuits, thus providing the background needed for the study of electrical machinery. Chapter 15 deals with the theory and operation of the transformer, which is a prerequisite to the study of a-c machinery. The first part of Chapter 16 is devoted to an analysis of the general expression for electromagnetic torque as derived from Ampere's law and an analysis of induced voltages as derived from Faraday's law. The treatment is kept as general as possible to stress that the same basic laws underly the operation of a-c and d-c machines. Once this objective is achieved, attention is directed to a description of the operation, performance, characteristics, and applications of the various categories of electrical machines: the three-phase induction motor, the three-phase synchronous motor, the d-c machines, and the single-phase induction motor. To make this treatment of machinery as useful as possible for the nonelectrical majors, particular attention is given to the ratings and applications of electric motors which are conveniently summarized in suitable tables. Controllers for these motors are also discussed quite extensively. When this text is used for a two-semester course for nonelectrical majors, the instructor must be selective in treating the subjects of Part IV, since such a great deal of material is covered.

As mentioned previously, *Principles of Electrical Engineering* may be used as an introduction to electrical engineering for electrical majors. Part I (Electric Circuit Theory) can very well serve as a prerequisite to a full course devoted to networks. Similarly, Part II taken in its entirety can precede a full course in electronics, before attention is given to advanced topics. A detailed study of Chapter 10–12 can readily be used as the preparation for a beginning graduate course in control systems. Finally, the material in Part IV was planned so that it could also meet the needs of a one-semester course in electric machinery for electrical majors.

It is with pleasure that I acknowledge gratitude to my colleagues and friends at The City College of New York for the fruitful technical discussions I have had with them. I want also to express my sincere appreciation to Miss Sadie Silverstein, administrative assistant of the Electrical Engineering Department, for her truly outstanding performance in assisting with the preparation of the manuscript.

principles of
electrical engineering

1

the fundamental laws of
electrical engineering

The science of electrical engineering is based on just a few experimentally established fundamental laws. The principles and concepts that underlie the operation and performance of many engineering devices are very often the same in spite of differences in appearance and arrangement. In the interest of stressing the importance of these basic laws, attention is focused on the historical frame of reference as well as the final experimentation which culminated in their strikingly simple formulations. Once the fundamental laws are studied and understood, a considerable amount of perspective will have been gained. In turn this will facilitate the understanding of those branches of engineering where the appropriate laws provide the foundation.

For example, it will be seen that the field of *electric circuit theory* emanates from the fundamental results achieved by Coulomb (1785), Ohm (1827), Faraday (1831), and Kirchhoff (1857). The dates denote the years in which the laws, which today bear their names, were first published. Likewise, in Part IV of this book, it will be seen that the whole subject matter of electromagnetic devices and electromechanical energy conversion can be treated and analyzed by applying just two of the fundamental laws—Ampere's law (1825) and Faraday's law of induction (1831). In electronics a similar situation prevails, although it is not quite as clear-cut. At the turn of the twentieth century the age of electronics was ushered in by the invention of the vacuum tube. In this field fundamental contributions were made by Richardson and Dushman, who were responsible for identifying the properties and characteristics of thermionic emission. This was accompanied by the significant efforts of Langmuir and Child in describing the space-charge behavior of these vacuum devices. A description of this basic work appears in Part II. As a result of the work of these experimenters it became possible to treat the vacuum device as a circuit element, the external characteristics of which are readily identifiable. In recent years, however, the vacuum tube has slipped to a position of secondary importance in electronics because of the advent of semi-conductor devices. Accordingly, considerable attention is devoted here to the principles of operation involved in these latter devices because an entirely new concept is encountered, namely, that of current flow by a diffusion process. The Boltzmann relation is fundamental in describing the external behavior of the semiconductor.

The subject matter of this chapter begins with a discussion of units. The fundamental laws are mathematical expressions of experimentally established conclusions. Hence means must be available for making measurements—and for this a consistent system of units is a prerequisite.

1-1 Units

Engineering is an applied science dealing with physical quantities. Accordingly, the establishment of a universally accepted system of units is indispensable if order, communication, and understanding among the engineers and scientists throughout the world are to be achieved. A *unit* of a physical quantity is a standard of measurement regarded as an undivided entity. Once a unit is agreed upon, its application to an unknown quantity merely requires determining how many units make up the whole. Thus, if the unit of length is the meter, then any unknown length can be measured by finding the number of times the unit is taken to make up the whole. To be useful all basic units must be permanent, reproducible, available for use, and adaptable to precise comparison.

The number of different physical quantities with which the electrical engineer must be concerned exceeds thirty in number. Although it is possible to establish a standard unit for each quantity, it is really unnecessary to do so because many of the quantities are functionally related through experiment, derivation, or definition. In the study of mechanics, for example, the units of only three quantities need to be declared arbitrarily as standards not specified in terms of anything else. All other quantities can be expressed in terms of the three arbitrary units by means of the experimental, derived, and defined relationships between the physical quantities. The three quantities of concern are called the *fundamental quantities* and in mechanics are identified as *length*, *mass* and *time*. The corresponding units selected for these quantities are referred to as the *fundamental units*.

Two considerations enter into the selection of the fundamental units. First, the basic relationship between the various quantities involved in the study of the given discipline should involve a minimum number of constants. Second, the measuring units should be of a practical size. In 1902 G. Giorgi proposed that the meter-kilogram-second (MKS) system of units best meets these needs, and today this comprehensive system is universally accepted.

It is worthwhile at this point to indicate briefly how other quantities in the study of mechanics can be described in terms of the three fundamental units. Consider, for example, velocity. It is referred to as a defined quantity because it is expressed as length per unit time, or, in terms of the fundamental units, as meters/second. Acceleration can be similarly treated. The force quantity can be described in terms of the fundamental units through the aid of an experimental relationship, namely, Newton's law. Thus

$$F = ma \tag{1-1}$$

where m is the mass in kilograms and a is the acceleration in meters/second^2. As a matter of convenience the unit for force in the MKS system is called the *newton*, but note that it is entirely expressible in terms of the fundamental units. Thus, from an inspection of the right side of Eq. (1-1) the dimensional relationship for force is $[MLT^{-2}]$. The *dimension* of a quantity shows the fundamental units which enter as factors and their exponents. *Work* is still another example of a defined quantity. Since work involves the product of force by length, it can be dimensionally expressed as $[ML^2T^{-2}]$. Note, too, that the simple form of Eq. (1-1), free of constant multiplying factors, resulted from the proper choice of fundamental units. Thus, if the centimeter had been chosen in place of the meter as the fundamental unit of length, and provided other units remain unchanged, Eq. (1-1) would have had to be written as $F = ma \times 10^{-2}$.

Although for the study of mechanics only three fundamental units are required, a fourth fundamental unit must be introduced for the study of electricity and magnetism. For practical reasons the additional fundamental quantity is chosen to be *current* and the corresponding fundamental unit is called the *ampere*. However, from a purely theoretical viewpoint the fourth basic quantity could be charge, having the fundamental unit of the coulomb. One is derivable from the other. The important consideration which led to the selection of the ampere is that the ampere serves as the link between electrical, magnetic, and mechanical quantities and is more readily measured. The definition of the ampere which appears below stresses the matter sufficiently.

With this background, attention may now be directed to the formal definitions of the four fundamental units used in the study of electrical engineering. They are as follows:

The *meter* was defined in 1960 by an international commission as 1,650,763.73 wavelengths of the radiation of the orange-red line of krypton 86. This is a definition in terms of the wavelength of light, which makes it more permanent than the definition previously used.

The *kilogram* is the mass of a platinum-iridium alloy of cylindrical shape kept at the International Bureau of Weights and Measures at Sevres, France.

It is approximately equal to the mass of a cube (measuring 0.1 meter on each side) of pure water at 4°C.

The *second* is now defined as 1/31,556,925.9747 of the tropical year 1900.

The *ampere* is that current which when flowing through two infinitely long straight wires of negligible cross-sectional area and placed one meter apart in a vacuum produces a force per unit length of 2×10^{-7} newtons/meter.

In recent years in electrical engineering, with the advent of very high-frequency oscillators, such terms as *pico* and *nano* have been occurring frequently in the literature. The following tabulation is presented as an explanation of these and other terms.

deci (d-, 10^{-1})	deka (dk-, 10^{1})
centi (c-, 10^{-2})	hekto (h-, 10^{2})
milli (m-, 10^{-3})	kilo (k-, 10^{3})
micro (μ-, 10^{-6})	mega (M-, 10^{6})
nano (n-, 10^{-9})	giga (G-, 10^{9})
pico (p-, 10^{-12})	tera (T-, 10^{12})

Thus a picosecond is one millionth of one millionth of a second, and a centimeter is one hundredth of a meter. Similarly a megaton is one million tons. A little thought reveals that the MKS system of units offers the additional flexibility of the use of prefixes, which is not so readily accomplished with a system of units such as the English system where inches, feet, and yards are used.

For convenience a table of units is given in Appendix A.

1-2 Electric Current

Charge has been associated with the idea of electricity from the very beginnings. As early as 600 B.C. it was found that if the fossil resin amber was rubbed with a dry substance, it was possible for the amber to exert a force of attraction on a light material such as a feather or straw. In fact it is from the Greek word for amber—*elektron*—that the word for electricity is derived. In this situation the amber is described as possessing *frictional electricity* through the accumulation of charge. Up to the end of the eighteenth century this was the only form of electricity known to man. In 1799, however, Allessandro Volta developed the copper-zinc battery, which he showed was capable of producing electricity in a continuously flowing state in wires. Initially this type of electricity was called *current electricity* to distinguish it from the frictional kind. But with further experimentation Volta was able to demonstrate that the two types were identical; he was able to produce the same results. Of course today our knowledge of the atomic structure of matter bears out the truth of Volta's work.

In terms of the atomic description we know that all matter is composed of atoms. The atom in turn consists of a central body called the *nucleus* and a number of smaller particles, called *electrons*, which move in approximately elliptical orbits around the nucleus. The electron is the smallest indivisible particle of electricity known to man. The American physicist R. A. Millikan in his renowned oil-drop experiment showed the charge of the electron to be 1.602×10^{-19} *coulomb*,† and it was arbitrarily called a *negative* charge. The nucleus of the atom is made up of two types of particles—the proton and the neutron. The *proton* has a mass 1837 times that of the electron and carries a positive charge equal to the sum of the electron charges of the atom. The *neutron* has the same mass as the proton but no charge. Thus we see that the role of positive and negative charges in the description of matter is an important one.

Electric current is defined as the time rate of change of charge passing through a specified area. The moving charges may be positive or negative; the area may be the cross-sectional area of a wire or some other suitable spatial area where charges are in motion. Expressed mathematically we can write

$$i = \frac{dq}{dt} \qquad (1\text{-}2)$$

In this equation i denotes the instantaneous electric current and q represents the net charge, which may be of the positive as well as negative kind. That is

$$q = N_1 p + N_2 e \qquad (1\text{-}3)$$

where $N_1 p$ denotes the total positive charge and $N_2 e$ the total negative charge. The positive direction of the current i is arbitrarily taken to be the direction of flow of positive charges.

Since the symbol e is used to denote the charge of an electron we may write

$$e = -1.602 \times 10^{-19} \quad \text{coulomb/electron}$$

From this result we see that there are 6.24×10^{18} electrons in one coulomb, which is a staggering number. Such a large number is used for the coulomb so that smaller numbers can be used for ordinary engineering calculations.

The ability to put charge in motion, i.e. to effect a current flow, is important in many ways in electrical engineering. For example, it is the means by which energy can be transferred from one point to another; as another example, control of the rate at which charge is put into motion makes it possible to transmit intelligence in such communication devices as radio, television, and telstar.

Circuit. An electric circuit is a closed path, composed of active and passive elements, to which current flow is confined. Figure 1-1 depicts a typical circuit containing one passive and one active element. An active element is one which supplies energy to the circuit. A passive element is one which receives energy and then either converts it to heat or stores it in an electric or magnetic field. The battery in Fig. 1-1 is the active element.

---

† The coulomb is that unit of charge which when placed one meter apart from an identical particle of charge is repelled by a force of $10^{-7}c^2$, where c is the velocity of light in m/sec.

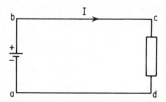

Fig. 1-1. Example of an electric circuit. The current *I* is shown pointed in the direction of conventional flow. Electrons flow in the opposite direction.

1-3 Coulomb's Law

Appearing in Fig. 1-2 are two point positive charges of value Q_1 and Q_2 separated by a distance of *r* meters. The distance between the charges is assumed to be large compared to the size of the mass on which the charges reside. Charles A. Coulomb was able to show experimentally that a force of repulsion exists

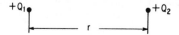

Fig. 1-2. Configuration for measuring the electrostatic force of repulsion by Coulomb's law.

between Q_1 and Q_2 which is directly proportional to the magnitude of the charges and inversely proportional to the distance between them. In terms of MKS units, Coulomb showed the magnitude of this force to be given by

$$F = \frac{Q_1 Q_2}{4\pi\epsilon r^2} \tag{1-4}$$

where Q_1 and Q_2 are expressed in coulombs and *r* in meters and ϵ is a constant related to the surrounding medium. The 4π is a proportionality constant which appears whenever Coulomb's law is expressed in rationalized MKS units. It is this equation which is known as Coulomb's law. However, Eq. (1-4) is also frequently written as

$$F = \mathscr{E} Q_2 \tag{1-5}$$

where

$$\mathscr{E} \equiv \frac{Q_1}{4\pi\epsilon r^2} \quad \frac{\text{newtons}}{\text{coulomb}} \text{ or volts/meter} \tag{1-6}$$

The quantity $\mathscr{E}$ is called the *electric field intensity* and has the units of newtons per coulomb or volts/meter. The former unit is obvious from Eq. (1-5); the latter is discussed presently.

Although Eq. (1-4) bears Coulomb's name, he is not generally credited with having originated the formulation of the law. Long before Coulomb began his experiments it was known that like charges repel and unlike charges attract. It was Benjamin Franklin who proposed inquiries about the behavior of static charges that finally culminated in Coulomb's successful measurements. While experimenting, Franklin one day discovered that when a conductor was in a charged state but not carrying current (i.e. not part of a complete circuit) all of the charge resided on the surface of the conductor. Franklin subsequently asked Joseph Priestley† to investigate the matter more thoroughly. In the process of doing so Priestley in 1766 was able to deduce the law which bears Coulomb's name. However, it was Coulomb's brilliant invention of a torsion balance which enabled him in 1785 to verify Priestley's deductions with significant accuracy. Coulomb's measurements led directly to Eq. (1-4).

Permittivity. A glance at Eq. (1-4) shows that through measurements it is possible to know all the factors in this equation with the exception of the proportionality factor ϵ, which is a property of the medium in which the experiment is performed. This quantity is called the *permittivity*. When the experiment is performed in a vacuum (free space), the value of the permittivity as obtained from Eq. (1-4) and expressed in MKS units comes out to be

$$\epsilon_0 = \frac{1}{36\pi \times 10^9} = 8.854 \times 10^{-12} \quad \text{farads/meter} \qquad (1-7)$$

By repeating Coulomb's experiment in oil for the same values of Q_1, Q_2 and r it is found that the resulting force is only about half as much as for air. Accordingly, this means the permittivity of oil is greater than that of air. A convenient way of expressing this difference is to introduce a quantity called *relative permittivity*, which is defined as

$$\epsilon_r = \frac{\epsilon}{\epsilon_0} \qquad (1-8)$$

Thus a value of ϵ_r of two reveals that for a given configuration, such as depicted in Fig. 1-2, the resulting force in the given medium is half the amount which is available in free space.

EXAMPLE 1-1   Find the force in free space between two like point charges of one coulomb each and placed one meter apart.

solution: From Eqs. (1-4) and (1-7) we have

$$F = \frac{1 \times 1}{4\pi \dfrac{1}{36\pi \times 10^9} \times 1} = 9 \times 10^9 \text{ newtons} \approx 1 \text{ megaton} \qquad (1-9)$$

† Priestley is better known for the discovery of oxygen.

Note the gigantic size of the force. This calculation again illustrates that one coulomb of electricity is an exorbitant quantity which is not available for ordinary engineering computations.

Potential Difference. Coulomb's law serves as the starting point for the study of a considerable part of electrical engineering. It is our purpose next to establish additional background that follows naturally from Coulomb's law and that will be useful later on in the book. Refer to Fig. 1-3. The charge Q_1 is assumed located at some fixed point in space. The charge Q_2 is assumed initially

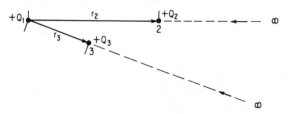

Fig. 1-3. Illustrating the law of potential difference.

to be on a horizontal path from Q_1 but infinitely far away. Hence the force of repulsion existing between Q_1 and Q_2 is zero. Let us consider next what happens as Q_2 is moved from infinity to a point where it is finally located r_2 meters from Q_1. It should be evident that as Q_2 is moved in this direct line from infinity to positions which are closer and closer to Q_1, greater and greater forces are required to accomplish this. Furthermore, since the force exerted on Q_2 acts through a distance which measures from infinity right up to r_2, it follows that work is being done on Q_2. As a matter of fact this work is imparted to Q_2 in the form of *potential energy*, which is recoverable in the form of kinetic energy immediately upon release of the force which keeps Q_2 fixed at a position r_2 meters from Q_1. Keep in mind that there is a force of repulsion between Q_1 and Q_2 which attempts to send Q_2 back to infinity. The fact that work must be done on Q_2 to bring it from infinity to r_2 may also be described in terms of the electric field intensity associated with charge Q_1 and expressed by Eqs. (1-6) and (1-5). Accordingly, owing to Q_1 an electric field of diminishing magnitude exists from r_2 to ∞. In moving Q_2 from ∞ to r_2 work must be done *against* this electric field. It is worthwhile at this point to use a description which stresses not so much the total work imparted but rather the total *work per unit charge*, which is called *voltage*. Viewed dimensionally the work imparted to Q_2 requires applying the length dimension to both sides of Eq. (1-5), which leads to

$$\text{newtons} \times \text{meters} = \left(\frac{\text{newtons}}{\text{coulomb}}\right) \times \text{coulombs} \times \text{meters} \qquad (1\text{-}10)$$

However, to deal in terms of work per unit charge it is necessary to divide both sides of Eq. (1-10) by the unit of charge Q_2. Thus

$$\frac{\text{newton-meters}}{\text{coulomb}} = \left(\frac{\text{newtons}}{\text{coulomb}}\right) \text{meters} = \text{voltage} \qquad (1\text{-}11)$$

It follows then that to find the work imparted per coulomb in moving Q_2 from infinity to r_2, it is necessary to perform an integration of electric field intensity $\mathscr{E}$(newtons/coulomb) with respect to length (meters). We may express this mathematically as

$$\boxed{\begin{aligned} \text{voltage} &= \frac{\text{work}}{\text{charge}} \\[4pt] V_2 &= \frac{W_2}{Q_2} \end{aligned}} \tag{1-12}$$

or, more fully,

$$\boxed{V_2 = \frac{W_2}{Q_2} = -\int_\infty^{r_2} \mathscr{E}_1\, dr} \tag{1-13}$$

The minus sign is arbitrarily inserted to indicate that work is done on the charge *against* the action of the electric field produced by Q_1. Since in this case $\mathscr{E}_1$ is given by Eq. (1-6), V_2 may be evaluated as

$$V_2 = -\int_\infty^{r_2} \frac{Q_1}{4\pi\epsilon r^2}\, dr = \frac{Q_1}{4\pi\epsilon r_2} \qquad \text{volts} \tag{1-14}$$

Thus Eq. (1-14) expresses the potential (or voltage) of point 2 in the configuration of Fig. 1-3 *with respect to a point at infinity*. Moreover, note that since voltage is work per unit charge, Eq. (1-14) is independent of Q_2.

In the interest of making clear to the reader the meaning of *potential difference* let us take another charge Q_3 and move it from infinity to r_3. Refer again to Fig. 1-3. Clearly, since r_3 is smaller than r_2, more work per unit charge must be done to place a charge at point 3. Reasoning as before, we may describe the potential of point 3 as

$$V_3 = -\int_\infty^{r_3} \frac{Q_1}{4\pi\epsilon r^2}\, dr = \frac{Q_1}{4\pi\epsilon r_3} \qquad \text{volts} \tag{1-15}$$

Because V_3 is greater than V_2, there exists a *potential difference* between these two points. It should be apparent then that when a unit charge moves from a point of higher potential (point 3) to one of lower potential (point 2) it gives up energy. Conversely, when a unit charge moves from a point of lower potential to one of higher potential it receives energy. This, then, is the meaning to be associated with such terms as potential difference or voltage rise or voltage drop. Return to Fig. 1-1 for an illustration of the application of these terms. Current is assumed flowing in a clockwise direction, which means that either positive charges are moving clockwise or negative charges are moving counterclockwise. As the positive charges move through the voltage rise from a to b they receive energy so that the potential of b is higher than a. However, as they move from c to d they undergo a voltage drop, thereby losing energy to the device appearing between these two points. It is interesting to note that the stream of charge gains energy in one part of the circuit (at the active element) and gives it up at the other part (the passive element). Of course the total energy remains unchanged in accordance with the law of conservation of energy.

Further Considerations. If the charge Q_1 in Fig. 1-3 is doubled, the electric field intensity by Eq. (1-6) correspondingly doubles. Therefore to move a unit positive charge from infinity to the same points 2 and 3 requires twice the energy. The potential of each of these points becomes twice as large as well, and so too does the voltage or potential difference. One can conclude from these observations that charge, Q, and voltage, V, are related through some proportionality factor which can be reasonably expected to depend upon the medium in which the charges find themselves as well as upon the geometrical configuration prevailing. Calling this proportionality the capacitance C we can write

$$Q = CV \qquad (1\text{-}16)$$

More is said about this equation and capacitance in Chapter 2.

Gauss's Law. This law is an important consequence of Coulomb's law and provides additional useful knowledge, which is needed in the work that follows in Chapter 2. In this connection consider that a sphere of radius r is put around a point charge Q as depicted in Fig. 1-4. By Eq. (1-6) we know that the value of the

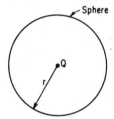

Fig. 1-4. Configuration for deriving Gauss's electric flux law.

electric field intensity at any point on the surface of the sphere is

$$\mathscr{E} = \frac{Q}{4\pi\epsilon r^2} \qquad (1\text{-}17)$$

Let us now form the product of $\mathscr{E}$ and the surface area of the sphere, which is

$$A = 4\pi r^2 = \text{area of sphere} \qquad (1\text{-}18)$$

Thus

$$\mathscr{E}A = \frac{Q}{4\pi\epsilon r^2}(4\pi r^2) = \frac{Q}{\epsilon} \qquad (1\text{-}19)$$

Rewriting Eq. (1-19) yields

$$Q = (\epsilon\mathscr{E})A = DA \qquad (1\text{-}20)$$

where

$$D \equiv \epsilon\mathscr{E} = \text{electric flux density} \qquad (1\text{-}21)$$

The quantity D is identified as an electric flux density because, as revealed by Eq. (1-20), it is combined with an area term A to yield charge, which is also called *electric flux.*

Equation (1-20) is known as Gauss's law and it states that the total electric flux through any closed surface surrounding charge is equal to the amount of charge enclosed. It is discussed here because if we compare Eq. (1-20) with Eq. (1-16), information about the nature of capacitance readily follows. This matter is treated in Sec. 2-4.

1-4 Ohm's Law

This law is perhaps one of the first things learned about electricity in any elementary course on the subject. It is not at all unlikely that the reader was introduced to Ohm's law in its simplest form in his physics course. This is mentioned by way of further stressing its importance in the study of electrical engineering.

George Simon Ohm (1787–1854) was a professor of mathematics and physics who devoted considerable time and effort to experiments dealing with voltaic cells and conductors.† These experiments included the effects of temperature on the resistivity of various metals. In spite of this extensive laboratory work, however, the law that bears his name was the result of a mathematical analysis of the galvanic circuit based on an analogy between the flow of electricity and the flow of heat. This work was described in a pamphlet published in 1827.‡ By following the formulation of Fourier's heat conduction equation and using electric field intensity as analogous to temperature gradient, Ohm was able to show that the current flow in a circuit composed of a battery and conductors can be expressed as

$$I = \frac{A}{\rho} \frac{dv}{dl} \tag{1-22}$$

where the derivative term denotes the electric field gradient. In the language used by Ohm v was called the electroscopic force by way of representing the volume density of electricity at a point in the conductor—a terminology consistent with the analogous situation in heat flow through a solid described in terms of the quantity of heat per unit volume. In the case where a conductor of uniform cross-sectional area is used, Eq. (1-22) may be written as

$$\boxed{I = \frac{A}{\rho} \frac{V}{L} = \frac{V}{R}} \text{ amps} \tag{1-23}$$

where V is the potential difference in volts appearing across the conductor of

† See William Francis Magie, *A Source Book in Physics*, 1st ed. (New York: McGraw-Hill Book Company, 1935).

‡ "Die galvanische Kette mathematisch bearbeitet," i.e., "The Galvanic Circuit Investigated Mathematically."

length L, A is the cross-sectional area, p is a property of the material called the *resistivity*, and R is the resistance of the conductor in ohms.

Equation (1-23) is a mathematical description of Ohm's law. It states that the strength of the current in a wire is proportional to the potential difference between its ends. Experimentation aimed at verifying Eq. (1-23) reveals the need for maintaining constant temperature—otherwise the results change accordingly. (More is said about this in Chapter 2.)

Ohm's law may be alternatively expressed as

$$\boxed{V = IR} \qquad (1\text{-}24)$$

In this form it states that for any given potential difference, the amount of current produced is inversely proportional to the resistance, which in turn is dependent upon the composition of the wire. It is interesting to note therefore that Ohm's law is not among those basic laws which are independent of materials.

The simplicity of Ohm's law is impressive. It almost commands disbelief; and as a matter of fact this is precisely what happened when Ohm first published his results. He was ridiculed and subsequently resigned from his professorship. Fourteen years later, however, he received the recognition due him, and all ended well.

During Ohm's time, experimental verification of his law was achieved exclusively by using direct-current sources, namely the voltaic cells. Since then further experiments have been made to show that Ohm's law is also valid when the potential difference across a linear resistor is time-varying. In this case Ohm's law is written as

$$\boxed{v = iR} \qquad (1\text{-}25)$$

where lower-case letters are used for potential difference and current to emphasize the nonconstant nature of these quantities. This matter is treated at considerable length in Chapter 6.

Devices are often encountered in the study of electrical engineering where the volt-ampere characteristic is not linear, i.e., the current is not related to potential difference by a *constant* proportionality factor. For example, for the incandescent lamp the volt-ampere curve looks as depicted in Fig. 1-5, which is clearly nonlinear.

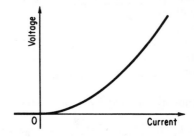

Fig. 1-5. Voltage vs current curve for incandescent lamp.

Does Ohm's law apply in this instance? An affirmative answer is possible provided that Ohm's law is expressed in differential form. Thus

$$dv = R\,di \qquad (1\text{-}26)$$

Note that R in this case is the slope of the curve of Fig. 1-5 corresponding to a given point on the curve. As this point is changed, so too is the value of resistance as it applies to that point in the differential sense.

The very simplicity of Ohm's law can lead to difficulty because of the tendency to oversimplify. It is important to know that although Ohm's law applies to many materials—copper, aluminum, iron, brass, bronze, tungsten, etc.—it does not apply to all. For example, it does not apply to nonlinear devices involving step discontinuities such as might be found in Zener diodes and voltage regulator tubes.

1-5 Faraday's Law of Electromagnetic Induction

Michael Faraday (1791–1867) distinguished himself as an outstanding experimentalist. He had very little formal academic training. He was born of humble parents and spent seven years as a bookbinder's apprentice. However, Faraday was an avid reader of scientific works and at the age of twenty, after listening to a lecture by Sir Humphrey Davy,† wrote to Davy requesting that he be taken on as an assistant. Following an interview, Davy agreed to the arrangement and so began a career for Faraday which culminated in fundamental contributions to the science of electricity. His crowning achievement was the establishment of the law of electromagnetic induction which related electricity and magnetism. This was accomplished in 1831 through a series of brilliant experiments.

In the beginning, upon joining Davy at the Royal Institution, Faraday directed his efforts to experiments in chemistry. In 1820, however, the whole scientific world was excited over the discovery by Hans Christian Oersted that a current-carrying wire could be made to deflect a freely suspended magnet. Faraday repeated Oersted's experiment and then went further to show that not only was it possible for a magnet to move round a current-carrying wire but also for a current-carrying wire to move round a magnet. The reason for the phenomena, however, remained a mystery. This mystery further deepened when in 1825 Andre M. Ampere demonstrated that a force also existed between two current-carrying wires. Although Oersted and Ampere both kept searching for the answer in the conductor or the magnet, Faraday did not. Rather he conceived the forces to be due to tension in the *medium* in which the conductors and/or magnets were placed. As a matter of fact it was this attitude which led Faraday to introduce the concept of lines of force. Since he was not a mathematician, he used descriptions which enhanced visualization.

In 1831 Faraday showed that electricity could be produced from magnetism.

† Davy developed the first electric arc lamp. He also isolated sodium and potassium by means of an electric current.

He demonstrated that induced currents could be made to flow in a circuit whenever (a) current in a neighboring circuit is established or interrupted, (b) a magnet is brought near a closed circuit, and (c) a closed circuit is moved about in the presence of a magnet or other closed current-carrying circuits. Figure 1-6(a) shows one of the experiments made by Faraday. By closing the switch he was able to observe a deflection in the galvanometer† connected to the second circuit.

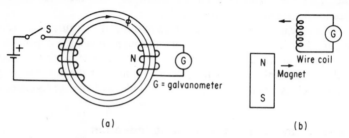

(a)   (b)

Fig. 1-6. Illustrating Faraday's law of induction.

Moreover, upon opening the switch, he again observed a deflection but this time in a reverse direction. Figure 1-6(b) shows schematically still another arrangement used by Faraday to show how electricity is produced from magnetism. If either the magnet was moved towards the circuit or the circuit towards the magnet, the galvanometer registered a deflection.

Faraday was a disciplined and meticulous experimenter who kept careful and complete records of his laboratory work. From these descriptions‡ it has been possible to formulate Faraday's law of induction mathematically as follows

$$e = -\frac{d\lambda}{dt} = -N\frac{d\phi}{dt} \quad \text{volts}$$

(1-27)

The quantity e denotes the electromotive force induced in a closed circuit having a flux linkage of λ weber-turns. In those instances where the magnetic flux ϕ penetrates all the turns of the coil as shown in Fig. 1-6(a), Faraday's law may be expressed by the second form of Eq. (1-27). The negative sign is due to Emil Lenz, who subsequent to Faraday's experiments pointed out that the direction of the induced current is always such as to oppose the action that produced it. This reaction is commonly called *Lenz's law*.

As pointed out in Chapter 16, Faraday's law as embodied in Eq. (1-27) is one of the two basic relationships upon which the entire theory of electromagnetic and electromechanical energy conversion devices is based. In fact, soon after Faraday's work of genius was published in 1831 explanations were at last possible for the phenomena observed by Oersted, Ampere, and other experimenters.

---

† A galvanometer is a sensitive instrument used to measure small currents and voltages.

‡ *Experimental Researches in Electricity*, 3 vols. (London, 1839).

The foundation was now established to facilitate the ensuing rapid development of electric motors and generators.

Faraday was also the first experimenter to identify the *emf of self-induction* which manifested itself whenever circuits carrying current in long wires or circuits wound with many turns were disconnected. The American inventor Joseph Henry also independently discovered the current of self-induction but not before Faraday. Both experimenters were able to demonstrate that a changing current produced an emf of self-induction in a coil of wire which varied directly with the time rate of change of current. Expressed mathematically

$$e = L \frac{di}{dt} \tag{1-28}$$

where L is a proportionality factor called the *coefficient of self-inductance* which is dependent upon the medium and some physical dimensions. (See Chapter 2.) This result is equivalent to Faraday's law of induction as expressed in Eq. (1-27). Moreover, Eq. (1-28) is extremely important in the development of electric circuit theory. It serves to identify one of the three basic circuit parameters—*inductance*. The other two parameters, resistance and capacitance, have already been identified with Ohm's and Coulomb's law, respectively.

1-6 Kirchhoff's Laws

By the middle of the nineteenth century a considerable amount of experience with electric circuits had been gained by the leading men of science of the time. However, it was Gustav Robert Kirchhoff (1824–1887) who published the first systematic formulation of the principles governing the behavior of electric circuits. He advanced no new experimental facts or concepts but merely restated familiar principles. His work was embodied in two laws—a current and a voltage law— which together are known as *Kirchhoff's laws*. It is upon these laws that electric circuit theory is based.

Kirchhoff's current law states that the sum of the currents entering or leaving a junction point at any instant is equal to zero. A junction point is that place in a circuit where two or more circuit elements are joined together. It is often called an *independent node.* In Fig. 1-7 *n* corresponds to the junction point or node. The

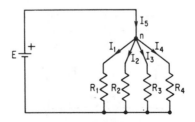

Fig. 1-7. Illustrating Kirchhoff's current law.

current law may be expressed mathematically as

$$\sum_{j=1}^{k} I_j = 0 \tag{1-29}$$

where k denotes the number of circuit elements connected to the node in question and $\sum$ is the Greek symbol used to indicate summation. For the circuit of Fig. 1-7 there are five circuit elements joined at node n—four passive elements (the resistors) and one active element (the battery). Applying Eq. (1-29) to node n for the situation where the currents are all assumed to be *entering* the node we have

$$I_5 - I_1 - I_2 - I_3 - I_4 = 0 \tag{1-30}$$

The minus signs are used because these currents are defined as leaving rather than entering the node. Equation (1-30) may be rewritten as

$$I_5 = I_1 + I_2 + I_3 + I_4 \tag{1-31}$$

and it states in mathematical terms that the current entering the node is equal to the sum of the currents leaving it.

In countless experiments performed over the past century Kirchhoff's current law has never been found to be invalid. This is understandable when it is realized that the current law is nothing more than a restatement of the *principle of conservation of charge*. This principle states that the number of electrons passing per second must be the same for all points in the circuit. Accordingly, if the summation of all the currents at a node were not to add up to zero, there would have to occur an accumulation of charge at the node. For the sake of argument suppose the sum of currents at node n were not zero but rather one ampere. Then charge would accumulate at the rate of one coulomb for each second. We have already learned, however, that by Coulomb's law a charge accumulation of this kind would produce explosive forces. But experiment shows no evidence of this whatsoever. Any accumulation of charge at the node means an accumulation of mass. Once again experiment fails to reveal that this happens. Therefore, it is proper to conclude that for every charged particle that enters the node there is another charged particle that leaves.

Kirchhoff's voltage law states that *at any time instant the sum of voltages in a closed circuit is zero*. Essentially this law is a restatement of the *law of conservation of energy*. To understand this, refer to Fig. 1-8 which depicts a 12-volt battery

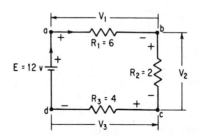

Fig. 1-8. Illustrating Kirrchoff's voltage law.

supplying energy to three series-connected resistors representing a total resistance of 12 ohms. By Ohm's law there exists a current of one ampere, which means that charge flows throughout the circuit at the rate of one coulomb per second. This current is denoted by I. The positive direction of current flow is a-b-c-d-a. The electrons, of course, are moving in the reverse sense.

The constant current which flows in this circuit means that the average rate of movement of charge is constant. However, as electrons traverse the circuit elements R_1, R_2, and R_3, they collide with the atoms in the space lattice configurations of these elements, thereby dissipating energy in the form of heat. Since this energy must be supplied by the battery, the law of conservation of energy permits us to write

$$W = W_1 + W_2 + W_3 \qquad (1\text{-}32)$$

where W denotes the energy supplied by the battery in a specified time interval, W_1 denotes the energy dissipated in R_1 in the same interval of time, and similarly for W_2 and W_3. Furthermore, if we denote by Q the total charge which moves throughout the circuit during this same interval of time, it follows that the work per unit charge may be written as

$$\frac{W}{Q} = \frac{W_1}{Q} + \frac{W_2}{Q} + \frac{W_3}{Q}$$

But because work per unit charge is voltage, this last equation can be expressed as

$$E = V_1 + V_2 + V_3$$

Transposing then yields

$$V_1 + V_2 + V_3 - E = 0 \qquad (1\text{-}33)$$

Note that Eq. (1-33) is a direct mathematical statement of Kirchhoff's voltage law. Due regard must be given to signs. As Eq. (1-33) is written, all voltage drops are treated as positive quantities. Conversely, then, all voltage rises encountered in the loop must be prefixed with a negative sign. A little thought reveals that the reverse convention is equally valid, i.e. voltage rises may be treated as positive quantities while voltage drops are taken as negative.

A generalized formulation of Kirchhoff's voltage law is written as

$$\boxed{\sum_{j=1}^{k} V_j = 0} \qquad (1\text{-}34)$$

where V_j represents the voltage drop of the jth element in any given closed circuit which is assumed to have k elements.

In Fig. 1-8 the source voltage was assumed to be a battery of fixed magnitude, thus producing a constant current. However, Kirchhoff's voltage law is general enough so that it applies equally as well when the source voltage is a time-varying quantity producing a time-varying current. Refer to Fig. 1-9. Appearing across d-a is the varying source voltage. Between points a-b is a resistor of value R similar to those used by Ohm in his experiments. If the instantaneous value of the current is denoted as i then the potential difference between the ends of the resistor is iR. Across points b-c appears a coil of the type used by Faraday

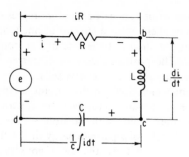

Fig. 1-9. Illustrating Kirchhoff's voltage law for a time varying source.

and across which is developed the emf of self-induction, $L(di/dt)$. Finally, between points *c-d* a capacitor appears. The voltage drop across the capacitor is readily found from Eqs. (1-16) and (1-2). Thus

$$V = \frac{1}{C} q = \frac{1}{C} \int i\, dt \tag{1-35}$$

Then by Kirchhoff's voltage law we can write

$$e = iR + L\frac{di}{dt} + \frac{1}{C} \int i\, dt \tag{1-36}$$

This is an equation concerning which a great deal is said in subsequent chapters.

1-7 Ampere's Law

Andre Marie Ampere (1775–1836) combined extraordinary experimental ability with a genius for mathematics to achieve outstanding contributions in the field of electricity and magnetism. Like all the other leading scientists of the day Ampere was so fascinated by Oersted's discovery of the effect of current in a wire on a magnetic needle that he forthwith directed all his attention to a study of this new phenomenon. A short time afterwards he not only succeeded in extending Oersted's experiments but also came up with a rigorous mathematical theory to describe them. Ampere's most famous experiment in this regard dealt with the force produced between two very long, parallel wires carrying currents. (It will be recalled that this is the configuration which is used today to identify the fundamental unit of the ampere.) Ampere's facility for mathematics enabled him to describe the force in terms of the following equation:

$$F = \frac{\mu I_1}{2\pi r} l I_2 \quad \text{newtons} \tag{1-37}$$

where I_1 and I_2 are the currents flowing in the two conductors, l is the length of the conductors, and r denotes the distance separating the wires at their center lines. The quantity μ is a property of the medium.

Ampere further showed that the current I_1 could be looked upon as producing a magnetic field given by†

$$B = \frac{\mu I_1}{2\pi r} \qquad (1\text{-}38)$$

This result was also observed by two other French experimenters of the day—J. B. Biot and F. Savart. Upon substitution of Eq. (1-38) into (1-37) there results that simplified form of Ampere's law which the reader very likely encountered for the first time in an elementary physics course. Thus

$$F = BlI_2 \qquad (1\text{-}39)$$

In order for Eq. (1-39) to be valid it is necessary for the wires carrying current I_1 and I_2 to lie in the same plane. If, for some reason, the conductor carrying I_2 is inclined by some angle θ with respect to a line perpendicular to the wire carrying I_1, then Ampere's law states that the force is given by

$$F = BlI_2 \sin \theta \qquad (1\text{-}40)$$

In the interest of covering all cases of this kind as well as those in which more than a single conductor is used to carry I_2, use is made of a generalized formulation of Ampere's law. In equation form it may be expressed as

$$\boxed{F = K_B B I_2 f(\theta) = K_\phi \phi Z_2 I_2 f(\psi)} \qquad (1\text{-}41)$$

Here B denotes the magnetic field produced by the circuit which carries current I_1, and $f(\theta)$ is some function of the angular displacement existing between the direction of the B-field and the direction in which I_2 flows. Often $f(\theta)$ is the sinusoidal function, $\sin \theta$. The quantity K_B accounts for such factors as the length of the conductor, the number of poles present in the field, and so forth. An alternate version of Ampere's law finds the force expressed in terms of the magnetic flux field, ϕ, the number of conductors Z_2 which carry I_2, and some function of the angular displacement between the flux and the physical distribution of the conductors Z_2. Often $f(\psi)$ is $\cos \psi$. Refer to Chapter 16 for more details.

On the basis of the information appearing in Eq. (1-41), Ampere's law may be described in words as follows. *To produce an average force (or torque) in an electromagnetic device, there must exist a flux field and an ampere-conductor distribution (which are stationary with respect to each other) and a favorable field pattern (i.e, $f(\psi) \neq 0$).* This statement applies even for those cases in which the flux and current are time-varying quantities. All that is required is that the frequency of variations be the same.

PROBLEMS

1-1   A negative point charge Q_1 of magnitude 2×10^{-12} colulomb is placed in a vacuous medium.

(a) Find the value of the electric field at a point one meter from the charge.

† See Sec. 12-1 for further explanation of B.

(b) Compute the work done in bringing an electron from a point very far removed from Q_1 to a point 0.5 meter away.

(c) What force is acting on the electron in part (b)?

1-2   A positive point charge of value 10^{-6} coulomb is fixed at a point in a vacuum. An identical point charge is then brought to a distance 10 cm away.

(a) Compute the potential energy of the second charge.

(b) How much work per unit charge must be done in moving charges about in the electric field created by the fixed charge source? Is this a constant quantity? Explain.

1-3   Two flat parallel plates, measuring 0.1 m by 0.2 m, are charged by transferring 10^{-6} colulomb from one plate to the other. The plates are separated by 2 cm and oil is placed between them. The relative permittivity of the oil is 2.

(a) Find the electric flux density between the plates.

(b) Determine the value of the electric field intensity.

(c) Compute the potential difference existing betwen the plates.

(d) What is the capacitance of the parallel plates?

1-4   For the circuit shown in Fig. P1-4 find the current that flows and also the potential difference across each resistor.

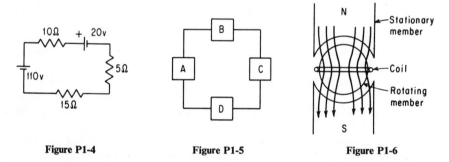

Figure P1-4   **Figure P1-5**   **Figure P1-6**

1-5   Four pieces of apparatus are connected as shown in Fig. P1-5. Free electrons circulate at a constant rate in the direction *a-b-c-d*. The quantity of electricity passing a reference point on the circuit in a given time is 50 coulombs. During this time, in A, the electrons receive energy in the amount of 600 joules. In B energy of 200 joules is given up by the electrons, and in *C*, 300 joules is supplied to the electrons. Redraw the circuit showing batteries or resistors instead of boxes A, B, C, and D. Mark polarities and compute the emf or voltage drop for each element.

1-6   In the configuration shown in Fig. P1-6 the coil has 100 turns and is attached to the rotating member which revolves at 25 revolutions per second. The magnetic flux is a radial uniform field and has a value of $\phi = 0.002$ weber. Compute the emf induced in the coil.

1-7   Determine the instantaneous polarity for the coil configurations shown in Fig. P1-7. The flux is increasing in the direction shown.

1-8   (a) Find the magnetic flux density in free space at a distance 5 cm from a long straight wire carrying a current of 50 amperes.

(b) Calculate the force produced on a second wire one meter long placed 5 cm from the first wire and carrying 100 amperes.

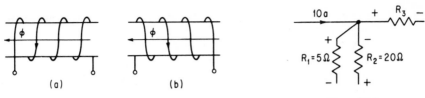

(a)   (b)

Figure P1-7   **Figure P1-9**

(c) Repeat (a) and (b) assuming the wires are imbedded in iron having a relative permeability of 1500. Use $\mu_0 = 4\pi \times 10^{-7}$.

1-9 A portion of a network has the configuration shown in Fig. P1-9. The voltage drops existing across the three resistors are known to be respectively 20, 40, and 60 volts having the polarities indicated. Find the value of R_3.

1-10 In Fig. P1-10 find the voltage drop across R_1 and R_2. The resistance R_3 is not specified.

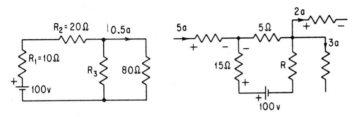

Figure P1-10   **Figure P1-11**

1-11 The voltage drop across the 15-ohm resistance in the circuit of Fig. P1-11 is 30 volts having the polarity indicated. Find the value of R.

1-12 A stationary magnetic field is produced by the magnets shown in Fig. P1-12. Between the magnets there is placed a cylindrical rotor which supports current-carrying conductors. The direction of current flow in these conductors is represented by dots (out of paper) and crosses (into the paper).

(a) If these current directions are assumed fixed, is there a torque on each conductor?

(b) Does the rotor experience a net torque different from zero, i.e., does the rotor turn?

(c) What recommendation can you make to improve the situation?

Figure P1-12

I

ELECTRIC CIRCUIT
THEORY

2

the circuit parameters

This is the first of four chapters devoted to a treatment of electric circuit theory. It is the purpose of *electric circuit theory* to furnish a means of describing the transfer of energy from a suitable source to appropriate engineering devices which are designed to achieve certain useful results. Through circuit theory, for example, we can readily analyze the generation, transmission, and application of electrical energy for home, commercial, and industrial uses. In like fashion the design, operation, and performance of radio and television circuits, telephone and telegraph circuits, and industrial electronic circuitry are easily determined through the techniques of circuit analysis.

It is important to understand at the outset, however, that *circuit analysis is really a simplified and approximate form of field analysis*. Recall that each of the basic experiments performed by Coulomb, Faraday and Ohm dealt with the medium of space, involving such quantities as electric field intensity (volts/meter), charge density (coulombs/meter³), and magnetic flux density (webers/meter²). The dependence of these quantities on length, volume, and area accordingly identified them as space or field variables. These variables also depend on time variations to which they may be subjected. If all problems in electricity and magnetism were to be solved as straight field problems, they would be quite complex and their solution would be laborious. Fortunately, however, under ordinary circumstances it is permissible to solve these problems by dealing with integrated effects. Thus, for example, instead of working with the space variable for electric field intensity we can work directly with voltage, which is the electric field intensity integrated with respect to distance. Similarly, in place of current density we can deal with current, which we obtain by integrating the current density over an appropriate area such as that of a conducting wire. This approach not only simplifies analysis but also facilitates measurements, because current and voltage are easily measured.

If we are to use circuit analysis rather than field analysis, the loss of energy through radiation must be negligible. Section 1-6 explains how charge receives energy from an active element in a circuit and delivers it to the passive elements. At the passive elements this energy may then be dissipated as heat or stored in magnetic or electric fields. Energy transfer requires motion of charges. Recall, too, that when a charge is accelerated by a field it gains energy, whereas when it is decelerated it gives up energy to the field. Now it can be shown that for engineering devices of dimensions small compared with a wavelength, the energy lost through radiation is indeed negligible unless exorbitant values of charge acceleration are achieved.† In turn, this means that throughout the closed path in which the charge particles move, the energy is confined solely to those devices present in the closed path. In such cases integrated effects may be used to describe and analyze problems which are determined in terms of the physical dimensions and properties of the devices as well as the voltage and current existing at a finite number of equipotential surfaces. These surfaces are called the terminals of the device. In Fig. 1-8, for example, it is possible to express the energy delivered to the device appearing across terminals a and b in terms of the integrated voltage between these points and the current flowing into terminal a. The subject matter of this book is restricted to those situations where there is negligible radiation loss. Hence circuit analysis (by application of Kirchhoff's laws) is always valid.

In Sec. 2-1 we shall see how the energy received by a device appearing between two terminals is expressible in terms of the associated voltage and current. The remaining sections describe the three important circuit elements of electric circuit theory which result from the fundamental investigations of Ohm, Faraday, and Coulomb. These circuit elements are respectively resistors, inductors, and capacitors.

2-1   Energy and Power

Our purpose here is to obtain a relationship between the energy delivered to a device and the integrated effects of the electric field intensity (voltage) and current density (current). Figure 2-1 depicts a battery delivering energy to a circuit element appearing between terminals c and d. It is assumed that the current flowing through the device is i and the potential difference across the terminals is v. From

† Of the order of 10^{16} ft/sec^2.

Eq. (1-12) we know that v denotes the amount of energy released by each unit charge as it moves through the circuit element from c to d. Therefore the total energy delivered to the device by the charge carriers is

$$w = vq \quad \text{coulomb-volt (or joules)} \qquad (2\text{-}1)$$

Moreover, the quantity of electricity q associated with a constant current i is

$$q = it \quad \text{coulombs} \qquad (2\text{-}2)$$

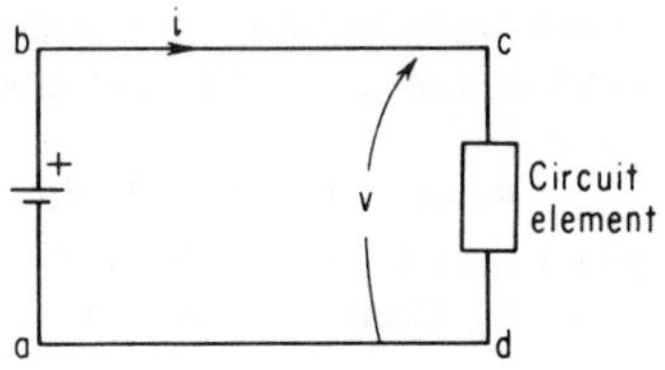

Fig. 2-1. Delivering energy to the circuit element by means of current i.

Inserting Eq. (2-2) into (2-1) yields the desired result. Thus

$$\boxed{w = vit} \quad \text{watt-secs (joules)} \qquad (2\text{-}3)$$

which shows that the energy absorbed is given by the product of the potential difference, the current, and the time. Although Fig. 2-1 depicts a situation where a constant current flows, Eq. (2-3) is equally applicable for time-varying quantities. Lower-case letters are used to stress this point.

Power is a defined quantity. It is the time rate of doing work. Letting p denote power it can be expressed mathematically as

$$\boxed{p \equiv \frac{w}{t}} \quad \text{joules/sec} \qquad (2\text{-}4)$$

Upon substituting Eq. (2-3), there results an alternative expression

$$\boxed{p = vi} \quad \text{watts (or joules/sec)} \qquad (2\text{-}5)$$

Note that power expresses a device's energy-delivering or energy-absorbing capability per unit of time. Accordingly, if we compare two devices and we find that one can absorb three times as much energy as the other can in any given second, we say that the first has three times the power of the second.

2-2 The Resistance Parameter

Resistance is one of three basic parameters of electric circuit theory. Strictly speaking, every circuit element exhibits some degree of resistance—but often, as in the case of inductors, capacitors, or ordinary wire connections between various circuit elements, it is of minor importance.

In Chapter 1 resistance is introduced as the proportionality factor in Ohm's law relating current to potential difference. In a general way resistance can be described as that property of a circuit element which offers opposition to the flow of current and in so doing converts electrical energy into heat energy. In the interest of furnishing a complete picture the resistance parameter is treated below from several viewpoints, each of which reveals some useful and interesting facet of the properties and characteristics of this important circuit element.

   Circuit Viewpoint. A circuit element is described from the circuit viewpoint when it is expressed in terms of an associated current and potential difference. Since by Ohm's law we have

$$R = \frac{v}{i} \qquad \text{ohms} \qquad (2\text{-}6)$$

it follows that this equation embodies the circuit-viewpoint description of resistance. Equation (2-6) is a linear algebraic expression when the proportionality

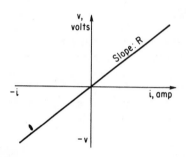

factor R is independent of the current. Moreover, Eq. (2-6) is valid for negative as well as positive values of i. These remarks are graphically depicted in Fig. 2-2. It is also worth noting that when Ohm's law correctly describes the relation between terminals c and d, the included circuit element is called a *resistor*.

   Experiment shows that the resistance of most metallic conductors varies with temperature. Specifically, if the resistance of a metal at temperature T_1 is R_1, then for normal range of temperatures the resistance at temperature T_2 is given by

Fig. 2-2. Graphical representation of Ohm's law.

$$R_2 = R_1[1 + \alpha_1(T_2 - T_1)] \qquad (2\text{-}7)$$

where α_1 is the temperature coefficient of resistance at T_1 and temperature is customarily measured in degrees centigrade. The values of α for various metals appear in Appendix C. Investigations reveal that a linear variation of resistance with temperature prevails over the range from approximately $-50°C$ to $200°C$. Consequently, a more convenient form of Eq. (2-7) is expressed approximately as

$$\frac{R_2}{R_1} = \frac{K + T_2}{K + T_1} \qquad (2\text{-}8)$$

For copper, the most commonly used conductor material, K takes on the value of 234.5. This value is obtained by extrapolating the linear curve to zero resistance as shown in Fig. 2-3. Hence Eq. (2-8) becomes

$$\frac{R_2}{R_1} = \frac{234.5 + T_2}{234.5 + T_1} \qquad (2\text{-}9)$$

This expression is useful in computing the effect of increased temperature on copper conductors, particularly because copper is almost always used in the temperature range for which the equation is valid. By way of illustration, Eq. (2-9) indicates that if the resistance of copper conductors in a given circuit at an ambient temperature of T_1 equal to 20°C is R_1, then the resistance at an operat-

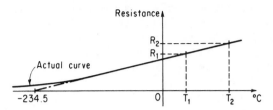

Fig. 2-3. Variation of resistance with temperature for copper.

ing temperature T_2 equal to 80°C is

$$R_2 = R_1 \left(\frac{234.5 + 80}{234.5 + 20}\right) = 1.235R_1 \qquad (2\text{-}10)$$

Thus there is a 23.5 per cent increase in resistance at the higher temperature. This increase in resistance occurs in many engineering devices such as motors, generators, and transformers when they operate at rated output power.

Energy Viewpoint. The resistance parameter may also be looked upon in terms of its characteristic property of converting electrical energy into heat energy. If Eqs. (2-5) and (2-6) are combined, it follows that the power absorbed by a resistor is

$$p = vi = (iR)i = i^2R \qquad (2\text{-}11)$$

This is called *Joule's law* and it states that the number of electrons colliding with the atoms of a circuit element to produce heat is proportional to the square of the current. The corresponding amount of energy converted to heat in the time interval $t_2 - t_1$ is readily found from

$$W = \int_{t_1}^{t_2} Ri^2\, dt \quad \text{joules} \qquad (2\text{-}12)$$

The integral form is applicable whenever the current i is a time-varying quantity. When i is a constant quantity I, Eq. (2-12) becomes

$$W = RI^2t \quad \text{joules} \qquad (2\text{-}13)$$

where

$$t = t_2 - t_1 \quad \text{sec} \qquad (2\text{-}14)$$

Accordingly, Eq. (2-13) indicates that resistance may also be expressed in terms of the dissipated energy per unit time per unit current squared. Thus

$$\boxed{R = \frac{W}{I^2t}} \quad \text{ohms} \qquad (2\text{-}15)$$

All current-carrying conductors and resistors involve a heat loss which occurs at a rate given by I^2R. At first this heat is stored in the body of the material, thereby bringing about a temperature rise. As soon as the body temperature exceeds the ambient temperature, however, heat is transferred to the surrounding medium. Eventually a point is reached where the heat is transferred at the same rate that it is produced. Consequently heat ceases to be stored in the material, and temperature no longer rises. The ability of a resistor to transfer heat necessarily depends on the area exposed to the surrounding medium. If the power-dissipating capability of a resistor or rheostat is inadequate for a specified rating it can easily burn out. To prevent such occurrences resistors and rheostats are given a specified safe power rating which is determined by test. Thus one may purchase a 2-watt, 100-ohm resistor or a 10-watt, 100-ohm resistor. Although the resistance values are the same, the physical dimensions of the 10-watt resistor are much larger.

EXAMPLE 2-1   A 100-ohm resistor is needed in an electric circuit to carry a current of 0.3 ampere. The following resistors are available from stock:

> 100 ohms,   5 watts
>
> 100 ohms, 7.5 watts
>
> 100 ohms, 10 watts

Which resistor would you specify?

solution: The power associated with a current of 0.3 ampere and 100 ohms is

$$P = I^2R = (0.3)^2(100) = 9 \quad \text{watts}$$

Hence select the 10-watt resistor to guard against overheating and possible damage.

Geometrical Viewpoint. The resistance parameter is fundamentally a geometric constant. This fact was discovered by Ohm in his original investigations. In his analogy to Fourier's heat-conduction equation Ohm showed that the resistance of a conductor of uniform dimensions was directly dependent upon the length, inversely proportional to the cross-sectional area, and also dependent upon the physical conduction properties of the material. This information appears in Eq. (1-23) and is repeated here for convenience. Thus

$$\boxed{R = \rho\,\frac{L}{A}} \quad \text{ohms} \tag{2-16}$$

where ρ denotes the resistivity of the material expressed in ohm-meters

L is the length of the conductor in meters

A is the cross-sectional area in meters squared

Equation (2-16) is easily verified experimentally. For example, it is found that for a material of fixed resistivity doubling the length doubles the resistance while doubling the area halves the resistance.

EXAMPLE 2-2   Find the resistance of a round copper conductor having a length of one meter and a uniform cross-sectional area of one square centimeter. The resistivity of copper is 1.724×10^{-8} ohm-meter.

solution: From Eq. (2-16) there results

$$R = \rho\,\frac{L}{A} = 1.724 \times 10^{-8}\,\frac{1\text{ m}}{10^{-4}\text{ m}^2} = 1.724 \times 10^{-4} \quad \text{ohm}$$

Often the resistivity of copper is expressed in units of microhm-cm in order to avoid the 10^{-8} factor. It is so expressed in Table C-1 of Appendix C. In this table note that the resistivity is also expressed in units of ohm-circular mils per foot. Reference to Eq. (2-16) reveals that this description represents the resistance of a piece of material which has a cross-sectional area of one circular mil and a length of one foot. The *circular mil* is a convenient unit which is used to describe the cross-sectional area of round conductors. By definition, a circular mil is a unit of area and specifically it denotes the area of a circle having a diameter of 0.001 inch or 1 mil. What then is the area of a round wire having a diameter of 0.10 inch expressed in units of circular mils? As is customary in dealing with measurements, we must find the number of times the circle of 1-mil diameter fits into the circle of 100-mil diameter. Since the area of a circle is proportional to the square of the diameter, it follows that the answer is $(100)^2 = 10,000$ circular mils. This clearly leads to the following simple rule: to find the area in circular mils multiply the diameter expressed in inches by 1000 and square the resulting number.

EXAMPLE 2-3   Find the resistance of a round copper wire having a diameter of 0.1 inch and a length of 10 feet.

solution:

$$R = \rho \frac{L}{A} = 10.37 \times \frac{10}{10,000} = 0.01037 \quad \text{ohm}$$

Commercial wire sizes are standardized in the United States in accordance with the area expressed in circular mils. Each wire size is given an AWG (American Wire Gage) number for identification and each has a maximum allowable current rating which varies with the type of insulation usde. Table C-2 in Appendix C lists this information.

Conductance is defined as the reciprocal of resistance. The symbol for conductance is G. Thus

$$G \equiv \frac{1}{R} \tag{2-17}$$

Hence

$$G = \frac{1}{\rho} \frac{A}{L} = \sigma \frac{A}{L} \tag{2-18}$$

The quantity σ denotes the conductivity and is the reciporcal of resistivity.

Ohm's law may be expressed in terms of the conductance of a circuit element. From Eqs. (2-6) and (2-17) we can write

$$i = \frac{v}{R} = Gv \quad \text{amps} \tag{2-19}$$

Thus by multiplying the potential difference by the conductance we obtain the corresponding current. Sometimes this is a preferred procedure in circuit analysis, as is demonstrated later in the book.

2-3   The Inductance Parameter

As described in Sec. 1-5 inductance was first discovered by Faraday in his renowned experiments of 1831. In a general way *inductance* can be characterized as that property of a circuit element by which energy is capable of being stored in a magnetic flux field. A significant and distinguishing feature of inductance, however, is that it makes itself felt in a circuit only when there is a *changing current*. Thus, although a circuit element may have inductance by virtue of its geometrical and magnetic properties, its presence in the circuit is not exhibited unless there is a time rate of change of current. This aspect of inductance is particularly stressed when we consider it from the circuit viewpoint.

Circuit Viewpoint. The current-voltage relationship involving the inductance parameter is expressed by Eq. (1-28) and is repeated here for convenience. Thus

$$v_L = L\frac{di}{dt} \tag{3-20}$$

In general both v_L and i are functions of time. Figure 2-4 depicts the potential difference v_L appearing across the terminals of the inductance parameter when

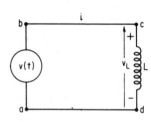

Fig. 2-4. Illustrating the inductance parameter.

a changing current flows into terminal c. Note that the arrowhead on the v_L quantity is shown at terminal c, indicating that this terminal is instantaneously positive with respect to terminal d. In turn this means that current is increasing in the positive sense, i.e., the slope di/dt in Eq. (2-20) is positive. Any circuit element that exhibits the property of inductance is called an *inductor* and is denoted by the symbolism shown in Fig. 2-4. In the ideal sense the inductor is considered to be resistanceless, although practically it must contain the wire resistance out of which the inductor coil is formed.

It follows from Eq. (2-20) that an appropriate defining equation for inductance is

$$\boxed{L = \frac{v_L}{(di/dt)}} \quad \frac{\text{volt-secs}}{\text{amp}} \text{ or henrys} \tag{2-21}$$

Thus by recording the potential difference at a given time instant across the terminals of an inductor and dividing by the corresponding derivative of the current time function we determine the inductance parameter. Note that the units of inductance are volt-second/ampere. For simplicity this is more commonly called the *henry*.

A *linear inductor* is one for which the inductance parameter is independent of current. As current flows through an inductor it creates a space flux. When this flux permeates air, strict proportionality between current and flux prevails so that the inductance parameter stays constant for all values of current. A plot of the potential difference across the coil as a function of the derivative of the current then appears as shown in Fig. 2-5, which is a plot of Eq. (2-21). Note the similarity to Fig. 2-2, which applies for the resistance parameter. Of course the abscissa is different in each case. When the flux is made to penetrate iron, however, it is possible for large currents to upset the proportional relationship between the current and the flux it produces. In such a case the inductor is called *nonlinear* and a plot of Eq. (2-21) will then no longer be a straight line.

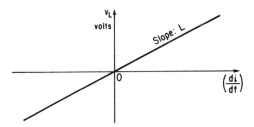

Fig. 2-5. Graphical representation of the inductance parameter L from the circuit viewpoint.

For the resistance parameter Ohm's law can be written to express either voltage in terms of current [see Eq. (1-24)] or current in terms of voltage [see Eq. (1-23)]. The same procedure may be followed for the inductance parameter. Equation (2-20) already expresses the voltage as a function of the current. However, to express the current in terms of the potential difference across the inductor, Eq. (2-20) must be transposed to read as follows:

$$di = \frac{1}{L} v_L \, dt \qquad (2\text{-}22)$$

In integral form this becomes

$$\int_{i(0)}^{i(t)} di = \frac{1}{L} \int_0^t v_L \, dt \qquad (2\text{-}23)$$

or

$$\boxed{i(t) = \frac{1}{L} \int_0^t v_L \, dt + i(0)} \qquad (2\text{-}24)$$

Equation (2-24) thus reveals that the current in an inductor is dependent upon the integral of the voltage across its terminals as well as the initial current in the coil at the start of integration.

An examination of Eqs. (2-20) and (2-24) reveals an important property of inductance: *the current in an inductor cannot change abruptly in zero time.* This is made apparent from Eq. (2-20) by noting that a finite change in current in zero time calls for an infinite voltage to appear across the inductor, which is physically impossible. On the other hand, Eq. (2-24) shows that in zero time the contribution

to the inductor current from the integral term is zero so that the current immediately before and after application of voltage to the inductor is the same. In this
sense, then, we may look upon inductance as exhibiting the property of inertia.

EXAMPLE 2-4   An inductor has a current passing through it which varies in time in
the manner depicted in Fig. 2-6(a). Find the corresponding time variation of the voltage
drop appearing across the inductor terminals, if it is assumed that the inductance of the
coil is 0.1 henry.

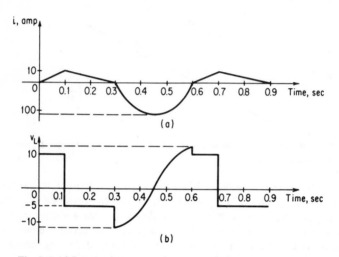

Fig. 2-6. (a) Input current wave shape to an inductor;
(b) corresponding voltage variation across inductor terminal.

solution: The solution appears in Fig. 2-6(b). Note that in the interval from 0 to 0.1
second, $di/dt = 100$ amp/sec. Hence the voltage across the coil is then a constant given by

$$v_L = L \frac{di}{dt} = (0.1)(100) = 10 \text{ volts} \quad \text{for } 0 < t < 0.1$$

In the time range from 0.1 to 0.3 seconds the slope of the current curve is -50. Hence
the voltage across the coil is -5 volts. Finally, in the interval from 0.3 to 0.6 seconds the
current wave slope is sinusoidal so that the corresponding voltage wave is a cosine. In
drawing the cosine wave it is assumed that the maximum slope of the sine wave exceeds
100 amp/sec.

It is interesting to note in Fig. 2-6 that unlike the current in an inductor, the voltage
across the inductor is allowed to change discontinuously.

Energy Viewpoint. Assume that an inductor has zero initial current. Then
if a current i is made to flow through the coil across which appears the potential
difference v_L, the total energy received in the time interval from 0 to t is

$$W = \int_0^t v_L i \, dt \quad \text{joules} \tag{2-25}$$

Inserting Eq. (2-20) leads to

$$W = \int_0^t \left(L \frac{di}{dt} \right) i \, dt = \int_0^i Li \, di \qquad (2\text{-}26)$$

or

$$\boxed{W = \frac{1}{2} Li^2} \quad \text{joules} \qquad (2\text{-}27)$$

Continuing with the assumption that the inductor has no winding resistance, Eq. (2-27) states that the inductor absorbs an amount of energy which is proportional to the inductance parameter L as well as the square of the instantaneous value of the current. Thus energy is stored by the inductor in a magnetic field. It is of finite value and retrievable. As the current is increased, so too is the stored magnetic energy. Note, however, that this energy is zero whenever the current is zero. Because the energy associated with the inductance parameter increases and decreases with the current, we can properly conclude that the inductor has the property of being capable of returning energy to the source from which it receives it.

A glance at Eq. (2-27) indicates that an alternative way of identifying the inductance parameter is in terms of the amount of energy stored in its magnetic field corresponding to its instantaneous current. Thus, in mathematical form we can write

$$\boxed{L = \frac{2W}{i^2}} \quad \text{henrys} \qquad (2\text{-}28)$$

This is an energy description of the inductance parameter.

There is one final point worthy of note. It has already been demonstrated that for a potential difference to exist across the inductor terminals the current must be changing. A constant current results in a zero voltage drop across the ideal inductor. This is not true, however, about the energy absorbed and stored in the magnetic field of the inductor. Equation (2-27) readily verifies this fact. A constant current results in a fixed energy storage. Any attempt to alter this energy state is firmly resisted by the effects of the initial energy storage. This again reflects the inertial aspect of inductance.

Geometrical Viewpoint. The voltage drop across the terminals of an inductor may be expressed from a circuit viewpoint by Eq. (2-20). However, this same voltage drop may be described by Faraday's law in terms of the flux produced by the current and the number of turns N of the inductor coil. Accordingly, we may write

$$v_L = L \frac{di}{dt} = N \frac{d\phi}{dt} \qquad (2\text{-}29)$$

It then follows that

$$L = N \frac{d\phi}{di} \qquad (2\text{-}30)$$

In those cases where the flux ϕ is directly proportional to current i for all values (i.e. for linear inductors), the last expression becomes

$$L = \frac{N\phi}{i} \quad \frac{\text{weber-turns}}{\text{amp}}$$

(2-31)

Here the inductance parameter has a hybrid representation because it is in part expressed in terms of the circuit variable i and in part in terms of the field variable ϕ. To avoid this we replace flux by its equivalent,[†] namely

$$\phi = \frac{mmf}{\text{magnetic reluctance}} = \frac{Ni}{\mathscr{R}}$$

(2-32)

where *mmf* denotes the magnetomotive force which produces the flux ϕ in the magnetic circuit having a reluctance $\mathscr{R}$. Appearing in Fig. 2-7 is an inductor consisting of N turns wound about a circular iron core. If the core is assumed to have a mean length of l meters and a cross-sectional area of A_m meters^2, then the magnetic reluctance can be shown to be

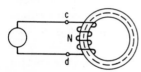

Fig. 2-7. Linear inductor with iron core.

$$\mathscr{R} = \frac{l}{\mu A_m}$$

(2-33)

where μ is a physical property of the magnetic material and is called the permeability. Note the similarity of Eq. (2-33) to Eq. (2-16).

Upon substituting Eqs. (2-32) and (2-33) into (2-31) there results the expression for the inductance parameter of the circuit of Fig. 2-7. Thus

$$L = \frac{N^2 \mu A_m}{l}$$

(2-34)

A study of Eq. (2-34) reveals some interesting and useful facts about the inductance parameter which are not readily available when this quantity is defined either from the circuit or the energy viewpoint. Most impressive is the fact that inductance, like resistance, is dependent upon the geometry of physical dimensions and the magnetic property of the medium. This is significant because it tells us what can be done to change the value of L. Thus for the inductor illustrated in Fig. 2-7, the inductance parameter may be increased in any one of four ways: increasing the number of turns, using an iron core of higher permeability, reducing the length of the iron core, and increasing the cross-sectional area of the iron core.

It is interesting to note that neither the circuit viewpoint nor the energy viewpoint could tell us these things, because essentially they deal with the effects associated with a given inductor geometry. It is emphasized that all three viewpoints are needed to complete the picture of circuit parameters and to give proper perspective.

† See Eq. (12-21) of Sec. 12-4.

2-4 The Capacitance Parameter

Attention is next directed to capacitance, the third basic parameter of electric circuit theory. In a general way capacitance can be characterized as that property of a circuit element in which energy is capable of being stored in an electric field. A significant and distinguishing feature of capacitance is that its influence in an electric circuit is manifested only when there exists a *changing potential difference* across the terminals of the circuit element. This aspect of capacitance is easily apparent when treated from a circuit viewpoint.

Circuit Viewpoint. In Sec. 1-3 capacitance is introduced as the proportionality factor relating the charge between two metal surfaces (or conductors) to the corresponding potential difference existing between them. From Eq. (1-16), therefore, we have

$$q = Cv_c \tag{2-35}$$

where q represents the charge and v_c denotes the potential difference. Lower-case letters are used here to stress the instantaneous nature of the quantities. To obtain a definition of capacitance from a circuit viewpoint it is necessary to introduce current into the formulation of Eq. (2-35). This is readily accomplished by substituting Eq. (2-35) into the general expression for current as given by Eq. (1-2). Thus

$$\boxed{i = \frac{dq}{dt} = C\frac{dv_c}{dt}} \quad \text{amps} \tag{2-36}$$

This expression shows the manner in which the current flowing through a capacitance parameter is related to the potential difference appearing across it. Any circuit element showing the property of yielding a current which is directly proportional to the rate of change of the voltage across its terminals is called a *capacitor*. A capacitor usually consists of large metal surfaces separated by small distances.

With the establishment of the current-voltage relationship of Eq. (2-36) the definition of capacitance from a circuit viewpoint readily follows. Thus

$$\boxed{C = \frac{i}{(dv_c/dt)}} \quad \text{farads} \tag{2-37}$$

Moreover, from the terms appearing on the right side of Eq. (2-37) it is seen that the unit of capacitance is amp-sec/volt or coulomb/volt. However, for convenience this quantity is defined as the *farad*. Hence the unit of capacitance is the farad. In accordance with Eq. (2-37) the capacitance of a circuit element may be found by taking at any instant the value of the current passing through it and dividing by the corresponding value of the derivative of the voltage appearing across its terminals. Note the similarity in form which this expression for capacitance bears to the corresponding expression for inductance. Refer to Eq. (2-21). The difference lies in the interchange of roles between current and voltage.

Equation (2-36) expresses the capacitor current in terms of the capacitor voltage. It is often necessary to express the capacitor voltage as a function of its current. To do this Eq. (2-36) must be transposed and integrated in the fashion indicated below. Thus

$$dv_c = \frac{1}{C} i \, dt \tag{2-38}$$

Integrating both sides we have

$$\int_{v_c(0)}^{v_c(t)} dv_c = \frac{1}{C} \int_0^t i \, dt \tag{2-39}$$

or

$$v_c(t) = \frac{1}{C} \int_0^t i \, dt + v_c(0) \tag{2-40}$$

The quantity $v_c(0)$ denotes the initial voltage appearing across the capacitor plates upon start of the integration process. When there is no initial voltage on the capacitor, Eq. (2-40) becomes more simply

$$\boxed{v_c = \frac{1}{C} \int_0^t i \, dt} \quad \text{volts} \tag{2-41}$$

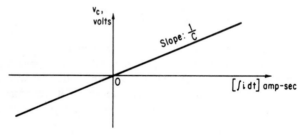

Fig. 2-8. Graphical representation of the capacitance parameter from the circuit viewpoint.

Figure 2-8 shows a graphical representation of this equation. Compare this with Figs. 2-2 and 2-5 and note the similarity in shapes and the difference in the abscissa quantities. Depicted in Fig. 2-9 is the circuit diagram for the capacitor circuit element showing the symbol used to denote the capacitor. As current enters the capacitor, Eq. (2-41) reveals that the potential difference increases, with the upper plate becoming more positive than the lower plate. Electrons are being transferred from the upper plate to the lower plate by the external circuit. To indicate this, the arrowhead on v_c is pointed upward. Of course, as i varies and becomes negative for a long enough period of time, it is entirely possible that the net result of integrating the current as called for in Eq. (2-41) will be a negative v_c.

A study of Eqs. (2-36) and (2-41) brings into evidence an important property of capacitance: *the voltage across a capacitor cannot change discontinuously.* Equation (2-36) shows that an abrupt change in capacitor voltage is not admissible because a finite change in v_c in zero time gives a value of infinity for dv_c/dt. Thus the capacitor current is infinite—a physical impossibility. On the other hand, Eq. (2-41) points out that for any finite change in capacitor current, however large, the integral contribution in

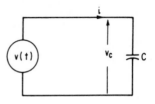

Fig 2-9. The capacitor circuit.

zero time must necessarily be zero. Hence the capacitor voltage cannot change instantaneously. It is interesting to note here that a step change in current through the capacitor is allowable.

EXAMPLE 2-5   Assume that the current waveshape of Fig. 2-6(a) is made to flow through the capacitor of Fig. 2-9. For a capacitance of 0.01 farad, find the capacitor voltage waveshape as a function of time.

solution: The solution readily follows from an application of Eq. (2-41). In the interval from 0 to 0.1 second the current has a variation given by $i = 100t$. Hence the corresponding voltage variation is

$$v_c = \frac{1}{C} \int_0^t 100t\, dt \quad 0 < t < 0.1$$
$$= 100 \times 50t^2 = 5000t^2 \tag{2-42}$$

Thus the voltage wave in this interval is parabolic and at $t = 0.1$ second has a value of 50 volts. The capacitor voltage can also be obtained by finding the area of triangle $0ab$ and multiplying by $1/C$. By either method the area under the current wave is obtained in the given time interval. This latter procedure was followed to find the key points shown in Fig. 2-10. Note that at time t' the sum of the positive and negative areas of the current wave is zero. Hence v_c is likewise zero. Then a short time later the capacitor voltage becomes negative. Note, too, that in the interval between 0.3 second and t' the capacitor voltage is positive in spite of the fact that the current is negative.

Energy Viewpoint. Assume that the capacitor of Fig. 2-8 has zero initial voltage across it and that a current i is allowed to flow in the circuit for a time interval t. The energy delivered to the capacitor during this time is given by

$$W = \int_0^t v_c i\, dt \tag{2-43}$$

Inserting Eq. (2-36) into the last expression yields

$$W = \int_0^t v_c \left(C \frac{dv_c}{dt} \right) dt = \int_0^v Cv_c\, dv_c \tag{2-44}$$

or

$$\boxed{W = \tfrac{1}{2} Cv_c^2} \quad \text{joules} \tag{2-45}$$

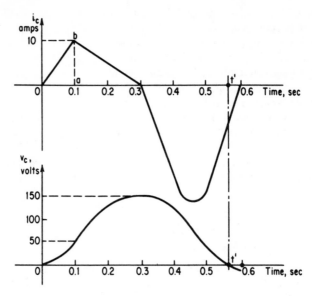

Fig. 2-10. Solution to Example 2-5.

Equation (2-45) states that the capacitor absorbs an amount of energy which is proportional to the capacitance parameter and the square of the instantaneous value of the voltage appearing across the capacitor. The absorbed energy in turn is stored by the capacitor in an electric field existing between its two plates. Note that as the capacitor voltage increases the energy increases and as v_c decreases in magnitude the associated energy decreases. Again it is reasonable to conclude therefore that like the inductor, the capacitor has the capacity of interchanging energy with the source. This stands in sharp contrast with the resistor, which can only dissipate energy in the form of heat.

With the derivation of Eq. (2-45) it is now possible to define capacitance from an energy viewpoint. Accordingly, we can write

$$\boxed{C = \frac{2W}{v_c^2}}\quad \text{farads} \tag{2-46}$$

Therefore capacitance may be identified in terms of the instantaneous value of the energy stored in its electric field and the corresponding value of the potential difference appearing across its terminals.

When the voltage across a capacitor is constant, there can be no current flow. This is required by Eq. (2-36). However, this does not mean that energy cannot be stored. In fact Eq. (2-41) states that for constant capacitor voltage there exists a finite and constant energy stored in its electric field. This situation is analogous to that which prevails in the inductor.

Geometrical Viewpoint. The amount of charge which accumulates on the plates of a capacitor is commonly expressed by Eq. (2-35) as $q = Cv_c$. However, by means of Gauss's flux theorem it is also possible to express this accumulated charge in terms of the electric field quantity $\mathscr{E}$. This description appears in Eq. (1-20) and is repeated here for convenience. Thus

$$q = \epsilon A \mathscr{E} \tag{2-47}$$

Recall that ϵ denotes the permittivity (or specific dielectric constant) of the material between the plates of the capacitor, and A represents the area of the plates. For the sake of illustration let us consider the capacitor to have a definite configuration, namely that of two flat, parallel plates separated by a distance of d meters. In such a case the electric field intensity, which has units of volts per meter, is simply described as

$$\mathscr{E} = \frac{v_c}{d} \quad \text{volts/meter} \tag{2-48}$$

Upon insertion of Eq. (2-48) into Eq. (2-47) the latter becomes

$$q = \epsilon A \frac{v_c}{d} \tag{2-49}$$

Note that now we have an expression for charge which involves the physical dimensions of the capacitor as well as the capacitor voltage.

By equating the two alternative forms, namely Eqs. (2-35) and (2-49), for the charge appearing on the plates of a capacitor for a voltage v_c we have

$$Cv_c = \epsilon \frac{A}{d} v_c \tag{2-50}$$

or

$$\boxed{C = \epsilon \frac{A}{d}} \quad \text{farads} \tag{2-51}$$

Equation (2-51) thus makes available a definition of capacitance expressed in terms of its geometrical configuration and the physical property of the material lying between its two metal surfaces. Although the foregoing expression is applicable to the specific arrangement of a parallel-plate capacitor, the conclusions relating to the factors upon which capacitance depends are general. Accordingly, it can be said that the capacitance parameter is as a rule directly proportional to the dielectric constant of the material, inversely proportional to the spacing between the metal surfaces, and directly proportional to the metal surface area.

It is interesting to note the striking resemblance which the geometrical descriptions of the resistance, inductance, and capacitance parameters of electric circuit theory bear to one another. Refer to Eqs. (2-16), (2-34), and (2-51). In each instance there is involved the physical property of the material. These are resistivity ρ for resistance, permeability μ for inductance, and permittivity ϵ for capacitance. Moreover, in each case there appears the ratio of length to area.

PROBLEMS

2-1   An electric heater draws 1000 watts from a 250-volt source. What power does it take from a 208-volt source? What is the value of resistance of the heater?

2-2   An electric broiler draws 12 amperes at 115 volts for a period of 3 hours.
(a) If electrical energy costs 5 cents per kilowatt-hour, determine the cost of operating the broiler.
(b) Find the quantity of electricity in coulombs which passes through the broiler.
(c) How many electrons are involved?
(d) What is the rate at which electrical energy is expended?

2-3   Determine the value of the resistance of the electric broiler of Prob. 2-2 in terms of the energy viewpoint. Check this value by finding the resistance also in terms of the circuit viewpoint.

2-4   A resistor is known to dissipate 27.8 kw-hr of energy in 30 minutes while drawing a current of 10 amperes.
(a) Obtain the value of resistance from an energy approach.
(b) Determine the resistance by proceeding in terms of a circuit approach.

2-5   A 10-ohm resistor has a voltage rating of 120 volts. What is its power rating?

2-6   A coil of standard copper wire having a resistance of 12 ohms at 25°C is imbedded in the core of a large transformer. After the transformer has been in operation several hours, the resistance of the coil is found to be 13.4 ohms. What is the temperature of the transformer core?

2-7   Find the resistance of a bus bar 20 feet long, $\frac{1}{3}$ inch by 3 inches in cross section, and composed of standard annealed copper. Assume temperature is 20°C.

3-8   Find the resistivity expressed in ohm-circular mils per foot of a material that has a resistivity of 4 microhm-cm.

2-9   What must be the length of No. 14 AWG round copper wire so that the resistance will be 0.2 ohm?

2-10   A 400-foot length of round copper wire has a resistance of 0.4 ohm at 20°C. What is the diameter of the wire in inches?

2-11   Find the inductance of a coil in which
(a) a current of 0.1 ampere yields an energy storage of 0.05 joule.
(b) a current increases linearly from zero to 0.1 ampere in 0.2 second producing a voltage of 5 volts.
(c) a current of 0.1 ampere increasing at the rate of 0.5 ampere per second represents a power flow of one-half watt.

2-12   The coil in the configuration of Fig. 2-7 is equipped with 100 turns. Moreover, the mean length of turn of the magnetic core is known to be 0.2 meter and the cross-sectional area is 0.01 square meter. The value of the permeability of the iron is 10^{-3}.
(a) Find the inductance of the coil.
(b) When a d-c voltage is applied to the inductor it is found to draw a current of 0.1 ampere. How much energy is stored in the magnetic field?

2-13 When a d-c voltage is applied to the coil of Prob. 2-12, the current is found to vary in accordance with $i = \frac{k}{10}(1 - \epsilon^{-5t})$ expressed in amperes. Moreover, it is found that after the lapse of 0.2 second, the voltage induced in the coil is 0.16 volt. Find the inductance of the coil.

2-14 Find the capacitance of a circuit element in which
(a) a voltage of 100 volts yields an energy storage in an electric field of 0.05 joule.
(b) a voltage increases linearly from zero to 100 volts in 0.2 second causing a current flow of 5 ma.
(c) two flat parallel plates are separated by a 0.1-mm layer of mica and have a total area of 0.113 square meter. Assume mica to have a relative permittivity of 10.

2-15 When a d-c voltage is applied to a capacitor, the voltage across its terminals is found to build up in accordance with $v_c = 150(1 - \epsilon^{-20t})$. After the lapse of 0.05 second, the current flow is equal to 1.14 ma.
(a) Find the value of the capacitance in microfarads.
(b) How much energy is stored in the electric field at this time?

2-16 Determine the current which flows through a 1-μf capacitor when
(a) the voltage increases linearly at the rate of 1000 volts/second and
(b) the energy storage remains fixed at 0.0005 joule.

2-17 A circuit element is placed in a black box. At $t = 0$ a switch is closed and the current flowing through the circuit element and the voltage across its terminals are recorded to have the waveshapes shown in Fig. P2-17. Identify the type of circuit element and its magnitude.

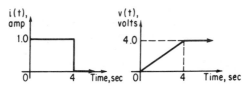

Figure P2-17

2-18 A voltage waveshape of the form $v(t) = K_1 t^n$ is applied to a de-energized circuit element at time $t = 0$. Determine the current through the element when the element is (a) a resistor, (b) an inductor, and (c) a capacitor.

2-19 Identify those characteristics which the circuit parameters resistance, inductance, and capacitance have in common when described in terms of their geometrical properties.

2-20 The voltage waveshape shown in Fig. P2-20 is applied to a coil having an inductance of $\frac{1}{2}$ henry and negligible resistance.
(a) Sketch the corresponding current waveshape as a function of time.
(b) Determine the instantaneous value of the current at the end of $t = 1$ second, $t = 2$ seconds, and $t = 4$ seconds.

2-21 A voltage wave having the time variation shown in Fig. P2-21 is applied to a pure inductor having a value of $2h$. Sketch carefully the variation of the current through the inductor over the first 10 seconds. Indicate all salient values in the current curve.

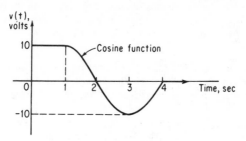

Figure P2-20

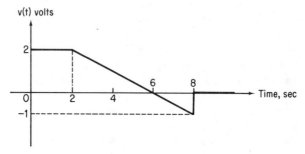

Figure P2-21

2-22   The voltage waveform shown in Fig. P2-22 is applied separately to a pure capacitor of 0.5 farad and to a pure inductor of 0.5 henry. Carefully sketch the current waveshape for each circuit element over the specified time interval.

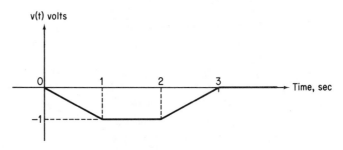

Figure P2-22

2-23   In the circuit of Fig. P2-23, if $v_{ab}(0) = 2$ volts and $i(t)$ has the variation shown, sketch the waveform $v_{ab}(t)$ on carefully labeled axes. Identify all salient points.

2-24   The voltage waveform shown in Fig. P2-24 is applied separately to a pure capacitor of 1 farad and to a pure inductor of 1 henry. Carefully sketch the current wave-shape for each circuit element over the specified time interval. (*Hint:* A graphical approach may be found easier than an analytical one.)

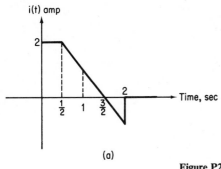

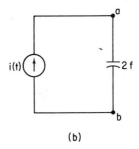

(a)   (b)

Figure P2-23

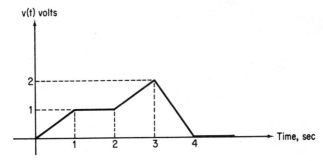

Figure P2-24

3

elementary network theory

The network theory treated in this chapter is used extensively in all branches of electrical engineering. It is customary, for example, in analyzing specific engineering devices (such as a transistor, or a vacuum tube, or a polyphase induction motor) to represent the device in terms of an appropriate equivalent circuit. Once this is accomplished, the performance of the device can easily be determined by applying the techniques of circuit theory. It is the purpose of this chapter to describe and develop these techniques.

The first three sections discuss the manner of handling the series and parallel combinations of resistances, capacitances, and inductances. In the remaining sections of the chapter, however, the various analytical techniques of network theory are developed in a framework of circuit elements consisting solely of resistors. This is done to achieve the maximum focusing of attention on the network theory itself by freeing the treatment of the clutter of complex numbers which must be involved† if inductors and capacitors are included. Such an approach in

† See Chapter 6.

no way causes a loss of generality. In Chapter 6 this same theory is shown to apply with equal validity for circuits involving all three circuit parameters in whatever combinations.

3-1 Series and Parallel Combinations of Resistances

Very often in circuit analysis it is necessary to deal with several elements in a closed-loop circuit which exhibit the property of dissipating heat. In the power-line circuits which supply electrical energy to homes and commercial establishments, for example, several resistive elements are frequently combined to carry the same current. Thus in a circuit which supplies electrical power to a lamp, three resistances are found: the internal resistance of the distribution transformer located beneath the street, the lamp resistance, and the resistance of the wires used to conduct the electrical power to the lamp. Similarly, radio and television circuits as well as industrial electronic circuitry employ series combinations of various resistors to achieve specific desirable objectives. To analyze such circuits properly we must know how to treat resistances in series.

By definition, circuit elements that carry the *same* current are said to be in *series*. It is not enough for the circuit elements to carry equal currents, for they can easily do this and yet be physically miles apart in entirely different circuits. The circuit parameters appearing in Fig. 3-1(a) are in series, for it is obvious here that the same current i flows through each circuit element.

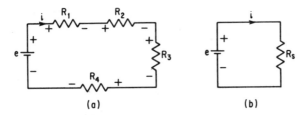

(a)   (b)

Fig. 3-1. Resistances in series:
(a) original configuration;
(b) equivalent circuit.

Applying Kirchhoff's voltage law, Eq. (1-34), to the circuit of Fig. 3-1(a), reveals a simple rule for handling resistances in series. Calling all voltage drops positive and voltage rises negative as the circuit is traversed in the assumed current-flow direction, we can write

$$iR_1 + iR_2 + iR_3 + iR_4 - e = 0 \qquad (3\text{-}1)$$

Rearranging yields

$$e = i(R_1 + R_2 + R_3 + R_4) \qquad (3\text{-}2)$$

The current i is factored out because it is common to each resistance. Consequently, the quantity in parentheses may be replaced by an equivalent resistance which is given by

$$R_s = R_1 + R_2 + R_3 + R_4 \tag{3-3}$$

Equation (3-2) may then be written simply as

$$e = iR_s \tag{3-4}$$

where R_s denotes the *equivalent series resistance* of the circuit. It follows, too, from this analysis that the original circuit configuration of Fig. 3-1(a) may be replaced by the equivalent circuit shown in Fig. 3-1(b), which is merely a circuit interpretation of Eq. (3-4).

In general, if there are n series-connected resistances in a circuit, *the equivalent series resistance is obtained by taking the sum of the individual resistances.* Expressed mathematically, we have

$$\boxed{R_s = R_1 + R_2 + \cdots + R_n = \sum_{j=1}^{n} R_j} \tag{3-5}$$

Circuit elements are also very frequently found in parallel combinations. In the home all electric light bulbs appear in parallel paths with respect to the source voltage. Other circuit elements such as the electric ironer and the electric broiler when used simultaneously are in parallel. By definition circuit elements are said to be in *parallel* when the *same* potential difference appears across their terminals. In accordance with this definition, the resistances R_1, R_2, and R_3 in Fig. 3-2(a) are in parallel.

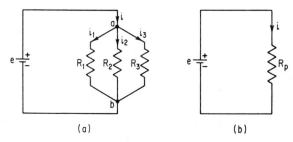

(a)   (b)

Fig. 3-2. Resistances in parallel:
(a) original configuration;
(b) equivalent circuit.

It is possible, again by means of circuit analysis, to treat this parallel combination of resistances in terms of an equivalent quantity. Kirchhoff's current law states that the current entering terminal b is equal to the sum of currents leaving this terminal. Expressed in equation form we have

$$i = i_1 + i_2 + i_3 \tag{3-6}$$

However, from Ohm's law as it relates to each resistance, Eq. (3-6) may be rewritten

$$i = \frac{e}{R_1} + \frac{e}{R_2} + \frac{e}{R_3} \tag{3-7}$$

Again by factoring out the common variable, which in this instance is the voltage e, there results

$$i = e\left(\frac{1}{R_1} + \frac{1}{R_2} + \frac{1}{R_3}\right) \tag{3-8}$$

The expression in parentheses may be replaced by an equivalent quantity defined as

$$\frac{1}{R_p} = \frac{1}{R_1} + \frac{1}{R_2} + \frac{1}{R_3} \tag{3-9}$$

where R_p denotes the equivalent resistance of the parallel combination of resistances. Upon substituting Eq. (3-9) into Eq. (3-8) we obtain a simplified equation for the circuit. Thus

$$i = \frac{e}{R_p} \tag{3-10}$$

Figure 3-2(b), which is the circuit representation of Eq. (3-10), may accordingly be considered as the equivalent circuit of the configuration of Fig. 3-2(a).

A general formulation of the foregoing procedure states that *the equivalent resistance of n parallel-connected resistances is the reciprocal of the sum of the reciprocals of the individual resistances.* Expressed in equation form this becomes

$$\boxed{\frac{1}{R_p} = \frac{1}{R_1} + \frac{1}{R_2} + \cdots + \frac{1}{R_n} = \sum_{j=1}^{n} \frac{1}{R_j}} \tag{3-11}$$

Equation (3-11) deals with the reciprocal of resistance. The unit for this quantity is *mhos*, a unit which has its origin in the fact that it represents inverse ohms. On the basis of the definition for conductance which was introduced in Chapter 2 [see Eq. (2-17)], Eq. (3-11) may also be expressed as

$$G_p = G_1 + G_2 \cdots G_n = \sum_{j=1}^{n} G_j \text{ mhos} \tag{3-11a}$$

Examples illustrating the use of Eqs. (3-5) and (3-11) to simplify circuit analysis are given in Sec. 3-4.

3-2 Series and Parallel Combinations of Capacitances

Series and parallel combinations of capacitances occur quite often in electronic circuitry—much more so than in the other fields of electrical engineering.

Appearing in Fig. 3-3(a) is a circuit involving two series-connected capacitors having capacitances C_1 and C_2 respectively. When a voltage is applied across the combination, the capacitors will have equal displacement of charge. But the potential difference across their terminals will be different provided that C_1 is not equal to C_2. However, by Kirchhoff's voltage law the sum of these two voltages must add up to the source voltage. Thus

$$e = v_1 + v_2 \tag{3-12}$$

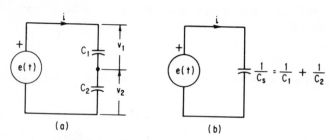

Fig. 3-3. Series-connected capacitances:
(a) original circuit;
(b) equivalent circuit.

But by Eq. (2-41) for zero initial capacitor voltage

$$v_1 = \frac{1}{C_1} \int_0^t i \, dt \quad \text{and} \quad v_2 = \frac{1}{C_2} \int_0^t i \, dt \tag{3-13}$$

Inserting Eqs. (3-13) into Eq. (3-12) yields

$$e = \frac{1}{C_1} \int_0^t i \, dt + \frac{1}{C_2} \int_0^t i \, dt \tag{3-14}$$

or

$$e = \left(\frac{1}{C_1} + \frac{1}{C_2} \right) \int_0^t i \, dt \tag{3-15}$$

This expression may now be simplified by introducing the substitution

$$\frac{1}{C_s} = \frac{1}{C_1} + \frac{1}{C_2} \tag{3-16}$$

where C_s denotes the equivalent capacitance of two series-connected capacitances. Accordingly, Eq. (3-15) becomes

$$e = \frac{1}{C_s} \int_0^t i \, dt \tag{3-17}$$

It follows then that Fig. 3-3(a) may be replaced by an equivalent circuit which is consistent with Eq. (3-17). This circuit is shown in Fig. 3-3(b).

A little thought indicates that a general formulation of the equivalent capacitance of n series-connected capacitances is

$$\boxed{\frac{1}{C_s} = \frac{1}{C_1} + \frac{1}{C_2} + \cdots + \frac{1}{C_n} = \sum_{j=1}^n \frac{1}{C_j}} \tag{3-18}$$

Alternatively,

$$\frac{1}{C_s} \equiv S_s = S_1 + S_2 + \cdots + S_n = \sum_{j=1}^n S_j \tag{3-19}$$

where S denotes stiffness expressed in *darafs*. Comparison of Eq. (3-18), which applies for *series-connected capacitances*, with Eq. (3-11), which applies for *parallel-connected resistances*, reveals the expressions to be identical in form. *Therefore, if*

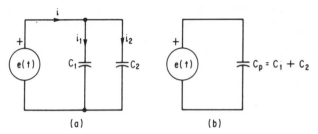

Fig. 3-4. Capacitances in parallel:
(a) original circuit;
(b) equivalent circuit.

the expression is in capacitances, series-connected capacitors are treated in the same manner as the resistances of parallel-connected resistors.

In the light of this conclusion it is reasonable to expect that capacitances in parallel may be treated in the same manner as resistances in series. The validity of this statement is readily established by applying Kirchhoff's current law to the circuit of Fig. 3-4(a). Therefore

$$i = i_1 + i_2 \tag{3-20}$$

But by Eq. (2-36)

$$i_1 = C_1 \frac{de}{dt} \quad \text{and} \quad i_2 = C_2 \frac{de}{dt}$$

Equation (3-19) then becomes

$$i = C_1 \frac{de}{dt} + C_2 \frac{de}{dt} \tag{3-21}$$

Collecting terms leads to .

$$i = (C_1 + C_2) \frac{de}{dt} \tag{3-22}$$

By introducing into Eq. (3-22) the expression

$$C_p = C_1 + C_2 \tag{3-23}$$

where C_p is the equivalent capacitance of the parallel combination, we obtain

$$i = C_p \frac{de}{et} \tag{3-24}$$

The circuit representation of this expression appears in Fig. 3-4(b), which is consequently the equivalent circuit of Fig. 3-4(a).

When n capacitances are connected in parallel it follows that the equivalent capacitance may be expressed as

$$\boxed{C_p = C_1 + C_2 + \cdots + C_n = \sum_{j=1}^{n} C_j} \tag{3-25}$$

Since Eq. (3-25) is identical in form to Eq. (3-5) we see that the conclusion stated at the outset is indeed valid.

It is interesting to observe in the foregoing analysis that out attention was focused solely on the manner in which the capacitance parameters combined for equivalence. We deliberately avoided discussing the influence of the time variation of current and voltage in the expressions of Eqs. (3-17) and (3-24). Consideration will be given to the effects of the integral of a time-varying current and the derivative of a time-varying voltage in Chapter 6, where the subject matter is more appropriate.

3-3   Series and Parallel Combinations of Inductances

Depicted in Fig. 3-5(a) is a circuit involving two inductors in series. The source voltage is assumed to be time-varying, causing a changing current to flow through the inductors. Then, by Kirchhoff's voltage law,

$$e(t) = L_1 \frac{di}{dt} + L_2 \frac{di}{dt} \tag{3-26}$$

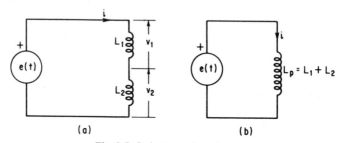

(a)   (b)

Fig. 3-5. Inductances in series:
(a) original circuit;
(b) equivalent circuit.

Factoring out the common term leads to

$$e(t) = (L_1 + L_2) \frac{di}{dt} \tag{3-27}$$

Letting

$$L_s = L_1 + L_2 \tag{3-28}$$

and substituting into Eq. (3-27) yields

$$e(t) = L_s \left(\frac{di}{dt} \right) \tag{3-29}$$

where L_s denotes the equivalent inductance for the series connection.

Figure 3-5(b) shows the equivalent circuit of Fig. 3-5(a) based on Eq. (3-29). Note that inductances in series are treated in the same fashion as resistances in series.

In a situation where there are n series-connected inductances in an electric circuit, the resultant equivalent inductance is the sum of the individual inductances. Thus

$$L_s = L_1 + L_2 + \cdots + L_n = \sum_{j=1}^{n} L_j \qquad (3\text{-}30)$$

A parallel combination of two inductances is shown in Fig. 3-6(a). By Kirchhoff's current law

$$i = i_1 + i_2 \qquad (3\text{-}31)$$

Then, in accordance with Eq. (2-24) and assuming zero initial current through each inductance

$$i_1 = \frac{1}{L_1} \int_0^t e(t)\,dt \qquad (3\text{-}32) \qquad \text{and} \qquad i_2 = \frac{1}{L_2} \int_0^t e(t)\,dt \qquad (3\text{-}33)$$

Inserting Eqs. (3-32) and (3-33) into Eq. (3-31) then yields

$$i = \left(\frac{1}{L_1} + \frac{1}{L_2}\right) \int_0^t e(t)\,dt \qquad (3\text{-}34)$$

or

$$i = \frac{1}{L_p} \int_0^t e(t)\,dt \qquad (3\text{-}35)$$

where

$$\frac{1}{L_p} = \frac{1}{L_1} + \frac{1}{L_2} \qquad (3\text{-}36)$$

The quantity L_p is the equivalent inductance of the parallel combination of the individual inductances and is depicted in Fig. 3-6(b).

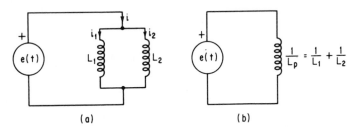

Fig. 3-6. Inductances in parallel:
(a) original circuit;
(b) equivalent circuit.

For the general case where there are n parallel-connected inductances, Eq. (3-36) becomes

$$\frac{1}{L_p} = \frac{1}{L_1} + \frac{1}{L_2} + \cdots + \frac{1}{L_n} = \sum_{j=1}^{n} \frac{1}{L_j} \qquad (3\text{-}37)$$

Equation (3-37) indicates that inductances in parallel are treated in the same way as resistors in parallel.

3-4   Series-Parallel Circuits

In many practical circuits in electrical engineering there occur situations
where a circuit element is in series with a parallel combination of other circuit
elements. Although these configurations may involve all three of the circuit
parameters, we shall, in the remainder of this chapter, confine our attention
exclusively to circuits involving only the resistance parameter. We follow this
procedure because it is simpler. Recall that where capacitance and inductance
are involved a complete description of the current-voltage relationships is dependent upon a specified variation of the voltage or current applied to each circuit
element. Reference to Eqs. (3-25) and (3-29) should make this apparent. The
presence of the derivative or integral in these expressions leads to an additional
complexity† which can only serve at this stage to obscure the development of the
theory as it applies to circuit analysis. There is no loss of generality in the application of the theory and principles herein described. A reading of Sec. 6-9 bears this
out. There it is seen that the very same theorems which are treated in this chapter
and illustrated solely in terms of resistive circuits are applied with equal validity
to circuit configurations involving all three circuit parameters.

The theoretical considerations which are required to solve series-parallel
combinations of the resistance parameter have already been studied. They are
embodied in Kirchhoff's current and voltage laws, Ohm's law, and Eqs. (3-5)
and (3-11) which are a consequence of these laws. The procedure for handling
series-parallel circuits is best illustrated by examples.

EXAMPLE 3-1   The circuit of Fig. 3-7(a) involves the combination of three parallel
resistors in series with the resistor R_1. For the resistance values shown find the value of
current flowing from the voltage source. All resistance values are expressed in ohms.

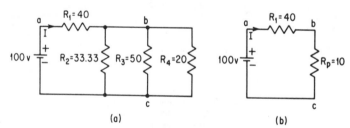

Fig. 3-7. Circuitry for Example 3-1:
(a) original configuration;
(b) equivalent circuit.

solution: As the first step in the procedure we find the equivalent resistance of the
parallel combination of R_2, R_3, and R_4. Thus, by Eq. (3-11),

$$\frac{1}{R_p} = \frac{1}{R_2} + \frac{1}{R_3} + \frac{1}{R_4} = \frac{1}{33.33} + \frac{1}{50} + \frac{1}{20} = \frac{3 + 2 + 5}{100} = \frac{1}{10} \qquad (3\text{-}38)$$

† By dealing with capacitance and inductance parameters at this point we would have to
work with complex numbers rather than real numbers alone. See Chapter 6.

Hence

$$R_p = 10 \text{ ohms}$$

The original circuit configuration can now be represented by the equivalent circuit of Fig. 3-7(b), which is merely a simple series circuit. Then, by Kirchhoff's voltage law

$$100 = IR_1 + IR_p = I(R_1 + R_p) \tag{3-39}$$

or

$$I = \frac{100}{R_1 + R_p} = \frac{100}{40 + 10} = 2 \text{ amp} \tag{3-40}$$

EXAMPLE 3-2   In the circuitry of Fig. 3-7 find the potential difference across terminals *bc*.

solution: Refer to Fig. 3-7(b). By Ohm's law the voltage drop across *bc* is

$$V_{bc} = IR_p = (2)(10) = 20 \text{ volts} \tag{3-41}$$

where *I* is given by Eq. (3-40).

A glance at Fig. 3-7(b) shows that the current *I* is common to both resistors. Accordingly it should be possible to obtain the solution without solving for the current specifically. This is readily demonstrated as follows. The voltage across *bc* can be generally written as

$$V_{bc} = IR_p \tag{3-42}$$

Furthermore, by Kirchhoff's voltage law

$$E = I(R_1 + R_p) \tag{3-43}$$

When we divide Eq. (3-42) by Eq. (3-43) the current term cancels, leaving as the expression for the voltage drop across *bc*

$$V_{bc} = \frac{R_p}{R_1 + R_p} E \tag{3-44}$$

Inserting the specified values of the parameters into Eq. (3-44) then gives

$$V_{bc} = \frac{10}{40 + 10}(100) = \frac{1000}{50} = 20 \text{ volts} \tag{3-46}$$

Equation (3-44) is used often in circuit analysis. It is worthwhile, therefore, to express it in more general terms as follows: In a circuit composed of *n* series-connected resistors and a source *E* the voltage drop V_k appearing across the terminals of resistor R_k is

$$\boxed{V_k = \frac{R_k}{R_1 + R_2 + \cdots + R_n} E} \tag{3-46}$$

Equation (3-46) is frequently referred to as the *voltage-divider theorem*.

EXAMPLE 3-3   Determine the equivalent series circuit of the arrangement of circuit elements appearing in Fig. 3-8(a).

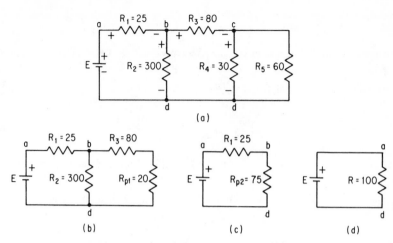

Fig. 3-8. Circuit configuration for Example 3-3:
(a) original;
(b) intermediate reduction;
(c and d) final form.

solution: To obtain the result we follow a process of *network† reduction* which basically
involves applying Eqs. (3-5) and (3-11) to series and parallel combinations of resistances
starting with the pair of terminals farthest removed from the source voltage. For the case
at hand terminals *cd* are the farthest removed. Since R_4 and R_5 are in parallel, Eq. (3-11)
yields

$$\frac{1}{R_{p1}} = \frac{1}{R_4} + \frac{1}{R_5} \tag{3-47}$$

or

$$R_{p1} = \frac{R_4 R_5}{R_4 + R_5} \tag{3-48}$$

Equation (3-48) is another useful expression to keep in mind because the combination
of two parallel resistances occurs often in circuit analysis. Substituting the values for R_4
and R_5 yields

$$R_{p1} = \frac{30(60)}{30 + 60} = 20 \text{ ohms} \tag{3-49}$$

By means of this computation Fig. 3-8(a) reduces to the form shown in Fig. 3-8(b). Note
there that R_3 and R_{p1} are in series, so that by Eq. (3-5) these may be combined into a single
resistance of value

$$R_s = R_3 + R_{p1} = 80 + 20 = 100 \text{ ohms} \tag{3-50}$$

A further reduction (or simplification) can be accomplished by again applying
Eq. (3-11) to the two parallel resistances, R_2 and R_s, which appear across terminals *bd*.
Thus

$$R_{p2} = \frac{R_2 R_s}{R_2 + R_s} = \frac{300(100)}{300 + 100} = 75 \text{ ohms} \tag{3-51}$$

† The word network is used synonymously with the word circuit.

Since R_{p2} is in series with R_1 it follows that Fig. 3-8(b) can now be represented by Fig. 3-8(c) or Fig. 3-8(d), which is the desired result.

EXAMPLE 3-4 In the circuit of Fig. 3-8(a) find the value of E which yields a power dissipation in R_5 of 15 watts.

solution: The solution is readily found by applying the laws of Ohm and Kirchhoff, but first it is necessary to find the current flowing through resistor R_5. Since power is energy per unit time, it follows from Eq. (3-15) that

$$I_5^2 R_5 = 15 \text{ watts} \tag{3-52}$$

where I_5 denotes the current flowing through R_5. Hence

$$I_5 = \sqrt{\frac{15}{R_5}} = \sqrt{\frac{15}{60}} = \frac{1}{2} \text{ amp} \tag{3-53}$$

Then by Ohm's law the voltage drop across terminals cd is

$$V_{cd} = I_5 R_5 = \tfrac{1}{2}(60) = 30 \text{ volts} \tag{3-54}$$

Because this same voltage appears across R_4, the current through R_4 is

$$I_4 = \frac{V_{cd}}{R_4} = \frac{30}{30} = 1 \text{ amp} \tag{3-55}$$

The current which flows through R_3 is the current entering junction c, which is equal to the two currents—I_4 and I_5—which leave junction c. Therefore

$$I_3 = I_4 + I_5 = 1 + \tfrac{1}{2} = \tfrac{3}{2} \text{ amps} \tag{3-56}$$

The corresponding potential difference associated with I_3 flowing through R_3 is then

$$V_{bc} = I_3 R_3 = \tfrac{3}{2}(80) = 120 \text{ volts} \tag{3-57}$$

To find the current which flows through R_2 we must first determine the voltage across it, V_{bc}. Applying Kirchhoff's voltage law to loop $dbcd$ in Fig. 3-8(a) we have

$$V_{bd} = V_{bc} + V_{cd} = 120 + 30 = 150 \text{ volts} \tag{3-58}$$

Hence

$$I_2 = \frac{V_{bd}}{R_2} = \frac{150}{300} = \frac{1}{2} \text{ amp} \tag{3-59}$$

Since the current flowing through R_1 and into junction b splits into two paths yielding currents I_2 and I_3, both of which are now determined, it follows that

$$I_1 = I_2 + I_3 = \tfrac{1}{2} + \tfrac{3}{2} = 2 \text{ amps} \tag{3-60}$$

The corresponding voltage drop caused by this current flowing through R_1 is therefore

$$V_{ab} = I_1 R_1 = 2(25) = 50 \text{ volts} \tag{3-61}$$

Finally, by applying Kirchhoff's voltage law to loop abd we obtain the desired solution.

$$E = V_{ab} + V_{bd} = 50 + 150 = 200 \text{ volts} \tag{3-62}$$

In recapitulation note that the solution of this problem was achieved by applying Ohm's law to get the voltage drop across an individual circuit element, Kirchhoff's current law at appropriate junction points such as b and c in Fig. 3-8(a) to obtain information about the current entering the junction, and also Kirchhoff's voltage law to appropriate loops in order to obtain the proper levels of potential differences.

EXAMPLE 3-5   Appearing in Fig. 3-9(a) is a series-parallel circuit which involves
two voltage sources. For the values of the parameters shown in the figure find the current
which flows through R_2, the 20-ohm resistor.

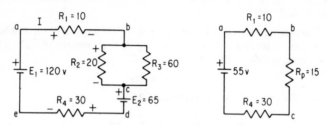

Fig. 3-9. Circuit for Example 3-5, to illustrate superposition:
(a) original configuration;
(b) equivalent circuit.

solution: First replace the parallel combination of resistances across *bc* by its equivalent value. Thus

$$R_p = \frac{R_2 R_3}{R_2 + R_3} = \frac{20(60)}{20 + 60} = 15 \text{ ohms} \tag{3-63}$$

Then apply Kirchhoff's voltage law to the loop. Assume that the loop current flows
clockwise. Calling voltage drops positive and voltage rises negative, the loop voltage
equation becomes

$$IR_1 + IR_p + E_2 + IR_4 - E_1 = 0 \tag{3-64}$$

Combining and rearranging terms leads to

$$I(R_1 + R_p + R_4) = E_1 - E_2 = 120 - 65 = 55 \tag{3-65}$$

The circuit representation of this last expression is shown in Fig. 3-9(b). The value of *I*
consequently is

$$I = \frac{55}{R_1 + R_p + R_4} = \frac{55}{10 + 15 + 30} = 1 \text{ amp} \tag{3-66}$$

To find the current through R_2 it is necesary first to compute the voltage across
terminals *bc*. Hence

$$V_{bc} = IR_p = 1(15) = 15 \text{ volts} \tag{3-67}$$

Then by Ohm's law the current through R_2 is

$$I_2 = \frac{V_{bc}}{R_2} = \frac{15}{20} = \frac{3}{4} \text{ amp} \tag{3-68}$$

which is the desired result.

3-5 The Superposition Theorem

This theorem has to do with the presence of two or more energy sources acting within a network. Accordingly, in the interest of establishing the proper
background, let us return temporarily to the circuit of Fig. 3-9(a) in which appear
two distinct voltage sources. By Eq. (3-64) the expression for the net current *I*

existing in this circuit can be written in terms of the effect which each voltage source produces. Thus

$$I = \frac{E_1}{R_1 + R_p + R_4} - \frac{E_2}{R_1 + R_p + R_4} \tag{3-69}$$

An examination of the first term on the right side of this equation reveals that $E_1/(R_1 + R_p + R_4)$ is precisely the current which flows in the circuit in the assumed direction of I when E_2 is removed and replaced with a short circuit between points c and d and E_1 is allowed to act alone. Similarly, the quantity $-E_2/(R_1 + R_p + R_4)$ is the current produced in the same circuit by the voltage source E_2 when E_1 is removed and replaced with a short circuit between points a and e. The minus sign indicates that the current produced by E_2 is in a direction *opposed* to that caused by E_1 acting alone. Furthermore, because the current produced by either voltage is *linearly* related to the voltage in accordance with Ohm's law, the two effects may be superposed to yield the net current I. Performing the operations called for individually in Eq. (3-69) we have as the net current

$$I = \frac{120}{55} - \frac{65}{55}$$
$$= 2.182 - 1.182 = 1.000 \text{ amp} \tag{3-70}$$

which is the same result as Eq. (3-66). Note that $I_1 = 2.182$ amp is the solution of the current in the circuit due to E_1 alone, and $I_2 = -1.182$ amp is the solution of the current (flowing opposite to the assumed positive direction of current flow for I) in the circuit due to E_2 acting alone.

Accordingly, on the basis of the foregoing discussion the *superposition theorem* may be stated as follows: If I_1 is the response to E_1 and I_2 is the response to E_2, then in any *linear* circuit the response to the combined forcing function $(E_1 + E_2)$ is $(I_1 + I_2)$. This theorem is not limited to circuit theory only. It is applicable in many fields. The sole requirement is that cause and effect bear a linear relationship to one another.

EXAMPLE 3-6   Find the power dissipated in the 10-ohm resistor R_1 in Fig. 3-9(a).

solution : The power is given by the equation

$$P_1 = I^2 R_1 = I^2(10) = 10 \text{ watts} \tag{3-71}$$

Let us investigate at this point whether this power can be computed by superposition. Proceeding for the moment on the assumption that such a procedure is valid, we write

$$\begin{aligned} P &= I_1^2 R_1 + (-I_2)^2 R_1 \\ &= (2.182)^2 10 + (-1.182)^2 10 \\ &= 47.6 + 13.95 \\ &= 61.55 \text{ watts} \end{aligned} \tag{3-72}$$

A comparison of the two results shows a vast inconsistency. In fact the result from superposition gives a power dissipation in R_1 alone which exceeds half the total power delivered by the E_1 source. By Eq. (3-5) this power is $E_1 I = 120(1) = 120$ watts. Clearly, then, superposition does not work in this case. A glance at Eq. (3-72) points out the reason. Note that the two variables involved, power and current, are *not linearly related* because current appears to the second power. Therefore superposition does not apply and so cannot be expected to give correct answers.

In the configuration of Fig. 3-9(a) there is little to be gained by using the principle of superposition to obtain the solution. The use of Kirchhoff's voltage law is indeed simpler. However, this is not true of the circuit arrangement of Fig. 3-10(a). This diagram is obtained from Fig. 3-9(a) by placing E_2 in the R_3 branch.

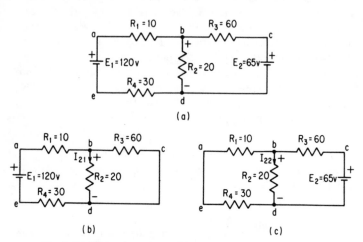

Fig. 3-10. Illustrating superposition (all resistances in ohms):
(a) original circuit;
(b) response due to E_1 alone;
(c) response due to E_2 alone.

EXAMPLE 3-7 By means of the superposition theorem find the current which flows through R_2 in the circuit of Fig. 3-10(a).

solution: Figure 3-10(b) depicts the circuit with E_1 acting alone. Note that this circuit is identical to Fig. 3-9(b) with the exception that the source voltage is 120 volts. The total current supplied by E_1 is therefore

$$I_1 = \frac{E_1}{55} = \frac{120}{55} = 2.182 \text{ amps} \tag{3-73}$$

The potential difference across R_2 due to I_1 is then equal to

$$V_{bd1} = I_1 R_p = 2.182(15) = 32.7 \text{ volts} \tag{3-74}$$

where the subscript 1 denotes the voltage across bd due to E_1. Hence the current which flows through R_2 as a result of E_1 is

$$I_{21} = \frac{32.7}{R_2} = \frac{32.7}{20} = 1.64 \text{ amps} \tag{3-75}$$

where the first subscript denotes that resistor R_2 is involved and the second subscript refers to source E_1.

Appearing in Fig. 3-10(c) is the circuit diagram which allows the effect of E_2 to be found. E_1 is removed. In arm *baed* the 10-ohm and the 30-ohm resistors may be combined into a 40-ohm resistor. Then between terminals *bd* there appears the parallel combination of the 40- and the 20-ohm resistors. Hence

$$R_p = \frac{40(20)}{40 + 20} = \frac{40}{3} \text{ ohms} \qquad (3\text{-}76)$$

With this quantity available the potential difference across R_2 caused by E_2 is readily obtained by using the voltage-divider theorem. Refer to Eq. (3-46). Thus

$$V_{bd2} = \frac{R_p}{R_p + R_3} E_2 = \frac{\frac{40}{3}}{60 + \frac{40}{3}} (65) = 11.8 \text{ volts} \qquad (3\text{-}77)$$

where the subscirpt 2 refers to source E_2. Note that the voltage drop across R_2 caused by E_2 is in the *same* direction as that caused by E_1. Therefore the current produced in R_2 by E_2 is

$$I_{22} = \frac{V_{bd2}}{R_2} = \frac{11.8}{20} = 0.59 \text{ amp} \qquad (3\text{-}78)$$

Accordingly, by superposition the total current flowing through R_2 in the configuration of Fig. 3-10(a) is

$$I_2 = I_{21} + I_{22} = 1.64 + 0.59 = 2.23 \text{ amps} \qquad (3\text{-}79)$$

The plus sign is used here because E_1 and E_2 both cause current to flow from terminal *b* to *d* in passing through R_2.

3-6 Network Analysis by Mesh Currents

Network reduction and superposition are only two of several methods which are available to the engineer for determining the response of a circuit to one or more energy sources acting within the circuit. In this section attention is focused on a third method which can be used to find the current flowing through any particular circuit element. However, before proceeding with the treatment, it is useful first to define a network terminology which makes it easier to discuss this method as well as the one described in the next section.

The word *network* is used synonymously with the term *circuit* and refers to any arrangement of passive and/or active circuit elements which form closed paths. Illustrated in Fig. 3-11 is a typical network. Note that this particular circuit is the same as Fig. 3-10(a) except that R_1 and R_4 are combined into a single 40-

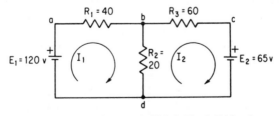

Fig. 3-11. A two-mesh network. This is Fig. 3-10(a) redrawn.

ohm resistor, which is permitted in accordance with Eq. (3-5). This network is composed of five circuit elements. There are three passive elements, namely resistances R_1, R_2, and R_3, and two active elements, namely the energy sources E_1 and E_2.

A *node* of a network is an equipotential surface at which two or more circuit elements are joined. Thus in Fig. 3-11 terminals *a*, *b*, *c*, and *d* are nodes.

A *junction* is that point in a network where three or more circuit elements are joined. In the network of Fig. 3-11 there are two junction points—*b* and *d*. The junction is sometimes referred to as an *independent node*.

A *branch* is that part of a network which lies between junction points. Accordingly, in Fig. 3-11 we see that there are a total of three branches. The branch *dab* consists of the two circuit elements E_1 and R_1. The second branch involves R_3 and E_2, and finally the third branch is merely R_2.

A *loop* is any closed path of the network. Examples of loops in Fig. 3-11 are *abda*, *abcd*, and *abcda*.

A *mesh* is the most elementary form of a loop. It is a property of a planar† network diagram and must be so identified that it cannot be further divided into other loops. In the circuit of Fig. 3-11 both loops *abda* and *abcd* qualify as meshes, but *abcda* cannot because it encloses the first two loops.

Fig. 3-12. The tree of the network of Fig. 3-11.

A *tree* is defined as that part of a network composed of those branches between the junction points which can be drawn without forming a closed path. A tree of the network of Fig. 3-11 is depicted in Fig. 3-12. It consists of the two junction points *b* and *d* and the branch R_2 between these points. Note that if either of the two remaining branches were drawn, a closed path would result. Hence these branches are not permitted and so are not part of the tree. The proper identification of the tree of a network is useful because it facilitates the determination of the minimum number of independent equations needed to solve the circuit completely.

We shall consider here the *mesh method* of solving problems in circuitry because it offers some advantages over the methods discussed so far. To illustrate, let us describe the method and then apply it to an example. For convenience the procedure is explained in conjunction with the circuitry of Fig. 3-11. Although in the solution of circuit problems we are often concerned with finding the current which flows in each branch of the network, by the use of loop currents we can often obtain this information with much less effort. As a general rule those loops should be chosen which are simplest in form. In other words we should work with meshes. Often we can readily identify the meshes to be used, once the tree of the network is drawn. It then simply requires introducing the remaining branches. Each branch so introduced identifies an appropriate mesh. Thus in Fig. 3-12, by adding branch *bad* to the tree a mesh results. Similarly by introducing branch *bad* a second mesh results. It is important to note here that as each mesh is formed, at least one branch is included in the newly formed mesh which was not included

† Two-dimensional.

as part of a loop previously formed. Consequently, by this process we are assured of forming the correct number and composition of the meshes needed to solve the problem. For the network of Fig. 3-11 we have two independent meshes, each of which leads to an independent mesh equation.

To write the mesh equations we must first introduce the mesh currents. By definition a *mesh current* is that current which flows around the perimeter of a mesh; for convenience, all mesh currents are drawn in a clockwise direction as shown in Fig. 3-11. Mesh currents may or may not have a direct identification with branch currents. For example, in mesh 1 of Fig. 3-11 the branch current through R_1 is identical to the mesh current. However, the branch current through R_2 is clearly the difference between the two mesh currents, $I_1 - I_2$, because I_1 and I_2 have opposite assumed positive directions. Thus branch currents have a physical identity, and they may be measured. Mesh currents, on the other hand, are fictitious quantities which are introduced because they allow us to solve problems in terms of a minimum number of unknowns. Hence in Fig. 3-11, although there are three unknown branch currents, we can readily find these by determining the two unknown mesh currents I_1 and I_2.

At this point in our study it is appropriate to refer to an expression which relates the number of independent mesh equations to the number of branches and junction points of the network. It reads as follows:

$$\boxed{m = b - (j - 1)} \qquad (3\text{-}80)$$

where m denotes the number of independent mesh equations
$\quad b$ denotes the number of branches
$\quad j$ denotes the number of junction points.

The -1 is included because some point in the network is needed as a reference point from which voltage measurements can be made, and a junction point is chosen for this purpose. Applying Eq. (3-80) to the circuitry of Fig. 3-11 yields

$$m = 3 - (2 - 1) = 2$$

Accordingly, this network can be completely analyzed by solving two independent equations. Of course this is a fact we had already learned in connection with forming the appropriate meshes from the tree. Equation (3-80), however, presents the information in precise mathematical form.

The independent mesh equations referred to in the foregoing discussion are those which are obtained by applying Kirchhoff's voltage law to each independent mesh. By traversing mesh 1 in the clockwise direction and calling voltage drops positive and voltage rises negative we obtain

$$I_1 R_1 + I_1 R_2 - R_2 I_2 - E_1 = 0 \qquad (3\text{-}81)$$

Note that the quantity $(-R_2 I_2)$ appears because of the effect which mesh current I_2 has in mesh 1 through the mutual resistance R_2. Since I_2 flows in a direction opposite to I_1 through R_2, it produces a rise in voltage for a clockwise traversal in mesh 1.

Rearranging Eq. (3-81) leads to

$$I_1(R_1 + R_2) - I_2R_2 = E_1 \tag{3-81a}$$

With this formulation the coefficient of mesh current I_1 is called the *self-resistance* of mesh 1 and is denoted by R_{11} (read as R one-one). A glance at Fig. 3-11 makes it apparent that R_{11} is simply the sum of all resistances found in mesh 1. Moreover, the coefficient of mesh current I_2 is called the *mutual resistance* of the network because it is common to meshes 1 and 2. It is represented by the symbol R_{12}. In terms of this standard notation, Eq. (3-81a) can be rewritten as

$$R_{11}I_1 - R_{12}I_2 = E_1 \tag{3-82}$$

Equation (3-82) is thus one of the two independent equations needed for a complete determination of the currents existing in the network. This equation involves the two unknown quantities I_1 and I_2. All else is known.

The second required mesh equation is found by applying Kirchhoff's voltage law in a similar manner to mesh 2. Thus

$$I_2R_2 + I_2R_3 + E_2 - R_2I_1 = 0 \tag{3-83}$$

Collecting terms yields

$$-R_2I_1 + (R_2 + R_3)I_2 = -E_2 \tag{3-84}$$

Again note the presence of a term which involves I_1. This comes about because of the mutual resistance R_2. The coefficient of I_2 is called the self-resistance of mesh 2 and is denoted R_{22}. Hence Eq. (3-84) may be rewritten as

$$-R_{21}I_1 + R_{22}R_2 = -E_2 \tag{3-85}$$

This is the second of the two required independent equations. A comparison of Eq. (3-85) with Eq. (3-82) shows each equation to be truly independent because R_{11} and R_{22} involve two different parameters of the network. That is, R_{11} involves R_1 but not R_3, and R_{22} involves R_3 but not R_1.

EXAMPLE 3-8 In the network of Fig. 3-11 find the magnitude and direction of each branch current by using mesh analysis.

solution: Inserting the specified values for R_1, R_2, R_3, E_1, and E_2 into Eqs. (3-82) and (3-84) gives

$$60I_1 - 20I_2 = 120 \tag{3-86}$$

and

$$-20I_1 + 80I_2 = -65 \tag{3-87}$$

Through the use of determinants we obtain directly

$$I_1 = \frac{\begin{vmatrix} 120 & -20 \\ -65 & 80 \end{vmatrix}}{\begin{vmatrix} 60 & -20 \\ -20 & 80 \end{vmatrix}} = \frac{83}{44} = 1.885 \text{ amps} \tag{3-88}$$

and

$$I_2 = \frac{\begin{vmatrix} 60 & 120 \\ -20 & -65 \end{vmatrix}}{4400} = -\frac{15}{44} = -0.341 \text{ amp} \tag{3-89}$$

The minus sign for I_2 indicates that the true direction of flow for I_2 is counterclockwise.

The branch currents expressed in terms of the computed mesh currents are then as follows:

$$I_{bd} = \text{current in branch } bd$$
$$= I_1 - I_2 = 1.885 - (-.341) = 2.23 \text{ amps} \tag{3-90}$$

$$I_{ab} = \text{current in branch } dab$$
$$= I_1 = 1.885 \text{ amp} \tag{3-91}$$

$$I_{cb} = \text{current in branch } dcb$$
$$= -I_2 = 0.341 \text{ amp} \tag{3-92}$$

Note that I_{bd} above is the same as the current found through R_2 by using the method of superposition. Refer to Eq. (3-79). It is interesting to note, too, that with the mesh method of solution the currents in the remaining two branches are available immediately as the mesh currents themselves, whereas to find these same branch currents by superposition additional calculations are needed.

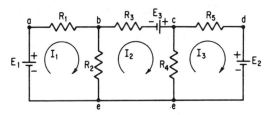

Fig. 3-13. A three-mesh network.

In the interest of gaining some additional experience with the mesh method of circuit analysis, let us apply the technique in a general way to the more complicated circuit of Fig. 3-13. An inspection of the network reveals that there are three junction points, b, c, and e. Furthermore, between these junction points there are a total of five branches. Therefore, by Eq. (3-80) it follows that there are three independent mesh equations.

$$m = b - (j - 1) = 5 - (3 - 1) = 3 \tag{3-93}$$

A tree of the network is depicted in Fig. 3-14. Although several other representations of the tree of this network are possible, the one shown in Fig. 3-14 is preferred because it leads directly to a convenient mesh formation. Thus by joining branch eab to be, mesh 1 is formed, and by introducing branch bc into place, mesh 2 is formed, and by inserting branch cde into place, mesh 3 is formed.

Although Kirchhoff's voltage law may be used in each mesh to obtain the three independent equations which permit a complete solution of the circuit, we shall do so instead by introducing self and mutual resistances. Thus, an examination of mesh 1 shows that the total

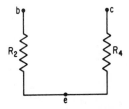

Fig. 3-14. A tree of the network of Fig. 3-13.

resistance in this mesh is

$$R_{11} = R_1 + R_2 \tag{3-94}$$

where R_{11} denotes the self resistance of mesh 1. It is total resistance encountered in traversing the perimeter of mesh 1. Furthermore, since R_2 is common to meshes 1 and 2

$$R_{12} = R_2 \tag{3-95}$$

where R_{12} denotes the mutual resistance between mesh 1 and mesh 2. Also, since there is no circuit element in mesh 1 which is common to mesh 3, it follows that the coupling term between meshes 1 and 3 is zero. Hence

$$R_{13} = 0 \tag{3-96}$$

Consequently, the voltage equation as it applies to mesh 1 is:

$$(R_1 + R_2)I_1 - R_2I_2 - 0 = E_1 \tag{3-97}$$

The rule for handling voltage sources that appear in a mesh is to use a plus sign for this quantity on the right side of the equation if it acts to produce a current which is in the same direction as the assumed mesh current, and to use a minus sign otherwise.

By the same reasoning which leads to Eq. (3-97) the voltage equation for the second mesh is

$$-R_2I_1 + (R_2 + R_3 + R_4)I_2 - R_4I_3 = E_3 \tag{3-98}$$

Note that mesh 2 is coupled to mesh current I_1 through R_2 and to mesh current I_3 through R_4.

Similarly the voltage equation for mesh 3 is

$$0 - R_4I_2 + (R_4 + R_5)I_3 = -E_2 \tag{3-99}$$

Equations (3-97), (3-98), and (3-99) involve three unknown mesh currents. Once these are found, then any one of the five branch currents can readily be determined. Mesh analysis has thereby simplified circuit analysis.

3-7 Circuit Analysis by Node-Pair Voltages

It is shown in the preceding section that finding the complete solution (i.e. all the branch currents) of the network of Fig. 3-11 is simpler by the mesh method than by the method of superposition. However, it should not be implied that there is no other, even simpler approach to the problem. For the network under discussion, in fact, there is another method of solution which makes available in a single independent equation the essential information needed to compute all branch currents.

A study of Fig. 3-11 shows why this is possible. For convenience Fig. 3-11 is repeated in Fig. 3-15 with notations which better suit our purposes here. As previously pointed out, the circuit of Fig. 3-15 contains two junction points which in Fig. 3-11 are called b and d but in Fig. 3-15 are identified as 1 and 0. Since the

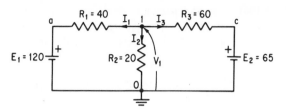

Fig. 3-15. Solution by nodal analysis.

voltages in a circuit must be measured between terminals, it is appropriate and useful to designate one terminal as a reference point. This is a point in the network with respect to which all other levels of potential are measured. In our work we assign that junction marked 0 as the reference point. Often this terminal is called the circuit *ground* and is represented by the symbol shown in Fig. 3-15. A glance at the circuit shows that node a is at a potential E_1 and node c is at a potential E_2 with respect to ground. The only potential which is unknown is that of junction 1. If this potential, V_1, can be found, each branch current can be determined because the voltage drop across each resistor will then be known.

The equation for finding V_1 is obtained by applying Kirchhoff's current law at junction 1. The usual procedure is to assume that all the branch currents leave the junction. Thus

$$I_1 + I_2 + I_3 = 0 \qquad (3\text{-}100)$$

where these currents are defined as shown in Fig. 3-15. But the voltage across R_1 is the difference in potential between points 1 and a. Hence we can write

$$I_1 = \frac{V_1 - E_1}{R_1} \qquad (3\text{-}101)$$

In writing the current expression the assumption is made that the node potential is always higher than the other voltages appearing in the equation. If, in fact, it turns out not to be so, a negative value for I_1 will result. This means that I_1 flows into the node and not away from it. The corresponding expressions for the currents in branches 2 and 3 are clearly

$$I_2 = \frac{V_1}{R_2} \qquad (3\text{-}102)$$

and

$$I_3 = \frac{V_1 - E_2}{R_3} \qquad (3\text{-}103)$$

Upon substituting these equations into Eq. (3-100) we obtain

$$\frac{V_1 - E_1}{R_1} + \frac{V_1}{R_2} + \frac{V_1 - E_2}{R_3} = 0 \qquad (3\text{-}104)$$

This last equation could be written directly from an examination of the circuit diagram. As the reader gains proficiency with these circuits he will learn to do so with ease.

Equation (3-104) is the independent equation which allows V_1 to be determined. All other quantities in the equation are known. For reasons which should be obvious the foregoing procedure is referred to as the *nodal method* of circuit analysis.

EXAMPLE 3-9   By using the values specified in Fig. 3-15 find the magnitude and direction of each of the branch currents.

solution: From Eq. (3-104) we have

$$V_1\left(\frac{1}{R_1} + \frac{1}{R_2} + \frac{1}{R_3}\right) = \frac{E_1}{R_1} + \frac{E_2}{R_3}$$

$$V_1\left(\frac{1}{40} + \frac{1}{20} + \frac{1}{60}\right) = \frac{120}{40} + \frac{65}{60} \tag{3-105}$$

$$\therefore \quad V_1 = 44.6 \text{ volts}$$

Inserting this value of V_1 into Eqs. (3-101), (3-102), and (3-103) yields the branch currents. Thus

$$I_1 = \frac{V_1 - E_1}{R_1} = \frac{44.6 - 120}{40} = -1.88 \text{ amps} \tag{3-106}$$

The minus sign means that I_1 flows into junction 1.

$$I_2 = \frac{V_1}{R_2} = \frac{44.6}{20} = 2.23 \text{ amps} \tag{3-107}$$

And

$$I_3 = \frac{V_1 - E_2}{R_3} = \frac{44.6 - 65}{60} = -0.341 \text{ amp} \tag{3-108}$$

This current too flows into the junction. Note that these answers compare favorably with those obtained by the mesh method. Note too the smaller amount of effort needed to obtain the solution.

We are now familiar with both the nodal and the mesh methods for finding the branch currents in electrical networks. At this point we may ask: when is one method to be preferred over the other? The answer is surprisingly simple. The general rule is this: Use that method which requires a smaller number of independent equations to obtain the solution. In the nodal method the number of independent node-pair equations needed is one less than the number of junctions in the network. In equation form

$$\boxed{n = j - 1} \tag{3-109}$$

when n denotes the number of independent node equations and j denotes the number of junctions. Of course the -1 accounts for the fact that one of the junctions is the reference. The number of independent mesh equations to solve a network problem is given by Eq. (3-80). In any given problem, therefore, start by computing m and n. If $m < n$, then the mesh method is generally preferred. Otherwise it is easier to use the nodal method. In the network of Fig. 3-15 the number of independent mesh equations needed to obtain a complete solution is

$$m = b - (j - 1) = 3 - (2 - 1) = 2 \tag{3-110}$$

whereas the number of independent node equations required is

$$n = j - 1 = 2 - 1 = 1 \tag{3-111}$$

Accordingly, the nodal method is the simpler approach. A comparison of Examples 3-8 and 3-9 bears this out. It should be pointed out, however, that the difference in effort is not overwhelming because we are dealing with a relatively simple problem. But as the complexity increases and the difference between the values of m and n gets large, the difference in effort between the two methods becomes significant. As a general rule the nodal method shows to advantage when the network has many parallel circuits.

It is possible to formulate the defining independent nodal equations of a network in the same manner as was done for the mesh method. To illustrate the procedure let us return to the network of Fig. 3-13, which is repeated in Fig. 3-16 with the modification that the junction points e, b, and c are replaced by the notation 0, 1, and 2 respectively. Here again we have a network for which the solution is easier by the nodal method than by the mesh method. In accordance with Eq. (3-80) there are three independent mesh equations, but by Eq. (3-109) there are only two independent node-pair voltage equations.

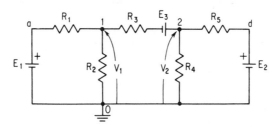

Fig. 3-16. A two node-pair network.

Applying Kirchhoff's current law at the independent node 1, and assuming that all branch currents are leaving the junction, we get

$$\frac{V_2 - E_1}{R_1} + \frac{V_1}{R_2} + \frac{V_1 + E_3 - V_2}{R_3} = 0 \tag{3-112}$$

The first two terms in this equation should be self-evident. The third term needs some comment. Keep in mind that the principle involved in the nodal method is to find the potential difference appearing across each circuit element in a branch. In the case of R_3 an inspection of the circuit diagram shows that the left side of R_3 is at a potential of V_1. The potential of the right side of R_3 is found by starting at the reference node and then summing all the voltages encountered in reaching the right side. Thus the potential of the right side of R_3 is $(+V_2 - E_3)$. Consequently, the potential difference across R_3 is $[V_1 - (V_2 - E_3)]$ or $(V_1 + E_3 - V_2)$. Division of this quantity by R_3 then gives the branch current through R_3.

By similar reasoning the current equation at independent node 2 can be written as

$$\frac{(V_2 - E_3) - V_1}{R_3} + \frac{V_2}{R_4} + \frac{V_2 - E_2}{R_5} = 0 \tag{3-113}$$

Thus there are two equations and two unknown quantities—V_1 and V_2.
Collecting terms and rearranging, Eq. (3-112) becomes

$$V_1\left(\frac{1}{R_1} + \frac{1}{R_2} + \frac{1}{R_3}\right) - V_2\left(\frac{1}{R_3}\right) = \frac{E_1}{R_1} - \frac{E_3}{R_3} \tag{3-114}$$

The coefficient of V_1 is a conductance; more specifically it is the *self-conductance* of node 1. It is found by forming the sum of the reciprocal of each of the resistances connected to node 1. It is denoted by the symbol G_{11}. That is

$$G_{11} = \frac{1}{R_1} + \frac{1}{R_2} + \frac{1}{R_3} = G_1 + G_2 + G_3 \tag{3-115}$$

The coefficient of V_2 is also a conductance. Called the mutual conductance, it is found by taking the reciprocal of the equivalent resistance that lies between nodes 1 and 2. It is denoted by G_{12} and in this case is

$$G_{12} = \frac{1}{R_3} \tag{3-116}$$

Accordingly, Eq. (3-114) can be written more compactly as

$$G_{11}V_1 - G_{12}V_2 = \left(\frac{E_1}{R_1} - \frac{E_3}{R_3}\right) \tag{3-117}$$

The minus sign always appears on the left side of this formulation because all independent node voltages are considered to be positive with respect to the reference node.

By performing the same operations on Eq. (3-113) we get

$$-G_{21}V_1 + G_{22}V_2 = \left(\frac{E_3}{R_3} + \frac{E_2}{R_5}\right) \tag{3-118}$$

where

$$G_{21} + G_{12} = \frac{1}{R_3} \tag{3-119}$$

and

$$G_{22} = \frac{1}{R_3} + \frac{1}{R_4} + \frac{1}{R_5} = G_3 + G_4 + G_5 \tag{3-120}$$

Note that the units of the right sides of Eqs. (3-117) and (3-118) are expressed in amperes and involve the source voltages. Hence one may look upon the expressions on the right side of these equations as equivalent current sources applied to nodes 1 and 2, respectively. For further discussion of this matter, refer to Sec. 3-9.

3-8 Thevenin's Theorem

Situations sometimes occur in electrical engineering in which it is desirable to find a particular branch current in a network as the resistance of that branch is varied while all other resistances and sources remain constant. For example, in the circuit configuration of Fig. 3-15 it may be desired to find the current which flows through R_2 for ten different values of R_2, assuming that E_1, E_2, R_1, and R_3

remain unchanged. We know that by repeatedly applying the superposition method or the mesh method or the nodal method to the network we can find the ten different values of branch current through R_2. We also know that perhaps the least effort is required in this case if we use the nodal method. In circumstances such as these, however, it is possible to resort to another technique of circuit analysis which appreciably simplifies the labor required to obtain the solution.

For the moment, so that we can better understand the procedure involved, let us leave Fig. 3-15 and turn attention to Fig. 3-17, which depicts a simple series circuit. The energy source is a battery having an emf E_b and an internal resistance R_i. The battery supplies energy to the load resistor R_L. When R_L is removed from the battery, i.e. the switch S is open, a voltmeter placed across terminals 01 reads an open-circuit voltage V_{oc} which is the battery emf. That is, $V_{oc} = E_b$. Furthermore, the resistance which is "seen" from the output terminals looking to the left into the source is clearly the internal resistance of the battery R_i. Therefore, as far as R_L is concerned, it "sees" appearing at terminals 01 a source voltage of value V_{oc} and an internal impedance R_i. As a matter of fact, even if that part of the circuit to the left of 01 were actually made up of series-parallel combinations of resistance and sources, nevertheless to an observer looking towards the left from the output terminals there appears an open-circuit voltage V_{oc} and an equivalent resistance associated with V_{oc} in exactly the same manner as depicted in the simple case of Fig. 3-17. The important thing to realize here is that at the output terminals with S open there can appear only one equivalent voltage V_{oc} and an equivalent resistance. Just how these quantities take on the values they do is irrelevant as far as the output terminals and the load resistor connected to these terminals are concerned. What really matters to R_L are the terminal quantities themselves—V_{oc} and R_i—irrespective of how these are arrived at. On this basis, when the switch S is closed the resulting current through the load resistor is readily identified as

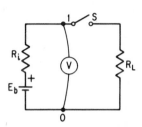

Fig. 3-17. Simple series circuit: battery supplying energy to a load resistor.

$$I_L = \frac{V_{oc}}{R_i + R_L} \qquad (3\text{-}121)$$

Let us now return to the circuit of Fig. 3-15 and rearrange it in the manner shown in Fig. 3-18(a). The drawing of the diagram is modified in order to bring it more into correspondence with Fig. 3-17. Note that R_2 of Fig. 3-18 corresponds to R_L of Fig. 3-17. With sources E_1 and E_2 and resistors R_1 and R_3 remaining unaltered in Fig. 3-18, it should be possible to determine the open-circuit voltage appearing across the output terminals 01 (with switch S open, of course) as well as the equivalent resistance looking into terminals 01 towards the left (or source) side. The open-circuit voltage is readily found as the potential difference between points 0 and 1 as determined by the flow of current in the closed path composed of E_1, R_1, R_3, and E_2. The equivalent internal resistance is merely the parallel

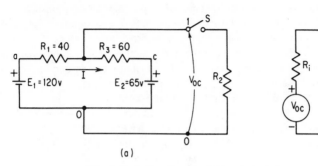

Fig. 3-18. Redrawing Fig. 3-15 to isolate R_2:
(a) original network redrawn;
(b) equivalent circuit.

combination of R_1 and R_3. Keep in mind that when we are finding the equivalent internal resistance the source voltages are set equal to zero. Figure 3-18(b) depicts the equivalent circuit of Fig. 3-18(a). The procedure is now illustrated by the following example.

EXAMPLE 3-10 In the circuit of Fig. 3-15 find the branch current which flows through R_2 when R_2 has the following values: 11, 19, 46, and 176 ohms.

solution: Refer to Fig. 3-18(a). Let us first find the open-circuit voltage $V_{oc} = V_{01}$. Thus we need to determine the potential of point 1 relative to 0, which can be considered the reference or ground point. Writing Kirchhoff's voltage law for circuit $ac0a$ yields

$$IR_1 + IR_3 + 65 - 120 = 0 \qquad (3\text{-}122)$$

where I is the current flowing in loop $ac0$ with switch S open. For the specified values of R_1 and R_3 this current becomes

$$I = \frac{55}{R_1 + R_3} = \frac{55}{40 + 60} = 0.55 \text{ amp} \qquad (3\text{-}123)$$

Therefore

$$V_{oc} = V_{01} = E_1 - IR_1 = 120 - (0.55)(40) = 98 \text{ volts} \qquad (3\text{-}124)$$

The equivalent resistance appearing between the output terminals 01 is then

$$R_i = \frac{R_1 R_3}{R_1 + R_3} = \frac{40(60)}{40 + 60} = 24 \text{ ohms} \qquad (3\text{-}125)$$

Having established V_{oc} and R_i, it then follows from Fig. 3-18(b) that the branch current through R_2 is

$$I_2 = \frac{V_{oc}}{R_i + R_2} = \frac{98}{24 + R_2} \qquad (3\text{-}126)$$

Accordingly

when $R_2 =$ 11 ohms,	$I_2 = \frac{98}{35} =$ 2.8 amps
when $R_2 =$ 19 ohms,	$I_2 = \frac{98}{43} =$ 2.28 amps
when $R_2 =$ 46 ohms,	$I_2 = \frac{98}{70} =$ 1.4 amps
when $R_2 =$ 176 ohms,	$I_2 = \frac{98}{200} =$ 0.49 amp

Note that once the equivalent source—V_{oc} and R_t—is determined, Eq. (3-126) allows the current through R_2 to be found with very little effort for each value of R_2. The same cannot be said for the superposition, mesh, and nodal methods.

A French engineer, M. L. Thevenin, was the first person to discover the equivalence in networks such as those depicted in Fig. 3-18. This technique of circuit analysis was first published by him in 1883. Known today as Thevenin's theorem, it may be stated as follows (for the resistive network case):

> Any two terminals of a network composed of linear passive and active circuit elements may be replaced by an equivalent voltage source and an equivalent series resistance. The voltage source is equal to the potential difference between the two terminal points caused by the active network with no external elements connected to these terminals. The series resistance is the equivalent resistance looking into the two terminal points with all power sources within the terminal pair inactive.

EXAMPLE 3-11 A network has the configuration shown in Fig. 3-19(a). All resistance values are expressed in ohms.

(a) Find the current through R_L when it takes on values of 10, 50, and 200 ohms.

(b) Determine the value of R_L corresponding to which there is a maximum power transfer to the load resistor. Compute this maximum power.

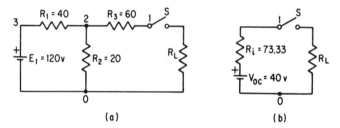

Fig. 3-19. Circuit configuration for Example 3-11:
(a) original network with resistances expressed in ohms;
(b) Thevenin equivalent circuit.

solution: Although the solution to part (a) can be found by applying the technique of network reduction, it is simpler to find the Thevenin equivalent circuit and then use Eq. (3-121) directly.

(a) The open-circuit voltage appearing at terminal 1 is the same as that at terminal 2. Applying the voltage-divider theorem to the closed path 032 we get

$$V_{oc} = V_{02} = V_{01} = \frac{R_2}{R_1 + R_2}E_1 = \frac{20}{40 + 20}(120) = 40 \text{ volts} \qquad (3\text{-}127)$$

Also the equivalent resistance looking into the network from terminal-pair 01 toward the left is R_3 in series with the parallel combination of R_1 and R_2. Thus

$$R_t = R_3 + \frac{R_1 R_2}{R_1 + R_2} = 60 + \frac{40(20)}{40 + 20} = 60 + \frac{40}{3} = 73.33 \text{ ohms} \qquad (3\text{-}128)$$

Thus the Thevenin equivalent circuit of Fig. 3-19(a) appears as shown in Fig. 3-19(b). Accordingly the corresponding load currents are

$$\text{for } R_L = 10, \quad I_L = \frac{40}{73.33 + 10} = 0.481 \text{ amp}$$

$$\text{for } R_L = 50, \quad I_L = \frac{40}{73.33 + 50} = 0.324 \text{ amp}$$

$$\text{for } R_L = 200, \quad I_L = \frac{40}{73.33 + 200} = 0.146 \text{ amp}$$

(b) The general expression for the power in the load resistor is

$$P_L = I_L^2 R_L = \left(\frac{V_{oc}}{R_i + R_L}\right)^2 R_L = \frac{V_{oc}^2 R_L)^2}{(R_i + R_L)^2} \tag{3-129}$$

In order to find the value of R_L for which P_L is a maximum, it is necessary to differentiate Eq. (3-129) with respect to R_L and set the result equal to zero. Thus

$$\frac{dP_L}{dR_L} = V_{oc}^2 \left[\frac{-2R_L}{(R_i + R_L)^3} + \frac{1}{(R_i + R_L)^2}\right] = 0$$

$$= \frac{-2R_L + R_i + R_L}{(R_i + R_L)^3} = 0 \tag{3-130}$$

From which it follows that

$$\boxed{R_i = R_L} \tag{3-131}$$

Hence for the problem at hand

$$R_L = R_i = 73.33 \text{ ohms}$$

The maximum power to R_L is then

$$P_{L \text{ max}} = \frac{V_{oc}^2}{4R_L^2} R_L = \frac{V_{oc}^2}{4R_L} = \frac{(40)^2}{4(73.33)} = 5.45 \text{ watts} \tag{3-132}$$

Equation (3-131) is a useful result. It states that the maximum transfer of power to a load resistor occurs when the load resistor has a value equal to the series equivalent or internal resistance of the source. Note that the efficiency at maximum power transfer is 50 per cent. This result brings into evidence another advantage of the Thevenin equivalent representation of a network. Once R_i is computed, it shows at a glance the condition for maximum power transfer. The Thevenin equivalent circuit conveys other useful information as well. Thus in Fig. 3-19(b) it is immediately apparent that the maximum voltage that can ever appear at the load resistor for the circuit configuration of Fig. 3-19(a) is 40 volts. This conclusion is not so obvious in the original circuit.

3-9 Norton's Theorem. Conversion of a Voltage Source to a Current Source

Let us look further into the description of the circuit of Fig. 3-17. Note that when R_L is made equal to zero, a current flows through the output terminals (from 1 to 0) which is the maximum possible current that can be supplied by the given voltage source. In this sense, then, one can treat the voltage source in terms of a source which is capable of supplying current up to a maximum limited by the series equivalent resistance R_i. In other words the voltage source may be considered in terms of a current source of magnitude $I = E_b/R_i$. However, in order for a description from this viewpoint to be completely consistent with the conditions as they prevail in Fig. 3-17 between the terminal-pair 01 at open circuit, the current source must appear across a resistor placed between points 0 and 1 and having such a value that the equivalent open-circuit voltage of the original circuit appears across it. Since we already know the value of the current source as $I = E_b/R_i$, we must now determine that value of shunt† resistance which when placed across the current source yields E_b. Expressed mathematically we have

$$IR_x = E_b \tag{3-133}$$

But

$$I = \frac{E_b}{R_i} \tag{3-134}$$

Therefore

$$R_x = R_i \tag{3-135}$$

Accordingly, as far as the network terminal-pair 01 is concerned at open-circuit, the part to the left of these terminals may be represented either by a voltage source and a series internal resistance or by a current source and a shunt resistance of value R_i. The current-source equivalent of the voltage source is illustrated in Fig. 3-20. The symbol used to denote it is a circle-enclosed arrow indicating the direction of current flow assumed as positive. In dealing with current sources it is

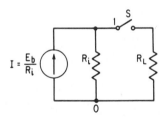

Fig. 3-20. Current source equivalent of the circuit of Fig. 3-19.

important to understand that the total current of the current source is supplied to the circuit elements placed across its terminals. If there is just one path available, as in Fig. 3-20 when the switch S is open, then that one circuit element draws the full current. If more than one circuit element is placed across the current source, then together they draw the full current of the source. *A current source always delivers constant current irrespective of the particular configuration of the circuit to which it is connected.* This is not unlike the behavior of a voltage source. Thus in the circuit of Fig. 3-17 the battery delivers its full voltage to the circuit

† This is called a shunt resistance because it is placed across the output terminals in parallel with the current source.

at all times. This continues to be true even if additional series-parallel circuit elements are placed in the circuit.

Why should it be at all desirable to work with equivalent current sources? What are the advantages? The biggest advantage is that the analysis of some very important engineering devices is much easier to perform. This is especially true of those engineering devices, such as the transistor and the vacuum-tube pentode, which are characteristically current-sensitive in their behavior and operation. Furthermore, in those instances of circuit analysis which can be most easily handled by the nodal method, the idea of a current-source equivalent of voltage sources is a natural consequence of the mathematical formulation associated with the method. This is readily borne out by referring to the right side of Eqs. (3-117) and (3-118). Here the quantities E_1/R_1, E_3/R_3, and E_2/R_5 are nothing more than the current-source equivalences of the corresponding voltage sources.

To help complete our understanding of the current-source equivalence let us investigate the behavior of the circuits shown in Figs. 3-17 and 3-20 when the switch S is closed. In Fig. 3-17, by Kirchhoff's voltage law, the expression for the output voltage appearing across terminal-pair 01 is

$$E_b - I_L R_i = I_L R_L = V_{01} \tag{3-136}$$

Dividing each term through by R_i yields

$$\frac{E_b}{R_i} - I_L = \frac{V_{01}}{R_i} \tag{3-137}$$

Inserting Eq. (3-134) and rearranging leads to

$$I_L = I - \frac{V_{01}}{R_i} \tag{3-138}$$

This last expression states that the load current is equal to the equivalent source current less the amount drawn by the shunt resistance R_i. It is interesting to note that Eq. (3-138) can be written immediately by referring to the current-source equivalent circuit of Fig. 3-20 and writing Kirchhoff's current law at node 1.

Another interesting result pertaining to the current-source equivalent circuit can be obtained by rewriting Eq. (3-138) and using the fact that $I_L = V_{01}/R_L$. Thus

$$I = I_L + \frac{V_{01}}{R_i} = \frac{V_{01}}{R_L} + \frac{V_{01}}{R_i} = V_{01}\left(\frac{1}{R_L} + \frac{1}{R_i}\right) \tag{3-139}$$

But the quantity in parenthesis is nothing more than the parallel combination of R_i and R_L in Fig. 3-20. Hence by setting

$$\frac{1}{R_p} = \frac{1}{R_L} + \frac{1}{R_i} \tag{3-140}$$

it follows that the expression for the voltage across the output terminals 01 is simply

$$\boxed{V_{01} = IR_p} \tag{3-141}$$

where I is the equivalent current source. Equation (3-141) provides a very simple and direct way of finding a terminal voltage, especially where many parallel elements are involved.

The current equivalent of a voltage source was first developed by E. L. Norton —an engineer employed by the Bell Telephone Laboratory. Today this equivalence is referred to as *Norton's theorem* and is described as follows:

> Any two terminals of a network composed of linear passive and active circuit elements may be replaced by an equivalent current source and a parallel resistance. The current of the source is the current measured in the short circuit placed across the terminal-pair. The parallel resistance is the equivalent resistance looking into the terminal-pair with all independent power sources inactive.

Note that this theorem does not require that the Thevenin equivalent circuit be found first. The short-circuit current and the parallel resistance may be found directly, as illustrated in the examples which follow.

EXAMPLE 3-12   By means of Norton's current-source equivalent circuit of the network shown in Fig. 3-21(a) find the current which flows through R_2 when it has a value of 11 ohms. All resistance values in Fig. 3-21 are expressed in ohms.

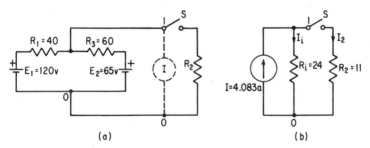

Fig. 3-21. Network configuration for Example 3-12:
(a) original circuit;
(b) Norton's current source equivalent.

solution: To find the current-source equivalent circuit associated with the network terminal pair 01, it is necessary to find the current which flows through a short circuit placed across the terminals, and the resistance looking into these same terminals when the short circuit is removed. In Fig. 3-21(a) the short circuit connection is represented by a broken line leading to a circle enclosing the letter I. This symbol can be thought of as depicting an instrument for measuring current, such as an ammeter. An examination of the circuit lying to the left of terminal-pair 01 indicates that the current I is composed of two components. There is a current contribution from source E_1 which takes a path through R_1, through the short circuit and then back to E_1. This component of I is thus $E_1/R_1 = 120/40 = 3$ amp. The contribution from source E_2 takes a path through R_3, then through the short circuit and back to E_2. The magnitude of this current is $E_2/R_3 = 65/60 = 1.083$ amp. Thus

$$I = \frac{E_1}{R_1} + \frac{E_2}{R_2} = 3 + 1.083 = 4.083 \text{ amps} \qquad (3\text{-}142)$$

Since the parallel resistance is the equivalent resistance looking into the network at the specified terminal-pair 01 with all source voltages set to zero, it is given by

$$R_i = \frac{R_1 R_3}{R_1 + R_3} = \frac{40(60)}{40 + 60} = 24 \text{ ohms} \tag{3-143}$$

Hence the Norton equivalent circuit takes on the form shown in Fig. 3-21(b).

When the switch S is closed, the source current I in Fig. 3-21(b) splits into two paths as I_i and I_2. Because R_i and R_2 are in parallel, the potential difference across each resistor is the same. Consequently we can write

$$I_2 R_2 = I_i R_i \tag{3-144}$$

But

$$I_i = I - I_2 \tag{3-145}$$

Therefore

$$I_2 R_2 = I R_i - I_2 R_i$$

or

$$\boxed{I_2 = \frac{R_i}{R_i + R_2} I} \tag{3-146}$$

Equation (3-146) is useful when circuit analysis is being performed with current sources. It is analogous to the voltage-divider theorem for voltage sources. It states that for two parallel branches the current in one branch is to the total current as the resistance of the second branch is to the sum of the resistances of the two branches. For the example at hand the current in branch I_2 is found to be

$$I_2 = \frac{R_i}{R_i + R_2} I = \frac{24}{24 + 11} (4.083) = 2.8 \text{ amps} \tag{3-147}$$

which checks with the value found by using Thevenin's theorem in Example 3-10.

EXAMPLE 3-13 The small-signal equivalent circuit of a vacuum-tube amplifier has the configuration depicted in Fig. 3-22(a). The indicated circuit parameters have the following values:

$$\mu = 99 \qquad\qquad R_1 = 147,000 \text{ ohms}$$
$$r_p = 66,000 \text{ ohms} \qquad R_2 = 100,000 \text{ ohms}$$

Find the value of the output voltage for $E_g = 0.4$ volt.

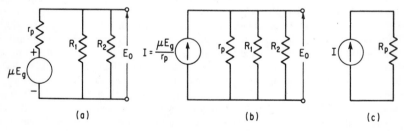

Fig. 3-22. Network for Example 3-13:
(a) original circuit;
(b) current source equivalent;
(c) simplified equivalent.

solution: The current which flows in the short circuit placed across the output terminals 01 is

$$I = \frac{\mu E_g}{r_p} = 1.5 \times 10^{-3} E_g \qquad \text{amp} \qquad (3\text{-}148)$$

Note that the short circuit across 01 removes R_1 and R_2 from any consideration in determining the current source I. Thus the current-source equivalent circuit may be represented as shown in Fig. 3-22(b). By finding the parallel combination of the three parallel resistances we can depict the equivalent circuit as shown in Fig. 3-22(c), where

$$\frac{1}{R_p} = \frac{1}{r_p} + \frac{1}{R_1} + \frac{1}{R_2} = \frac{1}{31,200} \qquad (3\text{-}149)$$

The output voltage is equal to the source current multiplied by R_p. Thus

$$E_0 = IR_p = \frac{\mu}{r_p} E_g R_p = 1.5 \times 10^{-3}(0.4)(31.2)10^{+3} = 18.72 \text{ volts} \qquad (3\text{-}150)$$

3-10   Network Reduction by Δ-Y Transformation

The techniques of series-parallel network reduction are not always applicable. A case in point is illustrated in Fig. 3-23. Examination of the diagram shows that although two parallel paths may be traced on leaving terminal a, yet the paths from b and c are neither of the series nor parallel kind. Of course one can argue here that the techniques of mesh and nodal analysis certainly lend themselves to a solution of the branch currents that flow in such a circuit configuration. A little thought, however, indicates that in either case it is necessary to solve three independent equations in order to arrive at a solution. In situations of the type under consideration the Δ-to-Y transformation simplifies the solution procedure considerably. As a matter of fact it often makes it unnecessary to solve simultaneous mesh or nodal equations by modifying the circuit in such a way as to permit the standard series-parallel network reduction technique to be applied directly.

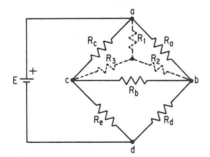

Fig. 3-23. The original circuit configuration appears drawn with solid lines. The Y-equivalent network of the abc delta network is drawn with broken lines.

The source of difficulty in applying the procedures of series-parallel reduction to the original configuration of Fig. 3-23 essentially lies in the three-terminal character of a circuit such as abc, which is often called a delta (Δ) network. It is significant to note that if this network could be replaced by an equivalent network which effectively removes the cross arm cb, then combinations of conventional series and parallel resistances could be readily identified and the familiar techniques of Sec. 3-4 applied. A network which possesses this capability is the wye (Y) network. It is depicted in Fig. 3-23 with a broken-line drawing. With such an equivalent circuit note that R_2 and R_d are in series and so, too, are R_3 and R_c. Moreover, these series combinations in turn are in parallel with each other. Accordingly, we find that the equivalent resistance between points *ad* can be written quite simply and directly as

$$R_{ad} = R_1 + \frac{(R_3 + R_e)(R_2 + R_d)}{R_3 + R_e + R_2 + R_d} \tag{4-151}$$

The current delivered by the source in the circuit of Fig. 3-23 is then easily determined.

We turn attention next to the procedure which describes how the equivalent Y network can be computed from the known arms of the given Δ network, which in Fig. 3-23 are R_a, R_b, and R_c. This part of the circuit is repeated for clarity's sake in Fig. 3-24. A study of Fig. 3-24 should make it apparent that for the Y and

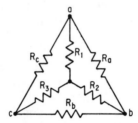

Fig. 3-24. The three-terminal Δ and Y networks.

Δ networks to be equivalent it is necessary that the resistance appearing between each of the three pairs of terminals—*ab*, *bc*, and *ca*—be the same. Thus, looking into the terminals *ab* the equivalent resistance for the Y arrangement is merely R_1 in series with R_2. On the other hand, for the Δ arrangement the equivalent resistance between *ab* is R_a in parallel with the series combination of R_b and R_c. Expressing this mathematically we can write

$$[R_{ab}]_Y = [R_{ab}]_\Delta \tag{3-152}$$

or

$$R_1 + R_2 = \frac{R_a(R_b + R_c)}{R_a + R_b + R_c} \tag{3-153}$$

Similarly for terminals *bc* we have

$$[R_{bc}]_Y = [R_{bc}]_\Delta \tag{3-154}$$

or

$$R_2 + R_3 = \frac{R_b(R_c + R_a)}{R_a + R_b + R_c} \tag{3-155}$$

And for terminals *ca*

$$[R_{ca}]_Y = [R_{ca}]_\Delta \tag{3-156}$$

or

$$R_3 + R_1 = \frac{R_c(R_a + R_b)}{R_a + R_b + R_c} \tag{3-157}$$

Equations (3-153), (3-155), and (3-157) represent three equations involving the three unknown resistances of the equivalent Y network. Solving these equations simultaneously leads to the following useful results:

$$R_a = \frac{R_1 R_2 + R_2 R_3 + R_3 R_1}{R_3}$$

$$R_b = \frac{R_1 R_2 + R_2 R_3 + R_3 R_1}{R_1}$$

$$R_c = \frac{R_1 R_2 + R_2 R_3 + R_3 R_1}{R_2}$$

$$R_1 = \frac{R_a R_c}{R_a + R_b + R_c} \tag{3-158}$$

$$R_2 = \frac{R_a R_b}{R_a + R_b + R_c} \tag{3-159}$$

$$R_3 = \frac{R_b R_c}{R_a + R_b + R_c} \tag{3-160}$$

A comparison of these equations with the circuitry of Fig. 3-24 reveals that any given arm of the Y network is found by taking the product of the two *adjacent* arms of the Δ network and dividing by the sum of the Δ-network arms. This is an easy rule for remembering the foregoing equations.

EXAMPLE 3-14 In the original circuit configuration of Fig. 3-23 assume that the resistances have the following values expressed in ohms:

$$R_a = 20 \qquad R_d = 24$$
$$R_b = 30 \qquad R_e = 5$$
$$R_c = 50$$

Find the current delivered by the source to the network. Assume that $E = 220$ volts.

solution: Replace the *abc* delta network by an equivalent wye which by Eqs. (3-158) to (3-160) has arm values as follows:

$$R_1 = \frac{R_a R_c}{R_a + R_b + R_c} = \frac{20(50)}{20 + 30 + 50} = 10 \text{ ohms}$$

$$R_2 = \frac{R_a R_b}{R_a + R_b + R_c} = \frac{20(30)}{20 + 30 + 50} = 6 \text{ ohms}$$

$$R_3 = \frac{R_b R_c}{R_a + R_b + R_c} = \frac{30(50)}{20 + 30 + 50} = 15 \text{ ohms}$$

Then by Eq. (3-151) the equivalent resistance in ohms between terminals *ad* is

$$R_{ad} = R_1 + \frac{(R_2 + R_d)(R_3 + R_e)}{R_2 + R_3 + R_d + R_c} = 10 + \frac{(6 + 24)(15 + 5)}{50} = 10 + 12 = 22$$

Hence

$$I = \frac{E}{R_{ad}} = \frac{220}{22} = 10 \text{ amps.}$$

PROBLEMS

3-1   Three resistances having values of 10, 20, and 30 ohms are connected in series to form an electric circuit. Find the equivalent series resistance.

3-2   The three resistances of Prob. 3-1 are placed in parallel across a voltage source. Find the equivalent parallel resistance.

3-3   Four resistances of values 50,000 ohms, 250 kilohms, 1 megohm, and 500 kilohms are placed in parallel. Compute the equivalent parallel resistance.

3-4   The capacitance values of three capacitors are 10 μf, 20 μf, and 40 μf. When these are placed in parallel across a 200-volt source find
(a) the equivalent capacitance,
(b) the total charge residing on the capacitors,
(c) the charge on each capacitor.

3-5   The three capacitors of Prob. 3-4 are placed in series across a 350-volt source.
(a) Compute the equivalent capacitance.
(b) Find the charge on each capacitor.
(c) Determine the voltage drop across each capacitor.

3-6   Three capacitors are connected across a 100-volt source in the manner shown in Fig. P3-6.
(a) Find the charge on each capacitor.
(b) Compute the total stored energy.
(c) Compute the energy stored in each capacitor.

3-7   In a configuration of three series-connected capacitors the voltage drops across their terminals are 20, 30, and 50 volts and the charge on each capacitor is 300 μcoulombs.
(a) Find the capacitance of each capacitor.
(b) What is the series equivalent capacitance?
(c) Compute the total stored energy.

3-8   The capacitors of Prob. 3-7 are disconnected in the charged condition and then placed in parallel. However, one of the capacitors is connected with its polarity reversed from that of the other two.
(a) Find the resultant voltage of the parallel combination.
(b) Determine the charge on each capacitor.
(c) Compute the stored energy.

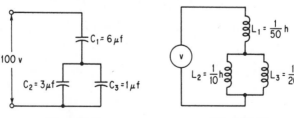

Figure P3-6   **Figure P3-10**

3-9 Three series-connected inductor coils have voltages of 20, 30, and 50 volts appearing across their terminals when the circuit current is changing at a rate of 100 amperes per second. Determine the equivalent series inductance.

3-10 Three inductors are arranged as shown in Fig. P3-10.
(a) Find the equivalent inductance as seen by the source.
(b) Determine the value of the emf of self-inductance in coil L_2 when the current in coil L_1 is changing at a rate of 1500 amperes per second.

3-11 The equivalent inductance of two inductors placed in parallel is 4 henrys. When series-connected, the equivalent inductance is 20 henrys. Find the values of the individual inductances.

3-12 Find the equivalent resistance appearing between terminals *ab* for the circuit configuration shown in Fig. P3-12. All resistance values are expressed in ohms.

3-13 In the circuit configuration of Fig. P3-13 determine the voltage drop across each circuit element as well as the power dissipated. All resistances are in ohms.

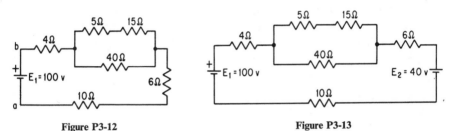

Figure P3-12   Figure P3-13

3-14 For the circuit of Fig. P3-14, find the current *I*. All resistance values are in ohms.

3-15 In the configuration of Fig. P3-15, find the value of *E* which permits a power dissipation of 180 watts in the 20-ohm resistor. All values are expressed in ohms.

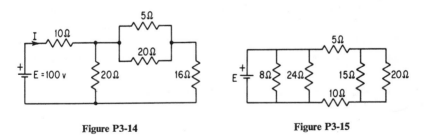

Figure P3-14   Figure P3-15

3-16 Determine the equivalent resistance appearing at the battery terminals in the circuit of Fig. P3-15.

3-17 Find the voltage drop appearing across the 18-ohm resistor in the network of Fig. P3-17. All values are in ohms.

3-18 Determine the equivalent resistance between points *ab* in the circuit of Fig. P3-18. All resistances are in ohms.

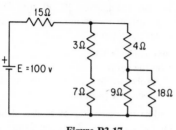

Figure P3-17   Figure P3-18

3-19   In the circuitry of Fig. P3-18, determine (a) the voltage needed across *ab* so that the voltage drop across the 15-ohm resistor is 45 volts, and (b) the corresponding voltage across the 8-ohm resistor for this condition.

3-20   Find the equivalent resistance between terminals *ab* for the circuit of Fig. P3-20. All resistance values are in ohms.

3-21   In Fig. P3-20 compute the voltage required between terminals *ab* so that a voltage drop of 45 volts occurs across the 15-ohm resistor.

3-22   Find the current *I* which flows through the 10-ohm resistor in the circuit of Fig. P3-22. All resistances are in ohms.

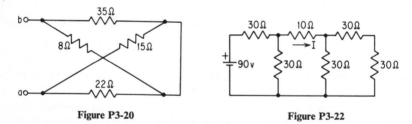

Figure P3-20   Figure P3-22

3-23   In the circuit of Fig. P3-23 find the value of R. All resistances are in ohms.

3-24   By means of superposition find the current which flows in $R_2 = 20$ ohms for the circuit of Fig. P3-24.

3-25   Find the current and the power dissipation in each resistor of the circuit shown in Fig. P3-25.

3-26   Determine the current which flows through each resistor in the circuit of Fig. P3-26. All resistance values are in ohms. Use superposition.

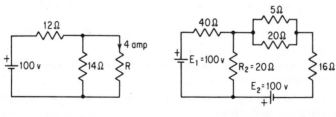

Figure P3-23   Figure P3-24

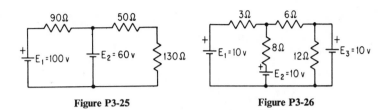

Figure P3-25   Figure P3-26

3-27 In the circuit configuration shown in Fig. P3-27 determine the number of (a) circuit elements, (b) nodes, (c) junction points, (d) branches, (e) meshes. Draw that tree of the network that leads directly to the mesh equations.

3-28 Repeat Prob. 3-27 for the circuit shown in Fig. P3-28.

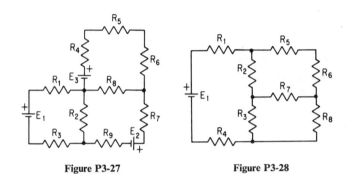

Figure P3-27   Figure P3-28

3-29 Repeat Prob. 3-27 for the network depicted in Fig. P3-29.

3-30 Find the current which flows through the 15-ohm resistor in the circuit of Fig. P3-30 by using the mesh method. Check your solution by using the superposition theorem.

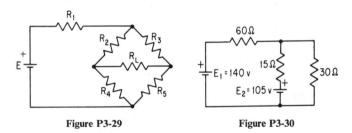

Figure P3-29   Figure P3-30

3-31 Solve Prob. 3-30 by using the nodal method of analysis.

3-32 In the network of Fig. P3-32 find the current through the 90-ohm resistor.

3-33 Determine the current which flows through the 30-ohm resistor in the circuit of Fig. P3-33.

3-34 In the circuit of Fig. P3-32 find the current through R_2 when it is made to take on the following values: $R_2 = 62.5, 106, 209,$ and 784 ohms. Use the method requiring the least effort to yield the answers.

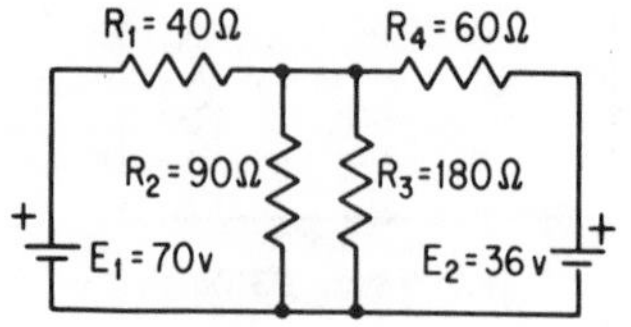

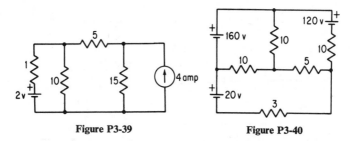

Figure P3-32 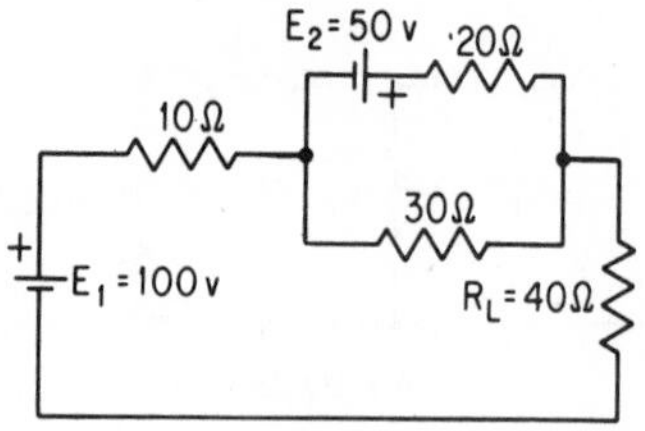   Figure P3-33

3-35 Assuming that R_L in Fig. P3-33 is the load resistor, find the Thevenin equivalent circuit.

3-36 Set up the equations which will allow you to determine all the branch currents in the circuits of Figs. P3-27 and P3-29. Give reasons to justify the method you have chosen to solve the problem.

3-37 Solve Prob. 3-25 by the mesh method.

3-38 Solve Prob. 3-26 by the nodal method.

3-39 Solve for the power delivered to the 10-ohm resistor in the circuit shown in Fig. P3-39. All resistances are in ohms.

3-40 Use Thevenin's theorem to find the power delivered to the 3-ohm resistance in the circuit of Fig. P3-40. All resistances are in ohms.

Figure P3-39   Figure P3-40

3-41 A circuit has an arrangement of circuit elements as depicted in Fig. P3-41.
(a) Find the Thevenin equivalent circuit considering R_4 as the variable load resistance.
(b) Find the current through R_4 when it has values of $\frac{4}{7}$ and $\frac{40}{7}$ ohms.

3-42 Use Thevenin's theorem to replace the three-loop equivalent circuit of Fig. P3-42 by a single-loop equivalent circuit in which the identity of R_L is preserved. All resistances are expressed in ohms.

3-43 In the circuit of Fig. P3-43 it is desirable to obtain information about each of the branch circuits.
(a) By using the mesh method write (but do not solve) the necessary independent equations which will allow all branch currents to be determined.
(b) By using the nodal method, write (but do not solve) the independent equations from which all the branch currents can be found.

3-44 In the circuit of Fig. P3-44 find the current delivered by the battery.

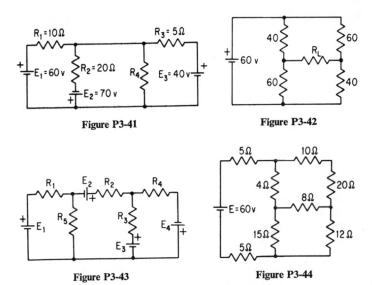

Figure P3-41

Figure P3-42

Figure P3-43

Figure P3-44

3-45 Derive the appropriate expressions which will permit converting a given wye arrangement of resistances to an equivalent delta arrangement.

3-46 Use Thevenin's theorem to replace the three-loop circuit of Fig. P3-46 by a single-loop equivalent circuit in which the identity of R_L is preserved.

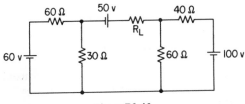

Figure P3-46

3-47 Refer to Fig. P3-47.
(a) Draw an appropriate tree of the network.
(b) How many equations must be solved to specify completely the current in each branch of the circuit, assuming all source and resistance values are known?

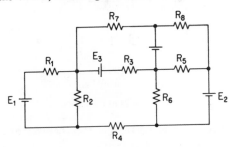

Figure P3-47

3-48 In the circuit shown in Fig. P3-48, find the current in the 90-ohm resistor using the mesh method.

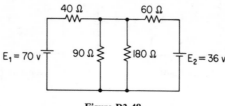

Figure P3-48

3-49 The circuit appearing in Fig. P3-49 is a typical potentiometer arrangement used to measure voltage.
(a) Find the voltage at point A with respect to the ground terminal.
(b) Find the Thevenin equivalent circuit for terminals $0A$.
(c) Calculate the galvanometer current.

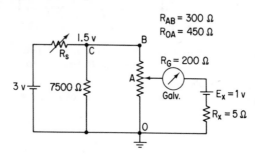

Figure P3-49

3-50 Refer to the circuit shown in Fig. P3-50. Consider that $0 \leq A \leq 1$.
(a) If I is a constant current source, find the output voltage E_0 in terms of I and the circuit parameters shown.
(b) Assume that the current source is ideal (i.e., has infinite resistance). Compute the resistance looking into terminals AB.
(c) If $R_L = 0$, compute the output current I_L.

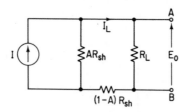

Figure P3-50

3-51 In the circuit of Fig. P3-51, find the Thevenin equivalent circuit existing between points a and b.

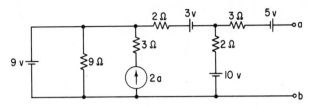

Figure P3-51

3-52 By direct application of Norton's theorem, find the Norton equivalent circuit as seen by R_L in the circuit of Fig. P3-52.

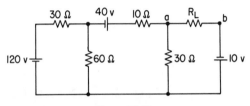

Figure P3-52

3-53 For the circuit shown in Fig. P3-53, find the Thevenin equivalent circuit as seen by the 20 ohm resistor appearing between points *a* and *b*.

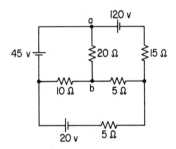

Figure P3-53

4

the laplace-transform
method of solving linear
differential equations

In dealing with the subject matter of electrical engineering, it is frequently necessary to determine the solutions of linear differential equations. In this way information is obtained about the dynamic and forced responses of a circuit, device, or system to a driving function. Moreover, the development of methods of analysis, design, and synthesis as well as the interpretation of associated results are very often influenced by the particular mathematical formulations used to generate these solutions. It is in the interest of furnishing the reader with the most consistent, systematic, and readily interpretable mathematical tool that attention is given in this chapter to the Laplace-transform method of solving linear differential equations. Certainly no more time is needed to learn the mechanics of the Laplace-transform method than the classical method. Moreover, the former has some very significant and important advantages to offer.

In the first place the Laplace-transform method of finding the solution to a defining differential equation involves a one-step systematic formulation which yields complete information about the sustained and transient solutions. This stands in sharp contrast to the three-part solution procedure of the classical method. In the latter, it is necessary first to find the particular solution associated with the forcing functions, then to find the complementary function which is the solution to the homogeneous differential equation, and finally to evaluate the constants of integration from the initial conditions. Although two aspects of the procedure remain the same, namely finding the roots of the characteristic equation and evaluating the initial conditions, it often happens that where the governing differential equation is of an order higher than the second the amount of labor required to get the solution by the Laplace-transform method is less.

However, by far the more important factor is the systematic formulation provided by the Laplace-transform method. It permits the treatment of forcing functions and disturbances in the same fashion as initial conditions. This means that the same basic procedure is used to evaluate the forced solution as is used to evaluate the transient solution. This feature is particularly useful because it allows a unifying approach in the analysis and design of circuits, devices, and systems which cannot be otherwise achieved. Thus, for example, in the study of electric circuits it is possible to treat the more general case of the transient response to deterministic inputs before the steady-state sinusoidal response. In addition, greater stress is placed on transform impedance as an operator which converts forcing function or initial condition to response.

Finally, the mathematical formulation of the Laplace transform permits a uniformity of analysis and of interpretation of results in situations which we find constantly recurring in all areas of electrical engineering and many other branches of engineering and science. Thus with this approach we will use the same techniques in arriving at answers to problems irrespective of whether they deal with circuits, amplifiers, feedback devices, control systems, or analog computers.

In the treatment which follows, the emphasis is put on the mechanics of the Laplace-transform method. No attempt is made to present a rigorous exposition. This is left to books devoted primarily to that objective.† Accordingly, a matter such as obtaining the final solution to a problem after applying the Laplace transform is accomplished through the use of tables rather than by means of integration in the complex plane, which is considered beyond the requirements for students of this book. When we accept the use of tables as a valid means of performing the inverse Laplace transformation, there is every reason to use the Laplace transform as a means of solving linear differential equations.

4-1   Nature of a Mathematical Transform

It is the nature of a transform to simplify the analytical procedure leading to the solution of a problem irrespective of whether the problem is concerned with numbers, functions, or the calculus of differential equations. Specifically the Laplace transform is a mathematical tool for transforming functions. In linear analysis it also serves to convert operations in time, such as differentiation and integration, into simpler algebraic functions of an intermediate variable such as multiplication and division.

† See M. F. Gardner and J. L. Barnes, *Transients in Linear Systems* (New York: John Wiley & Sons, Inc., 1942).

The idea of using transforms to simplify the procedure leading to a solution should not be new to the reader. It is assumed that he has used logarithms. Essentially the logarithm is a number transform which permits multiplication to be performed by means of the simpler operation of addition. An example readily illustrates the method. Let it be desired to find the product of

$$P = (35.6)\,(7.23)\,(181.47)$$

by using a number transform—the logarithm. The first step in the procedure is to find the transform of each number. The use of a logarithm table to the base 10 yields the results shown below.

original number		number transform
35.60	$\longrightarrow$	1.5514
7.23	$\longrightarrow$	0.8591
181.47	$\longrightarrow$	2.2588
46,700.	$\longleftarrow$	4.6693

The solution in the transformed domain involves next an addition yielding 4.6693. Finally, the solution in the original system of numbers is obtained by consulating the logarithm table. This last step is known as *performing the inverse transformation through the use of tables.* For the example the inverse operation yields the answer of 46,700. Thus multiplication, the desired operation, has actually been performed by addition, a simpler operation, in a transformed domain.

The Laplace transform achieves a similar result but on a different plane. Instead of simplifying the multiplication of numbers, it is concerned with functions and the simplification of operations of the calculus, such as differentiation and integration, into the easier operations of algebra such as multiplication and division. Note especially, however, that the method of procedure is identical. Thus to solve an integrodifferential equation the same basic steps are involved. (1) The Laplace transform is used to convert the integrodifferential equation into an algebraic equation. The algebraic equation is expressed in terms of a new (or transformed) variable. (2) The simpler operation of solving the algebraic equation is performed in the transformed domain. (3) The solution in the original domain is obtained by consulting appropriate tables.

4-2   The Laplace Transform: Definition and Usefulness

A typical form of the differential equations encountered in the study of electrical engineering is illustrated by Eq. (4-1):

$$a_2\frac{d^2i(t)}{dt^2} + a_1\frac{di(t)}{dt} + a_0i(t) - a_{-1}\int i(t)\,dt = f(t) \qquad (4\text{-}1)$$

where $f(t)$ refers to the applied forcing function, the coefficients a denote constants, and $i(t)$ is the solution or desired response. A test of the usefulness of the

Laplace transform is that it must succeed in converting this integrodifferential equation to a form which is easier to handle in finding the solution. Since Eq. (4-1) involves derivatives as well as an integral, and because of the importance of preserving the identity of the response function in spite of these operations of differentiation and integration, it can reasonably be concluded that the Laplace transform must in some way involve the exponential function. It will be recalled that the exponential function has the property that differentiation and integration always results in a new function which preserves the exponential character.

The direct Laplace transform of a function $f(t)$† is defined as follows

$$F(s) \equiv \mathscr{L}f(t) \equiv \int_0^\infty f(t)\epsilon^{-st}\,dt$$

(4-2)

where s is the intermediate or transformed variable. Note that s is part of the exponential function. Equation (4-2) states that to obtain the Laplace transform of a function $f(t)$, it must be multiplied by ϵ^{-st} and then integrated from $t = 0$ to $t = \infty$. Furthermore, in order that $F(s)$ be meaningful, it is necessary that the integral converge and that $f(t)$ be defined for $t > 0$ and equal to zero for $t \leq 0$. In general the variable s is complex. Accordingly, for most of the functions encountered in electrical engineering, convergence is assured by imposing the condition that the real part of s be positive, i.e. Re $[s] > 0$. Moreover, because t is used in this book exclusively to represent *time*, it follows that s must have the dimensions of inverse time, which is frequency. It is for these reasons, then, that the transformed variable is described as a *complex frequency*. An examination of Eq. (4-2) then reveals that after integration many time function can be expressed in terms of the complex frequency s, thus emphasizing that the Laplace transform is a mathematical tool which permits solving time-domain problems in the frequency domain.

To solve the integrodifferential equation by the Laplace transform, it is necessary to apply the Laplace integral to each term of the equation. This converts the original time-domain equation to one expressed entirely in terms of the intermediate variable s. In Eq. (4-1) the unknown response $i(t)$ is identified in the transform domain as $I(s)$, i.e., $\mathscr{L}i(t) = I(s)$. Also, a glance at Eq. (4-2) shows that $\mathscr{L}a_0i(t) = a_0I(s)$. Thus, one of the five terms of Eq. (4-1) is transformed. We next turn attention to the first derivative term.

How is the Laplace transform of a derivative term such as $a_1\,di(t)/dt$ found? The answer to this question is crucial in establishing whether or not the Laplace transform is useful in simplifying the solution of integradifferential equations. To demonstrate that the Laplace transform does, in fact, achieve this objective, let us start with the expression for the Laplace transform of $i(t)$. Thus

$$I(s) = \int_0^\infty i(t)\epsilon^{-st}\,dt$$

(4-3)

† The notation procedure in dealing with Laplace transform is to use lower-case letters to denote the time functions and capital letters for the corresponding Laplace-transformed function.

Although we do not as yet know the specific form of $i(t)$, we are assuming that it is a continuous function for $t > 0$ and that its Laplace transform does exist and is given by $I(s)$. The right side of Eq. (4-3) is readily recognized as the integral of the product of two variables. From the calculus we know that

$$\int u \, dv = uv - \int v \, du \tag{4-4}$$

Hence, letting

$$u = i(t), \qquad dv = \epsilon^{-st} \, dt$$
$$du = i'(t) \, dt, \qquad v = -\frac{1}{s}\epsilon^{-st} \tag{4-5}$$

and applying integration by parts yields

$$I(s) = -\frac{i(t)}{s}\epsilon^{-st}\bigg]_0^\infty + \frac{1}{s}\int_0^\infty i'(t)\epsilon^{-st} \, dt = \frac{i(0)}{s} + \frac{1}{s}\int_0^\infty i'(t)\epsilon^{-st} \, dt$$

Rearranging leads to

$$\int_0^\infty i'(t)\epsilon^{-st} \, dt = sI(s) - i(0) \tag{4-6}$$

An examination of the left side of Eq. (4-6) reveals that this is precisely the expression for the Laplace transform of the derivative of $i(t)$. Thus

$$\mathscr{L}\frac{di(t)}{dt} = \mathscr{L}i'(t) = \int_0^\infty i'(t)\epsilon^{-st} \, dt \tag{4-7}$$

Therefore Eq. (4-6) may be rewritten as

$$\boxed{\mathscr{L}i'(t) = sI(s) - i(0)} \tag{4-8}$$

Equation (4-8) is significant because it shows that the Laplace transform of a derivative in the original time domain carries over as the simpler operation of multiplication by the intermediate variable s in the transform domain. This result bears out the usefulness of the Laplace transform, for in transforming differentiation it preserves the transform of the original time function except for an algebraic operation. In addition, the term $i(0)$ describes in a direct and formal fashion the manner in which the initial condition is to be treated.

Extension of the Laplace transform to derivatives of higher order is readily accomplished by repeated application of the general procedure implied in Eq. (4-8). Thus the Laplace transform of the second derivative is

$$\mathscr{L}i''(t) = s[sI(s) - i(0)] - i'(0)$$
$$= s^2 I(s) - si(0) - i'(0) \tag{4-9}$$

For the nth derivative the Laplace transform is given by

$$\boxed{\mathscr{L}i^n(t) = s^n I(s) - s^{n-1}i(0) - s^{n-2}i'(0) - \cdots - i^{n-1}(0)} \tag{4-10}$$

Returning to Eq. (4-1) we see that we are now in a position to Laplace transform each term in the equation with the exception of the term involving the integral. By following a procedure similar to that used for differentiation it can be shown that the Laplace transform of an integral of time may be expressed as

$$\mathscr{L}i^{-1}(t) = \int_0^\infty \left[\int i(t)\, dt \right] \epsilon^{-st}\, dt = \frac{I(s)}{s} + \frac{i^{-1}(0)}{s} \tag{4-11}$$

where $i^{-1}(0)$ denotes the initial value of the integral term just before applying the forcing function. Note that integration in the time domain is transformed to the simpler operation of division by s in the frequency domain. Note too the presence of the initial-condition term associated with the energy-storing element which gives rise to the integral term.

For an n-order integral the result may be expressed as

$$\mathscr{L}i^{-n}(t) = \frac{I(s)}{s^n} + \frac{i^{-1}(0)}{s^n} + \frac{i^{-2}(0)}{s^{n-1}} + \cdots + \frac{i^{-n}(0)}{s} \tag{4-12}$$

Equations (4-10) and (4-12) illustrate another advantage of using the Laplace-transform method of solving differential equations. In situations were there are many energy-storing elements present (i.e. the order of the differential equation is high), this method indicates precisely how each initial-condition term enters into the expression for the derivatives and integrals.

With the results developed so far it is now possible to transform the integro-differential expression of Eq. (4-1) into a completely algebraic equation in the s domain. Thus, by means of Eqs. (4-3), (4-8), (4-9), and (4-11) the integro-differential equation becomes

$$a_2[s^2 I(s) - si(0) - i'(0)] + a_1[sI(s) - i(0)] + a_0 I(s)$$
$$+ a_{-1}\left[\frac{I(s)}{s} + \frac{i^{-1}(0)}{s} \right] = F(s) \tag{4-13}$$

where

$$F(s) = \mathscr{L}f(t) = \int_0^\infty f(t)\epsilon^{st}\, dt \tag{4-14}$$

The particular form of $F(s)$ depends upon the nature of $f(t)$. This matter is discussed in the next section.

It is significant to note that Eq. (4-13) is an algebraic expression which involves only one unknown quantity, namely, $I(s)$. Thus, the solution to the differential equation as it is found in the transform domain (also called the intermediate result) is

$$I(s) = \frac{F(s) + a_2 si(0) + a_2 i'(0) + a_1 i(0) - \dfrac{a_{-1} i^{-1}(0)}{s}}{a_2 s^2 + a_1 s + a_0 + \dfrac{a_{-1}}{s}} \tag{4-15}$$

Once $I(s)$ is found, the corresponding expression in the time domain is obtained from suitable tables as is done with logarithms. It is interesting to note in Eq. (4-15) that the initial-condition terms play a role similar to that of the driving function $F(s)$.

4-3   Laplace Transform of Common Forcing Functions

A formal solution to a linear integrodifferential equation is readily obtained in an intermediate form by Laplace transforming each term. This procedure leads to a result which is typically represented by Eq. (4-15). Here all the circuit parameters—a_1, a_2, and a_3—as well as all initial conditions are known. The only unknown term is $F(s)$. However, as Eq. (4-14) reveals, the specific form of $F(s)$ depends upon the nature of $f(t)$. Consequently, we next direct attention to the evaluation of $F(s)$ for those time functions which are commonly found in the study of electrical engineering. Once the Laplace transform $F(s)$ is found for a specific time function $f(t)$, it is tabulated for future reference just as is done for logarithms or integrals. Such a tabulation is particularly useful because, as can be shown, the Laplace transform of a time function, if it exists, is unique.

The Unit-Step Function u(t). This function is one of the most commonly used driving functions in electrical engineering. It frequently represents the closing of a switch which applies a constant voltage to a circuit or system. Mathematically, it is defined as

$$u(t) = \begin{cases} 0, & t \le 0 \\ 1, & t > 0 \end{cases} \tag{4-16}$$

and is plotted in Fig. 4-1. To determine the Laplace transform, we perform the integration called for by Eq. (4-2). Accordingly

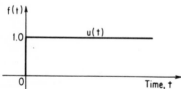

$$F(s) = \mathscr{L}u(t) = \int_0^\infty \epsilon^{-st}\, dt = \frac{1}{s} \tag{4-17}$$

Note that the discontinuous step function in the time domain is converted to a well-behaved analytic function in the s domain. The time function $u(t)$ and its corresponding Laplace transform $1/s$ constitute a unique transform-pair, and they are placed in Table 4-1 for ready reference. For situations where the step function has a magnitude F_0 different from one, it follows from Eq. (4-17) that the corresponding Laplace transform is F_0/s. That is, $\mathscr{L}F_0 u(t) = F_0/s$.

Fig. 4-1. Unit step function occurring at $t = 0$.

The Exponential Function, $f(t) = \epsilon^{-\alpha t}$. Direct application of Eq. (4-2) yields

$$F(s) = \mathscr{L}\epsilon^{-\alpha t} = \int_0^\infty \epsilon^{-\alpha t}\epsilon^{-st}\, dt = \frac{1}{s + \alpha} \tag{4-18}$$

This result is identified as the second Laplace-transform pair of Table 4-1. Keep in mind that t is restricted to positive values only.

The value of s in Eq. (4-18) for which the function becomes infinite is called a pole of $F(s)$. Hence it follows that $s = -\alpha$ is a pole of the Laplace transform of the exponential function. Poles are of particular importance to us because they characterize the nature of the time function associated with the solution in the s domain. This matter is discussed at length in Sec. 4-6.

The Sinusoidal Function, $f(t) = \sin \omega t$. For convenience in evaluating the Laplace integral the exponential form of $\sin \omega t$ is used. Thus

$$f(t) = \sin \omega t = \frac{\epsilon^{j\omega t} - \epsilon^{-j\omega t}}{2j} \tag{4-19}$$

Upon inserting into Eq. (4-2) we get

$$\mathscr{L} \sin \omega t = \frac{1}{2j} \int_0^\infty [\epsilon^{-(s-j\omega)t} - \epsilon^{-(s+j\omega)t}] \, dt$$

$$= \frac{1}{2j}\left[\frac{1}{s - j\omega} - \frac{1}{s + j\omega}\right] = \frac{\omega}{s^2 + \omega^2} \tag{4-20}$$

This appears as Laplace-transform pair 3 in Table 4-1. Here note that there are two poles of the Laplace transform function located at $\pm j\omega$ on the imaginary axis.

The Consinusoidal Function, $f(t) = \cos \omega t$. By recognizing that

$$\cos \omega t = \frac{1}{\omega}\frac{d}{dt}\sin \omega t$$

and applying the differentiation theorem we can write

$$\mathscr{L} \cos \omega t = \frac{1}{\omega}[sF(s) - f(0)]$$

where $F(s) = \mathscr{L} \sin \omega t = \omega/s^2 + \omega^2$. Therefore it follows that

$$\mathscr{L} \cos \omega t = \frac{1}{\omega}\left[\frac{s\omega}{s^2 + \omega^2} - 0\right] = \frac{s}{s^2 + \omega^2} \tag{4-21}$$

This result appears as transform pair 4 in Table 4-1.

The Ramp Function, $f(t) = t$. Applying the Laplace integral to the time function yields

$$F(s) = \int_0^\infty t\epsilon^{-st} \, dt \tag{4-22}$$

Integration by parts then leads to

$$F(s) = \frac{-t\epsilon^{-st}}{s}\Big]_0^\infty + \int_0^\infty \frac{1}{s}\epsilon^{-st} \, dt = \frac{1}{s^2} \tag{4-23}$$

Equation (4-23) reveals that the Laplace transform of the ramp function involves a double pole at the origin of the s domain.

The Parabolic Function, $f(t) = t^2$. Although a direct application of the Laplace integral can be used, it is more convenient here to use the integration theorem. However, it is first necessary to write t^2 in terms of an integral. Thus

$$\mathscr{L}t^2 = \mathscr{L}\int 2t \, dt \tag{4-24}$$

Table 4-1 Table of Laplace-Transform Pairs

$f(t)$	$f(t)$ vs. t, where $f(t) = 0$ for $t \leq 0$	$F(s)$	Pole location in s plane
1. $u(t)$		$\dfrac{1}{s}$	
2. $\epsilon^{-\alpha t}$		$\dfrac{1}{s + \alpha}$	
3. $\sin \omega t$		$\dfrac{\omega}{s^2 + \omega^2}$	
4. $\cos \omega t$		$\dfrac{s}{s^2 + \omega^2}$	
5. t		$\dfrac{1}{s^2}$	Double pole at $S = 0$
6. t^2		$\dfrac{2}{s^3}$	Triple pole at $S = 0$
7. t^n		$\dfrac{n!}{s^{n+1}}$	$(n+1)$ multiple pole at $S = 0$
8. $\epsilon^{-\alpha t} t^n$		$\dfrac{n!}{(s + \alpha)^{n+1}}$	$(n+1)$ multiple pole at $S = -\alpha$

Table 4-1 Table of Laplace-Transform Pairs (cont.)

$f(t)$	$f(t)$ vs. t, where $f(t) = 0$ for $t \leq 0$	$F(s)$	Pole location in s plane
9. $\epsilon^{-\alpha t} \sin \omega t$		$\dfrac{\omega}{(s+\alpha)^2 + \omega^2}$	
10. $\epsilon^{-\alpha t} \cos \omega t$		$\dfrac{s+\alpha}{(s+\alpha)^2 + \omega^2}$	
11. $\dfrac{\epsilon^{-\alpha t} - \epsilon^{-\beta t}}{\beta - \alpha}$		$\dfrac{1}{(s+\alpha)(s+\beta)}$	
12. $\sinh \beta t$		$\dfrac{\beta}{s^2 - \beta^2}$	
13. $\cosh \beta t$		$\dfrac{s}{s^2 - \beta^2}$	
14. $u(t - T)$		$\dfrac{\epsilon^{-sT}}{s}$	
15. $(t - T)u(t - T)$		$\dfrac{\epsilon^{-sT}}{s^2}$	Double pole at $S = 0$
16. $tu(t - T)$		$\dfrac{(1 + sT)\epsilon^{-sT}}{s^2}$	Double pole at $S = 0$
17. $\delta(t)$		1	
18. $\delta(t - T)$		ϵ^{-sT}	

Then by the integration theorem

$$\mathscr{L} \int 2t \, dt = \frac{F_1(s)}{s} + \frac{f_1^{-1}(0)}{s} = \frac{F_1(s)}{s} + 0 \tag{4-25}$$

where $F_1(s) = \mathscr{L}2t = 2/s^2$. Inserting this result into Eq. (4-25) gives

$$\mathscr{L}t^2 = \frac{2}{s^3} \tag{4-26}$$

It is interesting to note that the square function in time involves a cubic inverse function of s. As a matter of fact an extension of the foregoing analysis leads to the following general result:

$$\mathscr{L}t^n = \frac{n!}{s^{n+1}} \tag{4-27}$$

This result appears as item 7 in Table 4-1.

4-4 Laplace Transform of the Unit-impulse Function

The unit impulse can be of considerable importance in analyzing the transient behavior of systems. For example, the transient response of a system to an arbitrary forcing function can be determined once the system's response to the unit impulse is established. This is an especially useful tool in light of the fact that the unit-impulse response of linear systems is related in a simple way to its transfer function.

One mathematical definition[†] of the impulse function is

$$\delta(t) \equiv \lim_{\Delta t \to 0} \frac{u(t) - u(t - \Delta t)}{\Delta t} \tag{4-28}$$

which will be recognized as the derivative of the unit-step function. In other words,

$$\delta(t) \equiv \frac{d}{dt} u(t) \tag{4-29}$$

Strictly speaking, therefore, Eq. (4-29) has the value zero for $t > 0$ and infinity at $t = 0$. Now, because the infinity value at $t = 0$ is unrealistic, we introduce a new function

$$g(t) = 1 - \epsilon^{-\alpha t} \tag{4-30}$$

which can be made to represent the unit-step function very closely by choosing α very large (see Fig. 4-2). The use of $g(t)$ offers the advantage of dealing with a function which has a continuous first derivative. Specifically, its value is

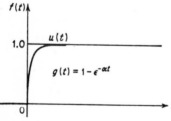

Fig. 4-2. Approximation of the unit-step function.

$$g'(t) = \frac{d}{dt} g(t) = \alpha \epsilon^{-\alpha t} \tag{4-31}$$

---

† For other definitions see S. Goldman, "Transformation Calculus and Electrical Transients," p. 101, Prentice-Hall, Inc., Englewood Cliffs, N. J., 1953.

An important property of this last equation is that

$$\int_0^\infty g'(t)\, dt = \int_0^\infty \alpha \epsilon^{-\alpha t}\, dt = 1 \tag{4-32}$$

regardless of the value of α. Equations (4-30) to (4-32) can now be used to lead us to a more realistic interpretation of the impulse function. It will be observed that, by making $\alpha \to \infty$, $g(t)$ approaches $u(t)$. Furthermore, by Eq. (4-31) the ordinate intercept approaches infinity, and by Eq. (4-32) the area enclosed by $g'(t)$ remains equal to 1. It therefore follows that, as the ordinate intercept approaches infinity, the abscissa intercept approaches zero with the enclosed area always equal to unity. The impulse function therefore refers to any quantity which has so large a value occurring in so negligibly small an interval of time that it causes a finite change in the state of the system.

The direct Laplace transform of the unit impulse is readily established from Eq. (4-31) as α is made to approach infinity. Thus,

$$\mathscr{L}\, \delta(t) = \lim_{\alpha \to \infty} \mathscr{L}\, g'(t) = \lim_{\alpha \to \infty} \mathscr{L}\, \alpha \epsilon^{-\alpha t} = \lim_{\alpha \to \infty} \frac{\alpha}{s + \alpha} = 1$$

The Laplace transform of a unit impulse may also be found by letting the height of a unit rectangular pulse approach infinity and the width approach zero in such a way as to maintain the enclosed area equal to unity as required by Eq. (4-32). This is illustrated in Fig. 4-3. The expression in the time domain for such a pulse is

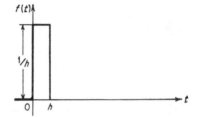

Fig. 4-3. Prelimit form of unit impulse in the form of a rectangular pulse.

$$f(t) = \frac{1}{h}[u(t) - u(t - h)] \tag{4-33}$$

The corresponding Laplace transform is

$$\mathscr{L}\, \text{pulse} = \frac{1}{h}\left(\frac{1}{s} - \frac{\epsilon^{-sh}}{s}\right) = \frac{1}{sh}(1 - \epsilon^{-sh}) \tag{4-34}$$

Inserting the power-series equivalent of ϵ^{-sh} yields

$$\mathscr{L}\, \text{pulse} = \frac{1}{sh}\left[1 - \left(1 - sh + \frac{s^2 h^2}{2!} - \frac{s^3 h^3}{3!} + \cdots + \right)\right]$$
$$= \left(1 - \frac{sh}{2!} + \frac{s^2 h^2}{3!} - \cdots\right)$$

Accordingly,

$$\mathscr{L}\, \text{unit impulse} = \mathscr{L}\, \delta(t) = \mathscr{L}\, \lim_{h \to 0} \text{pulse} = 1 \tag{4-35}$$

Consider next the evaluation of the Laplace transform of the unit impulse by yet another method—the use of Eq. (4-8). Since by Eq. (4-29) the impulse function is the derivative of the step function, application of Eq. (4-8) permits us to write

$$\mathscr{L}\, \delta(t) = sF(s) - u(0) \qquad \text{where } F(s) = \frac{1}{s} \tag{4-36}$$

Accordingly, we have

$$\mathscr{L}\,\delta(t) = s\frac{1}{s} - u(0) = 1 - u(0) \tag{4-37}$$

A little thought reveals a point of difficulty because of the ambiguity associated with the term $u(0)$. This difficulty indicates that a modification of the definition of the Laplace transform integral is clearly in order. A reasonable modification appears to be to change the lower limit in the integral definition of Eq. (4-2) to either 0^+ or 0^-. It is instructive to note that if 0^+ is used, then $u(0)$ in Eq. (4-37) becomes $u(0^+) = 1$. This leads to a result for the Laplace transform of the unit impulse contrary to that already derived by other means. On the other hand, if the lower limit of the integral is chosen to be 0^-, a perfectly consistent result ensues.

Consequently, it will be our practice from here on in this book to use an integral definition of the Laplace transform that employs 0^- as the lower limit. Thus, repeating the definition of Eq. (4-2), we now write

$$\boxed{F(s) \equiv \int_{0^-}^{\infty} f(t)\epsilon^{-st}\,dt} \tag{4-38}$$

Correspondingly, the Laplace transforms of all integral and derivative terms will now involve 0^- rather than 0. Thus

$$\mathscr{L}\,f'(t) = sF(s) - f(0^-) \tag{4-39}$$

and so on for higher derivative terms.

4-5 Useful Laplace-transform Theorems

There are four Laplace-transform theorems which when combined with the results of the preceding section provide a sufficient number of Laplace-transform pairs to take care of the vast majority of situations encountered in the study of electrical engineering. These are now described without proof.

Time-displacement Theorem (Real Translation). This theorem states that if a function $f(t)$ is Laplace-transformable and if the $\mathscr{L}f(t) = F(s)$, then

$$\mathscr{L}f(t - T) = \epsilon^{-sT}F(s) \tag{4-40}$$

The function $f(t - T)$ is the function $f(t)$ displaced by T. Hence by this theorem displacement in the time domain becomes multiplication by ϵ^{-sT} in the s domain. A useful application of this theorem is illustrated by the following example.

EXAMPLE 4-1 Find the Laplace transform of the pulse depicted in Fig. 4-4.

solution: Note that the pulse may be considered as composed of two delayed unit steps. Accordingly, we may write as the time expression for the pulse

$$\text{pulse} = u(t - T_1) - u(t - T_2)$$

then

$$\mathscr{L}(\text{pulse}) = \mathscr{L}u(t - T_1) - \mathscr{L}u(t - T_2) = \frac{\epsilon^{-sT_1}}{s} - \frac{\epsilon^{-sT_2}}{s} \tag{4-41}$$

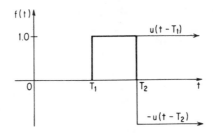

Fig. 4-4. Pulse as the superposition of two delayed steps.

Again note that the Laplace transform converts a discontinuous time function into an expression which is analytical† in the s domain.

Complex Translation. Of particular concern to us in our study throughout the book is the meaning attached to transform functions which involve a displacement in the complex variable s. We already know how to interpret any of the $F(s)$ functions treated in the preceding section. However, suppose now that for any one of these expressions the variable s is replaced by $(s + \alpha)$. How does this influence the corresponding time function? The answer is furnished by the complex translation theorem, which states that if $f(t)$ is Laplace-transformable and has the transform $F(s)$, then multiplication of $f(t)$ by the exponential time function $\epsilon^{-\alpha t}$ becomes a translation in the s domain and vice versa. Expressed mathematically we can write

$$\mathscr{L}\epsilon^{-\alpha t}f(t) = F(s + \alpha) \tag{4-42}$$

where $\mathscr{L}f(t)$ is $F(s)$.

By applying this theorem to known Laplace-transform pairs we may readily derive additional Laplace-transform pairs, as illustrated by the following two examples.

EXAMPLE 4-2 Find the Laplace transform of the exponentially damped cosine function.

solution: In accordance with the complex translation theorem this merely requires replacing s by $s + \alpha$ in the expression for the Laplace transform of the pure cosine function. Thus

$$\mathscr{L}\epsilon^{-\alpha t} \cos \omega t = \frac{s + \alpha}{(s + \alpha)^2 + \omega^2} \tag{4-43}$$

This result appears as item 10 in Table 4-1.

EXAMPLE 4-3 Find the Laplace transform of $\epsilon^{-\alpha t}t^n$.

solution: Applying the complex translation theorem to Eq. (4-27) yields

$$\mathscr{L}\epsilon^{-\alpha t}t^n = \frac{n!}{(s + \alpha)^{n+1}} \tag{4-44}$$

This is a very useful result as is demonstrated later in the book.

† That is, all derivatives of Eq. (4-41) exist in the s-plane except at $s = 0$.

Final-value Theorem. This theorem states that if $f(t)$ and its first derivative $f'(t)$ are Laplace-transformable and the $\mathscr{L}f(t) = F(s)$ and if the poles of $sF(s)$ lie *inside* the left half of the s-plane, then

$$\lim_{s \to 0} sF(s) = \lim_{t \to \infty} f(t) \qquad (4\text{-}45)$$

Equation (4-45) is useful in those situations where the transform solution of a problem is available and all that is needed is information about the final or steady-state solution. By performing the operation called for on the left side of Eq. (4-45), we obtain information about the final value without ever having to evaluate the complete solution in the time domain.

Initial-value Theorem. This theorem states that if $f(t)$ and $f'(t)$ are Laplace-transformable, $\mathscr{L}f(t) = F(s)$, and the $\lim_{s \to \infty} sF(s)$ exists, then

$$\lim_{s \to \infty} sF(s) = \lim_{t \to 0} f(t) \qquad (4\text{-}46)$$

Equation (4-46) permits us to evaluate the initial value of the time-domain solution $f(t)$ without ever having to determine $f(t)$ formally. One can work directly with the transform version of the solution to obtain initial values by performing the operation called for on the left side of Eq. (4-46).

4-6   Example. Laplace-transform Solution of First-order Equation

In the interest of bringing the material treated so far into sharper focus and to provide background for the next section, we turn attention next to the Laplace method of solving a linear, first-order differential equation. Refer to Fig. 4-5, which depicts a mechanical system comprised of a spring constant K and a viscous damper D. Mass is negligible. Let it be desired to describe as a function of time the velocity of the lower end of the spring, point p, when it is subjected to a constant downward force of magnitude F. Also, because initial conditions can be handled with ease in a direct and formal way by the Laplace-transform method, let us not hesitate to include this in the analysis by assuming that the spring is already stretched by an amount y_0 in the downward direction owing to a previous excitation. The equation which defines the velocity of point p is readily obtained by equating the applied force F to the sum of the opposing forces. Thus

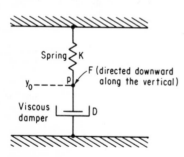

Fig. 4-5. Spring-damper arrangement. Force applied to lower end of spring at point p.

$$Fu(t) = Dv(t) + K \int v(t)\, dt \qquad (4\text{-}47)$$

Note that the equation has been written to bring into evidence the velocity $v(t)$ which is the variable of interest. The first term on the right side refers to the opposing torque associated with the viscous friction disk. The viscous friction coefficient D has units of force/velocity. The second term involves the integral of velocity which is displacement, and this displacement multiplied by the spring constant K, having units of force/displacement, yields the opposing force represented by the further stretching of the spring. The applied-force term F has the symbol $u(t)$ attached in order to emphasize that the force is applied at $t = 0$.

As the first step in the systematic Laplace-transform solution of this equation, it is necessary to identify the desired solution $v(t)$ in terms of its s-domain counterpart. Accordingly, we have

$$\mathscr{L}v(t) = V(s) \tag{4-48}$$

It is customary to use lower-case letters to denote the dependent variable in the time domain, and capital letters to denote the dependent variable in the s domain. Next we Laplace transform each term of Eq. (4-47). Thus

$$F\mathscr{L}u(t) = D\mathscr{L}v(t) + K\mathscr{L}\int v(t)\,dt \tag{4-49}$$

Then from item 1 in Table 4-1

$$F\mathscr{L}u(t) = \frac{F}{s} \tag{4-50}$$

Furthermore from Eq. (4-11) the Laplace transform of the integral term is

$$K\mathscr{L}\int v(t)\,dt = K\left[\frac{V(s)}{s} + \frac{\int_{-\infty}^{0} v_0\,dt}{s}\right] = K\left[\frac{V(s)}{s} + \frac{y_0}{s}\right] \tag{4-51}$$

where $\int_{-\infty}^{0} v_0\,dt$ denotes the initial displacement y_0 stored in the spring and brought about by an assumed previously applied velocity v_0.

Inserting Eqs. (4-48), (4-50), and (4-51) into Eq. (4-49) yields the Laplace-transformed differential equation. Thus

$$\frac{F}{s} = DV(s) + K\left[\frac{V(s)}{s} + \frac{y_0}{s}\right] \tag{4-52}$$

Since this equation involves only the single unknown quantity $V(s)$, the solution is immediately available in the s domain as

$$V(s) = \frac{F}{D}\left(\frac{1}{s + K/D}\right) - \frac{Ky_0}{D}\left(\frac{1}{s + K/D}\right) \tag{4-53}$$

It is important here to understand that Eq. (4-53) is the complete solution to the differential equation as it appears in the transform domain. To find the corresponding time solution, it is necessary to identify the meaning of terms such as $1/(s + K/D)$. If we continue to draw upon our experience of doing multiplication by logarithms, we can be led to expect that the inversion process of going from the s domain to the time domain can effectively be accomplished by the use of tables. Of course we understand that we must first prove that such a procedure is valid. We can accomplish this proof by means of the theory of functions of a complex variable; therefore Table 4-1 can be used in a bilateral fashion. On this basis, then, the term $1/(s + K/D)$ has a unique counterpart in the time domain, which by item 2 in Table 4-1 is identified as the exponential function $\epsilon^{-(K/D)t}$. It therefore follows that the time solution which corresponds to the s-domain solution of Eq. (4-53) is

$$v(t) = \frac{F}{D}\epsilon^{-(K/D)t} - \frac{Ky_0}{D}\epsilon^{-(K/D)t} \tag{4-54}$$

This is the complete solution to the problem described by Eq. (4-47) with the specified initial condition. Note that the initial condition is responsible for the second term on the right side of Eq. (4-54).

The foregoing example may also be used to illustrate the final-and initial-value theorems. A glance at Eq. (4-54) indicates that the final value (i.e. $t \rightarrow \infty$) of the velocity of point p is zero. This result can also be obtained by applying the final-value theorem to the transformed solution. Thus

$$\begin{aligned} v_{\text{final}} &= \lim_{s \to 0} s[V(s)] \\ &= \lim_{s \to 0} s\left[\frac{F}{D}\left(\frac{1}{s + K/D}\right) - \frac{Ky_0}{D}\left(\frac{1}{s + K/D}\right)\right] = 0 \end{aligned} \tag{4-55}$$

Note that there is no need to deal with the solution as a time function in order to obtain information about the final value of the solution. It can be done solely in terms of the s-plane expression, provided the specified conditions on the use of the theorem are satisfied.

Inspection of Eq. (4-54) also reveals that the initial value of the velocity immediately after application of the force is given by

$$v_{\text{initial}} = \frac{F}{D} - \frac{Ky_0}{D} \tag{4-56}$$

That this same result is obtained from the initial-value theorem is readily demonstrated as follows:

$$\begin{aligned} v_{\text{initial}} &= \lim_{s \to \infty} sV(s) = \lim_{s \to \infty} s\left[\frac{F/D}{s + K/D} - \frac{Ky_0/D}{s + K/D}\right] \\ &= \lim_{s \to \infty}\left[\frac{F/D}{1 + \frac{1}{s}\frac{K}{D}} - \frac{Ky_0/D}{1 + \frac{1}{s}\frac{K}{D}}\right] = \frac{F}{D} - \frac{Ky_0}{D} \end{aligned} \tag{4-57}$$

Although a comparison of the two methods of finding the initial value for the example at hand shows the initial-value theorem at a slight disadvantage in terms of the effort involved, this is not true when the s-domain solution is more complicated.

The process of finding the time solution by going from the s domain to the time domain through the use of appropriate tables can be described as *performing the inverse Laplace transformation*. In the next section we consider the matter of manipulating complicated expressions of transformed solutions into forms which are available in our listing of Laplace-transform pairs as presented in Table 4-1.

4-7 Inverse Laplace Transformation Through Partial-fraction Expansion

In Sec. 4-2 it is shown that Eq. (4-15) is the transformed solution of the integrodifferential equation (4-1). In the case where $F(s)$ in Eq. (4-15) corresponds to a unit-impulse function the expression for the transformed solution may be written as†

$$I(s) = \frac{a_2 s^2 + a_1 s + a_0}{s(s^3 + b_2 s^2 + b_1 s + b_0)} \tag{4-58}$$

Now if this form of the s-plane solution were available in a table of Laplace-transform pairs, the time solution could be written forthwith. A little thought reveals, however, that such a compilation is impractical because to cover all cases the table would have to be endless. Therefore it is customary to make available only a limited number of the basic forms and then to treat all other cases by reducing them through partial-fraction expansion into the basic forms. To accomplish this in the case of Eq. (4-58), it is necessary first to find the roots of the cubic equation by any of the standard methods of algebra.‡ For the sake of illustration assume these roots are found to be s_1, s_2, and s_3. Then Eq. (4-58) may be rewritten in terms of a partial-fraction expansion as

$$I(s) = \frac{a_2 s^2 + a_1 s + a_0}{s(s - s_1)(s - s_2)(s - s_3)} = \frac{K_0}{s} + \frac{K_1}{s - s_1} + \frac{K_2}{s - s_2} + \frac{K_3}{s - s_3} \tag{4-59}$$

Now since each term on the right side of this expression is identifiable in Table 4-1, a complete description of the time solution merely requires an evaluation of the K coefficients.

Before proceeding further with the partial-fraction expansion as a means of facilitating the evaluation of the inverse Laplace transformation, let us formulate the transformed solution of Eq. (4-59) in completely general terms. Accordingly we may write

$$I(s) = \frac{a_m s^m + a_{m-1} s^{m-1} + \cdots + a_1 s + a_0}{s^n + b_{n-1} s^{n-1} + \cdots + b_1 s + b_0} = \frac{N(s)}{D(s)} \tag{4-60}$$

Since in the vast majority of physical situations the order of the denominator polynomial is very often greater than that of the numerator, i.e. $n > m$, the partial-fraction expansion is directly applicable. In those rare cases where $n = m$ it is necessary to do longhand division in order that $I(s)$ be in the form of a proper fraction. Because of its infrequent occurrence this case will receive no further attention. Hence in the material that follows $I(s)$ is assumed to be a proper fraction so that partial-fraction expansion is immediately applicable. However, once this is done, the particular manner of evaluating the coefficients of the expansion is

---

† Note that the a-coefficients in Eq. (4-58) are not the same as those appearing in Eq. (4-15).

‡ This part of the solution procedure is also necessary when solving differential equations by the classical method.

dependent upon the character of the roots of $D(s)$ in Eq. (4-60). Attention is now directed to the three cases of interest which arise.

   $I(s)$ **Contains First-order Poles Only.** Assume that the roots of the denominator polynomial

$$D(s) = s^n + b_{n-1}s^{n-1} + \cdots + b_1s + b_0 = 0 \qquad (4\text{-}61)$$

are all real and distinct. Then Eq. (4-60) may be rewritten in terms of a partial-fraction expansion as

$$
\begin{aligned}
I(s) &= \frac{N(s)}{(s - s_1)(s - s_2)\cdots(s - s_n)} \\
&= \frac{K_1}{s - s_1} + \frac{K_2}{s - s_2} + \cdots + \frac{K_k}{s - s_k} + \cdots + \frac{K_n}{s - s_n}
\end{aligned}
\qquad (4\text{-}62)
$$

To evaluate any one of the coefficients it is necessary to isolate it in Eq. (4-62). This is readily accomplished by multiplying both sides of the equation by the root factor of the coefficient of interest. Thus, to find K_k requires multiplying through by $(s - s_k)$ and then setting s equal to s_k. This leads to the following general formulation for determining the coefficients.

$$K_k = \left[(s - s_k)\frac{N(s)}{D(s)}\right]_{s=s_k} \qquad (4\text{-}63)$$

Once the K_k coefficient is found, the contribution of the Kth term in the expansion is obtained by recognizing that the inverse Laplace transform of $1/(s - s_k)$ is $\epsilon^{s_k t}$. It follows then that the complete time solution corresponding to Eq. (4-62) is

$$i(t) = \mathscr{L}^{-1}I(s) = \sum_{k=1}^{n} K_k \epsilon^{s_k t} \qquad \text{for } t > 0 \qquad (4\text{-}64)$$

where K_k is given by Eq. (4-63) and $\mathscr{L}^{-1}$ is the symbol which denotes the inverse Laplace transformation.

EXAMPLE 4-4   Find the inverse Laplace transform of

$$I(s) = \frac{s + 3}{s(s + 1)(s + 2)}$$

solution: By a partial-fraction expansion we can write

$$I(s) = \frac{s + 3}{s(s + 1)(s + 2)} = \frac{K_1}{s} + \frac{K_2}{s + 1} + \frac{K_3}{s + 2}$$

Applying the operation called for in Eq. (4-63) yields

$$K_1 = [sI(s)]_{s=0} = \left[\frac{s + 3}{(s + 1)(s + 2)}\right]_{s=0} = \frac{3}{2}$$

$$K_2 = [(s + 1)I(s)]_{s=-1} = \left[\frac{s + 3}{s(s + 2)}\right]_{s=-1} = -2$$

$$K_3 = [(s + 2)I(s)]_{s=-2} = \left[\frac{s + 3}{s(s + 1)}\right]_{s=-2} = \frac{1}{2}$$

Hence

$$i(t) = \mathcal{L}^{-1}I(s) = \mathcal{L}^{-1}\frac{3}{2}\frac{1}{s} - 2\mathcal{L}^{-1}\frac{1}{s+1} + \frac{1}{2}\mathcal{L}^{-1}\frac{1}{s+2} = \frac{3}{2} - 2\epsilon^{-t} + \frac{1}{2}\epsilon^{-2t}$$

I(s) Contains a Pair of Complex Poles Plus One First-order Pole. In this case $I(s)$ may be written as

$$I(s) = \frac{N(s)}{(s - s_1)(s - s_1^*)(s - s_2)} = \frac{K_1}{s - s_1} + \frac{K_1^*}{s - s_1^*} + \frac{K_2}{s - s_2} \tag{4-65}$$

where s_1 and s_1^* are the complex conjugate roots and s_2 is the real root. Assume

$$s_1 = -\alpha + j\omega \quad \text{and} \quad s_1^* = -\alpha - j\omega$$

Then the coefficients K_1 and K_1^* are also, in general, complex conjugate. Hence we may write

$$K_1 = A + jB \quad \text{and} \quad K_1^* = A - jB$$

Inserting these quantities into Eq. (4-65) permits us to write it as

$$I(s) = \frac{A + jB}{(s + \alpha) - j\omega} + \frac{A - jB}{(s + \alpha) + j\omega} + \frac{K_2}{s - s_2} \tag{4-66}$$

Combining the first two terms on the right side and simplifying leads to

$$\boxed{I(s) = \frac{2A(s + \alpha) - 2B\omega}{(s + \alpha)^2 + \omega^2} + \frac{K_2}{s - s_2}} \tag{4-67}$$

The values of A and B are found by applying Eq. (4-63). Thus

$$\boxed{K_1 \equiv A + jB = \left[\frac{N(s)}{D(s)}(s - s_1)\right]_{s=-\alpha+j\omega}} \tag{4-68}$$

Since evaluation of the right side will in general be a complex number, it follows that

$$A \equiv \text{Re}(K_1) \quad \text{and} \quad B \equiv \mathcal{I}\text{m}(K_1) \tag{4-69}$$

where $\text{Re}(K_1)$ denotes the real part of K_1, and $\mathcal{I}\text{m}(K_1)$ denotes the imaginary part of K_1. Keep in mind, too, that K_2 is evaluated in the usual fashion.

Equation (4-67) is in a form which permits the corresponding time functions to be identified from the standard types appearing in Table 4-1. Thus

$$i(t) = 2A\mathcal{L}^{-1}\frac{s + \alpha}{(s + \alpha)^2 + \omega^2} - 2B\mathcal{L}^{-1}\frac{\omega}{(s + \alpha)^2 + \omega^2} + K_2\mathcal{L}^{-1}\frac{1}{s - s_2} \tag{4-70}$$

Then by transform-pairs 2, 9, and 10 the time solution becomes

$$\boxed{i(t) = 2A\epsilon^{-\alpha t}\cos\omega t - 2B\epsilon^{-\alpha t}\sin\omega t + K_2\epsilon^{s_2 t} \qquad t > 0} \tag{4-71}$$

I(s) Contains Multiple Poles Plus One First-order Pole. In the interest of illustrating the procedure more fully, a root of $D(s)$ of multiplicity three is assumed.

If we call this root s_0, the expression for $I(s)$ becomes

$$I(s) = \frac{N(s)}{(s - s_0)^3(s - s_1)} = \frac{K_{01}}{(s - s_0)^3} + \frac{K_{02}}{(s - s_0)^2} + \frac{K_{03}}{s - s_0} + \frac{K_1}{s - s_1} \qquad (4\text{-}72)$$

Note that when a multiple root is involved, the rules of the partial-fraction expansion require that there be as many coefficients associated with the multiple root as the order of the multiplicity. Moreover, these coefficients must be associated with fractions involving the multiple root which differ in power by one in the manner exhibited on the right side of Eq. (4-72). The general technique for evaluating the coefficients remains the same: it requires isolation through appropriate algebraic manipulation. In the case of K_{01} this merely necessitates multiplying both sides of Eq. (4-72) by $(s - s_0)^3$.

$$(s - s_0)^3 \frac{N(s)}{D(s)} = K_{01} + (s - s_0)K_{02} + (s - s_0)^2 K_{03} + (s - s_0)^3 \frac{K_1}{s - s_1} \qquad (4\text{-}73)$$

Upon insertion of $s = s_0$ all terms on the right side drop out with the exception of K_{01}. Hence

$$K_{01} = \left[(s - s_0)^3 \frac{N(s)}{D(s)}\right]_{s = s_0} \qquad (4\text{-}74)$$

To evaluate K_{02} requires that it, too, be made to stand alone. Inspection of Eq. (4-73) indicates that the most expedient way of accomplishing this is to differentiate both sides of the equation. Thus

$$\frac{d}{ds}\left[(s - s_0)^3 \frac{N(s)}{D(s)}\right]$$
$$= 0 + K_{02} + 2(s - s_0)K_{03} + 3(s - s_0)^2 \frac{K_1}{s - s_1} - (s - s_0)^3 \frac{K_1}{(s - s_1)^2} \qquad (4\text{-}75)$$

Again upon inserting $s = s_0$ we have

$$K_{02} = \left\{\frac{d}{ds}\left[(s - s_0)^3 \frac{N(s)}{D(s)}\right]\right\}_{s = s_0} \qquad (4\text{-}76)$$

A similar procedure makes available the expression for K_{03}. Differentiating Eq. (4-75) and setting $s = s_0$ yields

$$K_{03} = \left\{\frac{1}{2}\frac{d^2}{ds^2}\left[(s - s_0)^3 \frac{N(s)}{D(s)}\right]\right\}_{s = s_0} \qquad (4\text{-}77)$$

On the basis of Eqs. (4-76) and (4-77) a general expression for the coefficients associated with a multiple pole of $I(s)$ is readily seen to be

$$K_{0m} = \left\{\frac{1}{(m - 1)!}\frac{d^{m-1}}{ds^{m-1}}\left[(s - s_0)^p \frac{N(s)}{D(s)}\right]\right\}_{s = s_0} \qquad (4\text{-}78)$$

where p denotes the order of multiplicity and $m = 1, 2, \ldots, p$.

The last remaining coefficient K_1 in Eq. (4-72) is determined by Eq. (4-63). Accordingly the time solution may now be expressed as

$$i(t) = K_{01} \mathcal{L}^{-1} \frac{1}{(s - s_0)^3} + K_{02} \mathcal{L}^{-1} \frac{1}{(s - s_0)^2} + K_{03} \mathcal{L}^{-1} \frac{1}{s - s_0} + K_1 \mathcal{L}^{-1} \frac{1}{s - s_1}$$

$$\tag{4-79}$$

By use of transform-pairs 2 and 8 in Table 4-1 the final time solution becomes

$$\boxed{i(t) = \frac{K_{01}}{2} t^2 \epsilon^{s_0 t} + K_{02} t \epsilon^{s_0 t} + K_{03} \epsilon^{s_0 t} + K_1 \epsilon^{s_1 t}} \qquad t > 0 \qquad (4\text{-}80)$$

Therefore by means of a partial-fraction expansion, Eq. (4-78), and a table of Laplace-transform pairs the time solution for a transformed solution involving multiple poles is found in a direct and systematic fashion.

PROBLEMS

4-1 Obtain the Laplace transform of the following differential equations. Assume zero initial conditions except where noted.

(a) $\dfrac{d^2 m(t)}{dt^2} + \dfrac{dm(t)}{dt} = A \sin 2\omega t.$

(b) $\dfrac{d^3 m(t)}{dt^2} + a_2 \dfrac{d^2 m(t)}{dt^2} + a_1 \dfrac{dm(t)}{dt} + a_0 m(t) + a_{-1} \displaystyle\int m(t)\, dt = t \epsilon^{-\alpha t}.$

(c) $\dfrac{d^2 c(t)}{dt^2} + 3 \dfrac{dc(t)}{dt} + 2c(t) = t \left(\text{initial conditions:} \dfrac{dc}{dt} = 2,\, c = -1 \right).$

4-2 Find the Laplace transform of the following functions:

(a) $f(t) = \dfrac{\epsilon^{-\alpha t} + \alpha t - 1}{\alpha^2}.$

(b) $f(t) = \sin(\omega t + \theta).$

(c) $f(t) = \epsilon^{-\alpha t} \sin(\omega t + \theta).$

4-3 Find the Laplace transform of the time function depicted in Fig. P4-3.

4-4 Determine the Laplace transform of the time function depicted in Fig. P4-4.

4-5 Determine the Laplace transform of the time function shown in Fig. P4-5.

4-6 Find the inverse Laplace transform of $F(s) = 1/(s + 2)^3.$

4-7 The behavior of an engineering device is described by the differential equation

$$\frac{d^2 x}{dt^2} + 14 \frac{dx}{dt} + 40x = 5$$

It is desired to find the solution for x subject to the initial conditions $dx/dt = 2$ and $x = 1$.

(a) By using the Laplace transform find the s-domain solution for x.

(b) Determine the corresponding time solution for the result found in part (a).

4-8 The flow of current in an electric circuit is described by the differential equation:

$$\frac{d^3 i}{dt^3} + 14.1 \frac{d^2 i}{dt^2} + 41.4 \frac{di}{dt} + 4i = 10$$

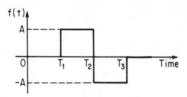

Figure P4-3

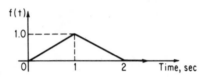

Figure P4-4

It is desired to find the solution for i subject to the initial condition that $i(0) = 0.725$ ampere.

    (a) Through use of the Laplace transform, obtain the s-domain solution for the current.

    (b) Determine the corresponding time solution for the result found in part (a).

4-9   Find the initial value (at $t = 0^+$) of the time function specified by

$$\mathscr{L}f(t) = F(s) = \frac{(2s + 5)(s + 4)}{(s + 6)^2(s + 2)}$$

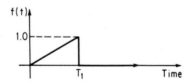

Figure P4-5

4-10   Determine the time function corresponding to

$$F(s) = \frac{s + 3}{(s + 1)(s + 2)^2}$$

4-11   Find the general form of the differential equation which leads to the $F(s)$ function specified in Prob. 4-10.

4-12   The Laplace transform of a time function is

$$F(s) = \frac{10}{s^2 + 4s + 8}$$

Find the time function.

4-13   Determine the time function whose Laplace transform is given by

$$F(s) = \frac{10}{s^2 + 6s + 8}$$

4-14   The Laplace transform function is given by

$$F(s) = \frac{3s + 8}{s^2 + 6s + 25}$$

What is the corresponding $f(t)$?

4-15   Obtain the time function whose Laplace transform is given by

$$F(s) = \frac{10}{s(s^2 + 6s + 8)}$$

4-16   Compute the initial value of the time function found in Prob. 4-15. Check this value by applying the initial-value theorem to the $F(s)$ function.

4-17   An engineering device contains two independent energy-storing elements. When subjected to a step forcing function, the governing differential equation becomes

$$\frac{dx}{dt} + 12x + 100 \int x \, dt = 300u(t)$$

Just prior to the application of the forcing function the initial value of x is known to be 10 and the intial value of the integrating device is known to be 6.4, i.e., $\int_{-\infty}^{0} x_0 \, dt = 6.4$.

   (a) By means of the Laplace transform, determine the response in the transformed domain.
   (b) Does the response have a steady-state value different from zero in the time domain? Why?
   (c) Find the expression for the complete solution in the time domain.

4-18   The Laplace-transformed solution for the current in a circuit that contains two energy-storing elements is given by

$$I(s) = \frac{105}{s(s^2 + 10s + 21)}$$

The initial conditions are known to be zero.

   (a) What type and what magnitude of forcing function was used?
   (b) What are the characteristic modes of the dynamic response?
   (c) What is the final value of the current?

4-19 Find the expression for the Laplace transform of $v(t)$ depicted in Fig. P4-19.

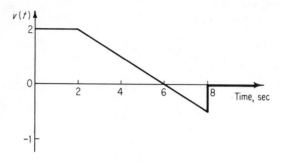

Figure P4-19

5

forced and transient responses to deterministic inputs

The study of the dynamic behavior of linear circuits and systems containing one or more energy-storing elements is of considerable importance to the engineer for two reasons. First, he often wants to know how long it takes for the circuit to respond to applied forcing functions. Recall that the energy in an energy-storing element such as an inductor or capacitor cannot change instantaneously. Accordingly, if an applied forcing function demands an increase in the amount of energy storage in a particular part of a circuit, this increase must occur gradually. The change cannot take place in a discontinuous manner, because then infinite forces would be needed. The period of adjustment during which the stored energy changes from some initial level to a new, commanded, final level is called the *settling time* of the circuit (or system). In many engineering applications it is important to keep the response time within tolerable limits.

Second, in situations where there are two or more energy-storing elements the engineer must be able to predict the occurrence of severe oscillations as the circuit (or system) changes from one energy state to another. In electric circuits such oscillations can readily cause ruinous voltages or currents. In electromechanical systems, such as a servomechanism, these oscillations can cause excessively high torques which in turn may damage mechanical parts. In fact, in some extreme

cases the interchange of energy during the transient state is such that the oscillations started by the application of the forcing function do not cease. Accordingly,
the new steady-state level is never reached. The system or circuit is then described
as being in a state of sustained oscillations. Whether or not this is a desirable
state of affairs depends upon the objective which the engineer has in mind. If his
intention is to build an oscillator, then his goal is achieved. However, if his intention is to build a follow-up control system which reaches its new steady-state
position after the elapse of a reasonable amount of time, then clearly a sustained
oscillation must be avoided at all costs.

It is the purpose of this chapter to furnish the background which will allow
the reader to determine the complete response of circuits and systems, when
subjected to conventional forcing functions. One commonly used forcing function is the step input of voltage or current or force or torque. The sinusoid is
another frequently used forcing function. In addition, there are the ramp function $[f(t) = t]$ and the parabolic function $[f(t) = t^2]$. It is interesting to note that a
Laplace transform exists for each of these forcing functions.

The complete response of a circuit to a deterministic input can always be
identified in two parts—a forced solution and a transient solution. In linear
systems and circuits the forced solution is readily recognizable because it has the
same form as the forcing function. Hence, if the forcing function is a sinusoid,
the forced response must itself be a sinusoid. However, the character of the transient solution remains the same irrespective of the type of forcing function used.
The transient solution makes possible a smooth transition from one energy level
to another in the circuit or system. The transient solution is a means of describing
the manner in which the circuit reacts to satisfy the boundary (or initial) conditions
upon application of a forcing function which calls for a change in energy state.
Thus the characteristic behavior of the transient solution is determined solely by
the circuit (or system) parameters.

The first necessary step in finding the complete solution of a circuit to a deterministic input is to identify correctly the differential equation for the circuit.
Once this is done, the Laplace transform method can be applied to it to find the
complete solution directly. This procedure is employed throughout this chapter.

5-1 Step Response of an *R-L* Circuit

Any circuit composed of resistance and inductance represents a situation
in which there occurs a dissipation of energy as well as a storage of energy in a
magnetic field. Since we now know how to deal with both voltage and current
sources, the step response of the *R-L* circuit is found for each source type. Attention is directed first to the step-voltage response.

Appearing in Fig. 5-1 is the circuit arrangement of an initially de-energized
inductor in series with a resistor. It is desired to find the complete solution for
the current when the switch S is closed. The governing differential equation results
upon applying Kirchhoff's voltage law to the circuit of Fig. 5-1. Thus

$$E = Ri + L\frac{di}{dt} \qquad (5\text{-}1)$$

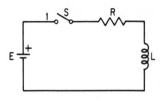

The current i is the solution being sought and E is the applied forcing function which causes the response i to exist. To emphasize that E exists only when the switch in Fig. 5-1 is closed, it is helpful to attach the unit step function $u(t)$ to the voltage source E in Eq. (5-1). Hence a more complete

Fig. 5-1. Series *R-L* circuit.

formulation of the governing differential equation of the circuit is

$$Eu(t) = Ri + L\frac{di}{dt} \qquad (5\text{-}2)$$

Equation (5-2) is a linear differential equation of first order. It contains a single energy-storing element as evidenced by the first derivative term.

The first step in solving Eq. (5-2) by the Laplace-transform method is to Laplace transform both sides of the equation. Thus

$$E\mathcal{L}u(t) = R\mathcal{L}i(t) + L\mathcal{L}\frac{di(t)}{dt} \qquad (5\text{-}3)$$

The desired solution in the time domain is identified as $i(t)$. The corresponding unknown function in the s domain is called $I(s)$. That is,

$$I(s) = \mathcal{L}i(t) \qquad (5\text{-}4)$$

Inserting Eqs. (5-4), (4-39), and (4-17) into Eq. (5-3) then yields the transformed algebraic counterpart of the governing differential equation (5-2).

$$\frac{E}{s} = RI(s) + L[sI(s) - i(0^-)] \qquad (5\text{-}5)$$

Here $i(0^-)$ is used to emphasize that information about the initial value of current immediately before application of the forcing function E is needed before a formal solution for the transformed response $I(s)$ can be written.

Because the switch in Fig. 5-1 is initially open, it follows that $i(0^-) = 0$. Hence Eq. (5-5) may now be explicitly written as

$$\frac{E}{s} = I(s)[R + sL] \qquad (5\text{-}6)$$

Therefore the transformed solution for the response becomes

$$I(s) = \frac{E/L}{s(s + R/L)} \qquad (5\text{-}7)$$

To obtain the time solution which corresponds to Eq. (5-7) we must perform the inverse Laplace transformation, which requires putting $I(s)$ into a form whose terms are readily identifiable in Table 4-1. This result is achieved by using a partial-fraction expansion. Accordingly

$$I(s) = \frac{E/L}{s(s + R/L)} = \frac{K_0}{s} + \frac{K_1}{s + R/L} \qquad (5\text{-}8)$$

Then by the application of the operation called for in Eq. (4-63) there results

$$K_0 = \left[\frac{E/L}{s + R/L} \right]_{s=0} = \frac{E}{R} \tag{5-9}$$

and

$$K_1 = \left[\frac{E/L}{s} \right]_{s=-R/L} = -\frac{E}{R} \tag{5-10}$$

The transformed solution may now be written more conveniently as

$$I(s) = \frac{E}{R} \frac{1}{s} - \frac{E}{R} \left[\frac{1}{s + R/L} \right] \tag{5-11}$$

From Table 4-1 we know the equivalent of $1/s$ to be a unit step in the time domain. Also the equivalent of $1/(s + R/L)$ is $\epsilon^{-(R/L)t}$ in the time domain. Hence the time solution which corresponds to the transformed solution of Eq. (5-11) is

$$i(t) = \mathscr{L}^{-1} I(s) = \frac{E}{R} \mathscr{L}^{-1} \frac{1}{s} - \frac{E}{R} \mathscr{L}^{-1} \frac{1}{s + R/L} \tag{5-12}$$

or

$$\boxed{i(t) = \frac{E}{R} - \frac{E}{R} \epsilon^{-(R/L)t}} \qquad \text{for} \quad t > 0 \tag{5-13}$$

In Eq. (5-12) the symbol $\mathscr{L}^{-1}$ is used to denote the inverse Laplace transformation. By means of Table 4-1 this operation is carried out by proceeding from the *s*-function column to the corresponding time-function column. Recall that the direct Laplace transformation involves going from the time-function to the *s*-function column.

Equation (5-13) is the complete solution of Eq. (5-2), which is the governing differential equation for the circuit of interest. Note that it consists of two parts. One part is E/R. This is called the *forced* solution. It depends directly upon the forcing function E and has the same form as E—that is, a constant. The forced solution is often called the *steady-state* solution because, after all transient terms have virtually disappeared, it is the only current which remains for the circuit of Fig. 5-1 with the switch closed. Mathematicians prefer to call this component of the solution the *particular* solution. The second part of the total solution is one which decays with time. Although it requires infinite time for this term to reach zero, in practice it takes a relatively short time for it to become negligibly small. As a matter of fact, when the power of ϵ has a value of five, this term is reduced to less than one per cent of its maximum value. In most situations this is equivalent to saying that the exponentially decaying term is for all intents and purposes equal to zero. As a result of its decaying nature this second part of the total solution is called the *transient* solution. An examination of Eq. (5-13) reveals that its presence in the solution is demanded by the need to satisfy the boundary conditions immediately upon application of the forcing function. That is, the forcing function E wants to establish the forced solution E/R, but the presence of the inductance prevents this from occurring instantly. As a result the circuit reacts in such a way as to assure a smooth transition from the initial to the final energy state. This is achieved through the generation of a component of the solution which not only satisfies the boundary conditions at $t = 0^+$ but also permits the forced solution to exist after a suitable period of adjustment. For the *R-L* circuit under discussion note that the equation for the transient solution is

$$i_t = -\frac{E}{R}\epsilon^{-(R/L)t} \tag{5-14}$$

At $t = 0^+ i_t$ has that value of $-E/R$ which it must have so that the total current at this time instant be zero as called for by the boundary condition, Equation (5-13) makes this obvious. However, note that as time elapses the transient term decays to zero, leaving just the forced solution. Of course as i_t decays, time is allowed for transferring energy to the inductance. When steady state is virtually reached, the energy stored in the inductor is

$$W = \frac{1}{2}L\left(\frac{E}{R}\right)^2 \quad \text{joules} \tag{5-15}$$

It is because this energy cannot be transferred instantaneously that the need for the transient term arises.

Inspection of Eq. (5-14) makes it clear that the decay of the transient term is solely dependent upon the circuit parameters—*R* and *L*—and entirely independent of the forcing function. This is certainly not unexpected when it is recalled that the transient term is really a description of the manner in which the circuit reacts to external disturbances, whatever their origin or nature. The only influence that the forcing function has on the transient term is in determining the magnitude of the coefficient of the exponential, but it in no way influences the rate of decay of the transient term. It can readily be shown that Eq. (5-14) is a solution to the homogeneous or force-free differential equation, which in this case takes the form

$$0 = Ri + L\frac{di}{dt} \tag{5-16}$$

For this reason the transient solution is often called the *source-free* solution. Mathematicians, however, prefer to call this the *complementary function*, because, when it is combined with the forced solution, a total function is obtained which satisfies the nonhomogeneous differential equation (5-2) for all time instants after switching.

Time Constant. A plot of Eqs. (5-13) and (5-14) appears in Fig. 5-2. Note that at $t = 0^+$ the steady-state and the transient solutions have equal and opposite values, thus yielding a zero value for the total current as called for by the boundary condition. Note also that as the transient term decays to zero, the total current reaches its forced value. In the interest of establishing a convenient measure of the duration of the transient, let us look more closely at the exponential function of Eq. (5-14). A glance at the power of the exponential, which must be a numeric, reveals that the quantity L/R must in turn bear units of time. Because of this fact

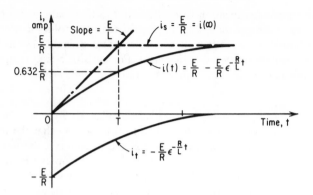

Fig. 5-2. The build-up of current in a series *R-L* circuit.

and because the quantity is determined solely in terms of the circuit parameters, it is called the *time constant* of the circuit and is denoted by T. Thus

$$T \equiv \frac{L}{R} \quad \text{sec} \tag{5-17}$$

A verification that L/R has units of seconds follows directly from the units for L and R. Thus

$$T = \frac{L\left(\dfrac{\text{volts} \times \text{sec}}{\text{amp}}\right)}{R\left(\dfrac{\text{volts}}{\text{amp}}\right)} = \frac{L}{R}\,(\text{sec}) \tag{5-18}$$

To understand how the time constant serves as a convenient measure of the duration of the transient, one need merely observe that at $t = T$ the value of the transient term from Eq. (5-14) is

$$i_t(T) = -\frac{E}{R}\epsilon^{-T/T} = -\frac{E}{R}\epsilon^{-1} = -0.368\frac{E}{R} \tag{5-19}$$

Therefore in a period equal to one time constant the transient term has decayed to 36.8 per cent of its initial value. Upon the elapse of a period equal to two time constants, the transient term has a value of

$$i_t(2T) = -\frac{E}{R}\epsilon^{-2} = -0.135\frac{E}{R} \tag{5-20}$$

Similarly

$$i_t(3T) = -\frac{E}{R}\epsilon^{-3} = -0.0498\frac{E}{R} \tag{5-21}$$

and

$$i_t(5T) = -\frac{E}{R}\epsilon^{-5} = -0.0067\frac{E}{R} \tag{5-22}$$

As the right side of Eq. (5-29) indicates, the transfer function indirectly conveys information about the transient response of the circuit. The transfer function involves just the circuit parameters.

EXAMPLE 5-1 In the circuit depicted in Fig. 5-4, the battery has been applied for a long time. For the values of the parameters indicated find the complete expression for the current after closing the switch which removes R_1 from the circuit.

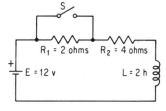

solution: The governing differential equation which applies when the switch is closed is

$$E\mu(t) = R_2 i + L\frac{di}{dt} \qquad (5\text{-}30)$$

Fig. 5-4. Circuit arrangement for Example 5-1.

Laplace transforming yields

$$\frac{E}{s} = R_2 I(s) + sLI(s) - Li(0^-) \qquad (5\text{-}31)$$

$$\frac{E}{s} + Li(0^-) = I(s)[R_2 + sL] \qquad (5\text{-}32)$$

Equation (5-32) is written with the forcing function and initial conditions on the left side and the transformed response and the operational impedance on the right side. This formulation indicates that the operational impedance may be singled out even in the presence of nonzero initial conditions.

From Eq. (5-32) the solution in the *s* domain may accordingly be written as

$$I(s) = \frac{E + sLi(0^-)}{s(R_2 + sL)} = \frac{E/L + si(0^-)}{s(s + R_2/L)} \qquad (5\text{-}33)$$

A comparison of Eq. (5-33) with Eq. (5-28) shows how an initial-condition term enters into the formulation of the solution. It is combined with the forcing function to give an altered magnitude.

By means of a partial-fraction expansion Eq. (5-33) becomes

$$I(s) = \frac{K_0}{s} + \frac{K_1}{s + R_2/L} \qquad (5\text{-}34)$$

Then

$$K_0 = \left[\frac{E/L + si(0^-)}{s + R_2/L}\right]_{s=0} = \frac{E}{R_2} = \frac{12}{4} = 3 \text{ amps} \qquad (5\text{-}35)$$

And

$$K_1 = \left[\frac{E/L + si(0^-)}{s}\right]_{s=-R_2/L} = -\frac{E}{R_2} + i(0^-) \qquad (5\text{-}36)$$

The current just before switching is given by

$$i(0^-) = \frac{E}{R_1 + R_2} = \frac{12}{2 + 4} = 2 \text{ amps} \qquad (5\text{-}37)$$

Inserting this result into Eq. (5-36) yields

$$K_1 = -\frac{E}{R_2} + i(0^-) = -\frac{12}{4} + 2 = -1 \qquad (5\text{-}38)$$

Accordingly, the expression for $I(s)$ from Eq. (5-34) becomes

$$I(s) = \frac{3}{s} - \frac{1}{s + R_2/L} = \frac{3}{s} - \frac{1}{s + 2} \tag{5-39}$$

Performing the inverse Laplace transformation by means of items 1 and 2 of Table 4-1 furnishes the desired response.

$$i(t) = 3 - \epsilon^{-2t} \tag{5-40}$$

As a check note that at $t = 0^+$ the value of $i(t)$ is 2 amperes and upon full decay of the transient it assumes a value of 3 amperes.

The time constant for this problem is

$$T = \frac{L}{R} = \frac{2 \text{ h}}{4 \text{ ohms}} = \frac{1}{2} \text{ sec} \tag{5-41}$$

Thus after 2.5 seconds the transient term has a value less than 1 per cent of its initial value of 1 ampere.

Step-current Response of an R-L Circuit. Attention is next turned to finding the complete response of an *R-L* circuit to which is applied a step-current forcing function. Because we are now involved with current sources, the resistance and inductance are assumed placed in parallel with one another as illustrated in Fig. 5-5. Recall that it is characteristic of a current source to deliver to the circuit elements connected to it the total value of its current. Accordingly, to simulate the application of a step current to the *R-L* combination, it is merely necessary to open switch S in Fig. 5-5. Clearly, when S is closed the short circuit draws all of the current from the source, leaving no current to flow through *R* and *L*.

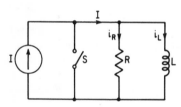

Fig. 5-5. Circuit for finding the step-current response of a parallel *R-L* circuit.

When switch S is opened, the source current flows through the parallel combination of resistance and inductance. By Kirchhoff's current law we then have

$$I = i_R + i_L \tag{5-42}$$

where i_R denotes the current which flows through the resistance and i_L denotes the current flowing through the inductance. Also, because of the parallel arrangement, the potential differences appearing across *R* and *L* are the same. Therefore we may further write

$$Ri_R = L\frac{di_L}{dt} \tag{5-43}$$

Inserting Eq. (5-42) into Eq. (5-43) leads to

$$RI = Ri_L + L\frac{di_L}{dt} \tag{5-44}$$

This expression is the governing differential equation for the circuit. Since *R* and *I* are both known, the left side of Eq. (5-44) is thereby known and effectively represents the forcing function. In fact the sole unknown in this equation is the current i_L. A comparison of Eq. (5-44) with Eq. (5-1) reveals them to be identical in form so that the solution may be taken to be Eq. (5-14) with the one modification that E is replaced by IR. Hence the complete solution can be written forthwith as

$$i_L = \frac{IR}{R}(1 - \epsilon^{-(R/L)t}) = I(1 - \epsilon^{-(R/L)t}) \tag{5-45}$$

The expression for the current through the resistor follows from Eq. (5-42)

$$i_R = I - i_L = I\epsilon^{-(R/L)t} \tag{5-46}$$

A study of Eq. (5-45) shows that at $t = 0^+$ there is no current flowing through the energy-storing element L, which is consistent with the boundary condition. Keep in mind that with S closed $i(0^-) = 0$. At the same time, however, the expression for i_R shows that all of the source current flows through R. This can take place instantly because there is no energy storage in a resistive element. As time progresses the current I transfers from a path through R to a path solely through L. This is again borne out by Eq. (5-45) and (5-46). After the elapse of five time constants, for all intents and purposes $i_L = I$ and $i_R = 0$. In the steady state the energy stored in the inductance is $\frac{1}{2}LI^2$.

5-2   Step Response of an *R-C* Circuit

Appearing in Fig. 5-6 is the circuit arrangement of a resistor in series with a capacitor. The capacitor is assumed to have an initial charge of value q_0 owing to the flow of a previously applied current. Let us now find the complete expression for the current after the switch S is closed. The governing differential equation is obtained upon applying Kirchhoff's voltage law to the closed circuit. Thus

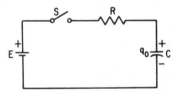

$$E\mu(t) = Ri + \frac{1}{C}\int i\,dt \tag{5-47}$$

Fig. 5-6. Series *R-C* circuit. The capacitor has an initial charge q_0.

Laplace transforming each term yields

$$E(s) = RI(s) + \frac{1}{C}\left[\frac{I(s)}{s} + \frac{i^{-1}(0)}{s}\right] \tag{5-48}$$

where

$$i^{-1}(0) = \int_{-\infty}^{0} i\,dt = q_0 \tag{5-49}$$

and represents the accumulation of charge on the capacitor brought about by the flow of current due to a previously applied action. This previous action is stressed in Eq. (5-49) by taking the limits of the integral from $-\infty$ to 0 where 0 refers to the instant of the closing of the switch in Fig. 5-6. Equation (5-48) may be rewritten as

$$E(s) = RI(s) + \frac{1}{C}\left[\frac{I(s)}{s} + \frac{q_0}{s}\right] \tag{5-50}$$

or

$$E(s) - \frac{q_0}{Cs} = I(s)\left[R + \frac{1}{sC}\right] \tag{5-51}$$

On the left side appear the forcing function and the effect of the initial condition. Appearing in brackets on the right side is the operational impedance of the R-C circuit.

For a step-voltage forcing function $E(s) = E/s$, so that Eq. (5-51) becomes

$$\frac{1}{s}\left(E - \frac{q_0}{C}\right) = I(s)\left[R + \frac{1}{sC}\right] \tag{5-52}$$

The quantity in parentheses in the last expression represents the effective voltage acting in the circuit to produce current. The transformed solution for the current is therefore

$$I(s) = \frac{E - \dfrac{q_0}{C}}{s\left(R + \dfrac{1}{sC}\right)} = \frac{E - \dfrac{q_0}{C}}{R\left(s + \dfrac{1}{RC}\right)} \tag{5-53}$$

The corresponding total solution in the time domain readily follows from item 2 of Table 4-1. Thus

$$\boxed{i(t) = \left(\frac{E}{R} - \frac{RC}{q_0}\right)\epsilon^{-t/RC}} \tag{5-54}$$

A plot of this equation as a function of time is depicted in Fig. 5-7. Note that at $t = 0^+$ there occurs a step change in current in the circuit. Its magnitude is equal to the difference between the applied voltage E and the initial voltage across the capacitor q_0/C divided by the resistance of the circuit. As time elapses this current gradually decays to zero.

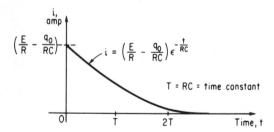

Fig. 5-7. Exponential decay of the current in an R-C circuit.

An inspection of the exponential term of Eq. (5-54) reveals that the time constant of an R-C circuit is given by

$$\boxed{T = RC} \quad \text{sec} \tag{5-55}$$

The complete current solution in the case of the simple R-C circuit consists merely of the transient term because the capacitor presents an open circuit to the battery source at steady state. Hence the forced solution is zero in this instance.

EXAMPLE 5-2 In the circuit of Fig. 5-8 the capacitor has an initial charge of such value that the voltage appearing across its plates is 40 volts. For the values of the parameters indicated find the expression for the current which flows when the switch is closed. Also, find the total energy dissipated in the resistor during the transient state and compare it with the energy initially stored in the capacitor.

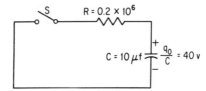

Fig. 5-8. Circuit configuration for Example 5-2.

solution: Taking voltage drops in a clockwise direction as positive, Kirchhoff's voltage law applied to the circuit with S closed yields

$$0 = Ri + \frac{1}{C}\int i \, dt \tag{5-56}$$

Upon Laplace transforming there results

$$0 = RI(s) + \frac{1}{C}\left[\frac{I(s)}{s} + \frac{q_0}{s}\right] \tag{5-57}$$

or

$$-\frac{q_0}{Cs} = I(s)\left(R + \frac{1}{sC}\right) \tag{5-58}$$

By the statement of the problem $q_0/C = 40$ volts. Hence

$$I(s) = \frac{-40}{s\left(R + \frac{1}{sC}\right)} = -\frac{40}{R}\left(\frac{1}{s + \frac{1}{RC}}\right) \tag{5-59}$$

From Table 4-1 the corresponding time solution is found to be

$$i(t) = -\frac{40}{R}\epsilon^{-t/RC} = -2 \times 10^{-4}\epsilon^{-t/2} \qquad \text{amp} \tag{5-60}$$

The negative sign indicates that the actual current flow is counterclockwise rather than clockwise as initially assumed.

The amount of energy dissipated in the resistor is readily found from

$$W_{\text{diss}} = \int_0^\infty i^2(t)R \, dt = \int_0^\infty (4 \times 10^{-8})2 \times 10^5\epsilon^{-t} \, dt \tag{5-61}$$
$$= -8 \times 10^{-3}[\epsilon^{-t}]_0^\infty = 8 \times 10^{-3} \text{ joules}$$

The amount of energy initially stored in the electric field of the capacitor is given by

$$W_e = \frac{1}{2}CV_0^2 = \frac{1}{2}10^{-5}(40)^2 = 8 \times 10^{-3} \text{ joules} \tag{5-62}$$

A comparison of Eqs. (5-61) and (5-62) shows that all of the capacitor energy is dissipated as heat after the switch is closed.

It does not always follow that all of the energy stored on a capacitor will be dissipated in a resistor appearing in series with it. This is illustrated by the following example.

EXAMPLE 5-3 In the circuit depicted in Fig. 5-9 assume that C_1 has an initial charge of such a magnitude that the potential difference appearing across its terminals is equal to

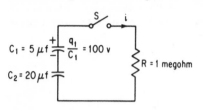

100 volts with a polarity which makes the upper plate positive relative to the lower plate. There is no initial charge on C_2. Determine the voltage which appears across each capacitor at steady state.

solution: Assuming a clockwise flow of current and writing Kirchhoff's voltage law for the circuit we have

Fig. 5-9. Circuit configuration for Example 5-3.

$$0 = Ri + \frac{1}{C_2} \int_0^t i\, dt + \frac{1}{C_1} \int i\, dt \qquad (5\text{-}63)$$

The last term is written as an indefinite integral to remind us that C_1 has an initial condition associated with it. The transformed version of Eq. (5-63) becomes

$$0 = RI(s) + \frac{1}{C_2}\left[\frac{I(s)}{s}\right] + \frac{1}{C_1}\left[\frac{I(s)}{s} - \frac{q_1}{s}\right] \qquad (5\text{-}64)$$

where q_1 denotes the initial charge on capacitor C_1. Also, the minus sign is used because the assumed positive flow of current through C_1 is in a direction which brings about a charge displacement on the plates of C_1 having a polarity opposed to that of the initial charge. Upon rearranging, the last expression becomes

$$I(s)\left(R + \frac{1}{sC}\right) = \frac{q_1}{sC_1} = \frac{100}{s} \qquad (5\text{-}65)$$

where

$$\frac{1}{C} = \frac{1}{C_1} + \frac{1}{C_2}$$

Inserting the specified values for C_1 and C_2 yields a value for C of $4\ \mu f$. The transformed solution for the current thus becomes

$$I(s) = \frac{100}{R}\left(\frac{1}{s + \frac{1}{RC}}\right) \qquad (5\text{-}66)$$

Then by Laplace-transform pair 2 of Table 4-1 the corresponding time solution is

$$i(t) = \frac{100}{R}\epsilon^{-t/RC} \qquad (5\text{-}67)$$

Accordingly, the voltage across C_1 is given by

$$
\begin{aligned}
v_1(t) &= -100 + \frac{1}{C_1}\int_0^t i(t)\, dt = -100 + \frac{1}{C_1}\frac{100}{R}\int_0^t \epsilon^{-t/RC}\, dt \\
&= -100 - \frac{1}{C_1}\frac{100}{R}(RC)[\epsilon^{-t/RC}]_0^t \\
&= -100 + \frac{C}{C_1}(100)(1 - \epsilon^{-t/RC}) \\
&= -100 + 80(1 - \epsilon^{-t/4})
\end{aligned}
\qquad (5\text{-}68)
$$

At steady state the voltage across C_1 is clearly -20 volts. That is,

$$v_2(\infty) = -100 + 80 = -20 \text{ volts} \qquad (5\text{-}69)$$

The minus sign is taken with respect to the clockwise direction of current. Therefore it means that there occurs a rise in voltage as C_1 is traversed in the direction of i. In other words the upper plate is 20 volts positive with respect to the lower plate.

The potential difference appearing across C_2 is expressed by

$$v_2(t) = \frac{1}{C_2} \int_0^t i(t)\, dt = \frac{100}{RC_2} \int_0^t \epsilon^{-t/RC}\, dt$$

$$= -\frac{C}{C_2}(100)[\epsilon^{-t/RC} - 1] \tag{5-70}$$

$$= \frac{C}{C_2}(100)[1 - \epsilon^{-t/RC}] = 20(1 - \epsilon^{-t/4})$$

At steady state the value of this voltage is

$$v_2(\infty) = 20 \text{ volts} \tag{5-71}$$

The plus sign here means that the bottom plate of C_2 is positive with respect to the upper plate.

Note that the potential difference across each capacitor is 20 volts with opposite signs. Consequently under these conditions the net potential difference across the resistor is zero, so that no further dissipation of the stored electric-field energy can take place. The energy is "trapped" by the action of the equal and opposite capacitor voltages.

Another interesting aspect of this example is revealed from a glance at the final expressions for $v_1(t)$ and $v_2(t)$. These equations show that the transient terms decay in accordance with an equivalent time constant which is determined by the circuit resistance and the equivalent capacitance as represented by Eq. (5-66).

Step-current Response of an R-C Circuit. Again in dealing with a current source the resistance and capacitance parameters are considered to be in parallel as illustrated in Fig. 5-10. As previously explained, a step-current forcing function is applied to the parallel combination by opening the switch. Then by Kirchhoff's current law we can write

$$I = i_R + i_C \tag{5-72}$$

Moreover the same potential difference appears across R and C. Hence

$$Ri_R = \frac{\int_0^t i_C\, dt}{C} \tag{5-73}$$

There can exist no initial charge on the capacitor because the switch S is initially assumed closed. Inserting Eq. (5-72) into Eq. (5-73) yields

$$RI = Ri_C + \frac{1}{C} \int_0^t i_C\, dt \tag{5-74}$$

The corresponding Laplace-transformed algebraic equation then becomes

$$\frac{RI}{s} = RI_C(s) + \frac{1}{sC} I_C(s) \tag{5-75}$$

or

$$I_C(s) = \frac{IR}{s\left(R + \frac{1}{sC}\right)} = I\left(\frac{1}{s + \frac{1}{RC}}\right) \tag{5-76}$$

By means of Laplace-transform pair 2 of Table 4-1 the time-domain solution for the capacitor current is found to be

$$i_C(t) = I\epsilon^{-t/RC} \tag{5-77}$$

Moreover, from Eqs. (5-77) and (5-72) the time solution for the current through the resistor is found to be

$$i_R(t) = I - i_C(t) = I(1 - \epsilon^{-t/RC}) \tag{5-78}$$

In the configuration of Fig. 5-10 the total source current initially flows entirely through the capacitor and then, as time elapses, it gradually transfers to the resistor. At steady state all of the source current flows through R and none flows through C.

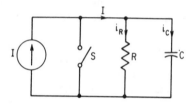

Fig. 5-10. Parallel R-C circuit with step-current forcing function.

5-3 Duality

Two circuits are said to be duals when the *mesh* equations that describe the behavior of one circuit are found to be identical in form to the *nodal* equations that describe the other. Duality, therefore, is a property of the circuit equations. There are times when the construction of the dual of a circuit is helpful in reducing the effort needed in analyzing simple, conventional circuits.

The origin of the principle of duality can be traced back to the laws of Ohm, Faraday and Coulomb. In the case of a voltage e applied to a series resistor R, we have, by Ohm's law, that

$$e = Ri \tag{5-79}$$

However, an alternative form of this equation states that the current i is expressed as the product of the conductance G and the voltage e appearing across G. That is,

$$i = Ge \tag{5-80}$$

A comparison of Eqs. (5-79) and (5-80) makes it plain that the roles of voltage and current are interchanged. Moreover, when the roles of current and voltage are interchanged, then the resistance R is replaced by the conductance G.

Similarly, by Faraday's law, we have the voltage and current of an inductor related by

$$e = L\frac{di}{dt} \tag{5-81}$$

For the capacitor the relationship is

$$i = C\frac{de}{dt} \tag{5-82}$$

Equation (5-81) is a mesh-oriented equation, since it is an expression for voltage. On the other hand, Eq. (5-82) is a node-oriented equation, because it is an expression for current. Again, note how one may proceed from the mesh orientation to the node orientation by merely interchanging the roles for voltage and current in the describing equations, and also by replacing the inductance parameter by the capacitance parameter. The reverse procedure is also valid. One useful feature of this duality property is that one can readily study, for example, the manner in which the voltage across a capacitor varies in response to a specific current source by simply investigating the variation of the current in an inductor subject to a voltage source of the same type as the current source.

Additional examples of the usefulness of duality can be cited. For example, it has already been demonstrated on several occasions *that the current through an inductor cannot change instantaneously*. A moment's thought should make it apparent that the dual statement is that *voltage across the capacitor cannot change instantaneously*. Here the dual words are *voltage* and *current*, *through* and *across*, *inductance* and *capacitance*. Similarly, we know that the voltage across an inductor can change instantly. The dual statement is that the current through a capacitor can change instantly—a result we are quite familiar with by now.

To consider the matter of duality further, refer to the simple series R-L circuit depicted in Fig. 5-11. The mesh equation for this circuit is clearly

$$e = Ri + L\frac{di}{dt} \tag{5-83}$$

Now, on the basis of the interchangeability of the roles of voltage and current, resistance and conductance, inductance and capacitance, we can write directly from Eq. (5-83) the current (or nodal) equation that identifies the dual circuit of Fig. 5-11. Thus

$$i = Ge + C\frac{de}{dt} \tag{5-84}$$

A circuit interpretation of this equation leads to the configuration appearing in Fig. 5-12. When the magnitude of G is selected equal to R, and the magnitude of C is chosen equal to L, an *exact dual* circuit results. The solution for e in Eq. (5-84) becomes exactly the solution for i in Eq. (5-83). If i in Eq. (5-83) has been previously found, then by the principle of duality the solution for the variation of e to a current source in the circuit of Fig. 5-12 is also known. To illustrate this point, let it be desired that the complete expression for the voltage e be found corresponding to the application of a step current $i = I$ in the circuit of Fig. 5-12.

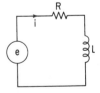

Fig. 5-11. Series R-L circuit.

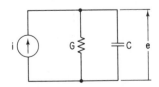

Fig. 5-12. Dual of the circuit shown in Fig. 5-11.

The solution for the current response in Fig. 5-11 to a step voltage $e = E$ was determined in Sec. 5-1 and is given by Eq. (5-14). By the principle of duality, then, the required solution may be written forthwith as simply

$$e(t) = \frac{I}{G}(1 - \epsilon^{-(G/C)t}) \tag{5-85}$$

The identification of the dual of a circuit need not proceed in the manner described in the foregoing. Rather, a direct graphical procedure can be employed. Thus, to draw the dual of the circuit of Fig. 5-11 directly, begin by drawing a broken line completely encircling the mesh circuit, as illustrated in Fig. 5-13a.

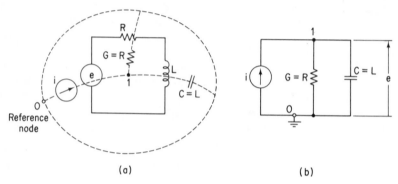

(a)   (b)

Fig. 5-13. (a) Illustrative of graphical procedure for identifying the dual circuit;
(b) The resulting dual circuit drawn in conventional form.

This closed loop represents the reference node of the nodal method. Next, place a dot at the center of each mesh of the planar network. Each dot denotes an independent node of the dual circuit. Obviously, for the circuit of Fig. 5-11, there is just one node required. Then identify appropriate branches between the node points by using a dual element for each element that appears in the original mesh circuit. Accordingly, the voltage source e is replaced by the current source i, the resistance R is replaced by the conductance G, and the inductance L is replaced by

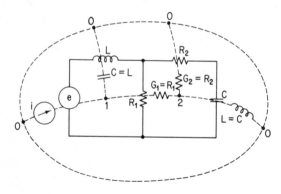

Fig. 5-14. Original two-mesh circuit of Example 5-4 together with graphical solution procedure for finding the dual circuit shown in broken lines.

the capacitance C. The resulting dual circuit is more clearly depicted in Fig. 5-13(b). The configuration is identical to that of Fig. 5-12.

EXAMPLE 5-4 Find the exact dual of the two-mesh circuit depicted in Fig. 5-14. Use the graphical procedure.

solution: Draw a broken line enclosing both meshes and call this the reference node 0. Place two dots in each mesh and call these nodes 1 and 2. Replace each element of the original circuit by its dual counterpart. The final result is shown in Fig. 5-15.

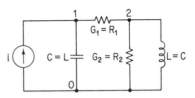

Fig. 5-15. The dual circuit of Fig. 5-14.

5-4 Pulse Response of the *R-C* Circuit

An interesting application of the theory of transients in *R-C* circuits occurs when a rectangular pulse is applied to a series combination of R and C, as depicted in Fig. 5-16. The capacitor is initially de-energized. Let it be desired to find the expressions describing the behavior of the current as well as the capacitor voltage for all time following the application of the pulse.

It is helpful at the outset to understand that any rectangular pulse can be considered as consisting of two step functions—one delayed with respect to the other. A glance at Fig. 5-17 makes this plain. The step function $e_1 = Eu(t)$ is assumed originating at time zero with magnitude E and existing to time infinity. The step function $e_2 = -Eu(t - T_1)$ is assumed originating at time $t = T_1$ with amplitude

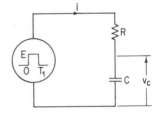

Fig. 5-16. Rectangular pulse applied to *R-C* circuit.

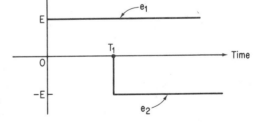

Fig. 5-17. Generation of the rectangular pulse by the superposition of two step functions—one delayed with respect to the other.

$-E$ and then persisting for all time thereafter. The superposition of these two functions yields a resultant function, which is a rectangular pulse having a magnitude E for the time interval $0 < t < T_1$, and zero for all other times.

To obtain the solution for the current in the circuit of Fig. 5-16, we begin by writing Kirchhoff's voltage law. Thus

$$e_1 + e_2 = \text{pulse} = iR + \frac{1}{C} \int i \, dt \qquad (5\text{-}86)$$

or

$$Eu(t) - Eu(t - T_1) = iR + \frac{1}{C} \int i \, dt \tag{5-87}$$

Laplace transforming both sides yields

$$E\left(\frac{1 - \epsilon^{-sT_1}}{s}\right) = I(s)\left(R + \frac{1}{sC}\right) = I(s)\frac{R}{s}\left(s + \frac{1}{RC}\right) \tag{5-88}$$

Thus the solution for the current in the transformed domain becomes

$$I(s) = \frac{E}{R}\left(\frac{1 - \epsilon^{-sT_1}}{s + (1/RC)}\right) \tag{5-89}$$

The corresponding time expression is found from Table 4-1 to be

$$i(t) = \frac{E}{R}[u(t)\epsilon^{-(t/RC)} - u(t - T_1)\epsilon^{-(t-T_1/RC)}] \tag{5-90}$$

This expression can be modified to make it apply specifically to various intervals of the time domain. For example, the analytical description of the current inside the interval from zero to T_1 is obtained from Eq. (5-90) by noting that $u(t - T_1) \equiv 0$ for $t < T_1$. Accordingly,

$$i(t) = \frac{E}{R}\epsilon^{-(t/RC)} \qquad \text{for} \quad 0 < t < T_1 \tag{5-91}$$

The corresponding expression for the voltage across the capacitor during this interval is readily obtained from

$$v_C(t) = \frac{1}{C}\int i \, d\tau = \frac{E}{RC}\int_0^t \epsilon^{-(\tau/RC)} \, d\tau = E(1 - \epsilon^{-(t/RC)}) \tag{5-92}$$

where τ is a dummy variable of integration and $0 < t < T_1$.

The expression for the current in the time interval beyond T_1 must plainly take into account the contribution from the second term appearing in Eq. (5-90). This contribution results from the presence of the negative step voltage occurring at $t = T_1^+$. It is helpful at this point to determine the value of the current at $t = T_1^+$, which is the time immediately following the drop of the pulse height to zero. Inserting $t = T_1^+$ into Eq. (5-90) thus yields

$$i(T_1^+) = \frac{E}{R}\epsilon^{-(T_1/RC)} - \frac{E}{R} = -\frac{E}{R}(1 - \epsilon^{-(T_1/RC)}) \tag{5-93}$$

Inspection of this expression discloses that in the interval from T_1^- to T_1^+, there is a step change in current by the amount $-E/R$, which is directly attributable to the negative delayed step function e_2 of Fig. 5-17. The net effect is that the current at T_1^+ has changed direction in the circuit of Fig. 5-16 and has the magnitude appearing in Eq. (5-93). Of course, as time is allowed to increase beyond T_1^+, the current expressed by Eq. (5-93) will decay in the customary exponential fashion in accordance with the circuit time constant RC. Thus the specific expression becomes

$$i(t') = -\left[\frac{E}{R}(1 - \epsilon^{-(T_1/RC)})\right]\epsilon^{-(t'/RC)} \tag{5-94}$$

where
$$t' = t - T_1$$
and
$$t > T_1$$

Note that the bracketed quantity is a constant. A graphical representation of the current variation over the entire time regime is depicted in Fig. 5-18 as the same as that of the voltage appearing across R. The variables $v_R(t)$ and $i(t)$ differ merely by a scale factor R.

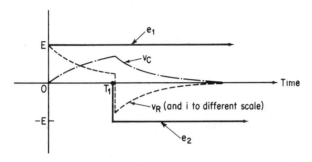

Fig. 5-18. Response of R-C circuit to rectangular pulse.

The voltage across the capacitor at time T_1 is obtained from Eq. (5-92) upon inserting $t = T_1$. Accordingly,

$$v_c(T_1) = E(1 - \epsilon^{-(T_1/RC)}) \qquad (5\text{-}95)$$

The sudden presence of the negative step voltage at T_1 in no way alters this quantity, because the voltage across a capacitor cannot change instantaneously in the presence of a finite force. As time elapses beyond T_1, however, this quantity, too, experiences an exponential decay consistent with the circuit time constant RC. Consequently, the expression for the capacitor voltage beyond T_1 becomes

$$v_c(t') = [E(1 - \epsilon^{-(T_1/RC)})]\epsilon^{-(t'/RC)} \qquad (5\text{-}96)$$

where $t' = t - T_1$ and $t > T_1$.

A plot of v_c also appears in Fig. 5-18. Note the absence of a discontinuity in the variation of the capacitor voltage in spite of the discontinuous character of the forcing function. The curves in Fig. 5-18 are drawn on the assumption that T_1 is comparable to the circuit time constant RC.

5-5 The Impulse Response

As a driving force, the impulse is a term often used to describe a situation in which an extraordinarily large forcing function is applied to a circuit or system over a very small period of time. The impulse can readily be derived as the limiting case of a pulse. In this connection, consider that the pulse depicted in Fig. 5-19(a)

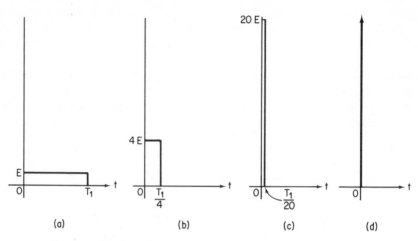

Fig. 5-19. Illustrating the impulse as the limiting case of the pulse:
(a) normal pulse;
(b) and (c) increased height and diminished pulse period for constant area;
(d) impulse function where height approaches infinity and duration approaches zero for constant area.

has its amplitude quadrupled and its period reduced by one quarter, as shown in Fig. 5-19(b). Note that the area is kept constant. Figure 5-19(c) shows the same pulse area, but the amplitude is now further increased by a factor of 5. In the limiting case, the height of the pulse is made to approach infinity, while the pulse duration is made to approach zero for constant area. Such a function is called the *impulse function* and is represented by the symbolism of Fig. 5-19(d). In each case illustrated in Fig. 5-19, the area of the pulse is given by

$$A = ET_1 \quad \text{volt-sec} \tag{5-97}$$

When this area is associated with the impulse function itself, it is often called the *strength* of the impulse. The *unit impulse function* results whenever A assumes a value of unity. On the basis of the foregoing description, the mathematical definition of the unit impulse function can be written as

$$\delta(t) = 0 \quad \text{for} \quad t \neq 0$$

and

$$\int_{-\infty}^{\infty} \delta(t)\, dt = \int_{0^-}^{0^+} \delta(t)\, dt = 1 \tag{5-98}$$

There are a number of areas in electrical engineering where waveshapes approaching the impulse function can be found serving practical purposes. This is particularly true in electronic circuits employing sweeps such as are found in television and radar applications. Similar high-force low-duration forces can be cited in other areas in engineering—for example, the hammer blow. However, knowledge of the impulse function and its properties is also useful from a purely theoretical point of view. One such application is the use of the impulse function to characterize the dynamical response of circuits and systems generally. This subject matter is too advanced to be treated in this book, but it is a point worth noting.

To focus attention on a specific problem, let it be desired to find the current response of a series R-C circuit to a driving force that is the impulse function of strength A. Refer to Fig. 5-20. The symbol used to denote the impulse function is $\delta(t)$. Since the impulse function is derived as a limiting case of the pulse function,

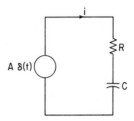

Fig. 5-20. Series R-C circuit with an applied impulse function.

we can reasonably expect that the results obtained for the pulse response can be employed, provided that appropriate modifications are made. Thus, the voltage appearing across the capacitor plates shortly after application of the impulse function should be derivable from Eq. (5-96). We need merely observe that T_1 in the case of the impulse is negligibly small compared to the circuit time constant $T = RC$. That is,

$$T_1 \ll T = RC \tag{5-99}$$

Repeating Eq. (5-96) for convenience, we have

$$v_c(T_1) = E(1 - \epsilon^{-(T_1/T)}) \tag{5-100}$$

At this point it is helpful to observe that when Eq. (5-99) prevails, the exponential term in the last equation may be replaced by

$$\epsilon^{-(T_1/T)} = 1 - \frac{T_1}{T} \tag{5-101}$$

Consequently, Eq. (5-100) becomes simply

$$v_c(T_1) = \frac{ET_1}{T} = \frac{\text{pulse area}}{T} = \frac{A}{T} \tag{5-102}$$

where, clearly, A carries the units of volt-seconds and T is in seconds.

A close examination of Eqs. (5-102) provides some interesting conclusions. For the case in which the pulse is modified to take on the properties of the impulse, T_1 obviously assumes an infinitesimal value, which we denote by 0^+. Certainly, in this instance, Eq. (5-101) is ever more valid. Accordingly, Eq. (5-102) may be simply rewritten as

$$v_c(0^+) = \frac{\text{pulse area}}{T} = \frac{A}{T} \tag{5-103}$$

By this result we must conclude that at an infinitesimal time after the application of the impulse, the voltage across the capacitor assumes the value A/T. At first this may appear to be a surprising result, for we learned previously that the voltage across a capacitor cannot change instantaneously. However, this statement had always been made subject to the condition "in the presence of a finite force," which is not true here. The impulse function has an amplitude approaching infinity, so it is capable of bringing about a sudden change.

It is now fair to speculate that the change in the capacitor voltage in 0^+ time must be the result of a sudden burst of current. In other words, there appears to exist an impulse of current that places the full charge on the capacitor plates consistent with the voltage A/T. To verify that this is indeed what happens, we begin by writing Kirchhoff's voltage law for the circuit of Fig. 5-20. Thus

$$A \, \delta(t) = Ri + \frac{1}{C} \int i \, dt \qquad (5\text{-}104)$$

Integrating both sides of the equation in the infinitesimal time interval from 0^- to 0^+ yields

$$A \int_{0^-}^{0^+} \delta(t) \, dt = R \int_{0^-}^{0^+} i \, dt + \frac{1}{C} \int_{0^-}^{0^+} \int_{0^-}^{0^+} i \, dt \qquad (5\text{-}105)$$

Examination of this last equation discloses that the term involving the double integral cannot possibly make any contribution to the equation balance. This follows from the fact that if the result of the first integration yields a finite value,[†] a second integration of this finite quantity over the infinitesimal time interval still gives a negligible sum. Hence, Eq. (5-105) reduces to

$$A \int_{0^-}^{0^+} \delta(t) \, dt = R \int_{0^-}^{0^+} i \, dt \qquad (5\text{-}106)$$

In view of the fact that A and R are both finite numbers, it must next be concluded that the only way for the impulse on the left side to be balanced on the right side is for a current impulse to exist on the right side. Accordingly, the waveshape for i in Eq. (5-106) must be one that has an amplitude approaching infinity and a duration approaching zero with the area fixed. Since the area of such a curve carries the units of amperes-seconds, which is the dimension of charge, we can properly replace the integrand on the right side of Eq. (5-106) by the impulse function $Q \, \delta(t)$, where Q denotes the area of the current pulse. Equation (5-106) thus becomes

$$A \int_{0^-}^{0^+} \delta(t) \, dt = R \int_{0^-}^{0^+} Q \, \delta(t) \, dt = RQ \int_{0^-}^{0^+} \delta(t) \, dt \qquad (5\text{-}107)$$

By the definition of the impulse function as described by Eq. (5-98), the last equation then reads simply as

$$A = RQ$$

or

$$Q = \frac{A}{R} \qquad (5\text{-}108)$$

---

† This result is possible only if the integrand is an impulse function.

Equation (5-108) represents the charge placed on the capacitor by the current impulse. The corresponding voltage is then

$$v_c(0^+) = \frac{Q}{C} = \frac{A}{RC} = \frac{A}{T} \qquad (5\text{-}109)$$

which agrees entirely with Eq. (5-103).

The solution for the current response after time 0^+ is straightforward. It is described simply by the action of "the initial condition" voltage on the capacitor discharging through the R-C circuit. Accordingly,

$$i(t) = -\frac{A}{RT}\epsilon^{-(t/T)} \qquad \text{for} \quad 0^+ < t < \infty \qquad (5\text{-}110)$$

The minus sign appears because the actual flow of current is opposite to the assumed positive direction shown in Fig. 5-20.

A more direct solution of the impulse response of the series R-C circuit is obtainable from a Laplace transform solution of Eq. (5-104). Thus, applying the Laplace transform to each term of Eq. (5-104) allows us to write

$$A = RI(s) + \frac{1}{C}\left[\frac{I(s)}{s} + \frac{\int_{-\infty}^{0^-} i\,dt}{s}\right] \qquad (5\text{-}111)$$

But the initial condition on the capacitor at time 0^- is zero. Accordingly, the last expression becomes simply

$$A = \left(R + \frac{1}{sC}\right)I(s) = \frac{R}{s}\left(s + \frac{1}{RC}\right)I(s) \qquad (5\text{-}112)$$

or

$$I(s) = \frac{A}{R}\left(\frac{s}{s + 1/RC}\right) = \frac{A}{R}\left[1 - \frac{1/RC}{s + 1/RC}\right] \qquad (5\text{-}113)$$

The bracketed version of the transformed solution is in the form of a proper fraction. Hence the inverse Laplace transform may be found by use of Table 4-1. Thus

$$i(t) = \mathscr{L}^{-1}I(s) = \mathscr{L}^{-1}\frac{A}{R} - \mathscr{L}^{-1}\frac{A}{RT}\frac{1}{s + 1/RC} \qquad (5\text{-}114)$$

which by transform pairs 17 and 2 yields the time expression for the current response of

$$i(t) = \frac{A}{R}\delta(t) - \frac{A}{RT}\epsilon^{-(t/T)} \qquad (5\text{-}115)$$

Then, by Eq. (5-108), this may be rewritten as

$$i(t) = Q\,\delta(t) - \frac{A}{RT}\epsilon^{-(t/T)} \qquad (5\text{-}116)$$

Examination of Eq. (5-116) discloses that the total current response consists of a current impulse of magnitude $Q = A/R$ occurring in the time interval from 0^- to 0^+, and a second component, $-(A/RT)\epsilon^{-(t/T)}$, which exists for all time from 0^+ to ∞. It is instructive to note the consistency of Eq. (5-116) with Eqs. (5-107) and (5-110).

5-6   Step Response of Second-order System (*R-L-C* Circuit)

The behavior of a circuit or system which contains two independent energy-storing elements is completely described by a second-order differential equation. Because the second-order system occurs frequently in engineering situations, considerable attention is given here to its analysis, especially with a view to putting the results in a universal form. In this way the results can be applied to second-order systems generally, independently of their particular composition. Thus the independent energy-storing elements may consist of inductance and capacitance or of mass and spring constant and so forth. The study of the second-order system is important also because the behavior of many higher-order systems can frequently be described in terms of an equivalent second-order system. Consequently the results developed here can frequently be applied to such systems with satisfactory results.

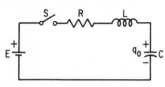

Fig. 5-21. Series *R-L-C* circuit with an initial charge on *C*.

Let it be desired to find the complete current response in the circuit of Fig. 5-21 after the switch is closed. Assume that the capacitor has an initial charge of q_0. The governing differential equation for the circuit is found upon applying Kirchhoff's voltage law. Thus

$$Eu(t) = Ri + L\frac{di}{dt} + \frac{1}{C}\int i\, dt \tag{5-117}$$

Equation (5-117) is a second-order nonhomogeneous differential equation because it involves a derivative as well as an integral term. Laplace transforming each term of the equation yields

$$\frac{E}{s} = RI(s) + sLI(s) + \frac{1}{C}\left[\frac{I(s)}{s} + \frac{q_0}{s}\right] \tag{5-118}$$

where

$$q_0 = i^{-1}(0^-) = \int_{-\infty}^{0^-} i\, dt$$

and represents the accumulation of charge on the capacitor owing to a previous current flow. Rearranging, we get

$$\left(\frac{E}{s} - \frac{q_0}{sC}\right) = I(s)\left[R + sL + \frac{1}{sC}\right] \tag{5-119}$$

The function in brackets on the right side is the operational impedance of the series *R-L-C* circuit. The expression for the current solution in the *s* domain then readily becomes

$$I(s) = \frac{\dfrac{E}{s} - \dfrac{q_0}{sC}}{\dfrac{L}{s}\left(s^2 + \dfrac{R}{L}s + \dfrac{1}{LC}\right)} = \frac{\dfrac{E}{L} - \dfrac{q_0}{LC}}{\left(s^2 + \dfrac{R}{L}s + \dfrac{1}{LC}\right)} \tag{5-120}$$

A glance at Eq. (5-120) shows that the denominator expression does not include an s factor standing alone as was the case with Eq. (5-7) for the R-L circuit. This means that a constant term in the steady-state solution does not exist. A glance at the circuit configuration verifies this conclusion because in the steady state the capacitor presents an open circuit to the battery so that the particular solution must be zero.

   It is significant to note that the quadratic expression in the denominator of Eq. (5-120) results from an algebraic manipulation of the operational impedance. Accordingly, it involves each of the three circuit parameters. Moreover, the expression is quadratic because of the presence of two energy-storing elements. From the experience we have gained so far in evaluating the inverse Laplace transform we know that the first step in the procedure is to identify the specific root factors of the poles of the transformed solution $I(s)$. This then permits a partial-fraction expansion to be made, each root factor being readily identifiable in Table 4-1. For each of the cases treated so far there was no need to find the root factors of the denominator of $I(s)$ because they automatically appeared in the desired form, as reference to Eqs. (5-7) and (5-53) indicates. This simplicity was a consequence of dealing with first-order circuits. Before we can proceed further in the solution procedure for $i(t)$ in the situation now under consideration we must first find the specific roots of the quadratic expression. That is, we must determine those values of s which satisfy the equation

$$s^2 + \frac{R}{L}s + \frac{1}{LC} = 0 \qquad (5\text{-}121)$$

This expression is called the *characteristic equation* of the second-order system. Keep in mind that it results from setting the operational impedance equal to zero and then performing an algebraic manipulation which serves to isolate the s^2 term. The roots of this equation, which are also the poles of $I(s)$ in this instance, depend solely upon the circuit parameters. In turn these roots determine entirely the nature of the transient response. To understand this better, consider that s_1 and s_2 are the roots of the characteristic equation. That is,

$$(s - s_1)(s - s_2) = s^2 + \frac{R}{L}s + \frac{1}{LC} \qquad (5\text{-}122)$$

Then Eq. (5-120) can be rewritten as

$$I(s) = \frac{\dfrac{E}{L} - \dfrac{q_0}{LC}}{\left(s^2 + \dfrac{R}{L}s + \dfrac{1}{LC}\right)} = \frac{\dfrac{E}{L} - \dfrac{q_0}{LC}}{(s - s_1)(s - s_2)} = \frac{K_1}{s - s_1} + \frac{K_2}{s - s_2} \qquad (5\text{-}123)$$

The formal expression for the corresponding time solution then becomes

$$i(t) = K_1 \epsilon^{s_1 t} + K_2 \epsilon^{s_2 t} \qquad (5\text{-}124)$$

Thus the characteristic manner in which the transient terms decay is solely dependent on the roots s_1 and s_2. For this reason the equation which yields these roots is called the characteristic equation. The characteristic equation, as we shall see, can often be used to obtain information about the nature of the dynamic response with a minimum of effort.

Although Eq. (5-124) presents a formal solution of the problem at hand, we are interested in writing this solution in a more useful and significant form. In this connection then let us return to Eq. (5-121) and write the specific expressions for the roots. Thus

$$s_{1,2} = -\frac{R}{2L} \pm \sqrt{\left(\frac{R}{2L}\right)^2 - \frac{1}{LC}}$$ (5-125)

Depending upon the expression under the radical the transient response can be any one of the following: (1) overdamped if $(R/2L)^2 > 1/LC$; (2) critically damped if $(R/2L)^2 = 1/LC$: (3) underdamped if $(R/2L)^2 < 1/LC$. Because the underdamped case is the most interesting as well as the most frequently encountered case, attention is confined to it throughout the remainder of this section. The term damping is an appropriate one to use in our description because, as previously demonstrated, it is characteristic of the resistance parameter to dissipate energy. Consequently, its presence serves to prevent an uninterrupted interchange of energy between the two energy-storing elements. Such an uninterrupted interchange would constitute a sustained (or undamped) oscillation. For the underdamped case then the expression for the roots of the characteristic equation becomes

$$s_{1,2} = -\frac{R}{2L} \pm j\sqrt{\frac{1}{LC} - \left(\frac{R}{2L}\right)^2}$$ (5-126)

This result is obtained by factoring out the minus sign under the radical and recalling that $j = \sqrt{-1}$.

The critical value of damping for fixed values of L and C corresponds to that value of R which makes the radical term go to zero. Hence

$$\left(\frac{R_c}{2L}\right)^2 = \frac{1}{LC}$$ (5-127)

or

$$R_c = 2\sqrt{\frac{L}{C}}$$ (5-128)

where R_c denotes the value of resistance which yields a critically damped transient response. In order to express the roots of the characteristic equation in a manner which makes the results applicable to all linear second-order systems as well as to provide a quick and convenient way of identifying the dynamic response, two figures of merit are introduced and defined. The first is the *damping ratio*, which is denoted by the Greek letter zeta (ζ) and defined as follows:

$$\boxed{\zeta \equiv \frac{\text{actual damping}}{\text{critical damping}} \equiv \frac{R}{2\sqrt{L/C}}}$$ (5-129)

The second figure of merit is the natural radian frequency of the circuit and is defined by

$$\omega_n \equiv \frac{1}{\sqrt{LC}} \tag{5-130}$$

Accordingly, the real part of the roots given by Eq. (5-126) may be expressed in terms of ζ and ω_n as

$$\frac{R}{2L} = \frac{\zeta R_c}{2L} = \frac{\zeta 2\sqrt{L/C}}{2L} = \zeta \frac{1}{\sqrt{LC}} = \zeta\omega_n \tag{5-131}$$

Also, by means of the results in Eqs. (5-130) and (5-131), the radical term may be expressed as

$$\sqrt{\frac{1}{LC} - \left(\frac{R}{2L}\right)^2} = \sqrt{\omega_n^2 - \zeta^2\omega_n^2} = \omega_n\sqrt{1 - \zeta^2} \tag{5-132}$$

Therefore Eq. (5-126) may be written

$$s_{1,2} = -\zeta\omega_n \pm j\omega_n\sqrt{1 - \zeta^2} \tag{5-133}$$

Introducing the substitution

$$\omega_d \equiv \omega_n\sqrt{1 - \zeta^2} = \text{damped frequency of oscillation} \tag{5-134}$$

simplifies Eq. (5-133) to

$$s_{1,2} = -\zeta\omega_n \pm j\omega_d \tag{5-135}$$

A study of the last expression reveals that both the real and the j parts of the complex roots have the units of inverse seconds, which is frequency. This in turn gives rise to the term *complex frequency*. It is shown presently that both parts of the complex frequency play important roles in establishing the character of the transient response.

By using the results appearing in Eq. (5-135) for s_1 and s_2 it is possible to write the characteristic equation entirely in terms of ζ and ω_n. Thus

$$s^2 + \frac{R}{L}s + \frac{1}{LC} = s^2 + 2\zeta\omega_n s + \omega_n^2 = (s + \zeta\omega_n)^2 + \omega_d^2 \tag{5-136}$$

Therefore, returning to Eq. (5-120), the expression for the solution in the s domain may be written

$$I(s) = \frac{\dfrac{E}{L} - \dfrac{q_0}{LC}}{s^2 + 2\zeta\omega_n s + \omega_n^2} = \frac{\dfrac{E}{L} - \dfrac{q_0}{LC}}{(s + \zeta\omega_n)^2 + \omega_d^2} \tag{5-137}$$

To put this equation in a form which is readily identifiable in Table 4-1, it is necessary to multiply the right side of Eq. (5-137) by ω_d/ω_d. Then

$$I(s) = \left(\frac{E - \dfrac{q_0}{C}}{\omega_d L}\right)\left[\frac{\omega_d}{(s + \zeta\omega_n)^2 + \omega_d^2}\right] \tag{5-138}$$

The quantity in parentheses is merely a constant, while that in brackets is seen to be the Laplace transform of an exponentially damped sine function as indicated by item 9 in Table 4-1. The corresponding solution in the time domain is therefore

$$i(t) = \frac{E - \dfrac{q_0}{C}}{\omega_d L} \epsilon^{-\zeta \omega_n t} \sin \omega_d t \qquad (5\text{-}139)$$

This expression is the specific version of the formal solution represented by Eq. (5-124). The total solution in this case involves only a transient term because, as already pointed out, the forced solution is zero.

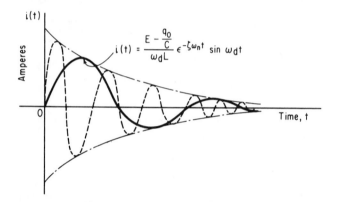

Fig. 5-22. Graphical representation of the current response in an R-L-C circuit for underdamped conditions. The broken-line curve is for a lower value of C.

A plot of the response function, Eq. (5-139), appears in Fig. 5-22. The response is a damped sinusoid which is confined to an envelope determined by the exponential function $\epsilon^{-\zeta \omega_n t}$. It is interesting to note that dealing in terms of the envelope alone it is possible to identify a time constant for the second-order system. Thus

$$\boxed{T = \frac{1}{\zeta \omega_n}} \qquad (5\text{-}140)$$

Accordingly, the meaning which can be attached to the real part of the complex frequency of Eq. (5-135) is that it gives a measure of the response time of the transient. It effectively describes the period of duration of the transient. Initially ω_n was introduced as a figure of merit. By its nature any figure of merit, if it is to be useful, must convey information about a specific performance feature. In the case of a circuit's natural frequency, it yields information about *settling time*,† i.e. the time needed to reach steady state, *The larger the ω_n of a circuit the smaller*

† This is also called the *response time*.

will be the settling time. Thus if two *R-L-C* circuits are compared and each has the same damping ratio but one has twice the natural frequency of the other, the transients in the circuit with the larger natural frequency will decay twice as fast. This conclusion follows from Eq. (5-140).

The meaning of the *j* part of the complex frequency of Eq. (5-135) is that it conveys information about the *actual* frequency at which the oscillations decay. This conclusion easily follows from Eq. (5-139).

For fixed values of *R* and *L* what is the effect on the transient response of decreasing the capacitance parameter *C*? Equation (5-140) tells us that the envelope is in no way affected by this change, since it is independent of *C*. In other words the time constant and therefore the settling time remain essentially unaltered. Equations (5-129) and (5-130), however, state respectively that the damping ratio decreases and the natural frequency increases. The resulting response is then characterized by a higher frequency of oscillation as well as a higher peak oscillation, as illustrated by the broken-line curve of Fig. 5-22. In Example 5-5 it is shown that the damping ratio is a figure of merit which yields information about maximum overshoot. The smaller the value of ζ the greater will be the peak values attained during the transient period.

Step-current Response of a Parallel G-L-C Circuit. The circuit configuration is depicted in Fig. 5-23. All initial conditions are assumed equal to zero. A step-current forcing function is applied to the parallel combination of the circuit elements by opening the switch. The lower of the two nodes of the circuit is taken as the reference node and is shown grounded to stress the point.

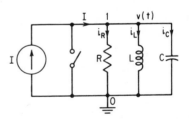

A little reflection here reveals that this parallel arrangement of circuit elements is the dual of the series *R-L-C* circuit. In fact, an exact dual occurs if it is assumed that $G = R$, $L = C$, and $C = L$. On this

Fig. 5-23. Parallel *G-L-C* circuit subjected to a step current forcing function.

assumption then, and by comparison with Eq. (5-118), the governing equation for the circuit of Fig. 5-23 is simply

$$I = Ge + C\frac{de}{dt} + \frac{1}{L}\int e\,dt \qquad (5\text{-}141)$$

Note that the roles of voltage and current have been interchanged. The corresponding solution for the output voltage *e* can be written forthwith from Eq. (5-139). Accordingly

$$e = \frac{I}{\omega_d C}\epsilon^{-\zeta\omega_n t}\sin\omega_d t \qquad (5\text{-}142)$$

where

$$\zeta = \frac{G}{2\sqrt{C/L}} \qquad (5\text{-}143)$$

and

$$\omega_n = \frac{1}{\sqrt{LC}} \tag{5-144}$$

Once the node voltage is determined, finding the expressions for the currents flowing through the individual elements is a routine matter. For example, the current through the resistor is simply given by

$$i_R = \frac{e(t)}{R} = \frac{I}{\omega_d RC} \epsilon^{-\zeta\omega_n t} \sin \omega_d t \tag{5-145}$$

To find the time function for the current through the inductor, more effort is required because it involves the integration of the product of two time functions. That is

$$i_L = \frac{1}{L} \int_0^t e(t)\, dt = \frac{I}{\omega_d LC} \int_0^t \epsilon^{-\zeta\omega_n t} \sin \omega_d t\, dt \tag{5-146}$$

After performing the integration and substituting the limits, the final expression becomes

$$i_L = \frac{I}{\omega_d LC} \left[\frac{\omega_d}{\omega_n^2} - \frac{\epsilon^{-\zeta\omega_n t}}{\omega_n} \sin(\omega_d t + \gamma) \right] \tag{5-147}$$

where

$$\gamma = \tan^{-1} \frac{\omega_d}{\zeta\omega_n} = \tan^{-1} \frac{\sqrt{1 - \zeta^2}}{\zeta} \tag{5-148}$$

A check at $t = \infty$ shows that the current through the inductor is equal to the source current. This is in accord with the physical situation, which indicates that at steady state the inductance behaves as a short circuit and thereby carries the total source current. The time solution for the capacitor current is found from

$$i_c = C \frac{de(t)}{dt} \tag{5-149}$$

where $e(t)$ is given by Eq. (5-145). It is left as an exercise for the reader to determine the final expression for i_c.

EXAMPLE 5-5   A step voltage of magnitude E is applied to an initially de-energized series R-L-C circuit for which the parameter values are

$$R = 40 \text{ ohms}, \qquad L = 0.2 \text{ henry}, \qquad C = 100 \,\mu f$$

Find the complete solution for the charge on the capacitor and show a plot of the response.

solution: The solution for the current is already determined and is given by Eq. (5-139). The only modification needed is to set q_0 equal to zero because of the statement of the problem. Thus

$$i(t) = \frac{E}{\omega_d L} \epsilon^{-\zeta\omega_n t} \sin \omega_d t \tag{5-150}$$

The time solution for the charge thus follows from

$$q(t) = \int_0^t i(t)\, dt = \frac{E}{\omega_d L} \int_0^t \epsilon^{-\zeta\omega_n t} \sin \omega_d t\, dt \tag{5-151}$$

After integrating, this becomes

$$q(t) = \frac{E}{\omega_d L}\left[\frac{\epsilon^{-\zeta\omega_n t}(-\zeta\omega_n \sin \omega_d t - \omega_d \cos \omega_d t)}{\omega_n^2}\right]_0^t \tag{5-152}$$

Substituting the upper and lower limits then yields

$$q(t) = \frac{E}{\omega_d L}\left[\frac{\omega_d}{\omega_n^2} - \frac{\epsilon^{-\zeta\omega_n t}}{\omega_n}\sin(\omega_d t + \phi)\right] \tag{5-153}$$

where

$$\phi = \tan^{-1}\frac{\omega_d}{\zeta\omega_n} \tag{5-154}$$

To put $q(t)$ in a more condensed from we must have the values of ζ, ω_d, and ω_n. These are readily obtained from the characteristic equation, which can be written as

$$s^2 + \frac{R}{L}s + \frac{1}{LC} = s^2 + \frac{40}{0.2}s + \frac{1}{0.2 \times 10^{-4}} = s^2 + 200s + 50{,}000 = 0 \tag{5-155}$$

or, more conveniently,

$$s^2 + 2\zeta\omega_n s + \omega_n^2 = s^2 + 200s + 50{,}000 = 0 \tag{5-156}$$

The left side is the general form of the characteristic equation expressed in terms of ζ and ω_n. A comparison of the constant terms in Eq. (5-156) readily shows that

$$\omega_n = \sqrt{50{,}000} = 100\sqrt{5} = 224 \text{ rad/sec} \tag{5-157}$$

Equating the coefficients of the s terms gives

$$\zeta\omega_n = 100 \tag{5-158}$$

or

$$\zeta = \frac{100}{\omega_n} = \frac{100}{100\sqrt{5}} = \frac{1}{\sqrt{5}} = 0.447 \tag{5-159}$$

The damped frequency of oscillation is then

$$\omega_d = \omega_n\sqrt{1 - \zeta^2} = 224\sqrt{1 - \tfrac{1}{5}} = 200 \text{ rad/sec} \tag{5-160}$$

Accordingly, Eq. (5-153) can now be rewritten

$$q(t) = E\left[C - \frac{C}{\sqrt{1 - \zeta^2}}\epsilon^{-100t}\sin(200t + \tan^{-1} 2)\right] \tag{5-161}$$

or

$$q(t) = EC[1 - 1.12\,\epsilon^{-100t}\sin(200t + 63.4°)] \tag{5-162}$$

A plot of the last equation appears in Fig. 5-24. Note that EC is the steady-state value of the charge that appears on the capacitor which is consistent with the potential difference E appearing across the capacitor terminals after the transient has decayed. Note too that the oscillations decay at a frequency of 200 rad/sec,

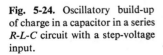

Fig. 5-24. Oscillatory build-up of charge in a capacitor in a series *R-L-C* circuit with a step-voltage input.

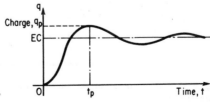

which is the value of ω_d. Inspection of Fig. 5-24 also shows that the first overshoot beyond the steady-state value represents the maximum overshoot associated with the transient. A quantitative measure of this quantity can be found by setting the derivative of $q(t)$ equal to zero. This yields the time at which the peak overshoot occurs. Specifically, it is given by the expression

$$t_p = \frac{\pi}{\omega_d} \tag{5-163}$$

where t_p denotes the time for the peak overshoot. The corresponding value of the peak overshoot expressed as a per cent of the steady-state value can be shown to be given by

$$\text{maximum per cent overshoot} = 100\frac{q_p - EC}{EC} = 100\epsilon^{-\zeta\pi/\sqrt{1-\zeta^2}} \tag{5-164}$$

The right side of Eq. (5-164) reveals that the maximum per cent overshoot is solely dependent upon the value of the damping ratio ζ. *It is for this reason that ζ is looked upon as a figure of merit which conveys information about maximum overshoot of the transient response.* Since for our example $\zeta = 0.447$, its insertion into Eq. (5-164) reveals that the maximum overshoot is 20 per cent of the steady-state value. Moreover, if it is assumed that steady state is reached after five time constants, that is, 5 ($\frac{1}{100}$), then clearly the settling time is 0.05 second. It is interesting to note that these two bits of information—settling time and maximum per cent overshoot—are really all that is needed to describe the dynamic response of the circuit. Knowledge of these quantities, which are directly obtainable from the characteristic equation and Eqs. (5-140) and (5-164), often makes it unnecessary to determine the formal solution of the response such as that represented by Eq. (5-162).

EXAMPLE 5-6   Refer to the circuit of Fig. 5-25. Assume it is initially de-energized.
   (a) When this circuit is disturbed by the application of the forcing function $e(t)$, determine whether or not the ensuing dynamic response will be a damped oscillation.
   (b) If the response in Part (a) is found to be oscillatory, compute the time it takes for

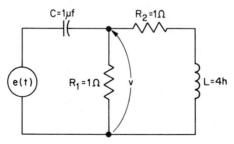

Fig. 5-25. Circuit of Example 5-6. Parameter values are chosen for convenience.

the transient response to decay to within 1 per cent of the initial value. (This is the setting time of the circuit.)

(c) What is the maximum percentage of overshoot?

solution: (a) The first step in arriving at a solution is to find the characteristic equation. Subsequent examination will then disclose the answers to the question. The characteristic equation can be found by obtaining the transformed solution of *any* of the circuit variables. The characteristic equation is the corresponding denominator expression set equal to zero.

To illustrate, let the nodal voltage $V(s)$ in Fig. 5-25 be the circuit variable for which the transformed solution is to be found. Kirchhoff's current law at the node allows us to write

$$\frac{V(s) - E(s)}{1/sC} + \frac{V(s)}{R_1} + \frac{V(s)}{R_2 + sL} = 0 \tag{5-165}$$

Inserting the specified parameter values and rearranging terms yields

$$V(s)\left[s + 1 + \frac{4}{1 + 4s}\right] = sE(s) \tag{5-166}$$

from which the transformed solution for the nodal voltage becomes

$$V(s) = \frac{s(s + \frac{1}{4})}{s^2 + \frac{5}{4}s + \frac{5}{4}}E(s) \tag{5-167}$$

It is instructive to note here that the identity of the Laplace transform of the source function is unnecessary, because the character of the dynamic response is determined by the circuit parameters and the configuration in which they are placed and not at all by the form of the source function. The characteristic equation is now found by setting the denominator expression of $V(s)$ equal to zero. Thus,

$$s^2 + \tfrac{5}{4}s + \tfrac{5}{4} = 0 \tag{5-168}$$

Comparing this equation with the standard formulation for a second-order system, that is,

$$s^2 + 2\zeta\omega_n s + \omega_n^2 = 0 \tag{5-169}$$

leads us to

$$\omega_n = \sqrt{\frac{5}{4}} = \frac{\sqrt{5}}{2} = 1.12 \tag{5-170}$$

and

$$\zeta\omega_n = \tfrac{1}{2}\left(\tfrac{5}{4}\right) = \tfrac{5}{8} \tag{5-171}$$

or

$$\zeta = \frac{\frac{5}{8}}{\omega_n} = \frac{\frac{5}{8}}{1.12} = 0.56 \tag{5-172}$$

Since the damping ratio ζ is less than unity, it follows that the ensuing response is a damped oscillation.

It is instructive to note here that if we had chosen to solve for any other variable, for example, the source current, the resulting transformed solution would lead to the same expression for the characteristic equation. This should not be a surprising result, for the reaction of the circuit to a disturbance anywhere in the circuit is the same.

(b) The settling time in this case corresponds to five time constants of the second-order system. Thus, from Eq. (5-140) we have

$$t_s = 5T = \frac{5}{\zeta\omega_n} = \frac{5}{\frac{5}{8}} = 8 \text{ secs} \tag{5-173}$$

Eight seconds after this circuit is disturbed—whether by the application of a source function, or the discharge of the capacitor, or whatever—the ensuing transients will disappear.

(c) Upon inserting the value of $\zeta = 0.56$ into Eq. (5-163), the maximum overshoot is found to be 12 per cent.

5-7 Complete Response of *R-L* Circuit to Sinusoidal Input

In this chapter up to this point attention has been confined to just one type of deterministic input, the step forcing function. In this section we now determine the total response of a series *R-L* circuit when a second type of deterministic input, the sinusoidal function, is applied to the circuit terminals. The circuit diagram is depicted in Fig. 5-26.

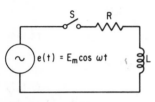

Since our concern is with finding the total solution, it follows that a steady-state solution as well as a transient solution must be determined. A point worthy of note here is that we shall be finding the steady-state solution to a sinusoidal forcing function in a linear *R-L* series circuit in spite of the fact that we have not yet discussed the sinusoidal steady-state theory of electrical networks.† That we can do this in a rather simple and direct fashion is attributable to the use of Laplace transforms as a means of solving linear nonhomogeneous differential equations. This is just another of the advantages the Laplace transform has to offer as a solution procedure.

Fig. 5-26. Series *R-L* circuit with sinusoidal forcing function.

The voltage source in the circuit of Fig. 5-26 is assumed to be varying in a cosinusoidal fashion. When the switch is closed the governing differential equation for the circuit becomes

$$E_m u(t) \cos \omega t = Ri + L\frac{di}{dt} \tag{5-174}$$

Since the switch is initially open, the initial value of the current through the inductor is obviously zero. With the use of Laplace-transform pair 4 of Table 4-1, the *s*-domain form of Eq. (5-174) is

$$\frac{E_m s}{s^2 + \omega^2} = I(s)[R + sL] \tag{5-175}$$

Hence the solution for the response in the *s* domain readily becomes available as

$$I(s) = \frac{E_m}{L}\left(\frac{s}{s^2 + \omega^2}\right)\left(\frac{1}{s + R/L}\right) \tag{5-176}$$

It is always impressive to see how easily the transformed solution is obtained with the Laplace transform. Also, note that the poles (or root factors of the denominator) of $I(s)$ are directly available. Therefore a partial-fraction expansion can be written forthwith.

† This is the subject matter of Chapter 6.

$$I(s) = \frac{s(E_m/L)}{(s^2 + \omega^2)(s + R/L)} = \frac{K_1}{s - j\omega} + \frac{K_1^*}{s + j\omega} + \frac{K_2}{s + R/L} \quad (5\text{-}177)$$

The cosinusoidal function is responsible for the presence of the purely imaginary complex conjugate poles. The associated coefficients for these poles in the partial-fraction expansion must themselves in general be complex conjugate numbers. To stress this point the symbol K_1^* is used to denote that this quantity is the complex conjugate of K_1. The evaluation of any coefficient in the partial-fraction expansion is accomplished by following the usual procedure as called for by Eq. (4-63) or Eq. (4-68). Thus for K_1 we have

$$\begin{aligned} K_1 &= \left[(s - j\omega)\frac{s(E_m/L)}{(s - j\omega)(s + j\omega)(s + R/L)}\right]_{s = +j\omega} \\ &= \left[\frac{E_m}{L}\frac{j\omega}{j2\omega(j\omega + R/L)}\right] = \left(\frac{E_m}{2}\right)\frac{1}{R + j\omega L} \\ &= \left(\frac{E_m}{2}\right)\frac{R - j\omega L}{R^2 + \omega^2 L^2} \end{aligned} \quad (5\text{-}178)$$

Since K_1^* must be the complex conjugate we can write forthwith

$$K_1^* = \left(\frac{E_m}{2}\right)\frac{R + j\omega L}{R^2 + \omega^2 L^2} \quad (5\text{-}179)$$

Recall that the rules for finding the complex conjugate is to replace all the j terms with opposite signs. For K_2 the determining expression is

$$\begin{aligned} K_2 &= \left[\left(s + \frac{R}{L}\right)\frac{s(E_m/L)}{(s^2 + \omega^2)(s + R/L)}\right]_{s = -R/L} \\ &= \left[\frac{E_m}{L}\frac{s}{s^2 + \omega^2}\right]_{s = -R/L} = -\frac{E_m R}{R^2 + \omega^2 L^2} \\ &= \frac{E_m}{\sqrt{R^2 + \omega^2 L^2}} \times \frac{R}{\sqrt{R^2 + \omega^2 L^2}} = -\frac{E_m}{\sqrt{R^2 + \omega^2 L^2}}\cos\theta \end{aligned} \quad (5\text{-}180)$$

where θ is defined by the right triangle shown in Fig. 5-27, i.e. $\theta = \tan^{-1}(\omega L/R)$. The introduction here of θ and of the algebraic manipulation which leads to its definition has the purpose of putting the final solution for the response in a form which allows easy comparison with similar results found by other means in Chapter 6. Actually, to reach a correct solution for the problem at hand it would suffice to work with that form of K_2 which does not involve the radical.

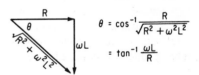

$\theta = \cos^{-1}\dfrac{R}{\sqrt{R^2 + \omega^2 L^2}}$

$= \tan^{-1}\dfrac{\omega L}{R}$

Fig. 5-27. Definition of θ for the R-L circuit of Fig. 5-26.

Inserting Eqs. (5-178), (5-179), and (5-180) into Eq. (5-177) yields an expression for $I(s)$ which is ready to be inverse Laplace transformed. Hence

$$\begin{aligned} I(s) &= \frac{E_m}{2(R^2 + \omega^2 L^2)}\left[(R - j\omega L)\frac{1}{s - j\omega} + (R + j\omega L)\frac{1}{s - j\omega}\right] \\ &\quad - \frac{E_m}{\sqrt{R^2 + \omega^2 L^2}}\cos\theta\left(\frac{1}{s + R/L}\right) \end{aligned} \quad (5\text{-}181)$$

The corresponding time solution is then

$$i(t) = \frac{E_m/2}{R^2 + \omega^2 L^2}[R - j\omega L)\epsilon^{j\omega t} + (R + j\omega L)\epsilon^{-j\omega t}]$$

$$- \frac{E_m}{\sqrt{R^2 + \omega^2 L^2}} \cos \theta [\epsilon^{-(R/L)t}] \tag{5-182}$$

Collecting terms yields

$$i(t) = \frac{E_m/2}{R^2 + \omega^2 L^2}[R(\epsilon^{j\omega t} + \epsilon^{-j\omega t}) + j\omega L(\epsilon^{-j\omega t} - \epsilon^{+j\omega t})]$$

$$- \frac{E_m}{\sqrt{R^2 + \omega^2 L^2}} \cos \theta [\epsilon^{-(R/L)t}] \tag{5-183}$$

Recalling that

$$\epsilon^{j\omega t} + \epsilon^{-j\omega t} = 2 \cos \omega t \quad \text{and} \quad (\epsilon^{-j\omega t} - \epsilon^{j\omega t}) = -2j \sin \omega t \tag{5-184}$$

simplifies Eq. (5-183) to

$$i(t) = \frac{E_m}{R^2 + \omega^2 L^2}[R \cos \omega t + \omega L \sin \omega t] - \frac{E_m}{\sqrt{R^2 + \omega^2 L^2}} \cos \theta [\epsilon^{-(R/L)t}] \tag{5-185}$$

Now since the cosine function leads the sine function by 90 degrees, the quantity in brackets in the last equation may be found with the aid of the right triangle in Fig. 5-27. If the horizontal line is assumed to represent the cosine having magnitude R, then the vertical downward line, which is 90 degrees displaced from the cosine line, represents the sine function having magnitude ωL. Clearly, then, the addition of these two quadrature terms is the hypotenuse of the right triangle the amplitude of which is $\sqrt{R^2 + \omega^2 L^2}$ and the position of which is θ degrees *behind* the horizontal line which denotes the cosine function. On the basis of this reasoning it follows that

$$R \cos \omega t + \omega L \sin \omega t = \sqrt{R^2 + \omega^2 L^2} \cos (\omega t - \theta) \tag{5-186}$$

where θ is defined in Fig. 5-27. The minus sign is used with θ to denote that the hypotenuse is in a position below the horizontal line.

Substituting the last expression into Eq. (5-185) yields the final form of the total solution for the current which flows in the R-L circuit when a sinusoidal forcing function is applied. Thus

$$i(t) = \frac{E_m}{\sqrt{R^2 + \omega^2 L^2}} [\cos (\omega t - \theta) - \epsilon^{-(R/L)t} \cos \theta] \tag{5-187}$$

A plot of the solution appears in Fig. 5-28.

An inspection of Eq. (5-187) and Fig. 5-28 reveals that the total solution is composed of a forced solution as well as a transient solution. The former is represented by the first term in brackets. That is,

$$i_s = \text{steady-state solution}$$

$$= \frac{E_m}{\sqrt{R^2 + \omega^2 L^2}} \cos (\omega t - \theta) \tag{5-188}$$

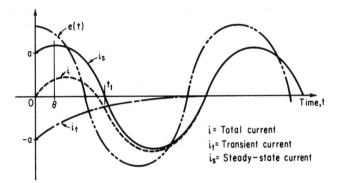

Fig. 5-28. The current response of an *R-L* circuit to a sinusoidal forcing function.

Because we are dealing with linear circuits, the forced solution has the same form as the forcing function, i.e. cosinusoidal. However, the response differs in two respects. Its amplitude is modified by the factor $1/\sqrt{R^2 + \omega^2 L^2}$, and its argument† is altered by the angle $-\theta$. Of course the transient solution behaves as expected. Initially it provides a magnitude of current which is equal and opposite to the instantaneous value of the steady-state current at the instant of switching. In Fig. 5-28 it is assumed that the switch is closed at the instant when the voltage has its positive maximum value. Note that the corresponding steady-state current is not zero at this instant. In Fig. 5-28 it is shown to have a value $0a$. Since the boundary condition demands that the current at this instant be zero, it is necessary for the transient term to have a value equal and opposite. This is depicted as $-0a$ in Fig. 5-28. As time elapses after switching, the transient term decays to zero at a rate determined by the circuit time constant, and the total current in Fig. 5-28 then becomes identical to the steady-state current.

The presence of a transient component in the total solution for the current is called for whenever the initial value of current in the *R-L* circuit is at variance with the value of the steady-state current at the instant of switching. As an illustration, if in the problem just discussed the initial current flowing through *L* were equal to $0a$ there would be no need for a transient term because the ensuing current would proceed directly into the steady state. As a result of the correspondence of the currents before and after switching, there is no force attempting to bring about an abrupt change in the amount of energy stored in the inductor. This behavior can be described in another way. Let t_1 in Fig. 5-28 denote the time when the steady-state current is zero. If the switch in Fig. 5-26 is closed at this instant, no violation of the boundary condition occurs; consequently there is no need for a transient current. In such a case the response proceeds directly into the steady state. Incidentally, note that although the steady-state current is zero the corresponding value of the applied voltage is not. This is characteristic of the *R-L* circuit for sinusoidal forcing functions.

† See p. 167 for a definition of this term.

EXAMPLE 5-7 A sinusoidal forcing function $e(t) = 141 \sin 377t$ is applied to an initially de-energized series R-L circuit in which $R = 100$ ohms and $L = \frac{1}{2}$ henry.

(a) If the switch which applied the voltage to the R-L circuit is closed at the instant when $e(t)$ is passing through zero with a positive slope, determine the initial value of the transient current.

(b) Write the complete expression for the transient solution.

(c) Write the expression for the complete solution of the current response.

(d) At what instantaneous value of the applied voltage will the closing of the switch result in no transient component of current?

solution: (a) The general form of the steady-state solution is

$$i_s = \frac{E_m}{\sqrt{R^2 + \omega^2 L^2}} \sin (\omega t - \theta) \qquad (5\text{-}189)$$

This follows directly from Eq. (5-188) except that the sine function is used in order to make it consistent with the assumed sinusoidal forcing function. Since $\omega = 377$ it follows that

$$\omega L = 377(\tfrac{1}{2}) = 188.5 \text{ ohms} \qquad (5\text{-}190)$$

and

$$\sqrt{R^2 + \omega^2 L^2} = \sqrt{100^2 + 188.5^2} = 213.5 \text{ ohms} \qquad (5\text{-}191)$$

Also

$$\theta = \tan^{-1} \frac{\omega L}{R} = \tan^{-1} 1.885 = 62° \qquad (5\text{-}192)$$

Therefore the value of the steady-state current at $t = 0^+$, which is the time immediately following application of the forcing function, is

$$i_s(0^+) = \frac{E_m}{\sqrt{R^2 + \omega^2 L^2}} \sin (-\theta) = \frac{141}{213.5} \sin (-62°) = -0.584 \text{ amp} \qquad (5\text{-}193)$$

However, in accordance with the boundary condition the initial value of the current through the inductor must be zero. That is,

$$i(0^+) = 0 = i_s(0^+) + i_t(0^+) \qquad (5\text{-}194)$$

or

$$0 = -0.584 + i_t(0^+)$$

Hence the initial value of the transient current is

$$i_t(0^+) = 0.584 \text{ amp} \qquad (5\text{-}195)$$

(b) The rate of decay of the transient term is determined by the time constant of the circuit, which here is

$$T = \frac{L}{R} = \frac{\frac{1}{2}}{100} = \frac{1}{200} \text{ sec}$$

Therefore the complete expression for the transient term becomes

$$i_t = 0.584\epsilon^{-200t} \qquad (5\text{-}196)$$

(c) By adding Eqs. (5-189) and (5-196) the equation for the total solution results.

$$i(t) = 0.661 \sin (377t - 62°) + 0.584\epsilon^{-200t} \qquad (5\text{-}197)$$

(d) In order that there be no transient current in the solution, it is necessary to close the switch at that time instant for which i_s is zero. A glance at Eq. (5-189) shows that this occurs whenever

$$(\omega t - \theta) = 0, \pi, 2\pi, \ldots \qquad (5\text{-}198)$$

Choosing the first value, it follows that

$$\omega t = \theta = 62° \qquad (5\text{-}199)$$

The corresponding instantaneous value of the forcing function is then

$$e = 141 \sin 62° = 124.2 \text{ volts} \qquad (5\text{-}200)$$

Therefore if the switch is closed when the applied voltage has an instantaneous value of 124.2 volts, there is no need for a transient term and so none exists.

5-8 Response of *R-L-C* Circuit to Sinusoidal Inputs

The treatment of this case presents nothing new in the way of concepts about transients; therefore the development of the solution is kept brief. The circuit configuration is shown in Fig. 5-29. With the switch closed, the differential equation for the circuit is

$$E_m u(t) \cos \omega t = Ri + L \frac{di}{dt} + \frac{1}{C} \int i\, dt \qquad (5\text{-}201)$$

Laplace transforming both sides gives

$$\frac{sE_m}{s^2 + \omega^2} = I(s)\left[s^2 + \frac{R}{L}s + \frac{1}{LC} \right]\frac{L}{s} \qquad (5\text{-}202)$$

Hence the expression for the transformed current response becomes

$$I(s) = \frac{E_m}{L} \frac{s^2}{(s^2 + \omega^2)(s^2 + 2\zeta\omega_n s + \omega_n^2)} \qquad (5\text{-}203)$$

where the generalized form of the characteristic equation is used. Note that here the expression for $I(s)$ contains two complex conjugates as well as two purely imaginary poles. It is assumed that the values of the circuit parameters are such that the dynamic response is underdamped. Obviously the presence of four poles in $I(s)$ means additional algebraic labor, but the method of solution remains the same. By means of a partial-fraction expansion Eq. (5-203) can be written as

$$I(s) = \frac{K_1}{s - j\omega} + \frac{K_1^*}{s + j\omega} + \frac{K_2}{s + \zeta\omega_n - j\omega_d} + \frac{K_2^*}{s + \zeta\omega_n + j\omega_d} \qquad (5\text{-}204)$$

The coefficients in this expansion are computed in the usual manner. The details are not shown here because of the unwieldy expressions which result when the analysis is carried out literally. However, the form of the solution can readily be identified from a study of Eq. (5-204).

The poles of $I(s)$ associated with the quadratic $(s^2 + \omega^2)$ are responsible for generating the steady-state or forced solution. This is entirely expected when it is recalled that these poles originated from

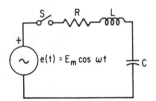

Fig. 5-29. Series *R-L-C* circuit with a sinusoidal forcing function. All initial conditions zero.

the expression for the Laplace transform of the cosinusoidal forcing function. This conclusion is also expected from our experience with the *R-L* circuit, which is described in the preceding section. After evaluating K_1 and K_1^* in Eq. (5-204) and performing some algebraic manipulation, it can be shown that the forced solution takes the form

$$i_s(t) = \frac{E_m}{\sqrt{R^2 + \left(\omega L - \dfrac{1}{\omega C}\right)^2}} \cos{(\omega t - \theta)} \tag{5-205}$$

where

$$\theta = \tan^{-1} \frac{\omega L - \dfrac{1}{\omega C}}{R} \tag{5-206}$$

Again note that the steady-state solution has the same form as the forcing function but it differs in magnitude and argument of the cosine.

Those poles of $I(s)$ associated with the quadratic which represents the characteristic equation are responsible for generating the transient solution. This we have already experienced several times before. Now from Eq. (5-204) we know that the general form which the transient solution must take is

$$i_t(t) = K_2\epsilon^{(-\zeta\omega_n + j\omega_d)t} + K_2^*\epsilon^{(-\zeta\omega_n - j\omega_d)t} = \epsilon^{-\zeta\omega_n t}(K_2\epsilon^{j\omega_d t} + K_2^*\epsilon^{-j\omega_d t}) \tag{5-207}$$

Therefore the form of the total solution is

$$i(t) = \frac{E_m}{\sqrt{R^2 + \left(\omega L - \dfrac{1}{\omega C}\right)^2}} \cos{(\omega t - \theta)} + \epsilon^{-\zeta\omega_n t}(K_2\epsilon^{j\omega_d t} + K_2^*\epsilon^{-j\omega_d t}) \tag{5-208}$$

which is obtained by adding Eqs. (5-205) and (5-207). A plot of $i(t)$ appears in Fig. 5-30.

Comparing Eq. (5-208), the current response of the *R-L-C* case, with Eq. (5-187), the current response for the *R-L* case, shows that the responses differ essentially in the character of the transient terms. Because the *R-L* configuration involves one energy-storing element and is described by a first-order differential equation there occurs just a simple exponential decay of the transient. The steady-

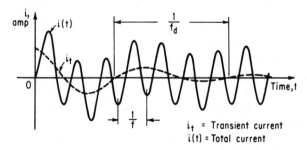

Fig. 5-30. Plot of current response to the configuration of Fig. 5-29. Here f_d is the damped frequency of oscillation in Hz of the transient term and f is the frequency of the applied forcing function.

state solution merely rides on this exponential decay as illustrated in Fig. 5-28. For the underdamped *R-L-C* configuration, however, the transient solution itself involves an exponentially damped oscillation whose frequency is ω_d. Here the steady-state term can be looked upon as riding on the damped oscillation as depicted in Fig. 5-30. It is basically in this respect that the two responses differ.

PROBLEMS

5-1 In the circuit of Fig. P5-1, determine the expression for the source current for all time after closing the switch. Assume zero current through the coil when S is closed.

5-2 A circuit containing two independent energy-storing elements is defined by the following differential equation:

$$\frac{d^2x}{dt^2} + 2\frac{dx}{dt} + 2x = f(t)$$

The circuit is initially de-energized so that $x(0^-) = \dot{x}(0^-) = 0$. Find $x(0^+)$, $\dot{x}(0^+)$, and $\ddot{x}(0^+)$ when $f(t) = u(t)$.

5-3 (a) In the circuit of Fig. P5-3 find the complete expression of the current which flows through the coil when switch S is closed.

(b) What is the final value of the coil current?

(c) How long does it take for the coil current to reach 95 per cent of its final value?

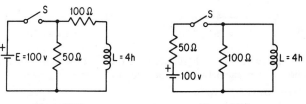

             Figure P5-1   **Figure P5-3**

5-4 A mass *M* is freely suspended and at rest in a viscous material having a viscous coefficient *D* equal to 20 newton-sec/m. Neglect gravity effects. A constant force of 10 newtons is applied to this mass. Determine:

(a) The expression for the mass velocity for all time after application of the force. Assume $M = 5$ newton-sec^2/m.

(b) The final velocity in meters/second.

(c) The time it takes for the mass to reach within 1 per cent of its final velocity.

5-5 The plot of the current response to a step forcing function of 100 volts has the variation depicted in Fig. P5-5. Find the values of resistance and inductance which apply for the circuit.

5-6 Assuming the coil initially de-energized and the current source suddenly applied to the circuit of Fig. P5-6, find the total expression for the current through the energy-storing element.

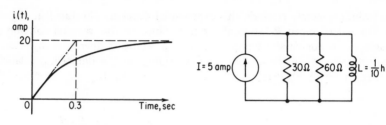

Figure P5-5   Figure P5-6

5-7   In the circuit of Fig. P5-7 switch S (denoted by the arrow) has been placed at *a* for a long time. It is then quickly moved to position *b* along the contact points shown as heavy lines.

(a) Find the expression for the current through the 30-ohm resistor.

(b) Compute the energy dissipated in this resistor.

(c) Compare the result found in (b) with the energy initially stored in the coil.

5-8   Determine the time constant at which the transients in the circuit of Fig. P5-8 decay.

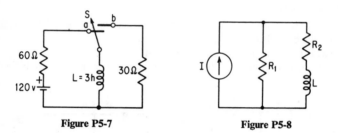

Figure P5-7   Figure P5-8

5-9   For the circuit of Fig. P5-9 determine the two time constants at which the transient terms decay when the network is subjected to a forcing function.

5-10   Refer to Fig. P5-10.

(a) Find the complete expression for the charging capacitor current when switch S is put to position *a*.

(b) After a long time switch S is placed at position *b*. Determine the expression for the current which flows through the 0.1 megohm resistor.

(c) Find the time constant of the circuit of part (b).

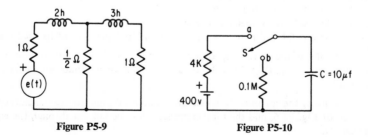

Figure P5-9   Figure P5-10

5-11 The forcing function in the circuit of Fig. P5-11(a) is the pulse depicted in Fig. P5-11(b).

(a) Find the expression for the current.

(b) Find the expression for the voltage which appears across the capacitor terminals.

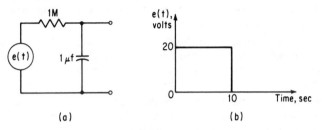

(a)   (b)

Figure P5-11

5-12 For the circuit shown in Fig. P5-12 determine (a) the characteristic equation, (b) the time constants at which transients decay in this network.

5-13 For the circuit of Fig. P5-13 determine whether oscillations occur when a forcing function is applied. If oscillations do occur find the value of the damping ratio and the damped frequency of response.

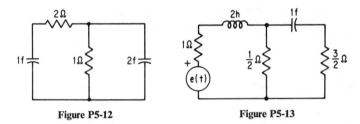

Figure P5-12   **Figure P5-13**

5-14 Refer to the circuit depicted in Fig. P5-14.

(a) Write the mesh equations for this circuit.

(b) From the results of part (a), write the corresponding equations that apply to the dual circuit.

(c) Employ the graphical procedure and find the exact dual circuit.

(d) Show that the results of parts (b) and (c) are consistent.

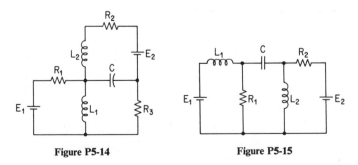

Figure P5-14   **Figure P5-15**

5-15 Repeat Prob. 5-14 for the circuit illustrated in Fig. P5-15.

5-16 A single rectangular pulse of magnitude E and duration T_1 is applied to a series R-L circuit. Find the expression for the current for (a) $0 \leq t \leq T_1$, and (b) $t \geq T_1$.

5-17 A single rectangular pulse of current having magnitude I and duration T_1 is applied to a parallel R-C circuit. Obtain the result for the voltage across the capacitor by employing the principle of duality.

5-18 A pulse having an amplitude of 10 volts and a period of duration of $T_1 = \pi/1000$ sec is applied to an initially de-energized series L-C circuit. Find the value of the coil current immediately upon the expiration of the pulse period. Assume that $L = 40$ millihenrys and $C = 100$ microfarads.

5-19 An impulse function of strength A volt-sec is applied to the series arrangement of a resistor R and inductor L, which is initially de-energized.
(a) Find the inductor current at time $t = 0^+$.
(b) Find $i(t)$ for $0^+ \leq t \leq \infty$.

5-20 An impulse function of strength Q amp-sec is applied to the parallel arrangement of a conductance G and capacitor C, which is initially de-energized.
(a) Find the capacitor voltage at time $t = 0^+$.
(b) Find the expression for the capacitor voltage for $0^+ \leq t \leq \infty$.

5-21 The circuit of Fig. P5-21 has been in the condition shown for a long time. The switch is then suddenly closed.
(a) What is the value of v before the switch is closed?
(b) What is the value of v immediately after the switch is closed?
(c) Find the complete expression for v after the switch is closed.
(d) What is the value of the time constant of the transient term?

5-22 The switch S is initially in position b with zero initial conditions on C and L. The switch is then put to terminal a moving along the contacts (heavy lines).
(a) What is the initial value of the voltage appearing across the capacitor when S is switched to a? Explain.
(b) Find the time expression for the current through the capacitor.
(c) How long does it take in seconds for the transient current to reach within 1 per cent of its initial value? An approximate answer is acceptable.
After a long time the switch S is quickly moved back to position b.
(d) Obtain the s-domain solution for the current which flows in the L-C circuit.

5-23 The configuration of an R-C circuit is as shown in Fig. P5-23. The initial-condition voltage on the capacitor is zero. Switch S is then put to terminal a.
(a) Find the expression for the current through the capacitor.

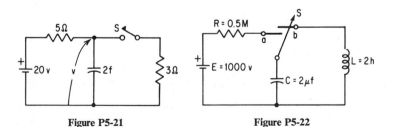

Figure P5-21   Figure P5-22

(b) What is the time constant of the charging circuit?

Ten seconds after switch S has been at terminal *a* it is then placed at terminal *b*.

(c) Find the current which flows through the 2-M (megohm) resistor.

(d) Compute the energy dissipated in this resistor after two seconds.

5-24 In the circuit of Fig. P5-24 switch S has been in position 1 for a long time.

(a) Find the complete solution for the current in the circuit when S is put to position 2.

(b) If it requires five time constants for the transient to disappear, find the time in seconds.

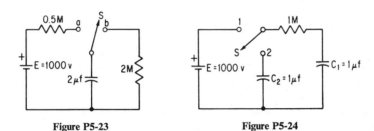

Figure P5-23   Figure P5-24

5-25 The circuit of Fig. P5-25 has been in steady state for a long time with the switch S open. Determine the complete time expression for the current when switch S is closed.

5-26 Assuming the circuit of Fig. P5-26 is in a steady-state condition with switch S open, compute the complete expression for the battery current when switch S is closed. What is the time constant of this circuit?

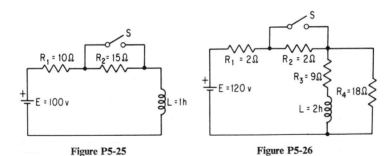

Figure P5-25   Figure P5-26

5-27 The circuit of Fig. P5-27 is initially de-energized. The switch is then closed.

(a) Find the expression for the current through R_2 as a function of time after the switch is closed.

(b) Repeat (a) for the current through C.

(c) What is the time constant of this circuit?

5-28 In the circuit of Fig. P5-28 the switch S is suddenly closed.

(a) Give the initial value of the current in each branch of the circuit.

(b) By replacing the portion of the circuit to the left of *cd* by its Thevenin equivalent, find the equation for the current in the inductance as a function of time.

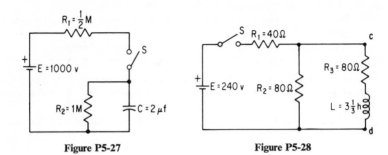

Figure P5-27   Figure P5-28

5-29 The response of a circuit to a step forcing function is known to be given by

$$i(t) = 10\epsilon^{-8t} \sin (6t + \theta)$$

Find the value of the damping ratio and the natural frequency of the current.

5-30 Refer to the circuit of Fig. P5-30. The switch is closed.
(a) Describe what happens at the capacitor.
(b) Find the battery current as a function of time.

5-31 After the circuit shown in Fig. P5-31 has been energized for a long time, switch S is suddenly opened.
(a) Find the expression for the field winding current as a function of time.
(b) What is the voltage across the $F_1 - F_2$ terminals at the instant the switch is opened?
(c) What would the voltage computed in part (b) be if R_1 were increased by ten times?

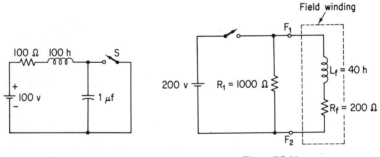

Figure P5-30   Figure P5-31

5-32 Refer to Fig. P5-32.
(a) With S_1 closed and S_2 open, find the time constant of the circuit.
(b) With S_2 open and S_1 suddenly closed, find the time it takes for the current to reach 0.05 amp.
(c) Assuming that S_1 has been closed a long time and that S_2 is then suddenly closed, find the relay current as a function of time.
(d) In part (c), how long will it take for the current in the relay to fall to 0.05 amp? Compare this with the result obtained in part (b).

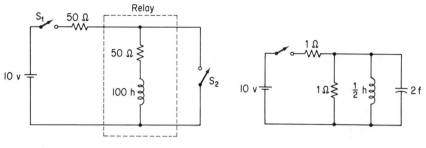

Figure P5-32   Figure P5-33

5-33 The inductor and capacitor in the circuit of Fig. P5-33 are initially de-energized.

(a) Find the time it takes for the battery current to reach within 99 per cent of its final value after the switch is closed.

(b) What is the maximum value of the coil current during the transient state?

5-34 In the circuit shown in Fig. P5-34, the switch has been at *a* for a long time.

(a) Derive the expression for the current in the inductor when the switch is placed from *a* to *b*.

(b) How long does it take the current to reach steady state?

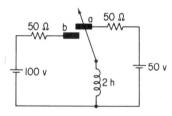

Figure P5-34

5-35 An engineering device contains two independent energy-storing elements. When subjected to a forcing function, the governing differential equation becomes

$$2\frac{di}{dt} + 36i + 450 \int i \, dt = 900u(t)$$

Just prior to the application of the forcing function the initial value of the response i is known to be 10, and the initial value of the integrating device is known to be 70, i.e., $\int_{-\infty}^{a} i \, dt = 70$.

(a) Does the response have a steady-state value different from zero in the time domain?

(b) Find the expression for the complete solution in the time domain.

5-36 The dynamic behavior of an electric circuit is described by

$$\frac{d^2i}{dt^2} + 5\frac{di}{dt} + 4i = \epsilon^{-t}$$

All initial conditions are zero. Find the complete solution for the current in response to the source function.

5-37   The behavior of a second-order system is described by

$$\frac{d^2c}{dt^2} + 6.4\frac{dc}{dt} + 64c = 160r$$

(a) What is the damping ratio of the system?

(b) Compute the per cent maximum overshoot.

(c) At what frequency does the system response oscillate during the transient state?

(d) How many seconds does it take for the response to settle within 5 per cent of its final value after the step forcing function r is applied? An approximate answer is acceptable.

5-38   A sinusoidal voltage $e(t) = 150 \cos 20t$ is applied at $t = 0$ to an initially de-energized series R-L circuit in which $R = 45$ ohms and $L = 3$ henrys.

(a) What is the expression for the forced current response?

(b) Calculate the maximum value of the transient term.

(c) Write the time expression for the transient current.

(d) Why does a transient-current term exist?

5-39   Repeat Prob. 5-38 for the case where the sinusoidal forcing function is described by $e(t) = 150 \cos (20t - 30°)$.

5-40   Repeat Prob. 5-38 for the case where $e(t) = 150 \cos (20t + 30°)$.

5-41   In the circuit of Fig. P5-41 switch S has been at position a for a long time. The switch is then put to position b at an instant when the a-c voltage is at its positive peak value. Determine the complete time expression for the current response when S is put to position b.

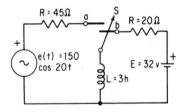

Figure P5-41

5-42   A sinusoidal voltage $e(t) = 100 \sin 20t$ is applied to a series R-L circuit. The magnitude of the transient term immediately upon switching at $t = 0$ is found to be 0.4 ampere. Moreover, the transient term is observed to diminish with a time constant of 0.5 second.

(a) Compute the values of R and L.

(b) Is the initial value of the transient current positive or negative? Explain.

(c) At what point in the voltage cycle must the switch be closed for the voltage across the inductance to be a maximum? What is the value of this voltage?

5-43   A sinusoidal voltage of $e(t) = E_m \cos \omega t$ is applied to an initially de-energized series combination of capacitance C and resistor R. Determine the complete expression for the current response.

6

sinusoidal steady-state
response of circuits

The theory of the sinusoidal steady-state response of circuits occupies a position of pre-eminence in electric-circuit theory. The analysis of many circuits and devices throughout all branches of electrical engineering is accomplished by the techniques embodied in the sinusoidal theory. Particularly impressive in this regard is the fact that the sinusoidal circuit theory is applicable not only in situations involving sinusoidal forcing functions but equally as well in those situations where the forcing functions are of a nonsinusoidal character.

It is not an accident that the bulk of the electric power generated in power plants throughout the world and distributed to the consumer appears in the form of sinusoidal variations of voltage and current. There are many technical and economical advantages associated with the use of sinusoidal voltages and currents. A significant appreciation of this statement will be gained upon the completion of the study of this book. In Chapter 15, for example, it will be learned that the use of sinusoidal voltages applied to appropriately designed coils results in a revolving magnetic field which has the capacity to do work. As a matter of fact, it is this principle which underlies the operation of almost all the electric motors found in home appliances and about 90 per cent of all electric motors found in commercial and industrial applications. Although other waveforms can be used in such devices, none leads to an operation which is as efficient and economical as that achieved through the use of sinusoidal functions.

In addition to these practical aspects, however, the sinusoidal function offers some very important and significant advantages in a mathematical sense. Recall that by Euler's theorem the sine as well as the cosine function can be represented quite simply by the exponential function. Thus, $\epsilon^{j\omega t} = \cos \omega t + j \sin \omega t$. Since as described in Chapter 5, the equations which govern the behavior of electric circuits frequently involve derivative and integral terms, this exponential character of the sinusoid is of prime importance. The reason is that it permits a simplification of mathematical analysis which cannot be achieved by any other function. The exponential function is the only mathematical function, the original form of which is preserved even though such operations as differentiation and integration are performed on it. Consequently, as described in this chapter, it is possible to treat sinusoidal time functions entirely in terms of complex numbers. A little thought indicates that the situation here is not unlike that encountered in Chapter 4 with the Laplace transform. In view of the involvement of the exponential function in both instances, the similarity is certainly to be expected.

Knowledge of the sinusoidal theory has another notable advantage in the mathematical sense. By means of the Fourier series it is possible to represent *any* periodic function of whatever form in terms of an infinite series of sinusoids. Accordingly, the steady-state response of electric circuits to nonsinusoidal forcing functions can be obtained through a repeated application of the sinusoidal theory followed by a summation of the individual sinusoidal responses.

6-1 Sinusoidal Functions—Terminology

In dealing with sinusoidal functions we must become familiar with the nomenclature before proceeding with the sinusoidal steady-state analysis of circuits. This makes it easier to describe and to interpret the results. Appearing in Fig. 6-1 are two sinusoids—one denoting voltage and the other current. The voltage sinusoid is a sine function which is represented mathematically as

$$v = V_m \sin \alpha = V_m \sin \omega t \quad \text{volts} \qquad (6\text{-}1)$$

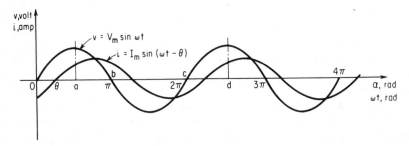

Fig. 6-1. Sinusoidal voltage and current waves of the same frequency. The current sinusoid has a phase lag of θ relative to the voltage wave.

where

$$\alpha = \omega t \quad \text{radians} \qquad (6\text{-}2)$$

and is called the *argument* of the sine function. The *amplitude* of the sine function is V_m and it denotes the *maximum* value of the sine function. Equation (6-1) is often referred to as the expression for the *instantaneous voltage* because by insertion of any particular value for t the corresponding value of v is determined. Note that as we substitute into Eq. (6-1) values of α that lie between 0 and 2π degrees, different and distinct values of the function result. However, as α is made to take on values between 2π and 4π there occurs a repetition of the first set of values obtained. Any complete set of positive and negative values of the function is called a *cycle*. Thus portion *oabc* and *abcd* of the voltage wave of Fig. 6-1 qualify as cycles of the time function.

Another characteristic of the sinusoid is its *frequency*, which is defined as the number of cycles of the function which are traversed in one second. Accordingly, the frequency is measured in units of *cycles per second* which has recently come to be called the hertz and abbreviatied Hz. In radio and radar transmission work where high frequencies are involved it is customary to use units of kilohertz, megahertz, and gigahertz. The time duration of one cycle is called the *period* of the function. It follows then that if f denotes the frequency of the periodic function, the period is related to the frequency by

$$\boxed{T = \frac{1}{f}} \quad \text{sec} \qquad (6\text{-}3)$$

Moreover, a glance at Fig. 6-1 indicates that each cycle spans 2π radians. Hence, if this quantity is divided by the period, there results the *angular velocity* (or angular frequency) of the sine function. It is denoted by ω and has units of rad/sec.

$$\boxed{\omega = \frac{2\pi}{T} = 2\pi f} \qquad (6\text{-}4)$$

The second form of ω is obtained from Eq. (6-3).

The equation to be used to describe the second sinusoid appearing in Fig. 6-1, i.e., the current wave, obviously cannot be identical to that used for the voltage wave. The reason is that the two sinusoids are displaced from one another by the angle θ and so cannot have the same instantaneous value. For example, at $t = 0$, $v = 0$ whereas the current has a finite negative value. A comparison of the two sine waves shows that the zero value of the current wave for positive slope occurs *later* in time by an amount $\alpha = \theta$. Accordingly, the current wave is said to be *lagging* behind the voltage wave by the relative phase angle θ. *Relative phase* is the term used to denote the angular displacement between two sinusoids of the same frequency. Thus in Fig. 6-1 the current sinusoid can be described as having a *phase lag* of θ degrees relative to the voltage sinusoid. Alternatively we can say that the voltage wave has a *phase lead* of θ degrees relative to the current wave.

The equation which describes the instantaneous value of the current wave must include the angular displacement (or phase angle) existing between the two sinusoids. A little thought reveals that the current expression is

$$i = I_m \sin(\omega t - \theta) \qquad (6\text{-}5)$$

Note that the phase angle is included as part of the argument of the sine function. A minus sign is used because there is a phase lag between the voltage and current sinusoids. This formulation assures that at $t = 0$ a finite, negative value for the current exists as called for by the plot of Fig. 6-1.

A word of caution is appropriate at this point concerning the units of the argument $(\omega t - \theta)$. The form of Eq. (6-5) demands that the unit for this total quantity be radians. However, in engineering usage of this equation it is customary to express ωt in radians and θ in degrees. The reason lies in the fact that in engineering calculations it is the phase angle which is important, not the total angular displacement. Therefore, when Eq. (6-5) is occasionally seen written as $i = I_m \sin(\omega t - 45°)$, the inconsistency in the units of the argument should be accepted in light of the foregoing comment.

There exist alternate forms with which to express sinusoids. These originate with Euler's identity, i.e.,

$$\epsilon^{j\omega t} = \cos \omega t + j \sin \omega t \qquad (6\text{-}6)$$

It should be apparent from this expression that a cosine and a sine function can be expressed in terms of the exponential notation as follows:

$$\cos \omega t = \text{Re} \left[\epsilon^{j\omega t} \right] \qquad (6\text{-}7)$$

$$\sin \omega t = \mathscr{I}\text{m} \left[\epsilon^{j\omega t} \right] \qquad (6\text{-}8)$$

where Re [ ] denotes the real part of the expression in brackets and $\mathscr{I}$m [ ] the imaginary part. This representation is used on various occasions throughout the book when it is convenient to do so. The sinusoids may also be expressed by the equations

$$\cos \omega t = \frac{\epsilon^{j\omega t} + \epsilon^{-j\omega t}}{2} \qquad (6\text{-}9)$$

$$\sin \omega t = \frac{\epsilon^{j\omega t} - \epsilon^{-j\omega t}}{2j} \qquad (6\text{-}10)$$

The insertion of Eq. (6-6) into the right side of either of the last two equations bears out the validity of the equivalence.

6-2 Average and Effective Values of Periodic Functions

The energy sources throughout the treatment of Chapter 3 are all of the non-varying, constant type—often called the *direct* kind. Thus when a voltage source of fixed magnitude is applied to a network, the forced solution is found to be a constant quantity too. This constant character of the response makes it a simple matter to identify the number of amperes flowing in the circuit and thereby to describe the energy-transferring capability of the circuit. Moreover, the computation of the power absorbed by each circuit element is accomplished in a direct manner through the use of Eq. (2-5) by inserting the constant values of voltage and current. The situation, however, is quite different when the voltages and currents associated with a circuit element are varying functions of time such as those appearing in Fig. 6-1. Note that the sinusoidal current is an *alternating current*, i.e., one which has positive and negative values. In such a case the manner of describing the energy-transferring capability of the current is not at all obvious, as it is when direct sources are used; in the latter case the average current flow is identical to the direct (or constant) value.

In view of the fact that the average current serves as a useful criterion in determining the energy transfer in circuits involving direct sources, let us investigate its usefulness in situations involving periodic driving functions. The term periodic is used rather than sinusoidal in order that the treatment be general. The sinusoid is only one example of a periodic function. Any function whose cycle is repeated continuously irrespective of waveform is called a periodic function.

A general definition of the average value of any function $f(t)$ over the specified interval between t_1 and t_2 is expressed mathematically as

$$F_{av} \equiv \frac{1}{t_2 - t_1} \int_{t_1}^{t_2} f(t)\, dt \qquad (6-11)$$

In the instance when $f(t)$ is a periodic function having a period of T sec, Eq. (6-11) becomes

$$F_{av} = \frac{1}{T} \int_0^T f(t)\, dt \qquad (6-12)$$

If the time function is expressed in radians through the use of Eq. (6-2), then an alternative form for the average value of the function results. Thus

$$F_{av} = \frac{1}{2\pi} \int_0^{2\pi} f(\omega t)\, d(\omega t) = \frac{1}{2\pi} \int_0^{2\pi} f(\alpha)\, d\alpha \qquad (6-13)$$

Average Value of a Sinusoid. By means of Eq. (6-13) let us now find the average value of the sinusoidal current variation of Fig. 6-1. Hence

$$I_{av} = \frac{1}{2\pi} \int_0^{2\pi} I_m \sin(\omega t - \theta)\, d(\omega t) = \frac{1}{2\pi} \int_0^{2\pi} I_m \sin(\alpha - \theta)\, d\alpha \qquad (6-14)$$

where $\alpha = \omega t$. Integrating and inserting limits yields

$$I_{av} = \frac{I_m}{2\pi} [-\cos(\alpha - \theta)]_{\alpha=0}^{\alpha=2\pi} = \frac{I_m}{2\pi} [-\cos(2\pi - \theta) + \cos(-\theta)] \equiv 0 \qquad (6-15)$$

Therefore the average value of a sinusoid over *one complete cycle* is identically equal to zero. A study of the plot of the sinusoid makes this conclusion obvious because for one cycle there is as much area above the abscissa axis as there is below. Accordingly, the net area is zero.

It is interesting to note that a finite average value can be found for the sinusoid for the *positive or negative half-cycle*. Because of the usefulness of this result in subsequent work, we determine the general expression for this quantity here. Hence

$$I_{\text{av}-1/2\,\text{cycle}} = \frac{1}{\pi} \int_{\theta}^{\pi+\theta} I_m \sin(\alpha - \theta)\, d\alpha \qquad (6\text{-}16)$$

The sinusoid being used is the one depicted in Fig. 6-1 and described by Eq. (6-5). The lower and upper limits of the integral are selected to include only the area above the abscissa axis. Upon integrating and substituting limits we obtain

$$I_{\text{av}-1/2\,\text{cycle}} = \frac{I_m}{\pi} \left[-\cos(\alpha - \theta)\right]_{\alpha=\theta}^{\alpha=\pi+\theta} = \frac{I_m}{\pi}\left[-\cos(\pi) + \cos 0\right] = \frac{2}{\pi} I_m \qquad (6\text{-}17)$$

or

$$\boxed{I_{\text{av}-1/2\,\text{cycle}} = \frac{2}{\pi} I_m = 0.636 I_m} \qquad (6\text{-}18)$$

Thus the average value of either the positive or negative half of a sine function can be found simply by multiplying the amplitude of the wave by 0.636. When taken over a full cycle the equal and opposite average values cancel out.

On the basis of the foregoing results it should be clear that, although the criterion of the average value of current works well in describing the energy-transferring capacity for direct sources, it is a meaningless criterion for sym-metrical† periodic functions because its value is always equal to zero. Therefore we must search for a more suitable criterion to measure the effectiveness of a periodic function. Preferably the chosen standard should in some way be related to the energy or power capability associated with the periodic function. Herein lies a clue about the procedure to follow in establishing such a criterion. Consider that the sinusoidal current of Eq. (6-5) is made to flow through a resistor having R ohms. Then by Joule's law the instantaneous power absorbed by the resistor and converted to heat is

$$p = i^2(t)R \quad \text{watts} \qquad (6\text{-}19)$$

Since the current in this expression varies usually from a positive maximum through zero to a negative maximum, the insertion of any one value of the current is of little usefulness in determining the actual power absorbed by the resistor. Rather a much more meaningful approach is to sum the instantaneous power consumption over one full cycle. Such a formulation cannot lead to a zero result because the power dissipation is real whether the current flows clockwise (positive) or counterclockwise (negative) in this circuit. Therefore a nonzero value must result. This conclusion is also borne out by the presence of the second power of the current in Eq. (6-19). Thus even if the current is negative the contribution to the power expression is positive. Proceeding on this basis, we find that the expression for the average power absorbed by the resistor becomes

$$P_{\text{av}} = \frac{1}{T}\int_0^T i^2(t)R\, dt = \left[\frac{1}{T}\int_0^T i^2(t)\, dt\right]R \qquad (6\text{-}20)$$

A study of the quantity in brackets reveals some interesting points. Note

---

† The term symmetrical is used to emphasize that we are here concerned only with those periodic functions with equal positive and negative areas.

first that this quantity is the average value of the function $f(t) = i^2(t)$. A comparison with Eq. (6-12) makes this obvious. We are still dealing with average effects—but with one important difference: instead of working with the periodic function $i(t)$ directly we are dealing with the *squared* function. This not only gives significance to the formulation in terms of a description involving the power capability of the current, but it also eliminates the possibility of dealing with a zero average value because the square of a periodic function always yields a positive contribution. Another point of interest is that the unit of the bracketed quantity is amperes squared. Accordingly we can interpret this quantity as the square of an *effective current*, which, when multiplied by R, yields the average power. Expressing this mathematically, we can write

$$I_{\text{eff}}^2 \equiv \frac{1}{T} \int_0^T i^2(t)\, dt = \text{average } i^2(t) \tag{6-21}$$

From this it follows that the effective current is the *root mean square* value. Thus

$$I_{\text{eff}} = I_{\text{rms}} = \sqrt{\text{average } i^2(t)} = \sqrt{\frac{1}{T} \int_0^T i^2(t)\, dt} \tag{6-22}$$

Although the effective current may be denoted by either of the subscript notations used in the last equation, it is frequently written without a subscript. The understanding is that reference is always to the effective value of a periodic function as defined in Eq. (6-22) unless otherwise specified.

Finally, as a third point of interest, let us make an analogy with the direct-current case. A direct voltage source applied to the resistor R causes an average power dissipation of I^2R to take place. Here I denotes the direct or average current. The flow of a periodic function of current through the same resistor yields an average power dissipation of I_{eff}^2R. Comparing this result with that of the direct-current case yields another interpretation of effective current: it is that current which produces the same heating effect as the direct current.

Effective Value of a Sinusoid. Let us now assume that the periodic function $i(t)$, referred to in the foregoing discussion, takes on the specific form of the sinusoid as depicted in Fig. 6-2. Then by Eq. (6-21) we have

$$I_{\text{eff}}^2 = \frac{1}{T} \int_0^T i^2(t)\, dt = \frac{1}{T} \int_0^T I_m^2 \sin^2 \omega t\, dt \tag{6-23}$$

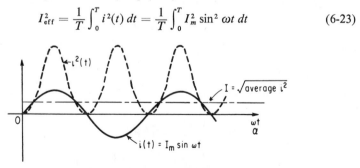

Fig. 6-2. Illustrating the computation of the effective value of a sinusoidal current.

Introducing the trigonometric identity

$$\sin^2 \omega t = \tfrac{1}{2} - \tfrac{1}{2}\cos 2\omega t \tag{6-24}$$

permits Eq. (6-23) to be written as

$$
\begin{aligned}
I_{\text{eff}}^2 &= \frac{I_m^2}{2T}\int_0^T (1 - \cos 2\omega t)\,dt \\
&= \frac{I_m^2}{2T}[T] - \frac{I_m^2}{4\omega T}[\sin 2\omega t]_0^T = \frac{I_m^2}{2}
\end{aligned}
\tag{6-25}
$$

because $T = 2\pi$.

Therefore

$$\boxed{I = I_{\text{eff}} = \frac{I_m}{\sqrt{2}} = 0.707\, I_m} \tag{6-26}$$

This is an important result which is used often in the study of sinusoidal steady-state theory. It is important to remember that it is valid only for sinusoidal functions.

EXAMPLE 6-1   Find the average value of the periodic function depicted in Fig. 6-3.

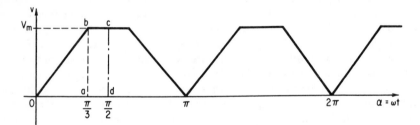

Fig. 6-3. Trapezoidal waveform for Example 6-1.

solution: A study of the waveshape shows that all the information about the curve is contained in the portion which lies from 0 to $\pi/2$. In this region two equations are needed to describe the function. Thus

$$v(t) = \frac{V_m}{\pi/3}(\omega t) = \frac{V_m}{\pi/3}\alpha \quad \text{for } 0 \le \alpha \le \frac{\pi}{3} \tag{6-27}$$

and

$$v(t) = V_m \qquad\qquad \text{for } \frac{\pi}{3} \le \alpha \le \frac{\pi}{2} \tag{6-28}$$

A direct application of Eq. (6-13) over the region from $\alpha = 0$ to $\alpha = \pi/2$ yields the average value. Thus

$$
\begin{aligned}
V_{\text{av}} &= \frac{1}{\pi/2}\left\{\int_0^{\pi/3}\frac{V_m}{\pi/3}\alpha\,d\alpha + \int_{\pi/3}^{\pi/2} V_m\,d\alpha\right\} \\
&= \frac{1}{\pi/2}\left\{\frac{V_m}{\pi/3}\left[\frac{\alpha^2}{2}\right]_0^{\pi/3} + V_m[\alpha]_{\pi/3}^{\pi/2}\right\} \\
&= \frac{V_m}{\pi/2}\left\{\frac{\pi}{6} + \frac{\pi}{6}\right\} = \frac{2}{3}\,V_m
\end{aligned}
\tag{6-29}
$$

We can check this result by finding the total area beneath the $v(t)$ curve over the portion from 0 to $\pi/2$ and dividing the result by $\pi/2$. Thus

$$\text{area under } v(t) \text{ curve} = \text{area } 0ab + \text{area } abcd$$

$$= \frac{1}{2} V_m \left(\frac{\pi}{3}\right) + V_m \left(\frac{\pi}{6}\right) = \frac{\pi}{3} V_m \tag{6-30}$$

$$\therefore \ V_{av} = \frac{\text{area}}{\pi/2} = \frac{(\pi/3) V_m}{\pi/2} = \frac{2}{3} V_m \tag{6-31}$$

6-3 Instantaneous and Average Power. Power Factor

Our interest in this section is to develop a general expression for the average power associated with a voltage and current in an a-c† circuit. The restrictive limitation of confining the treatment to resistive circuits is now dropped. In this connection then let $v(t) = V_m \sin \omega t$ represent the potential difference appearing across the branch terminals of a given circuit and let $i(t) = I_m \sin (\omega t - \theta)$ denote the corresponding current flowing through that branch. The relative phase angle is given by θ. The voltage and current sinusoids are shown in Fig. 6-4. It follows then that the expression for the instantaneous power is

$$p(t) = v(t)i(t) = V_m I_m \sin \omega t \sin (\omega t - \theta) \tag{6-32}$$

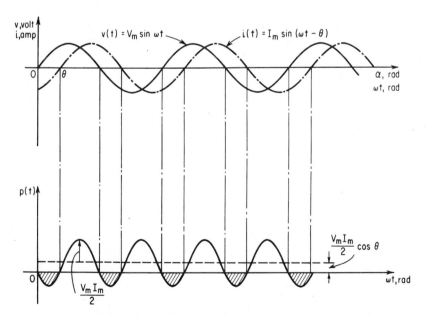

Fig. 6-4. Plot of instantaneous power as determined from the given voltage and current sinusoids.

† This notation is often used to denote circuits subjected to sinusoidal (i.e., alternating) forcing functions.

To put this in a more suitable form insert the identity

$$\sin(\omega t - \theta) = \sin \omega t \cos \theta - \cos \omega t \sin \theta \qquad (6\text{-}33)$$

into Eq. (6-32) to yield

$$p(t) = V_m I_m(\sin^2 \omega t \cos \theta - \sin \omega t \cos \omega t \sin \theta) \qquad (6\text{-}34)$$

To effect a further reduction introduce

$$\sin^2 \omega t = \tfrac{1}{2} - \tfrac{1}{2} \cos 2\omega t \qquad (6\text{-}35)$$

and

$$\sin \omega t \cos \omega t = \tfrac{1}{2} \sin 2\omega t \qquad (6\text{-}36)$$

into Eq. (6-34). Thus

$$p(t) = \frac{V_m I_m}{2} \cos \theta - \frac{V_m I_m}{2}(\cos \theta \cos 2\omega t + \sin \theta \sin 2 \omega t) \qquad (6\text{-}37)$$

Finally, by substituting the relation

$$\cos(2\omega t - \theta) = \cos 2\omega t \cos \theta + \sin 2\omega t \sin \theta \qquad (6\text{-}38)$$

into Eq. (6-37) the desired form of the expression for the instantaneous power is obtained. Thus

$$\boxed{p(t) = \frac{V_m I_m}{2} \cos \theta - \frac{V_m I_m}{2} \cos(2\omega t - \theta)} \qquad (6\text{-}39)$$

A plot of Eq. (6-39) appears in Fig. 6-4. Note that for a fixed θ the instantaneous power consists of two components—a constant part and a time-varying part. Note, too, that the varying part has a frequency which is twice that of the voltage and current sinusoids. The shaded portions of the plot of $p(t)$ refer to those time intervals when the power is negative. In effect this means that the circuit is returning power to the source during these intervals. It should be apparent, then, that the branch circuit under consideration contains at least one energy-storing element. A glance at Fig. 6-4 shows that the instantaneous power is negative whenever the voltage and current are of opposite sign. However, for the case plotted in Fig. 6-4 notice that the positive area under the $p(t)$ curve exceeds the negative area. Therefore the average power is positive and finite, and specifically it is equal to the constant term of Eq. (6-39). As θ is made smaller, i.e., as i is brought more nearly in phase with v, the negative areas of the $p(t)$ curve of Fig. 6-4 become smaller and so the average power increases. This is equivalent to raising the $p(t)$ curve higher above the abscissa axis. When $\theta = 0$ the current and voltage are in phase. There are no negative areas associated with the $p(t)$ curve; hence all the power is consumed between the circuit branch terminals. The circuit may then be called purely resistive. On the other hand, when θ is increased the negative areas become larger and so less power is consumed between the terminals and more returned to the source. At the extreme value of θ, i.e., $\theta = \pi/2$, the $p(t)$ curve is dropped to that position which makes the negative and positive areas equal. In this instance there is no power consumed between the circuit terminals.

The relative phase angle θ is determined by the values of the circuit parameters appearing between the circuit branch terminals across which $v(t)$ is assumed to exist. Because of the passive nature of these circuit parameters the value of θ is restricted to lie in the range expressed by $-\pi/2 \leq \theta \leq \pi/2$.

The general expression for the instantaneous power in an a-c circuit is described by Eq. (6-39). The really useful quantity in terms of the capability of the circuit to do work is the average value of the power over one cycle. Since each cycle is continuously repeated, whatever is done for one cycle applies equally as well for each succeeding cycle. It has already been stated in connection with the discussion of Fig. 6-4 that this average power is given by the constant term of Eq. (6-39). A mathematical verification now follows. From the general definition of the average value over one cycle we have

$$P_{av} = \frac{1}{T} \int_0^T p(t)\, dt \tag{6-40}$$

Inserting Eq. (6-39) for $p(t)$ yields

$$P_{av} = \frac{1}{T} \left\{ \int_0^T \frac{V_m I_m}{2} \cos \theta\, dt - \int_0^T \frac{V_m I_m}{2} \cos(2\omega t - \theta)\, dt \right\} \tag{6-41}$$

Since the second term on the right side involves the integration of a simple sine function over a time interval equal to two complete periods of the double frequency sine function, the value is always identically equal to zero. This leaves just the first term and, since θ is independent of t, it follows that the average power is

$$P_{av} = \frac{V_m I_m}{2} \cos \theta \quad \text{watts} \tag{6-42}$$

It is helpful at this point to rewrite the last equation as

$$P_{av} = \frac{V_m}{\sqrt{2}} \left(\frac{I_m}{\sqrt{2}} \right) \cos \theta \tag{6-43}$$

Recalling that the effective value of a sinusoidal quantity is the amplitude divided by $\sqrt{2}$, Eq. (6-43) can be expressed more significantly in terms of the corresponding effective values of the voltage and current sinusoids. Therefore

$$\boxed{P_{av} = V_{eff} I_{eff} \cos \theta = VI \cos \theta} \quad \text{watts} \tag{6-44}$$

Although it is customary to drop the subscripts entirely when writing this equation, one of our reasons for leaving them here is to emphasize better that *average power* is determined in terms of the *effective voltage and current* values. Equation (6-44) points out in a general and significant fashion the usefulness of the root mean square value of a periodic function as a criterion to measure its effectiveness. The effective values of voltage and current play key roles in measuring the ability of a circuit to do work.

In the interest of introducing another term of electric-circuit theory rewrite Eq. (6-44) as follows:

$$\cos \theta = \frac{P}{VI} \tag{6-45}$$

The quantity P is the average power and is expressed in watts—a unit which conveys the capability to do work. However, note that the denominator of Eq. (6-45) involves a quantity whose units are represented by the product of volts by amperes. When we are dealing with direct sources this product is called watts, because it is real power which can be entirely converted to work. The same is not true when sinusoidal quantities are involved. For example, in Fig. 6-4, corresponding to $\theta = \pi/2$ there is no useful (or work-producing) power in the circuit in spite of the large values which V and I may have. For this reason the product VI is called *apparent power*. This power is not always realizable in the circuit for doing work. The useful part depends upon the value of $\cos \theta$, and because of this $\cos \theta$ is called the *power factor* (abbreviated pf) of the circuit. Thus

$$\boxed{\text{pf} = \cos \theta = \frac{\text{average power}}{\text{apparent power}} = \frac{P}{VI}} \qquad (6\text{-}46)$$

where V and I are effective values.

EXAMPLE 6-2   A voltage $v(t) = 170 \sin (377t + 10°)$ is applied to a circuit. It causes a steady-state current to flow which is described by $i(t) = 14.4 \sin (377t - 20°)$. Determine the power factor and the average power delivered to the circuit.

solution: A comparison of the expressions for $v(t)$ and $i(t)$ reveals that the relative phase angle is 30° Hence.

$$pf = \cos \theta = \cos 30° = 0.866 \qquad (6\text{-}47)$$

Also

$$V = \frac{V_m}{\sqrt{2}} = \frac{170}{\sqrt{2}} = 120 \text{ volts} \qquad (6\text{-}48)$$

and

$$I = \frac{I_m}{\sqrt{2}} = \frac{14.14}{\sqrt{2}} = 10 \text{ amp} \qquad (6\text{-}49)$$

Therefore

$$P = P_{\text{av}} = VI \cos \theta = 120(10)(0.866) = 1040 \text{ watts} \qquad (6\text{-}50)$$

6-4   Phasor Representation of Sinusoids

Often in determining the sinusoidal steady-state response of circuits it is necessary to perform algebraic operations such as addition, subtraction, multiplication, and division on two or more sinusoidal quantities of the same frequency. Usually the sinusoids differ in amplitude and phase. Specifically consider the matter of adding two sinusoidal currents whose equations are

$$i_1 = I_{m1} \sin \omega t \qquad (6\text{-}51)$$

$$i_2 = I_{m2} \sin (\omega t + \theta_2) \qquad (6\text{-}52)$$

The current i_2 leads i_1 by the relative phase angle θ_2. The resultant current i_3 can obviously be written as

$$i_3 = I_{m1} \sin \omega t + I_{m2} \sin (\omega t + \theta_2) \tag{6-53}$$

Since the addition of two sinusoids of the same frequency always results in another sinusoid, it is desirable to express Eq. (6-53) in terms of an appropriate resultant amplitude and phase. Keep in mind that any sinusoid at a given frequency is completely specified once its amplitude and phase are known. Of course one obvious way of obtaining the result called for in Eq. (6-53) is to plot each sinusoid and then make a point-by-point summation of the two sine waves. The amplitude and phase of the resultant sinusoid can then be measured, thus allowing i_3 to be written in the more useful form

$$i_3 = I_{m3} \sin (\omega t + \theta_3) \tag{6-54}$$

where θ_3 is the phase angle measured with respect to the same reference point used for θ_2. Needless to say, such a procedure is laborious and time-consuming.

An alternative to the graphical solution is an analytical one which simplifies Eq. (6-53) to the form of Eq. (6-54) through the use of trigonometric identities. Accordingly, by introducing

$$\sin (\omega t + \theta_2) = \sin \omega t \cos \theta_2 + \cos \omega t \sin \theta_2 \tag{6-55}$$

into Eq. (6-53), it becomes

$$i_3 = (I_{m1} + I_{m2} \cos \theta_2) \sin \omega t + (I_{m2} \sin \theta_2) \cos \omega t \tag{6-56}$$

Moreover, the use of Eq. (6-55) in Eq. (6-54) allows Eq. (6-54) to be expressed alternatively as

$$i_3 = (I_{m3} \cos \theta_3) \sin \omega t + (I_{m3} \sin \theta_3) \cos \omega t \tag{6-57}$$

Since the last two equations for i_3 are to be identical, it follows that

$$I_{m3} \cos \theta_3 = I_{m1} + I_{m2} \cos \theta_2 \tag{6-58}$$

$$I_{m3} \sin \theta_3 = I_{m2} \sin \theta_2 \tag{6-59}$$

Equations (6-58) and (6-59) are obtained by equating the coefficients of like terms in Eqs. (6-56) and (6-57). By means of the last two equations the amplitude (I_{m3}) and the phase (θ_3) of the resultant sinusoid can be found in terms of the amplitudes and phase angles of i_1 and i_2, which are known.

Although this analytical procedure requires less effort than the graphical one, the method is still too cumbersome to be practical. This is particularly so when situations arise where more than two sinusoidal quantities must be summed. Furthermore, multiplication and division present additional complications. Clearly, then, we need a simpler and more direct method of treating sinusoidal quantities. In 1893 such a method was introduced, when Charles P. Steinmetz advanced the idea of using a constant-amplitude line rotating at a frequency ω to represent a sinusoid. Let us now see how such an idea is effective in simplifying the algebraic operations involving sinusoidal quantities.

Attention is first directed to the expression for i_1 as given by Eq. (6-51). By employing the notation of Eq. (6-8) we can write

$$i_1 = I_{m1} \sin \omega t = \mathscr{I}m \, [I_{m1} \epsilon^{j\omega t}] \tag{6-60}$$

Keep in mind that the exponential function $\epsilon^{j\omega t}$ may be treated as a rotational operator. Its amplitude is always unity, but the cosine and sine components vary as time progresses. This is illustrated in Fig. 6-5(a). As ωt moves through one full period of 2π radians (i.e., one complete cycle) the line OA makes one complete traversal of the circle in a counterclockwise direction. Line OA is fixed in value to the amplitude of the sine function it represents. Note that the vertical component of line OA is the sine function. As a matter of fact this is the meaning of the notation $\mathscr{I}\mathrm{m}\,[\quad]$—it refers to the values generated by taking the projections of a rotating line on a pre-established reference line (the vertical in this case). Accordingly, if we plot the vertical components of OA as it makes one complete revolution, the sine function shown in Fig. 6-5(b) is generated. When $\omega t = 0$ the position of OA is on the horizontal axis directed towards the right. Its vertical component at this instant is zero, as it should be for the sine function.

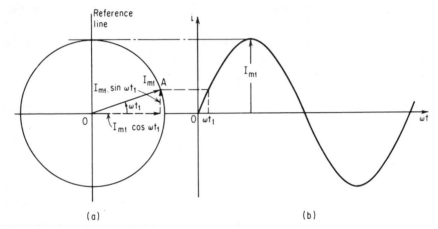

Fig. 6-5. Generating the sine function from the vertical component of the rotating line $I_{m1}\epsilon^{j\omega t} = OA$.

The current i_2 as given by Eq. (6-52) can be represented in a similar fashion. Thus in terms of the exponential notation we have

$$i_2 = I_{m2}\sin{(\omega t + \theta_2)} = \mathscr{I}\mathrm{m}\,[I_{m2}\epsilon^{j(\omega t+\theta_2)}] = \mathscr{I}\mathrm{m}\,[I_{m2}\epsilon^{j\theta_2}\epsilon^{j\omega t}] \qquad (6\text{-}61)$$

To simplify the notation we next define

$$\boxed{\mathbf{I}_{m2} \equiv I_{m2}\epsilon^{j\theta_2} = I_{m2}(\cos\theta_2 + j\sin\theta_2)} \qquad (6\text{-}62)$$

and denote it as line OB in Fig. 6-6(a). It is this quantity—$\mathbf{I}_{m2}$—which is called the *phasor* of the sinusoidal function of Eq. (6-61). In general the phasor can be represented by a *complex number* which is a result of locating a line in a plane. Thus the phasor OB can be located by specifying its magnitude, I_{m2}, and its displacement from the horizontal axis θ_2. The reader certainly recognizes these

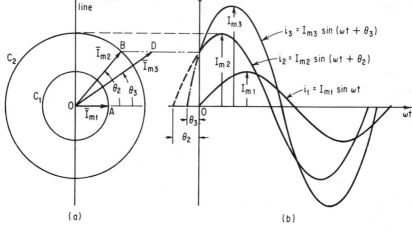

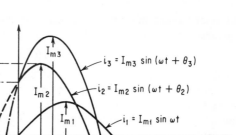

(a)   (b)

Fig. 6-6. Illustrating the phasor representation and addition of two sinusoids, $i_3 = i_1 + i_2$:
(a) position of the phasors for $t = 0$;
(b) sinusoidal function for increasing time.

quantities as the polar coordinates of line OB. However, note that OB can also be
located in terms of a horizontal (i.e., real-axis) component and a vertical (i.e.,
imaginary or j-axis) component as indicated by the second form of I_{m2} in Eq.
(6-62). For a given θ_2 the real part of the complex number is $I_{m2} \cos \theta_2$ and the
imaginary or j part is $I_{m2} \sin \theta_2$. It is important to understand that the position of
OB in Fig. 6-6(a) corresponds to time $t = 0$ in Eq. (6-61). Hence the corresponding
value of i_2 at this instant must be the projection of OB on the vertical reference
line, which clearly is $I_{m2} \sin \theta_2$.

The phasor of current i_1 is $I_{m1} = I_{m1}\epsilon^{j0°} = I_{m1}$. Hence it has no vertical (or j)
component and so initially (i.e., at $t = 0$) must lie along the horizontal axis as
shown by line OA in Fig. 6-6(a). Note too in this figure that the phasors I_{m1} and
I_{m2} are displaced from one another by θ_2. This is entirely consistent with Eqs.
(6-51) and (6-52). Now, as time elapses, both phasors I_{m1} and I_{m2} revolve at a
constant frequency of ω radians per second, and a continuous plot of the vertical
components of both phasors results in the sinusoids shown in Fig. 6-6(b). In this
way a sinusoid can be represented by a line fixed at one end and rotating at a
frequency equal to the angular velocity of the sinusoid. The magnitude of the
line must be equal to the amplitude of the sinusoid.

How does the phasor representation of sinusoids simplify the procedure for
adding two sinusoidal quantities? To understand this, let us find $i_3 = i_1 + i_2$
by rewriting Eq. (6-53) in exponential form. Thus

$$i_3 = i_1 + i_2 = \mathscr{I}\mathrm{m}\,[I_{m1}\epsilon^{j\omega t}] + \mathscr{I}\mathrm{m}\,[I_{m2}\epsilon^{j\omega t}] \qquad (6\text{-}63)$$

Keep in mind that attaching the factor $\epsilon^{j\omega t}$ to the phasors I_{m1} and I_{m2} has the effect of "animating" the phasors, i.e., it causes them to revolve at the angular frequency ω. From our knowledge of complex algebra we can rewrite Eq. (6-63) as

$$i_3 = \mathscr{I}\mathrm{m}\,[(I_{m1} + I_{m2})\epsilon^{j\omega t}] \qquad (6\text{-}64)$$

By factoring out $\epsilon^{j\omega t}$ in this last expression we are recognizing that both phasors are revolving at the same frequency ω. Therefore the only thing that needs to be done here is to perform the operation called for in parentheses—which, significantly, in no way involves ωt. Expressed in words, Eq. (6-64) states that *to add two sinusoidal quantities it is necessary merely to add the corresponding phasors.* Accordingly, if we call the phasor quantity of the resultant $I_{m3} = I_{m3}\epsilon^{j\theta_3}$, we may write

$$I_{m3} = I_{m1} + I_{m2} \qquad (6\text{-}65)$$

or

$$I_{m3}\epsilon^{j\theta_3} = I_{m1} + I_{m2}\epsilon^{j\theta_2} = (I_{m1} + I_{m2}\cos\theta_2) + jI_{m2}\sin\theta_2 \qquad (6\text{-}66)$$

The right side of the last equation involves all known quantities. Moreover, since it is a complex number having a horizontal and vertical component, it follows that the magnitude of the component is

$$I_{m3} = \sqrt{(I_{m1} + I_{m2}\cos\theta_2)^2 + (I_{m2}\sin\theta_2)^2} \qquad (6\text{-}67)$$

and the angle is

$$\theta_3 = \tan^{-1}\frac{I_{m2}\sin\theta_2}{I_{m1} + I_{m2}\cos\theta_2} \qquad (6\text{-}68)$$

The resultant phasor I_{m3} is depicted in Fig. 6-6(a) as line OD. Note that phasors are added in the same manner as vectors in mechanics. The sinusoid i_3 in Fig. 6-6(b) can then be considered as having been generated by the rotation of phasor I_{m3} in Fig. 6-6(a).

When the value of I_{m3} has been established, the corresponding time expression for i_3 readily follows from Eq. (6-64). Hence

$$\begin{aligned} i_3 &= \mathscr{I}\mathrm{m}\,[I_{m3}\epsilon^{j\omega t}] = \mathscr{I}\mathrm{m}\,[I_{m3}\epsilon^{j\theta_3}\epsilon^{j\omega t}] = \mathscr{I}\mathrm{m}\,[I_{m3}\epsilon^{j(\omega t + \theta_3)}] \\ &= \mathscr{I}\mathrm{m}\,[I_{m3}\cos(\omega t + \theta_3) + jI_{m3}\sin(\omega t + \theta_3)] \end{aligned} \qquad (6\text{-}69)$$

or

$$i_3 = I_{m3}\sin(\omega t + \theta_3) \qquad (6\text{-}70)$$

EXAMPLE 6-3 Two sinusoidal currents are described as follows:

$$i_1 = 10\sqrt{2}\,\sin\omega t \quad \text{and} \quad i_2 = 20\sqrt{2}\,\sin(\omega t + 60°)$$

Find the expression for the sum of these currents.

solution: The solution is found by performing phasor addition in the manner indicated by Eq. (6-65). In passing note that i_1 and i_2 have effective values of 10 and 20 amperes, respectively. The phasor quantities are

$$I_{m1} = I_{m1} = 10\sqrt{2}$$
$$I_{m2} = 20\sqrt{2}\,(\cos 60° + j\sin 60°)$$
$$= 20\sqrt{2}\left(\frac{1}{2} + j\frac{\sqrt{3}}{2}\right) = 10\sqrt{2} + j10\sqrt{6}$$

Hence

$$I_{m3} = I_{m1} + I_{m2} = (10\sqrt{2} + 10\sqrt{2}) + j10\sqrt{6}$$
$$= 20\sqrt{2} + j10\sqrt{6}$$

So that

$$I_{m3} = \sqrt{800 + 600} = 37.4$$

$$\theta_3 = \tan^{-1}\frac{10\sqrt{6}}{20\sqrt{2}} = \tan^{-1}\frac{\sqrt{3}}{2} = 41°$$

Therefore

$$I_{m3} = 37.4\epsilon^{j41°}$$

and

$$i_3 = 37.4\sin(\omega t + 41°)$$

It is pointed out in Secs. 6-2 and 6-3 that the most important and useful quantity of a sinusoidal function is its effective value. Although knowledge of the peak value of a sinusoid can be useful, it is not nearly so useful as the effective value. In Example 6-3 it is much more significant to speak in terms of the effective current than in terms of the peak value, because the former conveys information about its energy-transferring capability per cycle whereas the latter does not. Therefore in finding the sum of two currents the engineer is often really interested in the resultant effective current. Of course from Eq. (6-26) we know that the effective current can readily be obtained by dividing the peak value by $\sqrt{2}$. In view of the far greater importance of effective values it is therefore customary to use a phasor diagram in which the phasors are expressed as effective values rather than maximum values. Accordingly this procedure shall be followed in the remainder of the book.

To express the addition of two sinusoids in terms of the effective values of the phasors follows readily from Eq. (6-66) upon dividing each term in the expression by $\sqrt{2}$. Thus

$$\frac{I_{m3}}{\sqrt{2}} = \frac{I_{m1}}{\sqrt{2}} + \frac{I_{m2}}{\sqrt{2}} \qquad (6\text{-}72)$$

or, more simply,

$$I_3 = I_1 + I_2 \qquad (6\text{-}73)$$

We take note of one other convention which is used in phasor diagrams and the algebraic manipulations associated with them. In the interest of simplicity it is customary to replace $\epsilon^{j\theta}$ by $/\theta$. Accordingly the phasor quantity for a current such as i_2 of Eq. (6-52) can be expressed in terms of its effective value as

$$I_2 = I_2\epsilon^{j\theta_2} = I_2/\theta_2 \qquad (6\text{-}74)$$

where I_2 denotes the effective current and $/\theta_2$ denotes "at an angle θ_2 degrees counterclockwise." The angle θ_2 is always measured relative to the horizontal axis. For negative values of the angle the direction is taken as clockwise. Keep in mind too that

$$/\theta = \epsilon^{j\theta} = \cos\theta + j\sin\theta \qquad (6\text{-}75)$$

Example 6-3 is now repeated in order to illustrate the use of the foregoing modifications.

EXAMPLE 6-4   A sinusoidal current having an effective value of $10\underline{/0°}$ amperes is added to another sinuosidal curent of effective value $20\underline{/60°}$. Find the effective value of the resultant current.

solution: By Eq. (6-73) we have

$$I_3 = I_1 + I_2 = 10\underline{/0°} + 20\underline{/60°} = 10 + (10 + j10\sqrt{3})$$
$$= 20 + j10\sqrt{3} = 26.4\underline{/41°} \quad \text{amps} \tag{6-76}$$

To obtain the complete time expression we write

$$i_3 = \sqrt{2}(26.4)\sin(\omega t + 41°) = 37.4\sin(\omega t + 41°) \tag{6-77}$$

which checks with the result of Example 6-3.

Multiplication and Division of Complex Quantities. Up to now attention has been directed exclusively to the problem of adding sinusoidal quantities which are out of phase with one another. The use of phasor representation of the sinusoids simplified the procedure considerably. The method of solution was reduced to one of dealing with complex numbers as illustrated in Examples 6-3 and 6-4. In dealing with the sinusoidal steady-state response of electric circuits the need frequently arises to multiply and divide complex numbers. The complex numbers, however, do not always represent sinusoidal functions. In the interest of illustrating how the product of two complex numbers is obtained, consider the following two complex numbers. One is denoted by the phasor $I = I\epsilon^{j\theta}$; the other is represented by the operator† $\bar{Z} = Z\epsilon^{j\phi}$. The product is desired. As the first step in the procedure formulate the product in terms of the exponential form. The exponential form is preferred initially because as a legitimate part of the language of mathematics we are familiar with the rules governing its manipulation. Accordingly we can write

$$I\bar{Z} = I\epsilon^{j\theta}Z\epsilon^{j\phi} = IZ\epsilon^{j(\theta+\phi)} \tag{6-78}$$

When the notation of Eq. (6-75) is inserted this expression becomes

$$I\bar{Z} = IZ\underline{/\theta + \phi} \tag{6-79}$$

Therefore the product of two complex numbers is found by taking the product of their magnitudes and the sum of their angles.

The division of one complex quantity by another is treated in a similar fashion. To illustrate, let it be required to divide the complex quantity $\bar{Z} = Z\epsilon^{j\phi}$ into the phasor $\bar{V} = V\epsilon^{j\theta}$, which represents the sinusoid $v = \sqrt{2}\ V\sin(\omega t + \theta)$. We shall call the quotient I. Thus

$$I = \frac{\bar{V}}{\bar{Z}} = \frac{V\epsilon^{j\theta}}{Z\epsilon^{j\phi}} = \left(\frac{V}{Z}\right)\epsilon^{j(\theta-\phi)} \tag{6-80}$$

Expressed in terms of the shorthand notation of Eq. (6-75) we have

---

† This term will be better understood after a study of Sec. 6-5.

$$I = \frac{V\underline{/\theta}}{Z\underline{/\phi}} = \frac{V}{Z}\underline{/\theta - \phi} \tag{6-81}$$

Therefore the division of one complex number by another involves the division of their magnitudes to yield the magnitude of the quotient and the difference of their phase angles to yield the phase of the quotient. For the sake of completeness we give the corresponding time expression for the phasor of Eq. (6-81), which is

$$i = \sqrt{2}\,\frac{V}{Z}\sin(\omega t + \theta - \phi) \tag{6-82}$$

The angular frequency ω is the same as that for v.

Powers and Roots of Complex Numbers. Occasionally it becomes necessary to find the square or cube of a complex quantity. The manner of treatment again is revealed by using the exponential form. Hence to find the nth power of the complex quantity $\bar{Z} = Z\epsilon^{j\phi}$ we proceed as follows.

$$\bar{Z}^n = (Z\epsilon^{j\phi})^n = Z^n\epsilon^{jn\phi} = Z^n\underline{/n\phi} \tag{6-83}$$

Therefore the nth power of a complex number is a complex number whose magnitude is the nth power of the magnitude of the original complex number and whose angle is n times as large as that of the original complex number.

The root of a complex number can be found by making n a proper fraction in Eq. (6-83). However, one additional modification is necessary: the angle of the original complex number must be increased by $2k\pi$ (where k is an integer) in order to bring into evidence all those root values which satisfy Eq. (6-83). Thus to find the fourth power of $\bar{Z} = Z\epsilon^{j\phi}$ we proceed as follows. First replace ϕ by $\phi + 2k\pi$. Note that this in no way alters the value of the original complex number. Assign the value $1/4$ to n in Eq. (6-83). Then

$$\bar{Z}^{1/4} = [Z\epsilon^{j(\phi + 2k\pi)}]^{1/4} = Z^{1/4}\underline{\left/\frac{\phi}{4} + \frac{k\pi}{2}\right.} \tag{6-84}$$

Therefore the four different and distinct values which satisfy Eq. (6-84) are

$$\begin{aligned}
\bar{Z}_1^{1/4} &= Z^{1/4}\underline{\left/\frac{\phi}{4}\right.} && \text{for } k = 0 \\[4pt]
\bar{Z}_2^{1/4} &= Z^{1/4}\underline{\left/\frac{\phi}{4} + \frac{\pi}{2}\right.} && \text{for } k = 1 \\[4pt]
\bar{Z}_3^{1/4} &= Z^{1/4}\underline{\left/\frac{\phi}{4} + \pi\right.} && \text{for } k = 2 \\[4pt]
\bar{Z}_4^{1/4} &= Z^{1/4}\underline{\left/\frac{\phi}{4} + \frac{3\pi}{2}\right.} && \text{for } k = 3
\end{aligned} \tag{6-85}$$

Any further values assigned to k will yield results which are repetitions of those already listed in Eqs. (6-85).

EXAMPLE 6-5 The following three sinusoidal currents flow into a junction: $i_1 = 3\sqrt{2}\sin\omega t$, $i_2 = 5\sqrt{2}\sin(\omega t + 30°)$, and $i_3 = 6\sqrt{2}\sin(\omega t - 120°)$. Find the time expression for the resultant sinuosidal current which leaves the junction.

solution: The corresponding phasor expressions are

$$\bar{I}_1 = 3\underline{/0°}$$

$$\bar{I}_2 = 5\underline{/30°} = 5(\cos 30° + j \sin 30°) = 2.5\sqrt{3} + j2.5$$

$$\bar{I}_3 = 6\underline{/-120°} = 6[\cos(-120) + j \sin(-120)] = -3 - j3\sqrt{3}$$

Note that since we are concerned with addition it is the *rectangular form* of the complex number which is used. The resultant phasor quantity is therefore

$$\bar{I} = \bar{I}_1 + \bar{I}_2 + \bar{I}_3 = 2.5\sqrt{3} - j(3\sqrt{3} - 2.5) \qquad (6\text{-}86)$$

The phasor diagram representing the addition called for in the last equation is depicted in Fig. 6-7.

$$\bar{I} = 2.5\sqrt{3} - j2.7 = 5.1\underline{/-32°} \qquad (6\text{-}87)$$

The time expression for the resultant current is

$$i = \sqrt{2}(5.1) \sin(\omega t - 32°) = 7.22 \sin(\omega t - 32°) \qquad (6\text{-}88)$$

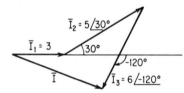

Fig. 6-7. Phasor diagram for Example 6-5 illustrating the addition of three sinusoidal currents by the use of phasors.

EXAMPLE 6-6   Find the quotient of $\bar{V}\bar{Z}_1/\bar{Z}_2$ where

$$\bar{V} = 45\sqrt{3} - j45$$

$$\bar{Z}_1 = 2.5\sqrt{2} + j2.5\sqrt{2} \qquad (6\text{-}89)$$

$$\bar{Z}_2 = 7.5 + j7.5\sqrt{3}$$

solution: Since we are not interested in addition but rather multiplication and division, it is simpler to work with the given complex quantities in polar form. Accordingly

$$\bar{V} = \sqrt{(45\sqrt{3})^2 + 45^2} \;\underline{/\tan^{-1}\dfrac{-45}{45\sqrt{3}}} = 90\underline{/-30°} \qquad (6\text{-}90)$$

$$\bar{Z}_1 = \sqrt{(2.5\sqrt{2})^2 + (2.5\sqrt{2})^2} \;\underline{/\tan^{-1} 1} = 5\underline{/45°} \qquad (6\text{-}91)$$

$$\bar{Z}_2 = \sqrt{7.5^2 + (7.5\sqrt{3})^2} \;\underline{/\tan^{-1}\sqrt{3}} = 15\underline{/60°} \qquad (6\text{-}92)$$

Hence

$$\frac{\bar{V}\bar{Z}_1}{\bar{Z}_2} = \frac{90\underline{/-30°} \times 5\underline{/45°}}{15\underline{/60°}} = \frac{450}{15}\underline{/-30 + 45 - 60} = 30\underline{/-45°} \qquad (6\text{-}93)$$

6-5   Sinusoidal Steady-state Response of Single Elements—*R, L, C*

The response of each circuit parameter to a sustained sinusoidal forcing function is found individually—for two reasons. First, it provides an opportunity to illustrate the manner in which the response can be found easily and directly by the use of the phasor representation of sinusoids. Second, it affords the opportunity to establish once and for all the phase-angle relationships existing between the current and voltage for each circuit parameter. It is seen that these relationships are fixed and must always be satisfied irrespective of whether a given circuit element is in a series or parallel arrangement with other circuit elements. Let us start by treating the simplest of the three parameters—resistance.

Resistive Circuit. The circuit of interest is depicted in Fig. 6-8. Assume a sinusoidal forcing function which can be described by

$$v = V_m \sin \omega t = \mathscr{I}\mathrm{m} \, [V_m \epsilon^{j\omega t}] \qquad (6\text{-}94)$$

Because the circuit is linear we know that the steady-state response must also be a sinusoid having the same frequency as the forcing function. However, in general, the response sinusoid (i.e., the current) differs in two respects—*amplitude and phase*. Therefore, on this basis we can say that the form of the response must be

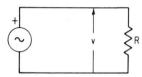

Fig. 6-8. Series resistive circuit with sinusoidal source.

$$i = I_m \sin (\omega t + \theta) = \mathscr{I}\mathrm{m} \, [I_m \epsilon^{j\theta} \epsilon^{j\omega t}] \qquad (6\text{-}95)$$

where I_m and θ are respectively the amplitude and phase of the response which must be determined to achieve a solution of the problem.

The equation which makes available the amplitude and phase information in this instance is Kirchhoff's voltage law applied to the circuit of Fig. 6-8. Thus

$$v = Ri \qquad (6\text{-}96)$$

Note that this expression is nothing more than Ohm's law with the modification that sinusoidal quantities are involved for the variables v and i. Inserting Eqs. (6-94) and (6-95) into Eq. (6-96) yields

$$\mathscr{I}\mathrm{m} \, [V_m \epsilon^{j\omega t}] = R \times \mathscr{I}\mathrm{m} \, [I_m \epsilon^{j\theta} \epsilon^{j\omega t}] \qquad (6\text{-}97)$$

The exponential form is preferred because it leads to the solution in a direct and easy fashion as is apparent from the material which follows. But first let us re-write the last equation in a simpler form by dropping the notation "Imaginary part of . . . ." *This is permissible because such equations reduce to an identity for all values of time whenever the coefficients of $\epsilon^{j\omega t}$ on one side equal the coefficients of $\epsilon^{j\omega t}$ on the other side.* There is a precaution to be observed, however, when using such a procedure. After completion of the algebraic manipulations which lead to the solution, it must be remembered that the actual solution is found by taking only the j part of the resulting exponential solution.

On this basis Eq. (6-97) can be written as

$$RI_m \epsilon^{j\theta} \epsilon^{j\omega t} = V_m \epsilon^{j\omega t} \qquad (6\text{-}98)$$

Since the time factor $\epsilon^{j\omega t}$ is common to both sides of the equation, it may be suppressed, thus leading to

$$I_m \epsilon^{j\theta} = \frac{V_m}{R} = \frac{V_m}{R} \epsilon^{j0°} \qquad (6\text{-}99)$$

On the left side of this equation appear the two unknown quantities I_m and θ, while on the right side appear the known quantities. Although in general both sides of this equation represent complex numbers, it is clear from the right side that in this case we have just a real number. This is characteristic of purely resistive circuits. A comparison of amplitudes and angles of the left and right sides of Eq. (6-99) leads to

$$I_m = \frac{V_m}{R} \quad \text{and} \quad \theta = 0° \tag{6-100}$$

Substituting this information into Eq. (6-95) yields the final expression for the solution. Hence

$$i = \frac{V_m}{R} \sin \omega t \tag{6-101}$$

Comparing Eq. (6-101) with Eq. (6-94) reveals that for a resistive circuit the current and voltage are in time phase—i.e., the peak values and the zero values occur at the same time instants. This is illustrated in Fig. 6-9.

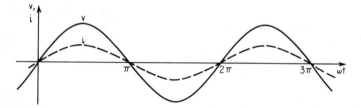

Fig. 6-9. Illustrating that i and v are in phase for the circuit of Fig. 6-8.

As noted in the discussion of phasor representation of sinusoids, it is more frequently desirable to deal in terms of the effective value of voltages and currents rather than their maximum values. Accordingly, to express the solution as represented by Eqs. (6-100) in terms of effective values, use is made of Eq. (6-26). Thus

$$I_m = \sqrt{2}\, I = \frac{\sqrt{2}\, V}{R} \tag{6-102}$$

where I and V denote effective values. Therefore, one can very simply describe the current response caused by a sinusoidal voltage of effective value V applied to a circuit of resistance R as

$$I = \frac{V}{R} \underline{/0°} \tag{6-103}$$

After experience has been gained in solving problems involving sinusoidal sources, it is usually unnecessary to go beyond this point in the solution. This is because the corresponding interpretation of Eq. (6-103) in the time domain is so well understood as to be self-evident. Recall from Sec. 6-4 that the procedure is to multiply the effective value of the response by $\sqrt{2}$ and add the computed phase angle to ωt to get the total argument. Applying this procedure to Eq. (6-103) yields

$$i = \frac{\sqrt{2}\,V}{R}\sin(\omega t + 0°) = \frac{V_m}{R}\sin\omega t \tag{6-104}$$

which obviously is identical to Eq. (6-101).

Appearing in Fig. 6-10 is the phasor diagram for the resistive circuit. The reference phasor is arbitrarily taken to be the applied voltage and is placed along the horizontal. By Eq. (6-103) the current phasor has an angle of 0° which means that it is in phase with the voltage. Hence it, too, is located on the horizontal axis. It is customary to show only effective values of voltage and current in the phasor diagrams. Note that the relationship depicted in Fig. 6-10 is consistent with that shown in Fig. 6-9, where time appears explicitly.

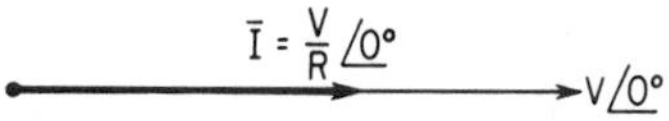

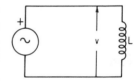

Fig. 6-10. Phasor diagram for resistive circuit.

Inductive Circuit. The same method of analysis is used to find the response of a purely inductive circuit to a sinusoidal forcing function in the steady state. Figure 6-11 shows the circuit. Assume that the potential difference appearing across the inductor terminals is given by Eq. (6-94). Because the circuit is assumed linear, the response can again be represented by Eq. (6-95). However, for the inductive circuit the defining equation is

$$v = L\frac{di}{dt} \tag{6-105}$$

Substituting Eqs. (6-94) and (6-95) gives

Fig. 6-11. Sinusoidal voltage applied to a pure inductor.

$$\mathscr{I}m\,[V_m\epsilon^{j\omega t}] = L\,\mathscr{I}m\,\frac{d}{dt}[I_m\epsilon^{j\theta}\epsilon^{j\omega t}] \tag{6-106}$$

where I_m and θ in general will be different from those values found for the resistive circuit. This is to be expected, since the defining equation involves a derivative of the current. Performing the differentiation and suppressing $\epsilon^{j\omega t}$ leads to

$$I_m\epsilon^{j\theta} = \frac{V_m}{j\omega L} \tag{6-107}$$

Again note that the term involving the unknown amplitude and phase angle is isolated to the left side of the equation. The quantities appearing on the right side are known. Examination of the right side in this instance shows that it is a number which is located along the vertical axis. In mathematical language this is described as the imaginary number. An alternative way of writing the right side follows upon recalling that the factor j is a rotational operator defined as

$$j \equiv \epsilon^{j(\pi/2)} = 1/90° \tag{6-108}$$

Any real number which is multiplied by j changes its position from the horizontal axis to the vertical axis. Thus $j10$ means that a line of 10 units (normally measured on the horizontal axis) is now measured on the vertical axis. Accordingly, Eq. (6-108) can be rewritten

$$I_m \epsilon^{j\theta} = \frac{V_m}{\omega L / 90°} = \frac{V_m}{\omega L} \underline{/-90°} \tag{6-109}$$

Therefore

$$\boxed{I_m = \frac{V_m}{\omega L} \quad \text{and} \quad \theta = -90°} \tag{6-110}$$

Expressed in terms of effective values, the solution for the response can be found conveniently by writing Eq. (6-109) as

$$I = \frac{\bar{V}}{j\omega L} \tag{6-111}$$

or

$$I\underline{/\theta} = \frac{V/0°}{\omega L/90°} = \frac{V}{\omega L} \underline{/-90°} \tag{6-112}$$

Therefore

$$\boxed{I = \frac{V}{\omega L} = \frac{V}{X_L} \quad \text{and} \quad \theta = -90°} \tag{6-113}$$

where

$$X_L \equiv \omega L \tag{6-114}$$

and is called the *inductive reactance* of the coil. The term reactance is used to distinguish it from resistance. A resistive circuit element causes no phase shift between v and i. However, an inductive element causes i to lag behind v by 90°. Any circuit element which exhibits this property in the sinusoidal steady state is said to have inductive reactance.

The phasor diagram for the purely inductive circuit is illustrated in Fig. 6-12(a). The location of the current phasor is drawn consistent with the results of Eq. (6-113), which calls for the current phasor to *lag* behind the voltage phasor by $\theta = -90°$. It is always true that for sinusoidal forcing functions *the current flowing through an inductor always lags behind the potential difference across the inductor terminals by 90°*. In the interest of completeness keep in mind the fact that the phasors of Fig. 6-12(a) are actually rotating counterclockwise (the assumed positive direction) at the angular frequency ω radians per second.

Equation (6-113) contains all the useful information needed for the solution of the problem. However, if the complete *time* solution for the current response is desired, it readily follows from Eq. (6-113) by recalling the meaning of phasor quantities. This leads to

$$i = \sqrt{2} \frac{V}{X_L} \sin(\omega t + \theta) = \frac{V_m}{X_L} \sin(\omega t - 90°) \tag{6-115}$$

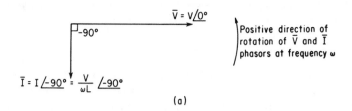

(a)

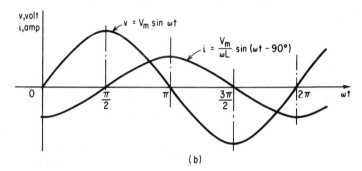

(b)

Fig. 6-12. Inductive circuit response to sinusoidal source:
  (a) phasor diagram;
  (b) time diagram.

A plot of Eq. (6-115) appears in Fig. 6-12(b). Note that the response sinusoid in the case of an inductive circuit is displaced 90° behind the potential difference across the inductor. Note also that just as Eq. (6-113) is a simpler and more direct way of representing the solution of Eq. (6-115) so too is Fig. 6-12(a) a simpler and more convenient way of expressing the results shown in Fig. 6-12(b).

Capacitive Circuit. In the circuit depicted in Fig. 6-13 the current which flows through the capacitor is related to the potential difference v by

$$i = C \frac{dv}{dt} \tag{6-116}$$

For an assumed sinusoidal forcing function of $v = V_m \sin \omega t$ the general form of the current response can be represented by $i_m = I_m \sin(\omega t + \theta)$. Inserting the exponential forms of v and i into the last expression yields

$$I_m \epsilon^{j\omega t} \epsilon^{j\theta} = C \frac{d}{dt} [V_m \epsilon^{j\omega t}] \tag{6-117}$$

Performing the differentiation leads to

$$I_m \epsilon^{j\theta} = j\omega C V_m \tag{6-118}$$

or

$$I_m \underline{/\theta} = \frac{V_m}{1/\omega C} \underline{/90°} \tag{6-119}$$

Fig. 6-13. A capacitive circuit with sinusoidal forcing function.

Equating the magnitudes and angles of the right and left sides then yields

$$\boxed{I_m = \frac{V_m}{1/\omega C} \quad \text{and} \quad \theta = +90°}$$ (6-120)

Once Eq. (6-119) is obtained by using the exponential forms of v and i in the governing differential equation, it can then be rewritten in terms of effective values. It is helpful at this point to employ effective values, because from here on the solution of capacitive circuits can be found by applying Ohm's law in the modified form dictated by Eq. (6-119). Accordingly we can write

$$I = \frac{\bar{V}}{(1/j\omega C)}$$ (6-121)

This states that the phasor current response I is equal to the phasor potential difference $\bar{V}$ across the capacitor divided by the quantity $1/j\omega C$. It is important to understand that the quantity $1/j\omega C$ arises from the exponential formulation as revealed by Eq. (6-118). Moreover, the units of $1/j\omega C$ are volts per ampere, which is ohms. This is evident from Eq. (6-121). Therefore $1/j\omega C$ may be looked upon as a kind of resistance, but it is not called this because it involves the rotational operator j. For this reason it is called a *reactance*; more specifically, it is called a *capacitive reactance* because a capacitor is involved. Equation (6-121) often appears in the form

$$I = \frac{\bar{V}}{X_c\underline{/-90°}} = \frac{\bar{V}}{X_c}\underline{/90°}$$ (6-122)

where

$$X_c \equiv \frac{1}{\omega C} \quad \text{capacitive reactance in ohms}$$ (6-123)

Also

$$I\underline{/\theta} = \frac{V}{X_C}\underline{/90°}$$ (6-124)

Therefore for the capacitive circuit

$$\boxed{I = \frac{V}{X_C} = V\omega C \quad \text{and} \quad \theta = 90°}$$ (6-125)

By the information contained in Eq. (6-125) the phasor diagram of a purely capacitive circuit is readily drawn. Figure 6-14(a) depicts the result. The corresponding time-domain description of the current response is illustrated in Fig. 6-14(b). The time-domain equation for the response is obtained in the usual way and found to be

$$i = \sqrt{2}\,\frac{V}{X_C}\sin(\omega t + \theta) = \frac{V_m}{X_C}\sin(\omega t + 90°)$$ (6-126)

A study of Eq. (6-125) as well as Fig. 6-14 reveals an important characteristic of the capacitive circuit: *the current flowing through a capacitor in the sinusoidal steady state always leads the potential difference across the capacitor by* 90°.

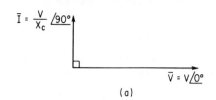

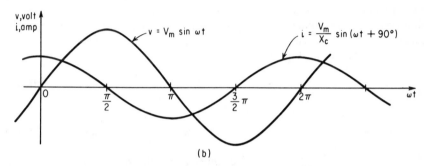

Fig. 6-14. Capacitive circuit response to sinusoidal source:
(a) phasor diagram;
(b) time diagram.

6-6   The Series *R-L* Circuit

In this section we direct attention to the sinusoidal steady-state analysis of a circuit configuration which involves the series connection of a resistive and an inductive element. This leads to the concepts of complex impedance and reactive power.

Appearing in Fig. 6-15 is a diagram of the series *R-L* circuit. Our objective is to find the steady-state current response assuming that *R*, *L*, and the forcing function are known. The differential equation for the circuit is

$$v = iR + L\frac{di}{dt} \qquad (6\text{-}127)$$

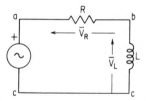

Fig. 6-15. Series *R-L* circuit.

If the applied forcing function is assumed to be $v = V_m \sin \omega t = \mathscr{I}\text{m}\,[V_m \epsilon^{j\omega t}]$, then the current response must be of the form $i = I_m \sin(\omega t + \theta)$ where I_m and θ are to be determined. Introducing these quantities into the last equation in exponential form gives

$$V_m \epsilon^{j\omega t} = RI_m \epsilon^{j\theta} \epsilon^{j\omega t} + j\omega L I_m \epsilon^{j\theta} \epsilon^{j\omega t} \qquad (6\text{-}128)$$

Suppressing $\epsilon^{j\omega t}$ and collecting terms

$$V_m = I_m \epsilon^{j\theta}(R + j\omega L) \qquad (6\text{-}129)$$

This expression can be written in terms of effective values by dividing both sides by $\sqrt{2}$. Accordingly

$$\bar{V} = \bar{I}(R + j\omega L) \tag{6-130}$$

Equation (6-130) is Kirchhoff's voltage law for the R-L circuit written in terms of phasor quantities. This expression can be written directly without going through the preceding steps. It is obtained by summing the voltage drops across each circuit element. However, it is important to attach the rotational operator j to the inductive reactance and $1/j$ to the capacitive reactance. This will be the procedure followed in the remainder of the book whenever dealing with such circuits. The analysis of the R-C circuit in the next section starts at this point.

Upon transposing Eq. (6-130) the solution for the current becomes

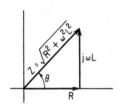

$$I = \frac{\bar{V}}{R + j\omega L} = \frac{\bar{V}}{\bar{Z}} \tag{6-131}$$

where

$$\bar{Z} \equiv R + j\omega L = Z \left/ \tan^{-1} \frac{\omega L}{R} \right. \tag{6-132}$$

Fig. 6-16. Complex impedance triangle.

This quantity $\bar{Z}$ is called the *complex impedance* of the R-L circuit, and can always be represented by a right triangle as shown in Fig. 6-16. An inspection of Eq. (6-131) indicates that effectively $\bar{Z}$ may be considered as a complex operator which when divided into the voltage phasor yields the desired current phasor. Information about the magnitude and relative phase angle of the response then readily follows when Eq. (6-131) is rewritten as

$$I \underline{/\theta} = \frac{V \underline{/0°}}{Z \left/ \tan^{-1} \dfrac{\omega L}{R} \right.} = \frac{V}{Z} \left/ -\tan^{-1} \frac{\omega L}{R} \right. \tag{6-133}$$

where

$$Z = \sqrt{R^2 + \omega^2 L^2} \tag{6-134}$$

Therefore

and

$$\boxed{\begin{array}{l} I = \dfrac{V}{Z} = \dfrac{V}{\sqrt{R^2 + \omega^2 L^2}} \\[3mm] \theta = -\tan^{-1} \dfrac{\omega L}{R} \end{array}} \tag{6-135}$$

Equation (6-135) furnishes all the information that is needed to identify the current response. Of course if the complete time expression is desired, it is found in the usual manner. In this case the result is

$$i = \frac{\sqrt{2}\,V}{Z} \sin(\omega t + \theta) = \frac{V_m}{\sqrt{R^2 + \omega^2 L^2}} \sin\left(\omega t - \tan^{-1} \frac{\omega L}{R}\right) \tag{6-136}$$

Comparing this equation with Eq. (5-188) shows the two to be equivalent. The difference in form lies in the fact that here the forcing function is assumed to be sinusoidal rather than cosinusoidal. However, note that the Laplace transform was not used in this instance to obtain the solution.

Examination of the solution for the relative phase angle reveals that for finite resistance the angle must be less than 90°. Note too that it is a *lag* angle, which means the current sinusoid is behind the voltage in time. These facts are represented

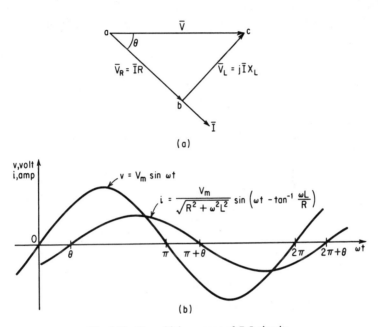

(a)

(b)

Fig. 6-17. Sinusoidal response of *R-L* circuit:
(a) phasor diagram;
(b) time diagram.

graphically in Figs. 6-17. Appearing in Fig. 6-17(a) is a graphical representation of Eq. (6-130). The potential difference across the resistive element is clearly

$$\bar{V}_R \equiv IR \qquad (6\text{-}137)$$

Because there is no phase shift associated with a resistor, it follows that the voltage drop IR must be in phase with I. Hence $\bar{V}_R$ is drawn on the same line as I. Equation (6-137) is represented by ab in the phasor diagram. Similarly the potential difference across the inductive element is shown by the second term of the right side of Eq. (6-110) to be expressed as

$$\bar{V}_L \equiv jI\omega L = jIX_L \qquad (6\text{-}138)$$

The presence of the rotational operator j conveys the meaning that the quantity IX_L is to be rotated in the positive (or counterclockwise) direction through 90°.

Accordingly, line bc in Fig. 6-17(a) represents Eq. (6-138). Of course the sum of $\bar{V}_R$ and $\bar{V}_L$ must add up to the line voltage ac. This too is borne out in the phasor diagram.

EXAMPLE 6-7   A 60-Hz sinusoidal voltage $v = 141 \sin \omega t$ is applied to a series R-L circuit. The value of resistance is 3 ohms and the inductance value is 0.0106 henry.

(a) Compute the effective value of the steady-state current as well as the relative phase angle.

(b) Write the expression for the instantaneous current.

(c) Compute the effective magnitude and phase of the voltage drops appearing across each circuit element.

(d) Find the average power dissipated by the circuit.

(e) Calculate the power factor.

solution: (a) The applied forcing function expressed in terms of its effective value is

$$\bar{V} = V\underline{/0^\circ} = \frac{V_m}{\sqrt{2}}\underline{/0^\circ} = 100 + j0 = 100 \text{ volts} \tag{6-139}$$

To find the current we must first compute the complex impedance. Thus by Eqs. (6-132) and (6-134) we have

$$\bar{Z} = R + j\omega L = 3 + j2\pi f L = 3 + j377(0.0106) = 3 + j4$$
$$= \sqrt{3^2 + 4^2} \underline{/\tan^{-1}\tfrac{4}{3}} = 5\underline{/53.1^\circ}$$

Therefore

$$I = \frac{\bar{V}}{\bar{Z}} = \frac{100\underline{/0^\circ}}{5\underline{/53.1^\circ}} = 20\underline{/-53.1^\circ} \tag{6-141}$$

The effective value of the steady-state current is 20 amperes and it lags behind the voltage by 53.1°.

(b) $$\qquad\qquad i = 20\sqrt{2} \sin (\omega t - 53.1^\circ) \tag{6-142}$$

(c) The voltage drop across the resistor is

$$\bar{V}_R = \bar{I}R = (20\underline{/-53.1})3 = 60\underline{/-53.1^\circ} \tag{6-143}$$

In Fig. 6-17(a) this quantity is denoted by line ab. Note that $\bar{V}_R$ is drawn in a lagging position of -53.1° with respect to the horizontal reference axis along which V is located. The voltage drop across the coil is

$$\bar{V}_L = j\bar{I}X_L = j(20\underline{/-53.1})4 = 1\underline{/90^\circ}(20\underline{/-53.1})4$$
$$= 80\underline{/90 - 53.1} = 80\underline{/36.9^\circ} \tag{6-144}$$

This quantity is represented by line bc in Fig. 6-17(a). Note that $\bar{V}_L$ makes an angle of 36.9° relative to the horizontal axis.

It is also worthwhile to note that the phasor sum of $\bar{V}_R$ and $\bar{V}_L$ yields the applied voltage $\bar{V}$. Thus

$$\bar{V} = \bar{V}_R + \bar{V}_L = 60\underline{/-53.1^\circ} + 80\underline{/36.9^\circ} = 100 + j0 \tag{6-145}$$

(d) From Eq. (6-49) we have

$$P_{av} = VI \cos \theta = 100(20) \cos 53.1^\circ = 2000(0.6) = 1200 \text{ watts} \tag{6-146}$$

Because the resistive element is the only element which dissipates power, the average power may also be found from

$$P_{\text{av}} = I^2 R = (20)^2 3 = 1200 \text{ watts} \tag{6-147}$$

(e) Power factor can be found by either of two methods:

$$\text{p f} = \frac{\text{average power}}{\text{apparent power}} = \frac{P_{\text{av}}}{VI} = \frac{1200}{100(20)} = 0.6 \tag{6-148}$$

or

$$\text{p f} = \cos \theta = \cos 53.1° = 0.6 \tag{6-149}$$

Reactive Power. From the expression for the average power dissipated in a circuit it should be apparent that power is dependent upon the in-phase component of the current. This point can be stressed by writing the expression for average power as

$$P_{\text{av}} = V[I \cos \theta] \tag{6-150}$$

and then noting that the quantity in brackets is the projection of the current phasor I onto the voltage phasor $\bar{V}$. Of course θ must be the angle between $\bar{V}$ and I as depicted in Fig. 6-17(a). When conditions in the circuit are such that the projection of the current phasor I onto $\bar{V}$ is zero, then no real power can be delivered or dissipated in the circuit. This condition is represented in Fig. 6-4 when θ equals 90°. An alternative way of expressing Eq. (6-150) is to introduce the following equivalences:

$$\cos \theta = \frac{R}{Z} \tag{6-151}$$

and

$$V = IZ \tag{6-152}$$

Equation (6-151) immediately follows from Fig. 6-16 and Eq. (6-152) states that the magnitude of the effective voltage is equal to the magnitude of the effective current multiplied by the magnitude of the complex impedance. Thus Eq. (6-150) becomes

$$P_{\text{av}} = V(I \cos \theta) = (IZ)I\frac{R}{Z} = I^2 R \text{ watts} \tag{6-153}$$

This equation states that the average power is equal to the power dissipated in the resistive portion of the circuit—a result already familiar to us.

A logical question to ask at this point is: What meaning, if any, can be given to the analogous equation involving the inductive reactance? That is, does the expression

$$P_X = I^2 X_L = I^2 \omega L \tag{6-154}$$

have any meaning? To seek out an answer, let us return to first principles—the expression for the instantaneous "power" across the inductance as expressed by the product of its current and potential difference. Thus

$$p_X = iv_L = iL\frac{di}{dt} \tag{6-155}$$

Inserting $i = I_m \sin(\omega t + \theta)$ we get

$$p_X = I_m^2 \omega L \sin(\omega t + \theta) \cos(\omega t + \theta) \tag{6-156}$$

Substituting the trigonometric identity

$$\sin (\omega t + \theta) \cos (\omega t + \theta) = \tfrac{1}{2} \sin 2(\omega t + \theta) \tag{6-157}$$

into the last equation yields

$$p_X = \frac{I_m^2}{2} (\omega L) \sin 2(\omega t + \theta) = I^2 \omega L \sin 2 (\omega t + \theta) \tag{6-158}$$

Equation (6-158) reveals that the product of the current and voltage of an inductor is a sinusoid having double the frequency of either the current or voltage sinusoids. It is important here to note that, unlike the case for resistance, the average value of p_X over one cycle yields a result which is identically equal to zero. This means that for part of the cycle energy is delivered to the inductance where it is stored in the magnetic field, but in the next half-cycle it is returned to the source. The net transfer of energy in a pure inductance is thereby zero.

A comparison of the amplitude of the double-frequency sinusoid of Eq. (6-158) with Eq. (6-154) reveals the meaning of the latter quantity: it denotes the amplitude of the energy which is interchanged between the source and the energy-storing inductive element. Although the units of Eq. (6-154) are volt-amperes, the units of P_X are not described in terms of watts. To distinguish P_X from real power, it is called *reactive power* expressed in units of *reactive voltamperes* (abbreviated *var*).

The reactive power P_x can also be expressed in terms of the effective values of current and voltage as well as the relative phase angle. From the impedance triangle of Fig. 6-16 we have

$$X_L = Z \sin \theta \tag{6-159}$$

Inserting this expression into Eq. (6-154) gives

$$P_X = I^2 X_L = I^2 Z \sin \theta = (IZ)I \sin \theta = VI \sin \theta \tag{6-160}$$

Accordingly, reactive power in a circuit may also be found by taking the product of V and the component of the current which is in quadrature with $\bar{V}$.

There remains one final item in our discussion of the R-L circuit. It is the treatment of the sinusoidal response of the parallel combination of resistance and inductance. The circuit arrangement is shown in Fig. 6-18. Because the same effective value of the sinusoidal voltage appears across each circuit element, the effective current through the resistor is

$$I_R = \frac{\bar{V}}{R} \tag{6-161}$$

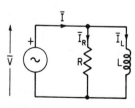

and that through the inductor is

$$I_L = \frac{\bar{V}}{jX_L} = \frac{\bar{V}}{j\omega L} \tag{6-162}$$

Fig. 6-18. The parallel R-L circuit; $G = 1/R$ and $B = (1/\omega L)$.

Accordingly, by Kirchhoff's current law the total current is

$$I = I_1 + I_2 = \bar{V}\left(\frac{1}{R} + \frac{1}{j\omega L} \right) = \bar{V}\bar{Y} \tag{6-163}$$

where $\bar{Y}$ is the *complex admittance*, and for the *R-L* case is obviously defined as

$$\bar{Y} \equiv \frac{1}{R} + \frac{1}{j\omega L} = \frac{1}{R} - j\frac{1}{\omega L} \tag{6-164}$$

The admittance too may be interpreted in terms of the role of an operator. It is that quantity which when multiplied by the phasor voltage yields the phasor current. In general $\bar{Y}$ is a complex number which often is denoted as

$$\bar{Y} \equiv G + jB \tag{6-165}$$

where G is the real part of the admittance and is called the *conductance*, and B is the quadrature component and is called *susceptance*. Upon comparing Eq. (6-165) with Eq. (6-164) it follows that for the parallel *R-L* case

$$G = \frac{1}{R} \quad \text{and} \quad B = -\frac{1}{\omega L} \tag{6-166}$$

Furthermore, the form of Eq. (6-164) makes it clear that the total current supplied by the source in steady state *lags* the voltage by the impedance angle, which for the parallel combination is

$$\theta = \tan^{-1}\frac{-1/\omega L}{1/R} = \tan^{-1}\frac{-R}{\omega L} \tag{6-167}$$

Therefore the same general result applies for the parallel arrangement of R and L as does for series combination: the resultant current *lags* the voltage.

In situations involving the sinusoidal steady-state analysis of circuits it is often helpful to replace a series combination of resistance and inductance by a corresponding parallel equivalent circuit which is expressed in terms of conductance and susceptance. The equivalent arrangement is depicted in Fig. 6-19. We consider this subject matter for two reasons. One, it is a useful technique to employ when three or more complex impedances appear in parallel because it is easier to add admittances than to deal with the reciprocal of impedances. Two, it is important to understand the distinc-

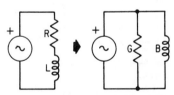

Fig. 6-19. The parallel equivalent of a series *R-L* circuit. $G = R/(R^2 + X_L^2)$ and $B = -X_L/(R^2 + X_L^2)$.

tion between the equivalent admittance (i.e., parallel equivalence) of a series *R-L* circuit and the admittance of an actual arrangement of R and L such as appears in Fig. 6-18. The admittance of the parallel R and L, of course, is represented by Eqs. (6-165) and (6-166).

Since the impedance of a series *R-L* circuit is $\bar{Z} = R + j\omega L$, the corresponding equivalent admittance is expressed as the reciprocal of this quantity. Thus

$$\bar{Y} \equiv \frac{1}{\bar{Z}} = \frac{1}{R + j\omega L} \tag{6-168}$$

To express this in terms of an equivalent conductance and susceptance Eq. (6-168) is rationalized as follows

$$\bar{Y} = \frac{1}{R + j\omega L} \times \frac{R - j\omega L}{R - j\omega L} = \frac{R}{R^2 + \omega^2 L^2} - j\frac{\omega L}{R^2 + \omega^2 L^2} \qquad (6\text{-}169)$$

Recalling that $\omega L = X_L$ and that $\bar{Y} = G + jB$, we may rewrite (6-169) as

$$G + jB = \frac{R}{R^2 + X_L^2} - j\frac{X_L}{R^2 + X_L^2} \qquad (6\text{-}170)$$

Equating the real and j parts then yields

$$G = \frac{R}{R^2 + X_L^2} \quad \text{and} \quad B = \frac{-X_L}{R^2 + X_L^2} \qquad (6\text{-}171)$$

where G is the conductance and B is the susceptance of the equivalent parallel combination of the series R-L configuration. A comparison of Eqs. (6-166) and (6-171) draws the distinction respectively of the circuit arrangements of Figs. 6-18 and 6-19.

6-7   The Series *R-C* Circuit

The principles underlying the method of solution are presented in detail in the two preceding sections. Accordingly, in the interest of brevity as well as to illustrate the method to be employed from here on in solving such problems, we proceed directly with Kirchhoff's voltage law for the circuit of Fig. 6-20 by using effective quantities throughout. Thus

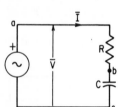

$$\bar{V} = IR + I\left(\frac{1}{j\omega C}\right) \qquad (6\text{-}172)$$

Fig. 6-20. The series *R-L* circuit.

The effective potential difference appearing across the capacitor is expressed always in terms of the effective current passing through the capacitor multiplied by its capacitive reactance, which is written in conjunction with the rotational operator j as called for by Eq. (6-121). Rewriting Eq. (6-172) gives

$$\bar{V} = I\left(R + \frac{1}{j\omega C}\right) = I\left(R - j\frac{1}{\omega C}\right) = I\bar{Z} \qquad (6\text{-}173)$$

where $\bar{Z}$ denotes the complex impedance for the R-C circuit and specifically is defined as

$$\bar{Z} \equiv R - j\frac{1}{\omega C} = \sqrt{R^2 + \frac{1}{\omega^2 C^2}} \left/ -\tan^{-1}\frac{1}{\omega RC}\right. \qquad (6\text{-}174)$$

Therefore the expression for the current response is

$$I\underline{/\theta} = \frac{V\underline{/0°}}{\sqrt{R^2 + \dfrac{1}{\omega^2 C^2}}} \left/ \tan^{-1}\frac{1}{\omega RC}\right. \qquad (6\text{-}175)$$

Thus

$$I = \frac{V}{\sqrt{R^2 + \dfrac{1}{\omega^2 C^2}}} \tag{6-176}$$

and

$$\theta = \tan^{-1} \frac{1}{\omega RC} \tag{6-177}$$

The corresponding time solution is then

$$i = I_m \sin(\omega t + \theta) = \frac{V_m}{\sqrt{R^2 + \dfrac{1}{\omega^2 C^2}}} \sin\left(\omega t + \tan^{-1} \frac{1}{\omega RC}\right) \tag{6-178}$$

The phasor diagram for the series *R-C* circuit is depicted in Fig. 6-21(a). The current now *leads* the voltage in time—a result which is consistent with Eq. (6-177) showing a positive value for the phase angle. Again note that the effective potential difference across the resistor is in phase with the current, while that across the capacitor is 90° *behind* the position of the current phasor in Fig. 6-21(a). This −90° rotation is brought about in the diagram by the attachment of $-j = /\underline{-90°}$ to the quantity *IX*. The corresponding time diagram of the current response is illustrated in Fig. 6-21(b).

The analysis of the parallel combination of resistance and capacitance is left as an exercise for the reader.

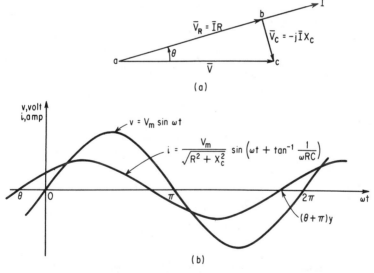

Fig. 6-21. (a) Phasor diagram for the *R-C* circuit of Fig. 6-20; (b) Time diagram for the *R-C* circuit of Fig. 6-20.

6-8   The *R-L-C* Circuit

The circuit configuration for the series combination of the three parameters is shown in Fig. 6-22; the terminals are lettered in order to facilitate the discussion

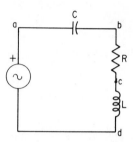

Fig. 6-22. The series *R-L-C* circuit.

of the phasor diagram. Writing Kirchhoff's voltage law expressed in terms of effective values we have

$$\bar{V} = I\left(\frac{1}{j\omega C}\right) + IR + I(j\omega L) = I\left(R + j\omega L + \frac{1}{j\omega C}\right) \qquad (6\text{-}179)$$

Calling the quantity in parentheses the complex impedance and denoting it as

$$\bar{Z} \equiv R + j\left(\omega L - \frac{1}{\omega C}\right) = \sqrt{R^2 + \left(\omega L - \frac{1}{\omega C}\right)^2}\left/\tan^{-1}\frac{\omega L - \frac{1}{\omega C}}{R}\right. \qquad (6\text{-}180)$$

enables us to write the expression for the current response

$$I\underline{/\theta} = \frac{\bar{V}}{\bar{Z}} = \frac{V\underline{/0^\circ}}{\sqrt{R^2 + \left(\omega L - \frac{1}{\omega C}\right)^2}}\left/-\tan^{-1}\frac{\omega L - \frac{1}{\omega C}}{R}\right. \qquad (6\text{-}181)$$

Therefore, the magnitude and phase of the effective current are

$$\boxed{I = \frac{V}{\sqrt{R^2 + \left(\omega L - \frac{1}{\omega C}\right)^2}}} \qquad (6\text{-}182)$$

and

$$\boxed{\theta = -\tan^{-1}\frac{\omega L - \frac{1}{\omega C}}{R}} \qquad (6\text{-}183)$$

The corresponding time solution is

$$i = \sqrt{2}\, I \sin(\omega t + \theta) = \frac{V_m}{\sqrt{R^2 + \left(\omega L - \frac{1}{\omega C}\right)^2}}\sin(\omega t + \theta) \qquad (6\text{-}184)$$

A study of Eq. (6-183) reveals that the current phasor may either lead or lag the voltage phasor, depending upon the relative values of the inductive and capacitive reactance ωL and $1/\omega C$. Whenever $\omega L > 1/\omega C$ the *R-L-C* circuit essentially behaves as an inductive circuit insofar as the current is concerned. It is interesting to note that this condition can be satisfied either by having a large inductance or else by operating at a high frequency. On the other hand, whenever $\omega L < 1/\omega C$ the current leads the voltage, thereby indicating that the *R-L-C* circuit behaves as a capacitive circuit as far as the current is concerned. However, there are some differences; these are discussed presently.

Appearing in Fig. 6-23 is the phasor diagram for the circuit of Fig. 6-22. It is drawn for the case where $\omega L > 1/\omega C$. Hence the current phasor must lag the voltage phasor. The component values of the effective potential difference appearing across each circuit element are also depicted. The voltage across the capacitor terminals $\bar{V}_c$ is represented by line *ab*. Note that the *I* leads $\bar{V}_c$ by 90° as it always must for sinusoidal forcing functions. The voltage drop across the resistor terminals $\bar{V}_R$ must be in phase with *I*. It is represented by line *bc*, which is parallel to phasor *I*. In phasor diagrams any line drawn parallel to another line means that the quantities represented by the two lines are in phase. Finally, note that line *cd* represents the effective potential difference appearing across the inductor terminals, $\bar{V}_L$. The current *I* lags $\bar{V}_L$ by 90° as expected.

Although the phasor diagram depicted in Fig. 6-23 is for an *R-L-C* circuit in which the inductive reactance predominates, note that the circuit does behave differently than the straight *R-L* circuit in two respects. One, the potential differences across the inductor contains a component (line *ce*) which is equal and opposite to the total voltage across the capacitor terminals. This leaves a net reactive voltage, as "seen" by the source, of amount *ed*. Two, the voltage across the inductor can be several times greater *in magnitude* than the source voltage $\bar{V}$, This cannot occur in the simple *R-L* circuit. The large value of $\bar{V}_L$

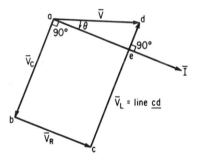

Fig. 6-23. Phasor diagram of the *R-L-C* circuit for $\omega L > 1/\omega C$.

should not be disturbing, however, because Kirchhoff's voltage law continues to be satisfied. Keep in mind that with a-c circuits it is the *phasor* sum that is important. An algebraic sum is meaningless except in those instances where only elements of the same type appear in the circuit.

The net reactive power which is interchanged between the source and the circuit continues to be given validly by Eq. (6-160), namely

$$P_X = VI \sin \theta = I(V \sin \theta) \qquad (6\text{-}185)$$

Here the quantity in parentheses is equivalent to line *ed* in Fig. 6-23, which is the net reactive voltage of the circuit.

6-9 Application of Network Theorems to Complex Impedances

In Chapter 3 the elementary network theory was illustrated by circuit configurations which included only resistive elements. This restriction was introduced in the interest of simplicity and also because we did not then know how to treat the voltage across a capacitor or an inductor for a time-varying current such as the sinusoid. We learned in Secs. 6-5 to 6-7 that for sinusoidal forcing functions the inductance and capacitance parameters exhibit an impediment to the flow current which is called reactance and which is characterized chiefly by the time phase displacement it causes between the response and the forcing function in the steady state. In the material that follows, the principal network theorems are applied to a-c circuits, i.e., to networks which include the three circuit parameters in various combinations. The methods of analysis in arriving at the desired solutions are the same as those used in Chapter 3. The difference lies merely in working with complex impedances rather than just resistances. Since the impedances as well as the phasor currents and voltages are expressed by complex numbers, the algebraic manipulations are more extensive. The application of the theory is best illustrated by means of examples.

EXAMPLE 6-8   *Series-parallel reduction of complex impedance.* In the circuit of Fig. 6-24 find the equivalent impedance which appears between points a and c. Assume that

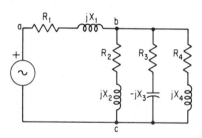

this circuit is subjected to a sinusoidal forcing function of frequency ω radians per second. At this frequency the impedances have the following values:

$$\bar{Z}_1 = R_1 + jX_1 = 10 + j10 = 14.14\underline{/45°}$$

$$\bar{Z}_2 = R_2 + jX_2 = 15 + j20 = 25\underline{/53.1°}$$

$$\bar{Z}_3 = R_3 - jX_3 = 3 - j4 = 5\underline{/-53.1°}$$

$$\bar{Z}_4 = R_4 + jX_4 = 8 + j6 = 10\underline{/36.9°}$$

Fig. 6-24. Network used to illustrate series-parallel reduction of complex impedances.

solution: Because there are more than two impedances in parallel between terminals b and c it is simpler to work first in terms of the admittance of the parallel combination. Thus

$$\bar{Y}_{bc} = \frac{1}{\bar{Z}_{bc}} = \frac{1}{\bar{Z}_2} + \frac{1}{\bar{Z}_3} + \frac{1}{\bar{Z}_4} = \frac{1}{25\underline{/53.1}} + \frac{1}{5\underline{/-53.1}} + \frac{1}{10\underline{/36.9°}}$$

$$= 0.04\underline{/-53.1} + 0.2\underline{/53.1°} + 0.1\underline{/-36.9}$$

$$= 0.024 - j0.032 + 0.12 + j0.16 + 0.08 - j0.06 \qquad (6\text{-}186)$$

Note that a quantity such as $\bar{Y}_2 = 1/\bar{Z}_2 = 0.024 - j0.032$ is the equivalent admittance of the impedance $\bar{Z}_2$. In fact these values can be found also by using Eqs. (6-171). Thus

$$\bar{Y}_2 = G_2 + jB_2 = \frac{15}{15^2 + 20^2} - j\frac{20}{15^2 + 20^2}$$

$$= \frac{15}{625} - j\frac{20}{625} = 0.024 - j0.032$$

A similar procedure may be employed in treating $\bar{Z}_3$ and $\bar{Z}_4$. Summing the real and j terms in Eq. (6-186) yields

$$\bar{Y}_{bc} = 0.224 + j0.068 = 0.234\underline{/16.9°}$$

$$\therefore\quad \bar{Z}_{bc} = \frac{1}{\bar{Y}_{bc}} = \frac{1}{0.234\underline{/16.9}} = 4.27\underline{/-16.9} = 4.09 - j1.24 \qquad (6\text{-}187)$$

Thus, by Eq. (6-187) the series equivalent impedance between b and c is reduced to an equivalent resistance of 4.09 ohms and an equivalent capacitive reactance of 1.24 ohms. The total impedance between a and c is then given by

$$\bar{Z}_{ac} = \bar{Z}_1 + \bar{Z}_{bc} = 10 + j10 + 4.09 - j1.24$$
$$= 14.09 + j8.76$$

Therefore

$$\bar{Z}_{ac} = 16.2\underline{/31.85°} \text{ ohms}$$

EXAMPLE 6-9 *Illustrating the mesh method of analysis.* Appearing in Fig. 6-25 is a two-mesh network with a sinusoidal source voltage appearing in each mesh. At a given frequency the impedance and phasor voltages are as follows:

$$\bar{Z}_1 = R_1 + jX_1 = 1 + j = \sqrt{2}\underline{/45°}$$
$$\bar{Z}_2 = R_2 - jX_2 = 1 - j = \sqrt{2}\underline{/-45°}$$
$$\bar{Z}_3 = R_3 + jX_3 = 1 + j2 = \sqrt{5}\underline{/63.5°}$$
$$\bar{V}_1 = 10\underline{/0°} \quad \text{and} \quad \bar{V}_2 = 10\underline{/-60°}$$

The phasor voltage $\bar{V}_1$ is taken as the reference. On this basis $\bar{V}_2$ lags $\bar{V}_1$ in time by $\omega t = 60°$. Find the phasor expression for the current which flows through branch $\bar{Z}_3$.

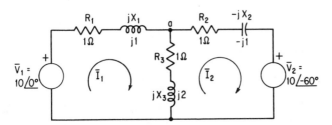

Fig. 6-25. Two-mesh network with complex impedances.

solution: The first step is to write the mesh equations in terms of phasor quantities. Thus for mesh 1 we have

$$I_1\bar{Z}_1 + I_1\bar{Z}_3 - I_2\bar{Z}_3 - \bar{V}_1 = 0$$

or
$$I_1(\bar{Z}_1 + \bar{Z}_3) - I_2\bar{Z}_3 = \bar{V}_1$$

$$(6\text{-}188)$$

For the second mesh Kirchhoff's voltage law yields

$$I_2\bar{Z}_2 + I_2\bar{Z}_3 - I_1\bar{Z}_3 + \bar{V}_2 = 0$$

or
$$-I_1\bar{Z}_3 + I_2(\bar{Z}_2 + \bar{Z}_3) = -\bar{V}_2$$

$$(6\text{-}189)$$

Equations (6-188) and (6-189) can be written more formally as

$$I_1 \bar{Z}_{11} - I_2 \bar{Z}_{12} = \bar{V}_1 \tag{6-190}$$

$$-I_1 \bar{Z}_{21} + I_2 \bar{Z}_{22} = -\bar{V}_2 \tag{6-191}$$

where

$$\bar{Z}_{11} = (R_1 + R_3) + j(X_1 + X_3) = 2 + j3$$

$$\bar{Z}_{12} = \bar{Z}_{21} = \bar{Z}_3 = 1 + j2$$

$$\bar{Z}_{22} = (R_2 + R_3) + j(X_3 - X_2) = 2 + j1$$

Applying determinants to the pair of equations (6-190) and (6-191) allows the expression for I_1 to be written as

$$I_1 = \frac{\begin{vmatrix} \bar{V}_1 & -\bar{Z}_{12} \\ -\bar{V}_2 & \bar{Z}_{22} \end{vmatrix}}{\begin{vmatrix} \bar{Z}_{11} & -\bar{Z}_{12} \\ -\bar{Z}_{21} & \bar{Z}_{22} \end{vmatrix}} = \frac{\bar{V}_1 \bar{Z}_{22} - \bar{V}_2 \bar{Z}_{12}}{\bar{Z}_{11} \bar{Z}_{22} - \bar{Z}_{12}^2} \tag{6-192}$$

$$= \frac{10(2 + j) - 10/\underline{-60}\,(1 + j2)}{(2 + j3)(2 + j1) - (1 + j2)^2} = \frac{20 + j10 - (5 - j5\sqrt{3})(1 + j2)}{1 + j8 + 3 - j4}$$

$$= \frac{20 + j10 - 22.32 - j1.34}{4 + j4} = \frac{8.97/\underline{105°}}{5.66/\underline{45°}}$$

$$= 1.58/\underline{60°} = 0.79 + j1.37$$

Similarly

$$I_2 = \frac{\begin{vmatrix} \bar{Z}_{11} & \bar{V}_1 \\ -\bar{Z}_{21} & -\bar{V}_2 \end{vmatrix}}{\bar{Z}_{11}\bar{Z}_{22} - \bar{Z}_{12}^2} = \frac{-\bar{V}_2 \bar{Z}_{11} + \bar{Z}_{21}\bar{V}_1}{5.66/\underline{45°}} \tag{6-193}$$

$$= \frac{-10/\underline{-60}\,(2 + j3) + (1 + j2)10}{5.66/\underline{45°}} = \frac{34.1/\underline{139.3°}}{5.66/\underline{45_°}}$$

$$= 6.03/\underline{94.3°} = -0.46 + j6.01$$

Therefore the current in branch $\bar{Z}_3$ is

$$I_{\text{branch 3}} = I_1 - I_2 = (0.79 + j1.37) - (0.46 + j6.01)$$

$$= 1.25 - j4.64 = 4.82/\underline{-75°} \text{ amps}$$

Hence the effective value of the current is 4.82 amperes and in the steady state this current lags behind the voltage $\bar{V}_1$ by 75 degrees.

EXAMPLE 6-10 *Illustrating the nodal method.* Find the current in the $\bar{Z}_3$ branch by employing the nodal method of solution.

solution: Inspection of the circuit reveals that there is only one independent node which is marked terminal a in Fig. 6-25. Upon assuming that all currents flow away from the node, the describing equation which results from Kirchhoff's current law is

$$\frac{\bar{V}_a - \bar{V}_1}{\bar{Z}_1} + \frac{\bar{V}_a}{\bar{Z}_3} + \frac{\bar{V}_a - \bar{V}_2}{\bar{Z}_2} = 0 \tag{6-194}$$

or

$$\bar{V}_a \left(\frac{1}{\bar{Z}_1} + \frac{1}{\bar{Z}_2} + \frac{1}{\bar{Z}_3} \right) = \frac{\bar{V}_1}{\bar{Z}_1} + \frac{\bar{V}_2}{\bar{Z}_2} \tag{6-195}$$

Let

$$\bar{Y} = \frac{1}{\bar{Z}_1} + \frac{1}{\bar{Z}_2} + \frac{1}{\bar{Z}_3} = \left(\frac{1}{2} - j\frac{1}{2}\right) + \left(\frac{1}{2} + j\frac{1}{2}\right) + \left(\frac{1}{5} - j\frac{2}{5}\right) \quad (6\text{-}196)$$
$$= 1.2 - j0.4 = 1.263\underline{/-18.4°}$$

Also

$$\frac{\bar{V}_1}{\bar{Z}_1} + \frac{\bar{V}_2}{\bar{Z}_2} = \frac{10\underline{/0°}}{\sqrt{2}\underline{/-45°}} + \frac{10\underline{/-60°}}{\sqrt{2}\underline{/-45°}} = 5 - j5 + 6.81 - j1.83 \quad (6\text{-}197)$$
$$= 11.81 - j6.83 = 13.65\underline{/-30°}$$

Inserting Eqs. (6-196) and (6-197) into Eq. (6-195) yields the phasor expression for the nodal voltage:

$$\bar{V}_a = \frac{13.65\underline{/-30°}}{1.263\underline{/-18.4°}} = 10.8\underline{/-11.6°} \text{ volts}$$

Therefore the branch current is simply

$$I_{\text{branch}} = \frac{\bar{V}_a}{\bar{Z}_3} = \frac{10.8\underline{/-11.6°}}{\sqrt{5}\underline{/63.5°}} = 4.82\underline{/-75°}$$

which checks with the result found by using the mesh method. Note, however, that considerably less effort is required to arrive at the solution.

EXAMPLE 6-11   *Solution by Thevenin's theorem.* Figure 6-25 is redrawn in Fig. 6-26 to separate branch $\bar{Z}_3$ from the remainder of the circuit. By using Thevenin's theorem find the current which flows through $\bar{Z}_3$.

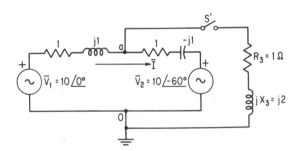

Fig. 6-26. The circuit of Fig. 6-25 is redrawn in preparation for applying Thevenin's theorem.

solution: The open-circuit voltage of the Thevenin equivalent circuit is equal to the potential difference appearing between points 0 and a in Fig. 6-26 when the switch is open. To determine this quantity let us first find the current $\bar{I}$ that flows in the circuit between the two sources when S is open. Kirchhoff's voltage law yields

$$-\bar{V}_1 + \bar{I}\bar{Z}_1 + \bar{I}\bar{Z}_2 + \bar{V}_2 = 0 \quad (6\text{-}198)$$

or

$$\bar{I} = \frac{\bar{V}_1 - \bar{V}_2}{\bar{Z}_1 + \bar{Z}_2} = \frac{10 - 10\underline{/-60}}{1 + j + 1 - j} = \frac{10 - 5 + j8.66}{2} = 5\underline{/60°}$$

Hence the voltage between points 0 and a is

$$\bar{V}_{0a} = \bar{V}_{\text{open-circuit}} = \bar{V}_1 - I\bar{Z}_1 = 10\underline{/0°} - 5\underline{/60°}\sqrt{2}\underline{/45°}$$
$$= 10 + 1.83 - j6.82 = 13.65\underline{/-30°} \tag{6-199}$$

Note that the open-circuit voltage has a larger magnitude than either source voltage. The equivalent impedance looking into the circuit at points 0 and a is

$$\bar{Z}_i = \frac{\bar{Z}_1\bar{Z}_2}{\bar{Z}_1 + \bar{Z}_2} = \frac{\sqrt{2}\underline{/45}\sqrt{2}\underline{/-45}}{1 + j + 1 - j} = 1\underline{/0°} \text{ ohm} \tag{6-200}$$

Therefore the branch current in Z_3 is given by

$$I_{\text{branch}} = \frac{\bar{V}_{oc}}{\bar{Z}_i + \bar{Z}_3} = \frac{13.65\underline{/-30}}{1 + 1 + j2} = 4.82\underline{/-75°} \text{ amps}$$

Again note the exact correspondence of this result with those obtained by the other methods.

The use of Norton's theorem in solving this problem is left as an exercise for the reeader.

6-10 Resonance

Resonance is identified with engineering situations which involve energy-storing elements subjected to a forcing function of varying frequency. Specifically resonance is the term used to describe the steady-state operation of a circuit or system at that frequency for which the resultant response is in time phase with the forcing function despite the presence of energy-storing elements. Resonance cannot take place when only one type of energy-storing element is present, e.g. inductance or mass. There must exist two types of independent energy-storing elements capable of interchanging energy between one another—for example, inductance and capacitance or mass and spring. Although attention here is confined to electric circuits, resonance is a phenomenon found in any system involving two independent energy-storing elements be they mechanical, hydraulic, pneumatic, or whatever. In the material which follows, the parallel as well as the series arrangement of the energy-storing elements—L and C—are treated as functions of frequency, with particular emphasis focused on the performance at resonance.

Series Resonance. The series arrangement of L and C along with resistance R is shown in Fig. 6-22. The expression for the effective current flow caused by a sinusoidal forcing function is given by Eq. (6-179) and is repeated for convenience. Thus

$$I = \frac{\bar{V}}{R + j\left(\omega L - \dfrac{1}{\omega C}\right)} = \frac{\bar{V}}{\bar{Z}} \tag{6-201}$$

What is the effect on I of increasing the frequency ω of the forcing function from zero to infinity? A glance at Eq. (6-201) indicates that a change in frequency means a change in the magnitude and phase angle of the complex impedance. Note that for ω close to zero the inductive reactance is almost zero but the capacitive reactance approaches infinity. Hence the current magnitude approaches zero. As ω increases the reactance part of $\bar{Z}$ decreases, thus causing an increase in current. As ω continues to increase, a point is reached where the reactance term is zero. Calling this frequency ω_0 we have

$$\omega_0 L - \frac{1}{\omega_0 C} = 0 \qquad (6\text{-}202)$$

or

$$\omega_0^2 = \frac{1}{LC} \qquad (6\text{-}203)$$

or

$$\boxed{\omega_0 = \frac{1}{\sqrt{LC}}} \qquad (6\text{-}204)$$

The frequency ω_0 is called the *resonant frequency* of the circuit. Its value is specified entirely in terms of the parameters of the two energy-storing elements of the circuit in the manner called for by Eq. (6-204).

At resonance the impedance of the circuit is a minimum, and specifically it is equal to R. Consequently, when a series R-L-C circuit is at resonance, the current is a maximum and is also in time phase with the voltage. The power factor is unity. The complete expression for the current phasor is then

$$I_0 = \frac{\bar{V}}{\bar{Z}} = \frac{V\underline{/0^\circ}}{R\underline{/0^\circ}} = \frac{V}{R}\underline{/0^\circ} \qquad (6\text{-}205)$$

where I_0 denotes the current at resonance.

For operation at a frequency below ω_0 the resultant j part of $\bar{Z}$ is capacitive so that the current leads the voltage. When $\omega > \omega_0$ the inductive reactance prevails, so that the current then lags the voltage. Depicted in Fig. 6-27 is the plot

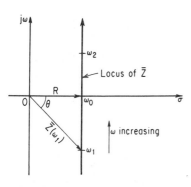

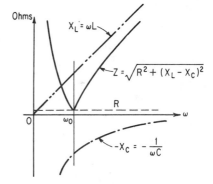

Fig. 6-27. Locus of the complex impedance $\bar{Z}$ as a function of frequency ω; $\bar{Z} = R + j(\omega L - 1/\omega C)$.

Fig. 6-28. Variation of impedance, inductive reactance and capacitive reactance as a function of frequency.

of the impedance in the complex plane. The magnitude of $\bar{Z}$ at a frequency ω_1 is obtained as the length of the line drawn from the origin to the point ω_1 on the heavy vertical line. This heavy vertical line denotes the value of the reactance portion of $\bar{Z}$. Note that at $\omega = \omega_0$ it has a zero value. An alternative way of representing the information of Fig. 6-27 is illustrated in Fig. 6-28. A plot of the variation of the magnitude and phase of the current with frequency for two values of the ratio $\omega_0 L/R$ is depicted in Fig. 6-29. Note that when the ratio of inductive reactance to resistance in a series R-L-C circuit is high, a very rapid rise of current to the maximum value occurs in the vicinity of the resonant frequency. A characteristic of this type is especially useful in radio and other communication applications.

Bandwidth. To describe the width of the resonance curve the term bandwidth is used. For the series R-L-C circuit bandwidth is defined as the range of frequency for which the power delivered to R is greater than or equal to $P_0/2$, where P_0 is the power delivered to R at resonance. From the shape of the resonance curve it should be clear that there are two frequencies for which the power delivered to R is half the power at resonance. For this reason these frequencies are referred to as those corresponding to the half-power points. The magnitude of the current at each half-power point is the same. Hence we can write

$$I_1^2 R = \tfrac{1}{2} I_0^2 R = I_2^2 R \qquad (6\text{-}206)$$

where subscript 1 denotes the lower half-power point and subscript 2 the higher half-power point. It follows then that

$$I_1 = I_2 = \frac{I_0}{\sqrt{2}} = 0.707 I_0 \qquad (6\text{-}207)$$

Accordingly the bandwidth may be identified on the resonance curve as that range of frequency over which the magnitude of the current is equal to or greater than 0.707 of the current at resonance. In Fig. 6-29(a) for $\omega_0 L/R = \tfrac{1}{2}$ the frequency at the lower half-power point is denoted ω_1 and that of the upper half-power point is denoted ω_2. Hence the bandwidth is $\omega_2 - \omega_1$.

In view of the fact that the current at the half-power points is $I_0/\sqrt{2}$, it follows that the *magnitude* of the impedance must be equal to $\sqrt{2}\,R$ to yield this current. This information can now be used to obtain an expression for the bandwidth in terms of the parameters of the series circuit. Calling the reactance at the lower half-power frequency X_1, we have

$$X_1 = \omega_1 L - \frac{1}{\omega_1 C} = -R \qquad (6\text{-}208)$$

The minus sign appears on the right side of the equation because below resonance the capacitive reactance exceeds the inductive reactance. Rearranging Eq. (6-208) leads to

$$\omega_1^2 + \frac{R}{L}\omega_1 - \frac{1}{LC} = 0 \qquad (6\text{-}209)$$

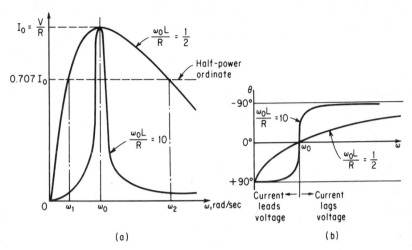

Fig. 6-29. Variation of current in a series R-L-C circuit with frequency:
(a) magnitude;
(b) phase angle.

The roots are therefore

$$(\omega_1)_{1,2} = -\frac{R}{2L} \pm \sqrt{\left(\frac{R}{2L}\right)^2 + \frac{1}{LC}} = -\alpha \pm \sqrt{\alpha^2 + \omega_0^2} \qquad (6\text{-}210)$$

where

$$\alpha \equiv \frac{R}{2L} \qquad (6\text{-}211)$$

Although Eq. (6-210) provides two solutions to Eq. (6-209), only one of these is physically realizable. The negative frequency is meaningless and so may be discarded. Hence the expression for the lower half-power frequency is

$$\omega_1 = -\alpha + \sqrt{\alpha^2 + \omega_0^2} \qquad (6\text{-}212)$$

The upper half-power frequency is found in a similar fashion. In this instance we have

$$X_2 = \omega_2 L - \frac{1}{\omega_2 C} = +R \qquad (6\text{-}213)$$

which leads to

$$\omega_2^2 - \frac{R}{L}\omega_2 - \frac{1}{LC} = 0 \qquad (6\text{-}214)$$

The expression for the useful ω_2 is then

$$\omega_2 = \alpha + \sqrt{\alpha^2 + \omega_0^2} \qquad (6\text{-}215)$$

Therefore the expression for the bandwidth becomes, from Eqs. (6-212) and (6-215),

$$\omega_{bw} = \omega_2 - \omega_1 = 2\alpha = \frac{R}{L} \qquad (6\text{-}216)$$

This expression is significant because it reveals that the bandwidth of the series *R-L-C* circuit depends solely upon the R/L ratio. Note that it is not the individual values of R or L but rather their ratio that is important. Note too that bandwidth depends not at all upon the capacitance parameter C.

Quality Factor Q_0 of the R-L-C Circuit. By forming the ratio of the resonant frequency to the bandwidth we obtain a factor which is a measure of the *selectivity* or *sharpness of tuning* of the series *R-L-C* circuit. This quantity is called the *quality factor* of the circuit and is denoted by Q_0. Thus

$$Q_0 \equiv \frac{\omega_0}{\omega_{bw}} \qquad (6\text{-}217)$$

Moreover, when Eq. (6-216) is inserted into the last equation an alternative expression results:

$$Q_0 = \frac{\omega_0}{R/L} = \frac{\omega_0 L}{R} = \frac{X_{L0}}{R} \qquad (6\text{-}218)$$

A glance at the curves appearing in Fig. 6-29(a) should make it clear as to why the quality factor Q_0 is used as a measure of the selectivity or sharpness of tuning. Note how much narrower the resonance curve is for $Q_0 = 10$ than for $Q_0 = \frac{1}{2}$. Equation (6-216) shows that the value of C in no way influences the bandwidth but it does alter the value of ω_0. Hence if C is changed, ω_0 may be made to occur at a different position along the frequency scale in Fig. 6-29(a) without in any way affecting the sharpness of tuning. This feature of the high-Q circuit is used often in communication networks. For example, in a radio the antenna may be considered in terms of an equivalent *R-L-C* circuit where C is adjusted by means of the dial tuning knob. If the dial is turned to the position which tunes in the frequency (ω_0) of a given radio transmitting station, the sharpness of tuning (i.e., small bandwidth) allows only signals from that station to produce large resonant current signals. The signals from other broadcasting stations, although present in equal strength at the antenna, produce little or no signal strength in the circuit because the dial is tuned to a frequency considerably off resonance relative to their broadcasting frequencies. Values of Q_0 of the order of 100 are typical in radio circuits.

By making use of the knowledge that Q_0 is very large in many resonant circuits, we can express the lower and upper half-power points in terms of the resonant frequency and the bandwidth. Returning to Eq. (6-212), we can rewrite it as

$$\omega_1 = -\frac{R}{2L} \pm \sqrt{\left(\frac{R}{2L}\right)^2 + \omega_0^2} = -\frac{R}{2L} + \sqrt{\left(\frac{R}{2L}\right)^2 + \frac{R^2}{L^2}\left(\frac{\omega_0^2 L^2}{R^2}\right)}$$

$$= -\frac{R}{2L} + \sqrt{\left(\frac{R}{2L}\right)^2 + Q_0^2 \frac{R^2}{L^2}} \qquad (6\text{-}219)$$

For values of Q_0 which are 5 or greater very little error is made by writing

$$\omega_1 = -\frac{R}{2L} + \frac{R}{L}Q_0 = -\frac{R}{2L} + \omega_0 \qquad (6\text{-}220)$$

Inserting Eq. (6-216) into Eq. (6-220) yields

$$\omega_1 = \omega_0 - \tfrac{1}{2}\omega_{bw} \qquad (6\text{-}221)$$

Similarly it can be shown that

$$\omega_2 = \omega_0 + \tfrac{1}{2}\omega_{bw} \qquad (6\text{-}222)$$

Another characteristic of a series R-L-C circuit at resonance is worth noting. It has to do with the magnitude of the voltage drop appearing across L and C. In terms of a phasor formulation the phasor voltage drop across L at resonance is given by

$$\bar{V}_{L0} = I_0(jX_L) = \frac{V}{R}\,\omega_0 L\underline{/90°} = Q_0 V\underline{/90°} \qquad (6\text{-}223)$$

Therefore the magnitude of the voltage across the terminals of the inductor is Q_0 times the rms value of the applied voltage. For high-Q circuits this represents a considerable amplification of the source voltage. By proceeding in a similar fashion for the capacitor we find

$$\bar{V}_{C0} = I_0\left(\frac{1}{j\omega C}\right) = \frac{V}{R}\left(\frac{1}{\omega_0 C}\right)\underline{/-90°}$$

$$= \frac{V}{R}(\omega_0 L)\underline{/-90°} = Q_0 V\underline{/-90°} \qquad (6\text{-}224)$$

Again note the resonant voltage-rise effect across the capacitor terminals. The phasor voltage diagram at resonance is depicted in Fig. 6-30.

On the basis of Eqs. (6-223) and (6-224) another definition of the quality factor Q_0 is possible. It represents the extent to which the voltage across L or C rises at resonance expressed as a multiple of the applied voltage. Stated mathematically, we have

$$Q_0 = \frac{V_{L0}}{V} = \frac{V_{C0}}{V} \qquad (6\text{-}225)$$

where V_{L0} and V_{C0} are both measured at resonance.

The definitions of Q_0 given by Eqs. (6-218) and (6-225) are restrictive in the sense that they apply specifically to the series R-L-C circuit. The definition of Eq. (6-217) is more generally applicable because it is based on the frequency-selectivity

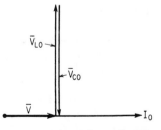

Fig. 6-30. Voltage phasor diagram of the series R-L-C circuit at resonance.

characteristic of the circuit. However, the most universally applicable definition is one that is expressed in terms of energy. In this connection, then, let us assume that the instantaneous expression for the forced solution of current at resonance is

$$i_0 = I_{m0} \cos \omega_0 t \qquad (6\text{-}226)$$

The total energy stored by the circuit in both L and C can be expressed as

$$W = \tfrac{1}{2} L i_0^2 + \tfrac{1}{2} C v_{c0}^2 \qquad (6\text{-}227)$$

Inserting the appropriate expressions for i_0 and v_{c0} yields

$$
\begin{aligned}
W &= \frac{1}{2} L I_{m0}^2 \cos^2 \omega_0 t + \frac{1}{2} C \left(\frac{1}{C} \int_0^t I_{m0}^2 \cos \omega_0 t \right)^2 \\
&= \frac{1}{2} L I_{m0}^2 \cos^2 \omega_0 t + \frac{I_{m0}^2}{2\omega_0^2 C} \sin^2 \omega_0 t \qquad (6\text{-}228) \\
&= \frac{1}{2} L I_{m0}^2 \cos^2 \omega_0 t + \frac{1}{2} L I_{m0}^2 \sin^2 \omega_0 t \\
&= \frac{1}{2} L I_{m0}^2 = L I_0^2
\end{aligned}
$$

where I_0 is the rms current at resonance. It is interesting to note that at resonance this energy is a constant quantity.

Returning to Eq. (6-218) and multiplying both numerator and denominator by I_0^2 we can write for Q_0

$$Q_0 = \frac{\omega_0 L I_0^2}{R I_0^2} = \frac{2\pi f_0 L I_0^2}{R I_0^2} = 2\pi \frac{I_0^2 L}{\left(\dfrac{I_0^2 R}{f_0} \right)} \qquad (6\text{-}229)$$

Now since $I_0^2 L$ represents the total stored energy at resonance and $I_0^2 R / f_0$ is the energy dissipated per cycle, Eq. (6-229) may be written more generally as

$$\boxed{Q_0 = 2\pi \frac{\text{total stored energy}}{\text{energy dissipated per cycle}}} \qquad (6\text{-}230)$$

Equation (6-230) is applicable to any resonant system regardless of its composition. As long as the quantities on the right side can be determined the quality factor can, in turn, be found.

Parallel Resonance. The circuit configuration for the study of parallel resonance appears in Fig. 6-31. The expression for the total current is

$$I = \bar{V}\bar{Y} = \bar{V}(G + jB) \qquad (6\text{-}231)$$

where I is a fixed, known quantity and $\bar{V}$ is the nodal voltage, which varies as $\bar{Y}$ varies with frequency. For the circuit of Fig. 6-31 the expression for the admittance is

$$\bar{Y} = G + jB = \frac{1}{R} + j\omega C + \frac{1}{j\omega L} = \frac{1}{R} + j\left(\omega C - \frac{1}{\omega L} \right) \qquad (6\text{-}232)$$

Hence

$$I = \bar{V}\left[\frac{1}{R} + j\left(\omega C - \frac{1}{\omega L}\right)\right] \quad (6\text{-}233)$$

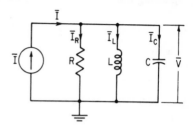

A study of this equation reveals that for small values of ω the susceptance B will be large. Then to keep the product of the right side constant and equal to I, the nodal voltage $\bar{V}$ must be correspondingly small. As ω increases, B decreases, and so $\bar{V}$ increases. When B equals zero, $\bar{V}$ has its maximum value. At this point the output nodal voltage is in time phase with

Fig. 6-31. Circuit configuration for studying parallel resonance.

the current source, and the circuit is said to be in parallel resonance. The frequency at which resonance occurs is found from

$$B = \omega_0 C - \frac{1}{\omega_0 L} = 0 \quad (6\text{-}234)$$

or

$$\omega_0 = \frac{1}{\sqrt{LC}} \quad (6\text{-}235)$$

which is identical to the expression which applies to the series $R\text{-}L\text{-}C$ circuit. As ω increases beyond ω_0, the susceptance gets ever larger, thus causing the nodal voltage to diminish towards zero. A sketch of $\bar{V}$ as a function of frequency is

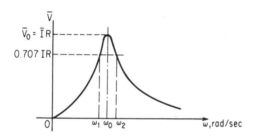

Fig. 6-32. Variation of the nodal voltage $\bar{V}$ of Fig. 6-31 as a function of frequency.

depicted in Fig. 6-32. The bandwidth is identified as the frequency range lying between ω_1 and ω_2. For the parallel case, the lower frequency is found from

$$\omega_1 C - \frac{1}{\omega_1 L} = -\frac{1}{R} \quad (6\text{-}236)$$

which leads to

$$\omega_1^2 + \frac{1}{RC}\omega_1 - \frac{1}{LC} = 0 \quad (6\text{-}237)$$

Ignoring the negative solution the desired expression is

$$\omega_1 = -\frac{1}{2RC} + \sqrt{\left(\frac{1}{2RC}\right)^2 + \frac{1}{LC}} = -\alpha + \sqrt{\alpha^2 + \omega_0^2} \qquad (6\text{-}238)$$

where

$$\alpha = \frac{1}{2RC} \qquad (6\text{-}239)$$

Similarly for the upper half-power frequency point

$$\omega_2 C - \frac{1}{\omega_2 L} = \frac{1}{R} \qquad (6\text{-}240)$$

This leads to

$$\omega_2 = \frac{1}{2RC} + \sqrt{\left(\frac{1}{2RC}\right)^2 + \frac{1}{LC}} = \alpha + \sqrt{\alpha^2 + \omega_0^2} \qquad (6\text{-}241)$$

Therefore the expression for the bandwidth is

$$\boxed{\omega_{bw} = \omega_2 - \omega_1 = 2\alpha = \frac{1}{RC}} \qquad (6\text{-}242)$$

Employing the definition of the quality factor given by Eq. (6-217) leads to

$$Q_0 \equiv \frac{\omega_0}{\omega_{bw}} = \frac{\omega_0}{1/RC} = \omega_0 RC \qquad (6\text{-}243)$$

It can be shown that Eq. (6-230) also applies with equal validity in determining Q_0.

Just as there occurs a resonant rise in voltage associated with L and C in the series-resonant case, so too there occurs a resonant rise in current in the L and C elements when the circuit is at parallel resonance. This is readily demonstrated. Recall that the value of the nodal voltage $\bar{V}_0$ at resonance is

$$\bar{V}_0 = IR\underline{/0°} \qquad (6\text{-}244)$$

Since this voltage appears across L we can write

$$I_{L0} = \frac{\bar{V}_0}{j\omega_0 L} = \frac{IR}{\omega_0 L}\underline{/-90°} = I\omega_0 RC\underline{/-90°} = IQ_0\underline{/-90°} \qquad (6\text{-}245)$$

For the capacitor

$$I_{C0} = \frac{\bar{V}_0}{1/j\omega_0 C} = jIR\omega_0 C = Q_0 I\underline{/90°} \qquad (6\text{-}246)$$

The subscript 0 denotes the resonant condition. An inspection of each of the last two expressions shows that the magnitude of the current through L or C is Q_0 times the source current I. It follows then that another way of expressing Q_0 in the parallel-resonance case is to write

$$Q_0 = \frac{I_{L0}}{I} = \frac{I_{C0}}{I} \qquad (6\text{-}247)$$

Equation (6-247) is analogous to Eq. (6-225) for the series-resonant case.

Appearing in Fig. 6-33 is the phasor diagram for the currents using the nodal voltage at resonance, $\bar{V}_0$, as the reference phasor. Note that I_{L0} lags and I_{c0} leads $\bar{V}_0$ by 90°, as called for by Eqs. (6-245) and (6-246). Of course the resultant of the three currents is the source current I.

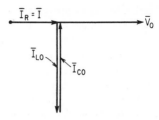

6-11   Balanced Three-phase Circuits

Fig. 6-33. Current phasor diagram for parallel resonance.

Almost all the electric power used in this country is generated and transmitted in the form of balanced three-phase voltage systems. The single-phase voltage sources referred to in the preceding sections originate as part of the three-phase system. The manner in which a three-phase voltage is generated is described in detail in Sec. 16-1. A balanced three-phase voltage system is composed of three single-phase voltages having the same amplitude and frequency of variation but time-displaced from one another by 120 degrees. A schematic representation is depicted in Fig. 6-34(a). The three single-phase voltages appear in a Y configuration; a Δ configuration is also possible. These single-

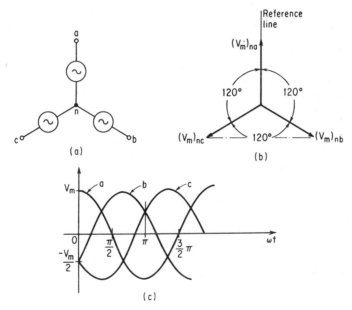

Fig. 6-34. Balanced 3-phase voltage system:
   (a) schematic diagram;
   (b) phasor diagram in terms of the maximum value of each phasor;
   (c) time diagram.

phase voltages are generated by a common rotating flux field in three identical windings which are separated from each other by 120 degrees inside the housing of the electric generator. When we join one end of each winding together to form terminal n, the Y connection results. Appearing in Fig. 6-34(b) is the phasor diagram of the three-phase voltage system. The corresponding time diagram for each single-phase voltage is shown in Fig. 6-34(c). For the time instant depicted in Fig. 6-34(b) [i.e., $(V_m)_{na}$ in phase with the vertical reference line] phase a is at its maximum value while simultaneously phases b and c have values of minus one-half the maximum value. Recall that in the phasor diagram instantaneous values are obtained by projecting the maximum-value phasor onto the reference line. Hence for $(V_m)_{nb}$ and $(V_m)_{nc}$ these projections are negative and equal to one-half V_m. Note that these instantaneous values exactly correspond to those obtained from Fig. 6-34(c) for $\omega t = 0$.

As the phasors in Fig. 6-34(b) revolve at the angular frequency ω with respect to the reference line in the counterclockwise (positive) direction, the complete time diagram of Fig. 6-34(c) is generated. The positive maximum value first occurs for phase a and then in succession for phases b and c. For this reason the three-phase voltage of Fig. 6-34 is said to have the *phase order abc*. The terms *phase sequence* and *phase rotation* are used synonymously. Phase order is important in certain applications For example, in three-phase induction motors it determines whether the motor turns clockwise or counterclockwise.

Current and Voltage Relations for the Y Connection. Every balanced three-phase system can be analyzed in terms of the procedures which apply to a single phase. Hence all the techniques developed in the preceding sections are applicable to symmetrical three-phase systems. The arrangement of the three single-phase voltages into a Y or Δ configuration calls for some modifications in dealing with total effects. For example the voltages appearing between any pair of the line terminals a, b, and c bear a different relationship in magnitude and phase to the voltages appearing between any one line terminal and the common terminal n called the *netural*. The former set of voltages—$\bar{V}_{ab}$, $\bar{V}_{bc}$, and $\bar{V}_{ca}$—are called the *line voltages*, and the latter set of voltages—$\bar{V}_{na}$, $\bar{V}_{nb}$, and $\bar{V}_{nc}$—are called the *phase voltages*. The relationships existing between the phase and line voltages are readily determined from an analysis of the phasor diagram in conjunction with Kirchhoff's voltage law. In this connection the phasor diagram of Fig. 6-34(b) is redrawn in Fig. 6-35 in terms of effective values rather than peak values. Moreover, V_{na} is made to coincide with the horizontal axis of the complex plane. The use of the double subscript notation is a convenience in treating three-phase circuits. If $\bar{V}_{na}$ is made to denote the voltage rise in going through the source from terminal n to a, then the quantity $-\bar{V}_{na} = \bar{V}_{an}$ represents the voltage drop in going from a to n.

The effective values of the phase voltages are shown in Fig. 6-35 as V_{na}, V_{nb}, and V_{nc}. Each has the same magnitude and each is displaced $120°$ from the other two phasors. To obtain the magnitude and phase angle of the line voltage from

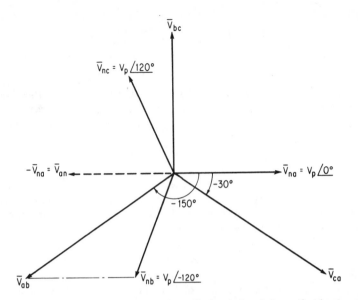

Fig. 6-35. Illustrating the phase and magnitude relations between the phase and line voltages of a Y connection.

a to b, i.e., $\bar{V}_{ab}$, we apply Kirchhoff's voltage law. Thus in Fig. 6-34(a)

$$\bar{V}_{ab} = \bar{V}_{an} + \bar{V}_{nb} \tag{6-248}$$

This equation states that the voltage existing from a to b is equal to the voltage from a to n (i.e., $-\bar{V}_{na}$) plus the voltage from n to b. Equation (6-248) can be rewritten as

$$\bar{V}_{ab} = -\bar{V}_{na} + \bar{V}_{nb} \tag{6-249}$$

Since for a balanced system each phase voltage has the same magnitude, let us set

$$V_{na} = V_{nb} = V_{nc} = V_p \tag{6-250}$$

where V_p denotes the effective magnitude of the phase voltage. Accordingly we may write

$$\bar{V}_{na} = V_p \underline{/0^\circ} \tag{6-251}$$

$$\bar{V}_{nb} = V_p \underline{/-120} \tag{6-252}$$

$$\bar{V}_{nc} = V_p \underline{/-240} = V_p \underline{/120} \tag{6-253}$$

Inserting Eqs. (6-251) and (6-252) into Eq. (6-249) yields

$$\bar{V}_{ab} = -V_p + V_p \underline{/-120^\circ}$$
$$= -V_p + V_p(\cos 120^\circ - j \sin 120^\circ) \tag{6-254}$$

$$= V_p \left(-\frac{3}{2} - j\frac{\sqrt{3}}{2} \right) = \sqrt{3}\, V_p \underline{/-150}$$

Similarly we obtain

$$\bar{V}_{bc} = \bar{V}_{bn} + \bar{V}_{nc} = \sqrt{3} \ V_p \underline{/-270^\circ} \tag{6-255}$$

and

$$\bar{V}_{ca} = \bar{V}_{cn} + \bar{V}_{na} = \sqrt{3} \ V_p \underline{/-30^\circ} \tag{6-256}$$

A study of the expressions for the line voltages shows that they constitute a balanced three-phase voltage system whose magnitudes are greater than the phase voltages by the $\sqrt{3}$. That is, the magnitude of the line voltage V_L is given by

$$\boxed{V_L = \sqrt{3} \ V_p} \quad \text{valid for Y connection} \tag{6-257}$$

Any current that flows out of line terminal a (or b or c) must be the same as that which flows through the phase source voltage appearing between terminals n and a (or n and b, or n and c). Therefore in the Y-connected three-phase generator the line current equals the phase current. That is,

$$\boxed{I_L = I_p} \tag{6-258}$$

where I_L denotes the effective value of line current and I_p denotes the effective value of the phase current.

Current and Voltage Relationships in a Δ-Connected Three-phase System. Assume next that the three single-phase sources of Fig. 6-34(a) are rearranged to form the Δ connection shown in Fig. 6-36. A three-phase load is shown connected

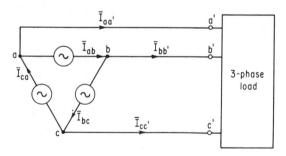

Fig. 6-36. A Δ-connected three-phase source.

to the three-phase source for the purpose of allowing line currents to flow. From inspection of the circuit diagram it should be obvious that the line and phase voltages have the same magnitude. Hence

$$\boxed{V_L = V_p} \quad \text{for Δ connection} \tag{6-259}$$

The phasor diagram appears in Fig. 6-37. It should also be obvious from Fig. 6-36 that the phase and line currents are not identical. Applying Kirchhoff's

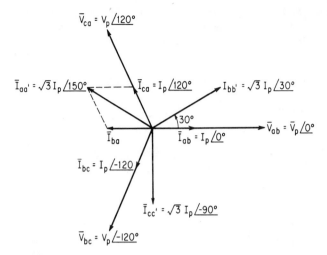

Fig. 6-37. Illustrating relation between phase and line currents in a Δ connection.

current law at one of the line terminals, e.g., *a*, emphasizes this point. Thus

$$I_{aa'} = I_{ca} + I_{ba} = I_{ca} - I_{ab} \tag{6-260}$$

where $I_{aa'}$ denotes the current which flows in the line joining *a* to *a'*. To illustrate the relationship existing between line and phase currents, let us consider that the nature of the load circuit is such that it causes the three phase currents to be expressed as

$$I_{ab} = I_p\underline{/0^\circ} \tag{6-261}$$

$$I_{bc} = I_p\underline{/-120^\circ} \tag{6-262}$$

$$I_{ca} = I_p\underline{/120^\circ} \tag{6-263}$$

As Fig. 6-37 indicates, these phase currents are assumed to be in phase with their respective phase voltages. Inserting Eqs. (6-261) and (6-263) into Eq. (6-260) then yields

$$I_{aa'} = I_p\underline{/120^\circ} - I_p = I_p\left(-\frac{3}{2} + j\frac{\sqrt{3}}{2}\right) = \sqrt{3}\,I_p\underline{/150^\circ} \tag{6-264}$$

In a similar fashion

$$I_{bb'} = I_{ab} + I_{cb} = \sqrt{3}\,I_p\underline{/30^\circ} \tag{6-265}$$

and

$$I_{cc'} = I_{bc} + I_{ac} = \sqrt{3}\,I_p\underline{/-90^\circ} \tag{6-266}$$

Note that a set of balanced phase currents yields a corresponding set of balanced line currents which differ in magnitude by the factor $\sqrt{3}$. Thus

$$\boxed{I_L = \sqrt{3}\,I_p} \quad \text{for Δ connection} \tag{6-267}$$

where I_L denotes the magnitude of any one of the three line currents.

Power in Three-phase Systems. It is shown in Sec. 6-3 and in Fig. 6-4 that the instantaneous power for a single-phase sinusoidal source varies itself sinusoidally at twice the frequency of the source. By Eq. (6-44) the expression for the instantaneous power is

$$p(t) = \frac{V_m I_m}{2} \cos \theta - \frac{V_m I_m}{2} \cos (2\omega t - \theta) \tag{6-268}$$

In terms of effective values this becomes

$$p(t) = VI \cos \theta - VI \cos (2\omega t - \theta) \tag{6-269}$$

Equation (6-269) can now be applied to each phase of the three-phase system. The only modification needed is introducing the 120° displacement that exists between phases. Accordingly, we may write

$$p_a(t) = V_p I_p \cos \theta - V_p I_p \cos (2\omega t - \theta) \tag{6-270}$$

$$p_b(t) = V_p I_p \cos \theta - V_p I_p \cos (2\omega t - \theta - 120°) \tag{6-271}$$

$$p_c(t) = V_p I_p \cos \theta - V_p I_p \cos (2\omega t - \theta - 240°) \tag{6-272}$$

where phase *a* is taken as the reference phase, V_p and I_p denote the effective values of the phase voltage and phase current, and θ denotes the impedance angle of the balanced three-phase load. As a result of the 120° displacement between p_a, p_b, and p_c, the total instantaneous power in the three-phase system is a constant equal to three times the average power per phase. This situation is depicted in Fig. 6-38. Expressed mathematically the total average three-phase power is

$$P = \frac{1}{T} \int_0^T [p_a(t) + p_b(t) + p_c(t)] \, dt \tag{6-273}$$

or

$$\boxed{P = 3V_p I_p \cos \theta} \tag{6-274}$$

For a Y-connected system the insertion of Eqs. (6-257) and (6-258) permits rewriting Eq. (6-274) as

$$P = 3 \frac{V_L}{\sqrt{3}} I_L \cos \theta = \sqrt{3} \, V_L I_L \cos \theta \tag{6-275}$$

For the Δ connection Eq. (6-274) becomes

$$P = 3V_L \frac{I_L}{\sqrt{3}} \cos \theta = \sqrt{3} \, V_L I_L \cos \theta \tag{6-276}$$

A comparison of the last two expressions thus indicates that the equation for the power in a three-phase system is the same either for a Y or a Δ connection when the power is expressed in terms of *line* quantities. However, it is important to remember that θ denotes the angle of the load impedance per phase and not the angle between V_L and I_L.

Before leaving this discussion on three-phase power, let us emphasize the fact that the total instaneous power for the three-phase system is a constant as illustrated by Fig. 6-38. This stands in sharp contrast to the single-phase case

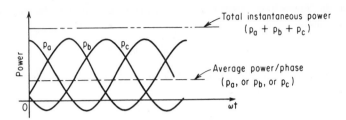

Fig. 6-38. Total instantaneous power in a three-phase system.

where the single-phase power pulsates at twice the line frequency. Herein, then, lies another significant advantage of the three-phase system over the single-phase system. Wherever large loads must be driven mechanically in commercial and industrial applications, the three-phase motor rather than the single-phase motor is used.

Three-phase Load Circuits. Appearing in each leg of the Y connection shown in Fig. 6-39 are identical impedances having the same values of resistance and

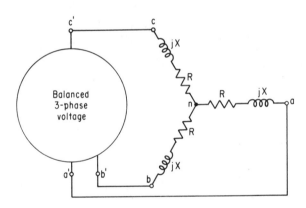

Fig. 6-39. A balanced three-phase system.

inductive reactances per phase. Such an arrangement is called a balanced three-phase load circuit. In the interest of demonstrating how to carry out the computation of line currents and power for such circuits, let us assume that a balanced three-phase voltage is applied to the load having an effective line voltage of 173.2 volts. At this point it is not necessary for us to know whether the generator is Y or Δ connected. Moreover, assume that each phase impedance is properly specified by

$$\bar{Z}_a = \bar{Z}_b = \bar{Z}_c = 7.07 + j7.07 = 10\underline{/45°} \qquad (6\text{-}277)$$

First find the current in line $a'a$. Because of the Y connection this line current is the same as the phase current I_{an}. To find I_{an}, however, we must first determine

$\bar{V}_{an}$. Since the line voltage is given as 173.2 volts, it follows that the magnitude of the phase voltage V_p is

$$V_p = \frac{V_L}{\sqrt{3}} = \frac{173.2}{\sqrt{3}} = 100 \text{ volts} \tag{6-278}$$

We can now select the voltage drop from a to n, i.e., $\bar{V}_{an}$, as our reference phasor and write it as

$$\bar{V}_{an} = V_p \underline{/0^\circ} = 100 \underline{/0^\circ} \tag{6-279}$$

Then because of the balanced nature of the system we can further write

$$\bar{V}_{bn} = V_p \underline{/-120^\circ} = 100 \underline{/-120^\circ} \tag{6-280}$$

and

$$\bar{V}_{cn} = V_p \underline{/120^\circ} = 100 \underline{/120^\circ} \tag{6-281}$$

Each of these line-to-neutral voltages is depicted in Fig. 6-40. The required line currents can then be specified as follows:

$$I_{a'a} = I_{an} = \frac{\bar{V}_{an}}{\bar{Z}_a} = \frac{100 \underline{/0^\circ}}{10 \underline{/45^\circ}} = 10 \underline{/-45^\circ} \tag{6-282}$$

$$I_{b'b} = I_{bn} = \frac{\bar{V}_{bn}}{\bar{Z}_b} = \frac{100 \underline{/-120}}{10 \underline{/45^\circ}} = 10 \underline{/-165^\circ} \tag{6-283}$$

$$I_{c'c} = I_{cn} = \frac{\bar{V}_{cn}}{\bar{Z}_c} = \frac{100 \underline{/120^\circ}}{10 \underline{/45^\circ}} = 10 \underline{/75^\circ} \tag{6-284}$$

These phasor quantities also appear in Fig. 6-40.

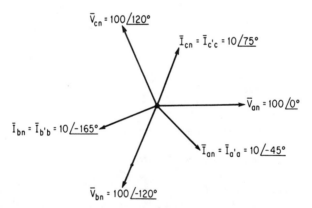

Fig. 6-40. Phasor diagram for the system of Fig. 6-30.

The power consumed in phase a is given by

$$P_a = V_p I_p \cos \theta \tag{6-285}$$

where

$$\theta = \tan^{-1} \frac{X}{R} = \tan^{-1} \frac{7.07}{7.07} = \tan^{-1} 1 = 45^\circ \tag{6-286}$$

Hence

$$P_a = V_p I_p \cos 45° = (100)(10)(0.707) = 707 \text{ watts} \qquad (6\text{-}287)$$

This power can also be computed from

$$P_a = I_{an}^2 R = (10)^2 (7.07) = 707 \text{ watts} \qquad (6\text{-}288)$$

Since the power consumed in each phase is the same, it follows that the total power delivered to the three-phase load by the source is

$$P_T = 3P_a = 2121 \text{ watts} \qquad (6\text{-}289)$$

The total power can also be calculated from a knowledge of the line quantities. Thus, by Eq. (6-275)

$$P_T = \sqrt{3}\ V_L I_L \cos \theta = \sqrt{3}\ (173.2)(10)(0.707) = 2121 \text{ watts} \qquad (6\text{-}290)$$

Note that the θ used in Eq. (6-290) is the very same one which is used in Eq. (6-287).

Let us now rearrange the given phase impedances $\bar{Z}_a$, $\bar{Z}_b$, and $\bar{Z}_c$ into a Δ connection as shown in Fig. 6-41(a). The source voltage is assumed to be the

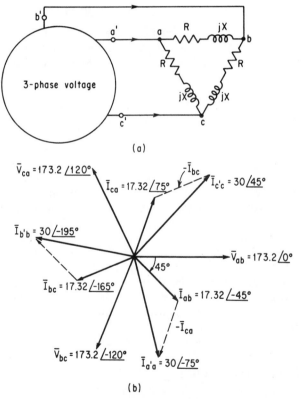

(a)

(b)

Fig. 6-41. A balanced three-phase delta-connected load:
(a) wiring diagram;
(b) phasor diagram.

same. Since in a Δ connection the line and phase voltages are the same, we arbitrarily select $\bar{V}_{ab}$ as the reference voltage and thereby draw the phasor diagram for the voltages as illustrated in Fig. 6-41(b). The expressions for the load phase currents become

$$I_{ab} = \frac{\bar{V}_{ab}}{\bar{Z}_a} = \frac{173.2\underline{/0°}}{10\underline{/45°}} = 17.32\underline{/-45°} \tag{6-291}$$

$$I_{bc} = \frac{\bar{V}_{bc}}{\bar{Z}_b} = \frac{173.2\underline{/-120°}}{10\underline{/45°}} = 17.32\underline{/-165°} \tag{6-292}$$

$$I_{ca} = \frac{\bar{V}_{ca}}{\bar{Z}_c} = \frac{173.2\underline{/120°}}{10\underline{/45°}} = 17.32\underline{/75°} \tag{6-293}$$

The line current $I_{a'a}$ is found upon applying the current law at terminal a. Thus

$$I_{a'a} = I_{ab} + I_{ac} = I_{ab} - I_{ca} = 17.32(0.448 - j1.673)$$
$$= 30\underline{/-75°} \tag{6-294a}$$

Similarly

$$I_{b'b} = 30\underline{/-195} \quad \text{and} \quad I_{c'c} = 30\underline{/45°} \tag{6-294b}$$

A comparison of the line current for a Δ-connected arrangement of three fixed impedances with a Y-connected arrangement shows that the Δ connection draws *three* times as much current. Compare Eq. (6-282) with Eq. (6-293). The total power delivered is also three times greater.

6-12   Fourier Series

Much is said in the preceding sections about the response of electric circuits to a periodic forcing function having a sinusoidal variation. This attention is merited, since the bulk of the electric power in the world is generated, transmitted, and consumed as a sinusoidally varying quantity at constant frequency. At this point, however, it is appropriate to ask: How can the response of electric circuits be obtained when the forcing function is not sinusoidally varying but is periodic? There are numerous applications in engineering where nonsinusoidal forcing functions such as rectangular, triangular, trapezoidal, or other waveshapes are used. Moreover, in communication engineering situations exist which involve the summation of sinusoidal signals of many frequencies, the resultant waveshape of which is extremely nonsinusoidal albeit periodic. This condition is illustrated in Fig. 6-42, which shows how it is possible to represent three entirely different waveforms by means of two sinusoidal components. Note that in each case one of the components is assumed to have a frequency three times greater that the other. For the condition depicted in Fig. 6-42(a) the resultant waveform may be expressed mathematically as

$$f(t) = f_1(t) + f_3(t) = F_1 \sin \omega t + F_3 \sin 3\omega t \tag{6-295}$$

where $f_1(t)$ is the *fundamental* sinusoidal component of frequency ω, and $f_3(t)$ is the *third harmonic* component of frequency 3ω. Both the fundamental and

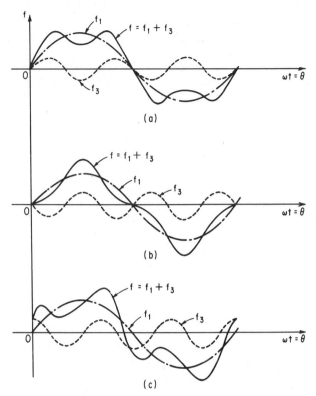

Fig. 6-42. (a) Addition of fundamental sine term plus positive third harmonic;

(b) addition of fundamental sine term plus negative third harmonic;

(c) addition of fundamental sine term and a third harmonic cosine term.

the third harmonic terms are called *periodic functions* because they satisfy the condition that

$$f(t) = f(t + T) \tag{6-296}$$

where T denotes the period of the fundamental wave. Observe that the period of the third harmonic is one-third that of the fundamental and that the resultant wave has the same period as the fundamental.

The equation which describes the waveform of Fig. 6-42(b) is

$$f(t) = f_1(t) - f_3(t) = F_1 \sin \omega t - F_3 \sin 3\omega t \tag{6-297}$$

By merely reversing the third harmonic term the resultant waveshape changes from one which is essentially flat-topped (Fig. 6-42a) to one which is peaked.

The resultant waveshape of Fig. 6-42(c) is obtained by shifting the third harmonic term of Fig. 6-42(a) 90° in the lead direction. The corresponding mathematical formulation is

$$f(t) = F_1 \sin \omega t + F_3 \sin \left(\omega t + \frac{\pi}{2} \right) \tag{6-298}$$

A little thought should make it apparent that even with only two sinusoidal components manipulated in the manner just described it is possible to represent a great number of different waveshapes.

A general formulation of this situation is provided by the Fourier series theorem,† which is stated here without proof. *A periodic function f(t) with period T can be represented by the sum of a constant term, a fundamental of period T, and its harmonics.* Expressed mathematically we have

$$\boxed{\begin{aligned} f(\theta) = \left(\frac{a_0}{2} \right) &+ a_1 \cos \theta + a_2 \cos 2\theta + a_3 \cos 3\theta + \cdots \\ &+ b_1 \sin \theta + b_2 \sin 2\theta + b_3 \sin 3\theta + \cdots \end{aligned}} \tag{6-299}$$

where

$$\theta \equiv \omega t$$

and

$$\omega = \frac{2\pi}{T}$$

where T is the period of the function $f(\theta)$. Thus Eq. (6-299) states that any periodic function of whatever shape can be replaced by the sum of a constant term $(a_0/2)$ and sine and/or cosine terms involving odd and/or even harmonics. The a- and the b- coefficients are evaluated from the shape of the $f(\theta)$ function. General expressions can readily be derived for evaluating these coefficients, as shown below. In this connection it is useful to keep in mind the following identities.

$$\sin^2 \theta = \tfrac{1}{2} - \tfrac{1}{2} \cos 2\theta \tag{6-300a}$$

$$\cos^2 \theta = \tfrac{1}{2} + \tfrac{1}{2} \cos 2\theta \tag{6-300b}$$

$$\int_0^{2\pi} \cos m\theta \sin n\theta \, d\theta = 0 \tag{6-301}$$

$$\int_0^{2\pi} \cos m\theta \cos n\theta \, d\theta = 0 \quad \text{for} \quad m \neq n \tag{6-302}$$

$$\int_0^{2\pi} \sin m\theta \sin n\theta \, d\theta = 0 \quad \text{for} \quad m \neq n \tag{6-303}$$

Consider first the evaluation of the constant term of Eq. (6-299). The waveform of $f(\theta)$ is assumed to be known for all values of θ and to be periodic. To evaluate a_0 it is necessary merely to integrate both sides of Eq. (6-299) over one period. Thus

$$\int_0^{2\pi} f(\theta) \, d\theta = \int_0^{2\pi} \frac{a_0}{2} \, d\theta + \int_0^{2\pi} a_1 \cos \theta \, d\theta + \cdots + \int_0^{2\pi} b_1 \sin \theta \, d\theta + \cdots \tag{6-304}$$

† Published in 1822.

Since the average value over one period of sine and cosine functions is zero, all terms on the right side of the last equation are zero with the exception of the first one. Hence we have

$$\int_0^{2\pi} f(\theta)\, d\theta = \frac{a_0}{2}(2\pi) \tag{6-305}$$

or

$$\boxed{a_0 = \frac{1}{\pi}\int_0^{2\pi} f(\theta)\, d\theta = \frac{2}{T}\int_0^{T} f(t)\, dt} \tag{6-306}$$

The strategy to be employed in evaluating a_1 is to manipulate Eq. (6-299) in such a way as to cause all terms to drop out with the exception of a_1. A little thought reveals that this is readily achieved by multiplying each term of Eq. (6-299) by $\cos\theta$ and then integrating over one period. Thus

$$\int_0^{2\pi} f(\theta)\cos\theta\, d\theta = \int_0^{2\pi}\frac{a_0}{2}\cos\theta\, d\theta + \int_0^{2\pi} a_1\cos^2\theta\, d\theta + \int_0^{2\pi} b_1\sin\theta\cos\theta$$
$$+ \int_0^{2\pi} a_2\cos\theta\cos 2\theta\, d\theta + \int_0^{2\pi} b_2\cos\theta\sin 2\theta\, d\theta + \cdots \tag{6-307}$$

All terms on the right side drop out but the one involving $\cos^2\theta$. Hence

$$\int_0^{2\pi} f(\theta)\cos\theta\, d\theta = \int_0^{2\pi} a_1(\tfrac{1}{2} + \tfrac{1}{2}\cos 2\theta)\, d\theta \tag{6-308}$$
$$= a_1\pi$$

Therefore

$$a_1 = \frac{1}{\pi}\int_0^{2\pi} f(\theta)\cos\theta\, d\theta = \frac{2}{T}\int_0^{T} f(t)\cos\omega t\, dt \tag{6-309}$$

To evaluate b_1 a similar procedure is used, but now since we are dealing with the coefficient of the sine term in the Fourier series it is necessary to multiply each term of Eq. (6-299) by $\sin\theta$ and then to integrate over T. Thus

$$\int_0^{2\pi} f(\theta)\sin\theta\, d\theta = \int_0^{2\pi}\frac{a_0}{2}\sin\theta\, d\theta + \int_0^{2\pi} a_1\sin\theta\cos\theta\, d\theta + \int_0^{2\pi} b_1\sin^2\theta\, d\theta$$
$$+ \int_0^{2\pi} a_2\sin\theta\cos 2\theta\, d\theta + \int_0^{2\pi} b_2\sin\theta\sin 2\theta\, d\theta + \cdots$$
$$\tag{6-310}$$

Performing the integration leads to

$$\int_0^{2\pi} f(\theta)\sin\theta\, d\theta = b_1\pi \tag{6-311}$$

or

$$b_1 = \frac{1}{\pi}\int_0^{2\pi} f(\theta)\sin\theta\, d\theta = \frac{2}{T}\int_0^{T} f(t)\sin\omega t\, dt \tag{6-312}$$

By following this procedure for the coefficients of each of the higher harmonic terms it can be shown that the general expressions for the a- and

b-coefficients for the nth harmonic are given by

$$a_n = \frac{1}{\pi} \int_0^{2\pi} f(\theta) \cos n\theta \, d\theta = \frac{2}{T} \int_0^T f(t) \cos n\omega t \, dt \qquad (6\text{-}313)$$

$$n = 0, 1, 2, 3, \cdots$$

and

$$b_n = \frac{1}{\pi} \int_0^{2\pi} f(\theta) \sin n\theta \, d\theta = \frac{2}{T} \int_0^T f(t) \sin n\omega t \, dt \qquad (6\text{-}314)$$

$$n = 1, 2, 3, \cdots$$

It is interesting to note that by writing the constant term in Eq. (6-299) as $a_0/2$ it is permissible to use Eq. (6-313) to evaluate a_0 as well as the coefficients of the cosine components in the wave.

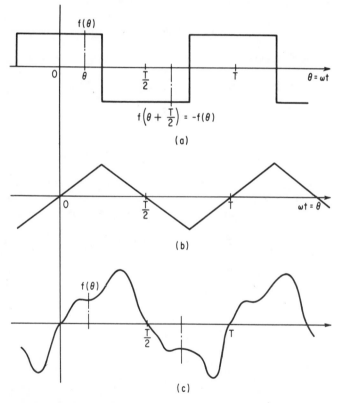

Fig. 6-43. Typical periodic wave forms illustrating symmetry properties:
(a) rectangular wave;
(b) triangular wave;
(c) arbitrary wave.

Symmetry Properties of Periodic Waves. Before proceeding with the solution of a problem illustrating the use of Eqs. (6-313) and (6-314), let us turn attention briefly to the simplification of these equations brought about by the presence of certain symmetry conditions which the $f(\theta)$ function may possess.

A periodic function is said to have *half-wave* symmetry when it satisfies the relation

$$f\left(\theta + \frac{T}{2}\right) = -f(\theta) \tag{6-315}$$

Each of the waveforms shown in Fig. 6-43 exhibits this property. Moreover, all periodic functions which obey Eq. (6-315) contain *no even harmonics*.

Sometimes periodic functions are encountered which contain either only sine functions or only cosine functions. For a periodic function to contain only sine terms it must satisfy the relation

$$f(-\theta) = -f(\theta) \tag{6-316}$$

The waveform depicted in Fig. 6-43(b) is an example of such a function. Because this waveform satisfies both Eqs. (6-315) and (6-316), its Fourier series representation includes only sine terms and in particular the fundamental sine function and the odd harmonics. For a periodic function to contain only cosine terms it must satisfy the equation

$$f(-\theta) = f(\theta) \tag{6-317}$$

The waveform of Fig. 6-43(a) meets this condition. Such functions are sometimes referred to as *even* functions because the value of the function is the same for a positive or negative value of the argument. Note that the waveshape of Fig. 6-43(c) satisfies neither Eq. (6-315) nor Eq. (6-316), hence it contains cosine as well as sine terms. However, it does not contain even harmonics.

EXAMPLE 6-12 Find the Fourier series representation of the waveform depicted in Fig. 6-44.

solution: A check of the symmetry conditions indicates that none is satisfied. Hence we can expect both even and odd harmonics to exist as well as sine and cosine terms. The average value of the function is found by evaluating Eq. (6-313) for $n = 0$. The periodic function in the first period may be expressed as

$$v(\theta) = \frac{V_m}{\pi}\theta, \quad 0 \leq \theta \leq \pi$$
$$v(\theta) = 0, \quad\quad \pi \leq \theta \leq 2\pi \tag{6-318}$$

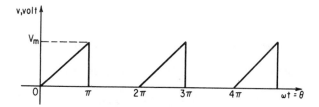

Fig. 6-44. Wave form for Example 6-12.

Accordingly we get

$$a_0 = \frac{1}{\pi} \int_0^{2\pi} f(\theta)\, d\theta = \frac{1}{\pi} \int_0^{\pi} \frac{V_m}{\pi} \theta\, d\theta = \frac{V_m}{2\pi^2} [\theta^2]_0^{\pi} = \frac{V_m}{2} \tag{6-319}$$

$$\therefore \quad \left(\frac{a_0}{2}\right) = \frac{V_m}{4} = \text{average value} \tag{6-320}$$

This average value of the function can also be found in this case by inspection.

The general expression for the coefficient a_n of the cosine terms is found as follows

$$a_n = \frac{1}{\pi} \int_0^{\pi} \theta \cos n\theta\, d\theta = \frac{V_m}{\pi^2}\left[-\theta\frac{\sin n\theta}{n} + \frac{\cos n\theta}{n^2}\right]_0^{\pi} = \frac{V_m}{\pi^2 n^2}(\cos n\pi - 1) \tag{6-321}$$

The limits of the integral for a_n are taken as zero to π rather than as zero and 2π because the contribution of the function from π to 2π is obviously zero. Inserting consecutive interger values for n yields

$$a_1 = \frac{V_m}{\pi^2}(-1 - 1) = -\frac{2V_m}{\pi^2}$$

$$a_2 = \frac{V}{4\pi^2}(1 - 1) = 0 = a_4 = a_6 = a_8 = \cdots$$

$$a_3 = \frac{V_m}{9\pi^2}(-1 - 1) = -\frac{2V_m}{9\pi^2}$$

$$a_5 = -\frac{2V_m}{25\pi^2}$$

$$\begin{array}{cc} \cdot & \cdot \\ \cdot & \cdot \\ \cdot & \cdot \end{array}$$

The evaluation of the coefficients of the sine terms follows in a similar fashion. Thus

$$b_n = \frac{1}{\pi} \int_0^{\pi} \frac{V_m}{\pi} \theta \sin n\theta\, d\theta = \frac{V_m}{\pi^2}\left[\frac{-\theta \cos n\theta}{n} + \frac{\sin n\theta}{n^2}\right]_0^{\pi} = -\frac{V_m}{\pi}\frac{\cos n\pi}{n} \tag{6-322}$$

Therefore

$$b_1 = -\frac{V_m}{\pi}(-1) = \frac{V_m}{\pi}$$

$$b_2 = -\frac{V_m}{2\pi}$$

$$b_3 = +\frac{V_m}{3\pi}$$

$$b_4 = -\frac{V_m}{4\pi}$$

$$\begin{array}{cc} \cdot & \cdot \\ \cdot & \cdot \\ \cdot & \cdot \end{array}$$

Hence the Fourier series representation of the waveform is

$$v(\theta) = v(\omega t) = \frac{V_m}{4} + \frac{V_m}{\pi}\left[\sin\theta - \frac{1}{2}\sin 2\theta + \frac{1}{3}\sin 3\theta - (-1)^n\frac{1}{n}\sin n\theta\right]$$

$$\qquad - \frac{2V_m}{\pi^2}\left[\cos\theta + \frac{1}{9}\cos 3\theta + \frac{1}{25}\cos 25\theta + \frac{1}{k^2}\cos k\theta\right] \tag{6-323}$$

where $n = 1,2,3,4,\cdots$, all integers
  $k = 1,3,5,7,\cdots$, odd integers

Equation (6-323) shows that a nonsinusoidal periodic function can be expressed entirely in terms of sines and cosines. It should be apparent then that if this waveform is used as the forcing function in a series *R-L* circuit, the corresponding current response can be found by a systematic application of the sinusoidal steady-state theory developed in the preceding sections. The current response at each frequency is found in the accustomed manner for each sine and cosine term as well as the constant term in $v(\theta)$ and then summed. The waveform of the current response in such a case, however, will be different than that of $v(\theta)$ because at each frequency the amount of phase lag caused by the inductance is different.

The Fourier series representation of a periodic time function has one other useful property which is worth noting. From the treatment so far it should be clear that an infinite number of terms in the series is required to get an exact representation of the original function. If we use only the first N terms of the series, then instead of an exact representation we have an approximation of the original time function yielding an error which can be expressed as

$$e(\theta) = f(\theta) - f_N(\theta) \tag{6-324}$$

where e denotes the error, $f(\theta)$ the original periodic function, and $f_N(\theta)$ the approximate function using the first N terms of the series. It is possible to identify an average value for this error by writing

$$E_{av} = \frac{1}{2\pi} \int_0^{2\pi} e(\theta)\, d\theta \tag{6-325}$$

However, this is not a very useful criterion of how well $f_N(\theta)$ represents $f(\theta)$ because large positive and negative errors may exist which could cancel each other, thereby giving a misleading result. A way of avoiding this difficulty is to use as a criterion the average value of the squared error. This is called the mean squared error (*MSE*). Thus

$$MSE = \frac{1}{2\pi} \int_0^{2\pi} e^2(\theta)\, d\theta \tag{6-326}$$

Here is the interesting aspect of this result: it can be shown that by using the coefficients of the Fourier series (i.e., a_n and b_n) for any given N the *MSE* is minimized. In other words any other choice of coefficients results in a larger mean squared error.

PROBLEMS

6-1   Find the average and effective values of each waveshape depicted in Fig. P6-1.

6-2   For each waveshape shown in Fig. P6-2 find the average and effective values.

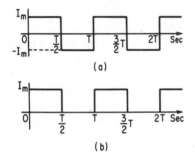

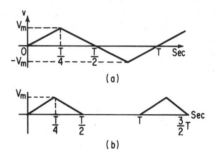

Figure P6-1   **Figure P6-2**

6-3   In each of the sketches of Fig. P6-3 the curves are sine waves or parts thereof. Find the average and effective values for each waveshape.

6-4   Compute the average and effective values for each waveshape shown in Fig. P6-4.

6-5   A sinusoidal source of $e(t) = 170 \sin 377t$ is applied to an R-L circuit. It is found that the circuit absorbs 720 watts when an effective current of 12 amperes flows.
   (a) Find the power factor of the circuit.
   (b) Compute the value of the impedance.
   (c) Calculate the inductance of the circuit in henrys.

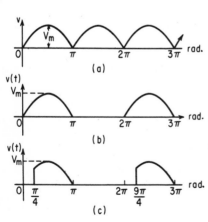

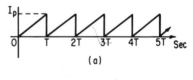

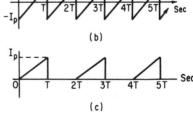

Figure P6-3   **Figure P6-4**

6-6 A voltage source of $e(t) = 141 \sin 377t$ is applied to two parallel branches. The time expression for the current in the first branch is

$$i_1(t) = 7.07 \sin \left(\omega t - \frac{\pi}{3}\right)$$

In the second branch it is

$$i_2(t) = 10 \sin \left(\omega t + \frac{\pi}{6}\right)$$

Compute the total power supplied by the source.

6-7 Refer to Prob. 6-6 and answer the following questions:
(a) Find the resultant current delivered by the source expressed in effective amperes.
(b) Write the expression for the instantaneous value of the resultant current.
(c) Compute the apparent power of the complete circuit.
(d) Find the resultant power factor of the circuit.
(e) Draw the phasor diagram showing the voltage and all current phasors.

6-8 A voltage of

$$e(t) = 141 \sin \left(377t + \frac{\pi}{3}\right)$$

is applied to a 20-ohm resistor.
(a) Find the effective value of the current in amperes.
(b) Write the expression for the instantaneous current.
(c) Compute the power supplied by the source.

6-9 A sinusoidal voltage

$$e(t) = 170 \sin \left(377t + \frac{\pi}{3}\right)$$

is applied to a 0.1-henry inductor.
(a) Find the effective value of the steady-state current in amperes.
(b) Write the expression for the instantaneous current.
(c) Draw a properly labeled phasor diagram. Use the rms value of the voltage phasor as reference.

6-10 A sinusoidal voltage

$$e(t) = 170 \sin \left(377t + \frac{\pi}{3}\right)$$

is applied to a 100-μf capacitor.
(a) Find the effective value of the steady-state current in amperes.
(b) Write the expression for the instantaneous current.
(c) Draw the phasor diagram. Use the rms value of the voltage phasor as reference.

6-11 A circuit is composed of a resistance of 9 ohms and a series-connected inductive reactance of 12 ohms. When a voltage $e(t)$ is applied to the circuit the resulting steady-state current is found to be $i(t) = 28.3 \sin 377t$.
(a) What is the value of the complex impedance?
(b) Determine the time expression for the applied voltage.
(c) Find the value of the inducatnce in henrys.

6-12 When a sinusoidal voltage of 120 volts rms is applied to a series R-L circuit, it is found that there occurs a power dissipation of 1200 watts and a current flow given by $i(t) = 28.3 \sin (377t - \theta)$.

(a) Find the the circuit resistance in ohms.

(b) Find the circuit inductance in henrys.

6-13   A sinusoidal voltage $e(t) = 141 \sin \omega t$ is applied to a series R-L circuit. In the steady state the effective value of the voltage measured across the terminals of the inductor is 60 volts. What is the voltage appearing across the resistor terminals?

6-14   In an R-C series circuit to which is applied a voltage of $170 \sin \omega t$ it is found that a steady-state current flows which leads the voltage by 30°. Find the effective voltage drops across the resistive and reactive elements.

6-15   A circuit is composed of a resistance of 6 ohms and a series capacitive reactance of 8 ohms. A voltage $e(t) = 141 \sin 377t$ is applied to the circuit.

(a) Find the complex impedance.

(b) Determine the effective and instantaneous values of the current.

(c) Compute the power delivered to the circuit.

(d) Find the value of the capacitance in farads.

6-16   A sinusoidal voltage is applied to three parallel branches yielding branch currents as follows:

$$i_1 = 14.14 \sin (\omega t - 45°), \quad i_2 = 28.3 \cos (\omega t - 60°), \quad i_3 = 7.07 \sin (\omega t + 60°)$$

(a) Find the complete time expression for the source current.

(b) Draw the phasor diagram in terms of effective values. Use the voltage as reference.

6-17   A sinusoidal voltage $V_m \sin \omega t$ is applied to three parallel branches. Two of the branch currents are given by

$$i_1 = 14.14 \sin (\omega t - 37°)$$

$$i_2 = 28.28 \cos (\omega t - 143°)$$

The source current is found to be

$$i = 63.8 \sin (\omega t + 12.8°)$$

(a) Find the effective value of the current in the thrid branch.

(b) Write the complete time expression for the instantaneous value of the current in part (a).

(c) Draw the phasor diagram showing the source current and the three branch currents. Use voltage as the reference phasor.

6-18   A voltage wave $e(t) = 170 \sin 120t$ produces a net current of $i(t) = 14.14 \sin 120t + 7.07 \cos (120t + 30°)$.

(a) Express the effective value of the current as a single phasor quantity.

(b) Draw the phasor diagram of part (a). Show the components of the current as well as the resultant.

(c) Find the power delivered by the voltage source of part (a).

6-19   A voltage wave has the variation shown in Fig. P6-19.

(a) Find the average and the effective values of the voltage.

(b) If the voltage of part (a) is applied to a 10-ohm resistance, find the dissipated power in watts.

6-20   A voltage of $e(t) = 150 \sin 1000t$ is applied across a series R-L-C circuit where $R = 40$ ohms, $L = 0.13$ henry, and $C = 10 \ \mu f$.

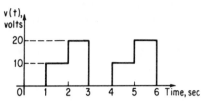

Figure P6-19

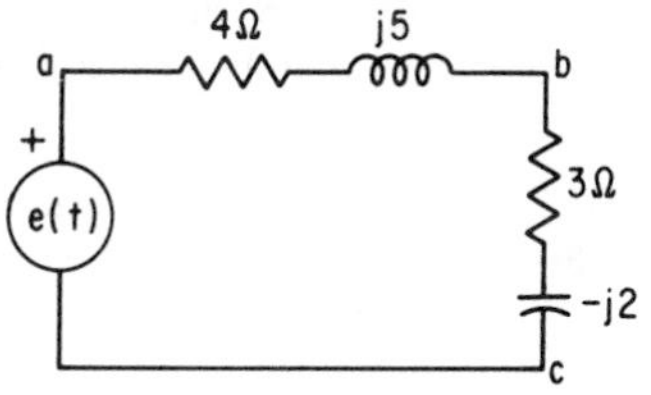

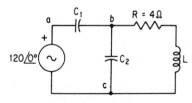

Figure P6-22

(a) Compute the rms value of the steady-state current.

(b) Find the expression for the instantaneous voltage appearing across the capacitor terminals.

(c) Determine the expression for the instantaneous voltage appearing across the inductor terminals.

(d) Compare the rms value of the voltages appearing across L and C with that of the applied voltage and comment.

(e) Draw the complete phasor diagram for the solution of this problem showing voltage components.

6-21 In the circuit of Prob. 6-20 determine (a) the power supplied by the source, (b) the reactive power supplied by the source, (c) the reactive power of the capacitor, (d) the reactive power of the inductor, (e) the power factor of the circuit.

6-22 In the circuit shown in Fig. P6-22 the applied forcing function is given by $e(t) = 141 \sin \omega t$.

(a) Express the voltage drops across terminals ab and bc in terms of phasor notation.

(b) Draw a phasor diagram showing $\bar{V}_{ab} + \bar{V}_{bc}$ and using current as the reference phasor.

6-23 An rms voltage of $100\underline{/0°}$ is applied to the series combination of $\bar{Z}_1$ and $\bar{Z}_2$ where $\bar{Z}_1 = 20\underline{/30°}$. The effective voltage drop across $\bar{Z}_1$ is known to be $40\underline{/-30°}$ volts. Find the reactive component of $\bar{Z}_2$.

6-24 An effective voltage of 100 volts is applied to the parallel combination of two impedances $\bar{Z}_1 = R_1 + jX_1$ and $\bar{Z}_2 = R_2 + jX_2$. Assuming that $R_1 = 3$ ohms and $R_2 = 4$ ohms and that the magnitudes of the two branch currents are the same, determine the values of X_1, X_2 and the resultant source current.

6-25 In the circuit shown in Fig. P6-25 the reactance of capacitor C_1 is 4 ohms, the reactance of C_2 is 8 ohms, and the reactance of L is 8 ohms. A sinusoidal voltage having an effective value of 120 volts is applied to the circuit.

(a) Find the effective value of the current delivered by the source.

Figure P6-25

(b) Write the expression for the instantaneous value of the current found in part (a).
(c) Draw a carefully labeled phasor diagram showing the source voltage, the source current, and voltages $\bar{V}_{ab}$ and $\bar{V}_{bc}$.

6-26   In the network configuration shown in Fig. P6-26 find the current which flows through the $\bar{Z}_3$ branch. Use the nodal method.

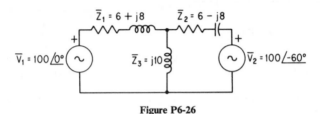

Figure P6-26

6-27   Solve Prob. 6-26 by using Thevenin's equivalent circuit.

6-28   In the circuit shown in Fig. P6-28 find the Thevenin and Norton equivalent circuits looking in at terminals *a-b*.

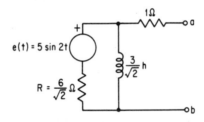

Figure P6-28

6-29   A circuit has the configuration depicted in Fig. P6-29.
(a) Find the equivalent impedance appearing to the right of points *ab*.
(b) Determine the value of the reactance X which puts the source current in phase with the source voltage.
(c) Should the reactance X of part (b) be inductive or capacitive? Find the required value of L or C.
(d) Compute the effective value of the source current for the condition described in part (b).

6-30   A sinusoidal forcing function having a frequency of 2 rad/sec is applied to the circuit of Fig. P6-30.
(a) For the parameter values specified find the value of the impedance appearing at the input terminals.
(b) Identify the type, location, and value of the circuit element which must be used with this circuit in order that the impedance presented to the source be entirely resistive.
(c) For the condition of part (b) find the phasor expression for the voltage across the 0.1-f capacitance when the applied voltage is $e(t) = 14.14 \sin 2t$.

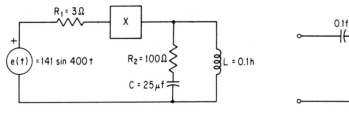

Figure P6-29   Figure P6-30

6-31   The inductive reactance in series with $\bar{Z}$ in the circuit of Fig. P6-31 has a value of 25 ohms. A voltmeter placed across $\bar{Z}$ registers a reading of 179 volts when 4 amperes flow through $\bar{Z}$. The power dissipated in the circuit is known to be 320 watts.

(a) Find the power factor of the circuit.

(b) What is the circuit resistance?

(c) Is it possible for the magnitude of the voltage across $\bar{Z}$ to be greater than the source voltage. Assume that $\bar{Z}$ contains an inductive reactance. Explain fully.

(d) Compute the reactive component of $\bar{Z}$.

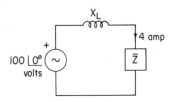

Figure P6-31

6-32   Refer to the circuit of Fig. P6-32.

(a) Find the equivalent reactance for the parallel branch.

(b) Find the rms line current.

(c) Determine the rms voltage across the parallel branch. Comment on this result.

(d) What is the power dissipated?

(e) What is the power factor?

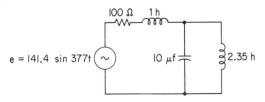

Figure P6-32

6-33   Refer to the circuit depicted in Fig. P6-33.

(a) What is the power factor?

(b) Find the power dissipated.

(c) Find the value of capacitance which when placed across the load will make the overall power factor unity.

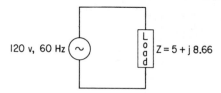

Figure P6-33

6-34 For the circuit shown in Fig. P6-34,
(a) Find I_1, I_2, I_3, I_4, and I_5.
(b) Compute $\bar{V}_{bc}$ and $\bar{V}_{cd}$.
(c) Draw the phasor diagram showing all currents and voltages.
(d) Compute the power supplied by the source.
(e) Find the line power factor.

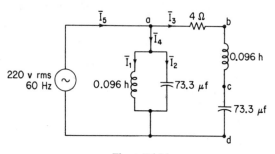

Figure P6-34

6-35 The circuit shown in Fig. P6-35 is in the sinusoidal steady state.
(a) Determine the phasor voltage $\bar{V}_{ab}$.
(b) To what value of radian frequency must the source voltage be set in order that $v_{cb}(t)$ and $i(t)$ be in phase?

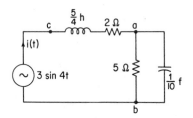

Figure P6-35

6-36 In the circuit of Fig. P6-36 the following relationships hold:

$$v_{ab}(t) = 4\sqrt{2}\ \sin{(\omega t + 135°)}$$
$$v_{bc}(t) = -4\sqrt{3}\ \sin{(\omega t + 60°)}$$
$$v_{cd}(t) = 4\cos{(\omega t - 150°)}$$

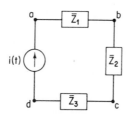

Figure P6-36

(a) Draw a clearly labeled phasor diagram for the voltages $v_{ab}(t)$, $v_{bc}(t)$, and $v_{cd}(t)$.
(b) Find the phasor voltage $\bar{V}_{ad}$.
(c) Write the expression for $v_{ad}(t)$.
(d) If $i(t) = 2 \sin(\omega t + 165°)$, find the average power delivered by the current source.

6-37 A series R-L-C circuit has the following parameter values: $R = 10$ ohms, $L = 0.01$ h, $C = 100$ μf.
(a) Compute the resonant frequency in radians per second.
(b) Calculate the quality factor of the circuit.
(c) What is the value of the bandwidth?
(d) Compute the lower and upper frequency points of the bandwidth.
(e) If a signal of $e(t) = 1 \sin 1000t$ is applied to this series R-L-C circuit, calculate the maximum value of the voltage appearing across the capacitor terminals.

6-38 Repeat Prob. 6-37 for circuit elements having the following values: $R = 10$ ohms, $L = 1$ henry, and $C = 1$ μf.

6-39 A current source is applied to the parallel arrangement of R, L, and C where $R = 10$ ohms, $L = 1$ henry, and $C = 1$ μf.
(a) Compute the resonant frequency.
(b) Find the quality factor.
(c) Calculate the value of the bandwidth.
(d) Compute the lower and upper frequency points of the bandwidth.
(e) If a signal of $i(t) = 1 \sin 1000t$ is applied to this parallel R-L-C circuit, calculate the maximum value of voltage appearing across the capacitor terminals.
(f) What is the capacitor current in part (e)?

6-40 In the circuit of Prob. 6-38 compute the value of the current which flows when the input signal is (a) $e(t) = 1 \sin 300t$ volts, (b) $e(t) = 1 \sin 1800t$ volts.

6-41 In the circuit of Prob. 6-39 compute the value of the voltage appearing across the parallel elements when the input signal is (a) $i(t) = 1 \sin 300t$ amp, (b) $i(t) = 1 \sin 1800t$ amp.

6-42 A voltage of $e(t) = 10 \sin \omega t$ is applied to a series R-L-C circuit. At the resonant frequency of the circuit the maximum voltage across the capacitor is found to be 500 volts. Moreover, the bandwidth is known to be 400 rad/sec and the impedance at resonance is 100 ohms.
(a) Find the resonant frequency.
(b) Compute the upper and lower limits of the bandwidth.
(c) Determine the value of L and C for this circuit.

6-43   The AM radio dial spreads over a frequency range of 570 kilohertz to 1560 kilohertz.

(a) Calculate the range of capacitance that is needed in series with a 20-microhenry inductance to tune over the entire frequency band.

(b) If the quality factor for a radio station operating at 570 kilohertz is 100, what is the bandwidth of the tuning circuit?

(c) What is the resistance of the tuning circuit?

(d) Determine the value of the quality factor at the upper end of the radio band.

6-44   A 220-volt, 3-phase voltage is applied to a balanced delta-connected three-phase load in the manner illustrated in Fig. P6-44. The rms value of the phase current measured between points a and b is $I_{ab} = 10/\underline{-30°}$.

(a) Find the magnitude and phase of the line current I. Draw the phasor diagram showing clearly the line voltages, phase currents, and line currents.

(b) Compute the total power received by the three-phase load.

(c) Find the value of the resistive portion of the phase impedance.

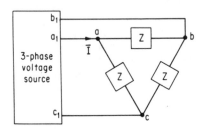

Figure P6-44

6-45   The delta-connected load of Fig. P6-44 consists of phase impedances each equal to $15 + j20$.

(a) Find the phasor current in each line.

(b) What is the power consumed per phase?

(c) What is the phasor sum of the three line currents? Why does it have this value?

6-46   A three-phase, 208-volt generator supplies a total of 1800 watts at a line current of 10 amps when three identical impedances are arranged in a wye connection across the line terminals of the generator. Compute the resistive and reactive components of each phase impedance.

6-47   A 208-volt, three-phase generator supplies power to both a delta- and a wye-connected load in the manner shown in Fig. P6-47. All the phase impedances are identical and specifically equal to $5 + j8.66$. Compute the total generator current which flows in line a.

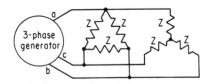

Figure P6-47

6-48 A balanced three-phase wye-connected load has an impedance of $4\underline{/60°}$ ohms from line to neutral. Moreover, from line a to neutral n the voltage in $\bar{V}_{an} = 20\underline{/30°}$.

(a) What is the current in phases b and c?

(b) What is the voltage from line b to neutral?

(c) What is the phasor expression for the voltage from line a to line c, i.e., $\bar{V}_{ac}$?

6-49 Determine the Fourier series for the waveforms shown in Fig. P6-1.

6-50 Find the Fourier series for the waveform of Fig. P6-2(a).

6-51 Determine the Fourier series of the rectified waveform shown in Fig. P6-3(a).

6-52 Find the Fourier series of the sawtooth waveform of Fig. P6-4(a).

6-53 A voltage wave has the form depicted in Fig. P6-53. Obtain the Fourier series representation of this periodic function.

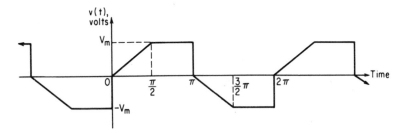

Figure P6-53

II

ELECTRONICS

7

electron control
devices—
semiconductor types

In Chapter 1 it is pointed out that electricity involves the flow of electrons no matter whether this flow takes place in a vacuum tube, in a semiconductor, or in a network consisting of circuit elements in whatever combination. In a broad sense, therefore, one may describe this electron flow as electronics. However, in common usage this term has a more restrictive meaning.

In accordance with the definition of the term published by the Institute of Electrical and Electronic Engineers "*electronics* is that branch of science and engineering which deals with electron devices and their utilization." Moreover, the I.E.E.E. goes on to define an *electron device* as "a device in which conduction by electrons takes place in a vacuum, gas, or semiconductor." Taken together, these two statements identify electronics not only in terms of the theory and design of electron devices such as the vacuum tube and the transistor, but also in terms of the vast variety of the circuitry built around them to do countless tasks in such fields as communication systems, computers, control systems, instrumentation, and telemetry.

Activity in electronics has been increasing at a phenomenal rate ever since World War II. Needless to say, national defense needs and space projects are important factors in this growth. An indication of the extent of this activity may be had by referring to some of the fields of interest of the organized professional technical groups of the I.E.E.E. as listed below:

Audio. The technology of communication at audio frequencies including acoustics and recordings as well as the reproductions from recordings.

Broadcasting. The design and use of broadcast equipment.

Nuclear Science. The application of electronic techniques and devices to the nuclear field.

Broadcast and Television Receivers. The design and manufacture of broadcast and television receivers and components, and related activities.

Space Electronics and Telemetry. The measurement and recording of data from remote points by electromagnetic media.

Navigational Electronics. The application of electronics to the operation and traffic control of aircraft and to the navigation of all craft.

Industrial Electronics and Control Instrumentation. Electronics pertaining to applied control, treatment, and measurement specifically directed to industrial processes.

Electronic Computers. Design and operation of electronic computers.

Biomedical Electronics. The use of electronic theory and techniques in problems of medicine and biology.

Communication Systems. Radio and wire telephony, telegraph and facsimile in marine, aeronautical, radio-relay, coaxial cable, and fixed-station services.

Automatic Control. The theory and application of automatic control techniques including feedback control.

Geoscience Electronics. Research and development in electronic instrumentation for geophysics and geochemistry especially regarding gravity, measurements, seismic measurements, space exploration, meteorology, and oceanography.

Aerospace. Theory and application of electronics and electricity for aerospace instrumentation and vehicle electric systems including electric propulsion and system application of electrically operated subsystems.

Although the foregoing list is only partial, the widespread activity in electronics is certainly apparent. This list is included to furnish the reader with a better perspective of what it is that electronics entails. The objective of Part II of this book is surely not to provide the detailed background needed to make useful contributions in any one of these fields of interest. Rather the goal is to furnish a general background, directing attention especially to a study of the theory and characteristics of the more important electron devices and their utilization in standard electronic circuitry. In this chapter and the next one therefore, a study is made of the vacuum-tube and solid-state electron devices, emphasizing their external characteristic and control capabilities. The next three chapters offer a detailed study of the operation, analysis, and performance description of the electronic circuitry commonly found in the various fields of interest in electronics.

The fundamental laws described in Chap. 1 provide the foundation on which the study of electric-circuit theory, electromagnetics, and electromechanical energy conversion are based. However, in the study of electronics, additional fundamental relationships are necessary. In high-vacuum electronics there are two important equations worthy of particular note. One is the Richardson-Dushman equation, which describes the thermionic electron emission capabilities of various metals; the other is the Langmuir-Child law, which identifies the space-charge limited current existing between two electrodes placed within a high-vacuum environment and across which there appears a suitable potential difference. These equations are significant because they depict the characteristics of the devices (e.g. diodes, triodes, tetrodes, and pentodes) used in vacuum-tube electronics and its associated circuitry. These matters are treated in Chap. 8. A similar situation prevails in semiconductor electronics except that different basic relationships are used in place of those which apply for the high-vacuum devices. Attention is directed first to semiconductor electronics in this chapter.

7-1   The Boltzmann Relation and Diffusion Current in Semiconductors

The availability as well as the flow of electrons in semiconductor materials such as germanium and silicon† depend upon processes which are distinctly different from those which occur in vacuum devices and in metallic conductors. As we study this distinction in the material which follows, we shall conclude with a description of the external characteristics of the semiconductor diode. This is basic to the understanding of semiconductor electronics. Once the theory of operation of the semiconductor diode is understood, other related devices such as the transistor can be readily treated because they are modifications of the fundamental device. For example, as is pointed out later, the transistor may be looked upon as consisting of two diodes connected back-to-back.

The properties of metal are quite different from those of semiconductor materials. For example, in a metal the valence electrons (i.e. those in the outermost orbit) are completely free and roam about from atom to atom, whereas in a semiconductor the motion of the valence electrons of one atom is coordinated with the motion of those of an adjacent atom, so that a covalent bond is imposed between them. The result is that under ordinary circumstances there are few, if any, free electrons in a pure semiconductor element. Accordingly, the conductivity property of the two elements is very different. Thus the application of just a fraction of a volt across the ends of a copper conductor can produce a large current flow, whereas the same potential difference appearing across a germanium material having the same dimensions results in negligible current flow.‡

Electronics is concerned with the control of electron flow in engineering devices. A primary consideration of any electron device, then, must be the ability and ease with which electrons can be made available and controlled. In Chap. 8 we shall see how in vacuum devices electrons are furnished through the application of heat energy and controlled by means of an electric field between electrodes. In a metal, electrons are available by virtue of the free-electron nature of the space

† Refer to column IV of the Periodic Table of Elements in Appendix B.

‡ A minute current flow can result because at room temperature the associated thermal energy can free some electrons from their covalent bonds, thus providing conduction.

lattice structure of the material, so that again the application of an electric field in the metal exerts control over the flow of the electrons. It should be apparent by now that since the pure semiconductor element lacks a source of electrons it can be only of limited use as an electronic device. To be useful on a scale comparable to that of the vacuum device, the pure semiconductor material must be altered so as to significantly increase its capability to transport charge carriers. Fortunately, this result can be achieved by the addition of small amounts of impurities to the composition of the pure semiconductor materials. Thus the addition of a small amount of pentavalent arsenic to a melt of pure tetravalent germanium makes one free electron per arsenic atom available for flow, and the impure germanium-arsenic mixture becomes a semiconductor. Because of the limited number of free electrons made available by this process, the conductivity of the semiconductor is less than that of a metal; however, the important point here is that there now exists a means of making available charge carriers for control purposes. It merely requires introducing impurities into a material characterized by covalent bonds. The higher the impurity content, the greater the conductivity of the semiconductor.

Essentially there are two types of impurities that can be added to a tetravalent element such as germanium or silicon—elements found in column IV of the periodic table. When the impurity has five valence electrons (pentavalent), it is referred to as a *donor* or *n-type impurity*. Depicted in Fig. 7-1 is a schematic representation of the convalent bonds existing between the valence electrons of the germanium atoms. Although germanium has a total of thirty-two orbital electrons, only the four in the outermost orbit are shown; for consistency the nucleus is shown carrying a corresponding charge of $+4$. Of course the total charge on the nucleus is $+32$. Note too that in the case of the arsenic† impurity

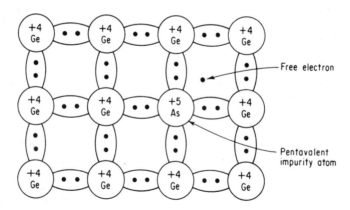

Fig. 7-1. Diagrammatic representation of the filled covalent bonds of the valence electrons. Note the free electron donated by the pentavalent impurity atom, arsenic.

† Other pentavalent impurities include phosphorus and antimony.

atom four of its five valence electrons have coordinated motion with electrons of adjacent germanium atoms, thus forming the covalent bonds. The fifth electron is unbonded and at room temperature has sufficient thermal energy so that it is not even electrostatically attracted to the positive charge at the nucleus of the arsenic atom. Hence this electron is free to roam through the material. It is for this reason that the arsenic atom is called a donor impurity: it donates one charge carrier in the form of an electron. The symbolism to be used to depict this situation is illustrated in Fig. 7-2. The circle around the plus sign emphasizes that the positive nucleus is an immobile charge and thus cannot be part of a current flow. On the other hand the free electron can.

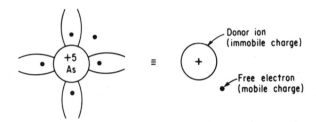

Fig. 7-2. Symbolism used to represent the donor impurity atom.

A second type of impurity semiconductor results when a trivalent element (such as indium, boron, gallium, or aluminum) is added to pure germanium. Since indium has only three valence electrons in the outer shell, the schematic representation of the covalent bonds between electrons of adjacent atoms appears as shown in Fig. 7-3. Note that only three covalent bonds can be filled. The vacancy that exists in the fourth bond is called a *hole*. The presence of this hole means that conditions are favorable for this impurity atom to accept an electron, thus filling

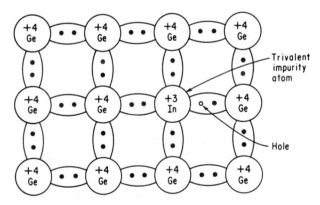

Fig. 7-3. Depicting the three filled covalent bonds of the trivalent impurity and the vacancy existing in the fourth bond.

the vacancy. For this reason this kind of impurity is called an *acceptor* or *p-type impurity*. Note that when an electron is received by the impurity atom it assumes a net negative charge that is immobile. The hole, which now no longer exists at this place in the material, may be considered to have moved elsewhere—for example, to the place from which the electron came. With such terminology one can look upon the hole as being a mobile charge carrier. This situation is depicted in Fig. 7-4, where the circle around the minus sign indicates that the trivalent impurity atom has accepted an electron, leaving the hole to appear elsewhere.

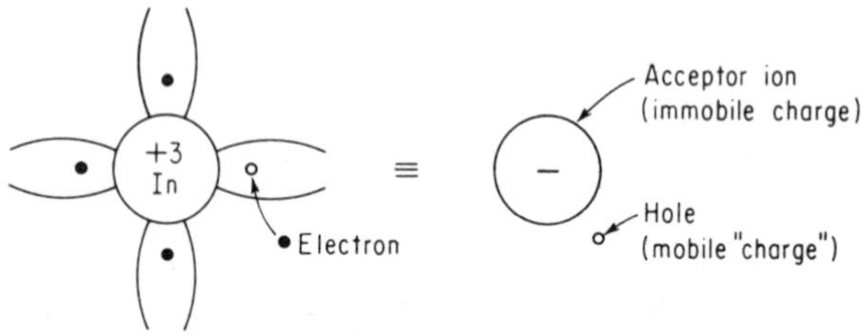

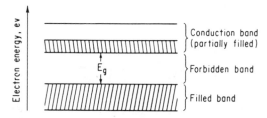

Fig. 7-4. Symbolism used to represent a trivalent impurity upon accepting an electron. When the electron fills the vacancy, the hole appears to go elsewhere. Hence the hole may be treated as a positive charge carrier.

It is useful at this point to consider the energy-band description of materials. It can be shown by quantum theory that in crystalline materials there are bands of allowed energy levels for the atoms separated by bands of forbidden energy levels. Although for a single atom the energy bands are discrete, it is found that the proximity of other atoms modifies the discrete energy levels of the individual atoms to the point where the energy distributions take on the appearances of bands. The energy-band description of a metallic conductor is illustrated in Fig. 7-5. The partially filled band refers to the free or valence electrons possessing the highest energy. Note that if small additional energy is added to the crystal these electrons are raised to higher energy levels without prohibition. This statement cannot be made for those electrons with energy levels that place them in the filled band. Moreover, if energy is applied to this material in the form of an externally applied electric field, conduction is found to occur. For this reason the uppermost band is also called the *conduction band*.

Fig. 7-5. Energy band description of a metallic conductor.

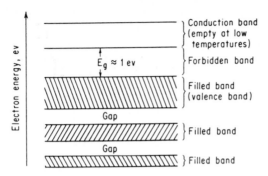

Fig. 7-6. Energy band structure of a semiconductor.

In a semiconductor material without impurities the energy-band description appears as shown in Fig. 7-6. The significant factor here is that the forbidden energy gap separating the filled valence band and the empty conduction band is about one electron-volt. Accordingly, as the temperature is increased, electrons whose energy levels are raised beyond the gap level will be freed of their covalent bonds and become available for conduction. At room temperature approximately one out of 10^{10} atoms has electrons so liberated.

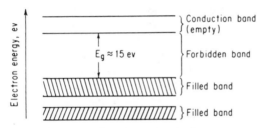

Fig. 7-7. Energy band structure of an insulator.

Figure 7-7 depicts the situation which prevails for insulators. In this case the gap energy is considerably larger than it is for semiconductors, with the result that at room temperature no covalent bonds are broken to provide free electrons in the conduction band. Even the application of extremely large electric fields fails to rupture any appreciable number of covalent bonds. Thus the conductivity of such materials is practically nil.

Semiconductor Types. We are now ready to consider the impurity-type semiconductors in terms of the energy-band concept. But first let us consider the *intrinsic germanium semiconductor*, which contains no impurities. The thermal energy associated with room temperature is sufficient to dislodge an electron from a covalent bond. As already mentioned, this happens to about one out of 10^{10} atoms. Figure 7-8(a) depicts this situation in terms of the lattice arrangement, and Fig. 7-8(b) does it by means of the energy-band diagram. Note the presence

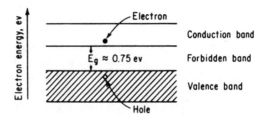

Fig. 7-8(a). Creation of electron-hole pair through thermal agitation.

Fig. 7-8(b). Intrinsic semiconductor at room temperature.

of the electron in the conduction band. As thermal energy increases, more electrons become available for conduction, which explains why a semiconductor material becomes a better conductor with rising temperature. In sharp contrast is the conductivity of metals, which decreases with temperature because of the greater impediment to the free flow of the electrons presented by the more violently vibrating atoms in the space lattice. Note too that current flow is now possible in two ways—the flow of electrons as well as the displacement of holes.

The band-structure description of the *n-type impurity semiconductor* is depicted in Fig. 7-9. Impurities are normally added in the ratio of about one part in 10^8. Accordingly, the energy diagram for the impurities remains discrete because of the relatively wide spacing between the impurity atoms. This situation is indicated in Fig. 7-9 by showing the individual impurity atoms. Furthermore, it is found that there are additional energy levels associated with pentavalent impurity atoms which are located just slightly below the lower level of the conduction band by approximately 0.01 ev. Because of this small difference in energy levels the thermal energy of room temperature supplies enough energy to each excess electron (i.e. the fifth valence electron) of the impurity atom to raise it into the conduction region. The impurity atoms are thus said to be fully ionized and are so depicted in Fig. 7-9 by the circles around the positive charges. Keep in mind, too, that the normal thermal agitation of the pure germanium atoms continues to make electron-hole pairs available. However, the number of electrons furnished from this source is small compared to that provided by the impurity atoms. A glance at Fig. 7-9

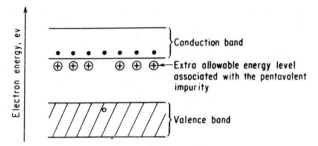

Fig. 7-9. Band structure of an *n*-type impurity semiconductor. Electrons are majority carriers; holes are minority carriers.

reveals that, like the intrinsic semiconductor, the *n*-type impurity semiconductor brings about current flow through the action of electrons and holes. But in the *n*-type impurity semiconductor the number of electrons (negative carriers) far exceeds the number of holes, which is why it is called "*n*-type." The holes constitute the minority carriers. In a pure semiconductor the number of holes is always equal to the number of electrons.

An analogous condition is found to occur when a trivalent impurity is added to germanium or silicon. The presence of the impurity allows an extra unfilled energy level to exist, which significantly is located a few hundredths of an electron-volt above the filled valence band as illustrated in Fig. 7-10. Thus enough electrons in the covalent bonds acquire the energy to be elevated to the additional energy level so that at room temperature all the holes of the impurity atoms are filled; the impurity is then said to be ionized 100 per cent. This ionization process, of course, leaves holes in the valence band which are free to move about to other locations. Thermal agitation of the pure germanium atoms also generates electron-hole pairs, thus adding to the total number of holes present in the valence band.

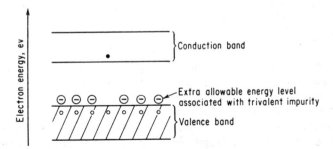

Fig. 7-10. Band structure of *p*-type impurity semiconductor. Holes are the majority carriers; electrons are minority carriers.

Note that when a trivalent impurity is used the minority carriers are the electrons whereas the holes are the majority charge carriers. For this reason such a semiconductor is referred to as a *p-type impurity semiconductor*. The *p* stands for "positive" charge carrier. Clearly, even where holes are involved current flow always occurs because of the movement of electrons, but the effect is that of a hole moving in the opposite direction. Therefore, electrically, one can say that the movement of a hole is equivalent to the movement of a positive ion having a mobility comparable to that of the electron. For convenience this description is used in all further work dealing with *p*-type devices.

The *p-n* Junction Diode with Forward Bias. The foregoing material was treated here for the purpose of providing the background for understanding the diffusion process which underlies the operation of the semiconductor diode. A *p-n* junction diode can be formed by growing a single crystal of semiconductor material in which the acceptor impurity is made to predominate in one part as the crystal is drawn out of a suitable melt and the donor impurity is made to predominate in the other part. A schematic diagram of the *p-n* junction diode appears in Fig. 7-11. Keeping

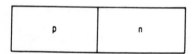

in mind that the *n* material has a relatively high density of electrons while the *p* material has a high density of holes (or electron vacancies), it is reasonable to expect that there will be a tendency for the electrons to *diffuse* over to the *p*-side and vice versa. As a matter of fact this is precisely what takes place initially, but the process does not continue unhindered. As electrons move across the junction from the *n*-side to the *p*-side, positive immobile charges are uncovered on the *n*-side and negative immobile charges are formed upon acceptance of an electron by the acceptor impurity on the *p*-side. When a sufficient number of these charges are uncovered, a potential energy barrier V_0 is created and it prevents any further diffusion from occurring. This situation is illustrated in some detail in Fig.

Fig. 7-11. A grown *p-n* junction.

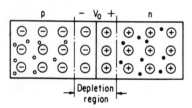

Fig. 7-12. The *p-n* junction diode showing the creation of the potential energy barrier V_0 through diffusion of mobile charge carriers across the junction.

7-12. Note that the figure depicts only the impurity atoms and one electron-hole pair resulting from the thermal agitation of the pure germanium. As a result of the absence of mobile charge carriers on either side of the junction, this region is called the *depletion region* and is of the order of 10^{-4} cm wide.

The potential energy barrier V_0 plays a role analogous to that of the potential energy barrier of metals. The magnitude of V_0 is of the order of a few tenths of a volt, and in any given situation its value may be found from the Boltzmann relation and a knowledge of the charge densities in the *p* and *n* regions. As an illustration, refer to Fig. 7-13 which depicts on a logarithmic scale the charge densities of the majority and minority carriers in the p as well as the n materials. Consider the equilibrium condition—switch S open. Note that the p material is more heavily doped with impurity than is the n material. However, it is a law of the semiconductor, impure or otherwise, that the product of holes and electrons in the p or n material is equal to a constant determined by the temperature. The Boltzmann relation is an equation which relates the density of particles in one region to those in an adjacent region when the densities involved are relatively sparse. The Boltzmann relation therefore is particularly useful, for example, in atmospheric studies. However, to a good approximation it is also applicable to the p-n junction diode because the density of holes in the valence band and the density of electrons in the conduction band for semiconductor materials is quite

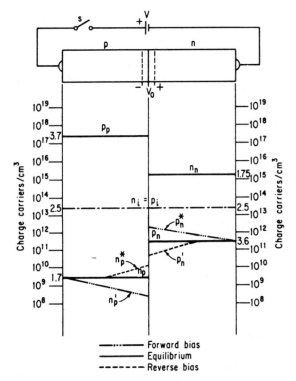

Fig. 7-13. Carrier densities in a p-n junction diode:

p_p = holes, or majority carriers in p material;

n_p = electrons, or minority carriers in p material;

n_n = electrons, or majority carriers in n material;

p_n = holes, or minority carriers in n material.

low. Expressed mathematically the Boltzmann relation is

$$N_1 = N_2 \epsilon^{V_{21}/E_T} \qquad (7\text{-}1)$$

where N_1 = density of particles in region 1
N_2 = density of particles in region 2
E_T = energy equivalent of temperature [see Eq. (8-4)]
V_{21} = the potential energy of region 2 with respect to region 1

Upon applying this equation to the *p-n* junction diode of Fig. 7-13 we have for the holes in the two regions

$$p_p = p_n \epsilon^{V_0/E_T} \qquad (7\text{-}2)$$

Inserting the values of p_n and p_p as specified in Fig. 7-13 as well as $E_T = 0.026$ volt for room temperature yields

$$\epsilon^{V_0/0.026} = \frac{3.7 \times 10^{17}}{3.6 \times 10^{11}} = 1.027 \times 10^6 \qquad (7\text{-}3)$$

and leads to

$$V_0 = 0.36 \text{ volt}$$

We are now at the point where we can begin to establish the volt-ampere characteristic of the semiconductor diode. Let us consider first the behavior of the diode when a forward bias voltage is applied. This merely requires closing switch S in Fig. 7-13. Note that a forward bias is applied by connecting the positive side of the battery to the *p* material and the negative side of the battery to the *n* material. Moreover, assume that the battery voltage V is less than the potential energy barrier V_0. Observe that, as a forward bias voltage, V is opposed to V_0. Consequently the density of holes in the *n* region at the junction must be altered in accordance with the modified Boltzmann relation, namely

$$p_p = p_n^* \epsilon^{(V_0-V)/E_T} \qquad (7\text{-}4)$$

where p_n^* is the new value of hole density in the *n* material. Upon rewriting Eq. (7-4) we get

$$p_p = p_n^* \epsilon^{V_0/E_T} \epsilon^{-V/E_T} \qquad (7\text{-}5)$$

Inserting Eq. (7-2) into Eq. (7-5) then yields

$$\boxed{p_n^* = p_n \epsilon^{V/E_T}} \qquad (7\text{-}6)$$

This is one of the most important relationships in junction theory. This equation states that upon applying a small forward bias there is an exponential increase in the minority charge carriers. Thus for $V = 0.102$ volt the new value of hole density in the *n* material at the junction is

$$p_n^* = p_n \epsilon^{0.102/0.026} \approx 50 p_n \qquad (7\text{-}7)$$

Thus there occurs a fiftyfold increase over the equilibrium value of the hole density.
Where does this large increase of holes come from? A little thought reveals

that the presence of the forward bias voltage V decreases the effective potential energy barrier from V_0 to $(V_0 - V)$, thereby allowing holes to move across the junction into the n region. It is interesting to note that although the hole density at the junction in the n material increases from 3.6×10^{11} to 180×10^{11}, this quantity is only a very small fraction of the hole density in the p material, which is 3.7×10^{17}. Furthermore, in accordance with the *law of charge neutrality*, every time a hole leaves the p material, it is replenished by one which enters the p material from the positive side of the battery.

What happens to this large increase in hole density after it is injected into the n material across the junction? Since we are dealing here with two contiguous regions having different particle densities there is a natural action of *diffusion* taking place. As the holes cross the junction into the n material the preponderance of negative charge carriers throughout the n material causes swift recombination to take place. Thus the charge density rapidly decreases as the distance from the junction increases. Again, by the law of charge neutrality for every electron in the n material which combines with a hole, there is another electron entering the n region from the negative side of the battery. It is important to understand that this change of charge density with distance in the semiconductor involves a transport of charge and thereby constitutes a current flow. But this is not a current flow in the usual sense that we have come to associate with metallic conductors. Rather in semiconductor diodes current flow takes place by *diffusion*. In fact the hole current may be generally written as

$$I_{pn} = -A_e D_p \frac{dp_n^*}{dx} \qquad \text{amps} \qquad (7\text{-}8)$$

where D_p = diffusion constant for holes, m^2/sec
     A = cross-sectional area of p-n junction, m^2
     e = electron charge in coulomb/electron
     I_{pn} = *hole* current in the n region

Equation (7-8) emphasizes the fact that the diffusion current is dependent upon the rate of change of charge density with distance. Diffusion current does not obey Ohm's law.

In the foregoing treatment attention was focused on the difference in the hole densities in the p and n regions. A similar treatment holds for the electron densities in the two regions. Accordingly, by the Boltzmann relation we can describe a simultaneous increase in the minority carriers of the p region expressed as

$$n_p^* = n_p \epsilon^{V/E_T} \qquad (7\text{-}9)$$

As these electrons diffuse through the p region they quickly recombine with holes, which are present in great numbers. Again there is a transport of charge with distance, leading to a second component of diffusion current but this time attributable to the electrons in the p region. Hence we can write

$$I_{np} = AeD_n\frac{dn_p^*}{dx} \quad \text{amps} \tag{7-10}$$

where D_n = diffusion constant for electrons, m^2/sec
$\quad A$ = cross-sectional area, m^2
$\quad e$ = electron charge
$\quad I_{np}$ = *electron* diffusion current in the p region

The total diffusion current is given by the sum of Eqs. (7-8) and (7-10) evaluated at the junction, i.e. $x = 0$. Thus

$$I = I_{pn}(0) + I_{np}(0) \tag{7-11}$$

It follows from Eqs. (7-6) and (7-8) as well as Eqs. (7-9) and (7-10) that the p-n junction diode diffusion current is exponentially related to the forward bias voltage. This is a consequence of the Boltzmann relation. Hence the external volt-ampere characteristic of the p-n junction diode appears as shown in Fig. 7-14 for positive values of V.

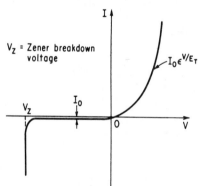

Fig. **7-14.** Volt-ampere characteristic of the p-n junction diode.

What happens to the p-n junction diode current when a reverse bias is applied (i.e. the positive side of the battery is connected to the n region and the negative side to the p region)? The Boltzmann relation reveals that under these conditions there is a marked decrease in holes at the junction of the n material. Thus we have $p_n' = p_n \epsilon^{-V/E_T}$ where p_n' is the hole density for reverse bias. This means that holes diffuse from the n region, where they are indeed scarce, into the p region. Similarly, the reverse bias causes a decrease of electrons at the junction in the p region. Effectively then, electrons diffuse from the p material, where they are relatively very few, into the n region. Because of the small numbers of charge carriers involved, the current flow in the reverse direction is correspondingly very small too. It is frequently of the order of 10^{-5} ampere and is called the *reverse saturation current*, denoted by I_0.

Accounting for both forward and reverse bias voltages, it is possible to write a single expression for the semiconductor diode current which is valid for positive as well as negative voltages up to V_z—the Zener breakdown voltage. Thus

$$\boxed{I = I_0(\epsilon^{V/E_T} - 1)} \tag{7-12}$$

where I_0 is the reverse saturation current. Note that for even small positive values of V the exponential term predominates so that the current variation is exponential, which is consistent with the Boltzmann relation. At $V = 0$ the current, of course, should be zero, and the form of Eq. (7-12) certainly bears this out. Finally, for negative V the exponential term approaches zero rapidly so that the only significant current flow is the reverse saturation current, $-I_0$.

It is important to keep in mind when using Eq. (7-12) that the exponential character of the variation for positive V is valid as long as V is less than the potential energy barrier V_0. When V exceeds V_0, then ohmic drops within the material must be accounted for.

A glance at Fig. 7-14 indicates that when a large negative voltage is applied to the diode, it is accompanied by a tremendous increase in current. Essentially, this condition is brought about by the complete disruption of all the covalent bonds, which releases huge numbers of electrons for conduction. In this state the semiconductor behaves as a metallic conductor.

7-2   The Semiconductor Diode

The *p-n* semiconductor junction diode with a forward bias is shown in Fig. 7-15. Recall that the *p* region is heavily doped with a trivalent impurity and the *n* region is doped with a pentavalent impurity. The application of a forward bias V, which is assumed smaller than the potential energy barrier V_0, causes an increase in the hole density in the *n* region at the junction. This increase is expressed by Eq. (7-6) and is repeated here. Thus

$$p_n^* = p_n\epsilon^{V/E_T} \tag{7-6}$$

where p_n^* denotes the new value of hole density in the *n* region

p_n denotes the equilibrium value of hole density

E_T denotes the energy equivalent of temperature

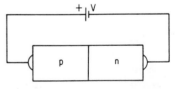

Fig. 7-15. The *p-n* junction diode with forward bias.

At the same time there is a similar action taking place which involves the electron density in the *p* region. The same forward bias causes an increase in the electron density of the *p* region which at the junction is described by Eq. (7-9). For convenience this equation too is repeated here:

$$n_p^* = n_p\epsilon^{V/E_T} \tag{7-9}$$

where n_p^* denotes the new value of electron density in the *p* region and

n_p denotes the equilibrium value

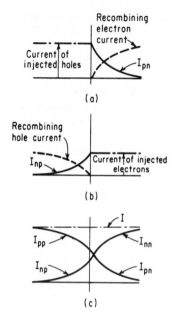

Fig. 7-16. Identifying the total diffusion current *I* in terms of the hole and electron current components.

The increased concentration of the hole density at the junction of the *n* region is followed by a diffusion of the charge carriers through the *n* material. Recombination with electrons then takes place rapidly because of the very high concentration of negative charge carriers in the *n* region. This action gives rise to a *diffusion current* which depends upon the change of charge density with distance from the junction. This hole (*p*) current in the *n* region, I_{pn}, is depicted in Fig. 7-16(a). Two things are worth noting here. One is the exponential decay in I_{pn} that occurs with increasing displacements from the junction. The other is the presence of a *recombining electron current* which is required by the law of charge neutrality. For every electron that recombines with a hole in the *n* region there enters into this region from the negative side of the battery another electron which preserves the charge neutrality of the *n* region.

A similar situation takes place in the *p* region. The increased electron density n_p^* brought about by the forward bias causes a diffusion of electrons away from the junction into the *p* region. However, because of the very high concentration of holes in this region, electrons rapidly recombine with holes. Since the decay in charge density from the junction can be shown to be an exponential decay, it follows that the electron (*n*) current in the *p* region, I_{np}, undergoes a similar decay as depicted in Fig. 7-16(b). In this instance, however, note the presence of a *recombining hole current*. Thus for every hole that recombines with an electron in the *p* region the positive side of the battery supplies an additional hole in order to preserve charge neutrality in the *p* region.

The total diffusion current is found by adding the effects depicted in Figs. 7-16(a) and (b). The total hole (*p*) current in the *p* region, I_{pp}, is composed of the recombining hole current of Fig. 7-16(b) plus the value of I_{pn} as *it exists at the junction*. It is important to understand that the increased hole density p_n^* originates from the positive side of the battery. The forward bias causes holes to be injected from the positive terminal through the entire *p* region and thence across the junction into the *n* region where diffusion occurs. Little or no recombination of the injected holes with electrons occurs in the *p* region because of the great scarcity of electrons there compared to holes. Accordingly, I_{pp} assumes the variation shown in Fig. 7-16(c). By a similar line of reasoning the total electron (*n*) current in the *n* region, I_{nn}, can be shown to vary as indicated.

Further examination of Fig. 7-16(c) reveals that the total diffusion current *I* is a constant throughout the body of the junction diode even though I_{nn} and

I_{pp} vary with the distance from the junction. A study of Fig. 7-16(c) makes it apparent that the diffusion current can be expressed in any one of three ways. Thus

$$I = I_{pp} + I_{np} \tag{7-13a}$$

or

$$I = I_{nn} + I_{pn} \tag{7-13b}$$

or

$$I = I_{pn}(0) + I_{np}(0) \tag{7-13c}$$

In this last equation $I_{pn}(0)$ refers to the hole current in the n region evaluated at the junction.

In dealing with the diffusion current it is important to understand the distinction between the two exponential functions with which it is related. An increase in forward bias voltage V causes an exponential increase in the diffusion current as described by Eq. (7-12). That is, the values of I_{np} and I_{pn} at the junction ($x = 0$) are increased exponentially with V. On the other hand, the values of these same currents as a function of the distance away from the junction decrease exponentially with x.

Semiconductor Diode Resistance. The volt-ampere characteristic of the semiconductor junction diode takes on the exponential character depicted in Fig. 7-17. If the forward bias is fixed at the value V_Q, there is associated with the diode an apparent forward resistance given by

$$R_f \equiv \frac{V_Q}{I_Q} \qquad \text{ohms} \tag{7-14}$$

When we are interested in the resistance of the diode as it appears to a signal source causing small changes in the bias voltage about the quiescent point Q, its value is determined as the reciprocal of the slope of the curve at point Q. This resistance is called the *incremental resistance*. An expression for it readily follows from Eq. (7-12), which mathematically describes Fig. 7-17. Thus

$$I = I_0(\epsilon^{V/E_T} - 1) \tag{7-15}$$

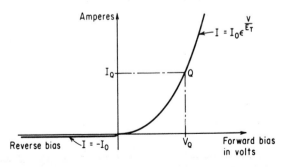

Fig. 7-17. Volt-ampere characteristic of *p-n* junction diode.

Differentiating yields

$$dI = \frac{I_0}{E_T}\epsilon^{V/E_T}\,dV \tag{7-16}$$

Therefore the forward incremental resistance is

$$r_f \equiv \frac{dV}{dI} = \frac{E_T}{I_0\epsilon^{V/E_T}} = \frac{E_T}{I + I_0} \tag{7-17}$$

At room temperature the value of E_T is 0.026 volt. Moreover, for a forward bias, i.e., V positive, Eq. (7-17) may be simplified to

$$r_f = \frac{E_T}{I} = \frac{0.026}{I} \tag{7-18}$$

Finally, if I is assumed to be expressed in milliamperes (ma), as is frequently the case, then Eq. (7-18) becomes

$$\boxed{r_f = \frac{26}{I\,(\mathrm{ma})}}\ \ \text{ohms} \tag{7-19}$$

This result indicates that the incremental forward resistance of the *p-n* junction diode is quite low. For example, when a current of 1 ma flows, this resistance has a value of 26 ohms.

The incremental backward resistance (i.e., the resistance with reverse bias) is ideally infinite because of the zero slope of the characteristic in this region. This conclusion also follows from Eq. (7-17) upon inserting $I = -I_0$ for negative values of bias voltage. Thus

$$r_b = \frac{E_T}{I + I_0} = \frac{0.026}{-I_0 + I_0} = \infty \tag{7-20}$$

However, the apparent backward resistance is not infinite because for any value of reverse bias there does exist a small current I_0. For germanium and silicon junction diodes this resistance is large compared to the apparent forward resistance. In germanium diodes the ratio of apparent backward to forward resistance is 400,000 : 1, while for silicon this ratio is 1,000,000 : 1.

The volt-ampere characteristic of an ideal semiconductor diode is depicted by Fig. 7-18, and the circuit symbols are shown in Fig. 7-19.

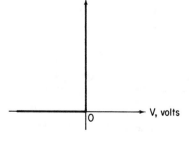

Fig. 7-18. Volt-ampere characteristic of an ideal diode, i.e. *rs* = 0.

Diffusion Capacitance. Appearing in Fig. 7-20 is the variation of the hole density injected into the *n* material of a *p-n* junction diode shown as a function of the distance from the junction for two values of forward bias voltage. At the junction the variation of the hole density with forward bias is an increasing exponential function. Thus for a forward bias voltage V_1 the hole density at the junction is

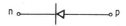

Fig. 7-19. Circuit symbol for the ideal diode. Conduction takes place from *p* to *n* only.

$$p_{n1} = p_n \epsilon^{V_1/E_T} \tag{7-21}$$

and for $V_2 > V_1$, it is

$$p_{n2} = p_n \epsilon^{V_2/E_T} \tag{7-22}$$

where in each case p_n denotes the equilibrium value of the hole density at the junction in the *n* material. When a forward bias is applied, however, as one moves away from the junction the hole charge density diminishes because of the combination which takes place. It can be shown that this decrease proceeds in accordance with an exponential decay ϵ^{-x/L_p} where L_p denotes the mean diffusion length for holes and x is the distance into the *n* material measured from the junction.

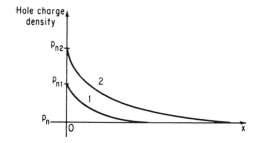

Fig. 7-20. Variation of hole charge density as a function of distance from the junction for two values of forward bias voltage.

It is important to note that the area beneath either curve in Fig. 7-20 represents the charge stored at the junction per unit cross-sectional area for each value of forward bias voltage. As the forward bias is changed from V_1 to V_2, operation goes from curve 1 to curve 2 in Fig. 7-20. Because the area beneath the charge-density curve increases with an increase in forward bias, it follows that the charge at the junction increases. Accordingly, there occurs at the junction a change in charge, ΔQ, per change in forward bias voltage $\Delta V = V_2 - V_1$, so that the ratio of the two, $\Delta Q/\Delta V$, can be interpreted in terms of an incremental capacitance. A similar analysis applies to the electron densities in the *p* region. The cumulative effect of the change in stored charge of the hole and electron densities at the junction corresponding to a change in forward bias voltage is called the *diffusion capacitance* and is denoted by C_D. Typical values of this quantity fall in the range of several thousand picofarads.

7-3 The Transistor (or Semiconductor Triode)

The addition of a third doped element to a semiconductor diode results in a triode which is capable of achieving amplification of signals in a fashion comparable and often superior to that realizable by the vacuum-tube triode. Figure 7-21(a) depicts one possible configuration of the semiconductor triode. It is a *p-n-p* type and is composed of two sections of germanium, heavily doped with a trivalent impurity, separated by a very thin section (of the order of 1 mil thickness) of germanium which is heavily doped with a pentavalent impurity. The semiconductor triode can also be formed by two *n* sections separated by a *p* section. Appearing in Fig. 7-21(b) is the symbol used to represent the semiconductor triode.

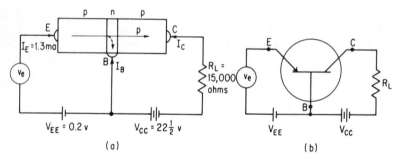

Fig. 7-21. Semiconductor triode in a common-base configuration:
    (a) circuit arrangement showing typical operating values for
    a *p-n-p* triode;
    (b) schematic representation.

How are control and amplification achieved in the semiconductor triode? This can be understood by investigating the manner in which the circuit of Fig. 7-21(a) operates. The thin center section of the semiconductor triode is called the *base* and the terminal connection associated with it is denoted by B. The input circuit consists of that part of the configuration associated with terminals E and B. Note that the bias voltage V_{EE} is connected in the input circuit in a way that assures *forward bias*. Thus when the input circuit is closed, V_{EE} causes holes to be injected into the p material. For this reason the second terminal of the semiconductor triode is denoted by E. The voltage V_{EE} is a fixed quantity which serves to establish an operating point about which changes may be effected by the action of the varying component of the signal v_e. The output circuit consists of that part of the configuration associated with terminals C and B. This section of the semiconductor triode circuit is *reverse-biased*, as a glance at Fig. 7-21(a) reveals. The third terminals of the triode is denoted by C in order to refer to the collector action that takes place at this terminal. Because the base terminal is common to both the input and output sections, the circuit configuration of Fig. 7-21(a) is called the *common-base* mode. In the representation of the semiconductor triode shown in Fig. 7-21(b) note that the emitter terminal has associated with it an arrowhead. When the arrowhead points towards the base terminal the triode is of the *p-n-p* type; when the arrowhead points away from the base terminal, it means that an *n-p-n* type is being used. Of course whenever the latter type is used, the voltages V_{EE} and V_{CC} must be connected in the reverse sense from that shown in Fig. 7-21.

Let us now analyze the circuit of Fig. 7-21 in terms of some typical operating figures. Assume that when the input circuit is quiescent (i.e., $v_e = 0$), the bias voltage V_{EE} causes an emitter current I_E of value 1.3 ma to flow. In other words a sufficient number of charge carriers (holes in this instance) are being injected from the forward bias battery into the *p* material to cause a current of 1.3 ma to flow. These injected holes then appear at the emitter-base junction, beyond which point a diffusion process should take place. However, the base region was made intentionally small with the expressed purpose of preventing the injected holes from recombining with the excess electrons existing in the *n* region. This result is achieved by making the thickness of the base region only a small fraction of the average diffusion length of a hole. Thus if the average diffusion length of a hole is L_p and the thickness of the base section is of the order of $0.05L_p$, then the chances of a hole recombining with an electron in the base region are small indeed. As a result the vast majority of the holes injected across the emitter-base junction never recombine but rather move on across the base-collector junction into the *p* material of the collector. Once the injected holes appear in this region, the effect of the voltage V_{CC} accelerates them towards the collector terminal and then through the load resistor R_L, through V_{CC}, and finally back to V_{EE}, thus completing the circuit. The remaining fraction of the injected holes, of course, take a path through the base terminal to the negative side of V_{EE}. In most semiconductor triodes operating in the common-base mode approximately 95 to 98 per cent of the injected holes take a path to the collector terminal and thence through the output circuit.

A figure of particular importance in semiconductor triodes is the ratio of the change of current appearing at the collector output terminal to a given current change appearing at the emitter input terminal. It is called the *current amplification factor* and is denoted by α. This quantity is also called the *current transfer ratio*.

Expressing this mathematically we can write

$$\alpha = \text{current transfer ratio} \equiv -\left(\frac{\partial i_c}{\partial i_e}\right)_{V_{CB} \text{ constant}} \tag{7-23}$$

This definition is valid for the common-base configuration. Moreover, α has a value which is always less than unity but frequently greater than 0.9. In this sense, then, it appears that use of the word "amplification" is misleading. However, it is used in order to keep clear the distinction between current amplification factor and current gain. The former is a property of a given semiconductor triode; the latter is influenced by the external circuitry connected to the triode.

In the interest of simplicity let us for the moment assume that α equals unity in the circuit of Fig. 7-21. Hence the entire emitter current of 1.3 ma flows to the collector and on through the 15,000-ohm resistance. Upon passing through this resistor, there occurs a voltage drop of $(1.3 \times 10^{-3})(1.5 \times 10^3) = 19.5$ volts. Consequently the voltage from collector to base becomes

$$V_{CB} = +19.5 - 22.5 = -3 \text{ volts} \tag{7-24}$$

The existence of this negative voltage from collector to base assures that the output circuit is in fact reverse biased.

Consider next the situation where a change in the emitter voltage is introduced by letting v_e assume a value of 1 millivolt; that is,

$$v_e = 0.001 \text{ volts} \tag{7-25}$$

To determine the corresponding change in the emitter current caused by v_e, it is necessary first to obtain the forward resistance of the *p-n* section of the triode appearing in the input circuit. This readily follows from Eq. (7-19). Thus

$$r_e = \frac{26}{I_E \text{(ma)}} = \frac{26}{1.3} = 20 \text{ ohms} \tag{7-26}$$

The subscript *e* indicates that the input change occurs at the emitter terminal. Therefore the change in emitter current is

$$\Delta I_E = \frac{v_e}{r_e} = \frac{0.001}{20} = 0.05 \text{ ma} \tag{7-27}$$

Continuing with the assumption that α is 1 yields a change in collector current of

$$\Delta I_C \approx \Delta I_E = 0.05 \text{ ma}$$

Accordingly, the change in output voltage, which is obtained across the 15,000-ohm resistor, is

$$v_c = \Delta I_C R_L = (0.05 \times 10^{-3})(15 \times 10^3) = 0.75 \text{ volts} \tag{7-28}$$

Here v_c denotes the change in voltage appearing at the collector terminal.

A comparison of Eq. (7-28) with Eq. (7-25) shows that there occurs a voltage gain A_v of 750. Thus

$$A_v \equiv \frac{v_c}{v_e} = \frac{0.75}{0.001} = 750 \tag{7-29}$$

The corresponding gain in power, G, is

$$G \equiv \frac{v_c \Delta I_C}{v_e \Delta I_E} = A_v \alpha = 750 \tag{7-30}$$

A study of the results shown in the last two equations makes it clear that the semiconductor triode may be used to amplify small voltage variations as well as power. It is important at this stage to avoid erroneous impressions regarding the power-amplifying capability of the circuit of Fig. 7-19. Although it is correct to say that one unit of power in the input circuit controls 750 times as many units in the output circuit, one must guard against thinking that the circuit is capable of generating any level of power called for by the input circuit. This is not so. Any power that is delivered to R_L must come from the source V_{CC}. Because of the nature of the circuit of Fig. 7-19, it is possible for a small signal in the input circuit to command V_{CC} to deliver large changes of power in the output circuit. In other words, the varying signals of the input circuit serve as valves controlling the output derived from V_{CC} at much higher voltage and power levels. The overall efficiency of this circuit, however, is considerably less than 100 per cent.

The key factor which is responsible for the amplifying capability of the semiconductor triode is the arrangement which causes the emitter input current to flow almost entirely through the output circuit. Thus easy control of the emitter current is obtained because the emitter-base junction is forward biased, and yet most of this current is made to flow through the output circuit because the average diffusion length of the charge carriers is many times greater than the thickness of the base region. By this scheme, therefore, there occurs a *trans*fer of the input signal current from a low-re*sistor* circuit (i.e., the forward-biased input circuit) to a high-resistor circuit (namely the output circuit containing R_L). By combination of parts of the two key words which describe the operation of Fig. 7-9, there is coined the new word *transistor*. Accordingly, the transistor is a semiconductor device which has a forward-biased input section and a reverse-biased output section. Moreover, because of the low-resistance forward-biased input section, the transistor is essentially a current-sensitive device.

A general formulation of the amplifying capabilities of the common-base transistor further bears out this transfer aspect from a low- to a high-resistance circuit. Corresponding to a change v_e in emitter voltage we can write

$$v_e = \Delta I_E r_e \qquad (7\text{-}31)$$

Moreover, the change in collector current is related to ΔI_E by

$$\Delta I_C = \alpha \Delta I_E \qquad (7\text{-}32)$$

Furthermore, the change in collector output voltage is described by

$$v_c = (\Delta I_C)R_L = \alpha \Delta I_E R_L \qquad (7\text{-}33)$$

Inserting the expression for ΔI_E from Eq. (7-31) leads to

$$v_c = \alpha \frac{R_L}{r_e} v_e \qquad (7\text{-}34)$$

Equation (7-34) indicates that any change in emitter input voltage v_e appears as a change in the collector output circuit amplified by $\alpha(R_L/r_e)$. Since α is very close to unity and R_L is very large compared to r_e, it follows that the input signal change is greatly amplified at the output terminal. Returning to the values used in the circuit of Fig. 7-19 we get

$$v_c = (1)\frac{15{,}000}{20}(0.001) = 0.75 \text{ volts} \qquad (7\text{-}35)$$

which agrees with Eq. (7-28).

Volt-ampere Characteristic of the Common-base Mode. The volt-ampere characteristic of the common-base transistor involves a plot of the variation of the output circuit current I_C versus the collector-to-base voltage V_{CB}. A family of curves results when these plots are made for various values of fixed emitter currents. Note that for the transistor the input current is the parameter while for the vacuum triode it is the grid voltage. As long as the reverse bias voltage across terminals C and B exceeds several tenths of a volt, Eq. (7-32) is valid, so that practically all the emitter current drifts through the base region to the collector terminal. Depicted in Fig. 7-22 is a set of typical common-base output characteristics for a *p-n-p* junction transistor. Note that the curves are almost straight horizontal lines. This is consistent with the theory that the emitter current flows almost entirely to the collector terminal.

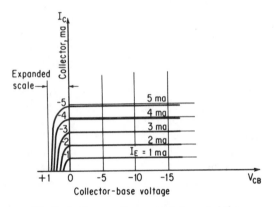

Fig. 7-22. Common-base output characteristics.

By convention a transistor current is said to be positive when it *enters* a terminal. Thus the positive direction of I_C is into terminal C in Fig. 7-19. However, if

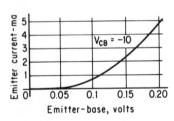

Fig. 7-23. Common-base input characteristics.

I_E is assumed to flow into terminal E, it then follows that almost all of this current flows out of terminal C. Because this is opposite to the assumed positive direction of I_C, the collector current is made to carry a minus sign as shown in Fig. 7-22.

Appearing in Fig. 7-23 are the corresponding input characteristics of the same junction transistor whose output characteristics appear in Fig. 7-22. Note that the exponential character of these curves is consistent with Eq. (7-12).

Transistor Parameters. Certain useful parameters may be identified for the junction transistor. Two of the parameters have already been identified. Equation (7-23) defines the current transfer ratio α, and Eq. (7-19) determines the forward-biased resistance r_e of the input circuit. The third useful parameter is the intrinsic transconductance, which is defined as the change in output collector current per unit change in emitter voltage. Expressing this mathematically we have

$$g_m \equiv \left(\frac{\partial i_c}{\partial v_e}\right)_{V_{CB} \text{ constant}} \quad \text{mhos} \tag{7-36}$$

Under the assumption that α is essentially constant, this expression may be rewritten as

$$g_m = \alpha\left(\frac{\partial i_c}{\partial v_e}\right)_{V_{CB} \text{ constant}} \tag{7-37}$$

A little thought reveals that the quantity in parentheses is the slope of the input characteristics of Fig. 7-23), which is the reciprocal of the forward-biased resistance r_e. Accordingly, Eq. (7-37) may be rewritten as

$$g_m = \alpha\left(\frac{1}{r_e}\right) \tag{7-38}$$

or

$$\boxed{\alpha = g_m r_e} \tag{7-39}$$

It is interesting to note the similarity which Eq. (7-39) for the current-sensitive transistor bears to Eq. (8-26) for the voltage-sensitive vacuum triode.

7-4 The Junction Field-Effect Transistor (JFET)

The conventional transistor described in Sec. 7-3 is chiefly characterized by its current sensitivity. The changes that are registered at the output terminals of the transistor are a direct consequence of changes that occur in the input current. The signal source that is applied to the input terminals must accordingly be capable of bringing about the required range of input current change if the transistor is to serve as a meaningful amplifier. There are many situations in electronics, however, in which the signal source is capable of effecting an appreciable change in voltage level but not an appreciable change in current. This happens, for example, whenever the source has a high internal impedance. In such instances, it is convenient to have available an amplifying device that possesses a very high input impedance. Such a device is the *field-effect transistor*, and it essentially provides features that, in the era of vacuum-tube electronics, were furnished by such popular devices as the vacuum triode and the pentode.

The field-effect transistor has been a well-established laboratory device since 1952. But it was not until a decade later that it began to receive wide acceptance. It had to await the perfection of thin-film and related technology to achieve with repeatable reliability configurations of thin, lightly doped layers of semiconductor material between more heavily doped layers of opposite type.

The composition of the junction field-effect transistor is best described by referring to Fig. 7-24, which depicts an n-channel JFET. The silicon body is lightly doped with a pentavalent impurity, thereby forming the n-channel. At the ends of the channel are placed two terminals—the drain D and the source S. On either side of the channel is deposited a much more heavily doped p-type impurity, as illustrated in Fig. 7-24. A terminal connection is provided for each section; these are called the gate terminals and denoted by G_1 and G_2. In use, both gate terminals are wired together.

For the n-channel JFET, the positive terminal of the supply voltage v_{DS} must

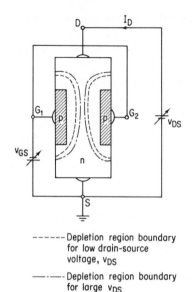

------ Depletion region boundary
for low drain-source
voltage, v_{DS}

—·—· Depletion region boundary
for large v_{DS}

Fig. 7-24. The JFET showing associated source and supply voltages.

be connected to the drain terminal of the JFET. The negative side of the drain source is connected to S. As a result, the supply voltage that appears entirely across the n channel is capable of producing a current flow from drain to source, I_D. At low values of v_{DS}, this current is limited essentially by the magnitude of v_{DS} and by the resistance of the n material. The latter in turn depends upon the length and cross-sectional area of the channel and the conductivity.

The success of the junction field-effect transistor depends upon the establishment of a substantial depletion region the width of which is readily and conveniently controlled. A glance at Fig. 7-24 discloses that a reverse-biased voltage is made to appear between each p section and the n-channel section about the drain terminal. The size of the voltage varies with the distance along the n channel because of the effect of the ohmic drop in the n channel. For points closer to the drain terminal, the reverse bias is larger than it is for points closer to the source terminal. This results in the creation of a depletion region surrounding each p section which is nonuniformly wide. It is also important here to understand that the depletion-region width that spreads into the n channel is many times greater than that appearing in the p material, simply because the n material is lightly doped, whereas the p material is much more heavily doped. This situation is emphasized in the diagram of Fig. 7-24 by showing that the spread of the depletion region takes place only in the n channel. At moderate values of reverse-biased voltage, the depletion region about the two p sections is denoted by the broken lines. At high values of reverse-biased voltage, brought about by increased values for v_{DS} and/or v_{GS}, the depletion region spreads to cover an area enclosed by the dash–dot lines. As the depletion region changes from one of less width to one of greater width, an effective control of the drain current, I_D, results. It is instructive here to recall that there exist only uncovered immobile charges in the depletion region, so the conductivity of this region is virtually neglible. Consequently, as the depletion region is made to increase in width, the path of the drain current in the n channel becomes more narrow. Accordingly, the path resistance increases and the drain current diminishes.

Volt-ampere Characteristic of the JFET. A typical set of drain characteristics for the 2N4222 JFET is depicted in Fig. 7-25. This set of curves very strongly resembles those of the conventional transistor in the common-base mode. Of course, the parameter here is the gate-source voltage, not the emitter-base current.

Let us direct attention first to the curve corresponding to $v_{GS} = 0$ in the configuration of Fig. 7-25. For positive values of the drain source voltage v_{DS} in the

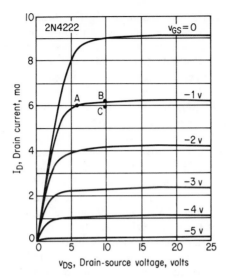

Fig. 7-25. Drain characteristics of the 2N4222 JFET.

vicinity of zero, the current is limited solely by the ohmic resistance of the n channel. By Ohm's law then, we have

$$i_D = \frac{v_{DS}}{l/\sigma A} \tag{7-40}$$

where l denotes the length of the n channel, A is the cross-sectional area, and σ denotes the conductivity. As v_{DS} increases, a reverse-biased voltage begins to appear, thus developing a depletion region that acts to decrease the effective cross-sectional area of the path of the drain current. Now, as v_{DS} increases further, the rate of increase of path resistance proceeds faster than the rate of increase in drain current owing to the increases in drain-source voltage. Consequently, the current curve begins to level off. Finally, a value of v_{DS} is reached beyond which no further significant increase in drain current occurs even for large changes in v_{DS}. In other words, the current saturates. The value of drain-source voltage at which current saturation begins is called the *pinch-off* voltage. This nomenclature is used to describe the squeezing effect on the drain current brought about by the two spreading depletion regions in the channel of the JFET. At the pinch-off voltage and in the vicinity of the drain terminal, the depletion regions have spread as far as they can go. It is not possible for the spread of the depletion regions to close off the channel completely. If this were to happen, the ohmic drop that is responsible for the nonuniform distribution of the depletion region would disappear, and the drain current would be re-established. Apparently, then, in the region beyond the pinch-off voltage, the JFET might best be described as being in a condition of dynamic balance.

What is the effect of varying the gate-source voltage when the drain-source voltage is kept fixed beyond the pinch-off value? It is helpful to keep in mind that the narrowest part of the drain-source current path occurs nearest the drain terminal for $v_{GS} = 0$. As v_{GS} is increased in the negative direction at fixed drain-source voltage, the reverse-biased voltage increases. But now the effect is to change the configuration of the depletion region about the source terminal S, where the depletion region is much less wide than it is near the drain terminal beyond pinch-off. The effect is an appreciable drop in drain current, as a glance at the drain characteristics of Fig. 7-25 readily discloses.

We have here a situation in which output current control is made possible by variation of the gate-source voltage. As a result, the JFET takes on the character of being a *voltage-sensitive* device with a very high input resistance. The input resistance is that which appears between the gate and source terminals. In view of the fact that these are reverse-biased terminals, negligible current flows. This explains why the value of the input resistance is of the order of 10^8 ohms.

The JFET is aptly named. As the gate-source voltage is varied to increase (or decrease) the reverse-biased voltage, especially as it appears between each p section and that part of the n channel about the source terminal, there occurs a further increase (or decrease) in the spread of the depletion region. Hence, more (or less) of the n channel becomes a region of uncovered charges and, thereby, reduced (or increased) conductivity. The drain current then drops (or rises). Throughout this action it is the *field effect* of the uncovered charges that influences the conductivity of the n channel and, therefore, the level of permissible drain current.

Although the treatment so far has made reference solely to an n-channel JFET, p-channel units are equally valid. The use of the latter requires reversing the polarity of the drain-source and gate-source voltages.

The useful portion of the volt-ampere characteristic of the JFET in amplifier applications is the region beyond the pinch-off voltages. Note, incidentally, that the pinch-off voltage diminishes with increasing values of the gate-source voltage. An expression that gives a good approximation of the manner in which the gate-source voltage determines the drain current is

$$I_{DS} = I_{DSS}\left(1 - \frac{V_{GS}}{V_{po}}\right)^2 \tag{7-41}$$

where I_{DSS} denotes the drain saturation current with the gate short circuited, and V_{po} is the value of the pinch-off voltage when v_{GS} is zero. Equation (7-41) describes the common-source transfer characteristic at constant drain-source voltage. A plot of this curve for a typical JFET is shown in Fig. 7-26.

The JFET Parameters. At this point we continue the policy that was established early in the chapter of identifying appropriate parameters for each new device as it is studied. This facilitates the analysis of those electronic circuits in which the device appears as a circuit element.

On the basis of the foregoing theory, it is certainly correct to state that generally the drain-source current is a function of both the gate-source voltage and the drain-source voltage. Expressed mathematically, we have

$$i_D = f(v_{GS}, v_{DS}) \tag{7-42}$$

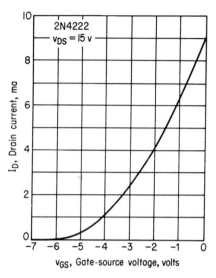

Fig. 7-26. Common-source transfer characteristic.

This equation involves the three variables i_D, v_{GS}, and v_{DS}. Accordingly, it is possible to identify three parameters. We need merely to allow any two variables to undergo change subject to the condition that the third variable be kept constant. To illustrate, consider the JFET to be operating initially at point A in the drain characteristics depicted in Fig. 7-25. At this operating point the drain current is 6 ma, the drain-source voltage is approximately 6 volts and the gate-source voltage equals -1 volt. In the interest of measuring the relative influence of a change in the drain-source voltage and the gate-source voltage on the drain current, consider first that with v_{GS} held momentarily fixed at -1 volt, the drain-source voltage is increased to 10 volts. This changes the operating point from A to B in Fig. 7-25. Note that the 4-volt increase in v_{DS} causes a slight increase in drain current. This increase may now by neutralized by allowing the gate-source voltage to increase slightly in the negative direction. It can be estimated from Fig. 7-25 that the new operating point moves from B to C by allowing a negative increase in the gate-source voltage of 0.1 volt. These figures thus point out that the change in the gate-source voltage is 40 times more effective than a change in the drain-source voltage. This ratio is called the *voltage amplification factor* of the junction field-effect transistor. It is readily computed for incremental changes from

$$\mu = -\left(\frac{\Delta v_{DS}}{\Delta v_{GS}}\right)_{i_D \text{ constant}} \tag{7-43}$$

A little thought about this calculation procedure should make it plain that the specific value of μ depends not only upon the size of the changes used, but also upon the particular initial operating point. For example, if A in Fig. 7-25 were selected initially at the ordinate line corresponding to a drain-source voltage of 15 volts, the value of the amplification factor would be found to be substantially greater than 40. In this connection, therefore, a more precise definition of the amplification factor is one that employs differentials rather than increments. Thus

$$\mu \equiv -\left(\frac{\partial v_{DS}}{\partial v_{GS}}\right)_{i_D \text{ constant}} \tag{7-44}$$

The minus sign appears because the changes in the quantities involved must move in opposite directions to maintain constant drain current. Since the JFET is used in the flat portions of the drain characteristics, high amplification factors of the order of several hundred are not uncommon.

A second parameter of the JFET can be identified by assuming operation to take place at constant gate-source voltage. The variables are then v_{DS} and i_D. Starting at the same operating point A of Fig. 7-25, it is clear that an incremental change of $+4$ volts in drain-source voltage brings about an increase of approximately 0.20 ma. Accordingly, we can identify a resistance parameter defined in incremental terms by

$$r_d = \left(\frac{\Delta v_{DS}}{\Delta i_D}\right)_{v_{GS} \text{ constant}} \tag{7-45}$$

This quantity is called the *incremental drain (or output) resistance*. For line AB of the drain characteristic of Fig. 7-25, this quantity has a value of approximately 20 kilohms. However, if operation were taking place beyond B, the quantity could easily take on values of several hundred thousand ohms. Typical values range anywhere from 0.1 to 1.0 megohms.

Again, a more precise definition is obtained by expressing Eq. (7-25) in terms of differentials. Thus

$$r_d \equiv \left(\frac{\partial v_{DS}}{\partial i_D}\right)_{v_{GS} \text{ constant}} \tag{7-46}$$

Note that this quantity at any operating point is conveniently identified in Fig. 7-25 as the reciprocal of the slope of the drain characteristic at fixed values of v_{GS}.

The third parameter of the JFET relates the drain current to the gate-source voltage at fixed drain-source voltage. Returning to Fig. 7-25, assume the operating point now to be B, corresponding to which $V_{DS} = 10$ volts, $I_{DS} = 6.2$ ma, and $V_{GS} = -1$ volt. Let the gate-source voltage increase negatively to -2 volts at a constant V_{DS} of 10 volts. The new drain current decreases to 4.2 ma. Accordingly, there occurs a change of 2 ma/gate-source volts. Since this quantity has the units of inverse ohms and since it relates an output quantity (drain current) to an input quantity (gate-source voltage), it is called the *mutual* or *transfer conductance*, or the *transconductance* for short. It is also frequently identified in manufacturers' literature as the *forward admittance*. In terms of incremental quantities, we can write

$$g_m = \left(\frac{\Delta i_D}{\Delta v_{GS}}\right)_{v_{DS} \text{ constant}} \tag{7-47}$$

Typical values range anywhere from 0.1 ma/v to 10 ma/v depending upon the operating point and the particular JFET used.

The differential form of Eq. (7-47) is expressed by

$$g_m \equiv \left(\frac{\partial i_D}{\partial v_{GS}} \right)_{V_{DS} \text{ constant}} \tag{7-48}$$

A glance at Fig. 7-26 reveals that this last expression is simply the slope of the transfer characteristic.

The foregoing three parameters of the JFET defined in terms of the partial derivatives of Eqs. (7-44), (7-46), and (7-48) are not independent, because they derive from a common expression, i.e., Eq. (7-42). The functional relationship existing between these parameters readily follows by writing the expression for the total differential of Eq. (7-42). Thus

$$\Delta i_D = \frac{\partial i_D}{\partial v_{GS}} \Delta v_{GS} + \frac{\partial i_D}{\partial v_{DS}} \Delta v_{DS} \tag{7-49}$$

Inserting Eqs. (7-46) and (7-48) yields

$$\Delta i_D = g_m \, \Delta v_{GS} + \frac{1}{r_d} \Delta v_{DS} \tag{7-50}$$

For convenience, if the condition is now imposed that Δv_{GS} and Δv_{DS} are so manipulated as to yield a Δi_D equal to zero, Eq. (7-50) may then be written as

$$0 = g_m \, \Delta v_{GS} + \frac{1}{r_d} \Delta v_{DS} \tag{7-51}$$

or

$$g_m r_d = - \left(\frac{\Delta v_{DS}}{\Delta v_{GS}} \right)_{i_D \text{ constant}} \tag{7-53}$$

Introducing Eq. (7-44) then leads to the simple result

$$\mu = g_m r_d \tag{7-53}$$

This last expression states that for any operating point on the drain characteristics, the product of the transconductance and the dynamic drain (or output) resistance is equal to the voltage amplification factor. It is instructive to note the similarity of Eq. (7-53), which applies to a voltage-sensitive device such as the JFET, with Eq. (7-39), which applies to a current-sensitive device such as the conventional transistor.

7-5   The Integrated-Gate FET (or MOSFET)

Recently, a modified version of the junction field-effect transistor has been developed that shows greater promise of achieving much wider commercial acceptance than the JFET. A much larger input resistance (about 100 times greater)

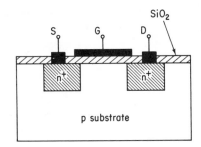

Fig. 7-27. Cross-sectional view of an integrated-gate FET illustrating composition.

and bigger transconductances are two of the reasons for this promise. A schematic diagram illustrating the composition of this device is depicted in Fig. 7-27. It is called the *integrated-gate field-effect transistor*, or IGFET.

The body of the IGFET consists of a lightly doped p substrate. Next, by means of masking and etching techniques, two highly doped isolated n regions are formed. These are called n^+ regions to emphasize the very large n-impurity content. This combination is then capped with a layer of silicon dioxide insulation by placing the substrate in a hot oven supplied with an oxygen environment. Finally, aluminum is deposited in the places indicated to establish the three terminals—D (drain), S(source), and G(gate). It is helpful to note that by this construction the metal gate terminal is insulated from the p and n^+ regions of the semiconductor by the silicon dioxide. For this reason this device is alternatively referred to as a metal-oxide-semiconductor FET or, in terms of its acronym, the MOSFET.

Figure 7-28(a) depicts the manner in which the MOSFET is wired to obtain the drain characteristics that appear in Fig. 7-29. The symbolic representation of

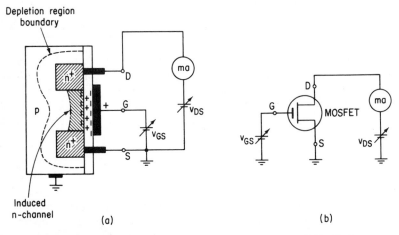

Fig. 7-28. (a) Diagram showing how the MOSFET is wired to obtain the drain characteristics depicted in Fig. 7-29; (b) symbolic representation.

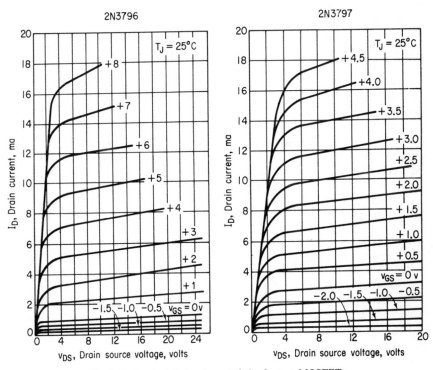

Fig. 7-29. Typical drain characteristics for two MOSFETs.

the MOSFET appears in Fig. 7-28(b). It is important to note that, unlike the JFET, the MOSFET is operated at positive gate-source voltages. Because of the lightly doped character of the substrate and the heavy dosage of n impurities in the n^+ sections, a widely spread depletion region is created that completely surrounds the n^+ sections, as shown in Fig. 7-28(a). For the moment, assume that the gate-source voltage is shorted, and let us recall what is taking place inside the depletion region. Besides the uncovered immobile charges that exist, electron-hole pairs are constantly being created as a result of thermal agitation of the pure silicon material of the substrate. Consider what happens next as the gate-source voltage is made to assume a positive potential. Immediately, this positive potential on the gate metal brings about a polarization of charges within the silicon dioxide insulator, as illustrated in Fig. 7-28(a). The positive charges take on a position of alignment along the inner surface of the insulator. Consequently, they attract the electrons in the depletion region to the region of the insulator, thereby creating an induced n channel that bridges the two n^+ sections of the MOSFET. If a positive drain voltage is now applied between the D and S terminals, n-channel electrons provide a conducting path between the two n^+ sections. This situation now resembles that of an n-channel JFET.

The magnitude of the gate-source voltage essentially determines how many thermally generated electrons from the depletion region will participate in the current flow from D to S. Hence drain current increases rapidly at first with increasing drain-source voltage, as depicted in Fig. 7-29. But a point of saturation is soon reached beyond which only small increases of drain current accompany large increases in drain-source voltage. To effect further appreciable increases in drain current, it becomes necessary to increase the gate-source voltage in order that additional electrons be drawn from the depletion region into the induced channel to partake in the current flow from drain to source.

An extension of the foregoing reasoning should make it apparent that when the gate-source voltage takes on negative values, the thermally generated electrons will be repelled from the region between the n^+ sections, thereby causing a decrease in conductivity. A glance at the drain characteristics of Fig. 7-29 for negative values of V_{GS} bears out this conclusion.

Investigators have shown that the drain characteristics of the MOSFET above the knee of the curves can be fairly well approximated by the expression

$$i_D = I_{po}\left(1 + \frac{v_{GS}}{V_{po}}\right)^2 \tag{7-54}$$

where I_{po} denotes the current beyond the knee of the drain characteristic for a zero gate-source voltage, and V_{po} is the corresponding value of the drain-source voltage. Appearing in Fig. 7-30 are the transfer characteristics of the MOSFETS whose drain characteristics appear in Fig. 7-29.

The parameters of the MOSFET can be identified and obtained in exactly the same manner as was done for the JFET. Hence, Eqs. (7-44), (7-46), (7-48), and (7-53) apply with equal validity.

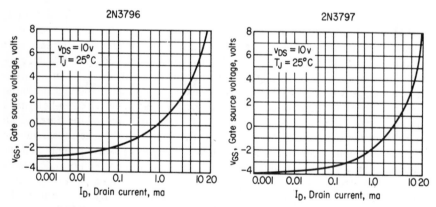

Fig. 7-30. Typical transfer characteristics for the MOSFETs whose drain characteristics appear in Fig. 7-29.

PROBLEMS

7-1 Indicate whether the following statements are true or false. When the statement is false, supply a reason to justify it:
(a) An example of an intrinsic semiconductor is pure Ge at zero degrees (Kelvin) absolute.
(b) An *n*-type impurity semiconductor is one in which the electron energy level of the impurity is selected close to the valence band of the intrinsic material.
(c) In a *p*-type impurity semiconductor, minority carriers exist by virtue of the presence of the impurity.
(d) A transistor is equipped with a filament heater supply voltage as is the vacuum-tube device.

7-2 A *p-n* junction diode has an equilibrium-value charge density in the *n* region of 2×10^{11} holes/cm^3. The electron density at equilibrium in the *p* region is 3×10^9 electrons/cm^3. The energy equivalent of ambient temperature is 0.026 v.
(a) Find the density of holes injected into the *n* region when a forward bias of 0.2 volt is applied to the junction diode.
(b) What is the density of injected electrons in part (a)?
(c) Sketch the variation of the injected charge densities in the *p* and *n* regions. Show also the corresponding variation of the diffusion currents.

7-3 Explain why Eq. (7-13a) and Eq. (7-13b) correctly represent the total diffusion current of the *p-n* junction diode.

7-4 A *p-n* junction diode has a reverse saturation current of 10^{-5} ampere. The diode is operated in an ambient temperature of 600°K. A forward bias of 0.2 volt is applied.
(a) Find the apparent diode resistance.
(b) Find the incremental resistance.

7-5 The volt-ampere characteristic of a junction diode is described by the equation $I = 10^{-5}(\epsilon^{38.5V} - 1)$. The diode is placed in series with an ohmic resistance of 5 ohms and a 1-v d-c source. Determine the diode current.

7-6 A junction diode at room temprature has a reverse saturation current of 5 μa. Find the maximum and minimum currents that flow in this diode when a voltage wave of 0.1 sin ωt is superimposed on a 0.2-v d-c source and applied to the diode.

7-7 The junction diode of Prob. 7-5 is placed in series with a resistor and a 1-v battery The resulting current is found to be 100 ma. What is the value of the series resistance?

7-8 A *p-n* junction diode has an equilibrium hole density of 2×10^{11} holes/cm^3. The average diffusion length of the holes is 0.1 cm and the length of the *n* section is 0.5 cm. The cross-sectional area at the junction is 0.001 cm^2. The diode is used at room temperature. Find the value of the diffusion capacitance due to holes associated with a change in forward bias from 0.1 v to 0.2 v.

7-9 A 2N104 transistor is used in the common-base configuration. The average collector characteristics can be found in any RCA semiconductor handbook.

(a) The emitter current is increased from 15 ma to 20 ma at fixed collector-to-base voltage. The corresponding increase in collector current is from 14.6 ma to 19.4 ma. What is the value of the current amplification factor?

(b) The transistor is operating at $V_{CB} = -5$ v and $I_E = 10$ ma. Find the value of the collector current when the emitter current is decreased by 5 ma and the collector-to-base voltage is doubled.

7-10   A 2N104 transistor is used in the circuitry of Fig. 7-21(a). Also $V_{CC} = 12$ v, $R_L = 400$ ohms, $\alpha = 0.97$, and $I_E = 10$ ma.

(a) Determine the value of the collector-to-base voltage.

(b) Find the change in the output voltage developed across R_L when the input voltage changes by 2.6 mv.

(c) What is the voltage gain in part (b)?

7-11   A 2N104 transistor is used in the circuitry of Fig. 7-21(a) where $R_L = 400$ ohms and $\alpha = 0.97$. For a specific change in v_e the voltage gain is found to be 125. What is the level of the collector current during this point of operation of the transistor?

7-12   Indicate whether the following statements are true or false. When the statement is found to be false, supply a reason to justify it.

(a) The field-effect transistor is so named because it produces an electric field that completely surrounds the body of the transistor.

(b) A p-channel JFET is used with the drain terminal connected to the positive side of the drain-source voltage supply.

(c) The channel of the JFET is more heavily doped than the regions immediately around the gate terminals.

(d) In a properly biased JFET, the depletion region spreads over a larger region as the magnitude of either the drain-source or the gate-source voltage is increased.

(e) At the pinch-off voltage, the drain-source current is cut off.

(f) Changes in the gate-source voltage primarily influence the depletion region in the vicinity of the drain terminal.

(g) The JFET is a current-sensitive device.

(h) The ohmic drop in the n channel of a properly biased JFET in no way influences the boundaries of the depletion regions.

(i) The conductivity of the depletion region is appreciable.

(j) Changes in the drain-source supply voltage do not affect the spread of the depletion regions.

7-13   The drain and transfer characteristics of the 2N4222 JFET are depicted in Figs. 7-25 and 7-26. This JFET is operated at a drain-source voltage of 15 and a gate-source voltage of -2. The incremental drain resistance is known to be 120 kilohms. Find the value of the forward admittance and amplification factor as they apply to the specified operating point.

7-14   The 2N3796 MOSFET is operated at a drain-source voltage of 16 volts and a gate source voltage of $+2$ volts.

(a) Find the drain-source current in milliamperes.

(b) Determine the incremental drain resistance.

(c) What is the value of transconductance at the specified operating point?

(d) Find the value of the amplification factor.

7-15   A MOSFET 3797 is operated at a drain-source voltage of 10 volts. Compare the value of the forward admittance prevailing at a gate-source voltage of -3 with that prevailing at $+3$ and comment.

8

electron control devices
—vacuum types

This chapter is devoted to a description of the input-output characteristics and the parameter representations of those electron control devices that are of the high-vacuum type. These devices were used extensively before the advent of the transistor. Although the semiconductor devices described in Chap. 7 have replaced the vacuum tubes almost completely, there are many pieces of industrial, commercial, and domestic equipment still functioning with vacuum tubes. Your television set is very likely still employing vacuum tubes in one form or another.

One of the chief shortcomings of the vacuum tube is the need to supply thermal energy to a cathode to make available electrons for control purposes. Moreover, upon switching, a warm-up time is always necessary before equipment employing vacuum tubes becomes operational. The transistor has no such disadvantage. Its source of electrons comes from the presence of impurities, and these are ready to go to work immediately upon switching on the equipment. Another serious drawback is size. For example, a transistor amplifier together with its supply voltage occupies only a very small fraction of the space required by its vacuum-tube counterpart. This advantage is even further exploited with the more recent advances in integrated circuitry. The relatively small space required to house giant computer centers stands as eloquent testimony to the advantages of the semiconductor devices.

A study of vacuum-tube electronics, however, is important from a historical viewpoint, as well as for the fact that equipment employing these devices will still be encountered, particularly in special applications involving high power at high frequencies. Moreover, it is also pedagogically gratifying to see just how the semiconductor units have come to replace the specific functions previously fulfilled by the various types of vacuum tubes. The similarities between the parameter representations of the field-effect transistors and the high-vacuum triode are indeed striking. The same is true of the volt-ampere output characteristics of the transistor and the plate characteristics of the high-vacuum pentode.

8-1 Richardson-Dushman Equation and the Langmuir-Child Law

Although the first vacuum diode was developed in 1904 by F. A. Fleming, it was not until two decades later that a complete mathematical description of this electronic device was formulated. Figure 8-1 depicts the essential features of the vacuum diode. It is composed of three elements placed inside a glass-enclosed high vacuum. A potential difference is placed across the two elements called the *plate* and *cathode* electrodes. The cathode emits electrons as a result of the action of the heater element and these electrons are then collected at the plate electrode by virtue of the electric field existing between plate and cathode.

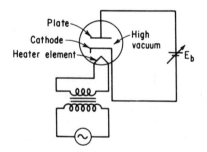

Fig. 8-1. Construction features of the vacuum diode.

To describe properly the amount of current that flows in such a situation, it is necessary to study two things: (1) the factors determining the number of electrons which can be emitted from the cathode material, and (2) the effect which these emitted electrons have on each other as well as the influence of the electric field acting upon the electrons. Knowledge about the first item was uncovered through the efforts of O. W. Richardson† and S. Dushman‡. On the basis of theoretical calculations using classical kinetic theory both investigators were able to show that electrons could be liberated from metals provided that sufficient *thermal energy* was imparted to the fastest-moving electrons so that they could overcome

† O. W. Richardson, *Phil. Mag.* (May 28, 1914), 633.

‡ S. Dushman, *Phys. Rev.*, **21** (1923), 623.

the surface-potential barrier. Expressed in terms of the current resulting from the liberated electrons, they showed that

$$I_{th} = SA_0 T^2 \epsilon^{-E_T/E_W} = SA_0 T^2 \epsilon^{-b_0/T}$$ (8-1)

where I_{th} = thermionic emission current, amp
   S = cathode emitter surface area, m²
   A_0 = a constant characteristic of the material and having units of amp/m² (°K)²
   T = temperature, °K
   E_W = work function, electron-volt
   E_T = $T/11,600$, the electron-volt equivalent of temperature
   $b_0 \equiv 11,600 E_W$, °K

The *electron-volt* is a unit of energy equal to 1.6 = 10^{-19} joule. Expressed differently, it is the energy associated with an electron when it experiences a one-volt change in potential. Equation (8-1) is known as the Richardson-Dushman equation and is often referred to as the *thermionic emission equation*.

Although Eq. (8-1) indicates that the thermionic emission current is directly dependent upon the square of the temperature, it is the exponential term which is the most influential. Accordingly, it is found that I_{th} is a very sensitive function of both E_W and T. The quantity E_W is called the *work function* because it represents the amount of energy which must be imparted to the fastest-moving electrons in a metal in order to free them from the bounds of the metal. Appearing in Fig. 8-2 is a graphical picture of this situation. The quantity E_B is the potential energy barrier existing at the surface of a metal, and E_f† is the *Fermi energy level*

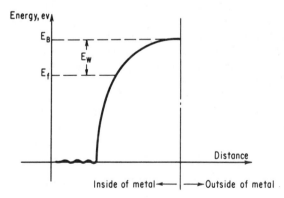

Fig. 8-2. Energy diagram showing how the boundary surface of a metal is described in terms of its potential distribution curve. The distances involved are infinitesimal.

† The Fermi energy level E_f for any given metal is readily calculable from a knowledge of the mass and charge of the electron, Planck's constant, and the atomic weight, specific gravity, and the number of free electrons of the metal.

representing the maximum kinetic energy which an electron can possess at zero degrees Kelvin. At this temperature no electrons can escape the metal because E_f is always less than E_B. However, as the temperature is increased from absolute zero and energy is imparted to the fastest-moving electrons a situation is reached where the total energy of these electrons exceeds the surface barrier potential E_B and so the electrons are liberated. It is important, of course, that these electrons be moving towards the surface and not towards the interior of the metal. The *work function* then is defined as the amount of energy that must be imparted to the fastest-moving electrons to give them a total energy equal to the barrier potential energy. Expressed mathematically

$$E_W = E_B - E_f \quad \text{ev} \tag{8-2}$$

For the cathode emitters used in practice, the values of the work function range from 1 to 4.5 electron-volts.

The electron-volt equivalent of temperature is defined by the relationship

$$eE_T \equiv kT \tag{8-3}$$

where k is the Boltzmann gas constant and is equal to 1.38×10^{-23} joule/°K. Upon substituting this value for k and the value for e in Eq. (8-3), we obtain

$$E_T = \frac{k}{e} T = \frac{T}{11,600} \quad \text{ev} \tag{8-4}$$

Equation (8-4) thus provides a rapid conversion from temperature to its energy equivalent expressed in units of electron-volts as called for in the Richardson-Dushman equation.

Notwithstanding the theoretical origin of the thermionic emission equation, there exists ample experimental evidence in support of this equation.[†] However, some modifications are necessary because b_0 in Eq. (8-1) is found to vary with temperature. For this reason the value of A and b_0 are determined experimentally in order that the computed and experimental results may be made to correspond. In this sense, then, it is interesting to note that the thermionic emission equation is an empirical equation the form of which was established by the theoretical work of Richardson and Dushman.

Information about the second item of importance relating to the description of the characteristics of high vacuum devices was provided by the work of I. Langmuir[‡] and C. D. Child.[§] These investigators showed that the actual current collected at the plate corresponding to a potential difference of E_b across the plate and cathode is given by

$$I_b = \left(2.33 \times 10^{-6} \frac{S_p}{d^2}\right) E_b^{3/2} \quad \text{amp} \tag{8-5}$$

where d denotes the distance between plate and cathode and S_p is the area of the

† S. Dushman, "Thermionic Emission," *Revs. Mod. Phys.*, **2** (Oct. 1930), 381–476.

‡ I. Langmuir, "The Effect of Space Charge and Residual Gases on Thermononic Currents in High Vacuum," *Phys. Rew.*, **2** (Dec. 1913), 450–486.

§ C. D. Child, "Discharge from Hot CaO," *Phys. Rev.*, **27** (May 1911), 392–511.

plate in square meters. Equation (8-5) is derived subject to the following assumptions: the electrodes are flat, parallel equipotential surfaces; only electrons having zero initial velocities are present in the interelectrode space, and the number of electrons emitted by the cathode exceeds the number which can be received by the plate for the given potential E_b. The important thing to note about Eq. (8-5) is the dependence of the current flowing through the diode upon the three-halves power of the voltage appearing across the electrodes. This expression states that for a given geometry of plate and cathode the current flow is solely dependent upon E_b. The only restriction is that $I_{th} \geq I_b$, since obviously no more electrons can be collected at the plate than are emitted by the cathode. The fact that I_b can be less than I_{th} is attributable to space-charge effects. That is, when the temperature of the cathode is raised to the point where more electrons are emitted than can be received by the plate, the excess electrons have a repelling effect upon each other, thereby causing a portion of the electrons to return to the cathode. For this reason Eq. (8-5), which is known as the Langmuir-Child law, is often referred to as the *space-charge equation*.

What happens to the form of Eq. (8-5) when the plate and cathode are arranged in a configuration which is different from that of flat, parallel surfaces? Under these circumstances Langmuir showed that the current collected at the plate continued to be functionally dependent solely upon the three-halves power of E_b. The only change was in the geometry factor. Accordingly, a general formulation of the Langmuir-Child law for the diode plate current is

$$\boxed{I_b = KE_b^{3/2}} \quad \text{amp} \tag{8-6}$$

where K is a geometry factor depending on the shape and separation of the cathode and plate.

A plot of the diode volt-ampere characteristic is depicted in Fig. 8-3. Note that as long as $E_b < E_b'$ the diode current is determined by Eq. (8-6). This is the space-charge limited current. However, when E_b is increased beyond E_b', then all the electrons which are emitted by the cathode are received by the plate. Then the plate current I_b becomes equal to the thermionic current, which in turn is determined by the assumed fixed cathode temperature. Because vacuum diodes are electron *control* devices it is important to understand that diodes are seldom operated in the saturated or temperature-limited region (i.e. in the flat portion of the curve); they are operated in the space-charge limited region (i.e. $0 \leq E_b < E_b'$). Another matter worthy of note here is that there is *zero current* (ideally) whenever the potential of the plate is made negative relative to the cathode. This is illustrated in Fig. 8-3 and is valid provided the plate cathode voltage is not made excessively large in the negative sense.

In 1906 L. De Forest introduced a third electrode into the vacuum tube in the form of a wire mesh placed close to and surrounding the cathode. This third electrode was called the *grid* and the resulting device a *triode*. De Forest discovered that he could obtain much greater control over the plate current by varying the negative potential applied to the grid than he could be varying the

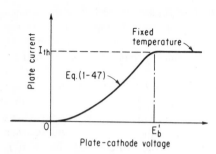

Fig. 8-3. Volt-ampere characteristic of the vacuum diode for a fixed cathode temperature.

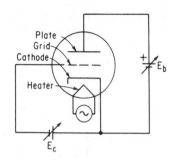

Fig. 8-4. Schematic diagram of the vacuum triode.

plate-to-cathode potential. So important was this discovery that it ushered in the electronics industry as we know it today. Appearing in Fig. 8-4 is the schematic representation for the triode.

As is the case with the diode, it is important to know the volt-ampere characteristic of the vacuum triode in order to deal with it properly in electronic circuitry. Experimental evidence reveals that the plate current of the triode may be represented approximately by the expression

$$I_b = K(\mu E_c + E_b)^n \quad \text{for} \quad (\mu E_c + E_b) > 0 \qquad (8\text{-}7)$$

where μ denotes a proportionality factor which accounts for the greater control that the grid voltage E_c bears on the flow of electrons because of its greater proximity to the cathode, K represents a proportionality factor which is dependent

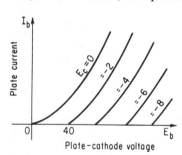

Fig. 8-5. Volt-ampere characteristic of the vacuum triode for various values of grid potential.

on physical dimensions, and n is an exponent which is approximately equal to 3/2. Depicted in Fig. 8-5 is a family of curves of space-charge limited plate current for the triode for various values of fixed grid potential. These curves have approximately a three-halves power relationship. The curve for $E_c = 0$ is nothing more than the diode characteristic referred to in Fig. 8-3. Moreover, note from Eq. (8-7) that if $\mu = 20$ and E_c is set at -2 then the expression for the plate current becomes

$$I_b = K(-40 + E_b)^n$$

Since the vacuum device is such that no plate current can flow for a negative value of the argument, it follows that the current will be zero until E_b exceeds 40 volts. Essentially, then, the effect of the grid potential is to cause a shift to the right of the diode characteristic, as depicted in Fig. 8-5.

The manner in which the diode and triode devices are used in electronics is the subject matter of Chap. 10. It is well to keep in mind that our concern here has been to describe the fundamental relationships which determine the characteristics of these devices. In Chap. 10 attention is devoted to the manner in which these electronic devices influence the behavior of circuits of which they are a part. By combining their external characteristics with a knowledge of circuit theory we shall readily accomplish the study of electronic circuitry.

8-2   The Vacuum Diode

In the preceding section it was shown that the temperature-limited current produced in a diode is very critically dependent upon the cathode material and its characteristics. The expression for this current, Eq. (8-1), is repeated here for convenience. Thus

$$I_{th} = SA_0 T^2 \epsilon^{-Ew/E_T} \quad \text{amp} \qquad (8\text{-}1)$$

where the letters have the meanings previously cited. It is important to note the dependence of this thermionic current upon the square of the temperature of the cathode material as well as upon its work function. Because of the exponential nature of this equation, it follows that the thermionic current is strongly influenced by the work function of the cathode material. The lower the work function, the higher the thermionic current. At the same time, however, the cathode material must be capable of withstanding high operating temperatures in order to take advantage of the square-law variation with temperature. For a material such as copper, which has the advantage of a low work function, it turns out that it cannot be used because it melts at 1083°K. Consequently, it vaporizes before it begins to emit electrons.

Tungsten is the most commonly used cathode material even though it has a slightly higher work function than copper. The important factor in its favor is that it melts at 3650°K. But often it is operated at about 2500°K, thus combining reasonable values of thermionic current with long life. In its pure form tungsten is used generally in applications involving potential differences exceeding 5000 volts. At lower voltages and power, it is combined with other elements in the interest of increasing the electron emission efficiency, which is defined as the emission in amperes per watts of heating power applied to the material. For example, in diode applications involving voltages from 750 to 5000 volts, *thoriated tungsten* is used as the cathode. If a thin layer of low-work-function material such as thorium is deposited on the tungsten, a tenfold increase in emission efficiency results. The upper voltage limit is placed on these diodes because at these potential levels sufficient energy can be imparted to the few gas ions (which are always present in the near-perfect vacuum) to cause them to deactivate the cathode layer. For applications below 750 volts oxide-coated cathode materials are used, such as *barium oxide* or *strontium oxide*. A further increase in cathode emission efficiency results with these materials. Since the majority of diode applications are in circuits involving voltages below 750 volts, most commercial diodes are equipped with oxide-coated cathodes.

Potential Profiles. Let us turn next to a consideration of the variation of the potential in the space between the cathode and plate electrodes of the vacuum diode. It is assumed that these electrodes are flat and placed parallel to each other. Moreover, the potential difference between the plate and cathode is assumed fixed and equal to E_b. Figure 8-6 depicts the situation. Note that the potential of the cathode is taken as the reference point and that the total distance between the electrodes is d. With the cathode unheated and a potential difference applied between plate and cathode, it follows that an electric field is established between these electrodes bearing the units of volts per meter. If one starts at the cathode and integrates the electric field over the distance d, the result is a potential at the plate of value E_b— a result which is entirely expected. A plot of the change in voltage that occurs in the space between cathode and plate as one moves from the former to the later is shown in Fig. 8-7(a). In the pictorial representation of Fig. 8-7(a′) electric field lines are depicted leaving the plate and terminating on the cathode. Note that for any vertical line drawn anywhere between cathode and plate the number of electric field lines is the same; in other words, the electric field is a constant. This result is also borne out in Fig. 8-7(a) by the straight line. The slope of this line is the electric field intensity.

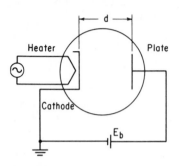

Fig. 8-6. Vacuum diode with parallel plate electrodes.

Now assume that the temperature is increased to some value T_1 degrees Kelvin.

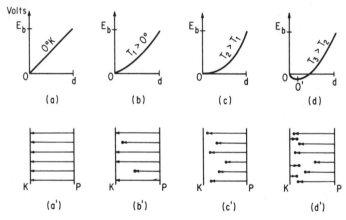

Fig. 8-7. Top row: Potential variation between cathode and plate in a vacuum diode as the temperature is increased.
Bottom row: Pictorial representation in terms of electric field lines. Here K denotes cathode, P denotes plate and ● denotes emitted electron.

How does this affect the curves of Figs. 8-7(a) and (a')? The increased temperature causes some electrons to be emitted from the cathode into the space immediately outside, thus causing an electron-cloud effect. If T_1 is assumed to be of moderate value, the electron density of this cloud will not be too high. Therefore most of the field lines from the plate continue to terminate on the cathode itself. However, some of the field lines will terminate on the emitted electrons as depicted in Fig. 8-7(b'). Now note that depending upon where one draws a vertical line in the space between plate and cathode, more or fewer electric field lines are encountered. Accordingly the value of the electric field intensity assumes different values as illustrated in Fig. 8-7(b) by the nonconstant slope of the curve.

If the cathode temperature is next assumed to be increased to that value which allows it to emit enough electrons so that all the field lines originating from the plate now terminate on electrons in the interelectrode space, none of the field lines terminates on the cathode itself. Consequently, the potential gradient at the cathode is zero and the potential distribution curve appears as shown in Fig. 8-7(c). Because of the greater energy associated with the higher temperature, there are some electrons that will have higher energies than others and so will be farther removed from the cathode, as illustrated in Fig. 8-7(c'). It is important to understand that the picture being described here is not a static one. Stating that an electric field line terminates on an electron does not imply that this electron remains suspended in space. On the contrary, it is immediately accelerated to the plate electrode, thus constituting part of the current flow. However, Fig. 8-7(c') can still be used to represent the situation, because as one electron is accelerated towards the plate there is another in the immediate vicinity to take its place.

Depicted in Figs. 8-7(d) and (d') is the situation where the temperature of the cathode has been increased to the point where more electrons are emitted than can be received by the plate for the given potential E_b. As a result there occurs a bunching of excess electrons at some point to the right of the cathode, viz. 0'. There occurs such a high concentration of electrons at 0' that its potential is found to be lower than that of the cathode. For this reason it is called a *virtual cathode*, because electrons appear to be originating from that point in the interelectrode space. Note that the potential distribution curve is negative between 0 and 0'. Thus the high concentration of electrons at the virtual cathode causes the electric field to be reversed in direction from what it was originally in this region of space. In this way some electrons are forced back into the cathode. An interesting aspect of this situation is that any further increase in temperature in no way changes the number of electrons arriving at the plate. That is, the plate current remains the same. This result is attributable to the repelling action which the excess electrons have upon each other. This action is commonly referred to as the *space-charge effect*, and the diode plate current as given by Eq. (8-1),

$$I_b = KE_b^{3/2} \quad \text{amp} \tag{8-6}$$

is called the space-charge limited current. As long as there are more electrons emitted than can be received by the plate, *control of the plate current can be effected only by adjusting the plate voltage E_b.*

Forward Resistance of the Vacuum Diode. In accordance with Eq. (8-6) the volt-ampere characteristic of the vacuum diode assumes a three-halves-power law variation as depicted in Fig. 8-8. For any point of operation Q corresponding to an applied plate-cathode voltage E_{b0}, there exists a current I_{b0}. Therefore one may look upon the diode as a vacuum device which has an *apparent forward resistance* given by

$$R_f \equiv \frac{E_{b0}}{I_{b0}} = \frac{E_{b0}}{KE_{b0}^{3/2}} = \frac{1}{KE_{b0}^{1/2}} \quad \text{ohms} \qquad (8\text{-}8)$$

This quantity is represented graphically as the reciprocal of the slope of the line $0Q$. It is called an apparent resistance because R_f is defined in terms of the *total* values of voltage and current. It is called a forward resistance because the plate is assumed to be positive with respect to the cathode.

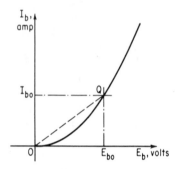

Fig. 8-8. Volt-ampere characteristic of the vacuum diode in the space-charge limited region.

In some applications small changes in E_b about the Q-point may be allowed to occur. In such a case the resistance seen by the changing component of the plate voltage is no longer R_f but rather the reciprocal of the slope of the volt-ampere curve at Q. This resistance is called the *dynamic* or *incremental plate resistance* and is defined as

$$r_p \equiv \frac{dE_b}{dI_b} \qquad (8\text{-}9)$$

Upon differentiating Eq. (8-6) and formulating dE_b/dI_b we get

$$r_p = \frac{2}{3}\left(\frac{1}{KE_b^{1/2}}\right) = \frac{2}{3}R_f \qquad (8\text{-}10)$$

Thus a comparison of Eqs. (8-8) and (8-10) reveals that for a diode which is operated in the space-charge limited region the incremental resistance is always less than the apparent resistance by a factor of 2/3. Typical values of r_p for various vacuum diodes vary from several hundred to about one thousand ohms.

Often the diode is used in circuits involving very high-resistance elements. In such cases the resistance of the diode may be neglected, which means that the diode may be considered to have the ideal volt-ampere characteristic shown in

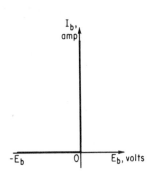

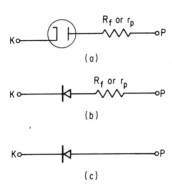

Fig. 8-9. Volt-ampere characteristic of an ideal diode; i.e., $R_f = 0$ and $R_b = \infty$.

Fig. 8-10. Circuit symbols used to represent the diode with a forward bias:
(a) practical vacuum diode;
(b) alternate symbolism;
(c) ideal diode.

Fig. 8-9. A little thought reveals that this characteristic has associated with it a zero forward resistance and an infinite backward resistance.

The circuit symbols used to represent both the practical and the ideal diode are shown in Fig. 8-10. The diode terminals are denoted by K and P, and that which appears between these terminals in the equivalent circuit of the diode.

8-3 The Vacuum Triode

Control of the tube current in the diode is achieved by adjusting the value of the plate-to-cathode voltage. However, in the triode far greater control of the tube plate current can be obtained by varying the negative potential of the grid. Recall that the grid is a wire mesh which is wrapped around the cathode in the fashion depicted in Fig. 8-11. Small variations of this gird voltage can be made to produce the same effects as very large variations in the plate-to-cathode voltage. Essentially this result is accomplished by having the negative potential on the grid behave as a valve controlling the flow of electrons from the cathode to the plate. When the grid electrode is made to assume large negative potentials, it creates a retarding electric field between the grid and the cathode which prevents any electrons from reaching the plate. In other words, the closer proximity of the grid to the cathode with its large negative potential predominates over the influence of the positive potential of the plate, which is farther removed from the cathode. In this state the vacuum tube is said to be operating *beyond cutoff*.

As the potential of the grid is made less and less negative relative to the cathode, more and more electrons find

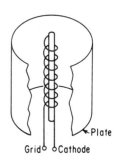

Fig. 8-11. Triode structure.

their way through to the plate electrode, the potential of which is assumed to remain fixed for simplicity. In this fashion control of the plate current in a triode is achieved.

To get a better understanding of the control action of the grid, refer to the potential profiles depicted in Fig. 8-12 for three values of negative grid potential. Appearing in Fig. 8-12(a) is the schematic diagram showing how the plate and grid voltages are applied to the three electrodes of the triode. The resultant grid-to-cathode voltage is denoted as e_c. The fixed plate-to-cathode voltage is called E_b. When e_c is made highly negative the tube is beyond cutoff so that the plate current is zero. Figures 8-12(b) and (b′) depict the electric field conditions prevailing within the triode for this condition. Note that the large negative grid potential causes the potential distribution curve between the cathode and the grid to be negative, thus producing a retarding field which prevents electrons from reaching

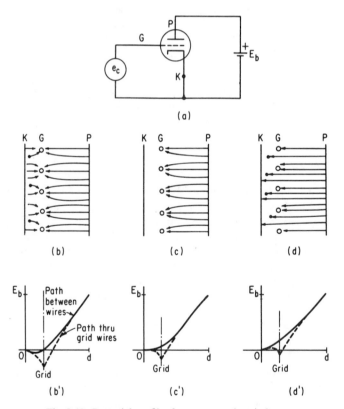

Fig. 8-12. Potential profiles for vacuum tube triode:
(a) Schematic diagram;
(b and b′) grid negative and beyond cutoff;
(c and c′) grid negative and at cutoff;
(d and d′) grid negative at one-half cutoff value.

the plate. If e_c is changed to that value of negative potential which causes the electric field at the cathode to be everywhere zero, the potential distribution curve then appears as shown in Fig. 8-12(c'). In this state the tube is said to be *at cutoff.* If the grid is made still less negative, the potential profile changes to the variation shown in Fig. 8-12(d'). Here note that for a path between the grid wires the potential distribution curve is always positive, which means that an accelerating electric field exists between the cathode and plate. Hence electrons readily reach the plate, thus establishing a current flow. Note too that in this case the electric field lines originating from the plate no longer terminate solely on the grid wire. Some now terminate on the cathode and others on electrons in the electron cloud outside the cathode.

If e_c is allowed to vary between zero and its cutoff value, considerable variation in the plate current can be obtained. Appreciable amplification of this varying grid voltage can be achieved if the varying plate current is caused to flow through a resistor placed in series with the plate electrode. This, in fact, is the principle of the voltage amplifier and is described further in Chap. 10. In such applications care is often taken to prevent the grid voltage e_c from becoming positive in order to avoid the flow of current in the grid-cathode circuit. In the sense described above, then, the triode may be looked upon as a *voltage-sensitive* control device—in contrast with the transistor, which is a current-sensitive control device.

Plate and Transfer Characteristics. On the basis of the foregoing discussion it should be clear that the plate current of the triode is a function of both the plate-to-cathode voltage and the grid-to-cathode voltage, with the latter exerting the greater influence. Expressing this mathematically we can write

$$i_b = K(\mu e_c + e_b)^n \tag{8-11}$$

where μ is a proportionality factor which accounts for the greater influence the grid voltage has in controlling i_b, K is a function of the geometry of the electrodes, and n is an exponent which is approximately equal to 3/2. Since i_b is a function of two variables, e_c and e_b, it is possible to fix either variable at different values and thereby obtain a family of curves describing the plate current as a function of the other variable. Thus, if e_c is fixed at various values of negative potential ranging from zero to -24 volts for a 6SN7 triode, the variation of the plate current i_b with plate voltage e_b is found to behave as depicted in Fig. 8-13. These curves are called the *average plate characteristics* of the triode because they relate the manner in which plate current changes with plate voltage for fixed grid potential.

Equation (8-11) can be used to describe the plate characteristics, but for each value of e_c it would require an adjustment in the parameters K, μ, and n. Information about these parameters in turn would have to come from experimental data taken on the given tube. Consequently, in dealing with electronic devices of this type it is customary to work directly with experimentally established and graphically represented average characteristics such as appear in Fig. 8-13. These characteristics are published and widely distributed by the vacuum

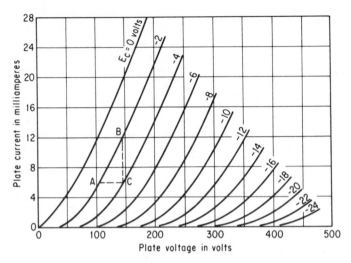

Fig. 8-13. Triode plate characteristic of a 6SN7 tube.

tube manufacturers. Since these are average characteristics, it is not unlikely that the actual characteristics of a specific tube used in a given application will differ from the published curves by as much as 5 or 10 per cent. Seldom, however, is this deviation of significant consequence. When it is important, corrective steps can readily be taken.

The plate characteristics of the tube contain the information concerning its performance capability. Accordingly, it should be possible by means of these curves to illustrate the predominating influence of the grid potential in controlling the plate current. In this connection, assume that in the circuit arrangement of Fig. 8-11(a) the grid voltage is fixed at E_c† = −2 volts and the plate voltage is fixed at E_b = 105 volts. This locates point *A* in Fig. 8-13 as the operating point to which there corresponds a plate current of 6 ma. Now let the plate potential increase to 145 volts, keeping the grid voltage fixed at −2 volts. This puts the operating point at *B*, thus yielding a plate current of 12 ma. If the plate voltage is then kept constant at 145 volts and the grid potential is made more negative by changing it from −2 to −4 volts, the point of operation goes from *B* to *C* in Fig. 8-13. At *C* the plate current is once again 6 ma. It is significant to note that a change of 2 volts in grid potential in the negative sense has completely nullified the effect of a 40-volt change in plate potential in the positive sense. Therefore it is said that in the 6SN7 triode the grid is 20 times more effective in controlling the plate current than is the plate potential. This effectiveness factor is commonly called the *voltage amplification factor* of the tube.

A mathematical description of the foregoing behavior follows from Eq. (8-11). Assume that e_c and e_b both undergo changes about some initial operating point

† Note that when e_c and e_b are assigned specific values, capital letters are used. Lower-case letters are reserved to represent instantaneous values.

having grid and plate voltages of value E_c and E_b, respectively. Then, in general, there will be a change in plate current of value ΔI_b. Thus we can write

$$I_b + \Delta I_b = K(\mu E_c + \mu \Delta E_c + E_b + \Delta E_b)^n \tag{8-12}$$

where I_b, E_b, and E_c identify the initial point. If the condition is now imposed that the grid and plate voltages are to be changed in such a manner that ΔI_b is to be zero, it follows that the sum of the incremental quantities on the right side of Eq. (8-12) must also be zero. Hence

$$\Delta E_b + \mu \Delta E_c = 0 \tag{8-13}$$

Equation (8-13) thus leads to the manner by which the proportionality factor μ can be identified. Specifically

$$\mu = -\frac{\Delta E_b}{\Delta E_c}\bigg]_{I_b \text{ constant}} \tag{8-14}$$

Using the numbers for the example illustrated in Fig. 8-13 we have

$$\Delta E_b = 145 - 105 = 40 \text{ volts} \tag{8-15}$$

$$\Delta E_c = -4 - (-2) = -2 \text{ volts} \tag{8-16}$$

Hence, the amplification factor of the tube is

$$\mu = -\frac{40}{-2} = 20 \tag{8-17}$$

It is worthwhile to note that the quantities $\mu \Delta E_c$ and ΔE_b are each represented by line AC in Fig. 8-13, subject to the condition that the plate current remain invariant. This observation provides a simple means of finding μ from the plate characteristics of a triode.

A plot of i_b versus e_c, with e_b held fixed at various suitable values, yields the family of curves depicted in Fig. 8-14. These curves are called the *transfer characteristics* of the triode for the reason that they relate i_b, which exists in the output section of the vacuum-tube circuit, to the grid voltage e_c, which appears in the input section. The transfer characteristics can be derived from the plate characteristics by plotting i_b versus e_c for fixed values of plate potential e_b. In some applications the transfer curves are more convenient to use than the plate characteristics. We shall consider this in more detail in Chap. 10.

Vacuum-tube Parameters. In the analysis of circuits which include the vacuum tube as a circuit element it is often useful to replace the vacuum device by suitable circuit parameters. Of course, the parameters must be so chosen that they provide an acceptable representation of the vacuum-tube characteristics. To show how this can be accomplished let us return to Fig. 8-13, and let it be proposed that we choose a suitable set of parameters to represent the operation of the triode in the region defined by area ABC. In general the plate current may be expressed as

$$i_b = f(e_c, e_b) \tag{8-18}$$

Since three variables are involved in this equation, it is possible to identify three parameters. As in turn one variable is held constant, the functional relationship

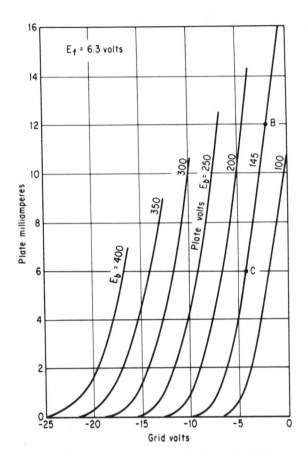

Fig. 8-14. Transfer characteristics of the 6J5 triode.

between the other two variables can be described in terms of a suitable parameter. Thus as the quantity i_b is held fixed, and e_c and e_b are allowed to vary, the proportionality factor between e_c and e_b is readily revealed. This result has already been described and is represented by Eq. (8-14). A more general formulation of this equation is

$$\text{amplification factor} = \mu \equiv -\left(\frac{\partial e_b}{\partial e_c}\right)_{i_b \text{ constant}} \qquad (8-19)$$

Accordingly, the parameter μ may be used to represent the operation of the triode along any constant-current line in region ABC of Fig. 8-13.

Next consider that e_c is held fixed and that a plot is made of i_b versus e_b. Clearly this leads to any one of the family of curves appearing in Fig. 8-13. If the grid voltage is assumed fixed at -2 volts and the plate-current variation is restricted to the range from 6 ma to 12 ma, then the line segment AB is the curve of interest. Because this line segment is a plot of plate current as a function of plate voltage, operation along AB can be represented by a parameter which is the slope of this curve. Since a plot of current versus voltage is involved, the slope has the units of the reciprocal of resistance. The resistance in turn is defined as

$$r_p \equiv \left(\frac{\partial e_b}{\partial i_b}\right)_{e_c \text{ constant}} \quad \text{ohms} \tag{8-20}$$

and is called the *dynamic plate resistance*. The prefix dynamic is used to emphasize that as different sections of the constant $-E_c$ curve are used, different slopes result, thus yielding somewhat altered values of r_p. In Fig. 8-13 a good approximation of Eq. (8-20) for the region under consideration is the ratio AC/BC. A study of the plate characteristics of Fig. 8-13 should make it evident that as long as i_b remains restricted to the range specified, the same value of r_p may be used even though e_c is allowed to vary considerably.

Finally consider that the plate voltage is held constant and that e_c is allowed to vary, say from -2 to -4 volts. This puts the operation of the triode along line BC. The variation of i_b with e_c is represented by line segment BC in Fig. 8-14. Here again the slope of line segment BC may be used to identify a suitable parameter to represent this mode of operation of the triode. Accordingly we may write

$$g_m \equiv \left(\frac{\partial i_b}{\partial e_c}\right)_{e_b \text{ constant}} \tag{8-21}$$

The quantity g_m is called the *transconductance* or *mutual conductance* of the triode. The word derives from the fact that the quantity involved is a conductance associated with the transfer characteristic.

The three tube parameters defined in terms of the partial derivatives of Eqs. (8-19), (8-20), and (8-21) cannot be independent because they derive from a single equation, Eq. (8-18). The functional relationship between these parameters can be obtained from the total differential of Eq. (8-18). Thus

$$\Delta i_b = \frac{\partial i_b}{\partial e_c}\Delta e_c + \frac{\partial i_b}{\partial e_b}\Delta e_b \tag{8-22}$$

Inserting Eqs. (8-20) and (8-21) yields

$$\Delta i_b = g_m \Delta e_c + \frac{\Delta e_b}{r_p} \tag{8-23}$$

If, for convenience, the condition is now imposed that Δe_c and Δe_b are so changed that Δi_b is zero, Eq. (8-23) may be written as

$$0 = g_m \Delta e_c + \frac{\Delta e_b}{r_p} \tag{8-24}$$

or

$$g_m r_p = \left(-\frac{\Delta e_b}{\Delta e_c}\right)_{i_b \text{ constant}} \tag{8-25}$$

Introducing Eq. (8-19) then leads to the simple result

$$\boxed{\mu = g_m r_p}$$ (8-26)

This last expression states that for any operating point on the plate characteristics the product of the transconductance and the dynamic plate resistance equals in value the voltage amplification factor.

EXAMPLE 8-1   Using the values of i_b, e_c, and e_b which define the region of operation ABC in Fig. 8-13, compute the corresponding three triode parameters.

solution: To find the amplification factor find the value in volts of $\Delta e_b = AC$ on the plate-voltage axis. Then divide this by the change in grid voltage needed to go from point A to C. Hence

$$\mu = -\frac{\Delta E_b}{\Delta E_c} = -\frac{40}{-2} = 20$$ (8-27)

The dynamic plate resistance is found conveniently as

$$r_p = \frac{AC}{BC} = \frac{40}{6 \times 10^{-3}} = 6667 \text{ ohms}$$ (8-28)

This value compares favorably with the manufacturer's published value of 6700 ohms for this region of operation.

The transconductance is found by noting that as the grid voltage is reduced from -2 to -4 volts the plate current decreases from 12 ma to 6 ma along line BC in Fig. 8-13. Accordingly

$$g_m = \frac{\Delta i_b}{\Delta e_c} = \frac{-6 \times 10^{-3}}{-2} = 3 \times 10^{-3} \text{ mho}$$ (8-29)

As a check on these computations note that the product of g_m and r_p does equal μ.

PROBLEMS

8-1   Three filaments each have the same dimensions. The first is made of tungsten, the second is made of thoriated tungsten, and third consists of a barium-oxide coated metal. The physical constants are:

	A_0	b_0	E_2
Tungsten	$60.2 \times 10^{+4}$	57,400	4.57 ev
Thoriated tungsten	$3 \times 10^{+4}$	30,500	2.63 ev
Oxide-coated	$1 \times 10^{+6}$	11,600	1.0 ev

Assuming each is operated at a temperature of 6000°K, determine which filament yields the greatest electron emission.

8-2   Two identical vacuum-tube diodes equipped with oxide-coated cathodes have a space-charge limited current of 16 ma at a plate voltage of 100 volts. Moreover, the temperature-limited current of the same type is 16 ma when operated at 2000°K.

(a) The two diodes are placed in parallel across a 25-volt d-c battery with proper polarity for both tubes to conduct. Assuming each tube is operated at 2000°K, find the total current supplied by the battery.

(b) In the circuit of part (a) the d-c battery is changed to 125 v and one diode is altered so that its saturation current is 25 ma. Find the current delivered by the battery.

8-3 The two diodes of Prob. 8-2(b) are arranged in series and placed across a d-c source of 250 volts with polarity for conduction. Determine the battery current and the voltage drop across each diode.

8-4 A test is performed on a 5BC3 vacuum diode for the purpose of obtaining the average plate characteristic. The data are as follows:

Plate current, ma	25	65	120	190	270	365
D-c plate voltage, volts	10	20	30	40	50	60

(a) Make a plot of plate milliamperes versus d-c plate voltage.

(b) What is the apparent forward resistance of this diode when the plate voltage is 40 v?

(c) Find the value of the incremental plate resistance at this voltage.

(d) Compute the ratio of the apparent to the incremental resistance. Compare with the result of Eq. (8-10) and comment.

8-5 The diode of Prob. 8-4 is placed in series with a 100-ohm load resistor and a 60-v battery.

(a) Find the current flow graphically.

(b) Check on the result of part (a) by using Ohm's law and the proper diode resistance.

(c) Explain your choice of the resistance used to represent the diode in part (b).

8-6 A 5Y3 vacuum diode has a plate characteristic described by the following data:

D-c voltage, volts	10	20	30	40	60
Plate current, ma	10	25	40	65	130

This diode is placed in series with a 5BC3 diode and a 60-v d-c supply. The plate characteristic of the 5BC3 diode is given in Prob. 8-4.

(a) Determine the current flow.

(b) Find the voltage drop across each diode.

(c) What is the apparent resistance of each diode?

8-7 The thermionic emission and space-charge limited characteristics of a vacuum diode are described respectively by the expressions

$$I_{th} = \frac{500}{T^2} \epsilon^{-16,000/T} \quad \text{and} \quad I_b = \frac{E_b^{3/2}}{500}$$

If the cathode is assumed operated at 2000°K, determine the threshold value of plate voltage beyond which operation is in the temperature-limited region.

8-8 From the plate characteristics of the 12AX7 triode (Appendix E) obtain the transfer characteristics for $E_b = 50$ v, 100 v, 150 v, 200 v, and 300 v.

8-9   From the plate characteristics of the 12AX7 triode (Appendix E) obtain the constant-current characteristics. Plot e_b versus e_c for $I_b = 0.5$ ma, 1.0 ma, and 2 ma.

8-10   A 6SN7 vacuum triode is operated at the following quiescent point: $E_b = 178$ v, $I_b = 10$ ma, and $E_c = -4$.

    (a) Find the value of the plate current when the plate voltage is decreased by 25 v and the grid voltage is descreased by 2 v.

    (b) The grid voltage is increased by 2 v. Determine the change in plate voltage needed to make the plate current 10 ma.

8-11   A vacuum triode has a voltage amplification factor of 20 and an $r_p = 7.7$ K. Find the change in plate voltage required for an increase of 4 ma plate current when the grid voltage is increased by 2 v.

8-12   The vacuum triode has a transconductance of 1.25 millimhos. There occurs a descrease in plate voltage of 20 v and a decrease in grid voltage of 0.5 v. The net decrease in plate current is 0.85 ma. Find the value of the dynamic plate resistance.

8-13   The plate current of a triode may be expressed approximately by the equation $i_b = 0.025(e_b + 20e_c)^{1.3}$ ma. The tube is operated at a plate potential of 250 v and a grid voltage of -8.

    (a) What is the voltage amplification factor of the tube?

    (b) What is the value of the plate current?

    (c) Find the value of the dynamic plate resistance at the specified operating point.

    (d) What is the value of the mutual conductance?

8-14   (a) The 12AT7 vacuum triode operates at the following point: $E_c = -1$ v, $E_b = 200$ v. Find the corresponding values of μ, r_b, and g_m.

    (b) Repeat part (a) for $E_c = -4$ and $E_b = 250$ v.

9

electronic circuits—
semiconductor types

In this chapter we study the composition, behavior, and performance of the main variety of circuits that use semiconductor diodes and triodes for control and amplifying purposes. These circuits are commonly referred to as *electronic circuits*. Although several chapters could be used to treat the subject matter dealt with here, this procedure is avoided in the interest of stressing that each section heading is an appropriate topic of electronic circuitry.

9-1 Graphical Analysis of Transistor Amplifiers

Common-emitter Configuration. Since a brief description of the transistor amplifier in the common-base mode is given in Sec. 7-3, in the interest of expanding our knowledge we shall direct our attention to the *common-emitter* arrangement which is depicted in Fig. 9-1. A glance at the diagram shows that the emitter section of the junction transistor is common to both the input and output circuits. This mode of operation is very frequently used because it offers an appreciable current gain as well as a sizable voltage gain.

How is a current gain exceeding unity achievable with the common-emitter circuit? To provide the answer to this question, let us first write Kirchhoff's

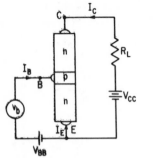

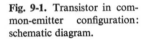

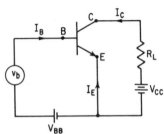

Fig. 9-1. Transistor in common-emitter configuration: schematic diagram.

Fig. 9-2. Transistor in common-emitter configuration: Circuit diagram.

current law for the transistor. In dealing with the transistor it is customary to consider the currents as positive when they flow into the terminals, as illustrated in Figs. 9-1 and 9-2. Thus

$$I_E + I_B + I_C = 0 \tag{9-1}$$

This equation applies irrespective of the circuit configuration used. Moreover, for the *n-p-n* transistor shown the collector current at a quiescent condition may be written as

$$I_C = I_{CO} - \alpha I_E \tag{9-2}$$

where I_{CO} is the reverse saturation current which exists in the reverse-biased section of the transistor and α is the common-base current amplification factor. The minus sign is used for the I_E term because in this case the conventional current flow in the input forward-biased section is *out* of the emitter terminal and not *into* the terminal, which is the assumed positive direction. When a number is introduced for I_E in Eq. (9-2), it must be negative, thus yielding the algebraic sum of the two terms involved. Equation (9-2) states that of the total number of electrons injected into the transistor at the emitter terminal constituting the current, $-I_E$, the portion arriving at the collector terminal is $-\alpha I_E$. Keep in mind that α lies between 0.9 and 1.0.

The current amplification factor for a common-base transistor can be found directly from Eq. (9-2) because it involves explicitly the input current, which is I_E, and the current in the output circuit I_C. However, note that for the common-emitter circuit under study, the input current is now I_B. Therefore to determine the relationship existing between the output and input currents for the common-emitter mode, it is necessary first to eliminate I_E from Eq. (9-2) in terms of I_B and I_C. This is readily accomplished by means of Eq. (9-1). Thus Eq. (9-2) becomes

$$I_C = I_{CO} + \alpha(I_B + I_C) \tag{9-3}$$

Solving for I_C yields

$$I_C = \frac{I_{CO}}{1-\alpha} + \frac{\alpha}{1-\alpha} I_B \tag{9-4}$$

Upon considering I_{CO} fixed and subjecting the input current to an incremental change ΔI_B, the corresponding change in the output current is found to be

$$\Delta I_C = \frac{\alpha}{1 - \alpha} \Delta I_B \qquad (9\text{-}5)$$

Consequently, the *common-emitter current amplification factor*, β, becomes

$$\boxed{\beta \equiv \frac{\Delta I_C}{\Delta I_B}\bigg|_{V_{CE}\ \text{constant}} = \frac{\alpha}{1 - \alpha}} \qquad (9\text{-}6)$$

where α denotes the common-base current transfer ratio. Typical values of β range from 20 to 80 depending upon α. Thus, if α equals 0.98, it follows that β is 49.

The common-emitter characteristics of the 2N241 transistor are depicted in Fig. 9-3. Unlike the common-base characteristics of Fig. 7-22, these curves are not horizontal but rather exhibit a positive slope. This comes about because the slight changes which occur in α with changes of the collector-to-emitter voltage appear in β with much greater sensitivity. Inspection of Eq. (9-6) makes this apparent since the denominator is a number appreciably less than unity.

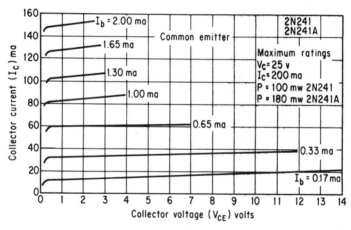

Fig. 9-3. Common-emitter characteristics of the 2N241 transistor.

Before proceeding with the analysis of the performance, it is worthwhile first to become familiar with the symbols to be used in connection with electronic circuits involving transistors. These are given in Table 9-1 for ready reference.

The circuitry for a common-emitter amplifier is depicted in Fig. 9-4. Use is made of an *n-p-n* 2N377 transistor.

How is a circuit such as the one depicted in Fig. 9-4 analyzed? Or, more specifically, how does one find the collector current i_c corresponding to a specified value of base current i_B? One step towards arriving at a solution would be to write Kirchhoff's voltage law for the output circuit. Thus

$$V_{CC} = v_C + i_C R_L \qquad (9\text{-}7)$$

Table 9-1. Transistor Symbols

	Base terminal	Emitter terminal	Collector terminal
Current Quantities:			
Instantaneous total value	i_B	i_E	i_C
Quiescent value	I_B	I_E	I_C
Instantaneous value of varying component	i_b	i_e	i_c
Effective value of varying component	I_b	I_e	I_c
Voltage Quantities:			
Instantaneous total value	v_B	v_E	v_C
Quiescent value†	V_B	V_E	V_C
Instantaneous value of varying component	v_b	v_e	v_c
Effective value of varying component	V_b	V_e	V_c
Supply voltage	V_{BB}	V_{EE}	V_{CC}

† Measured with respect to a ground reference.

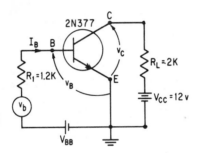

Fig. 9-4. A common-emitter transistor amplifier. Appearing at the output terminals is the varying component V_c of the collector-to-emitter voltage.

From this equation we then get an expression for the desired collector current, which is

$$i_C = \frac{V_{CC}}{R_L} - \frac{1}{R_L} v_C \tag{9-8}$$

Inspection of Eq. (9-8) quickly reveals that the collector-to-emitter voltage v_C is also unknown. We thus have a situation of a single equation involving two unknown quantities, i_C and v_C. Therefore, additional information is needed, and in particular it must relate i_C to v_C. Fortunately, once the semiconductor is specified, the required relationship is available graphically by means of the appropriate common-emitter characteristics, which for the 2N377 appear in Fig. 9-5(a).

A convenient and direct solution to the problem can now be obtained by plotting Eq. (9-8) on the same set of axes as are used for the common-emitter characteristics. Clearly, Eq. (9-8) is the equation of a straight line having a vertical intercept of V_{CC}/R_L and a slope of $-1/R_L$. This straight line is called the *d-c load line*. The point of intersection of the *d-c* load line with that curve of the common-

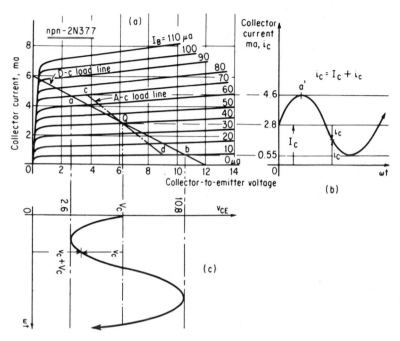

Fig. 9-5. Graphical analysis of transistor amplifier (these curves are drawn for an input signal of $V_b = 36 \times 10^{-3} \sin t$):

    (a) common emitter characteristics;

    (b) variation of plate current;

    (c) variation of collector-to-emitter voltage.

emitter characteristics corresponding to the specified value of base current provides the solution for the base current.

Let us illustrate this solution procedure for the circuit of Fig. 9-4. The manufacturers' published characteristics for the specified transistor appear in Fig. 9-5(a) drawn to scale. The *d-c* load line is readily plotted by identifying the end points of the straight line. Thus, from Eq. (9-8) it follows that when i_C is zero, v_C equals 12 volts. This locates one point on the load line—the one on the horizontal axis. Next, when v_C is set equal to zero, the corresponding value of collector current becomes $V_{CC}/R_L = 12/2000 = 6$ ma. This identifies the second point—the one on the vertical axis. The line drawn between these two points yields the *d-c* load line. Note that the slope is negative and has a value equal to $1/R_L$, all of which is in keeping with Eq. (9-8). If it is now desired to find the value of the collector current for a base current of 40 μa, a glance at Fig. 9-5(a) reveals the answer to be 2.8 ma. This point is identified as Q.

The base current of 40 μa is assumed to be produced by the voltage V_{BB} acting on the resistor R_1 of 1200 ohms in the input circuit of Fig. 9-4. The forward-bias resistance of the *p-n* section of the input circuit is neglected because generally

R_1 is many times larger. The signal voltage v_b is assumed to be zero. Consequently, because of the absence of a changing input signal, point Q is called the *quiescent operating* point of the circuit of Fig. 9-4. This is the operating point about which changes occur in the output circuit in response to changes in v_b. It is customary to identify the coordinates of the quiescent point as (I_C, V_C). In our example these quantities take on the values (2.8 ma, 6.2 volts). Thus, at the quiescent point it can be said for the circuit of Fig. 9-4 that the potential of the collector terminal is 6.2 volts corresponding to a collector current flow of 2.8 ma. Finally, note that the location of the quiescent point Q can be moved along the load line by changing the value of V_{BB}.

We next consider how the circuit of Fig. 9-4 behaves in the presence of a sinusoidal variation of $v_b = \sqrt{2}\ (25.4) \sin \omega t = 36 \sin \omega t$, where the amplitude is expressed in millivolts. The quantity $V_b = 25.4$ mv is the effective value of the specified sinusoidal input voltage. The corresponding change in plate current produced by this voltage is plainly

$$i_b = \frac{36}{1.2} \sin \omega t = 30 \sin \omega t \qquad \mu a \qquad (9\text{-}9)$$

where the amplitude here is expressed in microamperes (μa). The collector current and voltage for any value of i_b, as called for by Eq. (9-9), can be found in the same manner as was used for the quiescent point. Thus, if ωt is assumed to increase from 0 to $\pi/2$, i_B changes from 40 μa to 70 μa, which is consistent with

$$i_B = I_B + i_b \qquad (9\text{-}10)$$

Equation (9-10) describes the total instantaneous base current (i_B) in terms of the quiescent value (I_B) plus the instantaneous values of the varying component (i_b). In the plot of Fig. 9-5(a), this variation means that the *instantaneous operating point* moves along the load line from Q to a. The instantaneous operating point is confined to move along the load line because at every instant Kirchhoff's voltage law in the collector circuit that contains R_L must be satisfied. The value of the collector current corresponding to $i_B = 70$ μa, which is point a in Fig. 9-5(a), is read off the ordinate axis as 4.6 ma. A plot of the collector currents corresponding to the instantaneous operating points between Q and a on the d-c load line results in the variation shown in Fig. 9-5(b) between the points marked 2.8 (I_B) and a'. A corresponding plot of the collector voltage is depicted in Fig. 9-5(c). Note that the variation of the varying component of the collector voltage about the quiescent value is a negative sine wave. A comparison of this result with a plot of Eq. (9-9) brings out one of the distinguishing characteristics of an electronic amplifier: *phase inversion*. This means that the sign of varying collector voltage v_c is opposite to that of the varying input base current i_b. Finally, note that as i_b continues its variation through a complete sine wave, corresponding variations take place in the plots of i_c and v_c.

Does it follow that a sinusoidal variation of i_b results in a corresponding sinusoidal variation of the collector current and voltage? The answer depends on how the Q-point is selected and on how large a variation about the Q-point is allowed. If the Q-point is located in that region of the common-emitter characteristics where equal increments in base current are accompanied by equal spacing of the curves of the CE characteristics, then the answer is yes. The selection of Q in Fig. 9-5(a) was made with this in mind. Otherwise, the answer is no.

As indicated in Fig. 9-5(b), the instantaneous value of the varying component of the collector current is denoted by i_c. The amplitude, I_{cm}, of this current is readily found from the relationship

$$I_{cm} = \frac{I_{c_{max}} - I_{c_{min}}}{2} \tag{9-11}$$

where $I_{c_{max}}$ denotes the maximum value of the collector current corresponding to the positive maximum value of i_b, and $I_{c_{min}}$ denotes the minimum current corresponding to the negative maximum value of i_b. For the example illustrated in Fig. 9-5, this becomes

$$I_{cm} = \frac{4.6 - 0.55}{2} = 2.03 \text{ ma} \tag{9-12}$$

The rms or effective value then follows from

$$I_c = \frac{I_{cm}}{\sqrt{2}} = \frac{2.03}{\sqrt{2}} = 1.43 \text{ ma} \tag{9-13}$$

Inserting Eq. (9-11) into Eq. (9-12) leads to an expression for I_c related directly to the minimum and maximum values of collector current. Thus

$$\boxed{I_c = \frac{I_{c_{max}} - I_{c_{min}}}{2\sqrt{2}}} \tag{9-14}$$

In view of the sinusoidal character of the varying component of the collector voltage, a similar analysis leads directly to the following expression for the effective value of the varying component of the collector voltage:

$$V_c = \frac{V_{c_{max}} - V_{c_{min}}}{2\sqrt{2}} \tag{9-15}$$

Inserting the values found from Fig. 9-5 yields

$$V_c = \frac{10.8 - 2.6}{2\sqrt{2}} = 2.9 \quad \text{volts rms} \tag{9-16}$$

It is important to understand that this is the effective value of voltage developed across the load resistor in response to the varying component i_b in the input circuit. It should be apparent from Fig. 9-5 that if i_b were zero, V_c would also be zero.

We are now at a point in the analysis where we can compute the voltage gain of the amplifier. The *voltage gain* is defined as the ratio of the rms value of the varying component of the collector voltage to the rms value of the varying component of the input voltage. Expressed mathematically

$$\boxed{A_v \equiv \frac{V_c}{V_b}} \tag{9-17}$$

where A_v denotes the voltage gain, V_c denotes the rms value of v_c, and V_b denotes the rms value of v_b. For our example

$$A_v = \frac{V_c}{V_b} = -\frac{2.9}{0.0254} = -114 \qquad (9\text{-}18)$$

The minus sign signifies the phase reversal between input and output signals.

We can proceed in a similar fashion to find the current gain. The effective value of the varying component of the base current is found to be

$$I_b = \frac{I_{b_{max}}}{\sqrt{2}} = \frac{30}{\sqrt{2}} = 21.2 \qquad \mu a \qquad (9\text{-}19)$$

The *current gain* is then obtained as

$$A_c = \frac{I_c}{I_b} = \frac{1.43 \times 10^{-3}}{21.2 \times 10^{-6}} = 68.5 \qquad (9\text{-}20)$$

The listed current amplification factor β for this transistor is 70. In the presence of a finite load resistor, the current gain is always less than the current amplification factor.

The average value of a-c power that is delivered to the load resistor as a result of the varying signal at the input circuit is readily found to be

$$P_{\text{a-c}} = V_c I_c = 2.83(1.43)10^{-3}$$
$$= 0.0041 \text{ watt} = 4.1 \text{ mw} \qquad (9\text{-}21)$$

The power gain G of the transistor amplifier is the ratio of the average a-c power delivered to the load resistor to the average a-c power supplied by the signal source. Thus

$$G = \frac{V_c I_c}{V_b I_b} = 114(68.5) = 7809 \qquad (9\text{-}22)$$

Note that the power gain for the common-emitter configuration is considerably greater than for the common-base mode.

Since the transistor is a device that transfers a signal current from a low resistance circuit to a high resistance circuit and thereby brings about a voltage (and power) amplification, the voltage gain can also be computed by using an expression analogous to Eq. (7-34). A modification is necessary. The current gain A_c for the common-emitter circuit must replace the common-base current transfer ratio α. Accordingly, the alternative expression for the voltage gain becomes

$$A_v = -\frac{R_L}{R_1} A_c \qquad (9\text{-}23)$$

where R_1 refers to the total resistance of the input circuit. Inserting the numbers that apply for the circuit of Fig. 9-4 yields

$$A_v = -\frac{2000}{1200}(68.5) = -114 \qquad (9\text{-}24)$$

which agrees with Eq. (9-18). It should be noted, however, that if care is not exercized in using the graphical results leading to the determination of A_c and A_v, discrepancies will arise in these alternative computations of A_v.

Practical Voltage-amplifier Configuration: A-c Load Line. Practical considerations dictate the modification of the circuit of Fig. 9-4 to that shown in Fig. 9-6. An important change in this circuit is the insertion of the capacitor C_C between the collector load resistor at the collector and the input to a succeeding stage. The resistor R_2 represents the input resistance of a succeeding stage of amplification. By the proper selection of C_C, it can be made to pass the varying component of the collector voltage v_c practically undiminished, while completely blocking out the d-c component. To appreciate the importance of blocking out the d-c component, keep in mind that the a-c output of the circuit of Fig. 9-6 often serves as the input signal to a succeeding stage of amplification. If C_C were not used,

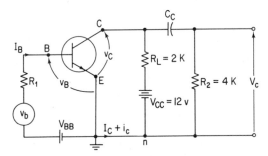

Fig. 9-6. A common-emitter transistor amplifier with capacitor coupling to a succeeding stage.

and if it were desired to operate the succeeding stage at an advantageous Q-point, it would be necessary to use a V_{BB} supply for the following stage, which would be large as well as inconvenient. Another reason for including C_C is that it prevents changes in the collector supply voltage V_{CC} from being interpreted by the succeeding stage as variations brought about by an input signal change. Usually, these changes occur very slowly so that C_C effectively blocks them out entirely. It is for this reason that the capacitor-coupled amplifier is said to be more stable than the direct-coupled amplifier (i.e., C_C omitted).

The presence of the coupling capacitor changes the a-c performance somewhat. As the varying component i_c of the collector current i_C leaves the emitter and flows toward terminal n, it no longer sees the single path through R_L in Fig. 9-6, but rather sees two paths. One is through R_L to C; the second is through R_2 and C_C to C. Of course, the d-c component (i.e., the quiescent current I_C) continues to flow through R_L to C, because C_C presents an open circuit to direct current. In view of the presence of the two paths for i_c, it follows that the effective load resistance seen by i_c changes from R_L to R_L in parallel with R_2. Accordingly, in the amplifier circuit of Fig. 9-6 the equivalent load resistance becomes

$$R'_L = \frac{R_L R_2}{R_L + R_2} \qquad (9\text{-}25)$$

When $R_2 \gg R_L$, very little effect on a-c amplifier performance results. Often, however, R_2 and R_L are comparable and adjustments become necessary. To illustrate, let us compute the a-c performance of the amplifier of Fig. 9-6 for the value of R_2 indicated. By means of Eq. (9-25), the equivalent load resistance becomes

$$R'_L = \frac{R_L R_2}{R_L + R_2} = \frac{2000(4000)}{6000} = 1333 \quad \text{ohms} \tag{9-26}$$

This value is sufficiently different from 2000 ohms to require the construction of a new load line. Since the quiescent point has not been changed, the new load line must also pass through point Q in Fig. 9-5(a). A convenient way of plotting the new load line is to write

$$R'_L = \frac{\Delta v_C}{\Delta i_C} \tag{9-27}$$

Then arbitrarily choose Δv_C as some convenient value measured from the Q-point. For example, choose $\Delta v_C = 1.8$ volts taken to the right of the Q-point. Equation (9-27) then yields the corresponding change in collector current of

$$\Delta i_C = \frac{\Delta v_C}{R'_L} = \frac{1.8}{1.333} 10^{-3} = 1.35 \quad \text{ma} \tag{9-28}$$

Thus, by moving from the Q-point by 1.8 volts to the right along the horizontal and then moving downward along the vertical by an amount equal to 1.35 ma, a second point on the new load line is established. The line drawn between this point and the Q-point locates the new load line, which is called the *a-c load line* for obvious reasons. In Fig. 9-5(a), the a-c load line is marked cQd. This line is sometimes called the *dynamic load line*.

As the base current varies in the manner described by Eq. (9-9), the instantaneous operating point for the circuit of Fig. 9-6 is now confined to the a-c load line. This leads to altered values of maximum and minimum collector voltages and current, thus resulting in a change in a-c performance. From Fig. 9-5(a), we have at point c,

$$I_{c_{max}} = 4.7 \text{ ma} \quad \text{and} \quad V_{c_{min}} = 3.6 \text{ volts}$$

and at point d,

$$I_{c_{min}} = 0.6 \text{ ma} \quad \text{and} \quad V_{c_{max}} = 9 \text{ volts}$$

Then by Eq. (9-14), the new value of the rms collector current is

$$I_c = \frac{4.7 - 0.6}{2\sqrt{2}} = 1.45 \quad \text{ma} \tag{9-29}$$

Hence the new current gain becomes

$$A_c = \frac{1.45 \times 10^{-3}}{21.2 \times 10^{-6}} = 68.2 \tag{9-30}$$

A comparison with Eq. (9-20) shows little difference occurs in this quantity. This is largely attributable to the horizontal nature of the collector volt-ampere characteristics. The corresponding rms value of the collector voltage is similarly found to be

$$V_c = \frac{9 - 3.6}{2\sqrt{2}} = 1.92 \text{ volts} \tag{9-31}$$

This leads to a voltage gain of

$$A_v = \frac{V_c}{V_b} = \frac{-1.92}{0.0254} = -75.6 \tag{9-32}$$

The minus sign denotes a phase reversal between input and output signals. Clearly, the voltage gain undergoes an appreciable reduction brought about by the shunting action of R_2. The related power gain then becomes

$$G = \left| \frac{V_c}{V_b} \right| \frac{I_c}{I_b} = 75.6(68.2) = 5160 \tag{9-33}$$

As expected, this quantity also diminishes by a good deal by operation on the a-c load line. However, we must not overlook the fact that this still represents a very healthy increase in signal power level. Vacuum-tube devices, for example, cannot match this performance.

The JFET Amplifier Configuration. In Chap. 7 it is pointed out that the field-effect transistor is a voltage-sensitive device. Accordingly, control of the output current is achieved by variations that take place in the *gate-source voltage*. In the interest of comparing the performance of an amplifier circuit employing a JFET to those already considered in this section, we focus attention on the simple circuit appearing in Fig. 9-7. This amplifier employs the silicon *n*-channel field-effect transistor 2N4220, whose drain characteristics appear in Fig. 9-8. The resistor R_g provides a path

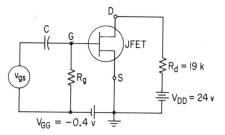

Fig. 9-7. Common-source amplifier employing the JFET 2N4220.

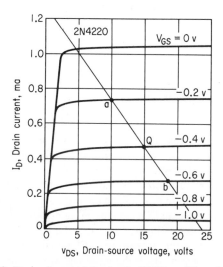

Fig. 9-8. Drain characteristics for the JFET amplifier of Fig. 9-7.

for maintaining a negative gate-source voltage even when the varying component of the gate-source voltage is removed.

For the specified value of drain load resistor R_d, the load line assumes the location depicted in Fig. 9-8. The Q-point is determined by the intersection of this load line with the drain characteristic corresponding to $V_{GS} = V_{GG} = -0.4$ volt. Assume next that the input signal voltage varies in accordance with $v_{gs} = 0.2 \sin \omega t$. The total instantaneous gate-source voltage can then be written as

$$v_{GS} = V_{GG} + v_{gs} = -0.4 + 0.2 \sin \omega t \tag{9-34}$$

Equation (9-34) states that as the varying component of the gate-source voltage swings through a complete cycle, operation on the load line goes from Q to a, back to Q, then to b, and back to Q again. From Fig. 9-8, it follows that the collector voltage and current at point a are

$$V_{d_{\min}} = 10 \text{ volts} \quad \text{and} \quad I_{d_{\max}} = 0.74 \text{ ma}$$

while at point b there results

$$V_{d_{\max}} = 18.8 \text{ volts} \quad \text{and} \quad I_{d_{\min}} = 0.27 \text{ ma}$$

Accordingly, we can compute the rms value of the varying component of the drain voltage as

$$V_d = \frac{V_{d_{\max}} - V_{d_{\min}}}{2\sqrt{2}} = \frac{18.8 - 10}{2\sqrt{2}} = 3.11 \quad \text{volts} \tag{9-35}$$

Similarly, the effective value of the varying component of the drain current is found to be

$$I_d = \frac{I_{d_{\max}} - I_{d_{\min}}}{2\sqrt{2}} = \frac{0.74 - 0.27}{2\sqrt{2}} = 0.166 \quad \text{ma} \tag{9-36}$$

The rms value of the varying component of the gate-source voltage is known to be

$$V_{gs} = \frac{0.2}{\sqrt{2}} = \frac{\sqrt{2}}{10} \tag{9-37}$$

Therefore, the voltage gain is

$$A_v = \frac{V_d}{V_{gs}} = \frac{-3.11}{\sqrt{2}/10} = -22 \tag{9-38}$$

and the corresponding a-c power is

$$P_{\text{a-c}} = V_d I_d = 3.11(0.166) = 0.516 \quad \text{mw} \tag{9-39}$$

It is instructive to note that the voltage gain of the JFET amplifier is lower than those encountered for the common-emitter amplifier configurations of the conventional transistors. This is typical of such amplifiers. Moreover, the field-effect transistor amplifiers also produce more distortion than the conventional transistor amplifiers. An examination of the characteristics of Fig. 9-8 makes this readily apparent. Note that equal increments in gate-source voltage do not produce equal increments in drain voltage and current. Or, stated graphically, distance Qa is not equal to Qb. For this reason, the JFET and MOSFET amplifiers are frequently used in situations in which the input signal variations produce a small excursion about the quiescent point. However, one should not lose sight of the fact that this amplifier type also has several outstanding advantages. Because the input circuit is reverse biased, the JFET amplifier draws negligibly small current. The input resistance is, thereby, extremely high. It can control power in the output circuit with negligibly small input power, thus yielding extremely high power gains.

9-2 Linear Equivalent Circuits

It is often advantageous, for reasons of accuracy as well as greater ease of analysis, to replace an engineering device by an equivalent circuit. Once this is done, all the tools and techniques of circuit theory can then be used in the analysis, design, and computation of performance of the device and the circuit or system of which it may be a part. This procedure is applied generally throughout the study of electrical engineering. In this section we apply it to the transistor.

Although the performance of a transistor amplifier can be conveniently determined by a graphical analysis for large variations of the input signal, a little thought should make it apparent that this approach leads to appreciable inaccuracies when the input signal varies over a very small range. In the transistor amplifier of Fig. 9-5 a variation of a few microamperes in base current cannot be treated graphically with any acceptable degree of accuracy. It is precisely in such instances that the equivalent-circuit representation of the device is useful. In essence the equivalent circuit is a means of replacing the transistor by a suitable arrangement of circuit parameters and voltage and/or current sources which are consistent with the characteristics of the original device. Moreover, since we are concerned with calculating the changes which occur about the operating point, the parameters which are selected for the equivalent circuit are those which apply for the given operating point. For small variations of the input signal insignificant changes take place in these parameters, so that even greater confidence may be put in the calculated results. In other words, as the input signal variations become smaller the results computed graphically become less accurate while those computed by the equivalent circuit become more accurate.

Transistor Equivalent Circuit in Terms of h Parameters. There are several types of equivalent circuits that may be used to represent the transistor. However, we treat here only the equivalent circuit that today has received the widest acceptance, the *h*-parameter equivalent circuit. The reasons for this preference are twofold: One, a relatively simple circuit results, which has the very desirable feature that the input and output circuits are completely isolated for the normal frequency range. Two, each of the parameters appearing in the equivalent is readily found either from the volt-ampere characteristic of the input circuit or the volt-ampere characteristic of the output circuit. A departure for this preference occurs only when the transistor amplifier is analyzed at very high frequencies. Under this condition the transition and diffusion capacitances of the transistor became important, and the easiest way to handle these quantities is to employ the hybrid-π equivalent circuit. More is said about this matter in Sec. 9-5.

Because of the vast importance of the common-emitter configuration, the equivalent circuit is derived specifically for this mode of operation. The procedure holds with equal validity, however, for the other operating modes—common collector and common base. It is important to keep in mind that we are exclusively concerned in this equivalent circuit derivation with a description of changes that take place about the quiescent point. There are four variables involved in describing these changes—two on the input side, v_B and i_B, and two on the output side, v_C and i_C. Since the transistor is a current-sensitive device on the input side, it is reasonable to choose i_B as one of the independent variables. Moreover, on the output side, it is really the voltage developed across the load resistor that is most important. Hence, v_C is arbitrarily chosen as the second independent variable. The remaining two variables—v_B and i_C—are then considered to be the dependent variables. On this basis we may accordingly write the following functional relationships:

$$v_B = f_1(i_B, v_C) \tag{9-40}$$

and

$$i_C = f_2(i_B, v_C) \tag{9-41}$$

From calculus, the corresponding expressions for the total differentials become

$$\Delta v_B = \left(\frac{\partial v_B}{\partial i_B}\right)_{V_C} \Delta i_B + \left(\frac{\partial v_B}{\partial v_C}\right)_{I_B} \Delta v_C \tag{9-42}$$

and

$$\Delta i_C = \left(\frac{\partial i_C}{\partial i_B}\right)_{V_C} \Delta i_B + \left(\frac{\partial i_C}{\partial v_C}\right)_{I_B} \Delta v_C \tag{9-43}$$

where the subscripts of the partial derivatives employing upper-case letters denotes that these quantities are held at fixed values. Now, recalling that the changes in total instantaneous quantities are the same as the changes in the varying component of these quantities, we can write as equivalencies

$$\Delta v_B = v_b \tag{9-44}$$

$$\Delta i_B = i_b \tag{9-45}$$

$$\Delta v_C = v_c \tag{9-46}$$

$$\Delta i_C = i_c \tag{9-47}$$

Inserting the last four equations into Eqs. (9-42) and (9-43) allows the latter to be written as

$$v_b = \left(\frac{\partial v_B}{\partial i_B}\right)_{V_C} i_b + \left(\frac{\partial v_B}{\partial v_C}\right)_{I_B} v_c \tag{9-48}$$

$$i_c = \left(\frac{\partial i_C}{\partial i_B}\right)_{V_C} i_b + \left(\frac{\partial i_C}{\partial v_C}\right)_{I_B} v_c \tag{9-49}$$

To further simplify these equations, let us take a closer look at each of the partial derivatives.

The coefficient of i_b in Eq. (9-48) clearly has the unit of resistance, because it is the ratio of the base voltage and the base current. More simply, it is the slope of the input characteristic. See Fig. 7-23. It is denoted by h_{ie}. Thus

$$h_{ie} \equiv \left(\frac{\partial v_B}{\partial i_B}\right)_{V_C} = \text{input resistance in common-emitter mode} \qquad (9\text{-}50)$$
$$\text{for constant collector voltage}$$

Typical values for amplifier applications range from a few hundred ohms to several kilohms.

The coefficient of v_c in Eq. (9-48) is a numeric. It represents that fraction of the collector (or output) voltage appearing at the input when the input is open circuited. Essentially it is a *reverse amplification factor*. An appreciation of how this quantity comes about and its order of magnitude can be had by referring to Fig. 9-1. In normal operation the *p-n* section from base to emitter is forward biased and, hence, of very low resistance. The *n-p* section from collector to base is reverse biased and so is characterized by extremely high resistance. Accordingly, for a given collector-to-emitter voltage there will be a portion appearing across the forward-biased *p-n* section between base and emitter terminals, but it will be only a very small fraction. Typical values are of the order of 10^{-4}. In most computations involving amplifier performance, little error results from neglecting this term entirely. However, it is included here for completeness and is denoted by h_{re}. Thus

$$h_{re} \equiv \left(\frac{\partial v_B}{\partial v_c}\right)_{I_B} = \text{reverse amplification factor} \qquad (9\text{-}51)$$

In Eq. (9-49), the coefficient of i_b is also a numeric. Now the ratio involves changes in current. Essentially, this coefficient describes the change that occurs in the collector current per unit change in base current at constant collector-emitter voltage. But this is precisely the definition of the common-emitter current amplification factor β. Refer to Eq. (9-6). In terms of the generally accepted hybrid notation, this ratio is denoted as

$$h_{fe} \equiv \left(\frac{\partial i_c}{\partial i_b}\right)_{V_C} = \beta = \text{forward current transfer ratio} \qquad (9\text{-}52)$$

Typical values range from 20 to 80.

The coefficient of v_c in Eq. (9-49) carries the unit of inverse ohms or conductance. This quantity relates to the output circuit, since it represents the change in collector current produced by a unit change in collector voltage at constant base current. The condition of fixed base current is equivalent to stating that the input terminals are short circuited. Because of the reverse-biased condition of the transistor in the collector output circuit, very little change in collector current is caused by a given change in collector voltage. Accordingly, this quantity assumes relatively small values and frequently can be neglected in computing amplifier performance without noticeably deteriorating accuracy. Expressed mathematically, we have

$$h_{oe} \equiv \left(\frac{\partial i_c}{\partial v_c}\right)_{I_B} = \text{output conductance with input short circuited} \qquad (9\text{-}53)$$

Equations (9-50) to (9-53) identify the four parameters that can be used to completely describe the operation of the transistor in small-signal applications.

Two parameters are numerics, one has the unit of resistance, and the other has the unit of conductance. In short, the parameters have *mixed* units. It is for this reason that these quantities are called the *hybrid* parameters or more simply the *h* parameters. The *e* subscript carried by each of these parameters merely serves to remind us that we are dealing with the common-emitter configuration.

The equations that describe the steady-state behavior of the transistor can now be written in terms of the *h* parameters. Thus

$$v_b = h_{ie}i_b + h_{re}v_c \qquad (9\text{-}54)$$

$$i_c = h_{fe}i_b + h_{oe}v_c \qquad (9\text{-}55)$$

The linear equivalent circuit of the transistor immediately follows from a circuit interpretation of Eqs. (9-54) and (9-55). The first equation is a Kirchhoff voltage equation. It states that the varying component of the base voltage is made up of a resistance drop involving the input resistance and the varying component of the base current plus a fractional portion of the varying component of the collector voltage. The circuit interpretation appears as the left side of the diagram of Fig. 9-9. The second equation is a Kirchhoff current equation. It states that the varying component of the collector current is an amplified version of the varying component of the base current plus a current component associated with the varying component of the collector voltage appearing across the output resistance ($1/h_{oe}$). Recall that the positive direction of collector current is defined as entering the collector terminal. Hence, the right side of the circuit of Fig. 9-9 depicts i_c entering the *C* terminal and then splitting into two paths, as called for by Eq. (9-55).

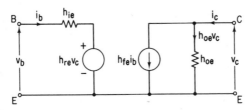

Fig. 9-9. The *h*-parameter equivalent circuit of the transistor in the common-emitter mode.

An equivalent circuit is a convenient way of describing the internal operation of a device as seen from the externally available terminals. For the transistor, these terminals are *B, C,* and *E.* Figure 9-9 is a means of representing this operation in terms of familiar elements so that performance can be readily determined by the powerful methods of circuit analysis studied in Chap. 3. Note that no d-c quantities such as V_{BB} and V_{CC} appear in the equivalent circuit, because the emphasis is on the *changes* that take place about the *Q*-point.

Depicted in Fig. 9-10 is an approximate form of the linear model of Fig. 9-9. Here both h_{re} and h_{oe} are set equal to zero. In many applications, entirely satisfactory results are obtained by the use of the approximate equivalent circuit.

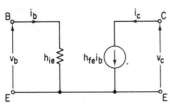

Fig. 9-10. The *h*-parameter approximate equivalent circuit of the common-emitter transistor.

Determination of h Parameters. The four h parameters described in the foregoing can be conveniently found once a set of input and output characteristics are available. These may be obtained either from the manufacturer or experimentally. A typical set of these curves is shown in Fig. 9-11.

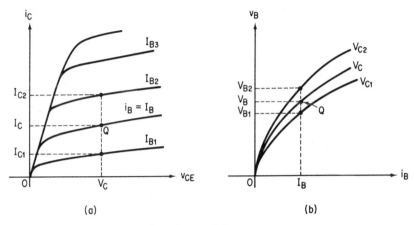

Fig. 9-11. Typical transistor characteristic curves:
   (a) the output characteristics from which h_{fe} and h_{oe} are obtained;
   (b) the input characteristics from which h_{ie} and h_{re} are obtained.

The output characteristics can be made to yield values for the parameters h_{oe} and h_{fe}. The output conductance h_{oe} is merely the slope of the curve at the quiescent point in Fig. 9-11(a) for fixed base current I_B. A typical variation with collector current is illustrated in Fig. 9-12(b). The current amplification factor h_{fe} can be found at constant collector voltage by measuring the change in collector current $(I_{C2} - I_{C1})$ corresponding to a change in base current from I_{B1} to I_{B2}. Thus

$$h_{fe} = \frac{I_{C2} - I_{C1}}{I_{B2} - I_{B1}} \tag{9-56}$$

It is important in making this calculation that the spreads $(I_{C2} - I_{C1})$ and $(I_{B2} - I_{B1})$ be as small as is consistent with reliable measurements. The parameter h_{fe} varies also with the location of the Q-point. Figure 9-12(a) depicts a typical variation.

Information about h_{ie} and h_{re} results from the input characteristics of Fig. 9-11(b). The slope of the curve at the Q-point at fixed V_C yields the value of h_{ie} directly. The manner in which this parameter varies with the collector current is shown in Fig. 9-12(c). The reverse voltage amplification factor can be determined from

$$h_{re} = \frac{V_{B2} - V_{B1}}{V_{C2} - V_{C1}} \tag{9-57}$$

where the quantities on the right side follow from Fig. 9-11(b). The corresponding variation with the Q-point at fixed collector-emitter voltage appears in Fig. 9-12(d).

h parameters

$(V_{CE} = 10 \text{ v}, f = 1 \text{ kHz}, T_A = 25°\text{ C})$

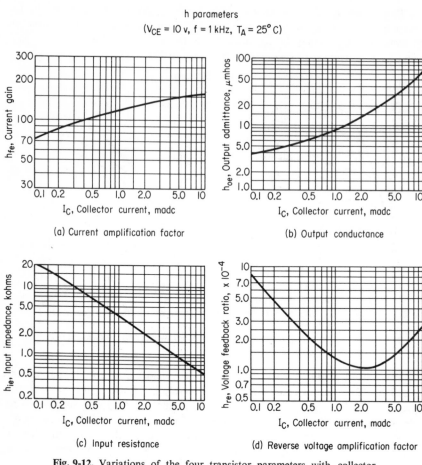

(a) Current amplification factor

(b) Output conductance

(c) Input resistance

(d) Reverse voltage amplification factor

Fig. 9-12. Variations of the four transistor parameters with collector current at fixed collector-emitter voltage. Transistor 2N3903.

Linear Model of the FET. The small-signal equivalent circuit of the field-effect transistor as it appears between its three external terminals is developed next. Since the FET is a voltage-sensitive device that draws negligible current on the input side, there are only three variables involved—the total instantaneous drain current i_D, the total instantaneous gate-source voltage v_{GS}, and the total instantaneous drain-source voltage v_D. Consequently, the derivation of the equivalent circuit is based on a single functional relationship,

$$i_D = f(v_{GS}, v_D) \tag{9-58}$$

In deriving the equivalent circuit, our objective is to replace the FET between terminals G, D, and S by a suitable arrangement of circuit elements. Although our interest at this point is solely with the FET, it is helpful in understanding the development to keep in mind the manner in which the FET is used in a typical amplifier circuit, such as the one depicted in Fig. 9-7. The equivalent circuit is useful in analyzing the behavior of such an electronic circuit whenever the input signal undergoes small variations. Again, because of this emphasis on the changes about the quiescent point, all d-c quantities are omitted from the equivalent circuit. It is the a-c performance that is important. Consistent with this theme, therefore, let us assume that a change Δv_{GS} occurs in the gate-source circuit. In general, this causes a change in drain voltage Δv_D and in drain current Δi_D. Since these variables are related by Eq. (9-58), we may write

$$\Delta i_D = \left(\frac{\partial i_D}{\partial v_{GS}}\right)_{V_D} \Delta v_{GS} + \left(\frac{\partial i_D}{\partial v_D}\right)_{V_{GS}} \Delta v_D \tag{9-59}$$

As a convenience, the changes in the total instantaneous quantities can be replaced by the varying components of the same quantities. Accordingly

$$\Delta i_D \equiv i_d, \qquad \Delta v_{GS} = v_{gs}, \qquad \Delta v_D = v_d \tag{9-60}$$

Inserting Eq. (9-60) into Eq. (9-59) yields

$$i_d = \left(\frac{\partial i_D}{\partial v_{GS}}\right)_{V_D} v_{gs} + \left(\frac{\partial i_D}{\partial v_D}\right) v_d \tag{9-61}$$

A further simplification results upon inserting Eqs. (7-46) and (7-48) into the last expression. This then brings into play two of the three parameters previously defined for the FET. Thus

$$\boxed{i_d = g_m v_{gs} + \frac{v_d}{r_d}} \tag{9-62}$$

Equation (9-62) is meaningful because it describes the functional relationship between the FET parameters and the changing quantities that take place between the gate-source and drain-source terminals. In view of the fact that Eq. (9-62) is an equation relating currents, a circuit inter- pretation naturally leads to the Norton equiv- alent circuit. This is shown in Fig. 9-13. Note that i_d is drawn flowing into the D terminal in accordance with the conventional positive direction established for it. The input circuit merely shows the varying component of volt- age appearing between the gate and source terminals. No closed path is provided for cur- rent flow on the input side since there is none.

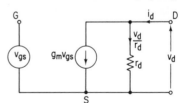

Fig. 9-13. Linear equivalent circuit of the FET, current-source version.

A linear model that emphasizes a voltage source and voltage drops rather than a current source and current paths is also possible. The expression for current of Eq. (9-62) is readily converted to an expression relating voltages by multiplying through by r_d. Thus

$$i_d r_d = g_m r_d v_{gs} + v_d \tag{9-63}$$

or

$$i_d r_d = \mu v_{gs} + v_d \tag{9-64}$$

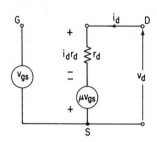

Fig. 9-14. Linear equivalent circuit of the FET, voltage-source version.

where μ is the voltage amplification factor described by Eq. (7-53). Upon rearranging terms, we have

$$v_d = i_d r_d - \mu v_{gs} \qquad (9\text{-}65)$$

The circuit interpretation of this expression is shown in Fig. 9-14. A comparison with the circuit of Fig. 9-13 should make it evident that the voltage-source equivalent circuit could be obtained directly from Fig. 9-13 by applying Thevenin's theorem. Both i_d and v_d in these figures are shown in their defined positive directions.

Equivalent Circuit of FET at Very High Frequencies. When the gate-source signal variations occur at very high frequencies, the equivalent circuit must be modified to take into account the capacitances that exist between the FET terminals. Keep in mind that at each pair of terminals voltages and charges appear. These in turn identify corresponding capacitances. Thus, a capacitance C_{gd} exists between the gate and drain terminals, another exists between the gate and source terminals, denoted by C_{gs}, and a third C_{ds} exists between the drain and source terminals. Usually, the drain-source capacitance is neglected entirely. The gate-drain capacitance has a value of the order of C_{ds}, but it is not neglected because its value is amplified for a reason to be pointed out in Sec. 9-5. Values of C_{gs} and C_{gd} are measured in picofarads (pf). Accordingly, it is only at very high frequencies (well above the audio range) that the associated reactances ($1/\omega C$) becomes small enough to be reckoned with.

The complete versions of the FET equivalent circuits depicting the foregoing capacitances appear in Fig. 9-15. The presence of capacitor C_{gd} indicates that a path exists that connects the output circuit at D to the input circuit at G. This

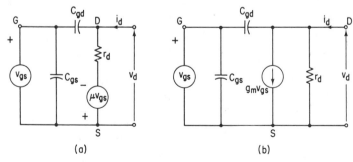

(a)   (b)

Fig. 9-15. The complete FET equivalent circuit including interterminal capacitances:

    (a) voltage-source equivalent;
    (b) current-source equivalent.

direct connection can cause oscillation difficulties if appropriate precautions are not taken to limit the amount of feedback that occurs between the output and input circuits. These capacitances also result in a decrease in voltage gain when the FET is used as an amplifier.

9-3 Biasing Methods for Transistor Amplifiers

In the transistor amplifier circuit of Fig. 9-4 the quiescent operating point was established with the aid of a battery source which was separate and distinct from the battery used in the output circuit. In the interest of simplicity and economy it is desirable that the amplifier circuit have a single source of supply—the one in the output circuit. We focus attention, therefore, in this section on the more commonly used methods of obtaining a fixed bias voltage or current in the input circuit without the use of a second power source.

Biasing Transistor Amplifiers. The treatment here is confined to the common-emitter configuration for two reasons. One, it is the most commonly used mode of transistor amplifier, and two, the high current amplification which occurs in this mode leads to a stability problem concerning the Q-point which must be reckoned with when selecting a bias method.

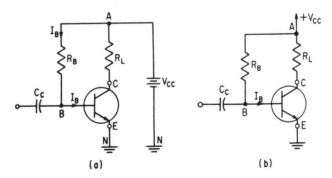

(a)   (b)

Fig. 9-16. Common-emitter transistor amplifier with fixed bias resistor R_B:
    (a) circuit diagram;
    (b) simplified form.

The simplest scheme for fixing the Q-point is to use the circuit shown in Fig. 9-16. A single source voltage V_{CC} is used. The quiescent base current is determined by the proper selection of the base resistor R_B. The closed circuit for this current is N-A-B-E-N. Applying Kirchhoff's voltage law to this circuit yields

$$V_{CC} - I_B R_B - V_{BE} = 0 \qquad (9\text{-}66)$$

where V_{BE} denotes the voltage drop between the base and emitter terminals. Since V_{BE} is at most several tenths of a volt, it can often be neglected with little error. It then follows from Eq. (9-66) that the required value of base resistance needed to yield the base current which for a specified load resistance establishes the Q-point is given by

$$R_B = \frac{V_{CC}}{I_B} \tag{9-67}$$

Keep in mind that V_{CC} is a known fixed quantity and I_B is chosen at some suitable value. Hence R_B can always be computed directly, and for this reason this procedure is called the *fixed-bias* method.

The circuit shown in Fig. 9-16(b) is a simplified form of Fig. 9-16(a). It is customary in drawing circuits of this kind to indicate only that side of V_{CC} which is connected to the load resistor R_L. It is assumed that the other side of the battery is connected to the ground (chassis) or reference point. The capacitor which appears in series with the base terminal is inserted for the purpose of blocking any d-c component in the signal from passing on through the base terminal.

On the basis of a superficial examination it would appear that the circuit of Fig. 9-16 should perform satisfactorily. However, experience shows that this is not so. The trouble lies in the inability of this circuit to maintain a fixed Q-point in the absence of signal inputs. Basically the reason for this behavior is that the reverse saturation current between the collector and emitter, I_{co}, is greatly influenced by the temperature. In fact it can be shown that the value of I_{co} is proportional to the cube of temperature. Thus

$$I_{co} = f(T^3) \tag{9-68}$$

When this fact is coupled with the current-amplifying feature of the common-emitter mode as reflected in Eq. (9-4), it is easy to understand why there occurs an instability of the Q-point. For convenience Eq. (9-4) is repeated here:

$$I_C = \frac{I_{co}}{1 - \alpha} + \frac{\alpha}{1 - \alpha} I_B \tag{9-69}$$

As the collector current flows in response to the base current in accordance with Eq. (9-69) heat is produced at the collector terminal, causing the temperature to rise. As the temperature rises, there occurs a rapid increase in I_{co} as called for by Eq. (9-68). Then, as revealed by Eq. (9-69), this causes a greatly magnified increase in collector current because I_{co} is divided by a very small number $(1 - \alpha)$. In turn the increased collector current causes a further increase in temperature which further increases the collector current. This behavior continues to that point where I_C is limited by the load resistor. This means that the Q-point has moved higher up along the load line to a position almost on the ordinate axis of the common-emitter characteristics. Any subsequent application of a sinusoidal input signal leads to an output signal that is greatly distorted. This instability of the Q-point is referred to as *thermal runaway* for obvious reasons.

A useful figure of merit which can be used to compare the relative instability of the various bias methods found in transistor amplifiers is the stability factor. It is simply defined as

$$S \equiv \frac{\partial I_C}{\partial I_{co}} \tag{9-70}$$

Applying Eq. (9-70) to Eq. (9-69) reveals that the stability factor for the fixed bias circuit of Fig. 9-16 is

$$S = \frac{\partial I_C}{\partial I_{co}} = \frac{1}{1 - \alpha} \qquad (9\text{-}71)$$

Accordingly, a value of $\alpha = 0.98$ yields a value of $S = 50$. This states that a per-unit change in the reverse saturation current results in a fiftyfold increase in collector current. Experience shows that values of S exceeding 25 result in unsatisfactory performance.

An improvement over the fixed-bias scheme of Fig. 9-16 is the use of the *collector-to-base-bias* method depicted in Fig. 9-17. In this arrangement note that the quiescent base current is determined not by V_{CC} but by the collector-to-base voltage. Consequently, any increase in I_C is partially offset by a decreased collector-to-base voltage swing due to the increased voltage drop in the load resistance R_L. The stability factor for this bias circuit is found to be

$$S = \frac{\partial I_C}{\partial I_{co}} = \frac{1}{1 - \alpha + \alpha \left(\dfrac{R_L}{R_L + R_B} \right)} \qquad (9\text{-}72)$$

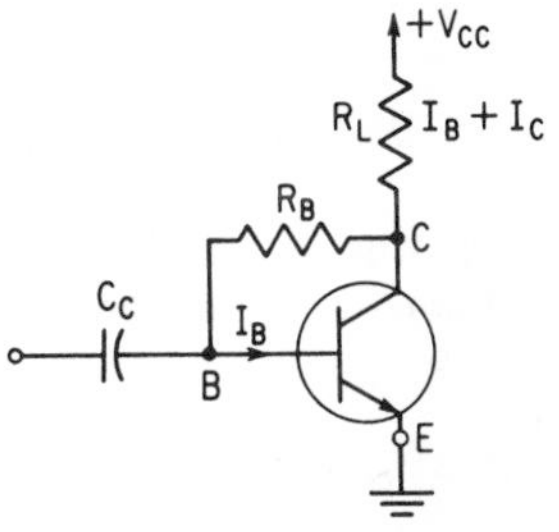

Fig. 9-17. Collector-to-base method of biasing a transistor amplifier.

Since the denominator term is increased the improvement in the stability of the Q-point is apparent.

The most frequently used bias method is the one illustrated in Fig. 9-18. It is called the *self-bias* or *emitter-bias* circuit, and it has a stability factor of the order of ten which yields very satisfactory results. To understand how this circuit brings about an effective stabilization of the Q-point, recall that the desired quiescent collector current I_C is determined by the voltage between the base-emitter terminals V_{BE}, which in turn determines the value of I_B in Eq. (9-69). This equation may be rewritten as

$$I_C = \frac{I_{co}}{1 - \alpha} + \frac{\alpha}{1 - \alpha} I_o \epsilon^{V_{BE}/E_T} \qquad (9\text{-}73)$$

where I_{co} is the reverse saturation current. Note too that V_{BE} is obtained as the difference of the voltages appearing across R_2 and R_E. That is

$$V_{BE} = V_{R2} - V_{RE} = V_{R2} - (I_B + I_C)R_E \qquad (9\text{-}74)$$

With this arrangement any increase in I_C due to a temperature rise causes the voltage drop across the emitter resistor R_E to increase. Since the voltage drop across R_2 is independent of I_C, it follows that V_{BE} diminishes [see Eq. (9-74)] which in turn causes I_B to decrease. The reduced value of I_B tends to restore I_C to its original value [see Eq. (9-69)].

A mathematical description of this behavior is given by the derivation of the stability factor. In the interest of simplifying the circuit analysis the d-c circuit to the left of the base terminal is replaced by its Thevenin equivalent circuit as depicted in Fig. 9-19. As far as the base terminal is concerned, looking towards

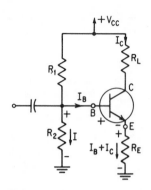

Fig. 9-18. Emitter-bias method of biasing an *n-p-n* transistor amplifier.

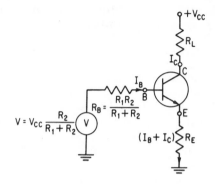

Fig. 9-19. Replacing the bias portion of the circuit of Fig. 9-18 by its Thevenin equivalent.

the left, there appears a voltage source of magnitude

$$V = \frac{R_2}{R_1 + R_2} V_{cc}$$

and of internal resistance R_B composed of the parallel combination of R_1 and R_2. Accordingly in the base-emitter circuit of Fig. 9-19 Kirchhoff's voltage law yields

$$V = I_B R_B + (I_B + I_C) R_E = I_B (R_B + R_E) + I_C R_E \qquad (9\text{-}75)$$

Hence the expression for the base current becomes

$$I_B = \frac{V}{R_B + R_E} - \frac{R_E}{R_B + R_E} I_C \qquad (9\text{-}76)$$

which upon insertion into Eq. (9-69) leads to an expression for the collector current given by

$$I_C \left(1 + \frac{\alpha}{1 - \alpha} \frac{R_E}{R_E + R_B} \right) = \frac{I_{co}}{1 - \alpha} + \frac{\alpha}{1 - \alpha} \left(\frac{V}{R_B + R_E} \right) \qquad (9\text{-}77)$$

Multiplying through on both sides by $(1 - \alpha)$ yields

$$I_C \left(1 - \alpha + \alpha \frac{R_E}{R_E + R_B} \right) = I_{co} + \alpha \frac{V}{R_B + R_E} \qquad (9\text{-}78)$$

Equation (9-78) is now in a form which leads directly to the stability factor. Thus

$$S = \frac{\partial I_C}{\partial I_{co}} = \frac{1}{1 - \alpha + \alpha \dfrac{R_E}{R_E + R_B}} \qquad (9\text{-}79)$$

Examination of this expression indicates that there are two ways in which the stability may be increased: (1) by decreasing R_B and (2) by increasing R_E. However, there is a limit on how much R_B can be decreased. Low values of R_B are obtained by making R_2 very small. In turn this has two drawbacks. It causes a shunting effect on the input signal, and it causes a heavy current drain on V_{CC}. Similarly, there is an upper limit on the value of R_E. For a fixed source V_{CC}, the larger R_E

is made, the smaller the allowable input signal must be to prevent distortion. On the other hand if it is desirable to maintain a suitably centered Q-point relative to the common-emitter characteristics, then the larger R_E, the greater V_{CC} must be. Typical values for the stability factor of the circuit of Fig. 9-18 are around ten.

EXAMPLE 9-1   The purpose of this example is to discuss the manner in which the resistors R_1, R_2, and R_E are selected in the circuit of Fig. 9-18, and then to consider how the actual Q-point can be found graphically.

Assume that the following information is known:

transistor type (see Fig. 9-20)
$R_L = 1000$ ohms
$V_{CC} = 12$ volts

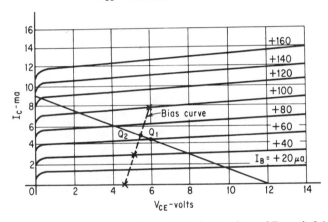

Fig. 9-20. Common-emitter characteristics for transistor of Example 9-1.

solution: Without R_E the d-c load line would be determined by R_L so that on the ordinate axis it would pass through the 12-ma point and on the abscissa axis through the 12-volt point. Refer to Fig. 9-20. With R_E the 12-volt point remains fixed but the ordinate-axis intercept is lowered. Since this lowering of the load line should not be excessive, a good rule of thumb is to choose R_E so that the new coordinate-axis intercept is about three-fourths of the value without R_E. Thus we may write

$$I_C\text{-axis intercept} = \frac{V_{CC}}{R_L + R_E} = \frac{3}{4}\left(\frac{V_{CC}}{R_L}\right) \tag{9-80}$$

or

$$R_E = \frac{R_L}{3} = 333 \text{ ohms} \tag{9-81}$$

Next construct the d-c load line and select a suitable Q-point at about mid range on the load line. This choice is identified as Q_1 in Fig. 9-20. Corresponding to this point read off the following pertinent quantities:

$$I_B = 60 \ \mu a, \quad V_{CE} = 6 \text{ volts}, \quad I_B = 4.4 \text{ ma} \tag{9-82}$$

To find R_2 use is made of Eq. (9-74) as it applies to Fig. 9-18. Thus

$$IR_2 = V_{BE} + (I_B + I_C)R_E \tag{9-83}$$

Inserting a typical value for V_{BE} of 0.2 volt as well as I_B and I_C from Eq. (9-82) yields

$$IR_2 = 0.2 + 4.46 \times 10^{-3} \times 333 = 1.69 \text{ volts} \tag{9-84}$$

In this expression neither I nor R_2 is known, so that we may arbitrarily select one and compute the other. In the interest of limiting the drain on the voltage source I is generally selected as a fraction of I_C (often one-fifth to one-tenth of I_C). In this case upon choosing $I \approx 0.5$ ma we get from the last equation

$$R_2 = \frac{1.69}{I} = \frac{1.69}{0.5} 10^3 = 2.38 \text{ K} \tag{9-85}$$

To find R_1 write Kirchhoff's voltage law for the loop containing V_{CC}, R_1, and R_2. Thus from Fig. 9-18 we have

$$V_{CC} = (I + I_B)R_1 + IR_2 \tag{9-86}$$

or

$$R_1 = \frac{V_{CC} - IR_2}{I + I_B} \tag{9-87}$$

Inserting the values for the quantities on the right yields

$$R_1 = \frac{12 - 1.69}{(0.5 + 0.06)10^{-3}} = 18.4 \text{ K} \tag{9-88}$$

Now that R_1, R_2, and R_E have specific values, the actual Q-point can be found so that it may be compared with the arbitrarily selected one. If sufficient disagreement occurs, a refinement on R_1 and R_2 will be introduced.

Kirchhoff's voltage law for the collector-emitter circuit of Fig. 9-19 leads to

$$V_{CC} = I_C R_L + V_{CE} + (I_C + I_B)R_E \tag{9-89}$$

Next insert the values for R_L and R_E espressed in kilohms. This permits I_B and I_C to be expressed in milliamperes. Thus

$$V_{CC} = 12 = V_{CE} + 1.333 I_C + 0.333 I_B \tag{9-90}$$

Similarly for the base-emitter loop we write

$$V_{CC} \frac{R_2}{R_1 + R_2} = I_B \frac{R_1 R_2}{R_1 + R_2} + (I_B + I_C)R_E \tag{9-91}$$

Inserting the values of R_1, R_2, and R_E just computed yields

$$1.86 = 0.333 I_C + 3.193 I_B \tag{9-92}$$

It is helpful at this point to eliminate I_C from Eqs. (9-90) and (9-92). This is readily done by multiplying Eq. (9-92) by 4 and then subtracting it from Eq. (8-80). The result is

$$V_{CE} = 4.56 + 12.44 I_B \tag{9-93}$$

where I_B must be expressed in ma.

Equation (9-93) is called the *bias equation* and involves two unknowns. The point of intersection of this equation with the d-c load line identifies the actual Q-point for the specified values of R_1, R_2, and R_E. Since Eq. (9-93) is a straight line, only two points are needed to plot it. Thus for

$$\begin{aligned} I_B &= 0, & V_{CE} &= 4.56 \text{ volts} \\ I_B &= 0.100 \text{ ma}, & V_{CE} &= 5.8 \text{ volts} \end{aligned} \tag{9-94}$$

This leads to the *bias curve* shown in Fig. 9-20. Note that it intersects the d-c load line at Q_2.

Since Q_2 is away from the assumed position of the Q-point, an adjustment is in order for R_1 and R_2. A little thought indicates that R_2 should be selected somewhat higher. Hence choose

$$R_2 = 4 \text{ K} \tag{9-95}$$

The adjusted value of I then becomes

$$I = \frac{1.69}{4} = 0.432 \text{ ma} \tag{9-96}$$

The new value of R_1 can still be found from Eq. (9-87), but the value of I_B corresponding to Q_2 must be used. This value is 0.065 ma. Hence

$$R_1 = \frac{10.31}{0.423 + 0.065} = 21.2 \text{ K} \tag{9-97}$$

We must now find the new bias curve so that the Q-point coresponding to the adjusted values of R_1 and R_2 can be found and compared to Q_2. Proceeding as before, the new bias equation becomes

$$V_{CE} = 4.4 + 13.48 I_B \tag{9-98}$$

A plot of this equation yields an intersection point with the load line which is almost identical to Q_1

Therefore the values of the resistors in the final design are:

$$R_E = 333 \text{ ohms}, \qquad R_1 = 4000 \text{ ohms}, \qquad R_2 = 21,000 \text{ ohms} \tag{9-99}$$

Although the introduction of R_E in the emitter leg helps to provide satisfactory stability of the quiescent point, the circuit configuration shown in Fig. 9-18 will cause a drop in gain unless suitably modified. This may be illustrated by assuming that a positive increment in base current occurs, which in turn causes a corresponding amplified increase in collector current. However, since this increased current causes the voltage drop across R_E to increase, thereby reducing the base-to-collector voltage, there occurs a subsequent drop in collector current.

The result is that the overall gain is less with R_E present than without it. This diminution in gain can be avoided by providing an alternate path from the emitter terminal through which the *changing* component of the emitter current may pass with little or no hindrance. The insertion of a low-impedance capacitor in parallel with R_E effectively accomplishes this purpose. See Fig. 9-21. The size of C_E is chosen so that at the frequency of the input signal the reactance $1/\omega C_E$ is small compared to R_E. Of course the presence of C_E does not upset the quiescent condition because the capacitor is an open circuit to d-c.

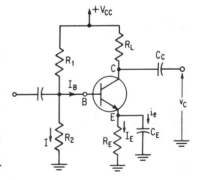

Fig. 9-21. Complete common-emitter amplifier circuit. The varying component of emitter current passes through C_E.

Biasing the FET. Reference to Fig. 9-7 shows that for a specified load resistance and junction field-effect transistor, the quiescent operating point is determined by the battery source V_{GG} in the gate-source circuit. As was true with the transistor

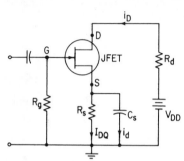

amplifier, it is desirable to eliminate this battery by using an appropriate biasing circuit. The most common procedure is to put a resistor R_s in the source leg, as illustrated in Fig. 9-22. Then at the quiescent condition the d-c drain current I_{DQ} flows through R_s, thus establishing a negative gate-source voltage given by

$$V_{GQ} = -I_{DQ}R_s \qquad (9\text{-}100)$$

Fig. 9-22. A JFET voltage amplifier with self-biasing source resistor R_s.

The minus sign appears because the voltage drop across R_s has the negative end on the gate side. Equation (9-100) is the bias equation and is analogous to Eq. (9-98) for the transistor case. The corresponding bias curve is found by plotting Eq. (9-100) on the drain volt-ampere characteristics for various assumed values of quiescent drain current I_{DQ}. The point of intersection of this bias curve with the load line then locates the actual quiescent point.

The source by-pass capacitor C_s is also included in the amplifier circuitry for the FET in order to prevent the drop in gain that results when the varying component of the drain current is made to flow through the self-bias source resistor. If C_s is selected sufficiently large so that its reactance at the lowest signal frequency is small compared to R_s, the voltage gain remains virtually intact.

EXAMPLE 9-2 A 2N4223 JFET is used in the amplifier circuit of Fig. 9-22. The drain supply voltage is 24 volts. It is desired that the quiescent point be located at $I_{DQ} = 3.4$ ma, $V_{DQ} = 15$ volts, and $V_{GQ} = 1$ volt. Find the values of self-bias source resistance R_s and drain load resistance R_d to assure this Q-point.

solution: The self-bias source resistance is found directly from Eq. (9-100). Thus

$$V_{GQ} = -I_{DQ}R_s$$
$$-1 = -3.4 \times 10^{-3}R_r$$

or

$$R_s = 293 \text{ ohms}$$

To find R_d, write Kirchhoff's voltage law for the ouput circuit at the quiescent point. Thus

$$V_{DD} = I_{DQ}R_d + V_{DQ} + I_{DQ}R_s \qquad (9\text{-}101)$$

Thus

$$R_d = \frac{V_{DD} - V_{DQ}}{I_{DQ}} - R_s = \frac{24 - 15}{3.4 \times 10^{-3}} - 293$$
$$= 2645 - 293 = 2352 \text{ ohms}$$

9-4   Computing Amplifier Performance

In this section attention is directed to the manner in which single-stage and multistage amplifier performance can be computed. In situations where the input signal variations are small the amplifier performance is determined through use of the equivalent circuit. Typical parameters are used for the transistors as obtained from the manufacturers' handbooks. The effects of the necessary biasing circuits on the overall amplifier performance are also treated. In those instances where the input signal goes through a wide excursion of values about a quiescent point, the performance is found more correctly by a graphical analysis. This is often necessary in the last stages of a multistage amplifier.

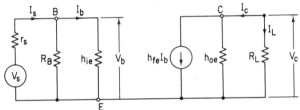

Fig. 9-23. Equivalent circuit of a transistor amplifier in the common-emitter mode.

General Expressions for Current and Voltage Gains. Appearing in Fig. 9-23 is the small-signal equivalent circuit of a transistor amplifier that includes the appropriate transistor parameters, the base bias equivalent resistor R_B, and the collector load resistor R_L. The source signal V_s and its internal resistance r_s are also shown. All symbols for voltages and currents are those that denote the effective values of the varying components of the quantities involved. Moreover, I_b, I_c, I_L, V_c, and V_b are shown in directions consistent with their assumed positive values. If actual current or voltage polarities are opposite to these, then the quantities are considered negative.

There are several current gains that can be identified with the configuration of Fig. 9-23. The first is the current amplification factor of the transistor itself—h_{fe}. This is an ideal current gain. It is realizable at the load resistor only when h_{oe} approaches zero and R_B approaches infinity. The second current gain is that which relates the collector (or load) current to the base current of the transistor. This quantity can be called the *transistor current gain* and can be expressed in terms of the load resistor and the transistor parameters. In the output circuit, the expression for load current becomes

$$I_L = -I_c = \frac{-1/h_{oe}}{1/h_{oe} + R_L} h_{fe} I_b = \frac{-h_{fe}}{1 + h_{oe}R_L} I_b \qquad (9\text{-}102)$$

The transistor current gain results upon formulating the ratio of output current to base current. Thus

$$A_I \equiv \frac{I_L}{I_b} = \frac{-I_c}{I_b} = -\frac{h_{fe}}{1 + h_{oe}R_L} \qquad (9\text{-}103)$$

The third current gain relates the load current to the source current. Accordingly, this is the current gain of the complete stage of the transistor amplifier. It is the most useful of the three current gains for obvious reasons. An asterisk notation is used to emphasize this fact. We have then

$$A_I^* \equiv \frac{I_L}{I_s} = \frac{I_L}{I_b} \times \frac{I_b}{I_s} \qquad (9\text{-}104)$$

The first part of the right side of Eq. (9-104) is available from the expression for the transistor current gain, Eq. (9-103). The second part is readily obtained by using the current-division principle as it applies to the parallel resistors R_B and h_{ie}. Thus

$$\frac{I_b}{I_s} = \frac{R_B}{R_B + h_{ie}} = \frac{1}{1 + h_{ie}/R_B} \qquad (9\text{-}105)$$

Inserting Eq. (9-103) and the last expression into Eq. (9-104) then yields the desired result.

$$\boxed{A_I^* = \frac{-h_{fe}}{(1 + h_{oe}R_L)(1 + h_{ie}/R_B)}} \qquad (9\text{-}106)$$

It is instructive to note here that the current gain of the complete amplifier stage becomes the current amplification factor only when $h_{oe} \to 0$ and $R_B \to \infty$. The more h_{oe} and R_B deviate from these ideal values, the greater will be the deterioration of the stage current gain from the ideal value of h_{fe}.

An analogous set of equations can be developed to express the voltage gain of the transistor amplifier. By definition, the voltage gain of the transistor (but not the complete stage) is

$$A_V \equiv \frac{V_c}{V_b} = \frac{-I_c R_L}{V_b} \qquad (9\text{-}107)$$

Here $-I_c$ can be replaced by its equivalent expression, $A_I I_b$ [see Eq. (9-103)]. Hence

$$A_V = \frac{A_I I_b R_L}{V_b} = \frac{A_I R_L}{V_b/I_b} = A_I \frac{R_L}{h_{ie}} \qquad (9\text{-}108)$$

It is significant to note here that the transistor voltage gain is determined by its current gain multiplied by the ratio of load resistor in the output circuit to the forward-biased resistor of the input circuit. This behavior was first encountered in Sec. 7-3. Refer to Eq. (7-34). Upon inserting Eq. (9-103) into the last equation, the transistor voltage gain may be written as

$$A_V = \frac{-h_{fe}}{1 + h_{oe}R_L}\left(\frac{R_L}{h_{ie}}\right) \qquad (9\text{-}109)$$

The role played by each of the transistor parameters and the load resistor in determining voltage gain is made clearly evident by this formulation.

The voltage gain of the entire stage is defined as

$$A_V^* \equiv \frac{V_c}{V_s} = \frac{V_c}{V_b} \times \frac{V_b}{V_s} = A_V \frac{V_b}{V_s} \qquad (9\text{-}110)$$

The first ratio on the right side is given by Eq. (9-109). The second ratio follows from an analysis of the input circuit of Fig. 9-23. Let the parallel arrangement of R_B and h_{ie} be represented by R_i, where the subscript stresses that this quantity is the input resistance of the amplifier. Thus

$$R_i = \frac{R_B h_{ie}}{R_B + h_{ie}} = \frac{h_{ie}}{1 + h_{ie}/R_B} \tag{9-111}$$

Then by voltage division

$$\frac{V_b}{V_s} = \frac{R_i}{R_i + r_s} = \frac{1}{1 + r_s/R_i} \tag{9-112}$$

The final expression for the overall voltage gain thus becomes

$$A_V^* = \frac{-h_{fe}}{1 + h_{oe}R_L}\left(\frac{R_L}{h_{ie}}\right)\left(\frac{1}{1 + r_s/R_i}\right) \tag{9-113}$$

Single-stage Transistor Amplifier. As stated in the introduction to this section, our interest next is to illustrate all of the computational features involved in computing the performance of a single-stage as well as a multistage amplifier. In this effort, a step-by-step analysis is made of a transistorized intercommunication system composed of four amplifying stages. Each stage is considered to be appropriately equipped with biasing circuits, by-pass capacitors, and coupling capacitors. A secondary objective of the analysis is to show how a succeeding amplifier stage "loads down" a preceding stage, thereby causing an overall reduction in performance.

Although the general expressions developed above could be applied directly to obtain the current and voltage gains, this procedure will not be used. Rather, in the interest of emphasizing first principles, each equivalent circuit as it arises is solved as a separate problem in circuitry. Besides offering an opportunity of exploiting our knowledge of circuit theory, such a procedure also makes us less dependent upon memorizing formulas.

As the first step in the development of the intercommunication system, consider the single-stage transistor amplifier depicted in Fig. 9-24. Assume the input signal to be a sinusoidal voltage having an effective value of 25 millivolts and an internal resistance of 1000 ohms. Let us find the performance of this amplifier circuit for the 2N104 transistor and the values of the circuit parameters shown in Fig. 9-24. Keep in mind that finding the performance of the amplifier generally means computing current gain, voltage gain, power gain, the input resistance as seen by the signal source, and the output resistance appearing at the output terminals.

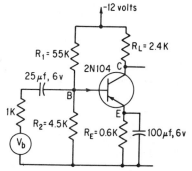

Fig. 9-24. Single-stage transistor amplifier. The input signal is a generator having an rms voltage V_b and internal resistance of 1K.

In view of the small magnitude of the input signal the performance is best determined by using an equivalent-circuit analysis. Also, because we are dealing here with an audio amplifier, the diffusion and transition capacitors in the complete equivalent circuit of the transistor may be neglected. These capacitors are important only at very high signal frequencies. Accordingly the equivalent circuit of Fig. 9-9 may be employed to replace the transistor in the circuit of Fig. 9-24, which in turn leads to the equivalent circuit of the transistor amplifier shown in Fig. 9-25. Here h_{re} is assumed equal to zero. Recall that R_B is the parallel combination of R_1 and R_2 of the base bias circuit which must be considered in the analysis. Depicted in Fig. 9-25 is the equivalent circuit showing the transistor manufacturer's published value of the equivalent circuit parameters.

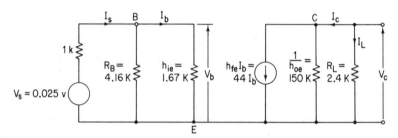

Fig. 9-25. The h-parameter equivalent circuit of the common-emitter amplifier of Fig. 9-23.

The amplifier performance is now readily calculated by applying the techniques of circuit analysis to the circuit of Fig. 9-25. To find the signal source current I_s, it is necessary first to find the parallel combination of the two resistances between terminals B and E. Thus

$$R_p = R_i = \frac{4.16(1.67)}{4.16 + 1.67}\,K = 1.19\,K \tag{9-114}$$

Then the effective value of the source current is

$$I_s = \frac{V_s}{1000 + R_p} = \frac{0.025}{2.19 \times 10^3} = 11.4\,\mu a \tag{9-115}$$

By the current-divider theorem the effective value of the varying component of the base current I_b is found to be

$$I_b = \frac{4.16}{5.825}I_s = 8.15\,\mu a \tag{9-116}$$

The current source generator in the output section of the equivalent circuit has the value

$$h_{fe}I_b = 44(8.15)\,\mu a = 0.358\,ma \tag{9-117}$$

This leads to an effective value of collector current I_c of

$$I_c = \frac{150}{152.4}(0.358) = 0.352\,ma \tag{9-118}$$

The current gain of the *complete amplifier stage* is then

$$A_I^* \equiv \frac{I_L}{I_s} = \frac{-I_c}{I_s} = -\frac{0.352 \times 10^{-3}}{11.4 \times 10^{-6}} = -30.9 \qquad (9\text{-}119)$$

This result is readily verified by direct application of Eq. (9-106). Thus

$$A_I^* = \frac{-44}{(1 + (2.4)(150))(1 + (1.67/4.16))} = -30.9 \qquad (9\text{-}120)$$

The transistor current gain, which is the ratio of the collector current to the base current, is given either by

$$A_I = \frac{I_L}{I_b} = \frac{-I_c}{I_b} = -\frac{352}{8.15} = -43.2 \qquad (9\text{-}121)$$

or computed directly from Eq. (9-103) as

$$A_I = \frac{-h_{fe}}{1 + h_{oe}R_L} = \frac{-44}{1 + (2.4/150)} = -43.2 \qquad (9\text{-}122)$$

The voltage gain for the complete amplifier stage is found to be

$$A_V^* \equiv \frac{V_L}{V_s} = \frac{V_c}{V_s} = \frac{-I_cR_L}{V_s} = -\frac{(0.352)(2.4)}{0.025} = -33.8 \qquad (9\text{-}123)$$

A direct application of Eq. (9-113) provides

$$A_V^* = \frac{-h_{fe}}{1 + h_{oe}R_L}\left(\frac{R_L}{h_{ie}}\right)\left(\frac{1}{1 + (r_s/R_i)}\right)$$

$$= \frac{-44}{1.016}\left(\frac{2.4}{1.67}\right)\left(\frac{1}{1 + (1/1.19)}\right) = -33.8 \qquad (9\text{-}124)$$

The corresponding value of the power gain is

$$G = |A_V^*||A_I^*| = 33.8(30.9) = 1042 \qquad (9\text{-}125)$$

The input resistance of the amplifier circuit is the resistance seen by the signal source between terminals B and E in Fig. 9-25. Accordingly

$$R_i = R_p = 1.19 \text{ K} \qquad (9\text{-}126)$$

The input resistance is of importance because it provides a measure of the extent to which the internal resistance of the source is significant in making available a signal variation to the transistor.

The output resistance R_0 of the amplifier circuit is defined here as the resistance appearing between the output terminals C and E in the equivalent circuit of Fig. 9-25 with R_L connected. The value is

$$R_0 = \frac{2.4(150) \text{ K}}{152.4 \text{ K}} = 2.36 \text{ K} \qquad (9\text{-}127)$$

Information about the output resistance of an amplifier stage is useful in determining matching characteristics when the first stage is followed by a second stage.

Two-stage Transistor Amplifier. Attention is next directed to the two-stage transistor amplifier depicted in Fig. 9-26. Note that the first stage is identical to

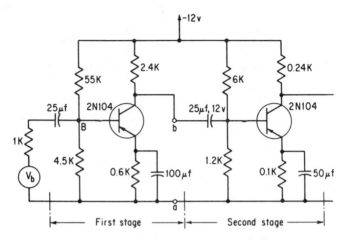

Fig. 9-26. Schematic diagram of a two-stage transistor amplifier.

that shown in Fig. 9-24. The input terminals to the second stage are identified as *ab*. As far as the first stage is concerned, it is helpful to note that it may be replaced by an equivalent current source having a value of $44I_b = 358$ μa and shunted by an equivalent resistance which is the output resistance of the first stage. This current source serves as the input signal to the second stage, the equivalent circuit of which is depicted in Fig. 9-27. Since the second stage also uses the 2N104 transistor, the same equivalent-circuit parameters are used as appear in Fig. 9-25. However, note that the equivalent resistance R_{B2} associated with the bias circuit is now equal to 1 K, which is the parallel combination of the 1.2-K and 6-K resistors of the second stage.

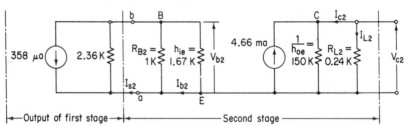

Fig. 9-27. The *h*-parameter equivalent circuit of the two-stage amplifier of Fig. 9-25.

The signal input current to the second stage is the current, I_{s2}, flowing into terminal *a* in Fig. 9-27. Its value is readily found by applying the current-divider theorem. But first it is helpful to obtain the equivalent resistance of the parallel combination of the 1-K and 1.67-K resistances. Thus

$$R_{p2} = \frac{1(1.67)}{1 + 1.67} = 0.625 \quad \text{K} \tag{9-128}$$

Then

$$I_{s2} = \frac{2.36}{2.36 + 0.625}(358) = 283 \quad \mu a \tag{9-129}$$

In order to obtain the current amplification as it appears in the output circuit of the second stage, we need the base current to the transistor in the second stage. Again by the current-divider theorem we have

$$I_{b2} = \frac{1}{1 + 1.67}I_{s2} = \frac{1}{2.67}(283) = 106 \quad \mu a \tag{9-130}$$

It is worthwhile at this point in the computations to check on the validity of proceeding with an equivalent-circuit approach for finding the performance. Because I_{b2} in Eq. (9-130) represents the effective value of the base current variation, it follows that the maximum input current swing is $\sqrt{2}(106) = 150 \ \mu a$. A glance at the average collector characteristics for the 2N104 transistor reveals that this swing is sufficiently small to allow computations to be made by means of the equivalent circuit rather than by a graphical analysis. However, we can well expect that for a third stage the swing will be so large that the equivalent-circuit representation is no longer proper. A graphical analysis will then be necessary; this matter is considered in the next phase of our study.

The current source appearing in the output section of the second stage is found to be

$$h_{fe}I_{b2} = 44(106) \ \mu a = 4.66 \quad ma \tag{9-131}$$

The output resistance of the second stage as seen between terminals C and E in Fig. 9-27 is for all practical purposes equal to 240 ohms. It also follows from this approximation that the collector current at the second stage is equal to the source current. Thus

$$I_{c2} \approx h_{fe}I_{b2} = -4.66 \quad ma \tag{9-132}$$

The minus sign indicates flow is opposite to the assumed positive direction.

We are now in a position to make an interesting comparison of the current gain of the first stage both with and without the second stage. The useful version of the current gain of the one-stage amplifier was found in Eq. (9-119) to be

$$A_I = -\frac{I_{c1}}{I_{s1}} = -\frac{352 \ \mu a}{11.4 \ \mu a} = -30.9 \tag{9-133}$$

When a second stage is made to follow the first stage, a loading effect occurs which serves to reduce the current gain. This comes about because the input resistance of the second stage is finite and of a value comparable to the output resistance of the first stage. A glance at the input section of Fig. 9-27 makes this apparent. Accordingly, a more significant expression for the current gain of the first stage of a two-stage amplifier is defined as

$$A'_I = \frac{I_{s2}}{I_{s1}} = -\frac{283 \ \mu a}{11.4 \ \mu a} = -24.8 \tag{9-134}$$

Note the substantial reduction that takes place from the unloaded current gain of Eq. (9-133).

The current gain for the second stage may be identified as

$$A_{I2}^* = \frac{I_{L2}}{I_{s1}} = -\frac{I_{c2}}{I_{s2}} = -\frac{4.66 \text{ ma}}{0.283 \text{ ma}} = -16.5 \qquad (9\text{-}135)$$

Keep in mind that this value is an unloaded current gain too. Therefore, in situations where the second stage is followed by a third stage, the same adjustment must be made for the second-stage gain as was done above for the first stage.

The total current gain of the complete two-stage amplifier (but with the second stage unloaded, i.e., as it appears in Fig. 9-27), is given by

$$\frac{I_{L2}}{I_{s1}} = -\frac{I_{c2}}{I_{s1}} = \frac{4.66 \text{ ma}}{0.0114 \text{ ma}} = 408 \qquad (9\text{-}136)$$

A check on this computation may be made by taking the product of the first-stage current gain, Eq. (9-134), and the second-stage current gain, Eq. (9-135). Thus

$$\frac{I_{L2}}{I_{s1}} = -\frac{I_{c2}}{I_{s1}} = \left(\frac{I_{s2}}{I_{s1}}\right)\left(-\frac{I_{c2}}{I_{s2}}\right) = 24.8(16.5) = 408 \qquad (9\text{-}137)$$

Continuing with the description of the performance of the two-stage amplifier we next compute the voltage gain. Clearly this is defined as the ratio of the a-c output voltage at the second stage divided by the input signal voltage. Thus

$$\frac{V_{c2}}{V_s} = -\frac{I_{c2}R_{L2}}{V_s} = \frac{4.66(0.24)}{0.025} = 44.74 \qquad (9\text{-}138)$$

The corresponding power gain for both stages then becomes

$$G = 408(44.74) = 18{,}254 \qquad (9\text{-}139)$$

Finally, the input resistance to the second stage is

$$R_{i2} = \frac{1(1.67)}{1 + 1.67} = 0.625 \text{ K} = R_{p2} \qquad (9\text{-}140)$$

and the output resistance of the second stage is

$$R_{o2} = \frac{(150 \text{ K})(0.24 \text{ K})}{150.24 \text{ K}} \approx 0.24 \text{ K} \qquad (9\text{-}141)$$

Three-stage Transistor Amplifier. The circuit configuration of a three-stage transistor amplifier appears in Fig. 9-28. It is formed by adding an appropriate third stage to the two-stage amplifier of Fig. 9-26. Before proceeding with the performance analysis of the three-stage amplifier, it is important first to determine how big a swing of the input current to the third stage occurs. Figure 9-29 shows the input section of the equivalent circuit of the third stage, which can be used to check on this swing. The input impedance as seen by the current source looking into terminals c and d of the third stage is

$$R_{i3} = \frac{R_{B3}h_{ie}}{R_{B3} + h_{ie}} \qquad (9\text{-}142)$$

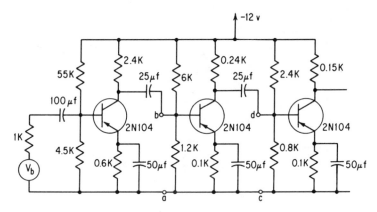

Fig. 9-28. Circuit diagram of a three-stage transistor amplifier.

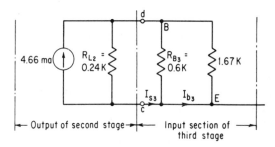

Fig. 9-29. Portion of input section of the equivalent circuit of the third stage which is used to check the maximum swing of the input signal.

where

$$R_{B3} = \frac{(0.8\ \text{K})(2.4\ \text{K})}{0.8\ \text{K} + 2.4\ \text{K}} = 0.6 \quad \text{K} \tag{9-143}$$

Hence

$$R_{i3} = \frac{0.6(1.67)}{0.6 + 1.67} = 0.441 \quad \text{K} \tag{9-144}$$

Therefore

$$I_{s3} = \frac{R_{i2}}{R_{i2} + R_{i3}} 4.66 = \frac{0.24}{0.681}(4.66) = 1.64 \quad \text{ma} \tag{9-145}$$

Accordingly, the effective value of the varying component of the base current to the transistor in the third stage is

$$I_{b3} = \frac{0.6}{0.6 + 1.67} I_{s3} = \frac{0.6}{2.27}(1.64) = 0.435 \quad \text{ma} \tag{9-146}$$

The maximum input current swing is thus $\sqrt{2}\,(435)\ \mu\text{a}$ or $615\ \mu\text{a}$. A glance at the average collector characteristics of the 2N104 depicted in Fig. 9-30 indicates

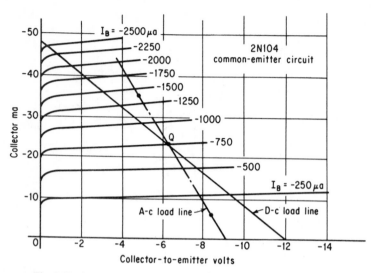

Fig. 9-30. Average collector characteristics of the 2N104 transistor.

that for a properly located quiescent point the base-current swing is so large as to require a graphical analysis.

To find the performance of the third stage graphically, it is necessary first to establish the quiescent point on the collector characteristics of Fig. 9-30. By following the procedure outlined in Sec. 7-3 for the transistor amplifier, the bias curve can be drawn and the Q-point located. For the third stage of the amplifier depicted in Fig. 9-28 the quiescent point is found to be

$$I_B = 750 \ \mu a, I_C = 28 \ \text{ma}, V_{CE} = -6.3 \qquad (9\text{-}147)$$

Therefore the maximum value of the base current in the third stage becomes

$$I_{b3_{max}} = 750 + 615 = 1365 \ \mu a \qquad (9\text{-}148)$$

Since capacitor C_E is a virtual short circuit to varying input signals, changes about the Q-point must be found on the a-c load line, which here has a slope equal to the reciprocal of the 150-ohm load resistance. The corresponding values of maximum collector current and collector-to-emitter voltage as obtained on the a-c load line in Fig. 9-30 are then

$$I_{c_{max}} = 35 \ \text{ma} \quad \text{and} \quad V_{CE_{min}} = -4.8 \ \text{volts} \qquad (9\text{-}149)$$

Similarly, the minimum value of the base current is

$$I_{b_{min}} = 750 - 615 = 135 \ \mu a \qquad (9\text{-}150)$$

which in turn yields

$$I_{c_{min}} = 6 \ \text{ma} \quad \text{and} \quad V_{CE_{max}} = -8.4 \ \text{volts} \qquad (9\text{-}151)$$

The effective value of the varying component of the collector current then becomes

$$I_{c3} = \frac{I_{c_{max}} - I_{c_{min}}}{2\sqrt{2}} = \frac{35 - 6}{2\sqrt{2}} = 10.2 \text{ ma} \tag{9-152}$$

This leads to a current gain for the three-stage amplifier of

$$\frac{I_{L3}}{L_{s1}} = -\frac{I_{c3}}{I_{s1}} = -\frac{10.2 \text{ ma}}{0.0114 \text{ ma}} = -898 \tag{9-153}$$

The effective value of the varying component of the collector voltage is given by

$$V_{c3} = \frac{8.4 - 4.8}{2\sqrt{2}} = \frac{3.6}{2\sqrt{2}} = 1.28 \text{ volts} \tag{9-154}$$

The corresponding overall voltage gain is then

$$\frac{V_{c3}}{V_s} = -\frac{1.28}{0.025} = -51.2 \tag{9-155}$$

Finally the overall power gain becomes

$$G = 898|51.2| = 45,978 \tag{9-156}$$

The Four-stage Transistor Amplifier. The addition of still another stage to the output terminals of the circuit of Fig. 9-28 leads to the four-stage amplifier appearing in Fig. 9-31. The selection of the transistor for the fourth stage must

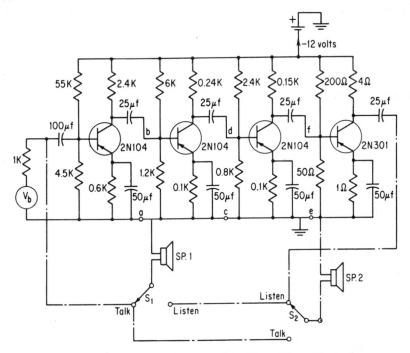

Fig. 9-31. Transistorized four-stage amplifier. The addition of the broken-line connections to the two speakers yields an intercommunication system.

be such that it can accommodate properly the wide swing in signal appearing at the output terminals of the third stage. The 2N301 transistor meets this condition.

In order to compute the output at the fourth stage, it is necessary to find the maximum and minimum values of the input base current to this stage. This is readily accomplished by representing the output of the third stage by the equivalent current source, 10.2 ma, and its shunt output resistance, $R_{L3} = 150$ ohms, and combining it with the input resistance of the fourth stage. Refer to Fig. 9-32.

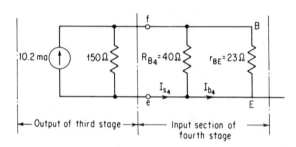

Fig. 9-32. Circuit used to obtain maximum swing of base current into fourth-stage transistor.

The manufacturer's transistor manual specifies the resistance between the base and emitter terminals for the 2N301 as 23 ohms. Hence the input resistance of the fourth stage, taking into account the effect of the bias circuit, becomes

$$R_{p4} = \frac{R_{B4}h_{ie}}{R_{B4} + h_{ie}} = \frac{40(23)}{40 + 23} = 14.6 \text{ ohms} \qquad (9\text{-}157)$$

Again by current division we find

$$I_{s4} = \frac{-150}{150 + 14.6} 10.2 = -9.12 \text{ ma} \qquad (9\text{-}158)$$

and

$$I_{b4} = \frac{-40}{40 + 23} 9.12 = -5.79 \text{ ma} \qquad (9\text{-}159)$$

Negative signs again indicate current flow is opposite to assumed positive directions. Thus the peak value of the varying component of the base current for the fourth stage becomes $\sqrt{2}\,(5.79) = 8.2$ ma.

Construction of the bias curve for the circuit configuration of the fourth stage locates the quiescent point at $I_B = 20$ ma, $I_C = 1.4$ amp, and $V_{CE} = -5$ volts. This is identified as Q in Fig. 9-33. Moreover, the a-c load line is drawn corresponding to the 4-ohm load resistance. The maximum value of the base current in this case then becomes

$$I_{b_{max}} = 20 + 8.1 = 28.1 \text{ ma} \qquad (9\text{-}160)$$

and is marked as point a† on the a-c load line of Fig. 9-33. The corresponding

---

† This point as well as b were found on a large version of Fig. 9-33 for greater accuracy.

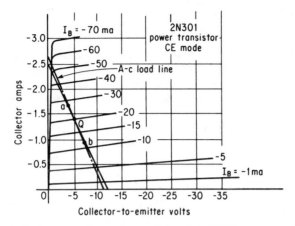

Fig. 9-33. Average collector characteristics of the 2N301 power transistor in the common-emitter mode.

values of the collector current and collector-to-emitter voltage are

$$I_{c_{max}} = 1.7 \text{ amps} \quad \text{and} \quad V_{CE_{min}} = -3.5 \text{ volts} \tag{9-161}$$

For the minimum value of base current, $I_{b_{min}} = 20 - 8.2 = 11.8$ ma, the collector current and voltage values are

$$I_{c_{min}} = 0.85 \text{ amp} \quad \text{and} \quad V_{CE_{max}} = -7.5 \text{ volts} \tag{9-162}$$

This point is identified as b on the a-c load line. It is worthwhile to note here that for the same swing above and below the Q-point of 8.2 ma in base current, the corresponding changes brought about in the collector current are quite different. The nonlinear nature of the spacing of collector characteristics is responsible for this effect.

The effective value of the magnitude of the varying component of the collector current is

$$|I_{c4}| = \frac{1.7 - 0.85}{2\sqrt{2}} = 0.3 \text{ amp} \tag{9-163}$$

This leads to an overall current gain of the four-stage amplifier of

$$\frac{I_{L4}}{I_{s1}} = -\frac{I_{c4}}{I_{s1}} = \frac{-(-0.3)}{11.4 \times 10^{-6}} = 26{,}300 \tag{9-164}$$

The value of I_{c4} is negative when expressed relative to I_{s1}. Proceeding in a similar fashion we find the rms value of the a-c component of the collector voltage to be

$$V_{c4} = \frac{7.5 - 3.5}{2\sqrt{2}} = \frac{2}{\sqrt{2}} = 1.4 \text{ volts} \tag{9-165}$$

and thereby yielding a four-stage voltage gain of

$$\frac{V_{c4}}{V_s} = \frac{1.4}{0.025} = 56 \tag{9-166}$$

The corresponding power gain is

$$G = 26.3 \times 10^3 \times 56 = 1.47 \times 10^6 \tag{9-167}$$

This completes the performance analysis of the four-stage amplifier.

As a matter of application interest it is useful to note that by the addition of two speakers wired in the manner shown by the broken-line connections depicted in Fig. 9-31, a transistorized intercommunication system results.

Single-stage FET Amplifier. The manner of computing the performance of a FET amplifier by the equivalent circuit can be illustrated by considering the FET amplifier configuration of Fig. 9-7. It is assumed that the varying input signal v_{gs} undergoes very small changes and that R_g is so large that it may be neglected from further consideration. Accordingly, the linear model for this amplifier may be drawn as shown in Fig. 9-34. We direct attention first to the current-source version. The effective value of the varying component of the drain voltage can be expressed as

$$V_d = I_L R_L = -I_d R_L \tag{9-168}$$

Here, too, the assumed positive directions for the load current and the drain current are opposite. This linear model of the amplifier states that the effective value of the voltage V_{gs} in the input circuit makes itself felt in the output circuit in the form of a current source of value $g_m V_{gs}$. This current then divides itself between r_d and R_L. By the current-divider principle, the drain current can be

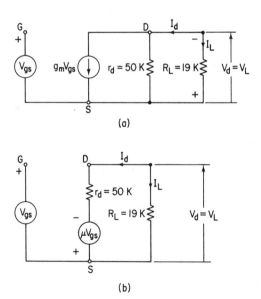

Fig. 9-34. Equivalent circuit of FET amplifier of Fig. 9-6:
(a) current-source version;
(b) voltage-source version.

expressed as

$$I_d = \frac{r_d}{R_L + r_d} g_m V_{gs} = \frac{g_m}{1 + R_L/r_d} V_{gs} \qquad (9\text{-}169)$$

The rms value of the varying drain voltage is thus related to the rms value of the varying gate-source voltage by

$$V_d = \frac{-g_m R_L}{1 + R_L/r_d} V_{gs} \qquad (9\text{-}170)$$

The voltage gain of the FET amplifier readily follows from Eq. (9-170) upon formulating the ratio of drain to gate-source voltage. Thus

$$\boxed{A_V \equiv \frac{V_L}{V_{gs}} = \frac{V_d}{V_{gs}} = \frac{-g_m R_d}{1 + R_L/r_d}} \qquad (9\text{-}171)$$

where g_m is the transconductance of the FET. The presence of the minus sign in this expression attests to the phase reversal that takes place in this amplifier. That is, instantaneously as the varying gate-source signal moves in a positive direction from the Q-point, the varying component of the drain-source voltage moves in a negative direction from the Q-point.

The 2N4220 FET used in the circuit of Fig. 9-7 has the following parameters at the specified Q-point:

$$g_m = 1200 \ \mu\text{mho}, \qquad r_d = 50 \text{ K}, \qquad \mu = 60$$

Accordingly, by Eq. (9-171) the specific value of the voltage gain for this amplifier is

$$A_V = \frac{-1.2 \times 10^{-3}(19 \times 10^3)}{1 + 19/50} = \frac{22.8}{1.38} = -16.5 \qquad (9\text{-}172)$$

An alternative expression for computing the voltage gain follows from an analysis of Fig. 9-34(b). Here the voltage appearing across the load is found by applying the voltage-division principle. Thus

$$V_d = V_L = \frac{-R_L}{R_L + r_d} \mu V_{gs} \qquad (9\text{-}173)$$

Rearranging terms leads to

$$\boxed{A_V = \frac{V_d}{V_{gs}} = \frac{-R_L}{R_L + r_d} \mu = \frac{-\mu}{1 + r_d/R_L}} \qquad (9\text{-}174)$$

Inserting the specified values yields

$$A_V = \frac{-60}{1 + \frac{50}{19}} = \frac{-60}{1 + 2.63} = -16.5 \qquad (9\text{-}175)$$

As expected, this result is identical to that of Eq. (9-172).

It is instructive to note that Eq. (9-174) could be derived from Eq. (9-171) by merely multiplying the right side of the latter by unity in the form of r_d/r_d and recalling that $\mu = g_m r_d$.

9-5   Frequency Response of *R-C* Coupled Transistor Amplifiers

A glance at the amplifier system depicted in Figs. 9-28 reveals that the coupling from one stage to the next is achieved by a coupling capacitor followed by a connection to a shunt resistor. Because of this arrangement such amplifiers are called *resistance-capacitance coupled amplifiers.* Since amplifiers are often designed to provide essentially uniform gain over some specified frequency range, it is important to know the factors which place upper and lower limits on this range. In this way appropriate steps may be taken in the amplifier design to assure satisfactory performance over the specified frequency range. For example, in the case of an audio amplifier, which is used to amplify speech and music sounds as detected by a microphone, it is essential that all the frequencies in the sound spectrum (15 Hz to 15,000 Hz) be uniformly amplified. Otherwise the amplified sound, emanating from the loudspeaker, will be a distorted version of that picked up by a quality microphone.

There are certain factors in *R-C* coupled amplifiers which will cause the amplifier not to provide the same gain at very low and very high frequencies. It is the purpose of this section to study what these factors are and how they may be controlled to yield a bandwidth which furnishes the desired fidelity of the output signals. Attention is directed first to the common-emitter transistor *R-C* coupled amplifier.

Transistor R-C Coupled Amplifiers. If an amplifier configuration contained no time-sensitive elements (such as capacitors), there would be no limit on the range of signal frequencies which could be amplified uniformly. That is, the bandwidth would be infinite. This situation, however, is not achievable in practice for two reasons. One, a coupling capacitor is generally employed between stages in order to eliminate transfer of the high d-c voltage levels to the input section of the following stage. This puts a limitation on the low-frequency transmission. Two, the amplifying devices themselves inherently possess capacitance, which although subject to diminution cannot be entirely eliminated. In transistors we must reckon with diffusion and transition capacitances, while in the vacuum tube it is the interelectrode capacitance which is important. These factors limit the high-frequency capability of the amplifier.

Three regions of the frequency spectrum are of concern to us in this study. One is the *midband frequency range.* This is defined as that frequency range over which the effects of all capacitors (coupling, diffusion, and transition) are negligible. The linear model of the common-emitter transistor amplifier as it applies to the midband frequency range takes on the form depicted in Fig. 9-35, which clearly is the very same one used in Sec. 9-4 to compute amplifier performance. For our purposes here, I_2 in Fig. 9-35 is considered to be the useful output current of the first stage (and, therefore, the input current of the second stage). The resistance R_2 denotes the net input resistance appearing at the base terminal of the second transistor. In this instance, it consists of the bias circuit resistance, R_{B2}, of the second stage in parallel with the h_{ie} parameter of the second transistor.

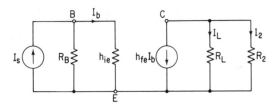

Fig. 9-35. Equivalent circuit of the common-emitter transistor amplifier as it applies to the midband frequency range.

The relationship of the current I_2 to the base current is found by applying the current-division principle to the output section of Fig. 9-35. Thus

$$I_2 = \frac{-R_L}{R_L + R_2} h_{fe} I_b = \frac{-h_{fe}}{1 + R_2/R_L} I_b \tag{9-176}$$

The minus sign simply recognizes that the actual current flow in this circuit is opposite to the assumed positive direction. Moreover, from the input circuit section we can write that

$$I_b = \frac{R_B}{R_B + h_{ie}} I_s = \frac{1}{1 + h_{ie}/R_B} I_s \tag{9-177}$$

Inserting the last equation into the preceding one yields

$$I_2 = \frac{-h_{fe}}{(1 + R_2/R_L)(1 + h_{ie}/R_B)} I_s \tag{9-178}$$

From Eq. (9-178), the current gain of the complete stage including the loading effect of the next stage becomes simply

$$A_{Im} \equiv \frac{I_2}{I_s} = \frac{-h_{fe}}{(1 + R_2/R_L)(1 + h_{ie}/R_B)} \tag{9-179}$$

This expression is the literal equivalent of Eq. (9-134). The subscript m is included to denote reference to the midband frequency range. In view the absence of any frequency-sensitive terms in Eq. (9-179), it follows that the gain in the midband frequency range remains invariant.

The second region of interest is the *low frequency range*. How should the equivalent circuit of Fig. 9-35 be modified so that it becomes applicable in this range? In this connection it is helpful to recall that the diffusion and transition capacitances (C_D and C_T) of the transistor are measured in picofarads, while the coupling capacitor C_C is measured in microfarads. Hence the reactances of the former two parameters are of the order of a million times larger than that of C_C. Recalling also that reactance is $1/\omega C$ and that ω is small at low frequencies, it follows that the reactance values of the diffusion and transition capacitors become so high as to appear as virtual open circuits. Hence, C_D and C_T may be omitted from the low-frequency model. The capacitor C_C, however, cannot be removed for, if it were, the transmission to the second stage would be zero. In fact, C_C is intentionally chosen large to assure satisfactory transmission even at low frequencies. Clearly, the larger the value of C_C, the smaller will be the frequency

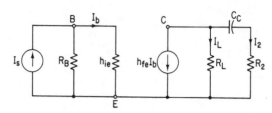

Fig. 9-36. Linear model of a common-emitter transistor amplifier that is applicable at low frequencies.

of the signal that can be successfully passed to the next stage. Of course, it is impossible to transmit a zero frequency (d-c) signal, because the capacitor then behaves as an open circuit. The applicable equivalent circuit at low frequencies is thus that shown in Fig. 9-36.

In this case, current division applied to the output section of the equivalent circuit yields

$$I_2 = \frac{-R_L}{R_L + R_2 + \dfrac{1}{j\omega C_C}} h_{fe} I_b = \frac{-R_L}{R_L + R_2} \left[\frac{1}{1 + \dfrac{1}{j\omega C_C(R_L + R_2)}} \right] \quad (9\text{-}180)$$

Inserting Eq. (9-177) for I_b and rearranging terms gives

$$I_2 = \frac{-h_{fe}}{\left(1 + \dfrac{R_2}{R_L}\right)\left(1 + \dfrac{h_{ie}}{R_B}\right)} \left[\frac{1}{1 - j\dfrac{1}{\omega(R_L + R_2)C_C}} \right] I_s \quad (9\text{-}181)$$

The quantity in brackets is a complex number, the value of which depends upon the frequency ω. Of particular interest is that value of frequency that makes the magnitude of the resulting complex number equal to $1/\sqrt{2}$ or 0.707. Examination of the bracketed term reveals that this magnitude prevails whenever the frequency has such a value as to make the coefficient of the j term unity. Hence, if we call this frequency ω_l, we have

$$\frac{1}{\omega_l(R_L + R_2)C_C} = 1$$

or

$$\omega_l = \frac{1}{(R_L + R_2)C_C} \text{ rad/sec} \quad (9\text{-}182)$$

Expressed in hertz this frequency becomes

$$\boxed{f_l = \frac{1}{2\pi(R_C + R_2)C_C}} \quad (9\text{-}183)$$

Upon inserting Eqs. (9-182) and (9-179) into Eq. (9-181) and formulating the ratio of currents, the current gain at low frequency for the entire stage under loaded conditions is obtained. Thus

$$A_{Il} = A_{Im} \left(\frac{1}{1 - j\frac{\omega_l}{\omega}} \right) \qquad (9\text{-}184)$$

This last expression shows that as the input signal frequency ω gets smaller, the magnitude of the low frequency gain gets correspondingly smaller by the factor

$$\frac{1}{\sqrt{1 + (\omega_l/\omega)^2}}$$

Of course, at $\omega = \omega_l$ there occurs about a 30 per cent decrease in gain from the midband value. The frequency ω_l identifies the lower end of the useful bandwidth of the amplifier.

The third and final region of interest is the *high frequency range*. By definition the high frequency range is that range of frequencies for which the reactance value of the diffusion capacitance† C_D and the transition capacitance‡ C_T are low enough to command inclusion in the equivalent circuit. However, to place these parameters, appropriate pairs of terminals must be identified across which to place C_D and C_T. In view of the fact that the diffusion capacitance is a property of the forward-biased section of the transistor, it would seem reasonable at first to place C_D between the base (B) and emitter (E) terminals. However, a more detailed study of the internal workings of the transistor discloses that the diffusion process of current flow begins at some internal point (B') of the base section after the current undergoes an *ohmic* drop. Accordingly, a more refined representation of the h_{ie} parameter is one that treats it in two parts: a base-spreading resistor $r_{BB'}$ that accounts for the ohmic drop, and the forward-biased resistor $r_{B'E}$ that relates the increase in diffusion current I_b to the corresponding increment of forward-biased applied voltage. This distinction is shown in the model appearing in Fig. 9-37. On this basis the correct location for the diffusion capacitance is across terminals B' and E. The value of base-spreading resistance lies in the vicinity of 20 per cent of h_{ie} and so needs to be reckoned with. The transition capacitance C_T is placed

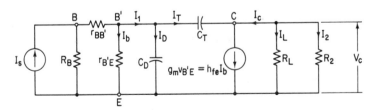

Fig. 9-37. Hybrid-equivalent circuit for computing high-frequency performance.

† See p. 263.

‡ C_T is associated with the charge distribution across the junction in the depletion region and the potential energy barrier V_0.

between terminals B' and C. This, too, is a correct location, because C_T arises from a reversed-bias state in the transistor existing between the collector terminal and the base section. It is important to note now that the input and output sections are no longer isolated. In fact, this bridging effect of C_T creates a π form, which is the reason why this version of the equivalent circuit is called the *hybrid-π* model. However, by means of an appropriate analytical procedure, it is possible to redraw the circuit of Fig. 9-37 so as to once again bring into evidence isolated input and output sections. Let us do this before proceeding with computing the high-frequency behavior.

Applying Kirchhoff's current law at the common junction point of C_D and C_T, we have

$$I_1 = I_D + I_T \tag{9-185}$$

The currents, I_D and I_T, may be expressed in terms of the product of the admittance of the circuit element carrying the current and the voltage appearing across its terminals. Thus

$$I_D = j\omega C_D v_{B'E} \tag{9-186}$$

and

$$I_T = j\omega C_T(v_{B'E} - V_c) \tag{9-187}$$

where V_c is the rms value of the collector voltage. Also, by the current law as it it applies to terminal C we can write

$$I_T - g_m v_{B'E} + I_c = 0 \tag{9-188}$$

where I_c is the current flowing through R_p, which is the equivalent resistance of the parallel combination of R_L and R_2.

The varying component of the collector-to-emitter voltage is given by

$$V_c = -I_c R_p \tag{9-189}$$

where

$$\frac{1}{R} = \frac{1}{R'_L} + \frac{1}{R_2} \tag{9-190}$$

By Eq. (9-188) we can then write

$$V_c = -g_m v_{B'E} R_p + I_T R_p \tag{9-191}$$

However, in practice I_T is very small compared to $g_m v_{B'E}$ so that the last equation may be written

$$V_c \approx -g_m v_{B'E} R_p \tag{9-192}$$

Introducing this equation into Eq. (9-187) then allows I_T to be expressed as

$$I_T = j\omega C_T(1 + g_m R_p) v_{B'E} \tag{9-193}$$

Finally, by substituting Eqs. (9-193) and (9-186) into Eq. (9-185) we get

$$I_1 = j\omega[C_D + C_T(1 + g_m R_p)]v_{B'E} = j\omega C v_{B'E} \tag{9-194}$$

where C is an equivalent capacitance as seen between terminals B' and E, and is defined by

$$C = C_D + C_T(1 + g_m R_p) \tag{9-195}$$

It is interesting to note here that the effect of C_T is amplified by the factor $(1 + g_m R_p)$. This behavior is readily explained as follows. When terminal B' (which is the left side of C_T) is driven positive in potential about the Q-point, simultaneously through the action of the current generator, $g_m v_{B'E}$, terminal C (the right side of C_T) is driven negative, thereby accounting for a larger potential difference across the capacitor terminals and hence a larger current. This effect is called the Miller effect.

By means of Eq. (9-194) the current I_1 in Fig. 9-37 may be looked upon as flowing into a terminal capacitor C. In this way the circuit of Fig. 9-37 may be replaced by the arrangement shown in Fig. 9-38(a). The high-frequency performance of the transistor amplifier may now be found by working with the equivalent circuit of Fig. 9-38(a).

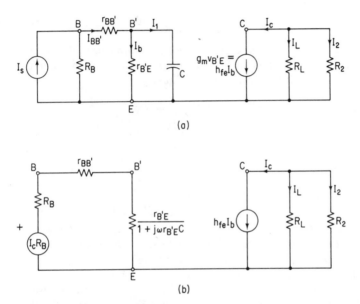

(a)

(b)

Fig. 9-38. The high-frequency equivalent circuit:
   (a) representation in terms of an isolated input and output section;
   (b) one-loop equivalent of the input section of (a).

The input current to the second stage I_2 is related to the base diffusion current I_b by

$$I_2 = \frac{-R_L}{R_L + R_2} h_{fe} I_b = \frac{-h_{fe}}{1 + R_2/R_L} I_b \qquad (9\text{-}196)$$

The parallel combination of $r_{B'E}$ and $1/j\omega C$ is expressed as $r_{B'E}/(1 + j\omega r_{B'E} C)$. Moreover, the input section of Fig. 9-38(a) can be replaced by the single-loop voltage-source equivalent circuit shown in Fig. 9-38(b). By the voltage-division principle, we can then write

$$v_{B'E} = \frac{\dfrac{r_{B'E}}{1 + j\omega r_{B'E}C}}{R_B + r_{BB'} + \dfrac{r_{B'E}}{1 + j\omega r_{B'E}C}} R_B I_s$$

$$= \frac{r_{B'E} R_B I_s}{R_B + h_{ie} + j\omega r_{B'E}(R_B + r_{BB'})C} \tag{9-197}$$

where $h_{ie} = r_{BB'} + r_{B'E}$. The diffusion base current thus becomes

$$I_b = \frac{v_{B'E}}{r_{B'E}} = \left(\frac{1}{1 + \dfrac{h_{ie}}{R_B}}\right)\left(\frac{1}{1 + j\dfrac{\omega}{\omega_h}}\right) I_s \tag{9-198}$$

where

$$\omega_h \equiv \frac{R_B + h_{ie}}{r_{B'E}(R_B + r_{BB'})C} \text{ rad/sec} \tag{9-199}$$

Substituting Eq. (9-198) into Eq. (9-196) and formulating the ratio of I_2 to I_s yields the desired expression for the complete stage current gain under loaded conditions for high-band frequencies. Thus

$$A_{Ih} = \frac{I_2}{I_s} = \frac{-h_{fe}}{\left(1 + \dfrac{R_2}{R_L}\right)\left(1 + \dfrac{h_{ie}}{R_B}\right)\left(1 + j\dfrac{\omega}{\omega_h}\right)}$$

or

$$\boxed{A_{Ih} = A_{Im}\left(\frac{1}{1 + j\dfrac{\omega}{\omega_h}}\right)} \tag{9-200}$$

Inspection of this result reveals that when the frequency of the input signal ω is allowed to become considerably larger than ω_h there is a sharp falloff of gain. For this reason the useful upper frequency limit of the amplifier is taken to be ω_h. It has already been pointed out that the useful lower frequency limit is ω_l. Accordingly, the useful range of frequencies over which the transistor amplifier performs effectively spreads from ω_l to ω_h. This range is called the *bandwidth*.

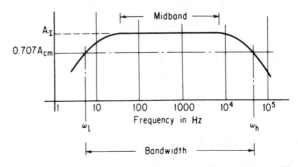

Fig. 9-39. Frequency response curve of an *R-C* coupled amplifier.

A plot of the current gain as a function of frequency on semilogarithmic paper appears as shown in Fig. 9-39. The behavior of the curve at low frequencies is dictated by Eq. (9-184). At the high frequencies the curve is described by Eq. (9-200). At both ends there is a serious dropoff of gain. Therefore the useful portion lies between ω_l and ω_h.

EXAMPLE 9-3   Refer to Fig. 9-31. The 2N104 transistor has the following parameters:

$$r_{BB'} = 0.29 \text{ K}, \qquad g_m = 0.032 \text{ mho}, \qquad h_{fe} = 44$$

$$r_{B'E} = 1.375 \text{ K}, \qquad C_T = 40 \text{ pf}$$

$$1/h_{oe} = 150 \text{ K}, \qquad C_D = 6900 \text{ pf}$$

(a) Compute the midband frequency current gain of the first stage of the circuit of Fig. 9-31. The first stage is to be taken from the base terminal of the first transistor to the base terminal of the second transistor.

(b) Find the bandwidth of the first-stage amplifier.

solution: Refer to the equivalent circuit shown in Fig. 9-40. In the midband frequency range C is removed and C_C is replaced with a short circuit. The resulting equivalent circuit then becomes the same as that of Fig. 9-35 upon recalling that $h_{ie} = r_{BB'} + r_{B'E}$. At low frequencies, C is removed and C_C is retained. At high frequencies, C is retained and C_C is replaced by a short circuit.

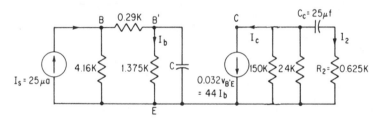

Fig. 9-40. Equivalent circuit for Example 9-3.

(a) The midband-frequency current gain for the complete stage follows directly from Eq. (9-179):

$$A_{Im} = \frac{-h_{fe}}{(1 + 0.625/2.36)(1 + 1.67/4.16)} = \frac{-44}{(1.265)(1.402)} = -24.8 \qquad (9\text{-}201)$$

Here the quantity 2.36 K is an equivalent load resistor that is the parallel resistance of R_L and $1/h_{oe}$. Note that this result is identical to that found in Eq. (9-134).

(b) To find the bandwith it is necessary to compute ω_l and ω_h. From Eq. (9-183), we get

$$f_l = \frac{1}{2\pi} \frac{1}{(R_L' + R_2)C_C} = \frac{1}{2\pi} \frac{1}{(2360 + 625)25 \times 10^{-6}}$$
$$= 2.14 \text{ Hz} \qquad (9\text{-}202)$$

Note that R_L' is used in place of R_L to denote the parallel resistance of R_L with $1/h_{oe}$. Similarly, from Eq. (9-199)

$$\omega_h = \frac{R_B + h_{ie}}{r_{B'E}(R_B + r_{BB'})C} = \frac{4.16 + 1.67}{1.375(4160 + 290)C} \qquad (9\text{-}203)$$

where

$$C = C_D + C_T(1 + g_m R_p) = 6900 \times 10^{-12} + 40 \times 10^{-12}(1 + 0.032 R_p) \qquad (9\text{-}204)$$

Now

$$R_p = \frac{R'_L R_2}{R'_L + R_2} = \frac{2.36(0.625)}{2.985} = 0.494 \text{ K} \qquad (9\text{-}205)$$

Thus

$$C = [6900 + 40(1 + 15.8)]10^{-12} = 7572 \times 10^{-12} \qquad (9\text{-}206)$$

Inserting Eq. (9-206) into Eq. (9-203) yields

$$\omega_h = 123{,}000 \text{ rad/sec} \qquad \text{or} \qquad f_h = 19{,}600 \text{ Hz} \qquad (9\text{-}207)$$

Thus, the bandwidth ranges from approximately 2 to 19,600 Hz, which gives excellent coverage of the audio range.

R-C Coupled FET Amplifiers. Analysis of the frequency response of the field-effect transistor amplifier follows in a manner identical to that of the vacuum-tube triode amplifier. The detailed derivation appears in Sec. 10-5. The results obtained there apply directly to the FET amplifier provided that the quantities listed in the triode column of Table 9-2 be replaced by the corresponding quantities appearing in the FET column.

Table 9-2. Analogous Quantities: Vacuum Tube vs. FET

Vacuum Triode Quantity	Symbol	Symbol	*FET* Quantity
Voltages and Currents:			
Effective value of varying plate voltage	E_p	V_d	Effective value of varying drain voltage
Effective value of varying grid-cathode voltage	E_g	V_{gs}	Effective value of varying gate-source voltage
Effective value of varying plate current	I_p	I_d	Effective value of varying drain current
Circuit Parameters:			
Dynamic plate resistance	r_p	r_d	Incremental drain (or output) resistance
Transconductance	g_m	g_m	Transconductance
Voltage amplification factor	μ	μ	Voltage amplification factor
Load resistance	R_L	R_L	Load resistance
Grid leak resistance	R_g	R_g	Gate-source resistance
Grid-cathode interelectrode capacitance	C_{gk}	C_{gs}	Gate-source interterminal capacitance
Grid-plate interelectrode capacitance	C_{gp}	C_{gd}	Gate-drain interterminal capacitance

The results are summarized here for completeness and ready reference. The equivalent circuit of the FET showing the coupling capacitance and the FET interterminal capacitance appear in Fig. 9-41. For the midband frequencies, C_{gs} and C_{gd} are removed and C_C is replaced by a short circuit. At the low frequencies, C_C is inserted and C_{gs} and C_{gd} continue to be left out. At the high frequencies, the

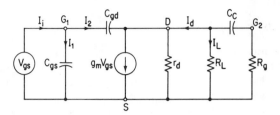

Fig. 9-41. The linear model of an R-C coupled FET amplifier showing the coupling and interterminal capacitances. The drain-to-source capacitance is neglected.

equivalent circuit is solved with C_{gs} and C_{gd} present, and C_C again replaced with a short circuit. The expression for the complete-stage voltage gain (gate of one stage to gate of next stage) at midfrequencies is

$$A_{Vm} = \frac{-g_m}{1/r_d + 1/R_L + 1/R_g} = \frac{-\mu}{1 + r_d/R_L + r_d/R_g} \qquad (9\text{-}208)$$

At the low end of the bandwidth, this equation is modified to

$$A_{Vl} = A_{Vm}\left(\frac{1}{1 - j\dfrac{\omega_l}{\omega}}\right) \qquad (9\text{-}209)$$

where

$$\omega_l = \frac{1}{R_l C_C} \qquad \text{and} \qquad R_l = R_g + \frac{r_d R_L}{r_d + R_L} \qquad (9\text{-}210)$$

At high frequencies, the expression becomes

$$A_{Vh} = A_{Vm}\left(\frac{1}{1 + j\dfrac{\omega}{\omega_h}}\right) \qquad (9\text{-}211)$$

where

$$\omega_h = \frac{1}{R_h C_i}, \qquad \frac{1}{R_h} = \frac{1}{r_d} + \frac{1}{R_L} + \frac{1}{R_g}$$

and

$$C_i = C_{gs} + (1 - A_{Vm})C_{gd} \qquad (9\text{-}212)$$

The bandwidth of the FET amplifier is equal to the range of frequencies from ω_l to ω_h. Expressed mathematically, we can write

$$\omega_{bw} = \omega_h - \omega_l = \frac{1}{R_h C_i} - \frac{1}{R_l R_C} \qquad (9\text{-}213)$$

Examination of Eq. (9-213) discloses that C_i plays the key role in establishing the upper frequency limit, and the coupling capacitor C_C exerts the key influence in fixing the lower limit of the bandwidth.

9-6   Integrated Circuits

In constructing the electronic amplifier circuits treated in the foregoing sections, use is made of discrete components, such as resistors, capacitors, diodes, and transistors, that are assumed to be suitably interconnected with appropriate wiring. In recent years, however, highly sophisticated technological advances have been achieved that allow many *complete* amplifier circuits to be formed simultaneously, including all the interconnections, on small silicon wafers. These circuits are called *integrated circuits*. The aim of this section is to give a brief description of the fabrication procedure involved in producing integrated components and then to show how a complete amplifier can be so formed.

The current great popularity of integrated circuits, especially in the area of digital computers, is attributable to several outstanding advantages. One is the tremendous saving in space. Many complete amplifier circuits can be formed on a wafer measuring only a small fraction of a square inch. Another is the considerable improvement in frequency response. The close spacing of the various diffused components and the necessarily short interconnections result in smaller capacitances, which in turn extend the frequency range over which the gain can remain constant. These short connections have also made possible appreciable improvements in the speeds of digital computers. A third noteworthy advantage is the increased reliability of such circuits. This attribute stems primarily from the fact that *all* circuit components (including the interconnections) are made at the same time. Finally, the cost of integrated circuits is lower than the more conventional types.

There are a few shortcomings, too. The current state of the art of fabrication limits production to essentially low power amplifiers. Moreover, the diffusion processes and other related procedures in the fabrication process are not good enough to permit a precise control of the parameter values for the circuit elements. However, control of the ratios is at a sufficiently acceptable level.

Fabrication Procedure. Attention is directed first to the forming of two diffused transistors on a single crystal silicon chip. The process begins by using a *p* substrate as a base having a thickness of approximately 6 mils. An epitaxial layer of about 25 mils is then applied to the *p* substrate by allowing a thin film of silicon in the gas phase to grow in such a manner as to trap *n*-type impurity atoms. This done, the chip is next placed in an oven with an oxygen atmosphere and heated to 1000°C, thereby permitting a silicon dioxide (SiO_2) layer of one-half micron to be placed over the entire top surface of the *n*-type epitaxial layer. The chip has now been prepared for the creation of isolated *n* regions. As will be seen presently, these *n* regions will serve as the collector sections of the diffused transistors. The silicon dioxide layer is useful because it forbids impurities to diffuse through it.

Let it now be desired to create *n* regions in the center of each half of the silicon chip. To achieve this result, the following steps are necessary. First, a photosensitive emulsion is placed on the chip covering the SiO_2 layer. This is followed by a mask (or stencil) with transparent windows appearing where the *n* regions are to

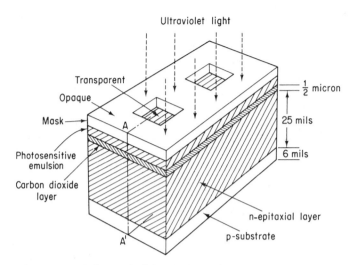

Fig. 9-42. Depicting the initial stages in the development of two diffused transistors for integrated circuits.

be located. The remainder of the mask is opaque. Refer to Fig. 9-42 for a pictorial description of the chip at this stage. Finally, the chip is subjected to ultraviolet light, which has the effect of polymerizing the photosensitive emulsion that appears beneath the transparent areas but not that hidden beneath the opaque sections. The mask is then removed and the chip is developed, which means that it is washed in a suitable chemical, thus allowing the unexposed portions of the chip to wash away. At this point the wafer has a *p*-type impurity applied to it with the result that what was originally an *n*-type epitaxial layer is now converted to a *p*-type layer *with the sole exception of those regions that remained protected by the layers of* SiO_2 *and photosensitive emulsion.* At this stage of the process, if a cross-sectional view of the chip were taken at a section such as *AA'* in Fig. 9-42, the result would appear as depicted in Fig. 9-43. Here the isolated *n* regions are clearly evident. It is instructive to note too that the *p* regions between the isolation islands are doped more heavily than the *p* substrate. In any integrated circuit in which the chip is used, the *p* substate is always connected to the most negative part of the circuit to assure that the *p-n* sections are all reverse biased. If the region

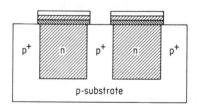

Fig. 9-43. Showing cross-sectional view through *AA'* of Fig. 9-42. Formation of isolated *n*-regions and p^+ doping.

between the *n* sections in Fig. 9-43 were not heavily doped, there would exist the danger that the depletion region of the reverse-biased *p-n* sections might spread sufficiently to touch, thereby connecting the two *n* regions. The heavier *p* doping guards against this and is represented by the p^+ notation.

Keep in mind that our aim is to develop two diffused transistors of the *n-p-n* type. Portions of the *n* regions already formed are to serve as the collector sections. The next step in the fabrication process is to develop successive diffusions for the purpose of creating the base and emitter sections. Before proceeding with the creation of the base section, it will be necessary to remove the remaining parts of the SiO_2 and mask, as shown in Fig. 9-43. This is readily accomplished by chemical solvents and mechanical abrasions. At this point, another layer of SiO_2 is formed over the entire top surface of the chip. This is then followed by suitable masking in conjunction with the photolithographic process to create new openings in accordance with the configuration of Fig. 9-44. The wafer is subjected to *p* doping to create the *p* sections shown within the *n* regions of this diagram. Because a portion of the newly created *p* sections will serve as the base of the diffused transistor, this part of the process is often referred to as the *base-diffusion* process.

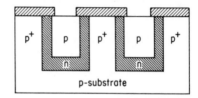

Fig. 9-44. Cross-sectional view illustrating openings for base diffusion within the isolated *n*-regions.

The foregoing process is now repeated once again for the purpose of creating isolated *n* sections in the newly formed isolated *p* regions. Accordingly, by proper application of the masking and photolithographic process, the openings depicted in Fig. 9-45 result. Then an *n*-type impurity is applied, thus creating the isolated n^+ regions that serve as the emitter sections of the transistors. Here again the $^+$ notation is used to denote a heavy amount of doping. It is common practice to use an especially heavy doping at those places of the isolated regions where metal contacts are to be inserted. This helps to reduce contact resistance.

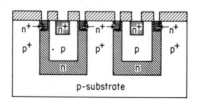

Fig. 9-45. Cross-sectional view illustrating openings for emitter diffusion within the isolated *p*-regions. Also illustrates n^+ diffusion for metal contacts for collector sections.

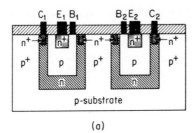

Fig. 9-46. Complete chip for two diffused transistors showing aluminum contacts (heavy shading):
  (a) cross-sectional view;
  (b) top view;
  (c) schematic representation.

The final stage of the fabrication process involves depositing aluminum contacts by employing the masking and etching techniques previously applied. The final result is depicted in Fig. 9-46. Typical dimensions of this double-diffused transistor wafer might measure 10 by 15 by 30 mils.

Diffused Circuit Elements. The diffusion process described in the foregoing pages to produce transistors has also been applied to produce microresistors, microcapacitors, and diodes. Resistive elements can be had by making use of the ohmic property of the p or n material. In this connection, refer to Fig. 9-44. It is instructive here to note that if metallic contacts were placed at the extreme ends of the opening inside the isolated p regions, a resistance would appear between the terminals, which is dependent upon the dimensions of the p region as well as its conductivity. Present technology permits values of resistance from 100 to 30,000 ohms to be realized by this scheme. The limits are dictated by practical factors associated with the fabrication procedure.

The microcapacitor can be formed by employing a diffusion procedure that leads to the configuration appearing in Fig. 9-47(a). There are two junctions in this arrangement. The junction between the *p* substrate and the isolated *n* region serves to provide isolation. The other junction between the *n* region and the isolated *p* region furnishes the desired capacitance, provided that this *p-n* junction is reverse biased. The need for a reverse bias represents a notable shortcoming of the *p-n* junction microcapacitor. Another disadvantage is the small size of capacitance that can be realized by this scheme owing to the limitations on the size of the microcircuit in which it appears. Typical values of capacitance are of the order of tens of picofarads.

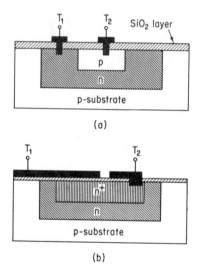

Fig. 9-47. The microcapacitor:
    (a) *p-n* junction type that must be kept reverse-biased;
    (b) nonpolarized MOS type.

An improvement over the *p-n* junction microcapacitor is the MOS microcapacitor illustrated in Fig. 9-47(b). Here the construction closely resembles that of the conventional capacitor. Terminal T_1 serves as one plate and the heavily doped isolated n^+ region serves as the other plate. The silicon dioxide fulfills the role of a thin-film dielectric. This arrangement offers the advantage of a nonpolarized capacitor and provides a slightly higher range of capacitance.

A diffused diode is obtained in integrated circuits from the transistor geometry. The two methods most preferred are either to use the base and emitter terminals of the diffused transistor, or to connect the collector and base terminals to form the anode of the diode and then to use the emitter terminal as the cathode.

The Integrated Transistor Amplifier. We are now in a position to illustrate how a complete transistor amplifier circuit including bias resistors can be formed on a silicon wafer employing the masking and photolithographic procedures previously described. The difference now is that the masking pattern must be laid out to include the appropriate interconnections as well. Appearing in Fig. 9-48(b) and (c) is the complete integrated circuit layout of the transistor amplifier shown in Fig. 9-48(a). The five circuit elements—one capacitor, three resistors,

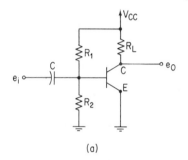

(a)

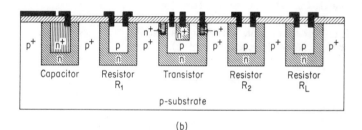

(b)

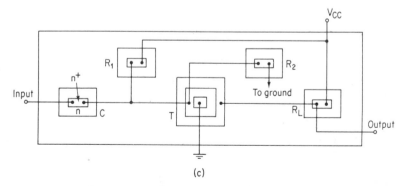

(c)

Fig. 9-48. Integrated transistor amplifier:
(a) schematic diagram;
(b) cross-sectional view showing each element;
(c) top view showing the interconnections.

and one transistor—and all the interconnections are created by the same masking, etching, and diffusion processes. In an actual integrated circuit, the circuit elements would not appear in the tandem arrangement shown in Fig. 9-48(b) and (c). Rather, the elements would be so placed as to make optimum use of the available space and to reduce the length of the interconnections to the minimum possible. The tandem arrangement is used here merely for convenience and clarification. The total area on the chip covered by this amplifier is a very small fraction of a square inch.

PROBLEMS

9-1   A transistor amplifier has the configuration shown in Fig. 9-6. The circuit elements and power sources are specified as follows:

$$2N109 \text{ transistor type,} \qquad V_{CC} = 12 \text{ volts}$$
$$R_L = 300 \text{ ohms,} \qquad V_{BB} = 0.2 \text{ volts}$$
$$R_2 = 900 \text{ ohms,} \qquad R_1 = 1000 \text{ ohms}$$

Neglect reactance of C_C.
(a) What is the value of the collector-to-emitter voltage at the quiescent state?
(b) Assume R_1 represents the total resistance of the forward-biased input section. If $v_b = 0.2 \cos \omega t$ volt, find the current gain.
(c) Determine the voltage gain in part (b).
(d) Compute the power gain in part (b).

9-2   A transistor amplifier has the circuit shown in Fig. P9-2.
(a) As v_b is allowed to increase in amplitude, will cutoff (i.e., zero collector current) be reached before saturation occurs? Explain.
(b) What must be the peak value of v_b to cause cutoff?
(c) What is the value of v_b beyond which saturation occurs?

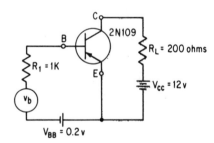

Figure P9-2

9-3   Repeat Prob. 9-2 for the case where the base bias battery V_{BB} is changed to 0.7 volt.

9-4   In the circuit of Fig. P9-4 the quiescent base current is 30 μa.
(a) Find the change in base input voltage needed to cause a 2-volt drop in the collector-to-emitter voltage.

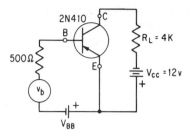

Figure P9-4

(b) Compute the ratio of the change in collector-to-emitter voltage to the change in input voltage for part (a).

(c) What is the value of the current gain?

(d) Compute the value of the voltage gain by using Eq. (9-23) and compare the result with that of part (b).

9-5 The 2N175 transistor, whose average collector characteristics appear in Appendix E, draws a quiescent current of 0.85 ma when the collector supply voltage is −6 volts and the base current is −10 μa.

(a) Find the value of load resistance.

(b) Assume that the output of the amplifier of part (a) is connected to a succeeding stage whose input resistance is 5100 ohms. Compute the total swing in collector current for a swing in base current of −5 to −15 μa.

9-6 In the JFET amplifier circuit of Fig. 9-7, the input signal is given by $v_{gs} = 0.4$ sin ωt.

(a) Compute the voltage gain and compare with the result given by Eq. (9-38). Comment.

(b) Make a rough sketch of the variation of the drain-source current as a function of time, and comment.

9-7 The voltage across the load resistor R_d in Fig. P9-7 is found to be 14 volts at the quiescent operating point. The average drain characteristics for this JFET appear in Fig. 9-8.

(a) Find the quiescent in drain-source current.

(b) Compute the value of R_d.

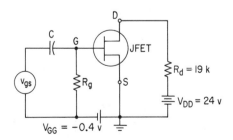

Figure P9-7

9-8   A manufacturer's handbook lists the following parameters for a 2N175 transistor as they apply to the quiescent point $v_{CE} = -4$ and $I_C = 0.5$ ma.

$$h_{ie} = 3570 \text{ ohms}$$
$$h_{fe} = 65$$
$$h_{oe} = 6.6 \times 10^{-6} \text{ mho}$$

(a) Draw the equivalent circuit showing all parameter values.
(b) A load resistor of 10 K is placed in the collector-to-emitter circuit of this transistor. If the base current is increased about its quiescent value by 5 μa, compute the change in collector current assuming a conventional common-emitter amplifier configuration. Neglect the effect of the bias circuit.
(c) Find the current gain in part (b).
(d) Repeat parts (b) and (c) for a load resistor of 1 K.
(e) Repeat parts (b) and (c) for a 100-ohm load resistor.

9-9   The equivalent circuit parameters of the 2N139 transistor, as they apply to the quiescent point $v_{CE} = -9$ volts and $I_C = -1.0$ ma, are as follows:

$$h_{ie} = 1325 \text{ ohms}$$
$$h_{fe} = 48.3$$
$$h_{oe} = 8.6 \times 10^{-6} \text{ mho}$$

This transistor is used in the common-emitter amplifier configuration having a load resistance of 2000 ohms. Neglect bias circuit effects.

Determine the change in base current that causes a 0.4-volt change across the terminals of the load resistor.

9-10   Prove the validity of Eq. (9-72).

9-11   The fixed-bias method illustrated in Fig. 9-16 is to be used for the 2N109 transistor operating with a supply voltage of $V_{CC} = 12$ v. The manufacturer's handbook lists the d-c base-to-emitter voltage as 0.15 v.

Find the value of base bias resistance which for a load resistance of 200 ohms yields a Q-point lying midway between cutoff and saturation.

9-12   In the circuit of Fig. P9-12 the quiescent point is to be at $V_{CE} = 8.8$ v, $I_c = 3$ ma, and $I_B = 200$ μa. Find R_B and R_E if $R_L = 3$ K.

9-13   The *n-p-n* germanium transistor shown in Fig. 9-17 has $\alpha = 0.98$, $I_{CO} = 4\mu$a, and $V_{BE} \approx 0$.

(a) Calculate R_B and R_L so that the quiescent point is placed at $I_E = 2$ ma and $V_{CE} = 6$ v.

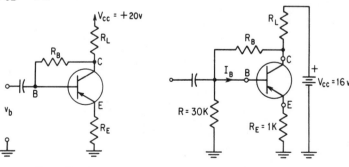

Figure P9-12   Figure P9-13

(b) Give a physical explanation of how this circuit provides d-c bias stabilization and compare it with other bias methods.

9-14 In the circuit of Fig. 9-18 the following data apply:

$$V_{CC} = 12 \text{ v}, \qquad R_1 = 6000 \text{ ohms}$$
$$R_L = 240 \text{ ohms}, \qquad R_2 = 1200 \text{ ohms}$$
$$R_E = 100 \text{ ohms}, \qquad 2\text{N}104 \text{ transistor type}$$

(a) Find the bias equation for this amplifier.
(b) Determine the Q-point.

9-15 An amplifier employing the 2N104 transistor has the configuration shown in Fig. 9-18. Also, $R_L = 150$ ohms, $R_E = 100$ ohms, and $V_{CC} = 12$ v.

Find the values of R_1 and R_2 which yield a quiescent point located at approximately the center of the load line.

9-16 A current source of 2 μa rms value is applied as the input signal of Fig. P9-16. The equivalent circuit parameters of the 2N175 transistor are listed in the statement of Prob. 9-8.

(a) Draw a completely labeled equivalent-circuit diagram of the transistor amplifier.
(b) What is the rms value of the base signal current?
(c) What is the rms value of the varying component of the collector current?
(d) Compute the transistor amplifier current gain (i.e., from signal current to collector current).

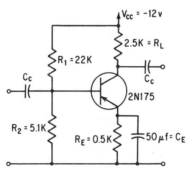

Figure P9-16

9-17 In the circuit of Prob. 9-16 a signal source having an rms voltage of 6 mv and an internal resistance of 3 K is applied to the input terminals.

(a) Find the stage amplifier current gain.
(b) Determine the value of voltage gain.
(c) Calculate the value of power gain.
(d) What is the value of the input resistance of the amplifier circuit?
(e) What is the value of output resistance?

9-18 A transistor amplifier has the circuit configuration shown in Fig. P9-16. However the values of the circuit elements are as follows:

$$R_1 = 42 \text{ K}, \qquad R_L = 3.2 \text{ K}$$
$$R_2 = 6.5 \text{ K}, \qquad R_E = 0.8 \text{ K}$$

Also, the transistor used is the 2N139, the equivalent-circuit parameters of which are given in Prob. 9-9.

Compute the rms value of a signal-source current which causes the varying component of the collector voltage to change by 0.8 v rms. Assume the reactance of C_E to be negligibly small.

9-19   A 2N410 transistor is used in the amplifier configuration shown in Fig. P9-19. The parameters for this transistor are:

$$h_{ie} = 1325 \text{ ohms}$$
$$h_{fe} = 48.3$$
$$h_{oe} = 8.6 \times 10^{-6} \text{ mho}$$

(a) Draw a clearly labeled equivalent circuit for this transistor amplifier.
(b) Compute the rms value of the output voltage for the specified input signal.
(c) Find the overall current, voltage, and power gains.

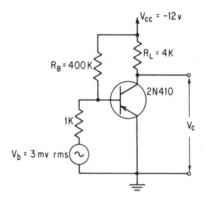

Figure P9-19

9-20   A two-stage transistor amplifier has the configuration shown in Fig. P9-20 The input signal is a sinusoid of 20-mv rms value. Assume the reactance of the coupling capacitors and the base-to-emitter voltage of the transistors to be negligibly small. The manufacturer's semiconductor handbook lists the following information:

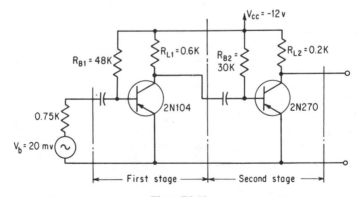

Figure P9-20

2N104 transistor: $h_{ie} = 1665$ ohms, $h_{fe} = 44$, $h_{oe} = 6.6 \times 10^{-6}$ mho
2N270 transistor: input resistance between base and emitter terminals
is 400 ohms

The average collector characteristics are available in Appendix E.

(a) Find the rms value of the current through R_{L1} when the second stage is disconnected.
(b) With the second stage connected, find the rms value of the base current in the 2N270 transistor.
(c) Compute the rms value of the varying components of the collector voltage and current at the output terminals of the second stage.
(d) Calculate the two-stage current gain.
(e) What is the total voltage gain?
(f) Find the overall power gain.

9-21 Answer the questions of Prob. 9-20 for the two-stage transistor amplifier circuit shown in Fig. P9-21. Assume the reactances of all capacitors are negligible. Note that the rms value of the input signal is 30 mv.

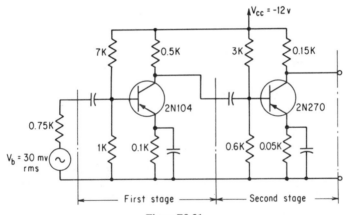

Figure P9-21

9-22 In the circuit of Fig. P9-20 design an appropriate third stage. Select a suitable transistor and load resistor which can be properly driven by the preceding stage. Compute the a-c performance of the third stage.

9-23 Design a suitable transistor amplifier stage which can be made to follow the two-stage amplifier of Fig. P9-21. Be sure to specify an appropriate transistor, a load resistor, and suitable bias-circuit resistors.

9-24 The transistor amplifier of Fig. P9-19 is coupled by a 25-μf capacitor to a second stage amplifier, the input resistance of which is 1200 ohms. Moreover, the diffusion capacitance of the 2N410 transistor is 1100 pf and the transition capacitance is 9.5 pf.

(a) Compute the midband-frequency current gain of the first stage. Consider the first stage as appearing between the base terminal of the first transistor and the base terminal of the second transistor.
(b) Find the lower limit of the bandwidth in Hz.

(c) Calculate the upper limit of the bandwidth in Hz.

(d) What is the magnitude and phase of the current which flows into the second stage?

9-25   In the circuit in Fig. P9-19 the coupling capacitor between the first and second stage has a capacitance of 25 μf. The diffusion capacitance of the 2N104 is 6900 pf and the transition capacitance is 40 pf. The input resistance of the 2N270 between base and emitter terminals is 400 ohms.

(a) Draw the complete equivalent circuit of the first stage showing all parameters clearly labeled.

(b) What is the midband-frequency current gain of the first stage?

(c) Find the bandwidth in Hz.

(d) What is the effect of decreasing the value of load resistance on the upper limit of the bandwidth?

(e) Repeat part (d) for the lower limit of the bandwidth.

9-26   Repeat Prob. 9-25 for the circuit shown in Fig. P9-21.

9-27   A solid-state amplifier employs the junction field-effect transistor whose parameters are as follows:

$$g_m = 3 \times 10^{-3} \text{ mho} \qquad C_{gs} = 6 \text{ pf}$$
$$r_d = 70 \text{ kilohms} \qquad C_{gd} = 2 \text{ pf}$$

In use, this amplifier is coupled to a succeeding stage by a 25-μf capacitor feeding into an input resistor of $R_g = 0.1$ M. The amplifier load resistor has a value of 10 K.

(a) Determine the midfrequency voltage gain.

(b) Find the frequency at which the amplifier circuitry causes the output signal to lead the input signal by 45 degrees.

(c) Find the frequency at which the amplifier circuitry causes the output signal to lag behind the input signal by 45 degrees.

(d) What is the bandwidth of this amplifier?

9-28   For the JFET amplifier of Prob. 9-27, compute the value of input capacitance as seen by the source voltage, and explain why this value is larger than the sum of the gate-source and gate-drain capacitances.

9-29   Design the layout for the integrated-circuit version of the amplifier depicted

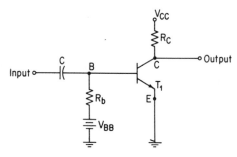

Figure P9-29

in Fig. P9-29. Show both a cross-sectional and a top view, and avoid overlapping connections.

9-30 Repeat Prob. 9-29 for the circuit amplifier circuit of Fig. P9-30.

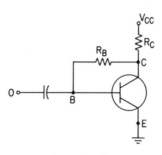

Figure P9-30

10

electronic circuits—vacuum types

The description of electronic circuits employing vacuum tubes follows in a manner similar to that of the semiconductor devices appearing in the circuits treated in Chap. 9. Attention is focused on computing amplifier performance by graphical techniques as well as through the use of the linear equivalent circuit. Frequency response is also treated.

10-1 Graphical Analysis of the Vacuum-tube Triode Amplifier

Appearing in Fig. 10-1 in its barest form is the electronic circuit for a vacuum-tube triode amplifier. As indicated, all voltages are measured with respect to the grounded cathode K. In series with the plate terminal of the triode is the load resistor R_L. It is across this resistor that the useful portion of the output voltage is obtained. The total plate current i_b is shown directed toward the plate terminal P. This is the only direction in which i_b can flow. The plate supply voltage E_{bb} is the source of power for the amplifier circuit.

Listed in Table 10-1 are the standard symbols used in the study of vacuum-tube circuits. Referrence to this table should make clear the meaning of each symbol in Fig. 10-1.

Table 10-1. Triode Symbols

	Grid voltage with respect to cathode	Plate voltage with respect to cathode	Current through load resistor
Instantaneous total value	e_c	e_b	i_b
Quiescent value	E_c	E_b	I_b
Instantaneous value of varying component	e_g	e_p	i_p
Effective value of varying component	E_g	E_p	I_p
Amplitude of varying component	E_{gm}	E_{pm}	I_{pm}
Supply voltage	E_{cc}†	E_{bb}†	

† These symbols refer to magnitudes only.

How is a circuit such as the one illustrated in Fig. 10-1 analyzed? To be more specific, how does one find the plate current i_b which corresponds to a specified value of the grid-to-cathode voltage e_c? A little thought discloses that one step towards arriving at a solution would be to write Kirchhoff's voltage law for the output circuit. Accordingly, we have

$$E_{bb} = e_b + i_b R_L \qquad (10\text{-}1)$$

From this equation we then can get an expression for the desired plate current, which is

$$i_b = \frac{E_{bb}}{R_L} - \frac{1}{R_L} e_b \qquad (10\text{-}2)$$

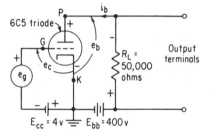

Fig. 10-1. Circuit diagram of a vacuum-tube amplifier in simplified form.

Inspection of Eq. (10-2), however, quickly reveals that the plate-to-cathode voltage e_b of the triode is also unknown. We thus have a situation of a single equation involving two unknown quantities—i_b and e_b. Therefore additional information is needed, and in particular it must relate i_b and e_b. Fortunately, once the vacuum tube is specified, this required relationship is available either analytically through the use of Eq. (8-11) or graphically by means of the appropriate plate characteristics. If the analytical approach is taken, it is necessary to solve simultaneously Eq. (10-2) and Eq. (10-3); the latter is Eq. (8-11), repeated here for convenience.

$$i_b = K(\mu e_c + e_b)^n \qquad (10\text{-}3)$$

In Eq. (10-3) the parameters K, μ, and n vary with the type triode used. Because the value of n often exceeds unity, the simultaneous solution of Eqs. (10-2) and (10-3) is not conveniently performed. Moreover, the parameters in Eq. (10-3) need to be adjusted in accordance with the values used for e_c and e_b.

A far more convenient and direct solution to the problem can be obtained by solving Eqs. (10-2) and (10-3) simultaneously in a graphical fashion. The graphical representation of Eq. (10-3) for various fixed values of e_c is nothing more than the plate characteristics, which are shown in Fig. 10-2(a) for the 6C5

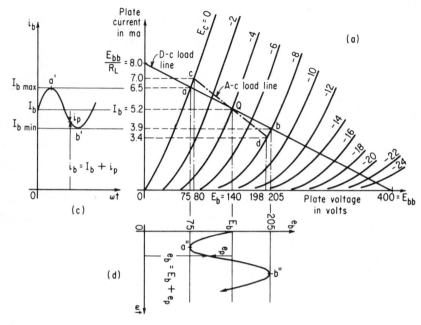

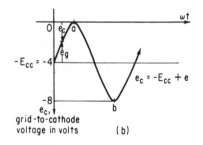

Fig. 10-2(a, c, d) Graphical analysis of voltage vacuum-tube amplifier
(these variations correspond to grid voltage changes shown in Fig. 10-2(b)):
>   (a) plate characteristics of 6C5 with d-c and a-c load
>   lines;
>   (c) plate current variation;
>   (d) plate voltage variations.

Fig. 10-2(b) Variation of grid-to-cathode voltage for a sinusoidal signal $e_s = 4 \sin \omega t$.

triode. Also appearing there is the graphical representation of Eq. (10-2), which is the equation of a straight line having a vertical intercept of E_{bb}/R_L and a slope of $-1/R_L$. This straight line is called the *d-c load line*. The point of intersection of the d-c load line with that curve of the plate characteristics corresponding to the specified value of grid-to-cathode voltage provides the solution for the plate current.

Let us illustrate this solution procedure for the circuit of Fig. 10-1. The manufacturer's published plate characteristics for the 6C5 triode appear in Fig. 10-2(a) drawn to scale. The d-c load line is readily plotted by identifying the two end-points of the straight line. Thus from Eq. (10-2) it follows that when i_b is zero, e_b is equal to $E_{bb} = 400$ volts. This locates one point on the load line—the one on the horizontal axis. Then when e_b is set equal to zero, the corresponding value of the plate current becomes $E_{bb}/R_L = 400/50,000 = 8$ ma.

This identifies the second point—the one on the vertical axis. The line drawn between these two points yields the d-c load line. Note that the slope is negative and has a value equal to $1/R_L$, all of which is consistent with Eq. (10-2). If it is now desired to find the value of the plate current for a grid-to-cathode voltage of -4, a glance at Fig. 10-2(a) reveals the answer to be 5.2 ma.

Applying Kirchhoff's voltage law to the input grid circuit of Fig. 10-1 leads to

$$e_c = -E_{cc} + e_s \qquad (10\text{-}4)$$

The minus sign is used for E_{cc} because the negative side of the battery is connected to the grid side of the input circuit. Inserting the specified value of E_{cc} then gives

$$e_c = -4 + e_s \qquad (10\text{-}5)$$

If the varying component of the input signal is considered to be zero (i.e., in a quiescent state) the resultant grid-to-cathode voltage then becomes

$$e_c = -4 = E_c$$

But this is precisely the value of grid-to-cathode voltage which led to point Q in the example just described. Consequently, because of the absence of any changing input signal, point Q in Fig. 10-2(a) is called the *quiescent operating point* of the circuit of Fig. 10-1. It is the operating point about which changes occur in the output circuit in response to changes in e_s. It is customary to identify the coordinates of the quiescent point as (I_b, E_b). In our example these quantities take on the values 5.2 ma and 140 volts, respectively. Thus at the quiescent point it can be said for Fig. 10-1 that the potential of the plate terminal P with respect to the cathode is 140 volts corresponding to a plate current flow of 5.2 ma. Finally, note that the quiescent operating point can be moved along the load line by changing the value of E_{cc}.

We next consider how the circuit of Fig. 10-1 behaves in the presence of a sinusoidal variation of $e_s = 4 \sin \omega t$ as illustrated in Fig. 10-2(b). Because the primary objective of the circuit of Fig. 10-1 is to produce at the output terminals a significantly amplified version of e_s, particular stress is put on identifying the varying components of the plate current and voltage rather than on their total values. Figure 10-2(b) is a plot of Eq. (10-4) which may be written more specifically as

$$e_c = -E_{cc} + e_s = -4 + 4 \sin \omega t \qquad (10\text{-}6)$$

The plate current and voltage for any value of e_c as called for by Eq. (10-5) can be found in the same manner as was done for the quiescent point. Thus, if ωt is assumed to increase from zero to $\pi/2$, e_c increases from -4 to zero sinusoidally. In terms of Fig. 10-2(a) this means that the *instantaneous operating point* moves along the load line from Q to a. Note on the plate characteristics that the grid-to-cathode voltage goes from -4 to 0, which is consistent with the variation depicted in Fig. 10-2(b). The instantaneous operating point is confined to move along the load line because at every instant Kirchhoff's voltage law in the plate circuit which contains R_L must be satisfied. The value of plate current corresponding to $e_c = E_c = 0$, which is point a in Figs. 10-2(a) and (b), is read off the ordinate axis as 6.5 ma. A plot of the plate currents corresponding to instantaneous operating points between Q and a on the d-c load line results in the variation shown in Fig. 10-2(c) between the points marked I_b and a'. A corresponding plot of the plate voltage is depicted in Fig. 10-2(d). A comparison of Figs. 10-2(b) and (d) reveals that corresponding to the maximum value of e_c there occurs a minimum value of plate voltage. In other words where the varying input signal e_s is driven in a positive direction, the plate voltage is driven negatively. This brings out one of the distinguishing characteristics of the triode amplifier: *phase inversion*. It means that the sign of the varying output plate voltage e_p is opposite to that of the varying input grid voltage e_s. Finally note that as the grid voltage e_s continues through a complete sine wave, corresponding variations take place in the plots of i_b and e_b.

Does it follow that a sinusoidal variation of e_s results in a corresponding sinusoidal variation of the plate current and voltage? The answer to this query depends upon how the Q-point is selected. If it is located in that region of the plate characteristics where equal increments in grid-to-cathode voltage are accompanied by equal spacing of the curves of the plate characteristics, then the answer is yes. The selection of Q in Fig. 10-2(a) was made with this in mind. Otherwise the answer is no.

As shown in Fig. 10-2(c), the instantaneous value of the varying component of the plate current is denoted by i_p. The amplitude, I_{pm}, of this current is readily found from the relationship

$$I_{pm} = \frac{I_{b_{max}} - I_{b_{min}}}{2} \tag{10-7}$$

where $I_{b_{max}}$ denotes the maximum value of the plate current corresponding to the positive maximum value of e_s, and $I_{b_{min}}$ denotes the minimum value of plate current corresponding to the negative maximum value of e_s. For the example illustrated in Fig. 10-2 this becomes

$$I_{pm} = \frac{6.5 - 3.0}{2} = 1.3 \text{ ma} \tag{10-8}$$

The rms or effective value then follows from

$$I_p = \frac{I_{pm}}{\sqrt{2}} = \frac{1.3}{\sqrt{2}} = 0.92 \text{ ma} \tag{10-9}$$

Inserting Eq. (10-7) into Eq. (10-9) leads to an expression for I_p related directly to the minimum and maximum values of plate current. Thus

$$\boxed{I_p = \frac{I_{b_{max}} - I_{b_{min}}}{2\sqrt{2}}} \tag{10-10}$$

In view of the sinusoidal character of the varying component of the plate voltage, a similar analysis leads directly to the following expression for the effective value of varying component of the plate voltage.

$$E_p = \frac{E_{b_{max}} - E_{b_{min}}}{2\sqrt{2}} \qquad\qquad (10\text{-}11)$$

Inserting the values found from Fig. 10-2 yields

$$E_p = \frac{205 - 75}{2\sqrt{2}} = 47 \text{ volts rms} \qquad\qquad (10\text{-}12)$$

It is important to understand that this is the effective value of the voltage developed across the load resistor in response to the varying component e_s in the grid circuit. It should be apparent from Fig. 10-2 that if e_s were zero, E_p would also be zero.

We are now at a point in the analysis where we can compute the voltage gain of the amplifier. The *voltage gain* is defined as the ratio of the rms value of the varying component of the plate voltage to the rms value of the varying component of the grid-to-cathode voltage. Expressed mathematically,

$$A_v \equiv \frac{E_p}{E_s} \qquad\qquad (10\text{-}13)$$

where A_v denotes the voltage gain
    E_s denotes the rms value of the grid signal
    E_p denotes the rms value of e_p

For our example,

$$E_g = \frac{E_{sm}}{\sqrt{2}} = \frac{4}{\sqrt{2}} = 2.82 \text{ volts} \qquad\qquad (10\text{-}14)$$

Therefore, the voltage gain is

$$A_v = \frac{E_p}{E_g} = -\frac{47}{2.82} = -16.7 \qquad\qquad (10\text{-}15)$$

The minus sign signifies the phase reversal that occurs between input and output.

The manufacturers' published value of the amplification factor μ for the 6C5 triode is 20. Note that the voltage gain is less than μ and in fact always is. Although the vacuum tube has the capability of amplifying voltages appearing at the grid terminals by a factor of 20, not all of this is realizable at the output terminals across R_L. Some of it appears as an internal voltage drop across the plate resistance of the triode. This can readily be illustrated with the figures at hand. Under ideal conditions an rms grid signal of $E_s = 2.82$ volts would appear at the output terminals μ times as large of $(E_p)_{\text{ideal}} = 20(2.82) = 56.4$ volts. The practical circuit of Fig. 10-1, however, yielded a value of 47 volts, thus indicating that the difference of 9.4 volts was consumed as an internal voltage drop within the tube itself. To verify this it is necessary merely to use the published value of r_p for the 6C5 triode, which is 10,000 ohms, and multiply it by our computed value of I_p, which is given in Eq. (10-9) as 0.92 ma. Accordingly, the rms value of voltage drop becomes $I_p r_p = (0.92 \times 10^{-3})10 \times 10^3 = 9.2$ volts which compares favorably with the aforementioned value of 9.4 volts.

The average value of the a-c power which is delivered to the load resistor as a result of the varying signal voltage in the grid is found to be

$$P_{\text{a-c}} = E_p I_p = 47(0.92 \times 10^{-3}) = 0.0432 \text{ watt} \qquad (10\text{-}16)$$

This power is supplied by the battery E_{bb} and does not come from the grid circuit. The grid circuit merely serves as the channel through which the flow of energy from the source E_{bb} to the load resistor R_L is controlled.

Practical Voltage-amplifier Configuration: A-c Load Line. Practical considerations dictate the modification of the circuit of Fig. 10-1 to that shown in Fig. 10-3. Note in Fig. 10-3 that if the signal source is assumed to be disconnected,

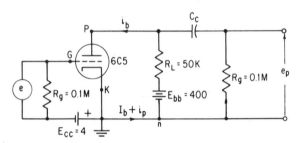

Fig. 10-3 Complete voltage amplifier circuit with capacitor coupling to output terminals. Note that only the a-c component of the plate voltage appears at the output terminals.

the grid-to-cathode circuit thereby becomes an open circuit. This means that the potential of the grid is no longer held fixed at $-E_{cc}$. Instead the grid assumes a potential determined by the electrostatic condition prevailing within the tube, which in turn is influenced by some of the high-velocity electrons arriving at the grid wire as well as by the positive ions originating from molecules of the residual gas in the tube. Since this condition generally leads to unfavorable operation, correction is achieved through the addition of the *grid-leak* resistor R_g connected as shown in Fig. 10-3. Thus R_g provides a path which allows the grid to be maintained at the potential determined by E_{cc}. Of course the high-energy electrons and positive gas ions will continue to be collected to some extent at the grid. Now, however, R_g furnishes a path through which this current may *leak* from the grid. Because the grid-leak current is only a fraction of a microampere, the selection of R_g in the range from 0.1 to megohm causes the effect of the grid-leak current on the grid-to-cathode voltage to be negligibly small.

A second important change in the circuit of Fig. 10-1 is the insertion of the capacitor C_c between the plate terminal and the output terminal. By the proper selection of C_c it can be made to pass the varying component of the plate voltage e_p practically undiminished while completely blocking out the d-c component. To appreciate the importance of blocking out the d-c component keep in mind that the a-c output of the circuit of Fig. 10-3 often serves as the input signal to a succeeding stage of amplification, such as depicted in Fig. 10-4. If C_c were not

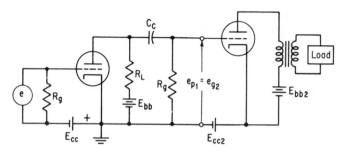

Fig. 10-4. Two-stage amplifier.

used in Fig. 10-4 and if it were desired to operate the second triode with a negative grid potential, then clearly the voltage E_{cc2} would have to be greater than
the average plate voltage of E_b, which in this case is 140 volts. Thus the presence
of C_c permits much smaller values of E_{cc} to be used in establishing a suitable
quiescent operating point for a given stage of amplification.

Another advantage of including C_c is that it prevents changes in the plate
supply voltage E_{bb} from being interpreted by the succeeding stage as changes
brought about by an input signal variation. Usually these changes are relatively
slow in character so that C_c effectively blocks them out. It is for this reason that
the capacitor-coupled amplifiers of Figs. 10-3 and 10-4 are said to be more stable
than direct-coupled amplifiers (i.e., C_c omitted).

The presence of the coupling capacitor changes the a-c amplifier performance
somewhat. As the varying component i_p of the plate current i_b leaves the cathode
and flows toward terminal n, it no longer sees the single path through R_L to the
plate terminal P in Fig. 10-3 but rather sees two paths. One is through R_L to P;
the second is through R_g and C_c to P. Of course the d-c component, I_b, of the
plate current continues to flow only through R_L to P because C_c presents an open
circuit to direct current. In view of the presence of two current paths for i_p, it
follows that the load resistor changes from R_L to R_L in parallel with R_g. Therefore
in the amplifier configuration of Fig. 10-3 the equivalent load resistance is

$$R'_L = \frac{R_L R_g}{R_L + R_g} \tag{10-17}$$

When R_g is considerably greater than R_L, very little difference exists between
R'_L and R_L. Often, however, R_g and R_L are comparable so that adjustments become necessary. In order to illustrate this, let us compute the a-c performance
of the circuit of Fig. 10-3 for the case where R_g is 0.1 megohm. All other quantities
are kept the same as in the circuit of Fig. 10-1. By means of Eq. (10-17) the equivalent load resistance becomes

$$R'_L = \frac{(50,000)(100,000)}{150,000} = 33,333 \text{ ohms} \tag{10-18}$$

This value is sufficiently different from 50,000 ohms to require the construction
of a new load line which applies for the modified circuit of Fig. 10-3. Since the

quiescent point has not been changed, the new load line must pass through point Q in Fig. 10-2(a). A convenient way of identifying the new load line is to write

$$R'_L = \frac{\Delta e_b}{\Delta i_b} \qquad (10\text{-}19)$$

Then arbitrarily choose Δe_b as some convenient value measured from the Q-point. For our example choose $\Delta e_b = 40$ volts taken to the left of the Q-point. Equation (10-19) then yields the corresponding change in plate current on the new load line as

$$\Delta i_b = \frac{\Delta e_b}{R'_L} = \frac{40}{33,333} = 1.2 \text{ ma} \qquad (10\text{-}20)$$

Thus by moving from the Q-point to the left by 40 volts along the horizontal axis in Fig. 10-2(a) and then upward along the vertical axis by an amount equal to 1.2 ma, a second point on the new load line is established. The line drawn between this point and the Q-point identifies the new load line which is called the *a-c load line* for obvious reasons. In Fig. 10-2(a) the a-c load line is marked as line cQd. This line is sometimes known as the *dynamic load line*.

As the input grid signal varies in the manner shown in Fig. 10-2(b), the instantaneous operating point for the circuit of Fig. 10-3 is now confined to the a-c load line. Clearly this leads to altered values of maximum and minimum plate current thus resulting in a change in a-c performance. By Eq. (10-10) the new value of the rms plate current is

$$I_p = \frac{7 - 3.4}{2\sqrt{2}} = 1.27 \text{ ma} \qquad (10\text{-}21)$$

The corresponding value of rms plate voltage is

$$E_p = \frac{198 - 80}{2\sqrt{2}} = 41.7 \text{ volts} \qquad (10\text{-}22)$$

Accordingly the voltage gain becomes

$$A_v = -\frac{41.7}{2.82} = -14.8 \qquad (10\text{-}23)$$

The a-c power delivered to the equivalent load resistance R'_L is

$$P_{\text{a-c}} = |E_p I_p| = 41.7(1.27)10^{-3} = 0.053 \text{ watt} \qquad (10\text{-}24)$$

A comparison of these results with those obtained for the circuit of Fig. 10-1 shows a slight drop in voltage gain for the modified circuit.

10-2 The Vacuum-tube Equivalent Circuit

To develop the equivalent circuit we start with the information contained in Eq. (8-18):

$$i_b = f(e_c, e_b) \qquad (10\text{-}25)$$

In deriving the equivalent circuit our objective is to replace the vacuum tube (see Fig. 10-5) between the terminals G, P, and K by a suitable arrangement of circuit elements. Although our interest at this point is solely with the tube, it is helpful in understanding the development to keep in mind the manner in which the tube is used in a typical amplifier circuit, such as depicted in Fig. 10-1.

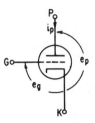

The equivalent circuit of a vacuum tube is especially useful in analyzing the behavior of an electronic circuit when the input signal undergoes small variations. Because of this emphasis on the changes about the quiescent point, all d-c quantities are omitted from the equivalent circuit. It is the a-c performance which is important. Consistent with this theme, therefore, let us assume that a change Δe_c occurs in the grid-to-cathode voltage of the triode. In general this brings about a change in plate voltage Δe_b as well as in the plate current Δi_b. Since all

Fig. 10-5. Vacuum-tube triode showing the varying quantities in the grid and plate circuits.

of these quantities are functionally related by Eq. (10-25) in accordance with the expression for the total differential, we may write

$$\Delta i_b = \left(\frac{\partial i_b}{\partial e_c}\right)\Delta e_c + \left(\frac{\partial i_b}{\partial e_b}\right)\Delta e_b \tag{10-26}$$

Introducing the notations for the varying quantities in the plate and grid circuits as specified in Table 10-1 we have

$$\Delta i_b \equiv i_p, \qquad \Delta e_b \equiv e_p, \qquad \Delta e_c = e_g \tag{10-27}$$

Inserting Eq. (10-27) into Eq. (10-26) yields

$$i_p = g_m e_g + \frac{e_p}{r_p} \tag{10-28}$$

where r_p and g_m are specified by Eqs. (8-20) and (8-21). Finally, rearrangement leads to

$$\boxed{i_p r_p - \mu e_g = e_p} \tag{10-29}$$

where $\mu = g_m r_p$ as given by Eq. (8-26).

Equation (10-29) specifies the functional relationship existing between the tube parameters and the changing quantities occurring between the grid-cathode and the plate-cathode terminals. Accordingly, a circuit interpretation of Eq. (10-29) leads directly to the equivalent-circuit representation of the vacuum tube. The result is shown in Fig. 10-6.

The varying grid voltage appears between the grid and cathode terminals. However, note that no physical connection is shown between the grid and plate terminals. This representation is valid as long as no grid current is allowed to flow and high-frequency effects are neglected. The modification in the equivalent circuit when e_g varies at very high frequencies is considered presently. The varying component of the plate voltage e_p appears across

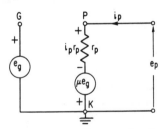

Fig. 10-6. The voltage-source equivalent circuit of the vacuum-tube of Fig. 10-5. No d-c quantities appear because only changes about the Q-point are represented.

the output terminals which are P and K. The voltage e_p is arbitrarily defined as positive when the upper terminal is at a higher potential relative to the quiescent voltage. Also, i_p is defined as a positive quantity when it flows into the P terminal. With this notation it then follows from Eq. (10-29) that as one traces the current i_p from the P to the K terminals, there occurs first the voltage drop $i_p r_p$ and then the rise $-\mu e_g$. Note that μe_g is the internally generated voltage of the vacuum tube corresponding to a grid voltage e_g. Note too that when the grid G is driven positive by e_g, the plate terminal will correspondingly undergo a drop in potential because the negative side of the μe_g generator is on the P side. (Keep in mind that the cathode terminal K serves as the voltage reference point.) This means

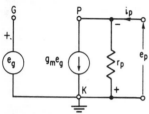

that e_p is negative whenever e_g is positive—a conclusion which agrees with the graphical analysis.

Often in the analysis of electronic circuits involving vacuum tubes it is advantageous to use a current-source version of the equivalent circuit. This is readily obtained by applying Norton's theorem to the circuit of Fig. 10-6. The result is shown in Fig. 10-7. The current generator has a value of $(\mu/r_p)e_g = g_m e_g$. It is also interesting to note that the circuit of Fig. 10-7 is really nothing more than a circuit interpretation of Eq. (10-28).

Fig. 10-7. Current-source equivalent circuit of the vacuum-tube triode.

When the grid signal variations occur at very high frequencies, the equivalent circuit must be modified to take into account the interelectrode capacitances. In accordance with the principles of electrostatics a capacitance exists between each pair of electrodes because the electrodes are conductors which are separated by a dielectric (a vacuum in this case) and across which different potentials appear. Thus a capacitance C_{gp} exists between the grid and plate terminals, another exists between the grid and cathode terminals denoted by C_{gk}, and a third C_{pk} exists between the plate and cathode terminals. Usually C_{pk} may be discarded entirely because its value is appreciably smaller than the others, owing to the shielding effect of the grid wire which is located between the plate and cathode electrodes. Typical values of C_{gp} and C_{gk} are measured in picofarads (pf) so that it is only at very high frequency (well above the audio range) that the reactance $1/\omega C$ becomes small enough to be reckoned with.

The complete form of the vacuum-tube equivalent circuit is depicted in Fig. 10-8. The presence of capacitor C_{gp} indicates that a path exists which connects the output circuit P to the input circuit G. This direct connection can cause oscillation difficulties if appropriate precautions are not taken to limit the amount of feedback that occurs between the output and input circuit. These capacitances also result in a diminution of the voltage gain when the tube is used as an amplifier. This matter is discussed in Sec. 10-5. Finally, it is interesting to note that in a strict sense the grid current is not zero. It must be equal to the sum of the currents flowing through the capacitances. Often, however, in triode applications this current is so small as to be truly negligible.

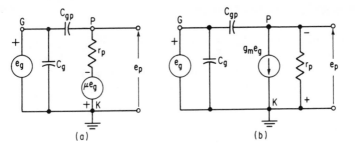

Fig. 10-8. The complete vacuum-tube equivalent circuit including inter-electrode capacitances:
(a) voltage-source equivalent;
(b) current-source equivalent.

10-3 Biasing Vacuum-tube Amplifiers

Reference to Figs. 10-1 to 10-3 shows that for a given load resistance and triode, the quiescent operating point is determined by the battery E_{CC} in the grid-cathode circuit. As was the case with the transistor amplifier, it is desirable to eliminate the need for this battery by using an appropriate biasing circuit. The most commonly used scheme for achieving this result is to put a resistor R_k in the cathode leg as illustrated in Fig. 10-9. Then under quiescent conditions the d-c

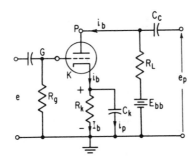

Fig. 10-9. Vacuum-tube voltage amplifier with self-biasing cathode resistor R_k.

plate current I_b flowing through R_k establishes a negative grid bias voltage given by

$$E_c = -I_b R_k \tag{10-30}$$

The minus sign appears because the voltage drop across R_k has the negative end on the grid side. Equation (10-30) is the bias equation and is analogous to Eq. (9-98) for the transistor. The corresponding bias curve is found by plotting Eq. (10-30) on the triode plate characteristics for various assumed values of I_b. The point of intersection of this bias curve with the load line then identifies the actual quiescent point.

The cathode by-pass capacitor C_k also is included in the circuitry for vacuum-tube amplifiers in order to prevent the decrease in gain that results when the varying component of the plate current is made to flow through the cathode resistor. If we select C_k sufficiently large so that its reactance at the signal frequency is small compared to R_k, the voltage gain remains virtually intact.

10-4   A Two-stage Vacuum-tube Amplifier

Depicted in Fig. 10-10 is the wiring configuration of a two-stage audio frequency amplifier employing the 12AX7 which is a twin triode housed in a single glass envelope and which can be used to provide two stages of voltage amplification.

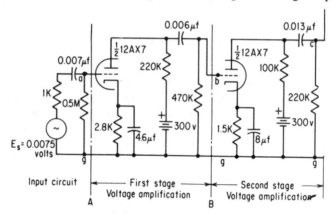

Fig. 10-10. A two-stage audio frequency vacuum-tube amplifier system.

Assume that the input signal to the first stage originates from a microphone. For the sake of illustration we further assume that the 1000-Hz frequency component has an rms value of 7.5 millivolts. Because of this small magnitude, an equivalent-circuit analysis of the first stage is appropriate. As shown in Fig. 10-10, the first stage is identified as appearing between lines A and B. This means that the loading effect of the 0.47-M grid resistor of the second triode is taken into account in determining the net input signal to the second stage. The a-c equivalent circuit for the first stage is shown in Fig. 10-11. The manufacturer's tube manual

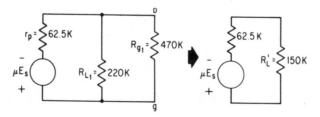

Fig. 10-11. Equivalent circuit for the first stage of the circuit of Fig. 10-10. R'_L is the equivalent parallel resistance of R_{L1} and R_{g1}.

lists the following values for the dynamic plate resistance and amplification factor:

$$r_p = 62,500 \text{ ohms}, \qquad \mu = 100 \tag{10-31}$$

Note that the reactance of the coupling capacitor has been neglected because it is small compared to the 470-K grid resistor. Moreover, note that the actual 220-K load resistance combines with the 470-K grid resistance to give an effective a-c load resistance of

$$R_L' = \frac{R_L R_g}{R_L + R_g} = \frac{220(470)}{220 + 470} = 150 \text{ K} \tag{10-32}$$

Calling the rms value of the a-c voltage developed across this resistance E_{p1}, it follows from the voltage-divider theorem that

$$E_{p1} = \mu E_s \frac{R_L'}{R_L' + r_p} \tag{10-33}$$

Rearranging then leads directly to the expression for the voltage gain of the first stage. Thus

$$\frac{E_{p1}}{E_s} = -\mu \frac{R_L'}{R_L' + r_p} = -\mu \frac{1}{1 + \dfrac{r_p}{R_L'}} \equiv A_v \tag{10-34}$$

where A_v denotes the voltage gain. Inserting the values of the parameters leads to

$$\frac{E_{p1}}{E_s} = -100 \frac{1}{1 + \dfrac{62.5}{150}} = -77.3 \tag{10-35}$$

When dealing with vacuum-tube devices, recall that the important performance characteristic is the voltage gain. The current gain is of no interest because the input grid current is assumed to be zero for all practical purposes.

As a preliminary to the study of the second stage, it is useful to compute the maximum value of the varying component of the voltage appearing at the grid of the second stage. Clearly this is $\sqrt{2}\, E_{p1}$, where E_{p1} is the varying component in the plate circuit of the first stage. Accordingly we find

$$E_{p1} = |A_v E_s| = 77.3(0.0075) = 0.58 \text{ volt} \tag{10-36}$$

and

$$\sqrt{2}\, E_{p1} = \sqrt{2}(0.58) = 0.82 \text{ volt} \tag{10-37}$$

The maximum swing of the input signal to the second stage is therefore ± 0.82 volt about the quiescent point.

Should the performance of the second stage be computed graphically or by means of the equivalent circuit? A glance at the published plate characteristics of the 12AX7 (see Fig. 10-12) reveals that an excursion about the Q-point of ± 0.82 volt involves such a wide swing that the graphical solution is to be preferred. Of course, to proceed graphically, the Q-point must first be established. The cathode bias curve for this second stage is given by Eq. (10-30), i.e., $E_c = -I_b R_k = -1.5 I_b$, where I_b is to be used in ma. The point of intersection of this straight line with the d-c load line yields the quiescent point. In this case the d-c load line is found

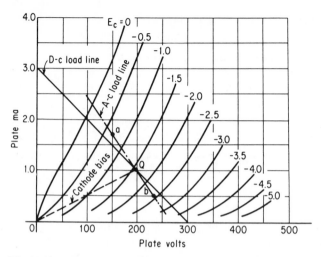

Fig. 10-12. Average plate characteristics of the 12AX7 vacuum tube.

in the usual fashion using the 0.1-M resistor of the second stage. The Q-point for this second stage occurs at $I_b = 1.05$ ma, $E_b = 195$ volts, $E_c = -1.55$ volts.

To determine the output voltage of the second stage, it is necessary to construct the dynamic or a-c load line. Recall that this is a line which passes through the Q-point with a slope equal to the reciprocal of the effective load resistance. For the second stage the effective load resistance is

$$R'_{L2} = \frac{R_{L2}R_{g2}}{R_{L2} + R_{g2}} \tag{10-38}$$

where R_{L2} = load resistance of second stage = 100 K
R_{g2} = grid resistance of succeeding stage = 220 K

Inserting values yields

$$R'_{L2} = \frac{100(220)}{320} = 68.8 \text{ K} \tag{10-39}$$

The a-c load line is identified as line aQb in Fig. 10-12. Point a on this line corresponds to the maximum positive value of the input signal for which the minimum value of the plate voltage is

$$E_{b_{\min}} = 150 \text{ volts} \tag{10-40}$$

Similarly, point b is reached when the varying input signal has its maximum negative value. The corresponding value of the plate voltage is then

$$E_{b_{\max}} = 230 \text{ volts} \tag{10-41}$$

Finally, by Eq. (10-11), the maximum value of the a-c voltage appearing across the output terminals of the second stage (g-c of Fig. 10-10) is

$$\sqrt{2}\, E_{p2} = \frac{E_{b_{\max}} - E_{b_{\min}}}{2} = \frac{230 - 150}{2} = 40 \text{ volts} \tag{10-42}$$

Note that E_{p2} denotes the rms value of this output voltage and specifically is equal to

$$E_{p2} = \frac{40}{\sqrt{2}} = 28.3 \text{ volts} \tag{10-43}$$

The voltage gain of the second stage is then

$$A_{v2} = \frac{E_{p2}}{E_{p1}} = -\frac{28.3}{0.58} = -48.8 \tag{10-44}$$

The combined voltage gain from the input to the output of the second stage accordingly is

$$A_v = -77.3(-48.8) = 3,772 \tag{10-45}$$

10-5. Frequency Response

Vacuum-tube amplifiers are subject to the same limitations of amplifying capability at high and low frequencies as are transistor amplifiers. At the low end of the frequency spectrum the dropoff in voltage gain is attributable also to the coupling capacitor. At the high frequencies the effect of the interelectrode capacitances looms important. The analysis proceeds in a manner very similar to that used for the transistor R-C coupled amplifiers.

For the sake of illustration let us consider the first stage of voltage amplification of the audio amplifier depicted in Fig. 10-10. The first stage is identified from the grid of the first triode to the grid of the second triode. Replacing the vacuum tube by its equivalent circuit (see Fig. 10-8), the complete equivalent circuit of the R-C coupled amplifier from grid G_1 to grid G_2 assumes the configuration shown in Fig. 10-13(a).

By appropriate mathematical manipulation this circuit may be redrawn with an input section separated from the output section. The input current I_i supplied by the signal source may be written as

$$I_i = I_1 + I_2 \tag{10-46}$$

where I_1 is the current flowing through the grid-to-cathode interelectrode capacitance and I_2 is the current flowing through the grid-to-plate capacitance.

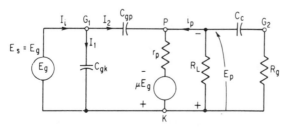

Fig. 10-13(a). The equivalent circuit of an R-C coupled vacuum-tube amplifier showing the coupling and interelectrode capacitances. The plate-to-cathode capacitance is neglected.

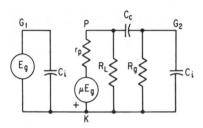

Fig. 10-13(b). Modified form of the equivalent circuit of Fig. 10-13(a).

It is interesting to note that when the interelectrode capacitances are negligibly small, the signal-source current is zero, so that the vacuum-tube amplifier behaves as a pure voltage-sensitive device. By applying Ohm's law we may rewrite the last equation as

$$I_i = j\omega C_{gk}E_g + j\omega C_{gp}(E_g - E_p) \quad (10\text{-}47)$$

Keep in mind that we are dealing throughout with the rms values of voltage and current. The quantity E_p is the rms value of the varying component of the plate voltage. It is related to the varying component of the grid voltage E_g by the relationship

$$E_p = A_{vm}E_g \quad (10\text{-}48)$$

where A_{vm} is the voltage gain in the midband frequency range (i.e., with negligible capacitance effects). Inserting Eq. (10-48) into Eq. (10-47) then yields

$$I_i = j\omega[C_{gk} + C_{gp}(1 - A_{vm})]E_g \equiv j\omega C_i E_g \quad (10\text{-}49)$$

where

$$C_i \equiv C_{gk} + C_{gp}(1 - A_{vm}) \quad (10\text{-}50)$$

Keep in mind that A_{vm} is a negative quantity.

An examination of Eq. (10-49) reveals that in essence the source voltage sees an input impedance which may be expressed as

$$Z_i \equiv \frac{E_g}{I_i} = \frac{1}{j\omega C_i} \quad (10\text{-}51)$$

Therefore the input section of the circuit of Fig. 10-13(a) may be redrawn to show that E_g feeds directly and solely into a capacitor having a capacitance of value C_i. This is depicted in Fig. 10-13(b). Note too that the expression for C_i again exhibits the Miller effect as was the case with the transistor amplifier. The effect of the grid-to-plate capacitance is greatly increased, and this is a factor which must be reckoned with especially at high frequencies. This is of particular importance in a multistage amplifier because the C_i of the second triode appears in parallel with the output terminals of the preceding stage and can thereby reduce

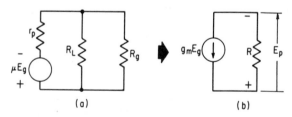

(a)　　　　　　　　　　　　　　(b)

Fig. 10-14. Mid-band frequency equivalent circuit:
(a) voltage source version;
(b) current source version

$$\left(\text{here } \frac{1}{R} = \frac{1}{r_p} + \frac{1}{R_L} + \frac{1}{R_g}\right)$$

the voltage gain appreciably. *The C_i appearing in the output section of the equivalent circuit of Fig. 10-13(b) represents the effective input capacitance of the following triode.*

The midband range of frequencies refers to that region for which C_C is virtually a short circuit and C_i is virtually an open circuit compared to R_L and R_g in the equivalent circuit of Fig. 10-13(b). Thus for the midband frequencies the equivalent circuit becomes that depicted in Fig. 10-14. From the current-source equivalent circuit it follows forthwith that the rms value of the varying component of the plate voltage is

$$E_p = -g_m E_g R = \frac{-g_m E_g}{\dfrac{1}{R}} = \frac{-g_m E_g}{\dfrac{1}{r_p} + \dfrac{1}{R_L} + \dfrac{1}{R_g}} \tag{10-52}$$

By rearranging terms in the last equation the expression for the midfrequency voltage gain, A_{vm}, becomes

$$\boxed{A_{vm} = \frac{E_p}{E_g} = \frac{-\mu}{1 + \dfrac{r_p}{R_L} + \dfrac{r_p}{R_g}}} \tag{10-53}$$

The high-frequency range is identified as that range for which the reactance of C_i is no longer negligible in comparison with R_L and R_g. Of course the reactance of C_C continues to be negligible. The equivalent circuit then takes the form illustrated in Fig. 10-15. The current-source equivalent circuit is shown in

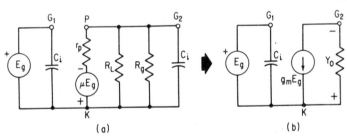

Fig. 10-15. High-frequency equivalent circuit of the *R-C* coupled vacuum-tube amplifier.

Fig. 10-15(b). Hence the rms output voltage may be written

$$E_p = -\frac{g_m E_g}{Y_o} \tag{10-54}$$

where

$$Y_o = \frac{1}{r_p} + \frac{1}{R_L} + \frac{1}{R_g} + j\omega C_i \tag{10-55}$$

Inserting Eq. (10-55) into Eq. (10-54) and formulating the ratio of output to input voltage yields, as the expression for the voltage gain at high frequencies,

$$A_{vh} = \frac{E_p}{E_g} = \frac{-g_m}{\dfrac{1}{r_p} + \dfrac{1}{R_L} + \dfrac{1}{R_g} + j\omega C_i} = \frac{-\mu}{1 + \dfrac{r_p}{R_L} + \dfrac{r_p}{R_g} + j\omega C_i r_p} \tag{10-56}$$

Rearranging terms leads to

$$A_{vh} = \frac{-\mu}{1 + \dfrac{r_p}{R_L} + \dfrac{r_p}{R_g}} \left[\frac{1}{1 + j\omega\dfrac{r_pC_t}{1 + \dfrac{r_p}{R_L} + \dfrac{r_p}{R_g}}} \right] \tag{10-57}$$

or

$$A_{vh} = A_{vm}\left(\frac{1}{1 + j\omega R_hC_t)}\right) \tag{10-58}$$

where A_{vm} is given by Eq. (10-53) and

$$R_h \equiv \frac{r_p}{1 + \dfrac{r_p}{R_L} + \dfrac{r_p}{R_g}} \tag{10-59}$$

Inspection of Eq. (10-58) reveals that as the frequency of the signal becomes greater, the voltage gain decreases. The frequency at which the gain drops to $1/\sqrt{2}$ of its midfrequency value is again readily determined by setting the coefficient of the j term in Eq. (10-58) equal to unity. Thus

$$\boxed{\omega_h = \frac{1}{R_hC_t}} \tag{10-60}$$

where R_h is given by Eq. (10-59) and C_t is given by Eq. (10-50). By substituting the last expression into Eq. (10-58) the voltage gain at high frequencies may be more conveniently expressed as

$$\boxed{A_{vh} = A_{vm}\left(\frac{1}{1 + j\dfrac{\omega}{\omega_h}}\right)} \tag{10-61}$$

Note that the foregoing results are completely analogous to those found for the transistor *R-C* coupled amplifier.

For any given *R-C* coupled amplifier the low-frequency range is identified as that portion of the frequency spectrum for which the reactance of C_c is sufficiently large that it can no longer be treated as a short circuit. At the same time the reactance of C_t is so enormous that it may be eliminated from the equivalent circuit. The low-frequency equivalent circuit then takes on the form depicted in Fig. 10-16.

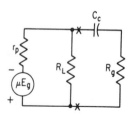

Fig. 10-16. Low-frequency equivalent circuit of the *R-C* coupled vacuum-tube amplifier.

To analyze this circuit for the purpose of finding the expression for the low-frequency voltage gain, one may use either mesh or nodal analysis. However, in this instance it is more convenient to use Thevenin's theorem in the interest of reducing the two-mesh network of Fig. 10-16 to a single mesh. When we apply the procedures outlined in Sec. 3-8 the Thevenin equivalent circuit becomes that shown in Fig. 10-17. The useful a-c output voltage is developed across R_g. The current which flows through this resistance is called I_g and is given by

$$I_g = \frac{\dfrac{\mu E_g R_L}{r_p + R_L}}{\dfrac{r_p R_L}{r_p + R_L} + R_g + \dfrac{1}{j\omega C_C}} \qquad (10\text{-}62)$$

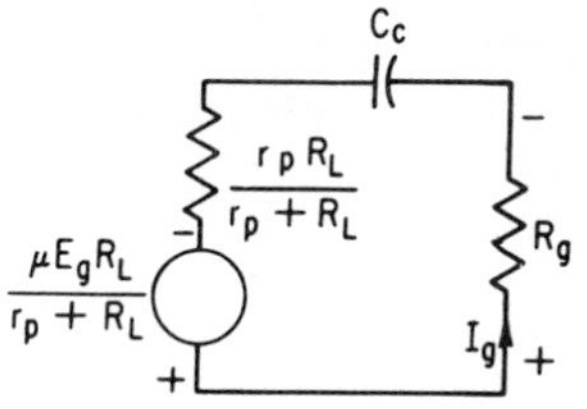

Hence the output voltage may be expressed as

Fig. 10-17. The Thevenin equivalent circuit of Fig. 10-16.

$$
\begin{aligned}
E_p = -I_g R_g &= \frac{-\dfrac{\mu E_g R_L R_g}{r_p + R_L}}{\dfrac{r_p R_L}{r_p + R_L} + R_g + \dfrac{1}{j\omega C_C}} \\[2mm]
&= \frac{-\mu E_g}{1 + \dfrac{r_p}{R_L} + \dfrac{r_p}{R_g} + \dfrac{r_p + R_L}{j\omega C_C R_L R_g}}
\end{aligned}
\qquad (10\text{-}63)
$$

The last term is obtained from the previous one by dividing the numerator and denominator by

$$\frac{R_L R_g}{r_p + R_L}$$

This is done so that the term

$$\left(1 + \frac{r_p}{R_L} + \frac{r_p}{R_g}\right)$$

may be readily identified in the equation. In turn Eq. (10-63) may be rearranged to read as

$$A_{vl} = \frac{E_o}{E_g} = \left(\frac{-\mu}{1 + \dfrac{r_p}{R_L} + \dfrac{r_p}{R_g}}\right)\left[\frac{1}{1 + \dfrac{r_p + R_L}{j\omega C_C\left(1 + \dfrac{r_p}{R_L} + \dfrac{r_p}{R_g}\right)R_L R_g}}\right] \qquad (10\text{-}64)$$

or

$$A_{vl} = A_{vm}\left(\frac{1}{1 - j\dfrac{1}{\omega R_l C_C}}\right) \qquad (10\text{-}65)$$

where A_{vm} is given by Eq. (10-53) and

$$R_l \equiv R_g + \frac{r_p R_L}{r_p + R_L} \qquad (10\text{-}66)$$

A glance at Eq. (10-65) indicates that as the signal frequency gets smaller the voltage gain diminishes from its value at the midband frequencies. The frequency at which there occurs a dropoff in gain by a factor of $1/\sqrt{2}$ is

$$\boxed{\omega_l = \frac{1}{R_l C_C}} \qquad (10\text{-}67)$$

where R_l is given by Eq. (10-66) and C_l is given by Eq. (10-50). When the last equation is inserted into Eq. (10-65), the low-frequency voltage gain may be more succinctly expressed as

$$A_{vl} = A_{vm}\left(\cfrac{1}{1 - j\cfrac{\omega_l}{\omega}}\right)$$

(10-68)

The bandwidth ω_{bw} of the *R-C* coupled vacuum-tube amplifier is equal to the range of frequencies from ω_l to ω_h. Expressing this mathematically we have

$$\omega_{bw} = \omega_h - \omega_l = \frac{1}{R_h C_t} - \frac{1}{R_l C_C}$$

(10-69)

Note that C_t plays the key role in establishing the upper frequency limit while the coupling capacitor C_C exerts the key influence in fixing the lower limit of the bandwidth. A plot of voltage gain as a function of frequency results in a curve which is identical in shape to that shown in Fig. 9-39 for the transistor *R-C* coupled amplifier.

10-6   Tetrodes and Pentodes

The use of the triode in electronic circuits involving signals whose frequencies are many times greater than those of the audio range (e.g. radio and video frequencies) leads to failure. This failure is a direct consequence of the interelectrode capacitance that exists between the grid (input section) and the plate (output section) which at high frequencies provides a path adequate to allow energy to be taken from the output circuit at the plate terminal and supplied to the input circuit at the grid terminal. In turn this causes the amplifier circuit to go into a state of sustained oscillation, thereby disrupting the amplifying function. Moreover, even in those situations where the circuit may not go into oscillations, there occurs a serious deterioration in amplifier voltage gain because of the shunting effect of the interelectrode capacitance C_t [refer to Fig. 10-15a]. For these reasons the design structure of the triode as a three-element electron-control device is in need of some modification if it is to be applied successfully at high frequencies.

The solution is to introduce a second grid wire between the control grid and the plate electrodes in the manner illustrated in Fig. 10-18. The triode now becomes a tetrode—a four-element device. By this scheme effective electrostatic shielding can be achieved between the output and input sections of the amplifier circuit. In most cases there is a reduction of grid-to-plate interelectrode capacitance by a factor of 1000. Because of the screening action exerted by the second grid, it is referred to as the *screen grid*. For the sake of distinction the first grid is then referred to as the *control grid*.

In normal use the screen grid has a fixed potential applied to it in the manner depicted in Fig. 10-19. Its value is less than the plate supply voltage. The potential

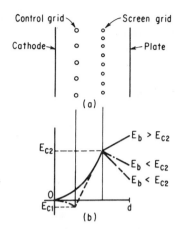

(a)

(b)

Fig. 10-18. (a) Physical arrangement of elements in a tetrode;
(b) the potential profile in the region between cathode and
plate electrodes (E_{c2} denotes the fixed screen grid potential).

profile of the tetrode, which shows the variation of the potential between the
cathode and the plate, is shown in Fig. 8-18(b). It is important to note here that
the slope of the potential profile curve between the screen grid and the plate may be
positive or negative depending upon whether the instantaneous value of the plate
potential is greater than or less than the fixed screen grid voltage. Recall that the
plate potential varies considerably with variation of the input control grid signal so
that it is possible to operate over a region involving negative as well as positive
slopes of the potential profile. This condition is described because the presence of
a negative potential profile slope between the plate and its adjacent electrode
(in this case the screen grid) introduces such a nonlinear effect in the plate characteristic of the tetrode (see Fig. 10-20) as to limit seriously its usefulness as an
electron-control device.

The kink in the plate characteristic of the tetrode is directly attributable to
the negative potential profile slope at the plate. This is another way of saying that

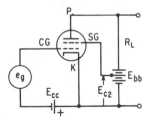

Fig. 10-19. Tetrode amplifier
circuit diagram.

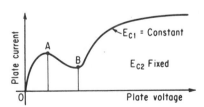

Fig. 10-20. Average tetrode plate characteristics. No secondary emission occurs in
the region from 0 to A because of low
plate voltages.

in the space around the plate there is a region of higher potential than the plate. When a cathode-emitted electron is accelerated towards the plate electrode, it moves at very high velocity. When it strikes the plate, the kinetic energy it releases may be capable of dislodging several other electrons from the plate. These are called *secondary-emission electrons*. In a triode such electrons are quickly drawn back to the plate because the plate is at the highest potential in the region. However, this is not necessarily true in the tetrode. In fact in the region between points A and B of Fig. 10-20 the plate potential is less than the screen grid potential, so that the secondary-emission electrons are collected by the screen grid instead of the plate. The dropping curve from A to B means that the plate now collects fewer electrons because more electrons are lost through secondary emission than arrive at the plate. Beyond B the potential profile at the plate becomes positive and so the secondary-emission electrons return to the plate. Below A the plate potential is so small that electrons do not acquire sufficient energy to cause secondary emission.

To remove the kink in the plate characteristic of the tetrode, a third wire called the *suppressor grid* is inserted in the region between the plate and the screen grid in the fashion shown in Fig. 10-21. The use of five elements now makes the device a *pentode*. If the suppressor grid is connected to the cathode, the space immediately surrounding the plate is always assured of a potential profile having a positive slope. This is clearly illustrated in Fig. 10-21(c). Typical plate characteristics of the pentode are shown in Fig. 10-22. Note the similarity of these characteristics with those of the transistor. Because of these nearby horizontal curves, the dynamic plate resistance r_p of the pentode is very large and is often measured in megohms.

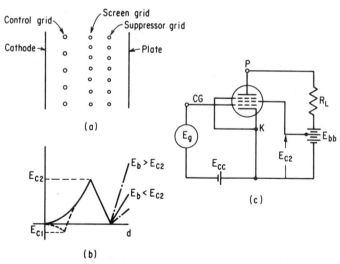

Fig. 10-21. (a) Physical arrangement of electrodes in a pentode;
 (b) the potential profiles for the case where the suppressor
 grid is connected to the cathode;
 (c) typical circuit connections involving the pentode.

This means that the amplification factor of the pentode appreciably exceeds that of the triode. In the pentode it requires much greater changes in the plate potential to offset given changes in the grid potential. It thus follows that the shielding of the plate electrode from the screen grid yields two significant results: it greatly reduces the grid-to-plate capacitance and it provides greater amplification factors.

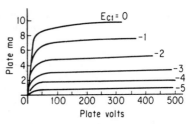

Fig. 10-22. Typical average plate characteristics of the 6SJ7 pentode.

PROBLEMS

10-1 A vacuum-tube amplifier employing a 12AT7 triode section (see Appendix E for curves) has the configuration shown in Fig. P10-1.

(a) Determine the plate current and the plate voltage at the quiescent state.

(b) Find the voltage amplification factor.

(c) Determine the change in plate current caused by a $+1$-volt change in grid-to-cathode voltage.

(d) Repeat part (c) for a -1-volt change in grid circuit.

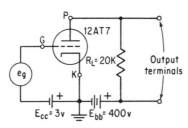

Figure P10-1

10-2 In the circuit of Fig. P10-1 a signal variation is applied to the grid circuit of $e_s = 2 \sin \omega t$.

(a) Sketch the waveform of the total instantaneous plate current.

(b) Compute the rms value of the varying component of current flowing through the load resistor.

(c) Calculate the rms value of the varying component of voltage appearing across R_L.

(d) What is the value of the a-c power delivered to the load resistor?

(e) What is the source of this power and how is its flow controlled?

(f) Find the value of the voltage gain and compare it with the amplification factor. Discuss the difference.

(g) Can you speak of a power gain for this circuit? Explain.

10-3 Repeat Prob. 10-2 for a case where $e_s = 3 \sin \omega t$ and comment on the differences computed.

10-4 In the voltage amplifier circuit of Fig. 10-3 the following components and sources are used:

12AT7 triode section, $E_{bb} = 400$ volts
$R_L = 20$ kilohms, $E_{cc} = 3$ volts
$R_g = 0.1$ megohm, $C_c = 0.1 \ \mu f$

A signal $e_s = 2 \sin 1000t$ is applied to the grid circuit.
  (a) Determine the voltage gain.
  (b) Determine the a-c power delivered to R_L.
  (c) What is the potential difference appearing across C_c in the quiescent state?

10-5 Repeat Prob. 10-4 for the case where $e_s = 3 \sin 1000t$ and comment on the differences computed. In particular discuss whether or not the output varies in direct proportion to the increased amplitude of the input signal.

10-6 A vacuum triode amplifier having the configuration of Fig. 10-1 and employing a 12AX7 triode section has the following quiescent point: $E_c = -1$ volt, $E_b = 110$ volts, and $I_b = 0.6$ ma. The supply voltage is 300 volts.
  (a) Find the value of load resistance.
  (b) A change in grid-to-cathode voltage from -1 to -0.5 volt in this circuit causes a current increase of 0.12 ma. If the voltage amplification factor is 100, find the value of the dynamic plate resistance.
  (c) What is the change in plate voltage in part (b)?
  (d) What is the largest rms value of a sinusoidal input signal that may be used in this circuit without driving the grid positive?
  (e) Find the minimum value of plate current corresponding to the input signal of part (d).

10-7 A triode amplifier employing a 6SN7 vacuum tube has the configuration shown in Fig. P10-7.
  (a) For the values shown find the quiescent operating point.
  (b) Sketch the variation of the total instantaneous current for an input signal e_g which has a rectangular waveform with peak values of ± 6 volts.
  (c) Compute the average value of the plate current in part (b) and compare with the quiescent value.
  (d) Compute the power delivered to the load resistor when the input signal is in the positive half of its cycle.

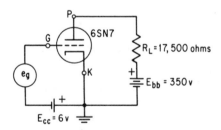

Figure P10-7

10-8 The 6EU7 vacuum triode which is commonly found in high-fidelity audio amplifiers operates at the following quiescent point: $E_b = 250$ volts, $E_c = -2$v, $I_b = 1.2$ ma. When the grid-to-cathode voltage is increased to -1.8 volts at constant plate potential, the plate current increases to 1.5 ma. Furthermore, it is found that to restore the current to 1.2 ma, a decrease in plate potential of 20 volts is needed at constant grid-to-cathode potential.

Draw the equivalent circuit of the vacuum tube with the values of all parameters clearly shown.

10-9 A 6SN7 triode is used in the circuit of Fig. 10-9. If $R_L = 17.5$ K, $E_{bb} = 350$ v, and $R_k = 1$ K, find the quiescent operating point.

10-10 In the circuit of Fig. 10-9 the load resistance is 17.5 K and $E_{bb} = 350$ v. Find the value of cathode resistor which places the quiescent operating point at $E_b = 205$ v, $I_b = 8$ ma, and $E_c = -6$ v.

10-11 A two-stage vacuum-tube amplifier has the configuration depicted in Fig. P10-11. Assume all capacitors have negligible reactance to varying signals. A sinusoidal input signal having an rms value of 0.63 v is applied to the grid-to-cathode circuit of the first stage. The manufacturer's tube handbook lists the following parameters for the 6SN7: $\mu = 20$, $r_p = 7.7$ K, and $g_m = 2.6 \times 10^{-3}$ mho.

    (a) Determine the quiescent operating point for the first stage.

    (b) Find the rms value of the varying component of the plate voltage of the first stage.

    (c) What is the rms value of the plate voltage of the second stage?

    (d) How much a-c power is delivered to R_L?

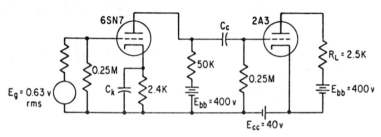

Figure P10-11

10-12 Refer to the circuit of Fig. 10-10. The vacuum-tube handbook lists the value of the grid-to-plate capacitance of the 12AX7 as 1.7 pf and the grid-to-cathode capacitance as 1.8 pf. Furthermore $\mu = 100$ and $r_p = 62.5$ K.

    (a) Draw the complete equivalent circuit of the first stage (see Fig. 10-13) showing all parameters clearly labeled.

    (b) Find the bandwidth of the first stage.

    (c) Is it possible to assign a new value to the plate load resistance in order to increase the upper half-power frequency by 25 per cent? Explain.

10-13 Determine the bandwidth of the first stage of the two-stage amplifier depicted in Fig. P10-11. The grid-to-plate capacitance of the 6SN7 is 4 pf and the grid-to-cathode capacitance is 3 pf. For the 2A3 triode these values are respectively 16.5 pf and 7.5 pf. Moreover, the 6SN7 has $\mu = 20$ and $r_p = 7.7$ K. Assume the coupling capacitance to be equal to 0.01 μf. Neglect the effect of the cathode capacitor.

10-14 In the circuit of Fig. P10-11 determine the magnitude and phase of the a-c output voltage of the first stage when a signal having a value of 0.63 v rms and a frequency of 100,000 Hz is applied to the input terminals. The parameters of the 6SN7 are $\mu = 20$, $r_p = 7.7$ K, $C_{gp} = 4$ pf and $C_{gk} = 3$ pf.

10-15 In the circuit of Fig. P10-11 what must be the value of the coupling capacitor in order that the lower frequency limit of the bandwidth be 10 Hz? Refer to Prob. 10-14 for the parameters of the 6SN7.

11

special topics
and applications

The subject matter that can be appropriately included in a chapter of this kind is almost limitless. In the interest of saving space, however, the selection of topics is restricted to those which are commonly found in electronic circuits and systems and, at the same time, are simple enough to be included in a book at this level.

11-1 Electronic-circuit Applications Involving Diodes

The diodes referred to in this section may be of either the semiconductor or the vacuum-tube type. Moreover, the diode is assumed to be ideal, which means it has zero forward resistance and infinite backward resistance. Thus the symbolism used to denote the diode is the same as that shown in Figs. 7-19 and 8-10(c).

The Half-wave Rectifier. The conversion of an alternating quantity to a unidirectional quantity is achieved by the circuitry depicted in Fig. 11-1. The alternating source e is connected across a load resistor R_L through an ideal diode D. Recalling that the diode conducts whenever its anode side (the arrowhead) is positive with respect to its cathode side (the quadrature line), it follows that during

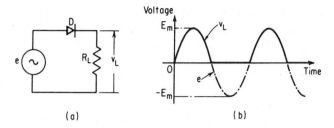

(a)   (b)

Fig. 11-1. (a) Half-wave rectifier circuit;
(b) unidirectional output voltage wave-shape.

the positive half-cycle of the voltage the entire value appears across R_L. When the voltage is in its negative half-cycle, the diode is open so that no voltage appears across the load. The resulting waveshape of the voltage across the load resistor, v_L, is shown in Fig. 11-1(b) as the solid curve. Expressed mathematically we can write

$$v_L = e = E_m \sin \omega t \quad \text{for} \quad e \text{ positive}$$
$$v_L = 0 \qquad\qquad\quad \text{for} \quad e \text{ negative} \tag{11-1}$$

Note the unidirectional aspect of the load voltage, which by virtue of the diode leads to an average different than zero. Specifically since the average value of a positive half sine wave is $(2/\pi)E_m$ and since there is zero contribution in the second half-cycle, the average value of the voltage appearing across R_L is clearly E_m/π.

The Peak Rectifier. If the output voltage of a half-wave rectifier is developed across a capacitor as shown in Fig. 11-2(a) rather than across a resistor, a non-pulsating unidirectional output voltage is obtained. Moreover, the magnitude of this voltage will be E_m rather than E_m/π. To understand how this is accomplished refer to Fig. 11-2 and assume that the source voltage is increasing from zero in the positive direction with the capacitor initially de-energized. As e increases positively, diode D conducts, thus permitting a charging current to flow through the capacitor. This builds up a voltage across the capacitor plates. At time t_1 in Fig. 11-2(b) the capacitor is charged to the peak value of the source voltage E_m. As

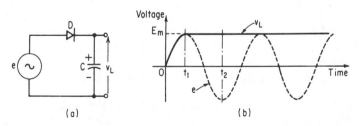

(a)   (b)

Fig. 11-2. (a) Peak rectifier circuit;
(b) resulting output waveshape.

time increases slightly beyond t_1, the potential of the anode side of the diode drops below E_m, but the potential of the cathode side remains at E_m because of the charge on the capacitor. Hence the diode opens. In fact the diode remains open for all time thereafter provided that there is no charge loss on the capacitor through leakage. Consequently the output voltage available at the capacitor terminals remains constant at E_m.

All diodes have a limit on the maximum reverse voltage they can reasonably be expected to withstand. For convenience this is called the *peak inverse voltage* (PIV) and must be checked when applying diodes so as to be sure not to exceed the value recommended by the manufacturer. The value of the PIV is readily found for the circuit of Fig. 11-2(a). Note that the peak negative voltage appearing across the diode terminals occurs when the cathode side is at the potential $+E_m$ and the plate side is at the potential $-E_m$. This occurs at time t_2 in Fig. 11-2(b). Thus the PIV is equal to $2E_m$. Accordingly in this application one must first compute the PIV and then select a diode capable of withstanding it. Otherwise the peak value of the source voltage must be correspondingly reduced.

In practical applications of the peak rectifier the load connected across the capacitor terminals is very often resistive. The circuit therefore takes on the configuration illustrated in Fig. 11-3(a). The presence of the load resistor across

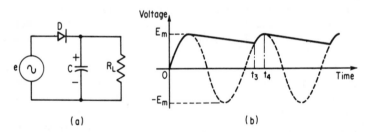

(a)   (b)

Fig. 11-3. (a) Peak rectifier circuit with connected load R_L;
(b) resulting waveshape.

the capacitor terminals provides a path for the capacitor to discharge during that part of the cycle when the source voltage has a magnitude less than the capacitor voltage or when it is negative. As a result the output voltage starts to diminish and the rate of decrease is dependent upon the time constant $R_L C$. Usually this time constant is very large compared to the period of the a-c source so that little loss of voltage occurs. The waveshape of the output voltage is shown in Fig. 11-3(b). Note that it is no longer nonpulsating. Also, in the interval between t_3 and t_4, the diode once again conducts, thus allowing the source to re-establish the output voltage to the E_m level at time t_4. This is repeated once in each cycle.

The Diode Clamper. An interesting variation of the peak rectifier circuit is obtained by taking the output across the diode terminals rather than the capacitor terminals in the manner depicted in Fig. 11-4. Assume that the voltage source has been applied for some time so that the magnitude of the voltage appearing across the capacitor terminals is E_m. Recalling that in this configuration the diode remains

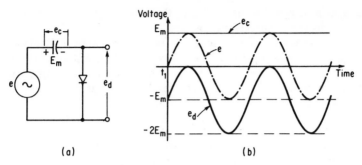

(a)   (b)

Fig. 11-4. (a) The diode clamper circuit;
(b) resulting waveshapes for $t_1 \gg 0$.

open for all time after the initial charging quarter-cycle,† the expression for the
voltage appearing across the diode terminals is

$$e_d = e - E_m \tag{11-2}$$

where e_d denotes the output voltage across the diode
e is the instantaneous value of the source voltage having a peak E_m.

A plot of Eq. (11-2) yields the variation shown in Fig. 11-4(b). Note that the only
difference between the input waveshape e and the output waveshape e_d is a downward shift of the latter relative to the former by the amount $-E_m$, which is the
fixed voltage appearing across the capacitor. This circuit is said to *clamp* the
positive peak value of the source voltage at zero volts at the output terminals.
The diode clamper finds wide application in television circuits because of the
need to provide fixed peak voltages at various points throughout the circuit.

The Voltage Doubler. If the diode clamper circuit is combined with the peak
rectifier circuit in the fashion shown in Fig. 11-5, a voltage-doubler circuit results.

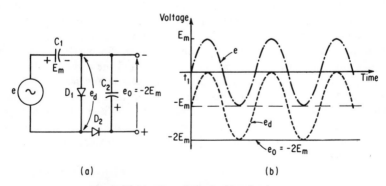

(a)   (b)

Fig. 11-5. (a) The voltage doubler circuit;
(b) resulting waveshapes for $t_1 \gg 0$.

† Assuming the charging process starts at the beginning of the positive half-cycle.

This arrangement is commonly found in the power-supply sections of electronic circuits such as tape recorders, radios, and so on.

The basic idea of the voltage doubler is to take the output of the diode clamper and apply it to a peak rectifier circuit. Since the output voltage at the $D1$ diode terminals is a sinusoidal voltage with a peak negative value of $-2E_m$, application of this voltage to peak rectifier circuit results in voltage appearing on C_2 equal to $2E_m$. The output waveshape is shown in Fig. 11-5(b).

The Diode Limiter. Diodes are frequently used in electronic circuits to generate nonlinear functions for simulation purposes. One example is the generation of the limit-stop action often found in mechanical systems. For example, in an aircraft the elevator, aileron, and rudder control surfaces can be deflected only through a restricted region, beyond which no further movement of the control surfaces is possible in spite of increasing input signals. This behavior is described as a *limiting* action and may be simulated by the electronic circuit shown in Fig. 11-6.

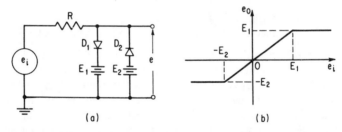

Fig. 11-6. (a) The diode limiter circuit;
(b) transfer characteristic relating output and input voltages.

The voltage E_1 is used to simulate the upper limit of the output quantity corresponding to positive inputs. Similarly the voltage E_2 is used to simulate the lower limit corresponding to negative inputs. By making E_1 and E_2 different, we can simulate limits of different values. To understand the operation of this circuit, first consider that the input signal e_i is increasing in the positive sense and is less in magnitude than E_1. For this condition then diode $D1$ is open because the anode side (arrowhead) is at potential $e_i < E_1$ and the cathode side is at the potential E_1. Moreover, diode $D2$ is also open because any positive value of e (with respect to ground) makes the voltage across $D2$ more negative. Consequently the input voltage e_i passes directly to the output terminals so that the output and input voltages are the same. Since there is no current flow in this instance, there is no loss of voltage across R. A straight line having unity slope graphically represents this situation.

Consider next that the input voltage e_i exceeds E_1. Now the voltage across $D1$ is positive and so it conducts. When this happens the voltage E_1 appears directly at the output terminals. Hence the output voltage remains fixed at this value irrespective of any further increase in e_i. The difference between e_i ($> E_1$) and E_1 appears across R. The greater this difference, the larger the voltage drop across R. A graphical description of the output for $e_i > E_1$ is thus a straight horizontal line as depicted in Fig. 11-6(b).

By similar reasoning it is found that for negative inputs diode $D1$ is always open while diode $D2$ is open only so long as the magnitude of this negative voltage is smaller than E_2.

11-2 The Emitter-Follower Amplifier

An amplifier configuration that has some very useful properties is depicted in Fig. 11-7. It differs from the circuitry of the conventional amplifier by the absence of a load resistor in the collector circuit as well as the lack of a by-pass capacitor for the emitter resistor. Another point of interest is that with respect to a-c signals the collector is at ground potential. For this reason, this configuration is sometimes called a common-collector amplifier.

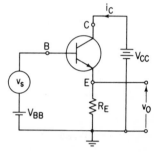

As the input signal v_s increases, the a-c plate current i_c also increases; and as i_c passes through the emitter resistor, it produces an increase in the a-c output voltage v_o. Similarly, as v_s decreases, there occurs a corresponding decrease in v_o. Since v_o is measured between the emitter and ground and because v_o follows the variations in v_s, this circuit is widely known as the *emitter-follower* amplifier.

Fig. 11-7. The emitter-follower amplifier.

In the interest of pointing out some of the useful features of this amplifier, it is helpful to analyze the emitter follower in terms of its equivalent circuit. As is customary, the transistor is replaced by its appropriate equivalent representation, as shown in Fig. 11-8. The simplified version of the equivalent circuit is used here and, of course, all d-c quantities are omitted. One notable characteristic of the emitter follower is that it possesses a high input resistance—considerably higher, in fact, than the common-emitter or common-base modes of transistor amplifiers. This fact can be readily demonstrated by redrawing the equivalent circuit of Fig.

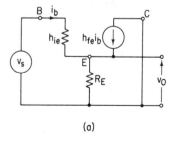

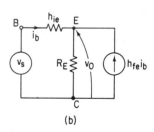

(a)   (b)

Fig. 11-8. Equivalent circuit of the emitter-follower amplifier:
(a) direct derivation from Fig. 11-7 with all d-c quantities omitted;
(b) rearrangement of the circuit elements in (a).

11-8(b) in the form shown in Fig. 11-9(a). Note that the shunt combination of R_E and the current source $h_{fe}i_b$ have been replaced by the Thevenin equivalent of a voltage source $h_{fe}R_E i_b$ and series resistor R_E. In turn, this voltage source may be replaced by the current i_b flowing through a resistor $h_{fe}R_E$ as illustrated in Fig. 11-9(b). It then follows from this circuit that the input resistance of the emitter follower is

$$R_i = \frac{v_s}{i_b} = h_{ie} + (1 + h_{fe})R_E \qquad (11\text{-}3)$$

Since h_{fe} is a large quantity (50 to 80 typically) and R_E is greater than h_{ie}, an appreciable increase in the level of input resistance occurs with the common-collector configuration.

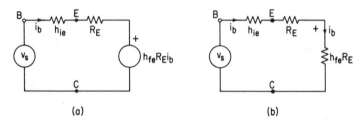

Fig. 11-9. (a) Replacing the current source of Fig. 11-8(b) by the Thevenin
equivalent voltage source;
(b) replacing the Thevenin voltage source by an equivalent
resistance drop.

In view of the high input resistance, it would be useful to know the value of the output resistance. This quantity is readily found by formulating the ratio of the open-circuit voltage to the short-circuit current as they appear at the emitter-collector terminals. The open-circuit voltage is found by removing R_E in Fig. 11-8(a) and measuring $v_o = v_{oc}$ for variations in v_s. A study of this circuit discloses that for the open-circuit condition, $i_b = 0$, and so $v_{oc} = v_s$. The short-circuit current is found by replacing R_E with a short circuit in Fig. 11-8(b) and measuring the current. An examination of the circuit reveals that both i_b and $h_{fe}i_b$ flow through short circuit, thus yielding a total current of $i_b + h_{fe}i_b$. Accordingly, the output resistance becomes

$$R_o = \frac{v_{oc}}{i_{sc}} = \frac{v_s}{i_b(1 + h_{fe})} \qquad (11\text{-}4)$$

But for the short-circuit condition

$$\frac{v_s}{i_b} = h_{ie} \qquad (11\text{-}5)$$

Therefore, we have, finally

$$R_o = \frac{h_{ie}}{1 + h_{fe}} \qquad (11\text{-}6)$$

Recall that h_{ie} is the forward-biased input resistance of the common-emitter mode and as such has a value of resistance that is relatively low. In the emitter-follower configuration, however, this quantity is even further reduced by a substantial factor, thus leading to a very low output resistance.

The combination of a very high input resistance and a very low output resistance makes the emitter follower a very popular circuit for providing *resistance transformation*. Very often in electronic circuits, the loads (i.e., the places where the amplified signals finally do their useful work) are characterized by low resistance, while the amplified signals are associated with circuits having high resistance levels. Without the use of an appropriate transformation device, such as the emitter follower, these amplified signals could not succeed in bringing about substantial changes in energy at the loads, because most of the signal would be consumed as an internal resistance drop. But with the emitter follower providing the coupling to the load, the load assumes the role of prime importance in the output circuit.

What is the voltage gain of the emitter-follower amplifier? The answer can be had by relating v_o to v_s in the configuration of Fig. 11-8. Writing Kirchhoff's current law at the emitter junction E in Fig. 11-8(b) yields

$$i_b - \frac{v_o}{R_E} + h_{fe}i_b = 0 \tag{11-7}$$

or

$$v_o = i_b(1 + h_{fe})R_E \tag{11-8}$$

Inserting the expression for i_b from Eq. (11-3) provides the relationship involving v_o and v_s. Thus

$$v_0 = (1 + h_{fe})R_E \left[\frac{v_s}{h_{ie} + (1 + h_{fe})R_E} \right] \tag{11-9}$$

Formulating the ratio of output voltage v_o to input signal v_s and manipulating for compactness leads to

$$\frac{v_o}{v_s} = \frac{1}{1 + \dfrac{h_{ie}}{(1 + h_{fe})R_E}} \tag{11-10}$$

A study of this last expression makes it quite clear that the voltage "gain" of an emitter-follower amplifier is never unity but comes extremely close. Values of 0.99 and 0.995 are typical. This is a significant conclusion, because it means that the high level of signal voltage often applied to the input of the emitter-follower amplifier appears at the output circuit almost undiminished. The important difference is that in the output circuit this voltage level is associated with a greatly amplified collector current. Hence, although there is no voltage gain, there is a very substantial gain in power delivered to the load as represented by the emitter resistor R_E.

11-3 The Push-Pull Amplifier

The push-pull amplifier is a power amplifier, and it is frequently found in the output stages of electronic circuits. It is used whenever high output power at high efficiency and little distortion is required. These are some of the advantages the

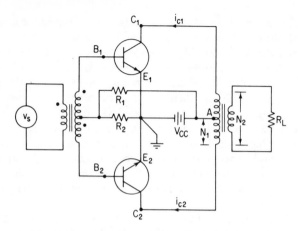

Fig. 11-10. Push-pull amplifier employing transistors.

push-pull amplifier has to offer over other arrangements, such as the parallel combination of two power amplifiers to provide the same output. Figure 11-10 shows the circuitry of a typical push-pull amplifier. Resistors R_1 and R_2 are used to furnish the proper bias for the circuit. The advantages of the push-pull arrangement, however, do not come without a price. Note the need in the base circuit of each transistor for signals of equal magnitude and opposite signs with respect to ground. Accordingly, it becomes necessary to apply the input signal to the push-pull amplifier through a transformer, as shown in Fig. 11-10, or else to include a driver stage to generate these signals.

In the push-pull circuit, the actual load resistor R_L is connected to the transistors by means of a transformer equipped with a center-tapped primary winding (point A in Fig. 11-10). The power supply voltage is connected between the emitter terminals and this center tap. The quiescent current of each transistor flows through each half of the primary winding in opposite directions so that no saturation of the magnetic core occurs.

When the base current of one transistor is being driven positive with respect to the Q-point, the collector current increases, thus causing a decrease in collector potential relative to ground. At the same time, however, a reverse action is taking place in the base circuit of the second transistor. Its base current is decreasing, thus causing a drop in the collector current with a consequent rise in collector potential with respect to ground. This means that the a-c current flowing through the transformer primary winding is in the *same* direction. As i_{c1} in Fig. 11-10 increases (i.e., pulls), the current i_{c2} decreases (i.e., pushes); hence the name push-pull amplifier.

In view of the fact that the load current in the secondary of the transformer is proportional to the *difference* of the collector currents, it follows that the harmonic content will be less than it is in either of the collector currents because of cancelation effects. Consequently, the push-pull arrangement yields much less distortion in the output.

11-4 Modulation

The term modulation is used to describe the process by which some characteristic of a carrier is varied in accordance with a modulating wave. A simple example of modulation can be illustrated by referring to the circuit of Fig. 11-11. If the switch is opened and closed, say in a periodic fashion, the voltage (or power) delivered to resistor R is altered. Thus when the switch is closed the d-c carrier is applied to the load; when the switch is opened the d-c carrier is reduced to zero. The opening and closing of the switch represents the modulating wave. Although in this example the carrier is a d-c quantity, in general it may be d-c, sinusoidal, or a series of pulses.

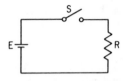

Fig. 11-11. A periodic opening and closing of the switch illustrates the modulation process.

One very important application of modulation occurs in radio transmission. To transmit radio signals successfully and in an efficient manner, it is necessary to radiate electric power. At d-c it is impossible to radiate electrical energy. At audio frequencies (15 to 15,000 Hz) radiation is not practicable because of the huge antenna sizes required; in addition, the efficiency of radiation is very poor. Radiation of electrical energy is practicable only at high frequencies—for example, above 20 kilohertz. It should be apparent then that if speech is to be transmitted properly, some means must be devised which will permit transmission to occur at high frequencies while it simultaneously allows carrying the intelligence contained in the audio-frequency signals. The solution lies in modulating the high-frequency signals (called the *carrier wave* and denoted by ω_c) by the audio-frequency signals (called the *modulating wave* and denoted by ω_m).

Although there are many modulation procedures possible, our attention is confined here to the two most important ones. Since it often happens that the wave to be modulated is sinusoidal, we may write for the unmodulated signal

$$i = A_c \sin(\omega_c t + \phi) \tag{11-11}$$

where A_c denotes the amplitude of the carrier signal and ω_c is its frequency in radians per second. A study of Eq. (11-11) reveals that the carrier wave may have either its amplitude A_c or its argument $\omega_c t$ altered. When the amplitude is modified in accordance with a much lower-frequency modulating wave, the alteration process is called *amplitude modulation* and abbreviated AM. When the argument is modified by changing the frequency ω_c, the process is described as *frequency modulation* and abbreviated FM. We shall discuss here only amplitude modulation.

It is helpful at this point to identify some of the distinguishing characteristics of an amplitude-modulated carrier wave. For such a wave the amplitude becomes time-dependent and may be expressed as

$$A_c(t) = A_c + A_m \sin \omega_m t \tag{11-12}$$

Introducing

$$m = \frac{A_m}{A_c} \tag{11-13}$$

Eq. (11-12) may be rewritten as

$$A_c(t) = A_c(1 + m \sin \omega_m t) \qquad (11\text{-}14)$$

The quantity m is called the *modulation index* and is restricted to values between zero and unity. The last expression indicates that the amplitude of the carrier signal varies sinusoidally at the modulating frequency ω_m between the limits $A_c(1 + m)$ and $A_c(1 - m)$. The complete time expression for the amplitude-modulated wave then becomes

$$i = A_c(1 + m \sin \omega_m t) \sin \omega_c t \qquad (11\text{-}15)$$

A normalized plot of Eq. (11-15) appears in Fig. 11-12(a). Note the maximum and minimum values of the amplitude. It is the variation in the amplitude of the carrier wave which contains the intelligence to be transmitted.

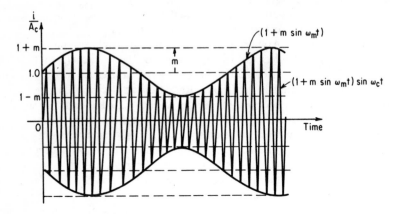

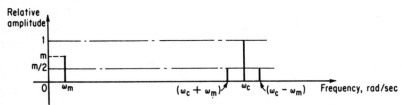

Fig. 11-12. (a) An amplitude modulated sine wave with $m < 1$;
(b) frequency spectrum of a sinusoidally modulated wave.

An alternative expression may be used to represent the modulated carrier wave which reveals additional information about the nature of amplitude modulation. From Eq. (11-15) we can write

$$\frac{i}{A_c} = \sin \omega_c t + m \sin \omega_m t \sin \omega_c t \qquad (11\text{-}16)$$

Inserting

$$\sin \omega_m t \sin \omega_c t = \tfrac{1}{2} \cos (\omega_c - \omega_m)t - \tfrac{1}{2} \cos (\omega_c + \omega_m)t \qquad (11\text{-}17)$$

into Eq. (11-16) yields

$$\frac{i}{A_c} = \sin \omega_c t + \frac{m}{2} \cos (\omega_c - \omega_m)t - \frac{m}{2} \cos (\omega_c + \omega_m)t \qquad (11\text{-}18)$$

Inspection of Eq. (11-18) reveals that the amplitude-modulated wave is equivalent to the summation of three sinusoids: one having a unity amplitude and frequency ω_c, the second having amplitude $m/2$ and frequency $(\omega_c - \omega_m)$, and the third having amplitude $m/2$ and frequency $(\omega_c + \omega_m)$. In practical radio transmission ω_c may be many times greater than ω_m. Hence the frequency of the second and third terms on the right side of Eq. (11-18) is generally close to the carrier frequency. Figure 11-12(b) represents this situation graphically on the frequency-spectrum plot. The frequency components contained in the amplitude-modulated wave are shown by vertical lines appropriately located along the frequency axis. The height of each vertical line is drawn proportional to its amplitude. The lower frequency component $(\omega_c - \omega_m)$ is called the *lower side frequency*. The upper frequency component $(\omega_c + \omega_m)$ is called the *upper side frequency*. Any amplitude-modulated wave must contain at least these frequencies.

How is the amplitude-modulation process implemented in electronic circuitry? Usually implementation is achieved by one of two methods: varying the quiescent operating point or using a square-law device. In either case the objective is to modify the single-frequency carrier wave so as to create new frequencies—the sideband frequencies. Appearing in Fig. 11-13 is an amplitude-modulated amplifier which achieves modulation by base injection. That is, the modulating signal develops a voltage across R_1, which in turn alters the quiescent point in the base circuit. Thus as the bias current is increased the signal at the collector terminal is correspondingly increased. Similarly as the bias current is decreased during the reversed part of the cycle of the modulating signal, the output at the collector is also decreased. Consequently, a plot of the collector output will be found to yield a variation similar to that shown in Fig. 11-12. The resistor R_E in Fig. 11-13 is the self-bias emitter resistor and C_E is its by-pass capacitor. Also, R_2 is a dropping resistor which limits the d-c voltage drop across R_1, which in

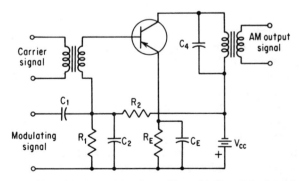

Fig. 11-13. Amplitude-modulated amplifier with base injection of the modulating signal.

turn establishes the initial quiescent base current. Capacitor C_2 is a by-pass capacitor for the carrier signal in the input circuit. In understanding the operation of this circuit it is helpful to keep in mind the fact that ω_m is very small compared to ω_c. Hence for any given cycle of the carrier frequency the bias in the base-to-emitter circuit appears to be fixed.

Another arrangement which permits the introduction of a time-varying element to create new frequencies is shown in Fig. 11-14. Here the modulating signal

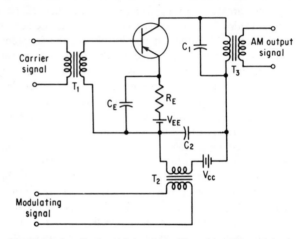

Fig. 11-14. Amplitude-modulated amplifier with collector injection.

is injected in the collector circuit. Accordingly the total supply voltage in the collector circuit is

$$\text{Total collector-supply voltage} = V_{CC} + kV_m \sin \omega_m t$$
$$= V_{CC}(1 + m \sin \omega_m t) \tag{11-19}$$

where

$$m = \frac{kV_m}{V_{CC}} \tag{11-20}$$

and k denotes a proportionality factor of transformer $T2$. For a given load impedance as the collector-supply voltage varies in accordance with Eq. (11-19), the load line moves parallel to itself up and down the transistor output characteristics, thereby changing the quiescent operating point. As the collector-supply voltage takes on its peak value of $V_{CC}(1 + m)$, the a-c current increases. Similarly as the collector-supply voltage diminishes to its minimum of $V_{CC}(1 - m)$, there occurs a corresponding decrease in a-c collector current. Once again the output appears as depicted in Fig. 11-12. Capacitor C_2 is used in Fig. 11-14 to by-pass the carrier signal in the collector circuit and C_1 is selected to provide series resonance with the primary winding of transformer $T3$.

The use of a square-law device to produce an amplitude-modulated signal can be described by the circuitry depicted in Fig. 11-15. Here it is assumed that

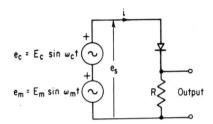

Fig. 11-15. A square-law diode modulator.

the current is describable by two terms of the Taylor series expansion

$$i = A_1 e_s + A_2 e_s^2 \tag{11-21}$$

where

$$e_s = E_c \sin \omega_c t + E_m \sin \omega_m t \tag{11-22}$$

Upon inserting the expression for e_s into Eq. (11-21) we obtain

$$i = A_1 E_c \sin \omega_c t + A_1 E_m \sin \omega_m t$$
$$+ A_2 [E_c^2 \sin^2 \omega_c t + 2 E_c E_m \sin \omega_c t + E_m^2 \sin^2 \omega_m t] \tag{11-23}$$

The quantity in brackets derives from the square term in Eq. (11-21). A little thought reveals that it is this term which gives rise to the product of the carrier and modulating sinusoids, which in turn yields the upper and lower frequency sidebands thereby permitting the amplitude-modulated signal to be identified. By introducing the trigonometric identity of Eq. (11-17) and $\sin^2 \omega_c t = \frac{1}{2} - \cos 2\omega_c t$, Eq. (11-23) may be rearranged to read as follows:

$$i = \frac{A_2}{2}(E_c^2 + E_m^2) + A_1 E_m \sin \omega_m t - \frac{A_2}{2} E_c^2 \cos 2\omega_c t - \frac{A_2}{2} E_m^2 \cos 2\omega_m t$$
$$+ A_1 E_c \sin \omega_c t + \frac{A_2 E_c E_m}{2} \cos (\omega_c - \omega_m)t - \frac{A_2 E_c E_m}{2} \cos (\omega_c + \omega_m)t \tag{11-24}$$

The last three terms of Eq. (11-24) are identical in form to those appearing in Eq. (11-18), so that together they yield in the output resistor the desired amplitude-modulated signal. The remaining four terms are of no use; they can be filtered out by appropriate band-pass filters which will pass ω_c and the two sideband frequencies but reject all else.

Frequency modulation is obtained in electronic circuits by varying the frequency of the carrier signal in a way which makes it dependent upon the instantaneous value of the modulating signal. Expressed mathematically the expression for ω_c in Eq. (11-11) becomes

$$\omega_c(t) = \omega_c + k_f E_m \sin \omega_m t \tag{11-25}$$

A frequency-modulated output is sometimes obtained by using the modulating signal to vary the gain of an oscillator, thereby changing its frequency of oscillation in accordance with Eq. (11-25). For further treatment of this subject matter the reader is referred to the bibliography at the end of the book.

11-5   Amplitude Demodulation or Detection

At the transmitting end of a radio station it is easier to radiate electrical energy at high frequencies (above 500 KHz). For this reason the intelligence contained in an audio signal (15 Hz to 15 KHz) is used to modify the amplitude of the high-frequency carrier wave. However, once the transmission has been achieved by being picked up by a local radio antenna and the AM signal suitably amplified, the radio receiver must also be equipped with electronic circuitry which will allow the recovery of the information appearing in the form of the amplitude variations of the carrier wave. The process of recovering the modulating wave from the carrier is called *amplitude demodulation* or, more simply, *detection*. Appearing in Fig. 11-16 is a circuit which is commonly used to accomplish this result. It is called a *diode detector*. The principle of operation is simple, since it essentially requires rectifying the carrier wave. As a matter of fact some form of rectification is involved in all demodulation circuits.

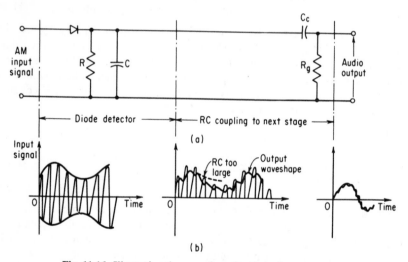

Fig. 11-16. Illustrating the operation of a diode detector:
(a) diode detector and coupling circuit;
(b) waveshapes at various points in the circuit of (a).

When the amplitude-modulated wave is applied to the input terminals of the diode detector, the diode conducts when the carrier signal is positive and cuts off when it is negative. Consequently, each positive half-cycle of the carrier wave appears across the parallel combination of R and C in the manner depicted in Fig. 11-16(b). The capacitor C charges to nearly the peak value of the positive half-cycle of the carrier wave. (A small voltage drop occurs across the diode.) During the negative half-cycle of the carrier the capacitor C discharges through the resistor R at a time constant RC. In choosing R and C care must be taken not to select too small a value for RC, otherwise the charge will be dissipated and the voltage reduced to zero during the negative half-cycle. This condition can be avoided by choosing RC large compared to the time it takes for the carrier to pass through one cycle. On the other hand, it is important that RC not be chosen excessively large, for then the capacitor will discharge so slowly that in the decreasing portion of the rectified output *diagonal clipping* occurs as indicated in Fig. 11-16(b). The best value for RC is that which causes the rate of discharge of the capacitor to follow the variations in the modulating signal.

The output signal of the diode detector has the somewhat jagged variation depicted in Fig. 11-16(b). It has a waveshape which very closely resembles that of the modulating signal, with the exception that it contains a d-c level. This, however, is readily removed by using capacitance-resistance coupling to the next stage as shown in Fig. 11-16(a). The output of this coupling network is a signal which very closely matches the original modulating signal. Since this signal may be thought to represent one of the many similar signals found in speech sounds, the combined process of modulation and demodulation thus performs a key role in the successful transmission of the audio frequencies generated by the human voice.

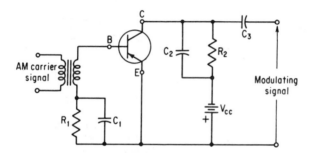

Fig. 11-17. Simplified transistor detector circuit.

It is possible to achieve amplitude demodulation by using a three-element electronic device. A typical example of such a configuration is illustrated in Fig. 11-17. This circuit is called a *transistor detector*. In addition to detection of the modulated carrier input signal, it also furnishes amplification of the modulating signal. Recalling that for the transistor shown in Fig. 11-17 conduction occurs only from emitter to base because of the diode action associated with these elements, it is easy to see that the amplitude-modulated wave is rectified in the emitter-base circuit in precisely the same manner as occurs in the diode detector of Fig. 11-16(a). In fact, resistor R_1 and capacitor C_1 serve the same function as do R and C in Fig. 11-16(a). Recall, too, that the voltage developed across R_1 in Fig. 11-17 varies in accordance with the modulating signal rate. Since this variation exists in the transistor input circuit, it causes a corresponding amplified variation to occur in the collector circuit—specifically across R_2. Capacitor C_2 is selected to by-pass the carrier-frequency signal and C_3 serves to block the d-c voltage level from passing to the next stage. As a result the voltage appearing across the output terminals of this circuit is an amplified form of the modulating portion of the input signal.

11-6   Transistor Logic Circuits

The transistor is used widely in electronic circuits to perform switching operations. One big area of application is in digital computers, where it is employed in so-called logic circuits to execute the basic operations of Boolean algebra. An understanding of how the transistor is used as a switching device can be obtained by examining the information depicted in Fig. 11-18. Appearing in Fig. 11-18(a)

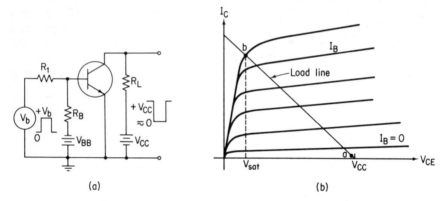

(a)   (b)

Fig. 11-18. (a) Common-emitter transistor amplifier used as a switching
device;
(b) average collector characteristic with load line. In absence
of input signal, V_{BB} holds transistor at cutoff (i.e. switch
OFF) thus yielding an output of $+V_{CC}$. In the presence of
$+V_b$ transistor conducts to saturation and output voltage
level drops to near zero (i.e. switch is ON).

is the common-emitter transistor configuration. Figure 11-18(b) shows the corresponding average collector characteristics with a load line superposed. When the input v_b is zero, the transistor is at *cutoff*—point *a* in Fig. 11-18(b). In this state the level of the voltage at the output terminals is $+V_{CC}$ for all practical purposes. The corresponding collector current is negligibly small since the transistor input is reverse biased. Upon admitting an input signal of the proper level, the transistor can readily be driven from a cutoff state to a state of saturation—point *b* in Fig. 11-18(b). In this state the level of the voltage at the output terminals is almost zero while the current is at a maximum value. Thus by allowing the input voltage v_b to vary from zero to some predetermined threshold value, the output voltage level of the transistor is made to *switch* from $+V_{CC}$ to zero. It is interesting to note that when the transistor is used as a switch in the manner just described, the two stable states are cutoff (point *a*) and saturation (point *b*). Operation along the load line in between these two states is referred to as the transient state. This situation stands in sharp contrast to that which prevails when the transistor is used as an amplifier.

The distinguishing feature of switching circuits is the existence of *two stable states*. These may be described either as *cutoff-saturation* or *off-on*. However, any device which can be made to assume either one of two possible states may also be used to simulate acts of logic. Thus the *on-off* states may be made to represent respectively the logical operations of *yes-no* or *true-false* or *something-nothing* or *closed-open* and so on. Mathematically, it is convenient to denote these two states by the numbers 1 and 0, comprising a binary number system. Moreover, the algebra which describes the rules for performing such basic operations as addition and multiplication in terms of the binary system is called Boolean algebra.† Our aim here is to describe briefly a few of the basic operations of this Boolean algebra and then to show how they can be implemented by using the switching properties of transistors.

OR Gate to Provide Addition. The basic operation of addition in the algebra of switching circuits may be denoted symbolically as

$$a + b = c \qquad (11\text{-}26)$$

i.e., "*a* plus *b* equals *c*." It is important to keep in mind, however, that this equation does not have the same meaning as it does in other number systems (e.g. the decimal system) because only two states are admissible. To illustrate this point consider that the quantities *a* and *b* can assume the states something-nothing. Then clearly the following results are possible for Eq. (11-26):

a	+	*b*	=	*c*
nothing		nothing		nothing
nothing		something		something
something		nothing		something
something		something		something

Another way of illustrating addition is to consider all possible ways by which a signal from one circuit may be transmitted to another circuit. A diagrammatic representation appears in Fig. 11-19. The variables *a* and *b* may assume either a closed or an open state. The results are then

a	+	*b*	=	*c*
open		open		open
open		closed		closed
closed		open		closed
closed		closed		closed

---

† Named after G. Boole, who was first to develop the laws of logic. Cf. G. Boole, *An Investigation of the Laws of Thought* (London, 1854), reprinted, Dover Publications, New York, 1951.

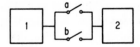

Fig. 11-19. Illustrating Boolean addition as a parallel operation. Signal transmission occurs when *a* OR *b* is closed.

Examination of this table reveals that transmission of the signal occurs whenever either or both switches are closed. The foregoing results are readily denoted in terms of the 0–1 mathematical representation as follows:

$$a + b = c$$

0	0	0
0	1	1
1	0	1
1	1	1

where 0 denotes the open state and 1 the closed state. This table is called the *truth table* for the operation called for by Eq. (11-19).

The implementation of addition in Boolean algebra can be achieved by employing circuitry similar to that appearing in Fig. 11-20. When both signals are zero, the transistor is at cutoff because of the reverse bias provided by V_B.

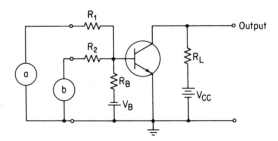

Fig. 11-20. NOR gate to implement Boolean addition. The presence of pulse signals *a* OR *b* provides enough forward bias to switch the output from $+V_{CC}$ to near zero.

When either signal appears at the input as a pulse of appropriate magnitude, the transistor will be sufficiently forward biased to be driven into saturation. The same result occurs when both signals appear simultaneously. Thus, the presence of either OR both signals means that the transistor as a switch is in the ON mode. An important distinction needs, however, to be made at this point. If information is assumed contained in the *level* of the voltages appearing in the circuitry, then care must be taken to assure that the desired correspondence called for in a truth table is in fact present in the voltage levels. Thus, for correspondence to exist in the foregoing truth table, the presence of a positive pulse at the input should command a change to a positive voltage level at the output terminals. But, as we have already pointed out, the circuitry of Fig. 11-20 does just the opposite. That is, a positive pulse at *a* or *b* produces a negative pulse at the output. For this reason the circuitry of Fig. 11-20 is called a *NOT-OR* or *NOR* gate. The circuit possesses the addition capability, but it does so with a phase inversion. This behavior is characteristic of transistor amplifiers, which are preferred in these amplifications precisely for the amplifying feature, which helps to preserve voltage levels in the presence of loading effects. It follows, then, that to obtain exact correspondence with the foregoing truth table an additional phase reversal would be necessary.

AND Gate to Provide Multiplication. Symbolically, multiplication in Boolean algebra is denoted by

$$ab = c \qquad (11\text{-}27)$$

However, again it is necessary to keep in mind that only two states are possible for the variables a and b. In terms of the something-nothing notation, the truth table for Eq. (11-27) becomes

a	$\cdot$	b	$=$	c
nothing		nothing		nothing
something		nothing		nothing
nothing		something		nothing
something		something		something

A better appreciation can be had of why multiplication is described as an ANDing process by resorting again to the analogy employing switches. Refer to Fig. 11-21. Note that circuits 1 and 2 are now joined by a series combination of the switching variables a and b. Clearly, if transmission is to be realized and multiplication achieved in turn, a AND b must be in the closed state. On the other hand, in the case of addition a OR b must be in the closed state to obtain transmission. Thus, it follows that Boolean multiplication is described in terms of a *series* arrangement of the two-state variables, addition is described in terms of a *parallel* arrangement of the two-state variables.

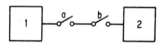

Fig. 11-21. Illustrating Boolean multiplication as a series operation. Signal transmission occurs when a AND b are closed.

Alternative representations of the truth table for Eq. (11-27) are as follows:

a	$\cdot$	b	$=$	c		a	$\cdot$	b	$=$	c
open		open		open		0		0		0
closed		open		open	$\Rightarrow$	1		0		0
open		closed		open		0		1		0
closed		closed		closed		1		1		1

where 0 denotes open and 1 denotes closed.

A typical electronic circuit implementation of Boolean multiplication appears in Fig. 11-22. The negative bias voltage in the base-emitter input circuits of the two transistors assures that both transistors are in the OFF (or cutoff) state. Hence in the absence of input signals the output voltage resides at about $+V_{cc}$. When input

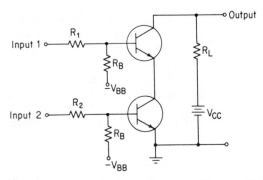

Fig. 11-22. Electronic circuit to implement Boolean multiplication. Because of the phase inversion, this is a NAND gate.

pulses of sufficient magnitude appear simultaneously, both transistors can be driven to the saturation (or ON) state. The output voltage then switches to near zero from $+V_{CC}$. Again note the presence of the phase inversion, thus making this circuit actually a NOT-AND or NAND gate. The series arrangement of the transistors assures implementation of the Boolean multiplication. If either input signal is lacking, the output voltage level will not switch.

The gates depicted in Figs. 11-20 and 11-22 are frequently referred to as RTL (resistance-transistor-logic) circuits, and they represent the simplest computer building block. An improvement over the RTL gate, especially with respect to speed of switching, is the DTL (diode-transistor-logic) circuit shown in Fig. 11-23.

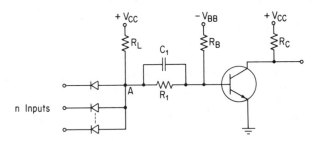

Fig. 11-23. Diode-transistor-logic (DTL) circuit that can qualify as either a NAND or a NOR gate.

Note the liberal use of diodes. This circuit can be made to function either as a NAND gate or a NOR gate. In the NAND mode, output switching occurs when all inputs appear simultaneously at the proper positive pulse heights. At this level, each input diode is opened and the effect of the supply voltage is transferred to the base-emitter circuit of the transistor, thus driving it into saturation. If any one input signal is missing, the diode of that input circuit will continue to conduct, thus keeping terminal A at essentially ground potential. The NOR feature can be realized by a suitable definition of voltage levels and proper choice of the resistors involved.

PROBLEMS

11-1 In the circuit of Fig. 11-1(a) assume that the sinusoidal source is replaced by a periodic symmetrical rectangular wave having a positive amplitude of 10 volts and a negative amplitude of 5 volts. Determine

(a) The waveform of the voltage developed across R_L.

(b) The average value of the load volage.

(c) The rms value of the load voltage.

11-2 A voltage of 100 sin ωt is applied to the circuit of Fig. P11-2.

(a) Sketch the variation of current as a function of time. Clearly indicate all points of interest.

(b) Find the average value of the current.

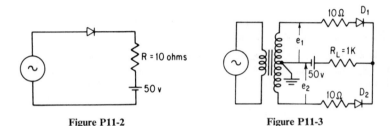

     Figure P11-2   **Figure P11-3**

11-3 In the circuit of Fig. P11-3 each diode is assumed to have a resistance of 10 ohms. With respect to ground the voltages e_1 and e_2 are equal and opposite. Specifically $e_1 = 100 \sin \omega t$.

(a) Sketch the variation of the battery current as a function of time.

(b) Find the average value of this current.

(c) What advantage does this circuit have over that depicted in Fig. P11-2?

11-4 The sinusoidal voltage $e = E_m \sin \omega t$ is applied to the circuitry shown in Fig. P11-4.

(a) Sketch the variation of the voltage across R for several cycles.

(b) If E_m is 100 v, what is the average value of the voltage across R?

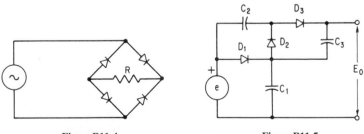

     Figure P11-4   **Figure P11-5**

11-5 If a sinusoidal source voltage e having an amplitude of E_m is applied to the circuit illustrated in Fig. P11-5, find the value of the output voltage in terms of E_m.

11-6 A sinusoidal voltage of amplitude E_m is applied to the circuit of Fig. P11-6.
(a) Identify the polarity and magnitude of voltage on each capacitor.
(b) What is the value of the output voltage?

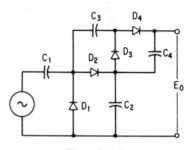

Figure P11-6

11-7 For the circuit shown in Fig. P11-7 plot the output voltage e_0 as a function of the input signal voltage e_i. Assume ideal diodes.

11-8 In the circuit of Fig. P11-8 determine the magnitude of the voltage appearing across R if the voltage source is $e = E_m \sin \omega t$. Assume that the resistance R has a very high value.

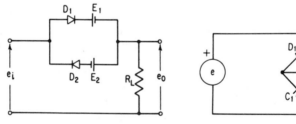

Figure P11-7   **Figure P11-8**

11-9 The transistor in the circuit shown in Fig. P11-9 has $h_{ie} = 3570$ ohms and $h_{fe} = 65$.
   (a) Obtain the expression for the voltage gain.
   (b) Find the value of the voltage gain when $R_L = 10$ K and $R_E = 1$ K.
   (c) Suggest a way of increasing the a-c output voltage while maintaining the same quiescent point, and compute the new value of gain.

11-10 Refer to the circuit of Fig. P11-9. Consider that $h_{ie} = 3570$ ohms, $h_{fe} = 65$, and let $R_L = 10$ K and $R_E = 0$.
   (a) Find the value of the input resistance and compare with the case in which $R_L = 10$ K and $R_E = 0$.
   (b) Compute the value of output resistance.
   (c) Compare the result of part (b) with that commonly encountered in the conventional common-emitter configuration, and comment.
   (d) Determine the value of the ouput voltage appearing across R_E when the signal voltage is sinusoidal and has an rms value of 5 volts.

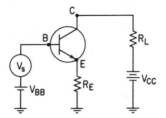

Figure P11-9

11-11 If an emitter-follower circuit is to be found in a transistor radio circuit, where is it likely to be located—immediately following the antenna pickup circuitry or at the other end where the loudspeaker is located? Explain.

11-12 A high-level signal source has an internal resistance of 50 K.
(a) Design a circuit employing a transistor having $h_{ie} = 1.7$ K and $h_{fe} = 44$ that makes the source signal feed into a total input resistance of 150 K.
(b) Find the percentage of the output voltage that appears across the load in the designed circuit of part (a).

11-13 A signal source having an internal resistance of 50 K must feed into a 2 K load resistor.
(a) Design a circuit employing a transistor that will permit the successful transmission of the signal to the load without exessive attenuation.
(b) Find the portion of the signal voltage that appears at the base-emitter terminals of the transistor.
(c) Compute the portion of the signal voltage that appears across the load resistor terminals.

11-14 An amplitude modulated wave is described by the equation

$$e_s = 10(1 + 0.4 \cos 1000t + 0.2 \cos 1800t) \cos 10^6 t$$

(a) List all the frequencies which exist in this wave.
(b) What is the modulation index of each frequency?
(c) Plot the frequency spectrum for this wave.

11-15 An amplitude-modulated wave having the form

$$e = E_m[1 + m \sin \omega_m t] \sin \omega_c t$$

is transmitted to a receiving station where the signal is demodulated by a square-law detector. If the volt-ampere characteristic of the detector is described by $i = ke^2$, find the frequencies which exist in the detector current. What are their relative amplitudes?

11-16 An amplitude modulated wave $e_s = (1 + 0.8 \sin 1000t) \sin 10^6 t$ is received by a square-law detector described by $i = 0.01e_s^2$. If the upper sideband frequency is eliminated from the transmitted wave, find the frequency components in the detector circuit. What are their relative amplitudes?

11-17 Repeat Prob. 11-17 for the case where the lower sideband frequency is removed from the transmitted wave.

11-18 In terms of the binary number system write the truth table for the OR operation represented by $a + b + c = d$.

11-19 Write the truth table in terms of the binary number system for that AND operation which expresses binary multiplication of three variables, i.e., $abc = d$.

11-20 The logic function represented by Fig. P11-20(a) can be written as the OR operation $1 + A = 1$, where 1 denotes the closed state. Write the corresponding logic functions for Figs. P11-20(b), (c), and (d). Consider that O represents the open state.

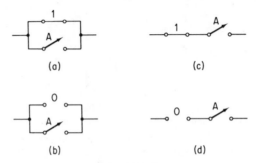

Figure P11-20

11-21 Write the logic function represented by Figs. P11-21(a) and (b). Are these representations identical?

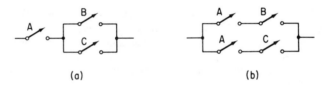

Figure P11-21

11-22 In Fig. P11-22, A, B, and C are inputs, and the output appears at D. This circuit is called a negative AND gate. Explain why and write the appropriate logic function.

11-23 The circuit depicted in Fig. P11-23 is called a positive NOR gate. The inputs are A and B, and the output is C. When the inputs are quiescent, the transistor is in the cutoff state. Explain the operation of this circuit and write the appropriate logic function. Use $\bar{C}$ to denote phase inversion and treat A and B as pulsed inputs.

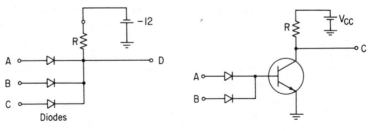

Figure P11-22   Figure P11-23

11-24 Refer to Fig. P11-24.

(a) Identify the logic function performed by this circuit.

(b) Develop the appropriate truth table.

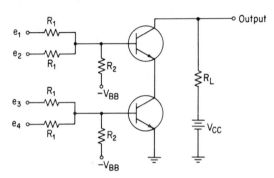

Figure P11-24

III

ELECTROMECHANICAL ENERGY CONVERSION

12

magnetic theory and circuits

An understanding of electromagnetism is essential to the study of electrical engineering because it is the key to the operation of a great part of the electrical apparatus found in industry as well as the home. All electric motors and generators, ranging in size from the fractional horsepower units found in home appliances to the 25,000-hp giants used in some industries, depend upon the electromagnetic field as the coupling device permitting interchange of energy between an electrical system and a mechanical system and vice versa. Similarly, static transformers provide the means for converting energy from one electrical system to another through the medium of a magnetic field. Transformers are to be found in such varied applications as radio and television receivers and electrical power distribution circuits. Other important devices—for example, circuit breakers, automatic switches and relays—require the presence of a confined magnetic field for their proper operation. It is the purpose of this chapter to provide the reader with background so that he can identify a magnetic field and its salient characteristics and more readily understand the function of the magnetic field in electrical equipment.

As has been previously pointed out, the science of electrical engineering is founded on a few fundamental laws derived from basic experiments. In the area of electromagnetism it is Ampere's law that concerns us, and, in fact, serves as the starting point of our treatment.† On the basis of the results obtained by Ampere in 1820, in his experiments on the forces existing between two current-carrying conductors, such quantities as magnetic flux density, magnetic field intensity, permeability, and magnetic flux are readily defined. Once this base is established, attention is then directed to a discussion of the magnetic properties of certain useful engineering materials as well as to the idea of a "magnetic circuit" to help simplify the computations involved in analyzing magnetic devices.

12-1 Ampere's Law—Definition of Magnetic Quantities

Appearing in Fig. 12-1 is a simplified modification of Ampere's experiment. The configuration consists of a very long conductor 1 carrying a constant-magnitude current I_1 and an elemental conductor of length l carrying a constant-magnitude current I_2 in a direction opposite to I_1.

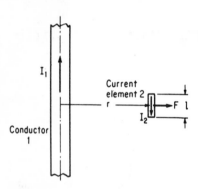

When taken together the elemental conductor and the current I_2 constitute a *current element* $I_2 l$. The elemental conductor 2 is actually part of a closed circuit in which I_2 flows, but for simplicity and convenience the details of the circuit are omitted except for the length l. Moreover, it is assumed that conductors 1 and 2 lie in the same horizontal plane and are parallel to each other. In accordance with Ampere's law it is found that with this configuration there exists a force on the elemental conductor directed to the right. Furthermore, the magnitude of the force is found to be directly proportional to I_1, I_2, l, and the medium surrounding the conductors as well as inversely proportional to the distance between them. In terms of MKS units the magnitude of this force can be shown to be given by

Fig. 12-1. Illustrating the force existing between a current element $I_2 l$ and a very long conductor carrying current I_1 as described by Ampere's law.

$$\boxed{F = \frac{\mu I_1}{2\pi r} I_2 l} \quad \text{newtons} \tag{12-1}$$

where I_1 and I_2 are expressed in amperes, l and r in meters, and μ is a property of the medium. A further interesting revelation about this experiment is that if the elemental conductor 2 is used as an exploring device to find those points in space where the force is of constant magnitude and outwardly directed, the locus is

---

† It should not be here inferred that this is the only starting point in developing a quantitative theory of the magnetic circuit. Faraday's law of induction is equally valid as a starting point, and is preferred when the goal is the development of an electromagnetic wave theory rather than a theory leading to the treatment of electromechanical energy conversion.

found to be a circle of radius r and centered along the axis of conductor 1. In other words it is possible to identify a field having constant *lines of force*. In this connection it is useful at this point to rewrite Eq. (12-1) as follows:

$$F = I_2 l B \tag{12-2}$$

where

$$B \equiv \frac{\mu I_1}{2\pi r} \tag{12-3}$$

From Eq. (12-2) it is obvious that the defined quantity B carries the units of force per current length. In rationalized MKS units this is expressed as

$$B \propto \frac{\text{newton}}{\text{amp} \times \text{meter}} \tag{12-4}$$

where $\propto$ denotes "is proportional to." A closer examination of this relationship reveals an interesting and very useful aspect of the significance of B. Recalling that

$$\text{newton} = \frac{\text{energy}}{\text{length}} = \frac{\text{watt-sec}}{\text{meter}} = \frac{\text{volt} \times \text{amp} \times \text{sec}}{\text{meter}} \tag{12-5}$$

and inserting into expression (12-4) allows B to be expressed alternatively as

$$B \propto \frac{\text{volt} \times \text{sec} \times \text{amp}}{\text{amp} + \text{meter}^2} = \frac{\text{volt} \times \text{sec}}{\text{meter}^2} \tag{12-6}$$

From Faraday's law ($d\phi = e\,dt$), we note also that the units of (volt $\times$ sec) is equivalent to that of flux. Therefore, we can properly conclude that B is in fact a *flux density*, since it has the units of flux per square meter. When the flux is expressed in webers, B then has the units of weber/meter², or *teslas*. Equation (12-3) indicates the factors that determine the magnitude of B, but it serves also as a means of identifying the manner in which the current I_1 influences the force field about the current element. It is significant to note that as long as I_2 is not zero, the force field and the magnetic field have the same characteristics—both have a circular locus and both are vector quantities possessing magnitude and direction. However, because of the way B is defined, the magnetic field exists as long as I_1 is not zero, irrespective of the value of I_2.

In our study of electrical machinery, which comes later, conductor 1 will be referred to as the *field winding* because it sets up the working magnetic field, whereas the total circuit of which the elemental conductor is a part is called the *armature winding*.

The direction of the magnetic field is readily determined by the *right-hand rule* which states that if the field winding conductor (1 in this case) is grasped in the right hand with the thumb pointing in the direction of current flow, the lines of flux (or flux density) will be in the direction in which the fingers wrap around the conductor.

Permeability. A glance at Eq. (12-1) shows that all the factors are known in this equation with the exception of the proportionality factor μ, which is a characteristic of the surrounding medium. Upon repeating the experiment of Fig. 12-1 in iron rather than air, it is found that the force is many times greater for the same values of I_1, I_2, l, and r. Therefore, it follows that μ may be defined from Eq. (12-1) for various media because it is the only unknown quantity. Moreover, because of the way magnetic flux density was defined, Eq. (12-3) indicates that the effect of the surrounding medium may be described in terms of the degree to which it increases or decreases the magnetic flux density for a specified current I_1. Thus when iron rather than free space is the medium it can be said that the iron provides a greater penetration of the magnetic field in a given region, i.e. there is a greater flux density. This property of the surrounding medium in which the conductors are imbedded is called *permeability*.

When the conductors of Fig. 12-1 are assumed placed in a vacuum (free space) and the force is measured for specified values of I_1, I_2, l, and r, the solution for the permeability of free space obtained from Eq. (12-1) and expressed in MKS units comes out to be

$$\mu_0 = 4\pi \times 10^{-7} \tag{12-7}$$

The unit of permeability also follows from Eq. (12-1). Thus

$$\mu_0 \propto \frac{\text{newtons}}{\text{ampere}^2}$$

But

$$\text{newton-meter} = \text{joule} = \text{volt} \times \text{amp} \times \text{sec}$$

or

$$\mu_0 \propto \frac{\text{volt} \times \text{amp} \times \text{sec}}{\text{amp}^2 \times \text{meter}} = \frac{\text{volt} \times \text{sec}}{\text{amp} \times \text{meter}}$$

However, volt $\times$ sec/amp is the unit of inductance expressed in henrys. Accordingly, permeability is expressed in units of henrys/meter.

In those cases where the surrounding medium is other than free space, the absolute permeability is again readily found from Eq. (12-1). A comparison with the result obtained for free space then leads to a quantity called *relative permeability*, μ_r. Expressed mathematically we have

$$\mu_r = \frac{\mu}{\mu_0} \tag{12-8}$$

Equation (12-8) clearly indicates that relative permeability is simply a numeric which expresses the degree to which the magnetic flux density is increased or decreased over that of free space. For some materials, such as Deltamax, the value of μ_r can exceed one hundred thousand. Most ferromagnetic materials, however, have values of μ_r in the hundreds or thousands.

Magnetic Flux ϕ. It is reasonable to expect that since B denotes magnetic flux density, multiplication by the effective area that B penetrates should yield the total *magnetic flux*. To illustrate this point refer to Fig. 12-2, which shows a coil of area ab lying in the same horizontal plane containing conductor 1. We already know that when a current I_2 flows through this conductor a magnetic field is

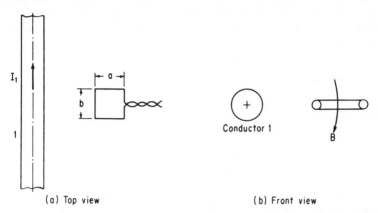

Fig. 12-2. Associating an area with a magnetic field B to identify a magnetic flux.

created in space and is specifically described by Eq. (12-3). To find the total flux penetrating the coil it is necessary merely to perform an integration of B over the surface area involved. Of course if B were a constant over the area of concern, the flux would be simply the product of B and the area ab. Next consider that the plane of the coil is titled with respect to the plane of conductor 1 by 60°, as depicted in Fig. 12-3. Clearly now the total flux penetrating the coil is reduced by a factor of one-half. If the coil is oriented to a position of 90° with respect to the horizontal plane, no flux threads the coil.

On the basis of these observations, then, the magnetic flux through any surface is more rigorously defined as the surface integral of the normal component of the vector magnetic field B. Expressed mathematically we have

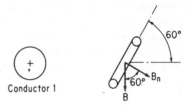

Fig. 12-3. Same as Fig. 12-2(b) except that the coil is tilted 60° relative to the horizontal plane.

$$\phi = \int_s B_n \, dA$$ (12-9)

where s stands for surface integral, A represents the area of the coil, and B_n is the normal component of B to the coil area. From expression (12-6) we know that magnetic flux must have the dimensions of *volt-seconds*. However, this is more commonly called *webers*. The volt-seconds unit of flux is better understood in terms of Faraday's law of induction.

Magnetic Field Intensity H. Often in magnetic circuit computations it is helpful to work with a quantity representing the magnetic field which is independent of the medium in which the magnetic flux exists. This is especially true in situations such as are found in electrical machinery where a common flux penetrates several different materials, including air. A glance at Eq. (12-3) discloses that division of B by μ identifies such a quantity. Accordingly, magnetic field intensity is defined as

$$H \equiv \frac{B}{\mu}$$ (12-10)

and has the units of

$$\frac{\dfrac{\text{newtons}}{\text{amp} \times \text{meter}}}{\dfrac{\text{newtons}}{\text{amp}^2}} = \frac{\text{amperes}}{\text{meter}}$$

Thus H is dependent upon the current that produces it and also on the geometry of the configuration but not the medium. For the system of Fig. 12-1 the value of the magnetic field intensity immediately follows from Eq. (12-3) and is given by

$$H = \frac{B}{\mu} = \frac{I_1}{2\pi r}$$ (12-11)

Because H is independent of the medium it is frequently looked upon as the intensity that is responsible for driving the flux density through the medium. Actually, though, H is a derived result.

More generally the units for H are ampere-turns/meter rather than amperes/meter. This is apparent whenever the field winding is made up of more than just a single conductor.

Ampere's Circuital Law. Now that the magnetic field intensity has been defined and shown to have dimensions of ampere-turns/meter, we shall develop a very useful relationship. Recall that H is a vector having the same direction as the magnetic field B. For the configuration of Fig. 12-1 H has the same circular locus as B. A line integration of H along any given closed circular path proves interesting. Of course the line integral is considered because H involves a per unit length dimension. Thus

$$\oint H\,dl = \int_0^{2\pi r} \frac{I_1}{2\pi r}\,dl = I_1 \text{ amp}$$ (12-12)

(Again keep in mind that the units here would be ampere-turns if more than one conductor were involved in Fig. 12-1.) Equation (12-12) states that the closed line integral of the magnetic field intensity is equal to the enclosed current (or ampere-turns) that produces the magnetic field lines. This relationship is called *Ampere's circuital law* and is more generally written as

$$\oint H\,dl = \mathscr{F}$$ (12-13)

where $\mathscr{F}$ denotes the ampere-turns enclosed by the assumed closed flux line path. The quantity $\mathscr{F}$ is also known as the *magnetomotive force* and frequently abbreviated as mmf. This relationship is useful in the study of electromagnetic devices and is referred to in subsequent chapters.

Derived Relationships. In the preceding pages the fundamental magnetic quantities—flux density, flux, field intensity and permeability—are defined starting with Ampere's basic experiment involving two current-carrying conductors. By the appropriate manipulation of these quantities additional useful results can be obtained.

Equation (12-10) is a vector equation describing the magnetic field intensity for a given geometry and current. If the total path length of a flux line is assumed to be L, then the total magnetomotive force associated with the specified flux line is

$$\mathscr{F} = HL = \frac{B}{\mu}L \tag{12-14}$$

Now in those situations where B is a constant and penetrates a fixed, known area A, the corresponding magnetic flux may be written as

$$\boxed{\phi = BA} \tag{12-15}$$

Inserting Eq. (12-15) into Eq. (12-14) yields

$$\mathscr{F} = HL = \phi\left(\frac{L}{\mu A}\right) \tag{12-16}$$

The quantity in parentheses in this last expression is interesting because it bears a very strong resemblance to the definition of resistance in an electric circuit. Refer to Eq. (2-16). Recall that the resistance in an electric circuit represents an impediment to the flow of current under the influence of a driving voltage. An examination of Eq. (12-16) provides a similar interpretation for the magnetic circuit. We are already aware that $\mathscr{F}$ is the driving magnetomotive force which creates the flux ϕ penetrating the specified cross-sectional area A. However, this flux is limited in value by what is called the *reluctance* of the magnetic circuit, which is defined as

$$\boxed{\mathscr{R} = \frac{L}{\mu A}} \tag{12-17}$$

No specific name is given to the dimension of reluctance except to refer to it as so many units of reluctance.

Equation (12-17) reveals that the impediment to the flow of flux which a magnetic circuit presents is directly proportional to the length and inversely proportional to the permeability and the cross-sectional area—results which are entirely consistent with physical reasoning.

Inserting Eq. (12-17) into Eq. (12-16) yields

$$\boxed{\mathscr{F} = \phi\mathscr{R}} \tag{12-18}$$

which is often referred to as the Ohm's law of the magnetic circuit. It is important to keep in mind, however, that these manipulations in the forms shown are permissible as long as B and A are fixed quantities.

Ampere's Law for Various Orientations of the Current Element. In Fig. 12-1 the assumption was made that the current element was located parallel to conductor 1 and lying in the same plane. Because this orientation was sufficient for the purpose at hand—to define the fundamental magnetic quantities—it was pursued as a matter of convenience. However, in the interest of furnishing a more complete picture of the experiment we shall now consider the effect on the force of placing the current element, $I_2 l$, in two additional different orientations. Consider first that the current element is no longer placed parallel to conductor 1 but continues to be located in the same horizontal plane. Refer to Fig. 12-4. The dots in this

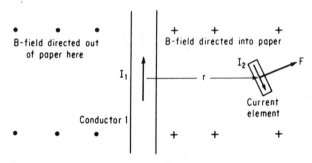

Fig. 12-4. Showing the direction of the force when the current element is no longer located parallel to conductor 1 but remains in the same plane.

figure indicate that the magnetic field is directed outward on the left side of conductor 1 and inward (with respect to the plane of the paper) on the right side of the conductor as indicated by the right-hand rule. The results of this experiment show that the magnitude of the force is the same as that found by using the configuration of Fig. 12-1.† This conclusion is not surprising because the value of the magnetic flux density as well as I_2 and l remain unchanged so that Eq. (12-2) is still valid in describing the force. The only change is the direction of the force. However, as Fig. 12-4 indicates, the force continues to be normal to the current element. It is worthwhile to keep this point in mind. Presently a general rule for establishing the force direction for all configurations will be described.

Next let us consider that orientation of the current element which places it parallel to conductor 1 but inclined at an angle $\theta = 30°$ with respect to the vertical. A side-view projection of the configuration is depicted in Fig. 12-5. Note that the magnetic field is directed downward along the vertical for this view. Actually, of course, the locus of B is circular, but in Fig. 12-5 we are looking at just that small portion of the B-field about the plane containing conductor 1.

----

† Because of the infinitesimal nature of the current element, the value of B is assumed to be constant about $I_2 l$ irrespective of its orientation.

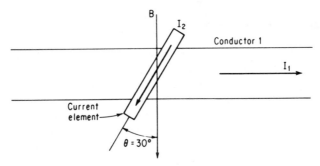

Fig. 12-5. Side-view projection of Fig. 12-1 but with the current element tilted relative to the horizontal plane. Force is directed out of paper.

With this configuration the force on the current element is found to have the same direction but one half the magnitude of that obtained with the orientation of Fig. 12-1. It follows, then, that the angle between the current vector I_2l and the flux density B affects the magnitude of the force. As a matter of fact, further experimentation reveals that the general expression for the force is

$$F = I_2lB \sin \theta \quad \text{newtons} \tag{12-19}$$

Equation (12-19) conveys information solely about the magnitude of the force and not its direction. It is possible, however, by employing the notation of vector analysis, to rewrite Eq. (12-19) so that information about magnitude as well as direction is present. This result is readily accomplished by the use of the *cross-product* notation between two vectors, yielding a third vector having magnitude and direction. Thus Eq. (12-19) is more completely expressed as

$$\bar{F} = I_2\bar{l} \times \bar{B} \quad \text{newtons} \tag{12-20}$$

The cross symbol must always be understood to involve the *sine* of the angle between the two vectors $\bar{B}$ and $\bar{l}$ (or the direction of I_2, which is determined by the orientation of $\bar{l}$). Moreover, wherever the cross product is involved, the direction of the resultant vector is always normal to the plane containing the vectors $\bar{B}$ and $\bar{l}$ in the sense determined by the direction of advance of a right-hand screw as $\bar{l}$ is turned into $\bar{B}$ through the smaller of the two angles made by the vectors. Accordingly, in the configuration of Fig. 12-5 the direction of the force is found by turning $I_2\bar{l}$ into $\bar{B}$ and then noting that this would cause a right-hand screw to advance out of the plane of the paper. Hence the force is directed outward, and this corresponds with the experimentally established result.

It is also possible to determine the direction of the force by means of another right-hand rule, which requires that the forefinger be put in the direction of the current and the middle finger in the direction of $\bar{B}$ and the two assumed lying in the same plane. The thumb of the right hand then points in the direction of the force when placed perpendicular to the other two fingers.

12-2 Theory of Magnetism

In order to understand the magnetic behavior of materials, it is necessary to take a microscopic view of matter. A suitable starting point is the composition of the atom, which Bohr described as consisting of a heavy nucleus and a number of electrons moving around the nucleus in specific orbits. Closer investigation reveals that the atom of any substance experiences a torque when placed in a magnetic field; this is called a *magnetic moment*. The resultant magnetic moment of an atom depends upon three factors—the positive charge of the nucleus spinning on its axis, the negative charge of the electron spinning on its axis, and the effect of the electrons moving in their orbits. The magnetic moment of the spin and orbital motions of the electron far exceeds that of the spinning proton. However, this magnetic moment can be affected by the presence of an adjacent atom. Accordingly, if two hydrogen atoms are combined to form a hydrogen *molecule*, it is found that the electron spins, the proton spins, and the orbital motions of the electrons of each atom oppose each other so that a resultant magnetic moment of zero should be expected. Although this is almost the case, experiment reveals that the relative permeability of hydrogen is not equal to one but rather is very slightly less than unity. In other words, the molecular reaction is such that when hydrogen is the medium there is a slight decrease in the magnetic field compared to free space. This behavior occurs because there is a precessional motion of all rotating charges about the field direction, and the effect of this precession is to set up a field opposed to the applied field regardless of the direction of spin or orbital motion. Materials in which this behavior manifests itself are called *diamagnetic* for obvious reasons. Besides hydrogen, other materials possessing this characteristic are silver and copper.

Continuing further with the hydrogen molecule, let us assume next that it is made to lose an electron, thus yielding the hydrogen ion. Clearly, complete neutralization of the spin and orbital electron motions no longer takes place. In fact when a magnetic field is applied, the ion is so oriented that its net magnetic moment aligns itself with the field, thereby causing a slight increase in flux density. This behavior is described as *paramagnetism* and is characteristic of such materials as aluminum and platinum. Paramagnetic materials have a relative permeability slightly in excess of unity.

So far we have considered those elements whose magnetic properties differ only very slightly from those of free space. As a matter of fact the vast majority of materials fall within this category. However, there is one class of materials—principally iron and its alloys with nickel, cobalt, and aluminum—for which the relative permeability is very many times greater than that of free space. These materials are called *ferromagnetic*† and are of great importance in electrical engineering. We may ask at this point why iron (and its alloys) is so very much more magnetic than other elements. Essentially, the answer is provided by the

† Derived from the Latin word for iron—*ferrum*.

domain theory of magnetism. Like all metals, iron is crystalline in structure with the atoms arranged in a space lattice. However, domains are subcrystalline particles of varying sizes and shapes containing about 10^{15} atoms in a volume of approximately 10^{-9} cubic centimeters. *The distinguishing feature of the domain is that magnetic moments of its constituent atoms are all aligned in the same direction.* Thus in a ferromagnetic material not only must there exist a magnetic moment due to a nonneutralized spin of an electron in an inner orbit, but also the resultant spin of all neighboring atoms in the domain must be parallel.

It would seem by the explanation so far that if iron is composed of completely magnetized domains then the iron should be in a state of complete magnetization throughout the body of material even without the application of a magnetizing force. Actually, this is not the case, because the domains act independently of each other, and for a specimen of unmagnetized iron these domains are aligned haphazardly in all directions so that the net magnetic moment is zero over the specimen. Figure 12-6 illustrates the situation diagrammatically in a simplified fashion. Because of the crystal lattice structure of iron the "easy" direction of domain alignment can take place in any one of six directions—left, right, up, down, out or in—depending upon the direction of the applied magnetizing force. Figure 12-6(a) shows the unmagnetized configuration. Figure 12-6(b)

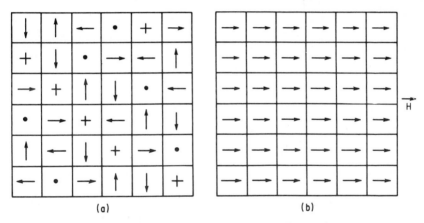

              (a)   (b)

Fig. 12-6. Representation of a ferromagnetic crystal:
        (a) unmagnetized and
        (b) fully magnetized by the field H.

depicts the result of applying a force from left to right of such magnitude as to effect alignment of all the domains. When this state is reached the iron is said to be *saturated*—there is no further increase in flux density over that of free space for further increases in magnetizing force.

Large increases in the temperature of a magnetized piece of iron bring about a decrease in its magnetizing capability. The temperature increase enforces the agitation existing between atoms until at a temperature of 750°C the agitation is so severe that it destroys the parallelism existing between the magnetic moments of the neighboring atoms of the domain and thereby causes it to lose its magnetic property. The temperature at which this occurs is called the *Curie point*.

12-3   Magnetization Curves of Ferromagnetic Materials

If the experiment of Fig. 12-1 is repeated with iron or steel as the medium for increasing values of the field winding current I_1 and the corresponding values of μ computed, it is found that the relative permeability varies considerably with the magnetizing force that establishes the operating flux density. A typical variation of μ_r for cast steel appears in Fig. 12-7. Here μ_r is plotted versus flux density rather than the magnetizing force because it is the onset of the realignment of more and more domains that brings about the change in permeability.

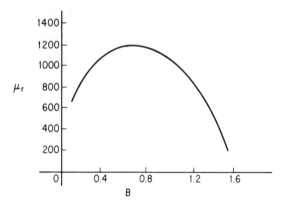

Fig. 12-7. Graph of relative permeability versus flux density.

Unfortunately, the state of development of the theory of magnetism is not so far advanced that it allows the prediction of the magnetic properties of a material on a purely theoretical basis even though the exact composition of the material is known. For example, with the present theory it is not possible to say exactly what the flux density will be in a given specimen of iron for a specified value of the magnetizing force. Rather, it is customary to obtain this information by consulting technical and descriptive bulletins where the measured magnetic properties of a representative sample of the specimen are published. These bulletins are made available to users by the manufacturers of magnetic steels and they include information on such varied shapes and forms as sheets, wires, bars, and even castings weighing up to hundreds of tons.

Usually the published magnetic characteristics of the various iron and steel samples are presented as plots of flux density B as a function of the magnetic field intensity H. In the interest of presenting a complete graphical picture of the functional relationship existing between these two quantities as well as to define

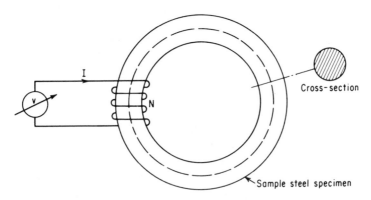

Fig. 12-8. Obtaining the magnetization curve of a sample steel specimen.

additional terms used in this connection, refer to Fig. 12-8. Assume that the steel specimen is initially unmagnetized and is in the form of a toroidal ring with a coil of N turns wrapped around it. Assume too that the coil can be energized from a variable voltage source capable of furnishing current flow in either direction in the coil. As the current I is increased from zero in the positive direction (current flowing into the top terminal of the coil), an increasing magnetic flux ϕ can be measured taking place within the body of the toroid in the clockwise direction. For any fixed value of I there is a specific value of flux. Then by Eq. (12-15) the corresponding flux density is determined since the toroidal cross-sectional area is known. Moreover, the magnetomotive force NI can be replaced by HL_m in accordance with Ampere's circuital law [Eq. (12-13)] where L_m is the mean length of path of the toroid (as shown by the broken line circle in Fig. 12-8). The two fundamental quantities involved in this arrangement then are B in teslas and H expressed in ampere-turns per meter. H is the quantity we want to deal with rather than magnetomotive force because for the same flux density, doubling the mean magnetic length will not change H but will require doubling the magnetomotive force. The conclusion to be drawn is that a plot of B versus H is a universal plot for the given material because it can be extended to any geometry of cross-sectional area and length. In contrast, a plot of ϕ versus NI is limited to a single geometrical configuration. Therefore, a plot of the magnetic characteristics of a material always involves plotting B versus H. For the virgin sample of Fig. 12-8 the graph of B versus H follows the curve Oa of Fig. 12-9 for field intensities up to H_a. Take note of the nonlinear relationship existing between these two quantities.

Another interesting characteristic of ferromagnetic materials is revealed when the field intensity, having been increased to some value, say H_a, is subsequently decreased. It is found that the material opposes demagnetization and, accordingly, does not retrace along the magnetizing curve Oa but rather along a curve located above Oa. See curve ab in Fig. 12-9. Furthermore, it is seen that when the field intensity is returned to zero, the flux density is no longer zero as was the case with the virgin sample. This happens because some of the domains

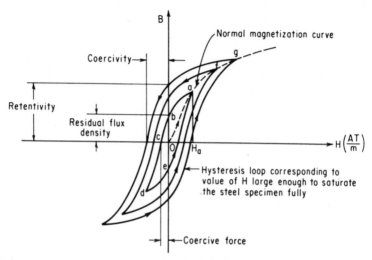

Fig. 12-9. Typical hysteresis loops and normal magnetization curve.

remain oriented in the direction of the originally applied field. The value of B that remains after the field intensity H is removed is called *residual flux density*. Moreover, its value varies with the extent to which the material is magnetized. The maximum possible value of the residual flux density is called *retentivity* and results whenever values of H are used that cause complete saturation.

Frequently, in engineering applications of ferromagnetic materials, the steel is subjected to cyclically varying values of H having the same positive and negative limits. As H varies through many identical cycles, the graph of B versus H gradually approaches a fixed closed curve as depicted in Fig. 12-9. The loop is always traversed in the direction indicated by the arrows. Since time is the implicit variable for these loops, note that B is always lagging behind H. Thus, when H is zero, B is finite and positive, as at point b, and when B is zero, as at c, H is finite and negative, and so forth. This tendency of the flux density to *lag behind* the field intensity when the ferromagnetic material is in a symmetrically cyclically magnetized condition is called *hysteresis*† and the closed curve *abcdea* is called a *hysteresis loop*. Moreover, when the material is in this cyclic condition the amount of magnetic field intensity required to reduce the residual flux density to zero is called the *coercive force*. Usually, the larger the residual flux density, the larger must be the coercive force. The maximum value of the coercive force is called the *coercivity*.

A glance at the hysteresis loops of Fig. 12-9 makes it quite evident that the flux density corresponding to a particular field intensity is not single-valued. Its value lies between certain limits depending upon the previous history of the ferromagnetic material. However, since in many situations involving magnetic

---

† Derives from the Greek *hysterein* meaning to be behind or to lag.

devices this previous history is unknown, a compromise procedure is used in making magnetic calculations by working with a single-valued curve called the *normal magnetization curve*. This curve is found by drawing a curve through the tips of a group of hysteresis loops generated while in a cyclic condition. Such a curve is *Oafg* in Fig. 12-9. Typical normal magnetization curves of commonly used ferromagnetic materials appear in Fig. 12-10.

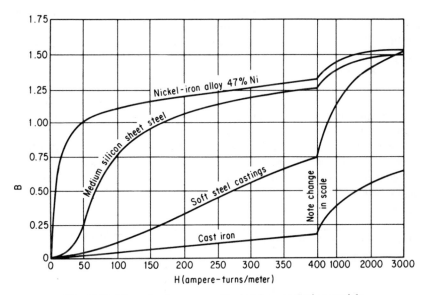

Fig. 12-10. Magnetization curves of typical ferromagnetic materials.

A final observation is in order at this point. By Eq. (12-10) the permeability of a material may be expressed as a ratio of B to H. Coupling this with the non-linear variation existing between B and H (see Fig. 12-9) bears out the variation of permeability with flux density as already cited in connection with Fig. 12-7. As a matter of fact, for a material in a cyclic condition the permeability is nothing more than the ratio of B to H for the various points along the hysteresis loop.

12-4 The Magnetic Circuit: Concept and Analogies

In general, problems involving magnetic devices are basically field problems because they are concerned with quantities such as ϕ and B which occupy three-dimensional space. Fortunately, however, in most instances the bulk of the space of interest to the engineer is occupied by ferromagnetic materials except for small air gaps which are present either by intention or by necessity. For example, in electromechanical energy-conversion devices the magnetic flux must permeate a stationary as well as a rotating mass of ferromagnetic material, thus making an air gap indispensable. On the other hand, in other devices an air gap may be intentionally inserted in order to mask the nonlinear relationship existing between *B* and *H*. But in spite of the presence of air gaps it happens that the space occupied by the magnetic field and the space occupied by the ferromagnetic material are practically the same. Usually this is because air gaps are made as small as mechanical clearance between rotating and stationary members will allow and also because the iron by virtue of its high permeability confines the flux to itself as copper wire confines electric current or a pipe restricts water. On this basis the three-dimensional field problem becomes a one-dimensional circuit problem and in accordance with Eq. (12-18) leads to the idea of a *magnetic circuit*. Thus we can look upon the magnetic circuit as consisting predominantly of iron paths of specified geometry which serves to confine the flux; air gaps may be included. Figure 12-11 shows a typical magnetic circuit consisting chiefly of iron. Note that the magnetomotive force of the coil produces a flux which is confined to the iron and to that part of the air having effectively the same cross-sectional area as the iron. Furthermore, a little thought reveals that this magnetic circuit may be replaced by a single-line equivalent circuit as depicted in Fig. 12-12. As suggested by Eqs. (12-17) and (12-18) the equivalent circuit consists of the magnetomotive force driving flux through two series-connected reluctances—$\mathcal{R}_i$, the reluctance of the iron, and $\mathcal{R}_a$, the reluctance of the air.

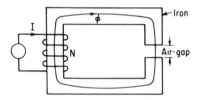

Fig. 12-11. Typical magnetic circuit involving iron and air.

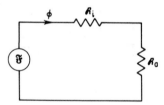

Fig. 12-12. Single-line equivalent circuit of Fig. 12-11.

This analogy of the magnetic circuit with the electric circuit carries through in many other respects. For the sake of completeness these details are presented below for the case of a toroidal copper ring and a toroidal iron ring having the same mean radius *r* and cross-sectional area *A*.

ELECTRIC CASE

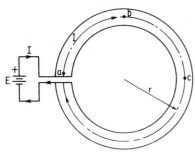

MAGNETIC CASE

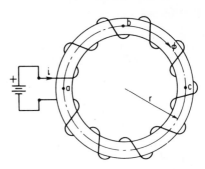

The toroidal copper ring is assumed open by an infinitesimal amount with the ends connected to a battery; a current of I amperes flows through the ring.

The toroidal iron ring is assumed wound with N turns of wire so that with a current i flowing through it the magnetomotive force creates the flux ϕ.

Driving Force

applied battery voltage $= E$

applied ampere-turns $= \mathcal{F}$

Response

$$\text{current} = \frac{\text{driving force}}{\text{electric resistance}}$$

or

$$I = \frac{E}{R}$$

$$\text{flux} = \frac{\text{driving force}}{\text{magnetic reluctance}}$$

or

$$\phi = \frac{\mathcal{F}}{\mathcal{R}} \qquad (12\text{-}21)$$

Impedance

Impedance is a general term used to indicate the impediment to a driving force in establishing a response.

$$\text{resistance} = R = \rho\frac{L}{A}$$

$$\text{reluctance} = \mathcal{R} = \frac{L}{\mu A} \qquad (12\text{-}22)$$

where $L = 2\pi r =$ mean length of turn of the toroid and A is the toroidal cross-sectional area.

where $L = 2\pi r =$ mean length of turn of the toroid and A is the toroidal cross-sectional area.

Equivalent Circuit

$$E = IR$$

$$\mathcal{F} = \phi\mathcal{R} \qquad (12\text{-}23)$$

Electric Field Intensity

With the application of the voltage E to the homogeneous copper toroid, there is produced within the material an electric potential gradient given by

$$\mathscr{E} \equiv \frac{E}{L} = \frac{E}{2\pi r} \quad \text{volts/meter}$$

This electric field must occur in a closed path if it is to be maintained. It then follows that the closed line integral of $\mathscr{E}$ is equal to the battery voltage E. Thus

$$\oint \mathscr{E}\, dl = E$$

Voltage Drop

If it is desired to find the voltage drop occurring between two points—a and b—of the copper toroid we may write:

$$V_{ab} = \int_a^b \mathscr{E}\, dl = \frac{E}{L} \int_a^b dl = \frac{IR}{L} l_{ab}$$
$$= \frac{I}{L} \rho \frac{L}{A} l_{ab} = I\rho \frac{l_{ab}}{A} = IR_{ab}$$

i.e.,

$$V_{ab} = IR_{ab}$$

where R_{ab} is the resistance of the copper toroid between points a and b.

Current Density

By definition, current density is the amount of amperes per unit area. Thus

$$J \equiv \frac{I}{A} = \frac{E}{AR} = \frac{\mathscr{E}L}{Ap\dfrac{L}{A}} = \frac{\mathscr{E}}{\rho}$$

or

$$\mathscr{E} = \rho J$$

This last expression is often referred to as the *microscopic* form of Ohm's law.

Magnetic Field Intensity

When a magnetomotive force is applied to the homogeneous iron toroid, there is produced within the material a magnetic potential gradient given by

$$H \equiv \frac{\mathscr{F}}{L} = \frac{\mathscr{F}}{2\pi r} \quad \text{AT/meter} \quad (12\text{-}24)$$

As already pointed out in connection with Ampere's circuital law, the closed line integral of H equals the enclosed magnetomotive force. Thus

$$\oint H\, dl = \mathscr{F} \quad (12\text{-}25)$$

mmf Drop

The letters *mmf* are used to represent magnetomotive force. The portion of the total applied *mmf* appearing between points a and b is found similarly:

$$\mathscr{F}_{ab} = \int_a^b H\, dl = \frac{F}{L} l_{ab} = \frac{\phi \mathscr{R}}{L} l_{ab}$$
$$= \frac{\phi}{L} \frac{L}{\mu A} l_{ab} = \phi \frac{l_{ab}}{\mu A} = \phi \mathscr{R}_{ab}$$

$$\mathscr{F}_{ab} = \phi \mathscr{R}_{ab} \quad (12\text{-}26)$$

where $\mathscr{R}_{ab}$ is the reluctance of the iron toroid between points a and b.

Flux Density

Flux density is expressed as webers per unit area. Thus

$$B = \frac{\phi}{A} = \frac{\mathscr{F}}{A\mathscr{R}} = \frac{HL}{A\dfrac{L}{\mu A}} = \mu H$$

or

$$H = \frac{B}{\mu} \quad (12\text{-}27)$$

It should not be inferred from the foregoing that electric and magnetic circuits are analogous in *all* respects. For example, there are no magnetic insulators analogous to those known to exist for electric circuits. Also, when a direct current is established and maintained in an electric circuit, energy must be continuously supplied. An analogous situation does not prevail in the magnetic case, where a flux is established and maintained constant.

12-5 Units for Magnetic Circuit Calculations

Magnetic circuit calculations can be carried out by use of any one of several different systems of units. These various systems arose initially because it was thought that the phenomena of electricity and magnetism were unrelated—thereby leading to the development of a separate system of units for each—and secondly, because of the desire to deal with practical values of the units once the relationship was discovered. Up to now attention has been given exclusively to the MKS (meter-kilogram-second) system of units as developed by Giorgi about the turn of the twentieth century. This policy is prompted by the acceptance in 1935 of the MKS system of units as the standard for scientific work by the International Electrotechnical Commission. However, a good part of the past literature is written in terms of the units of the CGS (centimeter-gram-second) system. Furthermore, many of the present-day computations are carried on in terms of the *mixed* system employing such units as ampere-turns/inch, maxwells/inch^2, and ampere-turns because of the convenience they offer in dealing with dimensions that are expressed in inches. For these reasons the units of all three systems are shown in Table 12-1. Note that two columns appear for the MKS system—the rationalized and the unrationalized. The rationalized MKS system is preferred. This matter of "rationalizing" merely involves the handling of the factor 4π. In the rationalized case the 4π is removed from the magnetomotive force and put instead with the permeability of free space. The end results, of course, remain unaffected, as a glance at the expression for ϕ in the table reveals.

The weber, which is the unit of flux in the MKS system, is equal to 10^8 *maxwells* (or lines) where the maxwell is the unit of flux in the CGS system. The *gilbert* is the CGS unit for mmf and is equal to 0.4π times the number of ampere-turns. The CGS unit for magnetic field intensity H is the *oersted* (or gilbert/cm) and the CGS unit for flux density B is the *gauss* (or lines/cm^2) and the MKS unit for B is the *tesla* (or wb/m^2). The relationships existing for the same quantity between the various systems of units are given in the last column of the table.

12-6 Magnetic Circuit Computations

Basically magnetic circuit calculations involving ferromagnetic materials fall into two categories. In the first the value of the flux is known and it is required to find the magnetomotive force to produce it. This is the situation typical of the

Table 12-1. Units

Quantity	Symbol	CGS Unit	CGS Relation	MKS unrationalized Unit	MKS unrationalized Relation	MKS rationalized Unit	MKS rationalized Relation	Mixed English system Unit	Mixed English system Relation	Conversion factors
mmf	$\mathscr{F}$	gilberts	$\mathscr{F} = 0.4\pi NI$	pragilbert	$\mathscr{F} = 4\pi NI$	amp-turn	$\mathscr{F} = NI$	amp-turn	$\mathscr{F} = NI$	pragilbert = 10 gilberts (AT) = 0.4π gilb = 1.257 gilb
Permeability: Free space Abs. nor. per.	μ_0 μ		$\mu_0 = 1$ $\mu = \mu_0\mu_r = \mu_r$		$\mu_0 = 10^{-7}$ $\mu = 10^{-7}\mu_r$		$\mu_0 = 4\pi10^{-7}$ $\mu = 4\pi10^{-7}\mu_r$		$\mu_0 = 3.19$ $\mu = 3.19\mu_r$	
Length	l	cm		meter		meter		inch		1 cm = 0.01 m 1 m = 39.4 in.
Area	A	cm²		m²		m²		in.²		1 m² = 1550 in.²
Reluctance	$\mathscr{R}$		$\mathscr{R} = \dfrac{l}{\mu_r A}$		$\mathscr{R} = \dfrac{l}{\mu_r 10^{-7}A}$		$\mathscr{R} = \dfrac{l}{4\pi10^{-7}\mu_r A}$		$\mathscr{R} = \dfrac{l}{3.19\mu_r A}$	
Flux	ϕ	maxwells = lines	$\phi = \dfrac{0.4\pi NI}{\frac{l}{\mu_r A}}$	weber	$\phi = \dfrac{4\pi NI}{\frac{l}{\mu_r 10^{-7}A}}$	weber	$\phi = \dfrac{NI}{\frac{l}{\mu_r 4\pi10^{-7}A}}$	maxwell lines	$\phi = \dfrac{NI}{\frac{l}{3.19\mu_r A}}$	1 weber = 10⁸ lines
Magnetic field intensity	H	glib/cm = oersteds	$H = \dfrac{0.4\pi NI}{l}$	pragilb/m = pra-oersteds	$H = \dfrac{4\pi NI}{l}$	amp-turn/meter	$H = \dfrac{NI}{l}$	amp-turn/inch	$H = \dfrac{NI}{l}$	1 oersted = 79.6 AT/m 1 praoersted = 1000 orsteds $1 \dfrac{AT}{in.} = \dfrac{0.4\pi}{2.54} = 0.495$ oers. $1 \dfrac{AT}{in.} = \dfrac{2.02}{1000}$ praoersteds
Flux density	B	lines/cm² = gausses	$B = \mu_r H$	webers/m² = teslas	$B = \mu_0\mu_r H$	teslas	$B = \mu_0 H$	lines/in.²	$B = 3.19\mu_r H$	1 gauss = 6.45 lines/in² 1 tesla = 64,500 lines/in² 1 tesla = 10,000 gausses

design of a-c and d-c electromechanical energy converters. On the basis of the desired voltage rating of an electric generator or the torque rating of an electric motor information about the required magnetic flux is readily obtained. Then with this knowledge and the configuration of the magnetic circuit the total mmf needed to establish the flux is determined straightforwardly. In the second case it is the flux for which we must solve, knowing the geometry of the magnetic circuit and the applied mmf. An engineering application in which this situation prevails is the magnetic amplifier, where it is often necessary to find the resultant magnetic flux caused by one or more control windings. Because the reluctance (or permeability) of the ferromagnetic material is not constant, the solution of this problem is considerably more involved than that of the first category as illustrated by the examples below.

EXAMPLE 12-1   A toroid is composed of three ferromagnetic materials and is equipped with a coil having 100 turns as depicted in Fig. 12-13. Material a is a nickel-iron alloy having a mean arc length L_a of 0.3 meter.
Material b is medium silicon steel and has a
mean arc length L_b of 0.2 meter. Material c is
of cast steel having a mean arc length equal
to 0.1 meter. Each material has a cross-
sectional area of 0.001 square meter.

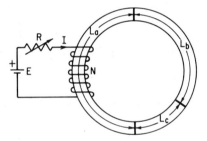

    (a) Find the magnetomotive force need-
ed to establish a magnetic flux of
$\phi = 6 \times 10^{-4}$ weber = 60,000 lines.
    (b) What current must be made to flow
through the coil?
    (c) Compute the relative permeability
and reluctance of each ferromagnetic
material.

Fig. 12-13. Toroid composed of three different materials.

solution: (a) To obtain the total mmf of the coil all we need to do is to apply Ampere's circuital law. Thus

$$\mathscr{F} = \mathscr{F}_a + \mathscr{F}_b + \mathscr{F}_c = H_a L_a + H_b L_b + H_c L_c$$

The unknown quantities here are H_a, H_b, and H_c. These can readily be found from a knowledge of the flux density, which here is the same for each section because the flux is common and the cross-sectional areas are the same. Hence

$$B_a = B_b = B_c = \frac{\phi}{A} = \frac{0.0006}{0.001} = 0.6 \text{ tesla}$$

Now H_a is found by entering the B–H curve of the nickel-iron alloy of Fig. 12-10 corresponding to $B_a = 0.6$. This yields

$$H_a = 10 \text{ AT/meter}$$

Similarly

$$H_b = 66 \text{ AT/m}$$
$$H_c = 324 \text{ AT/m}$$

Accordingly, the total required mmf is

$$\mathscr{F} = H_a L_a + H_b L_c + H_c L_b$$
$$= (10)(0.3) + 66(0.2) + 324(0.1)$$
$$= 3 + 13.2 + 32.4 = 48.6 \text{ AT}$$

Note that although the path length of cast steel is the smallest, it nonetheless requires the greatest portion of the mmf to force the specified flux through. This happens because of its much lower permeability as shown in part (c).

(b) In the rationalized MKS system the mmf is equal to the number of ampere-turns. Hence

$$I = \frac{\mathscr{F}}{N} = \frac{48.6}{100} = 0.486 \text{ amp}$$

(c) From Eq. (12-10)

$$\mu_a = \frac{B_a}{H_a} = \frac{0.6}{10} = 0.06 \text{ henry/m}$$

Also

$$\mu_{ra}\mu_0 = \mu_a$$

$$\therefore \mu_{ra} = \frac{\mu_a}{\mu_0} = \frac{0.06}{4\pi \times 10^{-7}} = 47{,}700$$

Furthermore, from Eq. (12-16) the reluctance is found to be

$$\mathscr{R}_a = \frac{\mathscr{F}_a}{\phi} = \frac{3}{6 \times 10^{-4}} = 5000 \text{ rationalized MKS units of reluctance}$$

Proceeding in a similar fashion for materials b and c leads to the following results:

$$\mu_{rb} = 7240, \qquad \mathscr{R}_b = 22{,}000$$
$$\mu_{rc} = 1470, \qquad \mathscr{R}_c = 54{,}000$$

Next we consider the more difficult problem: that of finding the flux in a given magnetic circuit corresponding to a specified mmf. The solution cannot be arrived at directly because, as a result of the nonlinear relationship between B and H, there are too many unknowns. The easiest way of finding the solution is to employ a cut-and-try procedure guided by a knowledge of the permeability characteristics of the materials such as appears in their magnetization curves. The following example illustrates the technique involved.

EXAMPLE 12-2   For the toroid of Example 12-1, shown in Fig. 12-13, find the magnetic flux produced by an applied magnetomotive force of $\mathscr{F} = 35$ ampere-turns.

solution: The solution cannot be determined directly because to do so we must know the reluctance of each part of the magnetic circuit, which can be known only if the flux density is known—which means that ϕ must be known right at the start. This is clearly impossible.

To obtain the solution by the cut-and-try procedure, we begin by first assuming that all of the applied mmf appears across the material having the highest reluctance. This yields an approximate value of ϕ which can subsequently be refined. A glance at Fig. 12-10 shows that the poorest magnetic "conductor" is cast steel. Hence by assuming the entire mmf to appear across material c we can find H, from which B follows, which in turn yields ϕ. Thus

$$H_c = \frac{\mathscr{F}_c}{L_c} = \frac{\mathscr{F}}{L_c} = \frac{35}{0.1} = 350 \text{ AT/m}$$

From Fig. 12-10

$$B_c = 0.65 \text{ tesla}$$

$$\therefore \phi_1 = B_c A_c = (0.65)(0.001) = 0.00065 \text{ weber}$$

This value represents the first approximation for the flux as indicated by the subscript. Also, since the cross-sectional area is the same for each material, it follows that

$$B_a = B_b = B_c = 0.65 \text{ tesla}$$

Reference to the nickel-iron magnetization curve reveals that the value of H_a corresponding to B_a is negligibly small compared to H_c. Hence for all practical purposes its effect can be neglected. However, note that for medium silicon steel the value of H_b is almost 90 AT/meter. This, coupled with the fact that $L_b = 2L_c$, indicates that material b takes about half as much mmf as material c in maintaining the flow of flux. In other words, at this point in our analysis we can make a refinement on our original assumption of assigning the entire mmf to material c. Now we see that about 50 per cent of that assigned to c should be assigned to b. Thus

$$\mathscr{F}_c + \mathscr{F}_b = \mathscr{F} \qquad \text{(assumed)}$$

but

$$\mathscr{F}_b = 0.5\mathscr{F}_c \qquad \text{(assumed)}$$

Hence

$$1.5\mathscr{F}_c = \mathscr{F} = 35$$

$$\therefore \mathscr{F}_c = 23.3 \text{ AT}$$

Accordingly a second approximation for the solution can be obtained. Therefore

$$H_c = \frac{23.3}{0.1} = 233 \text{ AT/m}$$

which in turn yields

$$B_c = 0.4 \text{ tesla}$$

so that the value of the flux now becomes

$$\phi_2 = B_c A_c = 0.0004 \text{ weber}$$

To determine whether or not this is the correct answer we must at this point compute the mmf drops for each material and add to see whether they yield a value equal to the applied mmf. If not, the foregoing procedure must be repeated until Ampere's circuital law is satisfied. Making this check for the second approximation we have

$$H_b = 60 \text{ AT/m} \qquad \text{corresponding to } B_a = 0.4$$

and

$$H_a = 7 \text{ AT/m} \qquad \text{for } B_b = 0.4$$

Accordingly

$$\begin{aligned} HL &= H_a L_a + H_b L_b + H_c L_c \\ &= (7)(0.3) + (60)(0.2) + 233(0.1) \\ &= 2.1 + 12 + 23.3 = 37.4 \text{ AT} \end{aligned}$$

Obviously this is too high by about 7 per cent. Hence, as a third try, reduce the biggest contributor to the mmf by a factor of 5 per cent. That is, assume that now

$$\mathscr{F}_c = 22 \text{ AT}$$

Then

$$H_c = 220 \text{ AT/m} \quad \text{and} \quad B_c = 0.375$$

$$\therefore \phi_3 = 0.000375 \text{ weber}$$

Corresponding to this flux we find

$$H_b = 57 \quad \text{and} \quad H_a = 6$$

Hence

$$\text{mmf} = (6)(0.3) + (57)(0.2) + 22 = 35.2 \text{ AT}$$

Since this summation of mmf's agrees with the applied mmf of 35 AT, the correct solution for the flux is

$$\phi = 0.000375 \text{ weber} = 37{,}500 \text{ lines}$$

When making magnetic circuit computations of the kind just illustrated, it is common practice to accept as valid any solution that comes within ± 5 per cent of the exact solution. The primary reason is that we are dealing with normal magnetization curves which neglect hysteresis and which are after all only *typical* of the material actually being used in the specified circuit. Deviations can and often do exist.

As a final example to illustrate magnetic circuit computations, we shall solve a problem involving parallel magnetic paths as well as the presence of an air gap. Moreover, we will consider only the first category type where the mmf needed to establish a specified flux is to be found. This is justified not only because the solution is straightforward but also because it is by far more representative of the kind of magnetic circuit problem the engineer is likely to be concerned with.

EXAMPLE 12-3   A magnetic circuit having the configuration and dimensions shown in Fig. 12-14 is made of cast steel having a thickness of 0.05 meter and an air gap of 0.002

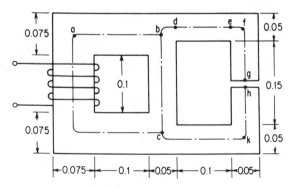

Fig. 12-14. Magnetic circuit for Example 12-3. All dimensions are in meters.

meter length appearing between points g and h. The problem is to find the mmf to be produced by the coil in order to establish an air gap flux of 4×10^{-4} weber (or 40,000 lines).

solution: The method of solution can be readily ascertained by referring to the equivalent circuit of this magnetic circuit as shown in Fig. 12-15. Knowledge of ϕ_g enables us to find the mmf drop appearing across b and c.

From this information the flux in leg bc can be determined and, upon adding it to ϕ_g, we find the flux in leg cab. In turn, the mmf needed to maintain the total flux in leg cab can be computed, and when we add it to the mmf drop across bc we obtain the resultant mmf.

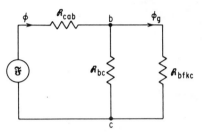

The computations involved for the various parts of the magnetic circuit are as follows.

Part gh. This is the air gap for which the flux is specified as 4×10^{-4} weber. The cross-sectional area of the gap is $(0.05)(0.05) =$

Fig. 12-15. Equivalent circuit of Fig. 12-14.

0.0025 m². Normally, however, this area is slightly higher because of the tendency of the flux to bulge outward along the edges of the air gap—which is often referred to as *fringing*. For convenience this effect is neglected. Thus, the air gap flux density is found to be

$$B_g = \frac{\phi_g}{A_g} = \frac{4 \times 10^{-4}}{0.0025} = 0.16 \text{ tesla}$$

Since the permeability of air is practically the same as that of free space, we have

$$H_g = \frac{B_g}{\mu_0} = \frac{0.16}{4\pi \times 10^{-7}} = 127,300 \text{ AT/m}$$

And

$$\mathscr{F}_g = H_g L_g = (127,300)(0.002) = 255 \text{ AT}$$

Part bg and hc. The length of ferromagnetic material involved here is

$$L_{bg} + L_{hc} = 2\left(L_{bd} + L_{de} + L_{ef} + \frac{L_{fk}}{2}\right) - L_g$$

$$= 2(0.025 + 0.1 + 0.025 + 0.1) - 0.002$$

$$= 0.498 \text{ m}$$

Moreover, corresponding to a flux density in the cast steel of 0.16 tesla, the field intensity H is found to be 125 AT/m. Hence

$$\mathscr{F}_{bg+hc} = 125(0.498) = 62.2 \text{ AT}$$

Part bc. Because path $bfkc$ is in parallel with path bc, the total mmf across path $bfkc$ also appears across path bc. Hence

$$\mathscr{F}_{bc} = 255 + 62.2 = 317.2 \text{ AT}$$

Also

$$L_{bc} = 0.1 + 0.075 = 0.175 \text{ m}$$

$$\therefore H_{bc} = \frac{317.2}{0.175} = 1810 \text{ AT/m}$$

And from Fig. 12-10 for cast steel the corresponding flux density is found to be

$$B_{bc} = 1.38 \text{ teslas}$$

Hence

$$\phi_{bc} = 1.38(0.0025) = 0.00345 \text{ weber}$$

Part cab. Accordingly, the total flux existing in leg *cab* is

$$\phi_{cab} = \phi_{bc} + \phi_g = 0.00345 + 0.0004 = 0.00385 \text{ weber}$$

Knowledge of this flux then leads to determination of the mmf needed in leg *cab* to sustain it. Thus

$$B_{cab} = \frac{0.00385}{0.00375} = 1.026 \text{ teslas}$$

From which

$$H_{cab} = 750 \text{ AT/m}$$

Hence

$$\mathscr{F}_{cab} = H_{cab}L_{cab} = (750)(0.5) = 375 \text{ AT}$$

Therefore the total mmf required to produce the desired air gap flux is

$$\mathscr{F} = \mathscr{F}_{cab} + \mathscr{F}_{bc} = 375 + 317.2 = 692.2 \text{ AT}$$

12-7   Hysteresis and Eddy-current Losses in Ferromagnetic Materials

The process of magnetization and demagnetization of a ferromagnetic material in a symmetrically cyclic condition involves a storage and release of energy which is not completely reversible. As the material is magnetized during each half-cycle it is found that the amount of energy stored in the magnetic field exceeds that which is released upon demagnetization. The background for understanding this behavior was provided in Sec. 12-3. There the hysteresis loop was identified as the variation of flux density as a function of the magnetic field intensity for a ferromagnetic material in a cyclic condition. The salient feature of the hysteresis loop is the delayed reorientation of the domains in response to a cyclically varying magnetizing force. A single hysteresis loop is depicted in Fig. 12-16. The direction of the arrows on this curve indicates the manner in which *B* changes as *H* varies from zero to a positive maximum through zero to a negative maximum and back to zero again, thus completing the loop. To appreciate the meaning of the various shaded areas shown in Fig. 12-16, let us look at the units associated with the product of *B* and *H*. Thus

$$\text{units of } (HB) = \frac{\text{amperes}}{\text{meter}} \times \frac{\text{newtons}}{\text{ampere-meter}} = \frac{\text{newtons}}{\text{meter}^2}$$

But

$$\text{newton-meter} = \text{joule}$$

Hence

$$\text{units of } (HB) = \frac{\text{joule}}{\text{meter}^3}$$

which is clearly recognized as an energy density. Therefore, in dealing with areas involving *B* and *H* in connection with a hysteresis loop we are really dealing with energy densities expressed on a per cycle basis because the hysteresis loop is repeatable for each cyclic variation of *H*.

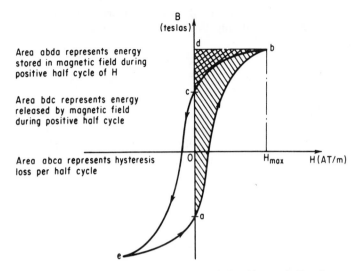

Area abda represents energy
stored in magnetic field during
positive half cycle of H

Area bdc represents energy
released by magnetic field
during positive half cycle

Area abca represents hysteresis
loss per half cycle

Fig. 12-16. Hysteresis loop and energy relationships per half-cycle.

The energy stored in the magnetic field during that portion of the cyclic variation of H when it increases from zero to its positive maximum value (assuming the material is already in a cyclic state) is given by

$$w_1 = \int_{B_a}^{B_b} H \, dB \qquad \text{joules/meter}^3 \qquad (12\text{-}28)$$

when rationalized MKS units are used; i.e., H must be expressed in AT/m and B in teslas. Note that the axes of Fig. 12-16 are so labeled. Moreover, during that portion of its cyclic variation when H decreases from its positive maximum value to zero (as it follows along curve bc of the hysteresis loop) energy is being released by the magnetic field and returned to the source, and this quantity can be represented as

$$w_2 = \int_{B_b}^{B_c} H \, dB \qquad \text{joules/meter}^3 \qquad (12\text{-}29)$$

In this equation, since $B_b > B_c$, the quantity w_2 will be negative, indicating that the energy is being released rather than stored by the magnetic field.

A graphical interpretation of Eq. (12-28) leads to the result that the energy absorbed by the field when H is increasing in the positive direction can be represented by the area $abdca$. Similarly the energy released by the field as H varies from $H_{\max}$ to zero can be represented by area $bdcb$. The difference between these two energy densities represents the amount of energy which is not returned to the source but rather is dissipated as heat as the domains are realigned in response to the changing magnetic field intensity. This dissipation of energy is called *hysteresis loss*. Keep in mind that Fig. 12-16 depicts this energy density loss for a one-half cycle variation of H. Hence area $abca$ represents the hysteresis loss per half-cycle. It certainly follows from symmetry that upon completion of the negative half-cycle variation of H an equal energy loss occurs. Therefore as H varies over one complete cycle the total energy loss per cubic meter is represented by the area of the hysteresis loop. More specifically this energy loss per cycle can be expressed mathematically as

$$w_h = \text{(area of hysteresis loop)} \quad \frac{\text{joules}}{\text{m}^3 \times \text{cycle}} \qquad (12\text{-}30)$$

where rationalized MKS units are used for H and B.

It is frequently desirable to express the hysteresis loss of ferromagnetic materials in watts—the unit of power. A little thought about the units of w_h in Eq. (12-30) shows how this can be directly accomplished. Thus

$$w_h = \frac{\text{energy}}{\text{vol} \times \text{cycles}} = \frac{\text{power} \times \text{sec}}{\text{vol} \times \text{cycles}} = \frac{\text{power}}{\text{vol} \times \text{cycles/sec}} \qquad (12\text{-}31)$$

Now let P_h = power loss in watts

v = volume of ferromagnetic material

f = cycles/sec = frequency of variation of H

Then Eq. (12-31) becomes

$$w_h = \frac{P_h}{vf} \qquad (12\text{-}32)$$

or

$$P_h = w_h vf \qquad (12\text{-}33)$$

where w_h—the energy density loss—is determined from Eq. (12-30).

To obviate the need of finding the area of the hysteresis loop in order to compute the hysteresis loss in watts from Eq. (12-33), Steinmetz obtained an empirical formula for w_h based on a large number of measurements for various ferromagnetic materials. He expressed the hysteresis power loss as

$$P_h = vf(K_h B_m^n) \qquad (12\text{-}34)$$

where B_m is the maximum value of the flux density and n lies in the range $1.5 \leq n \leq 2.5$ depending upon the material used. The parameter K_h also depends upon the material. Some typical values are: cast steel 0.025, silicon sheet steel 0.001, and permalloy 0.0001.

In addition to the hysteresis power loss, another important loss occurs in ferromagnetic materials that are subjected to time-varying magnetic fluxes— the *eddy-current* loss. This term is used to describe the power loss associated with the circulating currents that are found to exist in closed paths within the body of a ferromagnetic material, causing an undesirable heat loss. These circulating currents are created by the differences in potential existing throughout the body of the material owing to the action of the changing flux. If the magnetic circuit is composed of solid iron, the ensuing power loss is appreciable because the circulating currents encounter relatively little resistance. To increase significantly the resistance encountered by these eddy currents the magnetic circuit is invariably composed of very thin *laminations* (usually 14 to 25 mils thick) whenever the electromagnetic device is such that a varying flux permeates it in normal operation. This is the case with transformers and all a-c electric motors and generators. An empirical equation for the eddy-current loss is

$$P_e = K_e f^2 B_m^2 \tau^2 v \qquad \text{watts} \qquad (12\text{-}35)$$

where K_e denotes a constant dependent upon the material
     f denotes the frequency of variation of flux in Hz
     B_m is the maximum flux density
     τ denotes the lamination thickness
     v is the total volume of the material

A comparison of this equation with Eq. (12-34) reveals that eddy-current losses vary as the square of the frequency whereas the hysteresis loss varies directly with the frequency.

Taken together the hysteresis and eddy-current losses constitute what is frequently called the *core losses* of electromagnetic devices that involve time-varying fluxes for their operation. More than just passing attention is devoted to these losses here because, as will be seen, core losses have an important bearing on temperature rise, efficiency and rating of electromagnetic devices.

12-8   Relays—An Application of Magnetic Force

A *relay* is an electromagnetic device which can often be activated by relatively little energy, causing a movable ferromagnetic armature to open or close one or several pairs of electrical contact points located in another control circuit or in a main circuit handling large amounts of energy. A-c and d-c motor starters are equipped with relays designed to assure proper operation of motors during starting and running conditions. These devices are found in many applications in all fields of engineering, especially in situations where control of a process or machine is involved. Our objective in this section is to describe the principles that underlie the operation of these electromagnetic devices. Besides the knowledge it offers, this treatment gives the motivation for studying the theory of magnetic fields and circuits.

The derivation of a magnetic force equation is our primary interest because it shows how it is possible to do mechanical work—moving a relay armature—by abstracting energy from that stored in the magnetic field. Moreover, since the emphasis is on the principles involved, the simplifying assumptions of no saturation or no losses are imposed; i.e., linear analysis is used throughout. Accordingly, the magnetization curve of the ferromagnetic material is assumed to be a straight line as depicted in Fig. 12-17(a). (Only the curve corresponding to positive values of H is shown.) Now, by appropriately modifying the axes of the curve plotted in Fig. 12-17(a), a significant and useful thing happens. First recall that the area between the B-axis and the magnetization curve represents the energy absorbed

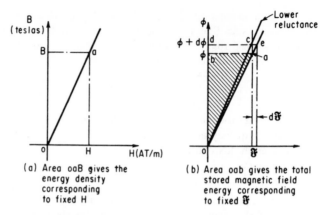

(a) Area oaB gives the energy density corresponding to fixed H

(b) Area oab gives the total stored magnetic field energy corresponding to fixed $\mathscr{F}$

Fig. 12-17. Linear magnetization curve of a relay.

from the source and stored in the magnetic field on a per unit volume basis. Equation (12-28) states this result mathematically. Repeating we have

$$w_f = \int_{B_a}^{B_b} H \, dB \qquad \text{joules/m}^3 \qquad (12\text{-}36)$$

In the simplified situation of Fig. 12-17(a), Eq. (12-36) becomes merely the area of triangle OaB. Thus

$$w_f = \tfrac{1}{2}BH \qquad \text{joules/m}^3 \qquad (12\text{-}37)$$

where H is assumed fixed at the value shown.

Moreover, since w_f is an energy density, the total energy stored in the magnetic field is found by multiplying Eq. (12-28) by the volume. Thus

$$W_f = w_f v = w_f AL \qquad \text{joules} \qquad (12\text{-}38)$$

where L is the length and A the cross sectional area of the magnetic circuit. Inserting Eq. (12-37) into Eq. (12-38) yields

$$\boxed{W_f = \tfrac{1}{2}(BA)(HL) = \tfrac{1}{2}\phi\mathscr{F}} \qquad \text{joules} \qquad (12\text{-}39)$$

Note that in this equation A is combined with B to identify the flux ϕ and H is combined with L to identify the magnetomotive force $\mathscr{F}$. A graphical representation of Eq. (12-39) appears in Fig. 12-17(b). It should be apparent that Fig. 12-17(b) derives from Fig. 12-17(a) by multiplying the ordinate axis by A and the abscissa axis by L, thus plotting ϕ versus $\mathscr{F}$. Then for a fixed H (or $\mathscr{F}$) area OaB of Fig. 12-17(a) gives the energy density whereas the corresponding area Oab of Fig. 12-17(b) gives the total energy stored in the magnetic field.

To understand how mechanical work can be done by the abstraction of energy stored in the magnetic field, consider the circuitry appearing in Fig. 12-18, which depicts the basic composition of an electromagnetic relay. It consists of an exciting coil placed on a fixed ferromagnetic core equipped with a movable element called the *relay armature*. The relay is energized from a constant voltage source through an adjustable resistor R. To begin with, consider that R is fixed at that

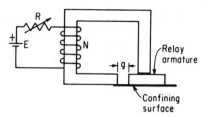

Fig. 12-18. Basic composition of an electromagnetic relay.

value which makes the coil mmf equal to $\mathscr{F}$ and producing the flux ϕ as shown in Fig. 12-17(b). Then adjust R to increase the mmf by $d\mathscr{F}$ and thereby the flux by $d\phi$, assuming the relay armature is held fast to keep the reluctance invariant. Figure 12-17(b) shows that an additional amount of energy is absorbed from the source and stored in the magnetic field virtually equal to the area of rectangle *abdc*. Next readjust R to make $d\mathscr{F}$ zero, and then, keeping the mmf fixed at $\mathscr{F}$, release the armature, thus allowing it to move in the direction to decrease the air gap. Hold the armature fast again when the decreased reluctance causes the flux to increase by the amount $d\phi$ obtained previously.

It is important to note here that, neglecting second-order effects, the new (lower-reluctance) position of the armature is also responsible for permitting the source to supply an additional amount of energy virtually equal to area *abdc* as before but with one significant difference. Whereas when the armature is held fixed, the additional energy supplied by the source is converted entirely to stored magnetic energy, on the other hand, when the armature is allowed to move towards a lower-reluctance position, only half of the same amount of additionally supplied energy is stored in the magnetic field. The other half is consumed in doing the mechanical work involved in moving the relay armature from the higher- to the lower-reluctance position. That the amount is one-half readily follows from a glance at Fig. 12-17(b) if we note that the area of triangle *Oac* is one-half the area of rectangle *abdc*. The slope of the magnetization curve is the permeability and so can be used as a measure of the reluctance. Hence the higher the slope of the magnetization curve, the higher μ and the lower the reluctance. For the magnetization curve in Fig. 12-17(b) to go from position *Oa* to *Oc*, it is necessary for the relay armature to be moved to a position corresponding to a smaller air gap. The mechanical work involved in accomplishing this is represented by area *Oac* in Fig. 12-17(b).

The conclusions described in the foregoing can now be expressed mathematically. It is important to keep in mind, however, that the interchange of energy between the magnetic field and the mechanical system (the relay armature) necessarily involves a *change in reluctance*. In other words the change in energy to do mechanical work, dW_m, is equal to the change in magnetic field energy associated with a change in reluctance $d\mathscr{R}$. Thus by Eqs. (12-18) and (12-39)

$$dW_m = -\tfrac{1}{2}\phi^2 d\mathscr{R} \tag{12-40}$$

where the negative sign emphasizes that mechanical work is done through a *decrease* in reluctance.

An expression for the magnetic force developed on the relay armature is readily obtained by recalling that

$$F \, dx = dW_m \qquad (12\text{-}41)$$

where F is the force in newtons. Inserting this expression into Eq. (12-40) then yields

$$\boxed{F = -\frac{1}{2}\phi^2 \frac{d\mathscr{R}}{dx}} \qquad (12\text{-}42)$$

Hence the magnitude of the instantaneous magnetic force is dependent upon the value of the flux as well as the rate of change of reluctance. Moreover, the direction of this force is always such as to bring about a decrease in reluctance as indicated by the minus sign.

EXAMPLE 12-4   In the relay circuit of Fig. 12-18 assume the cross-sectional area of the fixed core and the relay armature to be A, and the air gap flux to be ϕ. Neglecting the reluctance of the iron, find the expression for the magnetic force existing on the relay armature.

solution: We must find the rate of change of reluctance with distance along the sliding surface. Thus

$$\mathscr{R} = \frac{g}{\mu_0 A} \qquad (12\text{-}43)$$

where g = air gap length.

$$\frac{d\mathscr{R}}{dx} = \frac{1}{\mu_0 A}\frac{dg}{dx} \qquad (12\text{-}44)$$

But

$$dg = -dx \qquad (12\text{-}45)$$

Hence

$$\frac{d\mathscr{R}}{dx} = -\frac{1}{\mu_0 A} \qquad (12\text{-}46)$$

Inserting this last expression into Eq. (11-42) yields the desired result.

$$F = \frac{1}{2}\frac{\phi^2}{\mu_0 A} \qquad \text{newtons} \qquad (12\text{-}47)$$

EXAMPLE 12-5   The cross-sectional view of a cylindrical plunger magnet appears in Fig. 12-19. The plunger (or armature) is free to move inside a nonferromagnetic guide around which the coil is surrounded by a cylindrically shaped steel shell. The plunger is separated from the shell by an air gap of length g.

(a) Derive the expression for the magnetic force exerted on the plunger when it is in the position shown in Fig. 12-19. The length of the plunger is at least equal to that of the shell. Also it has a radius of a meters. Neglect the reluctance of the steel.

(b) Find the magnitude of the force when the mmf is 1414 ampere-turns and the plunger magnet dimensions are

$$x = 0.025 \text{ m}; \quad h = 0.05 \text{ m};$$
$$a = 0.025 \text{ m}; \quad g = 0.00125 \text{ m}$$

solution: Before proceeding with the calculations let us review the principle that gives rise to the force. For the direction of coil current assumed a typical flux path is that shown in Fig. 12-19. Note that the flux path must cross the air gap twice. It is particularly important to note too that the reluctance "seen" by the flux as it crosses the bottom gap is less than that seen by the same flux crossing the gap at the upper part of the plunger because the cross-sectional area of the magnetic circuit is smaller at the upper part than it is at the lower part of the plunger. As a matter of fact, at the lower part of the plunger the cross-sectional area seen by the flux is fixed for all positions of the plunger lying in the range $0 \leq x \leq h$, and

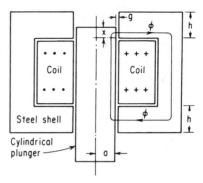

Fig. 12-19. Plunger magnet for Example 12-5.

specifically is equal to $2\pi a h$ (assuming g is small compared to a). Accordingly, a force is created on the plunger so directed as to decrease the reluctance. This means that the magnetic force acts to move the plunger upward.

(a) The solution of this part is obtained from Eq. (12-42), but first we need to identify the correct expression for the reluctance as seen by the flux. The reluctance of the steel is negligible. Hence the reluctance associated with the typical flux shown in Fig. 12-19 is merely the sum of the reluctances associated with each air gap. Thus

$$\mathscr{R} = \frac{g}{\mu_0 2\pi a h} + \frac{g}{\mu_0 2\pi a x} = \frac{g}{2\pi a \mu_0}\left(\frac{1}{h} + \frac{1}{x}\right) \tag{12-48}$$

Then

$$\frac{d\mathscr{R}}{dx} = \frac{g}{2\pi a \mu_0}\left(-\frac{1}{x^2}\right) \tag{12-49}$$

Also

$$\phi = \frac{\mathscr{F}}{\mathscr{R}} = \frac{\mathscr{F}}{g/\mu_0 2\pi a}\left(\frac{xh}{h+x}\right) \tag{12-50}$$

Inserting Eqs. (12-49) and (12-50) into Eq. (12-42) yields

$$F = \pi a \mu_0 \frac{\mathscr{F}^2 h^2}{g}\left(\frac{1}{x+h}\right)^2 = 3.94 \times 10^{-6}\frac{ah^2}{g}\mathscr{F}^2\left(\frac{1}{x+h}\right)^2 \quad \text{newtons} \tag{12-51}$$

(b) Inserting the specified dimensions into Eq. (12-51) gives the magnetic force as

$$F = 70 \text{ newtons} = 15.74 \text{ lb}$$

PROBLEMS

12-1 A long straight wire located in air carries a current of 4 amps. Assume the relative permeability of air is unity.

    (a) Find the value of the magnetic field intensity at a distance 0.5 m from the center of the wire.

    (b) A second long straight wire carrying a current of 2 amps is placed parallel to the first one at a distance of 0.5 m with the current flowing in the same direction. Find the direction and magnitude of the force per meter existing between the wires.

(c) Repeat part (b) for the case where the wires are imbedded in iron having a relative permeability of 10,000 and a spacing of 0.05 m.

12-2   The wires shown in Fig. P12-2 are long, straight, and parallel and are completely imbedded in an iron having a relative permeability of 1000. Each wire carries a current of 10 amps.

(a) Compute the magnitude and direction of the resultant force per meter on the wire in which I_2 flows.

(b) Compute the magnitude and direction of the force per meter on the wire in which I_1 flows.

(c) Repeat (a) for the third wire.

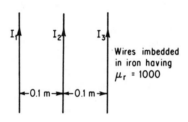

Figure P12-2

12-3   Repeat Prob. 12-2 for the case where the current I_2 flows opposite to I_1 and I_3.

12-4   A uniform magnetic field of 0.7 tesla in the iron is applied to the configuration of Fig. P12-2.

(a) Compute the direction and magnitude of the resultant force per meter when the magnetic field is directed perpendicularly into the plane of the paper.

(b) What is the value of this resultant force per meter when the magnetic field is applied in the plane of the wires and directed from right to left?

(c) Compute the resultant force per meter when the magnetic field is applied at an angle of 45° relative to the plane of the paper and directed into it from right to left.

(d) What is the resultant force per meter when the magnetic field is applied at an angle of 60° relative to the plane of the paper and directed into it from top to bottom?

12-5   In the configuration of Fig. P12-5 the current I_1 has a value of 40 amps. Find the value of I_2 that causes the magnetic field intensity at point P to disappear.

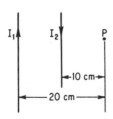

Figure P12-5

Figure P12-6

12-6   A circular loop of wire of radius r meters and consisting of a single turn carries the current I as shown in Fig. P12-6. Derive the expression for the magnetic field intensity at the center.

12-7   A magnetic circuit composed of silicon sheet steel has the square construction shown in Fig. P12-7.
(a) Find the mmf required to produce a core flux of 25×10^{-4} weber.
(b) If the coil has 80 turns, how much current must be made to flow through the coil?

12-8   The magnetic circuit of Prob. 12-7 has an air gap of 0.1 cm cut in the right leg. For a coil having 100 turns find the current that must be allowed to flow in order that the core flux be 0.0025 weber.

12-9   In the magnetic circuit of Fig. P12-9 determine the coil mmf needed to produce a flux of 0.0014 weber in the right leg. The thickness of the magnetic circuit is 0.04 m and is uniform throughout. Medium silicon steel is used.

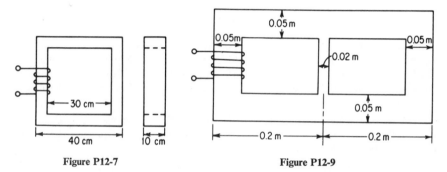

Figure P12-7   Figure P12-9

12-10   Repeat Prob. 12-9 for the case where the coil is placed on the center leg.

12-11   In the magnetic circuit of Prob. 12-7 find the core flux produced by a coil current of 200 ampere-turns.

12-12   In the magnetic circuit of Prob. 12-8 find the core flux produced by a coil current of 600 ampere-turns.

12-13   The core shown in Fig. P12-13 has a uniform cross-sectional area of 2 in.^{2} and a mean length of 12 in. Also, coil A has 200 turns and carries 0.5 amp. Coil B has 400 turns and carries 0.75 amp, and coil C carries 1.00 amp. How many turns must coil C have in order that the core flux be 120,000 lines? The coil currents have the directions indicated in the figure. The core is made of silicon sheet steel.

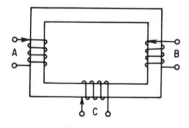

Figure P12-13

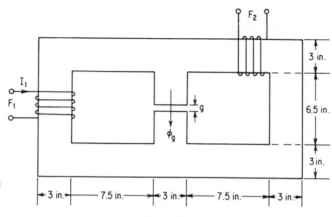

Figure P12-14

12-14   In the magnetic circuit shown in Fig. P12-14 the coil F_1 is supplied with 350 ampere-turns in the direction indicated. Find the direction and magnitude of the mmf required in coil F_2 in order that the air-gap flux be 180,000 lines. The core has an effective cross-sectional area of 9 in.^{2} and is made of silicon sheet steel. The length of the air gap is 0.05 in.

12-15   In the magnetic circuit of Fig. P12-14 the coil F_1 is supplied with 200 ampere-turns in the direction shown. Find the direction and magnitude of the mmf required of coil F_2 in order that the air-gap flux be 90,000 lines.

12-16   A sample of iron having a volume of 16.4 cm^3 is subjected to a magnetizing force sinusoidally varying at a frequency of 400 Hz. The area of the hysteresis loop is found to be 64.5 cm^2 with flux density plotted in kilolines per square inch and magnetizing force in ampere-turns per inch. The scale factors used are: 1 in. = 5 KL/in.^{2} and 1 in. = 12 AT/in. Find the hysteresis loss in watts.

12-17   A ring of ferromagnetic material has a rectangular cross section. The inner diameter is 7.4 in., the outer diameter is 9 in., and the thickness is 0.8 in. There is a coil of 600 turns wound on the ring. When the coil carries a current of 2.5 amps, the flux produced in the ring is 1.2×10^{-3} weber. Find the following quantities expressed in rationalized MKS units: (a) magnetomotive force; (b) electric field intensity; (c) flux density; (d) reluctance; (e) permeability; (f) relative permeability.

12-18   In plotting a hysteresis loop the following scales are used: 1 cm = 10 ampere-turns per inch, and 1 cm = 20 kilolines per square inch. The area of the loop for a certain material is found to be 6.2 cm^2. Calculate the hysteresis loss in joules per cycle for the specimen tested if the volume is 400 cm^3.

12-19   The flux in a magnetic core is alternating sinusoidally at a frequency of 500 cycles per second. The maximum flux density is 50 kilolines per square inch. The eddy-current loss then amounts to 14 watts. Find the eddy-current loss in this core when the frequency is 750 hertz and the flux density 40 kilolines per square inch.

12-20   The total core loss (hysteresis plus eddy current) for a specimen of magnetic sheet steel is found to be 1800 watts at 60 Hz. If the flux density is kept constant and the frequency of the supply increased 50 per cent, the total core loss is found to be 3000 watts. Compute the separate hysteresis and eddy-current losses at both frequencies.

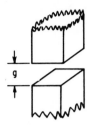

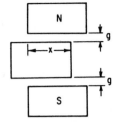

Figure P12-21   Figure P12-22

12-21   A flux ϕ penetrates the complete volume of the iron bars as shown in Fig. P12-21. When the bars are assumed separated by g meters, find the expression for the force existing between the two parallel plane faces. Neglect the reluctance of the iron.

12-22   A magnetic flux ϕ penetrates a movable magnetic core which is vertically misaligned relative to the north and south ploes of an electromagnet as depicted in Fig. P12-22. The depth dimension for the core and the electromagnet is b. Determine the expression for the force that acts to bring the core into vertical alignment. Express the result in terms of the air-gap flux density and the physical dimensions. Neglect the reluctance of the iron.

12-23   The magnetization curve of a relay is shown in Fig. P12-23.
(a) Find the energy stored in the field with the relay in the open position.
(b) Assume the relay armature moves rapidly under conditions of constant flux. Compute the work done in going from the open to the closed position.
(c) Calculate the force in newtons exerted on the armature in part (b).
(d) Assume the relay armature moves slowly at constant mmf. Compute the work done in going from the open to the closed position.
(e) Does the energy of part (d) come from the original stored field energy? Explain.

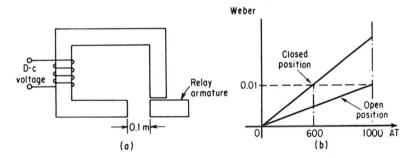

Figure P12-23

13

transformers

An understanding of the transformer is essential to the study of electromechanical energy conversion. Although electromechanical energy conversion involves interchanges of energy between an electrical system and a mechanical system while the transformer involves the interchange of energy between two or more electrical systems, yet the coupling device in both cases is the magnetic field and its behavior in each case is fundamentally the same. As a result we find that many of the pertinent equations and conclusions of transformer theory have equal validity in the analysis of a-c machinery and some aspects of d-c machinery. For example, the equivalent circuit of single-phase and three-phase induction motors is found to be identical in form to that of the transformer. Other examples can be cited, as a glance through Part III of the book readily reveals.

In addition to serving as a worthwhile prelude to the study of electromechanical energy conversion, an understanding of transformer theory is important in its own right because of the many useful functions the transformer performs in prominent areas of electrical engineering. In communication systems ranging in frequency from audio to radio to video, transformers are found fulfilling widely varied purposes. For example, *filament transformers* are used to supply heater power to the filaments of vacuum tubes. *Input transformers* (to connect a microphone output to the first stage of an electronic amplifier), *interstage transformers,* and *output transformers* are to be found in radio and television circuits. Transformers are also used in communication circuits as impedance transformation devices which allow maximum transfer of power from the input circuit to the coupled circuit. Telephone lines and control circuits are two more areas in which the transformer is used extensively. In electric power distribution systems the transformer makes it possible to convert electric power from a generated voltage of about 15 to 20 kv (as determined by generator design limitations) to values of 380 to 500 kv, thus permitting transmission over long distances to appropriate distribution points (e.g. urban areas) at tremendous savings in the cost of copper as well as in power losses in the transmission lines. Then at the distribution points the transformer is the means by which these dangerously high voltages are reduced to a safe level (208/120 volts) for use in homes, offices, shops, and so on. In short, the transformer is a useful device to be found in many phases of electrical engineering and therefore merits the attention given to it in this chapter.

13-1 Theory of Operation and Development of Phasor Diagrams

In the interest of clarity and motivation the theory of operation of the transformer is developed in five steps, starting with the ideal iron-core reactor and ending with the transformer phasor diagram at full-load. In the process the important transformer equations are developed along with a physical explanation of how the transformer operates. Emphasis on complete understanding of the physical behavior of a device is stressed, because once this is grasped, the theory as well as the associated mathematical descriptions readily follow.

In its simplest form a transformer consists of two coils which are mutually coupled. When the coupling is provided through a ferromagnetic ring (circular or otherwise), the transformer is called an *iron-core* transformer. When there is no ferromagnetic material but only air, the device is described as an *air-core transformer*. In this chapter attention is confined exclusively to the iron-core type. Of the two coils the one that receives electric power is called the *primary winding* whereas the other, which can deliver electric power to a suitable load circuit, is called the *secondary winding*.

Ideal Iron-core Reactor. As the first step in the development of the theory of transformers refer to Fig. 13-1 which depicts an ideal iron-core reactor energized from a sinusoidal voltage source v_1. The reactor is ideal because the coil resistance and the iron losses are assumed zero and the magnetization curve is assumed to be linear.

Our interest at this point is to develop the phasor diagram that applies to the circuitry of Fig. 13-1, because thereby much is revealed about the theory. As v_1 increases in

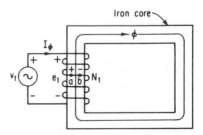

Fig. 13-1. Ideal iron-core reactor—zero winding resistance and no core losses. Polarities are instantaneous.

its sinusoidal variation, it causes a sinusoidal magnetizing current i_ϕ† to flow through the N_1 turns of the coil, which in turn produces a sinusoidally varying flux in time phase with the current. In other words, when the current is zero the flux is zero and when the current is at its positive maximum value, so too is the flux. Applying Kirchhoff's voltage law to the circuit we have

$$v_1 - e_1 = 0 \qquad (13\text{-}1)$$

where e_1 is the *voltage drop* associated with the flow of current through the coil and can be written in terms of i_ϕ as

$$e_{ab} = e_1 = L\frac{di_\phi}{dt} \qquad (13\text{-}2)$$

where L is the inductance of the coil and $i_\phi = i_{ab}$. Inserting Eq. (13-2) into Eq. (13-1) and rearranging yields

$$v_1 = e_1 = L\frac{di_\phi}{dt} = e_{ab} \qquad (13\text{-}3)$$

We also know that the magnetizing current is sinusoidal because the applied voltage is sinusoidal and the magnetization curve is assumed linear. Consequently we can write for the magnetizing current the expression

$$i_\phi = \sqrt{2}\, I_\phi \sin \omega t \qquad (13\text{-}4)$$

where I_ϕ is the rms value of the current.

Introducing Eq. (13-4) into Eq. (13-3) and performing the differentiation called for yields

$$v_1 = e_1 = \sqrt{2}\,(\omega L)I_\phi \cos \omega t = \sqrt{2}\,(\omega L)I_\phi \sin\left(\omega t + \frac{\pi}{2}\right) \qquad (13\text{-}5)$$

The maximum value of this quantity occurs when the $\cos \omega t$ has the value of unity. Then

$$E_{1\max} = \sqrt{2}\,(\omega L)I_\phi = \sqrt{2}\,E_1 \qquad (13\text{-}6)$$

where

$$E_1 \equiv \omega L I_\phi = \text{rms value of the inductive reactance drop} \qquad (13\text{-}7)$$

A comparison of Eqs. (13-4) and (13-5) shows that the magnetizing current lags the voltage appearing across the coil by 90 electrical degrees. The phasor diagram of Fig. 13-2(a) depicts this condition. Since all the quantities involved here are sinusoidal, the phasors shown represent the rms value of the quantities.

Another approach that can be used in arriving at the phasor diagram for the ideal iron-core reactor is worth considering because of its use later. The approach just described can be referred to as a *circuit viewpoint* because it deals solely with the electric circuit in which I_ϕ flows as well as the circuit parameter L. No direct use is made of the magnetic flux ϕ, which is a space (or field) quantity. In contrast the second approach starts with the magnetic field as it appears in Faraday's law of induction. This is one of the fundamental laws upon which the science of electrical engineering is based. Faraday's law states that the emf induced in a coil is proportional to the number of turns linking the flux as well as the time rate of change of the linking flux. In applying Faraday's law to the circuit of Fig. 13-1 the expression may be written either as a voltage drop (a to b) or as a voltage rise (b to a). By the former description we can write

$$e_{ab} = e_1 = +N\frac{d\phi}{dt} \tag{13-8}$$

Expressed as a voltage rise, the induced emf equation becomes

$$e_{ba} = -e_1 = -N\frac{d\phi}{dt} \tag{13-9}$$

In either case, the sign is attributable to Lenz's law. The minus sign states that the emf induced by a changing flux is always in the direction in which current would have to flow to oppose the changing flux. On the other hand, the plus sign in Eq. (13-8) denotes that the polarity of a is positive whenever the time rate of change of flux is positive as determined by the direction of current flow.

Equation (13-9) then represents the starting point for the second approach in analyzing the circuit of Fig. 13-1. Corresponding to the sinusoidal variation of the magnetizing current as expressed by Eq. (13-4), there is also a sinusoidal variation of flux which may be expressed as

$$\phi = \Phi_m \sin \omega t \tag{13-10}$$

where Φ_m denotes the maximum value of the magnetic flux. Putting this expression into Eq. (13-9) and performing the required differentiation leads to

$$e_{ba} = -e_1 = -N_1\frac{d\phi}{dt} = -N_1\Phi_m\omega \cos \omega t = E_{1\,max} \sin\left(\omega t - \frac{\pi}{2}\right) \tag{13-11}$$

where

$$E_{1\,max} \equiv N_1\Phi_m\omega \equiv \sqrt{2}\,E_1 \tag{13-12}$$

and

$$E_1 = \text{rms value of the induced emf}$$

A comparison of Eqs. (13-10) and (13-11) makes it clear that the reaction emf $- e_1$ *lags* the changing flux that produces it by 90 degrees. This situation is depicted in Fig. 13-2(b). Basically, the emf $- e_1$ is a voltage rise (or generated emf) which has such a direction that, if it were free to act, would cause a current to flow opposing the action of I_ϕ. To verify this solely in terms of the action-reaction law consider any two adjacent points such as a' and b' on coil N_1 in Fig. 13-1. If the flux is assumed increasing in the direction shown, then the reaction in the coil must be such that current would have to flow from point b' to a' in order to oppose the increasing flux. Recall that by the *right-hand rule* any current assumed to be flowing from b' to a' produces flux lines opposing the increasing flux. In order for this condition to prevail, it is therefore necessary for point a' to be positive with respect to b' for the instant being considered. This line of reasoning when extended to the full coil, leads to the polarity markings shown in Fig. 13-1. The interesting thing to note here is that on the basis of the flux viewpoint the emf $- e_1$ in Fig. 13-1 is treated as a reaction (or generated) voltage rise in progressing from b' to a'. On the other hand, from the circuit viewpoint the voltage

e_1, and its polarity when considered in the direction of flow of the current I_ϕ, appears as a voltage drop. The voltage e_1 remains the same; it is merely the point of view that changes.

Because I_ϕ and the flux ϕ are in time phase, it is customary to combine the two viewpoints into a single phasor diagram as depicted in Fig. 13-2(c). Since $E_{ab} = E_1$ represents the voltage drop, it follows that the same induced emf viewed on the basis of a reaction to the changing flux, being a voltage rise, is equal and opposite to E_1. That is, $E_{ba} = -E_1$. Another way of describing this is to say that the terminal voltage must always contain a component equal and opposite to the voltage rise.

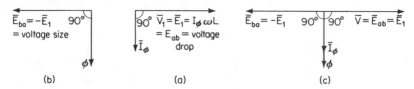

(b)   (a)   (c)

Fig. 13-2. Phasor diagrams of the ideal iron-core reactor: (a) circuit viewpoint; (b) field viewpoint; (c) combined diagram.

It should be apparent up to this point that the reaction voltage viewed from b' to a' in Fig. 13-1 does not succeed in actually establishing a reverse current to oppose the increasing flux. This never happens in the primary winding of a two-winding transformer, but it does happen in the secondary winding as is presently described.

In the work that follows in this chapter preference is given to treating the induced emf as a voltage drop. That is, in applying Faraday's law, the version described by Eq. (13-8) will be used. Due regard is given to Lenz's law in this equation by noting that, when the time rate of change of flux in the configuration of Fig. 13-1 is positive, the polarity of point a is also positive.

Practical Iron-core Reactor. As the second step in the development of transformer theory, let us consider how the phasor diagram of Fig. 13-2(a) must be modified in order to account for the fact that a practical reactor has a winding-resistance loss as well as hysteresis and eddy-current losses. A linear magnetization curve will continue to be assumed because the effect of the nonlinearity is to cause higher harmonics of fundamental frequency to exist in the magnetizing current—which, of course, cannot be included in the phasor diagram. Only quantities of the same frequency can be shown.

To develop the phasor diagram in this case we start with the flux phasor ϕ and then draw the induced emf E_1 90 degrees leading. This much is shown in Fig. 13-3(a). Next we must consider the location of the current. Keep in mind that the iron is in a cyclic condition, i.e., the flux is sinusoidally varying with time. Moreover, this cyclic variation takes place along a hysteresis loop of finite area because the core losses are no longer assumed zero. Now, as pointed out in Chapter 12, it is characteristic of ferromagnetic materials in a cyclic condition to

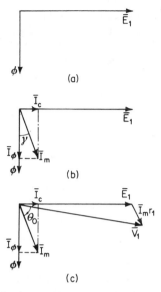

Fig. 13-3. Development of the phasor diagram of a practical iron-core reactor:

    (a) building on the flux phasor;
    (b) accounting for core losses;
    (c) accounting for core losses and winding resistance.

have the flux *lag* behind the magnetizing current by an amount called the *hysteretic angle* γ. See Fig. 13-3(b). Since the positive direction of rotation is counterclockwise in a phasor diagram, I_m is shown leading ϕ to bear out the physical fact. Note, too, that the magnetizing current I_m is now considered made up of two fictitious components I_ϕ and I_c. The quantity I_ϕ is the purely reactive component and is the current that would flow if the core losses were zero. The quantity I_c is that fictitious quantity which, when multiplied by the voltage E_1 produced by the changing flux, represents exactly the power needed to supply the core losses caused by the same changing flux. Thus we can write

$$I_c E_1 \equiv P_c \quad \text{watts} \tag{13-13}$$

or

$$I_c \equiv \frac{P_c}{E_1} \quad \text{amperes} \tag{13-14}$$

where P_c denotes the sum of the hysteresis and eddy-current losses. This technique for handling the core loss of an electromagnetic device is a useful one and is frequently applied in the study of a-c machinery.

To complete the phasor diagram for the practical iron-core reactor, it is only necessary to account for the winding-resistance drop. This is readily accomplished by applying Kirchhoff's voltage law to the circuit. Hence

$$\bar{V}_1 = I_m r_1 + \bar{E}_1 \quad \text{volts} \tag{13-15}$$

where r_1 is the winding resistance in ohms. Since r_1 is a constant, it follows that the $I_m r_1$ drop must be in phase with I_m or located parallel to I_m as depicted in Fig. 13-3(c). Note that the component of I_m that is in phase with the applied voltage $\bar{V}_1$ exceeds I_c, as well it should because this component must supply not only the core losses but the winding copper losses too. Thus, the input power may be expressed as

$$P_{in} = V_1 I_m \cos \theta_0 = I_m^2 r_1 + P_c \quad \text{watts} \tag{13-16}$$

where θ_0 denotes the power-factor angle.

The Two-winding Transformer. By placing a second winding on the core of the reactor of Fig. 13-1, we obtain the simplest form of a transformer. This is depicted in Fig. 13-4. The rms value of the induced emf E_1 appearing in the primary winding *ab* readily follows from Eq. (13-12) as

$$E_1 = \frac{N_1 \Phi_m \omega}{\sqrt{2}} = \frac{2\pi f}{\sqrt{2}} \Phi_m N_1 = 4.44 f \Phi_m N_1 \tag{13-17}$$

Keep in mind that this result derives directly from Faraday's law and is very important in machinery analysis. One significant result deducible from this equation is that for transformers having relatively small winding resistance (i.e., $E_1 \approx V_1$) the value of the maximum flux is determined by the applied voltage.

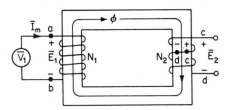

Fig. 13-4. The two-winding transformer.

The induced emf appearing across the secondary winding terminals is produced by the same flux that causes E_1. Hence the only difference in the rms values is brought about by the difference in the number of turns. Thus we can write

$$E_2 = 4.44 f \Phi_m N_2 \tag{13-18}$$

Dividing Eq. (13-17) by (13-18) leads to

$$\frac{E_1}{E_2} = \frac{N_1}{N_2} \equiv a = \text{ratio of transformation} \tag{13-19}$$

Another fact worth noting is that E_1 and E_2 are in time phase because they are induced by the same changing flux.

Transformer Phasor Diagram at No-load. When a transformer is at no-load, no secondary current flows. Figure 13-5 depicts the situation. A study of Fig. 13-5 should make it apparent that the development of the phasor diagram follows from Fig. 13-3(c) by making two modifications. One is to identify the secondary induced emf in the diagram. But, as already observed in the preceding step, the emf induced in the secondary winding must be in phase with the corresponding voltage E_1 of the primary winding. For convenience call this secondary induced emf E_2. Then E_2 and E_1 must both lead the flux phasor by 90 degrees as shown in Fig. 13-6. For convenience the turns ratio a is assumed to be unity. The second modification is concerned with accounting for the fact that not all of the flux produced by the primary winding links the secondary winding. The difference between the total flux linking the primary winding and the mutual flux linking both windings is called the *primary leakage flux* and is denoted by ϕ_{l1}. Note that some of the paths drawn to represent ϕ_{l1} encircle only several turns and not all N_1 turns. As long as such closed paths encircle an mmf different from zero, such a flux path does in fact exist. This is borne out by Ampere's circuital law.

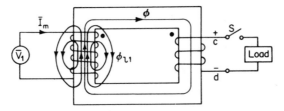

Fig. 13-5. Two-winding transformer at no-load; S open.

An important distinction exists between the mutual flux ϕ and the primary leakage flux ϕ_{l1}. It is this: The mutual flux exists wholly in iron and so involves a hysteresis loop of finite area. The leakage flux, on the other hand, always involves appreciable air paths, and, although some iron is included in the closed path, the reluctance experienced by the leakage flux is practically that of the air. Consequently the cyclic variation of the leakage flux involves no hysteresis (or lagging) effect. Hence in the phasor diagram the primary leakage flux must be placed in phase with the primary winding current as shown in Fig. 13-6. Now this leakage flux induces a voltage E_{l1} which must lead ϕ_{l1} by 90 degrees. This emf due to leak-

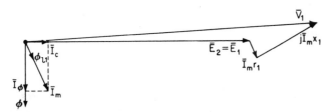

Fig. 13-6. Phasor diagram of transformer at no-load.

age flux may be replaced by an equivalent *primary leakage reactance drop*. Thus

$$I_m x_1 \equiv E_{l1}$$ (13-20)

This quantity must lead I_m by 90 degrees. The phasor addition of this drop and the primary winding resistance drop as well as the voltage drop $\bar{E}_1$ associated with the mutual flux yields the primary applied voltage $\bar{V}_1$ as depicted in Fig. 13-6.

The quantity x_1 of Eq. (13-20) is called the *primary leakage reactance*; it is a fictitious quantity introduced as a convenience in representing the effects of the primary leakage flux.

Transformer under Load. In this last step in the development of the theory of operation of the transformer, we direct attention first to the phasor diagram representation of Kirchhoff's voltage law as it applies to the secondary circuit. For simplicity consider that the load appearing in the secondary circuit of Fig. 13-7 is purely resistive and further that the flux has the direction indicated with

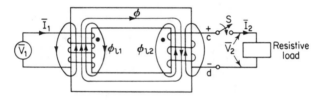

Fig. 13-7. Two-winding transformer under load; S closed.

the flux increasing. By Lenz's law there is an emf induced in the secondary winding which instantaneously makes terminal c positive with respect to terminal d. When switch S is closed, a current I_2 flows instantaneously from c through the load to d. For convenience the load resistor is assumed adjusted to cause rated secondary current to flow. $\bar{E}_2$ establishes the secondary load voltage besides accounting for other voltage drops existing in the secondary circuit. When equilibrium is established after the load switch is closed, the appropriate phasor diagram must show $\bar{V}_2$ and I_2 in phase as depicted in Fig. 13-8. Since it is physically impossible for the secondary winding to occupy the same space as the primary winding, a flux is produced by the secondary current which does not link with the primary winding. This is called the *secondary leakage flux*. The voltage E_{l2} induced by this secondary leakage flux leads ϕ_{l2} by 90 degrees. Again as a matter of convenience this secondary leakage flux voltage is replaced by a secondary leakage reactance drop. Thus

$$E_{l2} \equiv I_2 x_2$$ (13-21)

where x_2 is the fictitious secondary leakage reactance which represents the effects of the secondary leakage flux. Also, because leakage flux has its reluctance predominantly in air, Eq. (13-21) is a linear equation. Thus, doubling the current

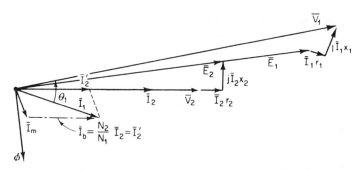

Fig. 13-8. Complete phasor diagram of transformer under load for the case where $a = (N_1/N_2) > 1$.

doubles the leakage flux, which doubles $\bar{E}_{l2}$, and by Eq. (13-21) this is represented by a doubled reactance drop. The phasor sum of the terminal voltage, the secondary winding resistance drop, and the secondary leakage reactance drop adds up to the source voltage $\bar{E}_2$ as demonstrated in Fig. 13-8.

When a phasor diagram shows $\bar{V}_2$ and I_2 in phase, the implication is that power is being supplied to the load equal to V_2I_2. Clearly this power must come from the primary circuit. But by what mechanism does this come about? To understand the answer to this question, consider the situation depicted in Fig. 13-7. With the flux increasing in the direction shown the secondary emf has the polarity indicated, so that when the switch is closed the action of the secondary current is to cause a decrease in the mutual flux. This tendency for the flux to decrease owing to the action of N_2I_2 causes the voltage drop E_1[or $e_1 = N_1(d\phi/dt)$] to decrease. However, since the applied voltage $\bar{V}_1$ is constant, the slight decrease in $\bar{E}_1$ creates an appreciable increase in primary current. In fact the primary current is allowed to increase to that value which, when flowing through the primary turns N_1, represents sufficient mmf to neutralize the demagnetizing action of the secondary mmf. Expressed mathematically we have

$$I_bN_1 - N_2I_2 = 0 \qquad (13\text{-}22)$$

where I_b denotes the increase in primary current needed to cancel out the effect of the secondary mmf to reduce the flux. I_b is often called the *balance* current. This is above and beyond the value I_m needed to establish the operating flux. Equation (13-22) reveals that N_1I_b must be oppositely directed to N_2I_2. In short, then, it is through the medium of the flux and its small attendant changes that power is transferred from the primary to the secondary circuit of a transformer. Of course, the total primary current is the phasor sum of I_b and I_m. In this case the primary leakage flux is in phase with I_1 rather than I_m, which is the case at no-load.

For transformers in which the winding resistance drop and the leakage reactance drop are negligibly small, the magnetizing mmf is the same at no-load as under load. In other words, the phasor sum of the total primary mmf N_1I_1

and the secondary mmf $N_2 I_2$ must yield $N_1 I_m$. Thus

$$\boxed{N_1 I_1 - N_2 I_2 = N_1 I_m} \tag{13-23}$$

Dividing through by N_1 leads to

$$I_1 - \frac{N_2}{N_1} I_2 = I_m \tag{13-24}$$

Inserting Eq. (13-22) then yields

$$I_1 - I_b = I_m = I_1 - I_2' \tag{13-25}$$

where I_2' is equivalent to I_b. These equations are readily verified in the phasor diagram of Fig. 13-8.

Finally, a comment regarding the construction features of practical transformers. Contrary to the schematic representation of Fig. 13-7 it should not be implied that the total number of primary winding turns are placed on one leg of the ferromagnetic core and all the turns of the secondary winding are placed on the other leg. Such an arrangement leads to an excessive amount of leakage flux, so that for a given source voltage the mutual flux is correspondingly smaller. The usual procedure in transformer construction is to put half the primary and secondary turns atop one another on each leg. This keeps the leakage flux within a few per cent of the mutual flux. Of course still greater reduction in leakage flux can be achieved by further subdividing and sandwiching the primary and secondary turns, but this is clearly obtained at an appreciable increase in the cost of assembly.

13-2   The Equivalent Circuit

Derivation. It is customary in analyzing devices in electrical engineering to represent the device by means of an appropriate equivalent circuit. In this way further analysis and design as well as computational accuracy is facilitated by the direct application of the techniques of electric circuit theory. This procedure is followed whenever new devices are investigated. Accordingly, equivalent circuits are derived not only for the transformer but for all a-c and d-c motors as well as important electronic devices such as the transistor and vacuum tubes. Generally, the equivalent circuit is merely a circuit interpretation of the equation(s) that describe the behavior of the device. For the case at hand there are two equations—Kirchhoff's voltage equations for the primary winding and for the secondary winding. Thus, for the primary winding we have

$$\bar{V}_1 = I_1 r_1 + j I_1 x_1 + \bar{E}_1 \tag{13-26}$$

and for the secondary winding

$$\bar{E}_2 = \bar{V}_2 + I_2 r_2 + j I_2 x_2 \tag{13-27}$$

Both equations are represented in the phasor diagram of Fig. 13-8, and the corresponding circuit interpretation appears in Fig. 13-9(a). The dot notation is used

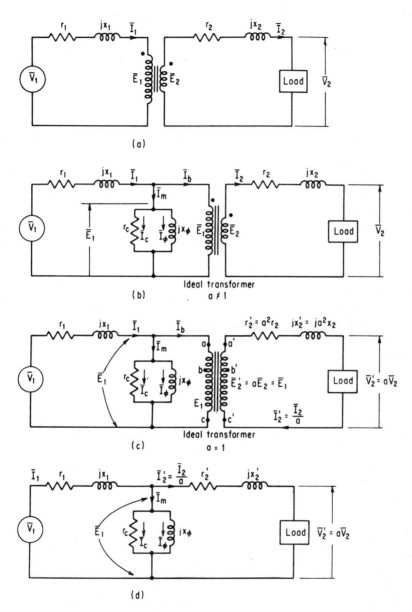

Fig. 13-9. Development of the exact equivalent circuit:
  (a) schematic diagram;
  (b) two-winding equivalent circuit;
  (c) replacing the actual secondary winding with an equivalent
      winding having N_1 turns;
  (d) final complete form.

471

to indicate which sides of the coils instantaneously carry the same polarity. The schematic diagram of Fig. 13-9(a) can be replaced by the equivalent circuit of Fig. 13-9(b) by recognizing that I_1 is composed of I_b and I_m. Moreover, since by definition I_m consists of two components, I_c and I_ϕ, it can be considered as splitting into two parallel branches. One branch is purely resistive and carries the current I_c, which was defined in-phase with $\bar{E}_1$. The other branch must be purely reactive because it carries I_ϕ, which was defined 90 degrees lagging $\bar{E}_1$ as depicted in Fig. 13-3(b). The ideal transformer is included in Fig. 13-9(b) to account for the transformation of voltage and current that occurs between primary and secondary windings. Thus, with the ratio of transformation $a = 2$, the ideal transformer conveys the information that $\bar{E}_2 = \frac{1}{2}\bar{E}_1$ and that $I_2 = 2I_b$. The fact that the actual transformer has primary winding resistance and leakage flux and must carry a magnetizing current is taken care of by the circuitry preceding the symbol for the ideal transformer in Fig. 13-9(b).

The resistive element r_c must have such a value that when the voltage E_1 appears across it, it permits the current I_c to flow. Thus

$$r_c \equiv \frac{E_1}{I_c} = \frac{P_c}{I_c^2} \tag{13-28}$$

The second form of this equation is obtained by replacing E_1 by the expression of Eq. (13-13). Hence, r_c may also be looked upon as that resistance which accounts for the transformer core loss. Similarly, the magnetizing reactance of the transformer may be defined as that reactance which when E_1 appears across it causes the current I_ϕ to flow. Thus

$$\boxed{x_\phi \equiv \frac{E_1}{I_\phi}} \tag{13-29}$$

A single-line equivalent circuit of the transformer is a desirable goal. However, as long as $E_2 \neq E_1$ this is impossible to accomplish because the potential distribution along the primary and secondary windings can never be identical. Only when $a = 1$ is this condition possible. Therefore, in the interest of achieving the desired simplicity in the equivalent circuit, let us investigate replacing the actual secondary winding by an equivalent winding having the same number of turns as the primary winding. A little thought should make it reasonable to expect that if an equivalent secondary winding having N_1 turns is made to handle the same KVA,† the same copper loss and leakage flux and the same output power as the actual secondary winding, then insofar as the applied voltage source is concerned it can detect no difference between the actual secondary winding of N_2 turns and the equivalent secondary winding of N_1 turns. That is, one winding can replace the other without causing any changes in the primary quantities.

Because the equivalent secondary winding is assumed to have N_1 turns, the corresponding induced emf must be different by the factor a. We have

$$\boxed{E_2' = aE_2 = E_1} \tag{13-30}$$

---

† KVA = kilovolt-amperes.

where the prime notation is used to refer to the equivalent winding having N_1 turns. For the KVA of the equivalent secondary winding to be the same as that of the actual secondary winding the following equation must be satisfied:

$$E'_2 I'_2 = E_2 I_2 \qquad (13\text{-}31)$$

Inserting Eq. (13-30) into the last equation shows that the current rating of the equivalent secondary winding must be

$$I'_2 = \frac{I_2}{a} \qquad (13\text{-}32)$$

Since in general $a \neq 1$ it follows that the winding resistance of the equivalent secondary must be adjusted so that the copper loss is the same in both. Accordingly

$$(I'_2)^2 r'_2 = I_2^2 r_2 \qquad (13\text{-}33)$$

This equation, together with Eq. (13-32), establishes the value of the winding resistance of the equivalent secondary. Thus

$$r'_2 = \left(\frac{I_2}{I'_2}\right)^2 r_2 = a^2 r_2 \qquad (13\text{-}34)$$

Often the quantity r'_2 is described as the secondary resistance referred to the primary. The phrase "referred to the primary" is used because we are dealing with an equivalent secondary winding having the same number of turns as the primary winding.

By proceeding in a similar fashion the secondary leakage reactance referred to the primary can be expressed as

$$x'_2 = a^2 x_2 \qquad (13\text{-}35)$$

Moreover, since the load KVA must be the same for the two windings we have

$$V'_2 I'_2 = V_2 I_2 \qquad (13\text{-}36)$$

from which it follows that the load terminal voltage in the equivalent secondary winding is

$$V'_2 = \frac{I_2}{I'_2} V_2 = a V_2 \qquad (13\text{-}37)$$

Appearing in Fig. 13-9(c) are all of the foregoing results in addition to an ideal transformer having unity turns ratio. Keep in mind that Fig. 13-9(c) is drawn with the equivalent secondary winding having N_1 turns whereas Fig. 13-9(b) is drawn with the actual secondary winding. As far as the primary is concerned it cannot distinguish between the two. For the ideal transformer of Fig. 13-9(c) note that joining point a to a', b to b', and c to c' causes no disturbance because instantaneously these points are at the same potential. Accordingly, this ideal transformer may be entirely dispensed with, in which case Fig. 13-9(c) becomes Fig. 13-9(d), which is called the *complete equivalent circuit* of the transformer. This is the single-line diagram we set out to derive.

Approximate Equivalent Circuit. In most constant-voltage, fixed-frequency transformers such as are used in power and distribution applications, the magnitude of the magnetizing current I_m is only a few per cent (2 to 5) of the rated winding current. This happens because high-permeability steel is used for the magnetic core and so keeps I_ϕ very small. Furthermore, steel with a few per cent of silicon added is often used and this helps to reduce the core losses, thereby keeping I_c small. Consequently, little error is introduced and considerable simplification is achieved by assuming I_m so small as to be negligible. This makes the equivalent circuit of Fig. 13-9(d) take on the configuration shown in Fig. 13-10. Note that the primary current I_1 is now equal to the referred secondary current I_2'. Also, we can now speak of an equivalent resistance referred to the primary

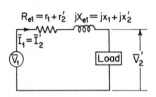

Fig. 13-10. Approximate equivalent circuit referred to the primary.

$$R_{e1} \equiv r_1 + r_2' = r_1 + a^2 r_2$$ (13-38)

and an equivalent leakage reactance referred to the primary

$$X_{e1} \equiv x_1 + x_2' = x_1 + a^2 x_2$$ (13-39)

The subscript 1 denotes the primary.

The corresponding phasor diagram drawn for the approximate equivalent circuit appears in Fig. 13-11. Note the considerable simplification over the phasor diagram of the exact equivalent circuit which appears in Fig. 13-8.

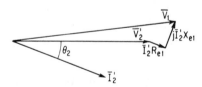

Fig. 13-11. Phasor diagram of the approximate equivalent circuit for a lagging power factor angle θ_2.

The analysis of the preceding pages concerns itself with replacing the actual secondary winding by an equivalent winding having the same number of turns as the primary winding. A little thought should make it apparent that a similar procedure may be employed to draw a single-line equivalent circuit with all quantities referred to the secondary winding. In other words, the actual primary winding can be replaced by an equivalent primary winding having N_2 turns. The resulting equivalent circuit then assumes the form depicted in Fig. 13-12. It is interesting to note that to refer the primary winding resistance r_1 to the secondary, it is necessary to divide by a^2. Thus

$$r_1' = \frac{r_1}{a^2}$$ (13-40)

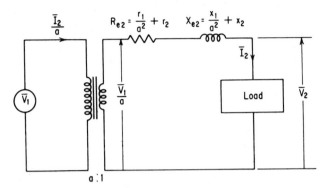

Fig. 13-12. Approximate equivalent circuit referred to the secondary.

Furthermore, the equivalent winding resistance referred to the secondary is

$$R_{e2} = r'_1 + r_2 = \frac{r_1}{a^2} + r_2 \qquad\qquad (13\text{-}41)$$

and the equivalent leakage reactance referred to the secondary is

$$X_{e2} = x'_1 + x_2 = \frac{x_1}{a^2} + x_2 \qquad\qquad (13\text{-}42)$$

Note too that the ideal transformer is included in Fig. 13-12 merely to emphasize the transformation that takes place from the primary side to the secondary side. Normally, the primary circuit is omitted entirely leaving only the single-line secondary circuit.

Meaning of Transformer Nameplate Rating. A typical transformer carries a nameplate bearing the following information: 10 KVA 2200/110 volts. What are the meanings of these numbers in terms of the terminology and analysis developed so far? To understand the answer refer to Fig. 13-13, which is really the approximate equivalent circuit drawn to include the ideal transformer in order to bring into evidence the transformation action existing between the primary and secondary circuits. The figure 110 refers to the rated secondary voltage. It is the voltage appearing across the load when rated current flows. We call this V_2. The 2200 figure refers to the primary winding voltage rating, which is obtained by taking the rated secondary voltage and multiplying by the turns ratio a between primary and secondary. In our terminology this is V'_2. By specifying the ratio $V'_2/V_2 = 2200/110$ on the nameplate, information about the turns ratio is given. Finally, the KVA figure always refers to the *output*

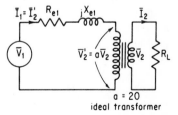

Fig. 13-13. Approximate equivalent circuit drawn with the ideal transformer.

KVA appearing at the secondary load terminals. Of course the input KVA is slightly different because of the effects of R_{e1} and X_{e1}.

Transfer Function. The transfer function of the transformer may be specified with respect either to voltage or to current. The voltage transfer function is defined as the ratio of the output voltage $\bar{V}_2$ to the input voltage $\bar{V}_1$ with the load connected. Thus

$$T_V = \frac{\bar{V}_2}{\bar{V}_1} = \frac{\bar{V}_2}{a\bar{V}_2 + I'_2 R_{e1} + jI'_2 X_{e1}} \approx \frac{1}{a} \qquad (13\text{-}43)$$

Equation (13-43) indicates that when the winding resistances and leakage fluxes are negligibly small the voltage transfer function T_V is approximately the reciprocal of the ratio of transformation. When these quantities cannot be neglected, T_V will be a complex number which varies with the load.

The current transfer function T_I is defined as the output current divided by the input current. For the approximate equivalent circuit of Fig. 13-13 we have

$$T_I = \frac{I_2}{I'_2} = a \qquad (13\text{-}44)$$

Here the current transfer function is exactly equal to a because the magnetizing current is assumed negligible. When I_m is not negligible, T_I will be somewhat smaller than a and slightly complex.

Input Impedance. The input impedance of a transformer is the impedance measured at the input terminals of the primary winding. In the case of the approximate equivalent circuit shown in Fig. 13-13 we have

$$\bar{Z}_i = \frac{\bar{V}_1}{I_1} = R_{e1} + a^2 R_L + jX_{e1} \qquad (13\text{-}45)$$

A glance at this result indicates that the input impedance is dependent upon the load impedance. In the case of power applications the input impedance is of no particular concern because the internal impedance of the source is negligibly small. In communication applications, however, where V_1 is obtained from a vacuum tube or transistor, input impedance is a useful quantity because it describes the effect that the transformer has on the voltage source.

Frequency Response. One of the things we are interested in, when referring to the frequency response of a transformer, is the manner in which the voltage transfer function varies for a fixed load as the frequency of the source voltage V_1 is varied. The amplitude of V_1 is assumed fixed. For power transformers frequency response is of no serious interest because these units are operated at a single fixed frequency—usually 60 Hz and sometimes 50 Hz. However, in communication circuits the frequency of the source voltage is likely to vary over a wide range. For example, in the case of an output transformer which couples the power stage of an audio amplifier to a loudspeaker the frequency may vary over the entire audio range. What, then, is the effect of this varying frequency on the voltage transfer function? The answer follows from an inspection of Fig. 13-9(d). Keep in mind that inductive reactance is directly proportional to the frequency. i.e.

$$x = \omega L = 2\pi f L \tag{13-46}$$

Hence at low frequencies the reactance x_ϕ, which normally is so high that it may be removed from the circuit, can now no longer be neglected. As a matter of fact this low magnetizing reactance can effectively shunt the fixed load impedance thereby causing a severe dropoff of output voltage as depicted in Fig. 13-15. Figure 13-14(a) shows the configuration of the equivalent circuit as it applies at very low frequencies. Note that x_1 and x_2 are so small at these frequencies that they may be omitted entirely from the diagram.

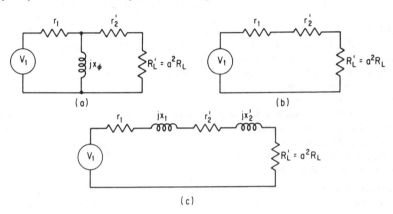

Fig. 13-14. Equivalent circuit for various portions of the frequency spectrum:

    (a) low frequencies;
    (b) intermediate frequencies;
    (c) high frequencies.

In the intermediate frequency range the design of these transformers for communication systems is such that x_1 and x_2 are still quite small while x_ϕ is sufficiently large that it may be omitted. The appropriate circuitry appears in Fig. 13-14(b). For a resistive load the equivalent circuit consists solely of resistive elements so that the transfer function remains constant over the band of intermediate frequencies. Refer to Fig. 13-15.

At very high frequencies x_1 and x_2 are no longer negligible and therefore must be included in the circuit. However, x_ϕ may continue to be omitted. The equivalent circuit now takes the form depicted in Fig. 13-14(c). As the frequency is allowed to get higher and higher, more and more of the source voltage V_1 appears across x_1 and x_2 and less and less appears across the fixed load resistance. Therefore again a dropoff occurs in the value of the voltage transfer function as depicted in Fig. 13-15. In communication transformers a suitable frequency response such as that shown in Fig. 13-15 is very often the most important characteristic of the device.

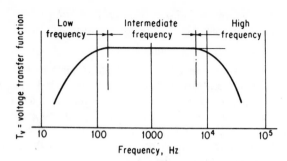

Fig. 13-15. Voltage transfer function V_2/V_1 versus frequency.

EXAMPLE 13-1   A distribution transformer with the nameplate rating of 100 KVA, 1100/220 volts, 60 Hz has a high-voltage winding resistance of 0.1 ohm and a leakage reactance of 0.3 ohm. The low-voltage winding resistance is 0.004 ohm and the leakage reactance is 0.012 ohm. The source is applied to the high-voltage side.

(a) Find the equivalent winding resistance and reactance referred to the high-voltage side and the low-voltage side.

(b) Compute the equivalent resistance and equivalent reactance drops in volts and in per cent of the rated winding voltages expressed in terms of the primary quantities.

(c) Repat (b) in terms of quantities referred to the low-voltage side.

(d) Calculate the *equivalent leakage impedances* of the transformer referred to the primary and secondary sides.

solution: From the nameplate data the turns ratio is $a = 1100/220 = 5$.

(a) Hence the equivalent winding resistance referred to the primary by Eq. (13-38) is

$$R_{e1} = r_1 + a^2 r_2 = 0.1 + (5)^2(0.004) = 0.2 \text{ ohm}$$

Also

$$X_{e1} = x_1 + a^2 x_2 = 0.3 + 25(0.012) = 0.6 \text{ ohm}$$

The corresponding quantities referred to the secondary are

$$R_{e2} = \frac{0.1}{25} + 0.004 = 0.008 \text{ ohm}$$

$$X_{e2} = \frac{0.3}{25} + 0.012 = 0.024 \text{ ohm}$$

These computations reveal that $r_1 = r_2'$ and $x_1 = x_2'$. This is a very common occurrence in well-designed transformers. A well-designed transformer is one that uses a minimum amount of iron and copper for a specified output. Note, too, that the resistance of the secondary winding is considerably smaller than that of the primary winding. This is consistent with the larger current rating of the former.

(b) The primary winding current rating in this case is

$$I_2' = \frac{100 \text{ KVA}}{1.1 \text{ KV}} = 91 \text{ amps}$$

Hence

$$I_2' R_{e1} = 91(0.2) = 18.2 \text{ volts}$$

Expressed as a percentage of the rated voltage of the high-tension winding we have

$$\frac{I_2' R_{e1}}{V_2'}(100) = \frac{18.2}{1100}(100) = 1.65 \text{ per cent}$$

Also

$$I_2' X_{e1} = 91(0.6) = 54.6 \text{ volts}$$

and

$$\frac{I_2' X_{e1}}{V_2'}(100) = \frac{54.6}{1100} = 4.96 \text{ per cent}$$

(c) The secondary winding current rating is

$$I_2 = a I_2' = 5(91) = 455 \text{ amps}$$

Hence

$$I_2 R_{e2} = 455(0.008) = 3.64 \text{ volts}$$

and

$$\frac{I_2 R_{e2}}{V_2}(100) = \frac{3.64}{220}(100) = 1.65 \text{ per cent}$$

Note that the percentage value of the winding resistance drop on the secondary side in this case is identical to the value on the primary side. Similarly for the leakage reactance,

$$I_2 X_{e2} = 455(0.024) = 10.9 \text{ volts}$$

$$\frac{I_2 X_{e2}}{V_2}(100) = \frac{10.9}{220}(100) = 4.96 \text{ per cent}$$

(d) The equivalent leakage impedance referred to the primary is

$$\bar{Z}_{e1} \equiv R_{e1} + jX_{e1} = 0.2 + j0.6 = 0.634\underline{/71.6°}$$

Referred to the secondary we have

$$\bar{Z}_{e2} = \frac{\bar{Z}_{e1}}{a^2} = 0.0253\underline{/71.6°}$$

13-3 Parameters from No-load Tests

The exact equivalent circuit of the transformer has a total of six parameters as Fig. 13-9(d) shows. A knowledge of these parameters allows us to compute the performance of the transformer under all operating conditions. When the complete design data of a transformer are available, the parameters may be calculated from the dimensions and properties of the materials involved. Thus, the primary winding resistance r_1 may be found from the resistivity of copper, the total winding length, and the cross-sectional area. In a similar fashion a parameter such as the magnetizing reactance x_ϕ may be determined from the number of primary turns, the reluctance of the magnetic path, and the frequency of operation. The calculation of the leakage reactance is a bit more complicated because it involves accounting for partial flux linkages. However, formulas are available for making a reliable determination of these quantities.

A more direct and far easier way of determining the transformer parameters is from tests that involve very little power consumption, called no-load tests.

The power consumption is merely that which is required to supply the appropriate losses involved. The primary winding (or secondary winding) resistance is readily determined by applying a small d-c voltage to the winding in the manner shown in Fig. 13-16. The voltage appearing across the winding must be large enough to cause approximately rated current to flow. Then the ratio of the voltage drop across the winding as recorded by the voltmeter, VD, divided by the current flowing through it as recorded by the ammeter, AD, yields the value of the winding resistance.

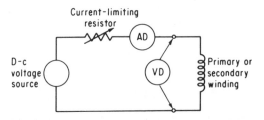

Fig. 13-16. Circuitry for finding winding resistance.

Open-circuit Test. Information about the core-loss resistor r_c and the magnetizing reactance x_ϕ is obtained from an *open-circuit test*, the circuitry of which appears in Fig. 13-17. Note that the secondary winding is open. Therefore, as Fig. 13-9(d) indicates, the input impedance consists of the primary leakage impedance and the magnetizing impedance. In this case the ammeter and voltmeter are a-c instruments and the second letter A is used to emphasize this. Thus AA denotes a-c ammeter. The wattmeter shown in Fig. 13-17 is used to measure the power drawn by the transformer. The open circuit test is performed by applying voltage either to the high-voltage side or the low-voltage side, depending upon which is more convenient. Thus, if the transformer of Example 13-1 is to be tested, the voltage would be applied to the low-tension side because a source of 220 volts is more readily available than 1100 volts. The core loss is the same whether 220 volts is applied to the winding having the smaller number of turns or 1100 volts is applied to the winding having the larger number of turns. The maximum value of the flux, upon which the core loss depends, is the same in either case as indicated by Eq. (13-17). Now for convenience assume that the instruments used in the open-circuit test yield the following readings:

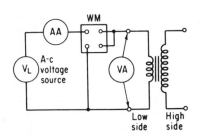

Fig. 13-17. Wiring diagram for transformer open-circuit test.

$$\text{wattmeter reading} = P$$

$$\text{ammeter reading} \; = I_m$$

$$\text{voltmeter reading} \; = V_L$$

The voltmeter reading carries the subscript L to emphasize that the test is performed with the source and instruments on the low-voltage side.

Neglecting instrument losses, the wattmeter reading may be taken entirely equal to the core loss, i.e. $P_c = P$. This is because the attendant copper loss is negligibly small. Moreover, since I_m is very small, the primary leakage impedance drop may be neglected, so that for all practical purposes the induced emf is equal to the applied voltage, i.e. $E_1 = V_L$. In accordance with these simplifications the open-circuit phasor diagram is represented by Fig. 13-3(b). The no-load power factor angle θ_0 is computed from

$$\theta_0 = \cos^{-1} \frac{P}{V_L I_m} \tag{13-47}$$

It then follows that

$$I_c = I_m \cos \theta_0 \tag{13-48}$$

and

$$I_\phi = I_m \sin \theta_0 \tag{13-49}$$

It is important to keep in mind that these currents are referred to the low-tension side. The corresponding values on the high-voltage side would be less by the factor $1/a$. The low-side core-loss resistor becomes

$$r_{cL} = \frac{P}{I_c^2} = \frac{P}{(I_m \cos \theta_0)^2} \tag{13-50}$$

The corresponding high-side core-loss resistor is

$$r_{cH} = a^2 r_{cL} \tag{13-51}$$

The magnetizing reactance referred to the low side follows from Eq. (13-29). Thus

$$x_{\phi L} = \frac{E_1}{I_\phi} = \frac{V_L}{I_m \sin \theta_0} \tag{13-52}$$

On the high side this becomes

$$x_{\phi H} = a^2 x_{\phi L} \tag{13-53}$$

Although the manner of finding the magnetizing impedance is treated in the foregoing, the motivation is not so much necessity as the desire for completeness. Invariably, the use of the approximate equivalent circuit is sufficient for all but a few special applications of transformers. Accordingly, the one useful bit of information obtained from the open-circuit test is the value of the core loss, which helps to determine the efficiency of the transformer.

Short-circuit Test. Of the six parameters of the exact equivalent circuit two remain to be determined: the primary and secondary leakage reactances x_1 and x_2. This information is obtained from a short-circuit test which involves placing a small a-c voltage on one winding and a short circuit on the other as depicted in Fig. 13-18(a). Appearing in Fig. 13-18(b) is the approximate equivalent circuit with the secondary winding short-circuited. Note that the input impedance is merely the equivalent leakage impedance.

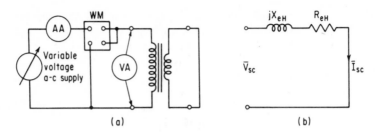

Fig. 13-18. Transformer short-circuit test:
(a) wiring diagram;
(b) equivalent circuit.

When performing the short-circuit test, a reduced a-c voltage is applied to the high side usually because it is more convenient to do so. From the computations of Example 13-1 recall that the leakage impedance drop is only about 5 per cent of the winding voltage rating. Hence 5 per cent of 1100 volts, which is 55 volts, is easier and more accurate to deal with than 5 per cent of 220 volts, which is 11 volts. Also, the wattmeter reading in the circuitry of Fig. 13-18(a) can be taken entirely equal to the winding-copper losses. This follows from the fact that the greatly reduced voltage used in the short-circuit test makes the core loss negligibly small. Calling the wattmeter reading P_{sc} and the ammeter reading I_{sc}, we can determine the equivalent winding resistance as

$$R_{eH} = \frac{P_{sc}}{I_{sc}^2} \tag{13-54}$$

The subscript H is used in place of L since it is known from the wiring diagram that the primary side is the high-voltage side. The equivalent resistance referred to the low side is then

$$R_{eL} = \frac{R_{eH}}{a^2} \tag{13-55}$$

where

$$a = \frac{N_H}{N_L} = \frac{\text{high-side turns}}{\text{low-side turns}} \tag{13-56}$$

Before the equivalent leakage reactance can be determined, it is first necessary to find the equivalent leakage impedance, which a glance at Fig. 13-18(b) reveals to be

$$\bar{Z}_{eH} = \frac{\bar{V}_{sc}}{\bar{I}_{sc}} \tag{13-57}$$

During the test the applied voltage, V_{sc}, is adjusted so that I_{sc} is at least equal to the rated winding current. The equivalent leakage reactance follows from Eqs. (13-54) and (13-57). Thus

$$X_{eH} = \sqrt{Z_{eH}^2 - R_{eH}^2} \tag{13-58}$$

It is important to note that this computation provides the *sum* of the primary and secondary leakage reactances. It gives no information about the individual breakdown between x_1 and x_2. Whenever the approximate equivalent is used in analysis such a breakdown is unnecessary. On the few occasions when the individual breakdown is needed, it is customary to assume that

$$x_1 = x_2' \tag{13-59}$$

This statement is based on the assumption that the transformer is well designed.

One final note. It should be apparent from the theory of transformer action that the establishment of rated current in the high-side winding requires a corresponding flow of rated current in the secondary circuit. It is for this reason that the *total* winding copper loss is measured in the short-circuit test.

EXAMPLE 13-2 The following data were obtained on a 50KVA, 2400/120-volt transformer.

> *Open-circuit test, instruments on low side:*
> > wattmeter reading = 396 watts
> > ammeter reading = 9.65 amps
> > voltmeter reading = 120 volts

> *Short-circuit test, instruments on high side:*
> > wattmeter reading = 810 watts
> > ammeter reading = 20.8 amps
> > voltmeter reading = 92 volts

Compute the six parameters of the equivalent circuit referred to the high and low sides.

solution: From the open-circuit test

$$\theta_0 = \cos^{-1}\frac{P}{V_L I_m} = \cos^{-1}\frac{396}{120(9.65)} = 70°$$

$$\therefore I_{cL} = I_m \cos\theta_0 = 9.65(0.342) = 3.3 \text{ amps}$$

$$I_{\phi L} = I_m \sin\theta_0 = 9.65 \sin 70° = 9.07 \text{ amps}$$

Hence

$$r_{cL} = \frac{396}{3.3^2} = 36.3 \text{ ohms}$$

This can also be found from

$$r_{cL} = \frac{E_1}{I_c} = \frac{V_L}{I_c} = \frac{120}{3.3} = 36.3 \text{ ohms}$$

On the high side this quantity becomes

$$r_{cH} = a^2 r_{cL} = 400(36.3) = 14,520 \text{ ohms}$$

The magnetizing reactance referred to the low side is

$$x_{\phi L} = \frac{E_1}{I_\phi} = \frac{V_L}{I_\phi} = \frac{120}{9.07} = 13.2 \text{ ohms}$$

The corresponding high-side value is

$$x_{\phi H} = a^2 x_{\phi L} = 400(13.2) = 5280 \text{ ohms}$$

From the data of the short-circuit test we get directly

$$Z_{eH} = \frac{92}{20.8} = 4.42 \text{ ohms} \qquad \therefore Z_{eL} = \frac{4.42}{400} = 0.011 \text{ ohm}$$

Also

$$R_{eH} = \frac{810}{(20.8)^2} = 1.87 \qquad \therefore R_{eL} = \frac{1.87}{400} = 0.0047$$

and

$$X_{eH} = \sqrt{Z_{eH}^2 - R_{eH}^2} = \sqrt{4.42^2 - 1.87^2} = 4 \text{ ohms}$$

$$\therefore X_{eL} = \frac{4}{400} = 0.01 \text{ ohm}$$

13-4   Efficiency and Voltage Regulation

The manner of describing the performance of a transformer depends upon the application for which it is designed. It has already been pointed out that in communication circuits the frequency response of the transformer is very often of prime importance. Another important characteristic of such transformers is to provide matching of a source to a load for maximum transfer of power. The efficiency and regulation is usually of secondary significance. This is not so, however, with power and distribution transformers, which are designed to operate under conditions of constant rms voltage and frequency. Accordingly, the material that follows has relevance chiefly to these transformers.

As is the case with other devices, the efficiency of a transformer is defined as the ratio of the useful output power to the input power. Thus

$$\eta = \frac{\text{output watts}}{\text{input watts}} \tag{13-60}$$

Usually the efficiency is found through measurement of the losses as obtained from the no-load tests. Therefore by recognizing that

$$\text{output watts} = \text{input watts} - \sum \text{losses} \tag{13-61}$$

and inserting into Eq. (13-60), a more useful form of the expression for efficiency results. Thus

$$\boxed{\eta = 1 - \frac{\sum \text{losses}}{\text{input watts}}} \tag{13-62}$$

where

$$\sum \text{losses} = \text{core loss} + \text{copper loss} = P_c + I_2^2 R_{e2} \tag{13-63}$$

Equation (13-62) also leads to a more accurate value of the efficiency.

Power and distribution transformers very often supply electrical power to loads that are designed to operate at essentially constant voltage regardless of whether little or full load current is being drawn. For example, as more and more light bulbs are placed across a supply line, which very likely originates from a distribution transformer, it is important that the increased current being drawn from the supply transformer does not cause a significant drop (more than 10 per cent) in the load voltage. If this should happen the illumination output from the bulbs is sharply reduced. Similar undesirable effects occur in television sets that are connected across the same supply lines. The reduced voltage means reduced picture size. An appreciable drop in line voltage with increasing load demands can also cause harmful effects in connected electrical motors such as those found in refrigerators, washing machines, and so on. Continued operation at low voltage can cause these units to overheat and eventually burn out. The way to prevent the drop in supply voltage with increasing load in distribution circuits is to use a distribution transformer designed to have small leakage impedance. The figure of merit used to identify this characteristic is the voltage regulation.

The *voltage regulation* of a transformer is defined as the change in magnitude of the secondary voltage as the current changes from full-load to no-load with the primary voltage held fixed. Hence in equation form,

$$\text{voltage regulation} = \frac{|\bar{V}_1| - |\bar{V}_2'|}{|\bar{V}_2'|} = \frac{\left|\dfrac{\bar{V}_1}{a}\right| - |\bar{V}_2|}{|\bar{V}_2|} \qquad (13\text{-}64)$$

where the symbols denote the quantities previously defined, absolute signs are used in order to emphasize that it is the *change in magnitude* that is important. Equation (13-64) expresses the voltage regulation on a per unit basis. To convert to percentage, it is necessary to multiply by 100. The smaller the value of the voltage regulation, the better suited is the transformer for supplying power to constant-voltage loads (constant voltage is characteristic of most commercial and industrial loads).

In determining the voltage regulation it is customary to assume that $\bar{V}_1$ is adjusted to that value which allows rated voltage to appear across the load when rated current flows through it. Under these conditions the necessary magnitude of $\bar{V}_1$ is found from Kirchhoff's voltage law as it applies to the approximate equivalent circuit. Refer to Figs. 13-10 and 13-11. For the assumed lagging power factor angle θ_2 of the secondary circuit the expression for $\bar{V}_1$ is

$$\bar{V}_1 = V_2'\underline{/0°} + I_2'\underline{/-\theta_2}\,Z_{e1}\underline{/\theta_z} \qquad (13\text{-}65)$$

where

$$\theta_z = \tan^{-1}\frac{X_{e1}}{R_{e1}} \qquad (13\text{-}66)$$

Equation (13-65) is written using the rated secondary terminal voltage as the reference.

EXAMPLE 13-3 For the transformer of Example 13-2, (a) find the efficiency when rated KVA is delivered to a load having a power factor of 0.8 lagging, (b) compute the voltage regulation.

solution: (a) The losses at rated load current and rated voltage are

$$\sum \text{losses} = 396 + 810 = 1206 \text{ watts} = 1.2 \text{ KW}$$

Also

$$\text{output KW} = 50(0.8) = 40 \text{ KW}$$
$$\text{input KW} = 40 + 1.2 = 41.2 \text{ KW}$$

Hence

$$\eta = 1 - \frac{\sum \text{losses}}{\text{input KW}} = 1 - \frac{1.2}{41.2} = 1 - 0.029 = 0.971$$

or

$$\eta = 97.1 \text{ per cent}$$

(b) From Eq. (13-65) we have

$$\bar{V}_1 = \bar{V}_H = 2400 + 20.8\underline{/-37°} \ 4.42\underline{/65°}$$

where

$$\theta_z = \tan^{-1}\frac{X_{eH}}{R_{eH}} = \tan^{-1}\frac{4}{1.87} = 65°$$

Therefore

$$\bar{V}_1 = 2400 + 92\underline{/28°} = 2400 + 81.1 + j43.1$$

$$V_1 \approx 2481.1$$

Note that the j-component is negligible in comparison to the in-phase component. This calculation indicates that in order for rated voltage to appear across the load when rated current is drawn, the primary voltage V_1 must be slightly different from the primary winding voltage rating of 2400 volts. The per cent voltage regulation is accordingly found to be

$$\text{per cent voltage regulation} = \frac{2481.1 - 2400}{2400} \times 100 = 3.38 \text{ per cent}$$

Thus there is approximately a 3.4 per cent change in voltage across the load as increased current is drawn from the transformer.

13-5　Mutual Inductance

Analysis of the two-winding transformer by the methods of electric circuit theory involves the use of self- and mutual-inductance. So far, Kirchhoff's voltage equation applied to the primary winding has led to Eq. (13-26). In the material that follows this result will be shown to be identical with that obtained by the classical approach.

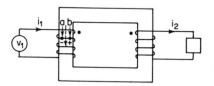

Fig. 13-19. Elementary transformer for analysis by the classical approach.

For simplicity assume the ferromagnetic material has a linear magnetization curve with no core losses. Refer to Fig. 13-19 where for the indicated current directions Kirchhoff's voltage equation becomes

$$v_1 = i_1 r_1 + L_1 \frac{di_1}{dt} - M \frac{di_2}{dt} \tag{13-67}$$

where lower-case letters for current and voltage denote instantaneous values. Here the quantity $L_1 \, di_1/dt$ denotes the voltage drop associated with the total flux produced by i_1 and linking the primary turns N_1. Similarly, the quantity $-M\,di_2/dt$ refers to the *voltage rise* associated with the flux produced by i_2 and linking the primary turns. The existence of a voltage rise in the primary circuit for the assumed direction of i_2 is easily verified. Note that the secondary mmf $N_2 i_2$, if allowed to act freely, produces a flux which threads counterclockwise in the configuration of Fig. 13-19. Then the reaction in the primary coil between two points such as a and b by Lenz's law requires that point b should be at a higher potential than a. Accordingly, this induced emf becomes a voltage rise with respect to the specified direction of i_1.

Assuming that the applied primary voltage is a sinusoid, Eq. (13-67) may be written in rms quantities as

$$\bar{V}_1 = I_1 r_1 + j\omega L_1 I_1 - j\omega M I_2 \tag{13-68}$$

Inserting the relationship

$$I_2 = a I_2' \tag{13-69}$$

wherein a is the ratio of transformation, we get

$$\bar{V}_1 = I_1 r_1 + j\omega L_1 I_1 - j\omega a M I_2' \tag{13-70}$$

But by Eq. (13-24)

$$I_2' = I_1 - I_m = I_1 - I_\phi$$

so that

$$\bar{V}_1 = I_1 r_1 + j\omega(L_1 - aM)I_1 + j\omega a M I_\phi \tag{13-71}$$

Equation (13-71) is now in a form where correspondence with Eq. (13-26) is possible after some algebraic manipulation.

The expression for the self-inductance of the primary winding may be written as

$$L_1 = N_1 \frac{d\phi_1}{di_1} = N_1 \frac{\phi_1}{i_1} \tag{13-72}$$

The equivalence of the differential form with the total quantities in this last expression is allowed by the linearity condition. Recalling that the total flux linking the primary coil consists of the mutual flux ϕ_m plus the primary leakage flux ϕ_{l1}, Eq. (13-72) can be rewritten as

$$L_1 = N_1 \frac{\phi_m + \phi_{l1}}{i_1} \tag{13-73}$$

Also,

$$\phi_m = \frac{N_1 i_1}{\mathscr{R}_m} \quad \text{and} \quad \phi_{l1} = \frac{N_1 i_1}{\mathscr{R}_{l1}} \tag{13-74}$$

where $\mathscr{R}_m$ is the reluctance experienced by the mutual flux and consists almost exclusively of iron and $\mathscr{R}_{l1}$ is the net reluctance experienced by the leakage flux of the primary in air. Accordingly, Eq. (13-73) becomes

$$L_1 = \frac{N_1^2}{\mathscr{R}_m} + \frac{N_1^2}{\mathscr{R}_{l1}} = L_m + L_{l1} \tag{13-75}$$

where L_m denotes an inductance associated with the mutual flux and L_{l1} is the inductance associated with the primary leakage flux. It is important to note here that L_m is not the mutual inductance.

The classical definition of mutual inductance is given in this case by

$$M_{21} = N_2 \frac{d\phi_{m1}}{di_1} = N_2 \frac{\phi_{m1}}{i_1} \tag{13-76}$$

Thus the mutual inductance is related to the number of turns of winding 2 and the change of flux with respect to the current in winding 1 which produces the mutual flux ϕ_{m1}. Upon inserting $N_1 i_1 = \phi_{m1} \mathcal{R}_m$ into Eq. (13-76), there results

$$M_{21} = \frac{N_2 N_1}{\mathcal{R}_m} \tag{13-77}$$

The expression for mutual inductance may also be found by considering a change in mutual flux corresponding to a change in secondary current i_2 linking N_1 turns. Thus

$$M_{12} = N_1 \frac{d\phi_{m2}}{di_2} = N_1 \frac{\phi_{m2}}{i_2} \tag{13-78}$$

But

$$\phi_{m2} \mathcal{R}_m = N_2 i_2 \tag{13-79}$$

Inserting this into Eq. (13-78) yields

$$M_{12} = \frac{N_1 N_2}{\mathcal{R}_m} = M \tag{13-80}$$

A comparison of Eq. (13-80) with Eq. (13-77) shows them to be the same, and from here on either expression will be denoted by M.

Returning to Eq. (13-71) we are now in a position to evaluate the significance of the term in parentheses. Thus, by Eqs. (13-75) and (13-77)

$$L_1 - aM = \frac{N_1^2}{\mathcal{R}_m} + \frac{N_1^2}{\mathcal{R}_{l1}} - \frac{N_1}{N_2} \frac{N_2 N_1}{\mathcal{R}_m} = \frac{N_1^2}{\mathcal{R}_{l1}} = L_{l1} \tag{13-81}$$

Clearly this quantity is the primary leakage inductance, and it represents the effect of the primary leakage flux. Of course the total quantity

$$\omega(L_1 - aM)I_1 = \omega L_{l1} I_1 = x_1 I_1 \tag{13-82}$$

is nothing more than the primary leakage reactance drop and is the counterpart of the second term of Eq. (13-26).

Let us now examine the last term of Eq. (13-71). Thus

$$\omega aMI_\phi = \omega \frac{N_1}{N_2} \frac{N_2 N_1}{\mathcal{R}_m} I_\phi = \omega N_1 \frac{N_1 I_\phi}{\mathcal{R}_m} \tag{13-83}$$

The fraction quantity on the right side of the last equation is the expression for mutual flux. Specifically, however, since it involves the reluctance and the rms value I_ϕ of the magnetizing current, this expression represents the magnitude of the maximum value of the mutual flux divided by $\sqrt{2}$. That is,

$$\frac{N_1 I_\phi}{\mathcal{R}_m} = \frac{\Phi_m}{\sqrt{2}} \tag{13-84}$$

where Φ_m denotes the maximum value of the mutual flux. Accordingly, Eq. (13-83) becomes

$$\omega a M I_\phi = \omega N_1 \frac{\Phi_m}{\sqrt{2}} = \frac{2\pi}{\sqrt{2}} f N \Phi_m = E_1 \qquad (13\text{-}85)$$

It therefore follows from Eq. (13-85) that the mutual inductance parameter so commonly encountered in linear circuit analysis is implied in the expression for the induced emf. Clearly both the induced emf and the mutual inductance are related by the same mutual flux.

PROBLEMS

13-1 A reactor coil with an iron core has 400 turns. It is connected across the 115-volt 60-Hz power line.
(a) Neglecting the resistance voltage drop, calculate the maximum value of the operating flux.
(b) If the flux density is not to exceed 75 kilolines per square inch, what must be the cross-sectional area of the core?

13-2 A fixed sinusoidal voltage is applied to the circuit shown in Fig. P13-2. If the voltage remains connected and the shaded portion of the iron removed, state what happens to the maximum value of the flux and the maximum value of the magnetizing current. Justify your answer. Assume negligible leakage impedance.

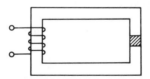

Figure P13-2.

13-3 Repeat Prob. 13-2 assuming the voltage applied to the core is a fixed d-c quantity. Assume d-c current is limited by an external resistor.

13-4 A transformer coil rated at 200 volts and having 100 turns is equipped with a 0.5 tap. If 50 volts are applied to half the number of turns, find the change in the maximum value of the flux compared to the rated value.

13-5 In an open-circuit test of a 25-KVA, 2400/240-volt transformer made on the low side, the corrected readings of amps, volts, and watts are respectively 1.6, 240, and 114. In the short-circuit test the low side is short-circuited and the current, voltage, and power to the high side are measured to be respectively 10.4 amp, 55 volts, and 360 watts.
(a) Find the core loss.
(b) What is the full-load copper loss?
(c) Find the efficiency for a full-load of 0.8 power factor leading.
(d) Compute the per cent voltage regulation for part (c).

13-6 A 50-KVA, 2300/230-volt, 60-Hz transformer has a high-voltage winding resistance of 0.65 ohm and a low-voltage winding resistance of 0.0065 ohm. Laboratory tests showed the following results:

> Open-circuit test: $V = 230$ v, $I = 5.7$ amps, $P = 190$ w
> Short-circuit test: $V = 41.5$ v, $I = 21.7$ amps, $P =$ no WM used

(a) Compute the value of primary voltage needed to give rated secondary voltage when the transformer is connected *step-up* and is delivering 50 KVA at a power factor of 0.8 lagging.

(b) Compute the efficiency under the conditions of part (a).

13-7   The following test data were taken on a 15-KVA, 2200/440-volt, 60-Hz transformer:

> Short-circuit test: $P = 620$ w, $I = 40$ amps, $V = 25$ v
> Open-circuit test: $P = 320$ w, $I = 1$ amp, $V = 440$ v

(a) Calculate the voltage regulation of this transformer when it supplies full-load at 0.8 pf lagging. Neglect the magnetizing current.

(b) Find the efficiency at the load condition of part (a).

13-8   The following test data were taken on a 110-KVA, 4400/440-volt, 60-Hz transformer:

> Short-circuit test: $P = 2000$ w, $I = 200$ amps, $V = 18$ v
> Open-circuit test: $P = 1200$ w, $I = 2$ amps, $V = 4400$ v

Calculate the voltage regulation of this transformer when it supplies rated current at 0.8 pf lagging. Neglect the magnetizing current.

13-9   A 50-KVA, 2300/230-volt, 60-Hz distribution transformer takes 360 watts at a power factor of 0.4 with 2300 volts applied to the high-voltage winding and the low-voltage winding open-circuited. If this transformer has 230 v impressed on the low side with no load on the high side, what will be the current in the low-voltage winding? Neglect saturation.

13-10   A 200/100-volt, 60-Hz transformer has an impedance of $0.3 + j0.8$ ohm in the 200-volt winding and an impedance of $0.1 + j0.25$ ohm in the 100-volt winding. What are the currents on the high and low side, if a short circuit occurs on the 100-volt side with 200 volts applied to the high side?

13-11   A 60-Hz, three-winding transformer is rated at 2300 volts on the high side with a total of 300 turns. Of the two secondary windings, each designed to handle 200 KVA, one is rated at 575 volts and the other at 230 volts. Determine the primary current when rated current in the 230-volt winding is at unity power factor and the rated current in the 575-volt winding is at 0.5 pf lagging. Neglect all leakage impedance drops and magnetizing current.

13-12   An "ideal" transformer has secondary winding tapped at point *b*. The number of primary turns is 100 and the number of turns between *a* and *b* is 300 and between *b* and *c* is 200. The transformer supplies a resistive load connected between *a* and *c* and drawing 7.5 KW. Moreover, a load impedance of $10/45°$ ohms is connected between *a* and *b*. The primary voltage is 1000 volts. Find the primary current.

13-13   Draw a neat phasor diagram of a transformer operating at rated conditions. Assume $N_1/N_2 = 2$ and

$$I_1 r_1 = 0.1 E_1, \qquad I_2 r_2 = 0.1 V_2, \qquad I_m = 0.3 I_2'$$
$$I_1 x_1 = 0.2 E_1, \qquad I_2 x_2 = 0.2 V_2, \qquad I_c = 0.1 I_2'$$

Consider the load power factor to be 0.6 leading. Use V_2 as the reference phasor and show all currents and voltages drawn to scale.

13-14 A 10-KVA, 500/100-volt transformer has $R_{eH} = 0.3$ ohm and $X_{eH} = 5.2$ ohms, and it is used to supply power to a load having a lagging power factor. When supplying power to this load an ammeter, wattmeter, and voltmeter placed in the high-side circuit read as follows:

$$I_1 = 20 \text{ amps}, \qquad V_1 = 500 \text{ volts}, \qquad \text{WM} = 8 \text{ KW}$$

For this condition calculate what a voltmeter would read if placed across the secondary load terminals. Assume the magnetizing current to be negligibly small.

13-15 The following data are taken on a 30-KVA, 2400/240-volt, 60-Hz transformer:

Short-circuit test: $V = 70$ v, $I = 18.8$ amps, $P = 1050$ w
Open-circuit test: $V = 240$ v, $I = 3.0$ amp, $P = 230$ w

(a) Determine the primary voltage when 12.5 amps at 240 volts are taken from the low-voltage side supplying a load of 0.8 pf lagging.
(b) Compute the efficiency in part (a).

13-16 A 30-KVA, 240/120-volt, 60-Hz transformer has the following data:

$$r_1 = 0.14 \text{ ohm}, \qquad r_2 = 0.035 \text{ ohm}$$
$$x_1 = 0.22 \text{ ohm}, \qquad x_2 = 0.055 \text{ ohm}$$

It is desired to have the primary induced emf equal in magnitude to the primary terminal voltage when the transformer carries the full-load current. Neglect the magnetizing current. How must the transformer be loaded to achieve this result?

13-17 The two windings of a 2400/240-volt, 48-KVA, 60-Hz transformer have resistances of 0.6 and 0.025 ohm for the high- and low-voltage windings, respectively. This transformer requires that 238 volts be impressed on the high-voltage coil in order that rated current be circulated in the short-circuited low-voltage winding.

(a) Calculate the equivalent leakage reactance referred to the high side.
(b) How much power is needed to circulate rated current on short circuit?
(c) Compute the efficiency at full-load when the power factor is 0.8 lagging. Assume that the core loss equals the copper loss.

13-18 A coil located on an iron core has a constant 60-Hz voltage applied to it. The total inductance is found to be 10 henries. A second identical coil is placed in parallel with the first one and is so wound that it produces aiding flux. What is the total inductance of the parallel combination? Neglect the winding resistance and leakage flux of each winding.

13-19 Show the derivation that allows Eq. (13-68) to be written from Eq. (13-67) for a sinusoidal forcing function and no saturation.

14

electromechanical

energy conversion

Electromechanical energy conversion involves the interchange of energy between an electrical system and a mechanical system through the medium of a coupling magnetic field. The process is essentially reversible except for a small amount which is lost as heat energy. When the conversion takes place from electrical to mechanical form the device is called a *motor*. When mechanical energy is converted to electrical energy, the device is called a *generator*. Moreover, when the electrical system is characterized by alternating current the devices are referred to as *a-c* motors and *a-c* generators, respectively. Similarly, when the electrical system is characterized by direct current the electromechanical conversion devices are called *d-c* motors and *d-c* generators.

The same fundamental principles underly the operation of both a-c and d-c machines, and they are governed by the same basic laws. Thus, in the computation of the developed torque of an electromechanical energy-conversion device, one basic torque formula (which derives directly from Ampere's law) applies whether the machine is a-c or d-c. The ultimate forms of the torque equations appear different for the two types of machines only because the details of mechanical construction differ. In other words, starting with the same basic principles for the production of electromagnetic torque, the final forms of the torque equations differ to the extent that the mechanical details differ. These comments apply equally in the matter of generating an electromotive force in the armature winding of a machine be it a-c or d-c. Once again a simple basic relationship (Faraday's law) governs the voltage induced. The final forms of the voltage equations differ merely as a reflection of the differences in the construction of the machines. It is important that this be understood at the outset, because a prime objective of the treatment of the subject matter as it unfolds in this chapter is to impress upon the reader the fact that a-c machines are not fundamentally different from d-c machines. They differ merely in construction details; the underlying principles are the same.

This chapter is arranged to give still greater emphasis to this theme. Attention first is given to a basic analysis of the production of electromagnetic torque as well as the generation of voltages starting with the classical fundamental laws. Then the construction features of the various types of electric machines are examined with the purpose of indicating how the conditions for the production of torque and voltage are fulfilled as well as to point out why differences are necessary. Once this background is established, the fundamental equations for developed torque and induced voltage are modified to make these expressions consistent with the particular construction features of the machines involved. In this way the equations are put in a more useful form for purposes of analysis and design. Attention is then directed in succeeding chapters to an analysis of the four major types of electromechanical energy-conversion devices to be found performing innumerable tasks throughout industry and in homes, shops, and offices everywhere.

14-1 Basic Analysis of Electromagnetic Torque

Since electromechanical energy conversion involves the interchange of energy between an electrical and a mechanical system, the primary quantities involved in the mechanical system are torque and speed, while the analogous quantities in the electrical system are voltage and current, respectively. Figure 14-1 represents the situation diagrammatically. *Motor action* results when the electrical system causes a current i to flow through conductors that are placed in a magnetic field. Then by Eq. (12-2) a force is produced on each conductor so that if the conductors are located on a structure that is free to rotate, an electromagnetic torque T results which in turn manifests itself as an angular velocity ω_m. Moreover, with the development of this motor action the revolving conductors cut through the magnetic field, thereby undergoing an electromotive force e, which is really a reaction voltage analogous to the induced emf in the primary winding of a transformer. Note that the coupling field is involved in establishing

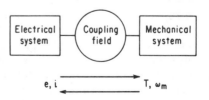

Fig. 14-1. Block representation of electromechanical energy conversion.

the electromagnetic torque T as well as the reaction induced emf e. In the case of *generator action* the reverse process takes place. Here the rotating member—the *rotor*—is driven by a prime mover (steam turbine, gasoline engine, etc.), causing an induced voltage e to appear across the armature winding terminals. Upon the application of an electrical load to these terminals, a current i is made to flow, delivering electrical power to the load. The flow of this current through the armature conductors interacts with the magnetic field to produce a reaction torque opposing the applied torque originating from the prime mover.

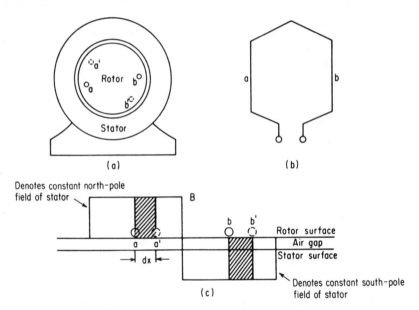

Fig. 14-2. Showing some construction features of an electromechanical conversion device:
    (a) stator and rotor;
    (b) full-pitch coil;
    (c) air-gap field distribution.

In the interest of emphasizing further the character of this motor and generator action and its relationship to the coupling magnetic field, a mathematical description is now undertaken. Consider that we have an electromechanical energy-conversion device which consists of a stationary member—the *stator*—and the rotating member, the rotor. Refer to Fig. 14-2(a). Assume that the stator is equipped with a coil which produces a uniformly distributed pair of magnetic poles and that the rotor is equipped with a total of Z conductors, only two of which (a and b) are depicted in Fig. 14-2(c). It is also assumed that conductor a is connected to conductor b to form a coil as shown in Fig. 14-2(b). By Faraday's law displacement of the coil by a differential distance dx to the new position $a'b'$

causes a change in the flux linking the coil equal to twice the shaded area.† Thus

$$d\phi = 2B \, dA \tag{14-1}$$

The factor 2 accounts for the two areas involved; B is the value of the flux density at the points where conductors a and b lie, and dA is the differential area through which the B-field penetrates. A little thought reveals this area to be equal to the axial length of the rotor l times dx. Thus

$$dA = l \, dx \tag{14-2}$$

However, if it is further assumed that dx results from a linear velocity v imparted to the conductor for the differential time dt, it follows that

$$dx = v \, dt \tag{14-3}$$

Insertion of Eqs. (14-2) and (14-3) into Eq. (14-1) then leads to

$$d\phi = 2Blv \, dt \tag{14-4}$$

The corresponding emf induced in coil ab becomes, by Faraday's law,

$$e = \frac{d\phi}{dt} = 2Blv \tag{14-5}$$

The factor 2 appears in Eq. (14-5) because the coil involves two conductors. It then follows that if the rotor is equipped with a total of Z conductors, which are all series-connected, the total induced voltage can be expressed as

$$e = ZBlv \tag{14-6}$$

When a current i flows through conductors a and b in the configuration of Fig. 14-2(c), the direction of i and the direction of B are 90 degrees apart.‡ Accordingly, a force exists on each conductor which is given by Eq. (14-7). Thus

$$F_c = Bli \tag{14-7}$$

where F_c denotes the force developed on an individual conductor. Since there are Z conductors on the rotor surface, it follows that the total developed force is§

$$F = ZBli \tag{14-8}$$

Moreover, if this force is made to act through a moment arm r, which is the radius of the rotor, the resulting developed torque can be expressed as

$$T = Fr = ZBlri \tag{14-9}$$

A glance at Eqs. (14-6) and (14-9) shows that both the induced voltage and the developed electromagnetic torque are dependent upon the coupling magnetic

---

† In the initial position ab the net flux linked by coil ab is zero, because it encloses one-half of the north-pole flux as well as one-half of the south-pole flux.

‡ Keep in mind that i flows through the conductors and that the conductors by the geometry of the machine will always cut the B-field at right angles.

§ This equation is valid provided that the conductors are paired to form series-connected coils that have a span of one *pole pitch*. A pole pitch is equal to 180 electrical degrees.

field. The factor that determines whether the induced voltage e and the developed torque T are actions or reactions depends upon whether generator or motor action is being developed.

The two mechanical quantities T and ω_m and the two electrical quantities e and i, which are coupled through the medium of the magnetic field B, are related further by the law of conservation of energy. This is readily demonstrated by dividing Eq. (14-6) by Eq. (14-9). Thus

$$\frac{e}{T} = \frac{ZBlv}{ZBlir} = \frac{\omega_m}{i} \tag{14-10}$$

where

$$\omega_m = \frac{v}{r} = \text{mechanical angular velocity of the rotor} \tag{14-11}$$

Rearranging Eq. (14-10) leads to

$$ei = T\omega_m \tag{14-12}$$

which states that the developed electrical power is equal to the developed mechanical power. This statement is valid for generator as well as motor action.

. Finally, on the basis of the foregoing discussion, it should be apparent that a conventional electromechanical energy-conversion device must involve two components. One is the *field winding*, which by definition is that part of the machine which produces the coupling field B. The other is the *armature winding*, which by definition is that part of the machine in which the "working" emf e and the current i exist.

Developed Torque for Sinusoidal B and Ni (the Situation of A-C Machines). The expression for electromagnetic torque as it appears in Eq. (14-9) has limited application because B is assumed constant and all conductors are assumed to produce equal torques in the same direction—a situation which rarely occurs in practical machines. In fact in most a-c machines the field winding is intentionally designed to produce an almost sinusoidally distributed flux density, and the armature winding is similarly arranged to yield an ampere-conductor distribution which is also almost sinusoidally distributed along the periphery of the rotor structure. For this reason we shall now consider the development of an electromagnetic torque equation that applies for such cases.

In differential form Eq. (14-1) can be expressed as

$$dF = Bl\,di \tag{14-13}$$

Here $l\,di$ is an assumed differential current element and its orientation is one of quadrature with the B-field. Moreover, because the field distribution is essentially sinusoidal for a-c machines, we can write for B that

$$B = B_m \sin \theta \tag{14-14}$$

where B_m denotes the amplitude of the flux density wave and θ is a displacement angle measured in electrical degrees. The differential current can be expressed in terms of a spatial distribution as

$$di = J \, d\theta = J_m \sin(\theta + \psi) \, d\theta \tag{14-15}$$

where J is expressed in units of amperes per radian. Of course, J_m denotes the peak value of J. Equation (14-15) essentially states that the current distribution in the armature winding of a-c machines also assumes a sinusoidal distribution. A derivation of this expression appears in Appendix D for the convenience of the interested reader. In general, the armature-winding current distribution is displaced from the field distribution by the angle ψ. See Fig. 14-3. The sine wave

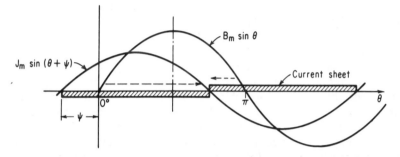

Fig. 14-3. Configuration for development of the torque equation showing the sinusoidally distributed flux density produced by the field winding and the current sheet which represents the armature winding. The broken-line arrows denote relative magnitude and direction of the electromagnetic torque developed per pole. The magnetic field and the current sheet are stationary relative to each other.

representing the current distrubution always has the same number of poles as the field distribution, because the latter induces the former. Inserting Eqs. (14-14) and (14-15) into Eq. (14-13) allows the differential force produced on the rotor of the machine to be expressed as

$$dF = Bl \, di = B_m J_m l \sin \theta \sin(\theta + \psi) \, d\theta \tag{14-16}$$

The expression for the differential torque then becomes

$$dT = r \, dF = B_m J_m lr \sin \theta \sin(\theta + \psi) \, d\theta \tag{14-17}$$

The total developed torque follows readily from this last equation by taking the summation of all torques associated with the current elements beneath one pole and then multiplying by the number of poles of the machine. The result is

$$T = \pi \frac{p}{2} B_m J_m lr \cos \psi \tag{14-18}$$

The details of the derivation appear in Appendix D.

The flux per pole, Φ, is a very useful and important quantity in machine analysis. Consequently, it is frequently desirable to express the developed electromagnetic torque in terms of this quantity. Obtaining the flux per pole is an easy matter if the field distribution is known, as we have certainly here assumed by Eq. (14-14). It merely requires finding the area of the B curve. Such a computation for the sine wave is straightforward, as a glance at Appendix D discloses. The final result for the developed electromagnetic torque thus becomes

$$T = \frac{\pi}{8} p^2 \Phi J_m \cos \psi \qquad \text{n-m} \qquad (14\text{-}19)$$

A close examination of Eq. (14-19) points out clearly the three conditions that must be satisfied for the development of torque in conventional electromechanical energy-conversion devices. First, there must be a field distribution. This is represented by Φ. Second, there must be a current distribution, which here is denoted by J_m. Actually, a better way to describe this is to say that there must be an *ampere-conductor distribution* that has the same number of poles as the field distribution and is stationary relative to it. After all, the current must flow in conductors and the conductors† are the elements that are distributed over the surface of the rotor, thereby becoming subject to the influence of the flux density. Finally, there must exist a favorable space displacement angle between the field distribution and the ampere-conductor distribution. One should note that it is possible to have both a field and an ampere-conductor distribution and still have zero net torque. This happens whenever the field pattern is such that equal and opposite torques beneath each pole are developed on the conductors. In other words, $\psi = 90°$.

It is also worthwhile to state at this point that no really new concepts have been used to derive Eq. (14-19). This result is merely an extension of Ampere's law.

Torque for Nonsinusoidal B and Uniform Ampere-Conductor Distribution (*the Situation of D-C Machines*). The approach in the derivation here is the same as in the preceding case. Again we start with the expression for the differential force as given by Eq. (14-13). In the case of d-c machines, however, B is nonsinusoidal, as a glance at Fig. 14-4 discloses. Moreover, an analytical expression for

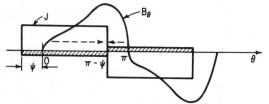

Fig. 14-4. Current-sheet and flux-density curves characteristic of d-c machines.

this field is virtually impossible to obtain because of the saturated state of the sheet steel that this flux penetrates. For our purposes at the moment, we therefore introduce

$$B = B_\theta \qquad (14\text{-}20)$$

where the θ subscript reminds us that because of the nonsinusoidal variation, the value of the flux density for a particular θ would have to come from a field plot.

---

† The magnitude of the currents existing in successive conductors, however, varies sinusoidally.

The differential current, on the other hand, is given simply as

$$di = J\, d\theta \tag{14-21}$$

where J is a constant quantity. This particular characteristic of the d-c machine is attributable to the presence of conducting brushes placed on a commutator. More is said about this in Sec. 14-3.

The differential force appearing on a differential current element located beneath a pole of the field distribution thus becomes

$$dF = B_\theta J l\, d\theta \tag{14-22}$$

and the related torque is

$$dT = r\, dF = B_\theta J l r\, d\theta \tag{14-23}$$

Integrating Eq. (14-23) over one pole and multiplying by the number of poles yields the total resultant electromagnetic torque. Thus

$$T = pJ \int_0^\pi B_\theta l r\, d\theta \qquad \text{n-m} \tag{14-24}$$

Note that an expression without the integral is not possible in this case as it was with the a-c machine, because an analytical expression for B_θ is not known. Of course, the integration called for in Eq. (14-24) could be performed graphically once the field plot for B_θ is known. However, this inconvenience can be avoided entirely by working rather in terms of the flux per pole of the machine, as described in Appendix D. Moreover, information about the flux per pole is available from other considerations, such as the power rating and voltage rating of the machine. Proceeding in this fashion, then, the expression for the developed electromagnetic torque for the d-c machine becomes simply

$$\boxed{T = \frac{p^2}{2} J \Phi} \qquad \text{n-m} \tag{14-25}$$

Here again it is worthwhile to note that the electromagnetic torque is dependent upon a field distribution represented by Φ, an ampere conductor distribution represented by J, and the angular displacement between the two distributions. The quantity ψ does not appear in Eq. (14-25), because for simplicity (and also because it is common practice) the current sheet was assumed in phase with the flux-density curve, thus assuring that all elemental parts of the current sheet experience a unidirectional torque. A glance at Fig. 14-4, however, should make it plain that if ψ is changed from 0° to 90°, the resultant electromagnetic torque becomes zero.

14-2　Analysis of Induced Voltages

The electromechanical energy-conversion process involves the interaction of two fundamental quantities which are related through the law of conservation of energy and the coupling magnetic field. These are, of course, electromagnetic torque and induced voltage. The manner in which electromagnetic torque is developed is described in the preceding section. Here we shall consider the generation of voltages in the armature winding brought by changing flux linkages. Whether or not this generated voltage is a reaction voltage depends upon whether or not the conversion device is behaving as a motor or generator. Our objective now is to obtain a general result for the armature-winding induced emf that can be applied to either a-c or d-c type electromechanical energy-conversion devices.

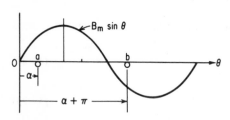

Consider the configuration shown in Fig. 14-5. The flux-density curve is assumed to be sinusoidally† distributed along the air gap of the machine, and the armature winding is assumed to consist of a full-pitch coil having N turns.

Fig. 14-5. Configuration for derivation of induced-emf equation.

A full-pitch coil is one that spans π electrical radians. Just as Ampere's law serves as the starting point in the development of the torque relationships, so Faraday's law serves as the starting point in establishing the induced-voltage relationships. Equation (14-5) is not the result we are seeking, because it applies for the case where the flux-density curve is uniform over the pole pitch. It is more generally useful to deal rather with a sinusoidal distribution of the flux density, as explained more fully later. Since the energy-conversion process involves relative motion between the field and the armature winding, we know that the flux linking the coil changes from a positive maximum to zero to a negative maximum and back to zero again. This cycle repeats every time the coil passes through a pair of flux-density poles. It should be clear from Fig. 14-5 that maximum flux penetrates the coil when it is in a position corresponding to $\alpha = 0°$ or $\alpha = 180°$ or multiples thereof. Minimum flux linkage occurs when $\alpha = 90°$ or $270°$ or multiples thereof. It should be apparent too that the maximum flux that links the coil is the same as the flux per pole, which for sinusoidal distributions is given by Eq. (D-16) of Appendix D. Moreover, as the coil moves relative to the field, it follows that the instantaneous flux linking the coil may be expressed as

$$\phi = \Phi \cos \alpha = \frac{4}{p} B_m lr \cos \alpha \qquad (14\text{-}26)$$

The $\cos \alpha$ factor is included because of the sinusoidal distribution. Furthermore, since α increases with time in accordance with

$$\alpha = \omega t \qquad (14\text{-}27)$$

where ω is expressed in electrical radians per second, the expression for the instantaneous flux linkage becomes

$$\phi = \frac{4}{p} B_m lr \cos \omega t = \Phi \cos \omega t \qquad (14\text{-}28)$$

---

† For a nonsinusoidal field distribution, Fig. 13-5 represents the fundamental component.

Then by Faraday's law the expression for the instantaneous induced voltage becomes

$$e = -N\frac{d\phi}{dt} = \omega N\Phi \sin \omega t \qquad (14\text{-}29)$$

It is important to keep in mind here that Φ represents the *maximum flux linkage*, which is the same as the flux per pole.

Equation (14-29) is applicable to d-c as well as a-c machines. In a later section the induced emf per phase in an a-c machine and the induced voltage appearing between brushes in a d-c machine are derived starting with Eq. (14-29). The fact that in the d-c machine the flux density curve is nonsinusoidal, as Fig. 14-4 shows, is relatively unimportant; what is of primary importance is the fundamental component of the nonsinusoidal distribution. It is for this reason that Eq. (14-29) is derived for sinusoidal distributions. Recall that by means of the Fourier series any nonsinusoidal periodic wave may be expressed entirely in terms of sinusoids.

14-3 Construction Features of Electric Machines

The energy-conversion process usually involves the presence of two important features in a given electromechanical device. These are the field winding, which produces the flux density, and the armature winding, in which the "working" emf is induced. In this section the salient construction features of the principal types of electric machines are described to show the location of these windings, as well as to demonstrate the general composition of such machines.

The Three-phase Induction Motor. This is one of the most rugged and most widely used machines in industry. Its stator is composed of laminations of high-grade sheet steel. The inner surface is slotted to accommodate a three-phase winding. In Fig. 14-6(a) the three-phase winding is represented by three coils, the axes of which are 120 electrical degrees apart. Coil aa' represents all the coils assigned to phase a for one pair of poles. Similarly coil bb' represents phase b coils, and coil cc' represents phase c coils. When one end of each phase is tied to the others, as depicted in Fig. 14-6(b), the three-phase stator winding is said to be Y-connected. Such a winding is called a three-phase winding because the voltages induced in each of the three phases by a revolving flux density field are out of phase by 120 electrical degrees—a distinguishing characteristic of a balanced three-phase system.

The rotor also consists of laminations of slotted ferromagnetic material, but the rotor winding may be either the squirrel-cage type or the would-rotor type. The latter is of a form similar to that of the stator winding. The winding terminals are brought out to three slip rings as depicted in Fig. 14-7(a). This allows an external three-phase resistor to be connected to the rotor winding for the purpose of providing speed control. As a matter of fact it is the need for speed control which in large measure accounts for the use of the wound-rotor type

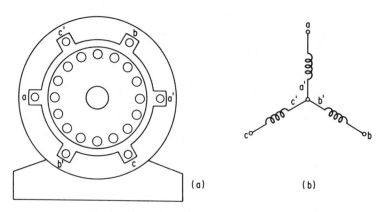

Fig. 14-6. Three-phase induction motor:
(a) showing stator with three-phase winding and squirrel-cage rotor;
(b) schematic representation of three-phase Y-connected stator winding.

induction motor. Otherwise the squirrel-cage induction motor would be used. The squirrel-cage winding consists merely of a number of copper bars imbedded in the rotor slots and connected at both ends by means of copper end rings as depicted in Fig. 14-7(b). (In some of the smaller sizes aluminum is used.) The squirrel-cage construction is not only simpler and more economical than the wound-rotor type but more rugged as well. There are no slip rings or carbon brushes to be bothered with.

In normal operation a three-phase voltage is applied to the stator winding at points a-b-c in Fig. 14-6. As described in Sec. 15-1 magnetizing currents flow in each phase which together create a revolving magnetic field having two poles. The speed of the field is fixed by the frequency of the magnetizing currents and the number of poles for which the stator winding is designed. Figure 14-6 shows the configuration for two poles. If the pattern a-c'-b-a'-c-b' is made to span only 180 mechanical degrees and then is repeated over the remaining 180 mechanical degrees, a machine having a four-pole field distribution results. For a p-pole machine the basic winding pattern must be repeated $p/2$ times within the circumference of the inner surface of the stator.

The revolving field produced by the stator winding cuts the rotor conductors, thereby inducing voltages. Since the rotor winding is short-circuited by the end rings, the induced voltages cause currents to flow which in turn react with the field to produce electromagnetic torque—and so motor action results.

Accordingly, on the basis of the foregoing description, it should be clear that for the three-phase induction motor the field winding is located on the stator and the armature winding on the rotor. Another point worth noting is that this machine is singly excited, i.e. electrical power is applied only to the stator winding. Current flows through the rotor winding by induction. As a consequence both the

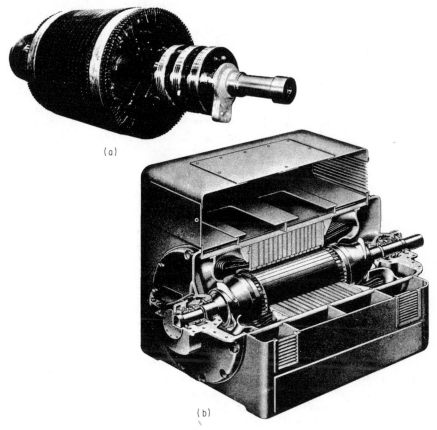

Fig. 14-7. (a) Wound rotor for three-phase induction motor. (Courtesy
of General Electric Company);
(b) Cutaway view showing bars in rotor of squirrel-cage
induction motor. (Courtesy of General Electric Company)

magnetizing current, which sets up the magnetic field, and the power current,
which allows energy to be delivered to the shaft load, flow through the stator
winding. For this reason, and in the interest of keeping the magnetizing current
as small as possible in order that the power component may be correspondingly
larger for a given rating, the air gap of induction motors is made as small as
mechanical clearance will allow. The air-gap lengths vary from about 0.02 inch for
small machines to 0.05 inch for machines of higher rating and speed.

Synchronous Machines. The essential construction features of the synchronous machine are depicted in Fig. 14-8. The stator consists of a stator frame,
a slotted stator core, which provides a low-reluctance path for the magnetic flux,
and a three-phase winding imbedded in the slots. Note that the basic two-pole

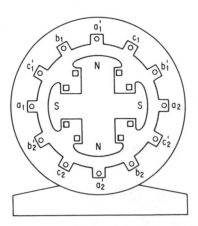

Fig. 14-8. Salient pole synchronous machine.

pattern of Fig. 14-6(a) is repeated twice, indicating that the three-phase winding is designed for four poles. The rotor either is cylindrical and equipped with a distributed winding or else has salient poles with a coil wound on each leg as depicted in Fig. 14-8. The cylindrical construction is used exclusively for turbogenerators, which operate at high speeds. On the other hand the salient-pole construction is used exclusively for synchronous motors operating at speeds of 1800 rpm or less.

When operated as a generator the synchronous machine receives mechanical energy from a prime mover such as a steam turbine and is driven at some fixed speed. Also, the rotor winding is energized from a d-c source, thereby furnishing a field distribution along the air gap. When the rotor is at standstill and d-c flows through the rotor winding, no voltage is induced in the stator winding because the flux is not cutting the stator coils. However, when the rotor is being driven at full speed, voltage is induced in the stator winding and upon application of a suitable load electrical energy may be delivered to it.

A little thought should make it apparent that for the synchronous machine the field winding is located on the rotor; the armature winding is located on the stator. This conclusion is valid even when the synchronous machine operates as a motor. In this mode a-c power is applied to the stator winding and d-c power is applied to the rotor winding for the purpose of energizing the field poles. Mechanical energy is then taken from the shaft. Note, too, that unlike the induction motor, the synchronous motor is a doubly fed machine, i.e., energy is supplied to the rotor as well as the stator winding. In fact it is this characteristic which enables this machine to develop a nonzero torque at only one speed—hence the name synchronous. This matter is discussed further in Sec. 16-3.

Because the magnetizing current for the synchronous machine originates from a separate source (the d-c supply), the airgap lengths are larger than those found in induction motors of comparable size and rating. However, synchronous machines are more expensive and less rugged than induction motors in the smaller horsepower ratings because the rotor must be equipped with slip rings and brushes in order to allow the direct current to be conducted to the field winding.

D-c Machines. Electromechanical energy-conversion devices that are characterized by direct current are more complicated than the a-c type. In addition to a field winding and armature winding a third component is needed to serve the function of converting the induced a-c armature voltage into a direct voltage. Basically the device is a mechanical rectifier and is called a *commutator*.

Appearing in Fig. 14-9 are the principal features of the d-c machine. The

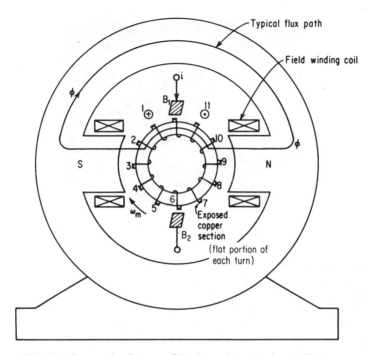

Fig. 14-9. Construction features of the d-c machine showing the Gramme-ring armature winding.

stator consists of an unlaminated ferromagnetic material equipped with a pro-truding structure around which coils are wrapped. The flow of direct current through the coils establishes a magnetic field distribution along the periphery of the air gap in much the same manner as occurs in the rotor of the synchronous machine. Hence in the d-c machine the field winding is located on the stator. It follows then that the armature winding is on the rotor. The rotor is composed of a laminated core, which is slotted to accommodate the armature winding. It also contains the commutator—a series of copper segments insulated from one another and arranged in cylindrical fashion as shown in Fig. 14-10. Riding on the commutator are appropriately placed carbon brushes which serve to conduct direct current to or from the armature winding depending upon whether motor or generator action is taking place.

In Fig. 14-9 the armature winding is depicted as a coil wrapped around a toroid. This is merely a schematic convenience. An actual winding is wound as shown in Fig. 14-10; that is, no conductors are wasted by placing them on the inner surface of the rotor core where no flux penetrates. In Fig. 14-9 those parts of the armature winding which lie directly below the brush width are assumed to have the insulation removed, i.e. the copper is exposed. This allows current to be conducted to and from the armature winding through the brush as the rotor revolves. In a practical winding each coil is made accessible to the brushes

Fig. 14-10. Armature winding of a d-c machine showing the coils connected to the commutator.

by connecting the coils to individual commutator segments as indicated in Fig. 14-10 and then placing the brushes on the commutator.

For motor action direct current is made to flow through the field winding as well as the armature winding. If current is assumed to flow into brush B_1 in Fig. 14-9, then note that on the left side of the rotor for the *outside* conductors current flows into the paper while the opposite occurs for the conductors located on the outside surface of the right side of the rotor. By Eq. (14-2) a force is produced on each conductor, thereby producing a torque causing clockwise rotation. Now the function of the commutator is to assure that as a conductor such as 1 in Fig. 14-9 revolves and thus goes from the left side of brush B_1 to the right side, the current flowing through it reverses, thus yielding a continuous unidirectional torque for the entire armature winding. Recall that a reversed conductor current in a flux field of reversed polarity keeps the torque unidirectional. The reversal of current comes about because the commutator always allows current to be conducted in the same directions in either side of the armature winding whether or not it is rotating.

Another point of interest in Fig. 14-9 concerns the location of the brushes. By placing the brushes on a line perpendicular to the field axis all conductors contribute in producing a unidirectional torque. If, on the other hand, the brushes were placed on the same line as the field axis, then half of the conductors would produce clockwise torque and the other half counterclockwise torque, yielding a zero net torque. In the notation of Fig. 14-4 this situation would correspond to a value of ψ equal to 90 degrees.

14-4   Practical Forms of Torque and Voltage Formulas

Our objective here is to modify the basic expressions for induced voltage and electromagnetic torque to a form that is more meaningful in terms of the specific data of a particular electromechanical energy-conversion device be it a-c or d-c.

A-c Machines. If N denotes the total number of turns per phase of a three-phase winding, the instantaneous value of the emf induced in any one phase can be represented by Eq. (14-29), which is repeated here for convenience. Thus

$$e = \omega N\Phi \sin \omega t \qquad (14\text{-}30)$$

Keep in mind that Φ denotes the total flux per pole and ω represents the relative cutting speed in electrical radians per second of the winding with respect to the flux-density wave. It is related to the frequency f of the a-c device by

$$\omega = 2\pi f \qquad (14\text{-}31)$$

where f is expressed in cycles per second or Hz.

The maximum value of this a-c voltage occurs when $\sin \omega t$ has a value of unity. Hence

$$E_{\max} = \omega N\Phi \quad \text{volts} \qquad (14\text{-}32)$$

The corresponding rms value is

$$E \equiv \frac{E_{\max}}{\sqrt{2}} = \sqrt{2}\,\pi f N\Phi = 4.44 f N\Phi \quad \text{volts} \qquad (14\text{-}33)$$

A comparison of this expression with Eq. (13-18) reveals that the equations have an identical form. There is a difference, however, and it lies in the meaning of Φ. In the transformer, Φ_m is the maximum flux corresponding to the peak magnetizing current consistent with the magnitude of the applied voltage. In the a-c electro-mechanical energy-conversion device Φ is the maximum flux per pole that links with a coil having N turns and spanning the full pole pitch.

In any practical machine the total turns per phase are not concentrated in a single coil but rather are distributed over one-third of a pole pitch (or 60 electrical degrees for each of the three phases). In addition the individual coils that make up the total N turns are intentionally designed not to span the full pole pitch but rather only about 80 to 85 per cent of a pole pitch. Such a coil is called a *fractional-pitch coil*. The use of a *distributed* winding employing *fractional-pitch* coils has the advantage of virtually eliminating the effects of all harmonics that may be present in the flux-density wave while only slightly reducing the fundamental component.[†] The reduction in the fundamental component can be represented by a winding factor denoted by K_w. Usually K_w has values ranging from 0.85 to 0.95. Accordingly, the final practical version of the induced rms voltage equation for an a-c machine is

$$\boxed{E = 4.44 f N K_w \Phi}\quad \text{volts} \qquad (14\text{-}34)$$

We turn attention next to the practical forms of the torque equations. There are two useful forms to which the basic expression for electromagnetic torque in a-c machines as expressed by Eq. (14-19) may be converted. In one form the ampere-conductor distribution of the armature winding is emphasized. In the other it is the conservation of energy which is stressed.

[†] Consult M. Liwshitz-Garik and C. C. Whipple, *Electric Machinery*, vol. II (Princeton, N. J.: D. Van Nostand Co., Inc., 1946), Chapter 5.

The quantity J_m is Eq. (14-19) is related to the mmf per pole by

$$J_m = \mathscr{F}_p \qquad (14\text{-}35)$$

Consult Appendix D for the derivation. Furthermore, for any given a-c machine the quantity $\mathscr{F}_p$ is entirely determinable from the design data as they appear in the following equation:[†]

$$\mathscr{F}_p = \frac{2\sqrt{2}}{\pi} q \frac{N_2 K_{w_2}}{p} I_2 = 0.9 q \frac{N_2 K_{w_2}}{p} I_2 \qquad (14\text{-}36)$$

where q = number of phases of the armature winding
 N_2 = number of turns per phase of the armature winding
 K_{w_2} = armature winding factor
 I_2 = armature winding current per phase
 p = number of poles

Hence the expression for electromagnetic torque as described by Eq. (14-19) becomes

$$T = \frac{\pi}{8} p^2 \Phi \mathscr{F}_p \cos \psi = \frac{\pi}{8}(0.9) p \Phi N_2 K_{w_2} q I_2 \cos \psi \qquad (14\text{-}37)$$

Moreover, qN_2 is the *total* number of turns on the armature surface. This can be expressed in terms of the total number of conductors, Z_2, by making use of the fact that it requires two conductors to make one turn. Thus

$$N_2 = q \frac{Z_2}{2} \qquad (14\text{-}38)$$

Inserting this into Eq. (14-37) yields

$$\boxed{T = 0.177 p \Phi (Z_2 K_{w_2} I_2) \cos \psi} \quad \text{n-m} \qquad (14\text{-}39)$$

The quantity in parentheses emphasizes the role played by the ampere-conductor distribution of the armature winding. Of course in order to compute the torque by this equation we need information about the space displacement angle ψ in addition to the design data, viz. p, Φ, Z_2, K_{w_2}. Normally ψ can be determined by the same data that lead to the determination of I_2, which is described in the next chapter.

A second approach to obtaining a useful form of the basic torque equation is to replace Φ by Eq. (14-34) and J_m by Eqs. (14-35) and (14-36) in Eq. (14-19). Thus

$$T = \frac{\pi}{8} p^2 \left[\frac{E_2}{\sqrt{2}\,\pi f N_2 K_{w_2}} \right]\left[\frac{2\sqrt{2}}{\pi} q \frac{N_2 K_{w_2}}{p} I_2 \right] \cos \psi$$

or

$$T = \frac{p}{4\pi f} p E_2 I_2 \cos \psi \qquad (14\text{-}40)$$

---

† Consult V. Del Toro, *Electromechanical Devices for Energy Conversion and Control Systems* (Englewood Cliffs, N. J.: Prentice-Hall, Inc., 1968), Appendix C.

Moreover, since the mechanical angular velocity ω_m is related to the electrical angular velocity ω by

$$\omega_m = \frac{2}{p}\omega = \frac{2}{p}(2\pi f) = \frac{4\pi f}{p} \tag{14-41}$$

it follows that Eq. (14-40) may be written as

$$T = \frac{1}{\omega_m}qE_2I_2\cos\psi \qquad \text{n-m} \tag{14-42}$$

where E_2 and I_2 denote the induced voltage and current per phase of the armature winding.

It can be shown that the *space* displacement angle ψ in this equation is identical to the time displacement angle θ_2†, which is the phase angle existing between the two time phasors E_2 and I_2. Accordingly, the expression for the electromagnetic torque may be written as

$$\boxed{T = \frac{1}{\omega_m}qE_2I_2\cos\theta_2} \qquad \text{n-m} \tag{14-43}$$

Note the power balance, which states that the developed mechanical power is equal to the developed electrical power.

D-c Machines. The useful version of the expression for the emf induced in the armature winding of a d-c machine and appearing at the brushes also derives from Eq. (14-29). The presence of the commutator, however, makes a difference in the manner in which this equation is to be treated to obtain the desired result. We are interested in the voltage as it appears at the brushes. If we follow coil 1 in Fig. 14-9 as it leaves brush B_2 and advances to brush B_1 we notice that the emf induced in this coil is always directed out of the paper beneath the south-pole flux. In fact this is true of any coil that moves from B_2 to B_1. Having established the direction of this voltage for the coils on the left side of the armature, we then note that upon tracing through the winding starting in the direction called for by the sign of the emf induced in coil 1 we terminate at brush B_2. When coil 1 rotates to a position on the right side of the armature winding such as position 11, the direction of the induced emf is out of the paper. If we start with this indicated direction and trace the armature winding we again terminate at brush B_2. Therefore it is reasonable to conclude that brush B_1 is at a fixed polarity and brush B_2 is at a reversed fixed polarity. In essence this means that, if we were to ride along as observers on coil 1 as it traverses through one revolution beginning at brush B_2, the fundamental component of the induced emf would exhibit the variation shown in Fig. 14-11. Note that *relative to brush B_1* the voltage induced in the coil as it moves from B_1 to B_2 appears *rectified* because the action of the commutator in conjunction with the brushes is to fix the armature winding in space in spite of its rotation. Accordingly, whether we look at a coil moving from B_2 to B_1 on the left side or from B_1 to B_2 on the right side, the directions of the induced emf's

† See p. 531

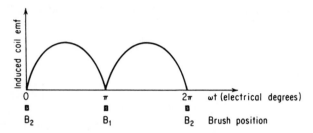

Fig. 14-11. Fundamental component of the induced emf in a coil of the armature winding of a d-c machine.

are such as to make B_1 assume one fixed polarity and B_2 the opposite fixed polarity.

For a single coil having N_c turns the average value of the unidirectional voltage appearing between the brushes is obtained by integrating Eq. (14-29) over π electrical degrees and dividing by π. Thus

$$E_c = \frac{1}{\pi} \int_0^\pi \omega N_c \Phi \sin (\omega t)\, d(\omega t) = 4f N_c \Phi \qquad (14\text{-}44)$$

It is customary to express the frequency in terms of the speed of the armature in *rpm* which is given by the relationship†

$$\boxed{f = \frac{pn}{120}} \qquad (14\text{-}45)$$

where p denotes the number of poles and n the speed in *rpm*. Inserting Eq. (14-43) into Eq. (14-44) and rearranging leads to

$$E_c = p\Phi(2N_c)\frac{n}{60} = p\Phi z \frac{n}{60} \qquad (14\text{-}46)$$

where z denotes the number of conductors per coil.

Equation (14-44) is the average value of the induced emf for a single coil. If many such coils are placed to cover the entire surface of the armature, the total d-c voltage appearing between the brushes can be considerably increased. If the armature winding is assumed to have a total of Z conductors and a parallel paths, then the induced emf of the armature winding appearing at the brushes becomes

$$\boxed{E_a = p\Phi\frac{Z}{a}\frac{n}{60} = \frac{pZ}{60a}\Phi n = K_E \Phi n} \qquad (14\text{-}47)$$

† Rotating a coil in a two-pole field at a rate of one revolution per second results in a frequency of one Hz. For a four-pole field each revolution per second yields two Hz. For a p-pole distribution the frequency generated in Hz is

$$f = \frac{p}{2}(\text{rps}) = \frac{p}{2}\left(\frac{\text{rpm}}{60}\right) = \frac{pn}{120}$$

where K_E is a winding constant defined as

$$K_E \equiv \frac{pZ}{60a} \tag{14-48}$$

It is worth keeping in mind that the induced-emf equation for the a-c machine and that for the d-c machine originate from the same starting point. To further emphasize this common origin let it be said that we can derive Eq. (14-45) from Eq. (14-34). It merely requires introducing the appropriate winding factor as it applies to the d-c armature winding. In fact it is the quasi-annihilating effect of the winding factor on the harmonics that allows the derivation of Eq. (14-47) to proceed in terms of the fundamental in spite of the nonsinusoidal field distribution characteristic of d-c machines. It is assumed throughout, however, that the fundamental component of flux is virtually the same as the total flux per pole.

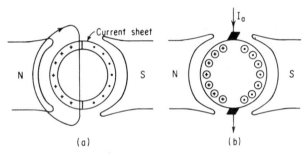

Fig. 14-12. D-c machine:
    (a) armature with current sheet;
    (b) armature with finite ampere-conductor distribution.

A practical version of the electromagnetic torque formula readily follows from Eq. (14-25) if we replace the current-density sheet J by an equivalent expression involving the design data of the machine. In this connection refer to Fig. 14-12(a). Applying Ampere's circuital law to the typical flux path shown leads to

$$\mathscr{F} = \pi \,(\text{rad})\, J\left(\frac{\text{amp-turn}}{\text{rad}}\right) \tag{14-49}$$

Hence the mmf per pole is

$$\mathscr{F}_p = \frac{\mathscr{F}}{2} = \frac{\pi J}{2} \tag{14-50}$$

This expression is applicable whenever the armature winding is represented by a current sheet. In a practical situation the armature winding is always represented by a finite ampere-conductor distribution as depicted in Fig. 14-12(b). In such cases the armature mmf per pole may be expressed in terms of the total current I_a entering or leaving a brush and the total conductors Z on the armature surface. Thus

$$\mathscr{F}_p = \left(\frac{I_a}{a}\right)\left(\frac{Z}{2}\right)\frac{1}{p} = \text{mmf/pole} = \text{AT/pole} \tag{14-51}$$

Upon equating the last two expressions we obtain the equation for J. Thus

$$J = \frac{I_a Z}{\pi p a} \tag{14-52}$$

Insertion of Eq. (14-50) into Eq. (14-25) then yields the desired expression for the electromagnetic torque developed by the d-c machine. Hence

$$T = \frac{p^2}{2} J\Phi = \frac{pZ}{2\pi a} \Phi I_a \qquad \text{n-m} \tag{14-53}$$

or

$$\boxed{T = K_T \Phi I_a} \tag{14-54}$$

where

$$K_T \equiv \frac{pZ}{2\pi a} = \text{torque constant} \tag{14-55}$$

The foregoing form of the torque formula stresses the ampere-conductor distribution taken in conjunction with the flux field. It is analogous to Eq. (14-39) for the a-c machine. Of course in Eq. (14-54) ψ does not appear because it was made zero by placing the brush axis in quadrature with the field axis. It is important to note, however, that the fundamental quantities for the production of torque are there. The equations differ only to the extent that the mechanical details of construction differ.

An alternative expression for the electromagnetic torque results upon substituting in Eq. (14-53) the expression for Φ as obtained from Eq. (14-47). Thus from Eq. (14-47)

$$\Phi = \frac{60a}{pZn} E_a \tag{14-56}$$

so that

$$T = \frac{pZ}{2\pi a} \left[\frac{60a}{pZn} \right] E_a I_a = \frac{60}{2\pi n} E_a I_a \tag{14-57}$$

But

$$\omega_m = \frac{2\pi n}{60} \tag{14-58}$$

Therefore

$$\boxed{T = \frac{1}{\omega_m} E_a I_a} \tag{14-59}$$

A glance at this result again points out the power-balance relationship that underlies the operation of electromechanical energy-conversion devices. Equation (14-59) for the d-c machine is analogous to Eq. (14-43) for the a-c machine.

PROBLEMS

14-1 The instantaneous voltage generated in a coil revolving in a magnetic field can be calculated either by the flux-linkage concept or by the "*Blv*" concept. With this in mind consider the following problem. A square coil 20 cm on a side has 60 turns and is so located that its axis of revolution is perpendicular to a uniform magnetic field in air of 0.06 weber per square meter. The coil is driven at a constant speed of 150 rpm. Compute:
   (a) The maximum flux passing through the coil.
   (b) The maximum flux linkage.
   (c) The time variation of the flux linkage through the coil.
   (d) The maximum instantaneous voltage generated in the coil using both concepts referred to above. Indicate by a sketch the position of the coil at this instant.
   (e) The average value of the voltage induced in the coil over one cycle.
   (f) The voltage generated in the coil when the plane of the coil is 30° from the vertical. Compute by both methods.

14-2 A square coil of 100 turns is 10 cm on each side. It is driven at a constant speed of 300 rpm.
   (a) The coil is placed so that its axis of revolution is perpendicular to a *uniform* (see Fig. P14-3(b)) field of 0.1 weber/m². At time $t = 0$, the coil is in a position of maximum flux linkages. Derive an expression for the instantaneous voltage generated. Sketch the voltage waveshape for one cycle.
   (b) The coil is now placed in a *radial* field (see Fig. P14-3(a)). All other data remain the same as for part (a). Sketch the voltage waveshape for one cycle showing the numerical value of the maximum voltage.

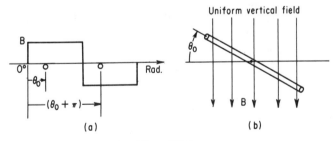

Figure P14-3

14-3 In Figs. P14-3(a) and (b) the flux per pole is 0.02 weber. The coil is revolving at 1800 rpm and consists of 2 turns.
   (a) For the configuration of Fig. P14-3(a) derive an expression for the flux linking the coil in terms of the flux per pole and θ_0.
   (b) What is the instantaneous value of the coil voltage in Fig. P14-3(a).
   (c) Compute the maximum value of the voltage induced in the arrangement of Fig. P14-3(b).
   (d) Assuming the coil of Fig. P14-3(b) is connected to a pair of commutator segments, find the d-c value of this voltage.

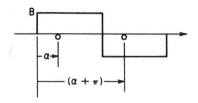

Figure P14-4

14-4   An electric machine has a field distribution as shown in Fig. P14-4. The coil spans a full 180° and has N turns. The machine has two poles and an axial length l and rotor radius r.

(a) Obtain the expression for the flux linkage per pole in terms of α expressed in electrical degrees.

(b) By using Faraday's law find the expression for the induced emf in terms of the flux per pole.

(c) Use the Blv form for the induced emf and verify the result of part (b).

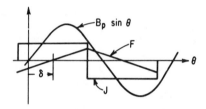

Figure P14-5

14-5   A p-pole machine has the sinusoidal field distribution $B_p \sin \theta$, where θ is in electrical degrees. The armature consists of a uniform current sheet of value J, which has assoicated with it a triangular mmf having an amplitude per pole of F_A. The machine has an axial length l and a radius r.

(a) Determine the expression for the maximum flux pole.

(b) Obtain the expression for the developed electromagnetic torque in terms of B_p, J, and δ.

(c) Express the torque found in part (b) in terms of F_A and the flux per pole Φ. Assume $\delta = \pi/2$.

(d) Is the torque found in part (b) constant at every point in the air gap? Explain.

14-6   A 4-pole a-c induction motor is characterized by a sinusoidal rotor mmf as well as a sinusoidal flux density. Moreover, the flux per pole is 0.02 weber. When operating at a specified load condition, the space displacement angle is found to be 53.3°. What is the amplitude of the rotor mmf wave required to produce a torque of 40 newton-meters?

14-7   A 60-Hz, 8-pole a-c motor has an emf of 440 volts induced in its armature winding, which has 180 effective turns. When this motor develops 10 KW of power, the amplitude of the sinusoidal field mmf is equal to 800 ampere-turns.

(a) Find the value of the resultant air-gap flux per pole.

(b) Find the angle between the flux wave and the mmf wave.

14-8 The induced emf per phase of a 60-Hz, 4-pole induction motor is 120 volts. The number of effective turns per phase is 100. When this motor develops a torque of 60 newton-meters the space displacement angle is 30°. Find the value of the total armature ampere-turns.

14-9 The three-phase armature winding of a 4-pole, 60-Hz, a-c induction motor is equipped with 320 effective turns per phase. When the motor develops 10 KW it has an armature current of 40 amps, a space displacement angle of 37°, and a speed of rotation of 1700 rpm.

(a) Find the amplitude of the armature mmf per pole.

(b) What is the value of the developed torque?

(c) Compute the induced emf in the armature winding per phase.

(d) Find the flux per pole.

15

the three-phase induction motor

One distinguishing feature of the induction motor is that it is a *singly excited* machine. Although such machines are equipped with both a field winding and an armature winding, in normal use an energy source is connected to one winding alone, the field winding. Currents are made to flow in the armature winding by induction, which creates an ampere-conductor distribution that interacts with the field distribution to produce a net unidirectional torque. The frequency of the induced current in the conductor is affected by the speed of the rotor on which it is located; however, the relationship between the rotor speed and the frequency of the armature current is such as to yield a resulting ampere-conductor distribution that is stationary relative to the field distribution of the stator. As a result the singly excited induction machine is capable of producing torque *at any speed below synchronous speed*.† For this reason the induction machine is placed in the class of *asychronous machines*. In contrast, *synchronous machines* are electromechanical

† Synchronous speed is determined by the frequency of the source applied to the field winding and the number of poles for which the machine is designed. These quantities are related by Eq. (14-45). Thus,

$$\text{synchronous speed} = \frac{120f}{p} = n_s$$

energy-conversion devices in which a net torque can be produced at only one† speed of the rotor. The distinguishing characteristic of the synchronous machine is that it is a *doubly excited* device except when it is being used as a reluctance motor.

The salient construction features of the three-phase induction motor are described in Sec. 13-3. Because the induction machine is singly excited, it is necessary that both the magnetizing current and the power component of the current flow in the same lines. Moreover, because of the presence of an air gap in the magnetic circuit of the induction machine, an appreciable amount of magnetizing current is needed to establish the flux per pole demanded by the applied voltage. Usually, the value of the magnetizing current for three-phase induction motors lies between 25 and 40 per cent of the rated current. Consequently, the induction motor is found to operate at a low power factor at light loads and at less than unity power factor in the vicinity of rated output.

We are concerned in this chapter with a description of the theory of operation and of the performance characteristics of the three-phase induction motor. Our discussion begins with an explanation of how a revolving magnetic field is obtained with a three-phase winding. After all it is this field that is the driving force behind induction motors.

15-1 The Revolving Magnetic Field

The application of a three-phase voltage to the three-phase stator winding of the induction motor creates a rotating magnetic field, which by transformer action induces a "working" emf in the rotor winding. The rotor-induced emf is called a working emf because it causes a current to flow through the armature winding conductors. This combines with the revolving flux-density wave to produce torque in accordance with Eq. (14-39). Consequently, we can view the revolving field as the key to the operation of the induction motor.

The rotating magnetic field is produced by the contributions of space-displaced phase windings carrying appropriate time-displaced currents. To understand this statement let us turn attention to Figs. 15-1 and 15-2. Appearing in Fig. 15-1 are the three-phase currents which are assumed to be flowing in phases *a*, *b*, and *c* respectively. Note that these currents are time-displaced by the equivalent of 120 electrical degrees. Depicted in Fig. 15-2 is the stator structure and the three-phase winding. Note that each phase (normally distributed over 60 electrical degrees) for convenience is represented by

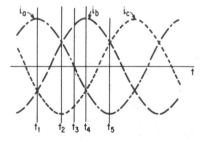

Fig. 15-1. Balanced three-phase alternating currents.

† Theoretically there are two rotor speeds at which a net torque different from zero can exist, but at the second speed enormous currents flow, which makes operation impractical.

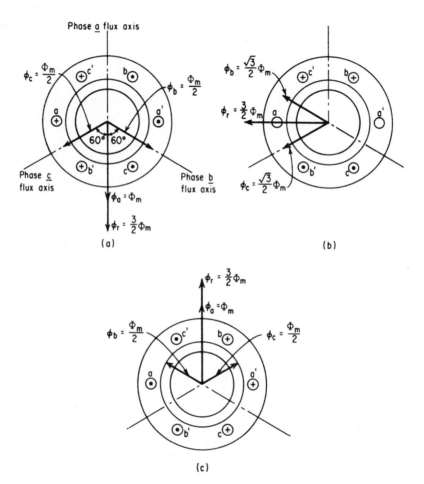

Fig. 15-2. Representing the rotating magnetic field at three different instants of time:

(a) time t_1 in Fig. 15-1;

(b) time t_3;

(c) time t_5.

a single coil. Thus coil *a-a'* represents the entire phase *a* winding having its flux axis directed along the vertical. This means that whenever phase *a* carries current it produces a flux field directed along the vertical—up or down. The right-hand flux rule readily verifies this statement. Similarly, the flux axis of phase *b* is 120 electrical degrees displaced from phase *a*, and that of phase *c* is 120 electrical degrees displaced from phase *b*. The unprimed letters refer to the beginning terminal of each phase.

Let us consider the determination of the magnitude and direction of the resultant flux field corresponding to time instant t_1 in Fig. 15-1. At this instant the current in phase a is at its positive maximum value while the currents in phases b and c are at one-half their maximum negative values. In Fig. 15-2 it is arbitrarily assumed that when current in a given phase is positive, it flows into the paper with respect to the unprimed conductors. Thus since, at time t_1, i_a is positive, a cross is used for conductor a. See Fig. 15-2(a). Of course a dot is used for a' because it refers to the return connection. Then by the right-hand rule it follows that phase a produces a flux contribution directed downward along the vertical. Moreover, the magnitude of this contribution is the maximum value because the current is a maximum. Hence $\phi_a = \Phi_m$ where Φ_m is the maximum flux per pole of phase a. It is important to understand that phase a really produces a sinusoidal flux field with the amplitude located along the axis of phase a as depicted in Fig. 15-3. However, in Fig. 15-2(a) this sinusoidal distribution is conveniently represented by the vector ϕ_a.

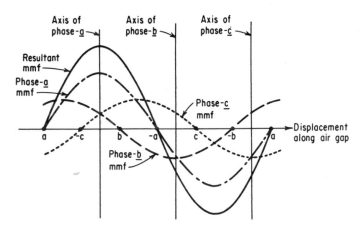

Fig. 15-3. Component and resultant field distributions corresponding to t_1 in Fig. 15-1.

In order to determine the direction and magnitude of the field contribution of phase b at time t_1, we note first that the current in phase b is negative with respect to that in phase a. Hence the conductor that stands for the beginning of phase b must be assigned a dot while b' is assigned a cross. Hence the instantaneous flux contribution of phase b is directed downward along its flux axis and the magnitude of phase b flux is one-half the maximum because the current is at one-half its maximum value. Similar reasoning leads to the result shown in Fig. 15-2(a) for phase c. A glance at the space picture corresponding to time t_1, as illustrated in Fig. 15-2(a), should make it apparent that the resultant flux per pole is directed downward and has a magnitude 3/2 times the maximum flux per pole of any one phase. Figure 15-3 depicts the same results as Fig. 15-2(a) but does so in terms of sinusoidal flux waves rather than flux vectors. Keep in mind that the resultant flux vector in Fig. 15-2 shows the direction in which flux crosses the air gap. Once across the air gap the flux is confined to the iron in the usual fashion.

Next let us investigate how the situation of Fig. 15-2(a) changes as time passes through 90 electrical degrees from t_1 to t_3 in Fig. 15-1. Here phase a current is zero, yielding no flux contribution. The current in phase b is positive and equal to $\sqrt{3}/2$ its maximum value. Phase c has the same current magnitude but is negative. Together phases b and c combine to produce a resultant flux having the same magnitude as at time t_1. See Fig. 15-2(b). It is important to note, too, that an elapse of 90 electrical degrees in time results in a rotation of the magnetic flux field of 90 electrical degrees.

A further elapse of time equivalent to an additional 90 electrical degrees leads to the situation depicted in Fig. 15-2(c). Note that again the axis of the flux field is revolved by an additional 90 electrical degrees.

On the basis of the foregoing discussion it should be apparent that the application of three-phase currents through a balanced three-phase winding gives rise to a rotating magnetic field that exhibits two characteristics: (1) it is of constant amplitude and (2) it is of constant speed. The first characteristic has already been demonstrated. The second follows from the fact that the resultant flux traverses through 2π electrical radians in space for every 2π electrical radians of variation in time for the phase currents. Hence for a two-pole machine, where electrical and mechanical degrees are identical, each cycle of variation of current produces one complete revolution of the flux field. Hence this is a fixed relationship which is dependent upon the frequency of the currents and the number of poles for which the three-phase winding is designed. In the case where the winding is designed for four poles it requires two cycles of variation of the current to produce one revolution of the flux field. Therefore it follows that for a p-pole machine the relationship is

$$f = \frac{p}{2}\,(\text{rps}) = \frac{p}{2}\left(\frac{n}{60}\right) \tag{15-1}$$

where f is in Hz and rps denotes revolutions per second. Note that Eq. (15-1) is identical to Eq. (14-45).

An inspection of the ampere-conductor distribution of the stator winding at the various time instants reveals that the individual phases cooperate in such a fashion as to produce a solenoidal effect in the stator. Thus in Fig. 15-2(a) the directions of the currents are such that they all enter on the left side and leave on the right side. The right-hand rule indicates that the flux field is then directed downward along the vertical. In Fig. 15-2(b) the situation is similar except that now the cross and dot distribution is such that the resultant flux field is oriented horizontally towards the left. Therefore, it can be concluded that the rotating magnetic field is a consequence of the revolving mmf associated with the stator winding.

In the foregoing it is pointed out that the flow of balanced three-phase currents through a balanced three-phase winding yields a rotating field of constant amplitude and speed. If neither of these conditions is exactly satisfied it is still possible to obtain a revolving magnetic field but it will not be of constant amplitude nor of constant linear speed. In general for a q-phase machine a rotating field of constant amplitude and constant speed results when the following two conditions are satisfied: (1) there is a *space* displacement between balanced phase windings of $2\pi/q$ electrical degrees, and (2) the currents flowing through the phase windings are balanced and *time*-displaced by $2\pi/q$ electrical degrees. For the three-phase machine $q = 3$ and so the now familiar 120° figure is obtained. The only exception to the rule is the two-phase machine. Because the two-phase situation is a special case of the four-phase system, a value of q equal to 4 must be used.

One final point is now in order. The speed of rotation of the field as described by Eq. (15-1) is always given relative to the phase windings carrying the time-varying currents. Accordingly, if a situation arises where the winding is itself revolving, then the speed of rotation of the field relative to inertial space is different than it is with respect to the winding.

15-2 The Induction Motor as a Transformer

The three-phase induction motor may be compared with the transformer because it is a singly-energized device which involves changing flux linkages with respect to the stator and rotor windings. In this connection assume that the rotor is of the wound type and Y-connected as illustrated in Fig. 15-4. With the rotor winding open-circuited no torque can be developed. Hence the application of a three-phase voltage to the three-phase stator winding gives rise to a rotating magnetic field which cuts both the stator and rotor windings at the line frequency f_1.

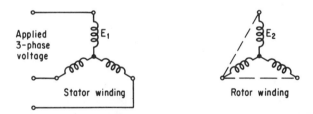

Fig. 15-4. Schematic representation of three-phase wound rotor induction motor. Broken line indicates short-circuit links for normal operation.

The rms value of the induced emf per phase of the rotor winding is given by Eq. (14-34) as

$$E_2 = 4.44 f_1 N_2 K_{w_2} \Phi \qquad (15-2)$$

where the subscript 2 denotes rotor winding quantities. Note that the stator frequency f_1 is used here because the rotor is at standstill. Hence E_2 is a *line*-frequency emf. Of course the flux Φ is the flux per pole, which is mutual to the stator and rotor windings.

A similar expression describes the rms value of the induced emf per phase occurring in the stator winding. Thus

$$E_1 = 4.44 f_1 N_1 K_{w_1} \Phi \tag{15-3}$$

From Eqs. (15-2) and (15-3) we can formulate the ratio

$$\frac{E_1}{E_2} = \frac{N_1 K_{w_1}}{N_2 K_{w_2}} \tag{15-4}$$

Note the similarity of this expression to the voltage transformation ratio of the transformer. The difference lies in the inclusion of the winding factors of the motor, necessitated by the use of a distributed winding for the motor as contrasted to the concentrated coils used in the transformer. In essence, then, the induction motor at standstill exhibits the characteristics of a transformer wherein the stator winding is the primary and rotor winding is the secondary.

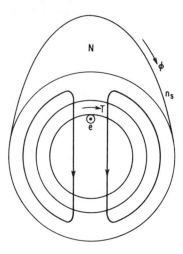

Next let us consider the behavior of the induction motor under running conditions—again with the intention of pointing out similarities to the transformer. To produce a starting torque (and subsequently a running torque) it is necessary to have a current flowing through the rotor winding. This is readily accomplished by short-circuiting the winding in the manner indicated by the broken line in Fig. 15-4. Initially the induced emf E_2 causes a rotor current per phase I_2 to flow through the short circuit, producing an ampere-conductor distribution which acts with the flux field to produce the starting torque. The sense of this torque is always such as to cause the rotor to travel in

Fig. 15-5. Showing the direction of induced voltage and developed torque on a typical rotor conductor.

the same direction as the rotating field. An examination of Fig. 15-5 makes this apparent. Assume that the flux field is revolving clockwise at a speed corresponding to the applied stator frequency and the poles of the stator winding. This speed is called the *synchronous speed* and is described by Eq. (15-1). Thus

$$\boxed{n_s = \frac{120 f_1}{p}} \quad \text{rpm} \tag{15-5}$$

By the $\bar{v} \times \bar{B}$ rule the induced emf in a typical conductor lying beneath a north-pole flux is directed out of the paper as indicated in Fig. 15-5. Then for this direction of current the $I_2 \times \bar{B}$ rule reveals the torque to be directed clockwise. Therefore the rotor moves in a direction in which it tries to catch up with the stator field.

As the rotor increases its speed, the rate at which the stator field cuts the rotor coils decreases. This reduces the resultant induced emf per phase, in turn diminishing the magnitude of the ampere-conductor distribution and yielding less torque. In fact this process continues until that rotor speed is reached which yields enough emf to produce just the current needed to develop a torque equal to the opposing torques. If there is no shaft load, the opposing torque consists chiefly of frictional losses. It is important to understand that as long as there is an opposing torque to overcome—however small or whatever its origin—*the rotor speed can never be equal to the synchronous speed.* This is characteristic of singly excited electromechanical energy-conversion devices. Since the rotor (or secondary) winding current is produced by induction, there must always be a difference in speed between the stator field and the rotor. In other words, transformer action must always be allowed to take place between the stator (or primary) winding and the rotor (or secondary) winding.

This speed difference, or *slip*, is a very important variable for the induction motor. In terms of an equation we may write

$$\text{slip} \equiv n_s - n \quad \text{rpm} \tag{15-6}$$

where n denotes the actual rotor speed in *rpm*. The term slip is used because it describes what an observer riding with the stator field sees looking at the rotor— the rotor appears to be slipping backward. A more useful form of the slip quantity results when it is expressed on a per unit basis using synchronous speed as the reference. Thus the slip in per unit is

$$\boxed{s = \frac{n_s - n}{n_s}} \tag{15-7}$$

For the conventional induction motor the values of s lie between zero and unity.

It is customary in induction-motor analysis to express rotor quantities (such as induced voltage, current, and impedance) in terms of line-frequency quantities and the slip as expressed by Eq. (15-7). For example, if the rotor is assumed to be operating at some speed $n < n_s$, then the actual emf induced in the rotor winding per phase may be expressed in terms of the line-frequency quantity E_2 as sE_2. This formulation has definite advantages, as described in the next section. In a similar fashion it is possible to express the rotor-winding impedance per phase as

$$z_2 = r_2 + jsx_2 \tag{15-8}$$

where z_2 denotes the rotor phase impedance, r_2 is the rotor resistance per phase, and x_2 is the line-frequency leakage reactance per phase of the rotor winding. Of course the effective value of this reactance when the rotor operates at a speed n (or slip s) is only s times as large. Keep in mind that the frequency of the currents in the rotor is directly related to the relative speed of the stator field to the rotor winding. Accordingly we may write

$$f_2 = \frac{p(\text{slip rpm})}{120} = \frac{p(n_s - n)}{120} \tag{15-9}$$

where f_2 is the frequency of the rotor emf and current. By means of Eq. (15-7) it is possible to rewrite Eq. (15-9) as

$$f_2 = \frac{psn_s}{120} = s\frac{pn_s}{120} = sf_1$$ (15-10)

which indicates that the rotor frequency f_2 is obtained by merely multiplying the stator line frequency by the appropriate per unit value of the slip. For this reason f_2 is often called the *slip frequency*.

15-3   The Equivalent Circuit

It is desirable to have an equivalent circuit of the three-phase induction motor in order to direct the analysis of operation and to facilitate the computation of performance. From the remarks made in the preceding section it should not come as a surprise that the equivalent circuit assumes a form identical to that of the exact equivalent circuit of the transformer. The derivation proceeds in a similar fashion with necessary modifications introduced to account for the fact that the secondary winding (the rotor) in this instance revolves and thereby develops mechanical power.

The Magnetizing Branch of the Equivalent Circuit. All the parameters of the equivalent circuit are expressed on a per phase basis. This applies whether the stator winding is Y- or Δ-connected. In the latter case the values refer to the equivalent Y connection. Appearing in Fig. 15-6(a) is the portion of the equivalent circuit that has reference to the stator (or primary) winding. Note that it consists of a stator phase winding resistance r_1, a stator phase winding leakage reactance x_1, and a magnetizing impedance made up of the core-loss resistor r_c and the magnetizing reactance x_ϕ. There is no difference in form between this circuit and that of the transformer. The difference lies only in the magnitude of the parameters. Thus the total magnetizing current I_m is considerably larger in the case of the induction motor because the magnetic circuit necessarily includes an air gap. Whereas in the transformer this current is about 2 to 5 per cent of the rated current, here it is approximately 25 to 40 per cent of the rated current depending upon the size of the motor. Moreover, the primary leakage reactance for the induction motor also is larger because of the air gap as well as because the stator and rotor windings are distributed along the periphery of the air gap rather than concentrated on a core as in the transformer. The effects of the actions that take place in the rotor (or secondary) winding must reflect themselves at the proper equivalent-voltage level at terminals *a-b* in Fig. 15-6(a). We next investigate the manner in which this comes about.

The Actual Rotor Circuit per Phase. For any specified load condition that calls for a particular value of slip *s*, the rotor current per phase may be expressed as

$$I_2 = \frac{s\bar{E}_2}{r_2 + jsx_2}$$ (15-11)

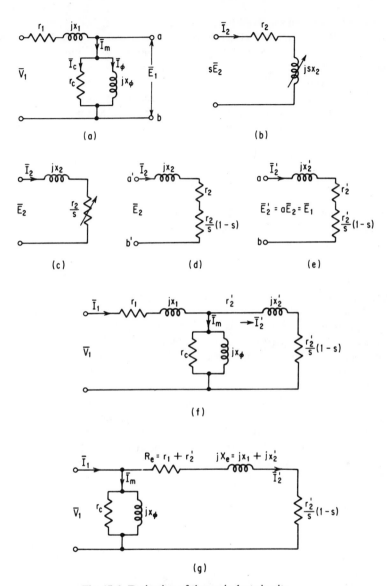

Fig. 15-6. Derivation of the equivalent circuit:
  (a) stator winding section;
  (b) actual rotor circuit;
  (c) equivalent rotor circuit;
  (d) modified equivalent rotor circuit;
  (e) stator-referred equivalent rotor circuit;
  (f) exact equivalent circuit;
  (g) approximate equivalent circuit.

where $\bar{E}_2$ and x_2 are the standstill (or line frequency) values. The circuit interpretation of Eq. (15-11) is depicted in Fig. 15-6(b). It illustrates that I_2 is a slip-frequency current produced by the slip-frequency induced emf $s\bar{E}_2$ acting in a rotor circuit having an impedance per phase of $r_2 + jsx_2$. In other words, this is the current that would be "seen" by an observer riding with the rotor winding. Furthermore, the amount of real power involved in this rotor circuit is the current squared times the real part of the rotor impedance. In fact, this power represents the rotor copper loss per phase. Hence the total rotor copper loss may be expressed as

$$\boxed{P_{\text{Cu2}} = qI_2^2 r_2} \qquad (15\text{-}12)$$

where q denotes the number of rotor phases.

The Equivalent Rotor Circuit. By dividing both the numerator and denominator of Eq. (15-11) by the slip s we get

$$I_2 = \frac{\bar{E}_2}{\dfrac{r_2}{s} + jx_2} \qquad (15\text{-}13)$$

The corresponding circuit interpretation of this expression appears in Fig. 15-6(c). Note that the magnitude and phase angle of I_2 remain unaltered by this operation. However, there is a significant difference between Eqs. (15-11) and (15-13). In the latter case I_2 is considered to be produced by a line-frequency voltage $\bar{E}_2$ acting in a rotor circuit having an impedance per phase of $r_2/s + jx_2$. Hence the I_2 of Eq. (15-13) is a *line-frequency* current, whereas the I_2 of Eq. (15-11) is a *slip-frequency* current. It is important that this distinction be understood.

Division of Eq. (15-11) by s has enabled us to go from an actual rotor circuit characterized by constant resistance and variable leakage reactance [see Fig. 15-6(b)] to one characterized by variable resistance and constant leakage reactance [see Fig. 15-6(c)]. Moreover, the real power associated with the equivalent rotor circuit of Fig. 15-6(c) is clearly

$$P = I_2^2 \frac{r_2}{s} \qquad (15\text{-}14)$$

Hence the total power for q phases is

$$\boxed{P_g = qI_2^2 \frac{r_2}{s}} \qquad (15\text{-}15)$$

A comparison of this expression with Eq. (15-12) indicates that the power associated with the equivalent circuit of Fig. 15-6(c) is considerably greater. For example, in a large machine a typical value of s is 0.02. Hence P_g is greater than the actual rotor copper loss by a factor of fifty.

What is the meaning of this power discrepancy? The answer lies in the fact that by Eq. (15-13) I_2 is a *line-frequency* current. This means that the point of reference has changed from the rotor (where slip-frequency quantities exist) to the stator (where line-frequency quantities exist). In accordance with the circuit representation of Fig. 15-6(c) the observer changes his point of reference from the rotor to the stator. This shift is significant because now, upon looking into the rotor, the observer "sees" not only the rotor copper loss *but the mechanical power developed as well.* The latter quantity is included because with respect to a stator-based observer the rotor speed is no longer zero as it is relative to a rotor-based observer. As a matter of fact Eq. (15-15) gives the total power input to the rotor. It is the power transferred across the air gap from the stator to the rotor. We can rewrite Eq. (15-15) in a manner that stresses this fact:

$$P_g = qI_2^2 \frac{r_2}{s} = qI_2^2 \left[r_2 + \frac{r_2}{s}(1 - s) \right]$$ (15-16)

In other words, the variable resistance of Fig. 15-6(c) may be replaced by the actual rotor winding resistance r_2 and a variable resistance R_m, which represents the mechanical shaft load. That is,

$$R_m \equiv \frac{r_2}{s}(1 - s)$$ (15-17)

This expression is useful in analysis because it allows any mechanical load to be represented in the equivalent circuit by a resistor. Figure 15-6(d) depicts the modified version of the rotor equivalent circuit. Finally, it should be apparent, on the basis of the foregoing remarks, that the rotor equivalent circuit is equivalent only insofar as the magnitude and phase angle of the rotor current per phase are concerned.

The Stator-referred Rotor Equivalent Circuit. The voltage appearing across terminals *a-b* in Fig. 15-6(a) is a line-frequency quantity having $N_1 K_{w1}$ effective turns. The voltage appearing across terminals *a'-b'* in Fig. 15-6(d) is also a line-frequency quantity but has $N_2 K_{w2}$ effective turns. In general $\bar{E}_1 \neq \bar{E}_2$, so that *a'-b'* in Fig. 15-6(d) cannot be joined to *a-b* in Fig. 15-6(a) to yield a single-line equivalent circuit. To accomplish this it is necessary to replace the actual rotor winding with an equivalent winding having $N_1 K_{w1}$ effective turns as was done with the transformer. In other words, all the rotor quantities must be referred to the stator in the manner depicted in Fig. 15-6(e). The prime notation is used to denote stator-referred rotor quantities.

The Complete Equivalent Circuit. The voltage appearing across terminals *a-b* in Fig. 15-6(e) is the same as that appearing across terminals *a-b* in Fig. 15-6(a). Hence these terminals may be joined to yield the complete equivalent circuit as it appears in Fig. 15-6(f). Note that the form is identical to that of the two-winding transformer.

The Approximate Equivalent Circuit. Considerable simplification of computation with little loss of accuracy can be achieved by moving the magnetizing branch to the machine terminals as illustrated in Fig. 15-6(g). This modification is essentially based on the assumption that $V_1 \approx E_1 = E_2'$. All performance calculations will be carried out using the approximate equivalent circuit.

15-4 Computation of Performance

When the three-phase induction motor is running at no-load, the slip has a value very close to zero. Hence the mechanical load resistor R_m has a very large value, which in turn causes a small rotor current to flow. The corresponding electromagnetic torque, as described by Eq. (14-39), merely assumes that value which is needed to overcome the *rotational losses* consisting of friction and windage. If a mechanical load is next applied to the motor shaft, the initial reaction is for the shaft load to drop the motor speed slightly, thereby increasing the slip. The increased slip subsequently causes I_2 to increase to that value which, when inserted into Eq. (14-39), yields sufficient torque to provide a balance of power to the load. Thus equilibrium is established and operation proceeds at a particular

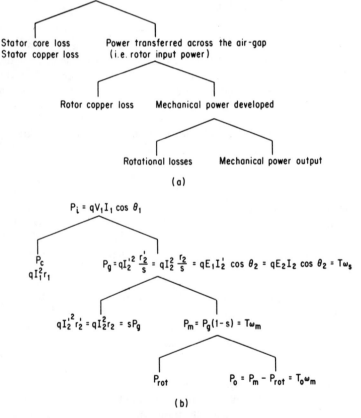

Fig. 15-7. Power-flow diagram:
(a) statement form;
(b) equation form.

value of s. In fact for each value of load horsepower requirement there is a unique value of slip. This can be inferred from the equivalent circuit, which shows that once s is specified then the power input, the rotor current, the developed torque, the power output, and the efficiency are all determined.

The use of a power-flow diagram in conjunction with the approximate equivalent circuit makes the computation of the performance of a three-phase induction motor a straightforward matter. Depicted in Fig. 15-7(a) is the power flow in statement form. Note that the loss quantities are placed on the left side of a flow point. Appearing in Fig. 15-7(b) is the same power-flow diagram but now expressed in terms of all the appropriate relationships needed to compute the performance. It should be clear that to calculate performance one must first compute the currents I_2 and I_1 from the equivalent circuit and then make use of the pertinent relationships depicted in Fig. 15-7(b).

EXAMPLE 15-1 A three-phase, four-pole, 50-hp, 440-volt, 60-Hz, Y-connected induction motor has the following parameters per phase:

$$r_1 = 0.10 \text{ ohm,} \qquad x_1 = 0.35 \text{ ohm}$$
$$r_2' = 0.12 \text{ ohm,} \qquad x_2' = 0.40 \text{ ohm}$$

It is known that the stator core losses amount to 1200 watts and the rotational losses equal 950 watts. Moreover, at no-load the motor draws a line current of 18 amps at a power factor of 0.089 lagging.

When the motor operates at a slip of 2.5 per cent find: (a) input line current and power factor, (b) the developed electromagnetic torque in newton-meters, (c) the horsepower output, (d) the efficiency.

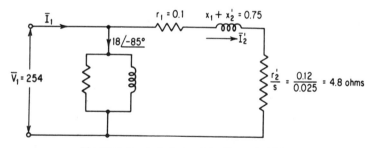

Fig. 15-8. Equivalent circuit for Example 15-1.

solution: (a) The computations are carried out on a per phase basis. Hence the phase voltage is $440/\sqrt{3}$ or 254 volts, and the equivalent circuit is depicted in Fig. 15-8. The stator-referred rotor current then follows from

$$\bar{I}_2' = \frac{\bar{V}_1}{r_1 + \dfrac{r_2'}{s} + j(x_1 + x_2')} = \frac{254}{4.9 + j0.75}$$
$$= 51.3\underline{/-8.7^\circ} = 50.9 - j7.8$$

For all practical purposes the magnetizing current may be taken equal to the no-load current because the corresponding rotor current is negligibly small. Thus

$$I_m = 18\underline{/-85°} = 1.6 - j17.95$$

Hence, the input line current is

$$I_1 = I_m + I_2' = (50.9 + 1.6) - j(17.95 + 7.8)$$

$$I_1 = 52.5 - j25.75 = 58.2\underline{/-26.2°} \tag{15-18}$$

and

$$\text{pf} = \cos\theta_1 = \cos 26.2° = 0.895 \text{ lagging}$$

(b) The developed torque is found from

$$T = \frac{P_g}{\omega_s} \tag{15-19}$$

Also

$$\omega_s = \frac{2\pi n_s}{60} = \frac{2\pi(1800)}{60} = 60\pi \quad \text{rad/sec}$$

and

$$P_g = qI_2'^2 \frac{r_2'}{s} = 3(51.3)^2 4.8 = 37,820 \text{ watts}$$

Therefore

$$T = \frac{37,820}{60\pi} = 201 \text{ newton-meters}$$

(c) From the power-flow diagram the power output is

$$P_o = P_m - P_{rot} = P_g(1 - s) - P_{rot} = 36,875 - 950 = 35,925 \text{ watts}$$

The horsepower output is therefore

$$HP_0 = \frac{35.925}{0.746} = 48.15$$

Note that this is slightly less than the rated horsepower of 50. Rated hp occurs at a slip somewhat larger than 2.5 per cent.

(d) It is more accurate to find the efficiency from the relationship

$$\eta = 1 - \frac{\sum \text{losses}}{P_i} \tag{15-20}$$

rather than from the output-to-input ratio. The tabulation of the losses is as follows:

P_c = core loss = 1200 watts

stator copper loss = $qI_1^2 r_1 = 3(58.2)^2(0.1) = 1030$ watts

rotor copper loss = $qI_2'^2 r_2' = sP_g = (0.025)(37,820) = 945$ watts

P_{rot} = rotational loss = 950 watts

$\sum$ losses = 4125 watts

Also, the input power is

$$P_i = \sqrt{3}(440(58.2)(0.895) = 39,700 \text{ watts}$$

Hence

$$\eta = 1 - \frac{4125}{39,700} = 1 - 0.104 = 0.896$$

The efficiency is 89.6 per cent.

15-5   Torque-speed Characteristic. Starting Torque and Maximum Developed Torque

The variation of torque with speed (or slip) is an important characteristic of the three-phase induction motor. The general shape of the curve can be identified in terms of the basic torque equation [Eq. (14-39)] and knowledge of the performance computation procedure. When the motor operates at a very small slip, as at no-load, the rotor current is practically zero so that only that amount of torque is developed which is needed to supply the rotational losses. As the slip is allowed to increase from nearly zero to about 10 per cent, Eq. (15-11) shows that the rotor current increases almost linearly. This is because the j part of the impedance sx_2 is small compared to r_2. Furthermore, it can be shown that the space-displacement angle ψ in Eq. (14-39) is identical to the rotor power-factor angle θ_2. That is,

$$\psi = \theta_2 = \tan^{-1} \frac{sx_2}{r_2} \qquad (15\text{-}21)$$

Therefore, for values of s from zero to 10 per cent, ψ varies over a range of about zero to 15 degrees. This means that the $\cos \psi$ remains practically invariant over the specified slip range, and so the torque increases almost linearly in this region. Of course the quantity Φ in Eq. (14-39) is essentially fixed since the applied voltage is constant.

As the slip is allowed to increase still further, the current continues to increase but much less rapidly than at first. The reason lies in the increasing importance of the sx_2 term of the rotor impedance. In addition, the space angle ψ now begins to increase at a rapid rate which makes the $\cos \psi$ diminish more rapidly than the current increases. Since the torque equation now involves two opposing factors, it is entirely reasonable to expect that a point is reached beyond which further increases in slip culminate in decreased developed torque. In other words, the rapidly decreasing $\cos \psi$ factor predominates over the slightly increasing I_2 factor in Eq. (14-39). As ψ increases, the field pattern for producing torque becomes less and less favorable because more and more conductors that produce negative torque are included beneath a given pole flux. A glance at Fig. 14-3 makes this statement self-evident. Accordingly, the composite torque-speed curve takes on a form similar to that shown in Fig. 15-9.

The *starting torque* is the torque developed when s is unity, i.e. the speed n is zero. Figure 15-9 indicates that for the case illustrated the starting torque is somewhat in excess of rated torque, which is fairly typical of such machines. The starting torque is computed in the same manner as torque is computed for any value of slip. Here it merely requires using $s = 1$. Thus the magnitude of the rotor current at standstill is

$$I_2' = \frac{V_1}{\sqrt{(r_1 + r_2')^2 + (x_1 + x_2')^2}} \qquad (15\text{-}22)$$

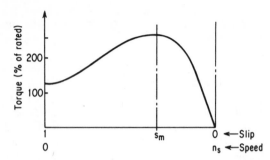

Fig. 15-9. Typical torque-speed curve of a three-phase induction motor.

The corresponding gap power is then

$$P_g = qI_2'^2 r_2' = \frac{qV_1^2 r_2'}{(r_1 + r_2')^2 + (x_1 + x_2')^2} \tag{15-23}$$

It is interesting to note that higher starting torques result from increased rotor copper losses at standstill.

At unity slip the input impedance is very low so that large starting currents flow. Equation (15-22) makes this apparent. In the interest of limiting this excessive starting current motors whose ratings exceed 3 hp are usually started at reduced voltage by means of line starters. This matter is discussed further in Sec. 15-8. Of course, starting with reduced voltage also means a reduction in the starting torque. In fact if 50 per cent of the rated voltage is used upon starting, then clearly by Eq. (15-23) it follows that the starting torque is only one-quarter of its full-voltage value.

Another important torque quantity of the three-phase induction motor is the maximum developed torque. This quantity is so important that it is frequently the starting point in the design of the induction motor. The maximum (or breakdown) torque is a measure of the reserve capacity of the machine. It frequently has a value of 200 to 300 per cent of rated torque. It permits the motor to operate through momentary peak loads. However, the maximum torque cannot be delivered continuously because the excessive currents that flow would destroy the insulation.

Since the developed torque is directly proportional to the gap power it follows that the torque is a maximum when P_g is a maximum. Also, P_g is a maximum when there is a maximum transfer of power to the equivalent circuit resistor r_2'/s. Applying the maximum-power-transfer theorem to the approximate equivalent circuit leads to the result that

$$\frac{r_2'}{s_m} = \sqrt{r_1^2 + (x_1 + x_2')^2} \tag{15-24}$$

That is, maximum power is transferred to the gap power resistor r_2'/s when this resistor is equal to the impedance looking back into the source. Accordingly the slip s_m at which the maximum torque is developed is

$$s_m = \frac{r_2'}{\sqrt{r_1^2 + (x_1 + x_2')^2}} \tag{15-25}$$

Note that the slip at which the maximum torque occurs may be increased by using a larger rotor resistance. Some induction motors are in fact designed so that the maximum torque is available as a starting torque, i.e. $s_m = 1$.

With the slip s_m known, the corresponding rotor current can be found and then inserted into the torque equation to yield the final form for the breakdown torque. Thus

$$T_m = \frac{1}{\omega_s} q I_2'^2 \frac{r_2'}{s_m} = \frac{1}{\omega_s} \frac{q V_1^2}{2[r_1 + \sqrt{r_1^2 + (x_1 + x_2')^2}]} \tag{15-26}$$

An examination of (Eq. 15-26) reveals the interesting information that the maximum torque is independent of the rotor winding resistance. Thus increasing the rotor winding resistance increases the slip at which the breakdown torque occurs but it leaves the magnitude of this torque unchanged. Figure 15-10 shows the effect of increasing the rotor resistance on a typical torque-speed curve.

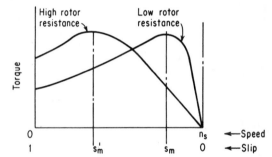

Fig. 15-10. Showing the effect of increased rotor resistance on torque-speed curve.

15-6   Parameters from No-load Tests

The performance computations of a three-phase induction motor presuppose knowledge of the parameters of the equivalent circuit. This information may be available either from design data or from appropriate tests. In the case where the design data are not known, a simple no-load test yields information about the magnetizing current as described in Example 15-1. It merely requires measuring the input line current and power. Note that specific knowledge about the core-loss resistor r_c and the magnetizing reactance x_ϕ is usually unnecessary. However, these parameters can readily be determined by proceeding in a fashion similar to that used in finding these same quantities for the transformer. A slight modification is necessary, though, because of the presence of the rotational losses.

To obtain information about the winding resistances and the leakage reactances a rotor-blocked test is necessary. This test is analogous to the short-circuit test of the transformer. It requires that the rotor be blocked to prevent rotation and that the rotor winding be short-circuited in the usual fashion. Furthermore, since the slip is unity the mechanical load resistor R_m is zero and so the input impedance is quite low. Hence, in order to limit the rotor current in this

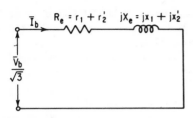

Fig. 15-11. Induction motor equivalent circuit for the rotor-blocked test.

test to reasonable values, a reduced voltage must be used—usually about 10 to 25 per cent of the rated value. Also, operation at such a reduced voltage renders the core loss as well as the magnetizing current negligibly small. Accordingly, the equivalent circuit in this test takes on the configuration shown in Fig. 15-11.

Assume now that the following instrument readings are taken in performing a rotor-blocked test on a Y-connected three-phase induction motor:

P_b = total wattmeter reading in watts
I_b = line current of Y-connection
V_b = line voltage of Y-connection

From these measurements it follows that the equivalent winding resistance is given by

$$R_e \equiv r_1 + r'_2 = \frac{P_b}{3I_b^2} \tag{15-27}$$

where R_e is the equivalent resistance per phase and the factor 3 accounts for the three phases. The equivalent phase impedance is obtained from

$$Z_e = \frac{V_b}{\sqrt{3}\,I_b} \tag{15-28}$$

Finally, the equivalent leakage reactance is determined from

$$X_e = \sqrt{Z_e^2 - R_e^2} = x_1 + x'_2 \tag{15-29}$$

Note that as long as computations are carried on in connection with the approximate equivalent circuit, it is sufficient to deal directly with X_e without a further breakdown into x_1 and x'_2. In fact it is the chief purpose of the blocked-rotor test to make X_e available.

15-7 Ratings and Applications of Three-phase Induction Motors

Now that the theory of operation, the characteristics, and the performance of the three-phase induction motor are understood, we can study their standard ratings and typical applications. Of course, before it is possible to specify a particular motor for a given application, the characteristics of the load must be known. This includes such items as horsepower requirement, starting torque, acceleration capability, speed variation, duty cycle, and the environment in which the motor is to operate. Once this information is available it is often possible to select a general-purpose motor to do the job satisfactorily. Table 15-1 is a list of such motors which are readily available and standardized in accordance with generally accepted criteria established by the National Electrical Manufacturers Association (NEMA). The table is essentially self-explanatory. The primary distinguishing features of the various classes of squirrel-cage motors are the construction details of the rotor slots. For example, the Class A motor uses a low-resistance squirrel cage with slots of medium depth. Classes B and C use much deeper bars in order to reduce the amount of full-voltage starting current and to obtain higher starting torque (for Class C). The Class D motor uses very high-resistance bars for the squirrel cage. This yields very high starting torque but also poor operating efficiency. Figure 15-12 shows the differences in the torque-speed curves.

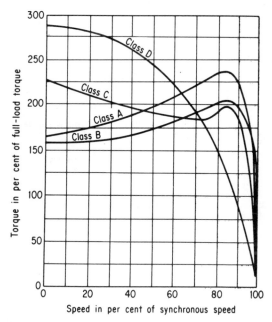

Fig. 15-12. Typical torque-slip curves for various classes of induction motors.

15-8 Controllers for Three-phase Induction Motors

After the right motor is selected for a given application, the next step is to select the appropriate controller for the motor. The primary functions of a controller are to furnish proper starting, stopping, and reversing without damage or inconvenience to the motor, other connected loads, or the power system. However, the controller fulfills other useful purposes as well, especially the following:

(1) It limits the starting torque. Some connected shaft loads may be damaged if excessive torque is applied upon starting. For example, fan blades can be sheared off or gears having excessive backlash can be stripped. The controller supplies reduced voltage at the start and as the speed picks up the voltage is increased in steps to its full value.

(2) It limits the starting current. Most motors above 3 hp cannot be started

Table 15-1. Characteristics and Applications

Type classification	HP range	Starting torque (%)‡	Maximum torque (%)‡	Starting current (%)‡
General-purpose, normal torque and starting current, NEMA Class A	0.5 to 200	Poles-Torque 2–150 4–150 6–135 8–125 10–120 12–115 14–110 16–105	Up to 250 but not less than 200	500–1000
General-purpose, normal torque, low starting current, NEMA Class B	0.5 to 200	Same as above or larger	About the same as Class A but may be less	About 500-550, less than average of Class A
High torque, low starting current, NEMA Class C	1 to 200	200 to 250	Usually a little less than Class A but not less than 200	About same as Class B
High torque, medium and high slip, NEMA Class D	0.5 to 150	Medium slip 350 High slip 250–315	Usually same as standstill torque	Medium slip 400–800, high slip 300–500
Low starting torque, either normal starting current, NEMA Class E, or low starting current, NEMA Class F	40 to 200	Low, not less than 50	Low, but not less than 150	Normal 500–1000, low 350–500
Wound-rotor	0.5 to 5000	Up to 300	200–250	Depends upon external rotor resistance but may be as low as 150

† By permission from M. Liwschitz-Garik and C. C. Whipple, *Electric Machinery*, vol. II (Princeton, N. J.: D. Van Nostrand Co., Inc., 1946.)

‡ Figures are given in per cent of rated full-load values.

Slip (%)	Power factor (%)	Efficiency (%)	Typical applications
Low, 3–5	High, 87–89	High, 87–89	Constant-speed loads where excessive starting torque is not needed and where high starting current is tolerated. Fans, blowers, centrifugal pumps, most machinery tools, woodworking tools, line shafting. Lowest in cost. May require reduced voltage starter. Not to be subjected to sustained overloads, because of heating. Has high pull-out torque
3–5	A little lower than Class A	87–89	Same as Class A—advantage over Class A is lower starting current, but power factor slightly less
3–7	Less than Class A	82–84	Constant-speed loads requiring fairly high starting torque and lower starting current. Conveyors, compressors, crushers, agitators, reciprocating pumps. Maximum torque at standstill
Medium 7–11, high, 12–16	Low	Low	Medium slip. Highest starting torque of all squirrel-cage motors. Used for high-inertia loads such as shears, punch presses, die stamping, bulldozers, boilers. Has very high average accelerating torque. High slip used for elevators, hoists, etc. on intermittent loads
1 to 3½	About same as Class A or Class B	About same as Class A or Class B	Direct-connected loads of low inertia requiring low starting torque, such as fans and centrifugal pumps. Has high efficiency and low slip
3–50	High, with rotor shorted same as Class A	High, with rotor shorted same as Class A, but low when used with rotor resistor for speed control	For high-starting-torque loads where very low starting current is required or where torque must be applied very gradually and where some speed control (50%) is needed. Fans, pumps, conveyors, hoists, cranes, compressors. Motor with speed control more expensive and may require more maintenance.

directly across the three-phase line because of the excessive starting current that flows. Recall that at unity slip the current is limited only by the leakage impedance, which is usually quite a small quantity, especially in the larger motor sizes. A large starting current can be annoying because it causes light to flicker and may even cause other connected motors to stall. Reduced-voltage starting readily eliminates these annoyances.

(3) It provides overload protection. All general-purpose motors are designed to deliver full-load power continuously without overheating. However, if for some reason the motor is made to deliver, say, 150 per cent of its rated output continuously, it will proceed to accommodate the demand and burn itself up in the process. The horsepower rating of the motor is based on the allowable temperature rise that can be tolerated by the insulation used for the field and armature windings. The losses produce the heat that raises the temperature. As long as these losses do not exceed the rated values there is no danger to the motor, but if they are allowed to become excessive, damage will result. There is nothing inherent in the motor that will keep the temperature rise within safe limits. Accordingly, it is also the function of the controller to provide this protection. Overload protection is achieved by the use of an appropriate time-delay relay which is sensitive to the heat produced by the motor line currents.

(4) It furnishes undervoltage protection. Operation at reduced voltage can be harmful to the motor, especially when the load demands rated power. If the line voltage falls below some preset limit, the motor is automatically disconnected from the three-phase line source by the controller.

Controllers for electric motors are of two types—manual and magnetic. We shall consider only the magnetic type, which has many advantages over the manual type. It is easier to operate. It provides undervoltage protection. It can be remotely operated from one or several different places. Moreover, the magnetic controller is automatic and reliable whereas the manual controller requires a trained operator, especially where a sequence of operations is called for in a given application. The one disadvantage is the greater initial cost of the magnetic type.

Appearing in Fig. 15-13 is the schematic diagram of a magnetic full-voltage

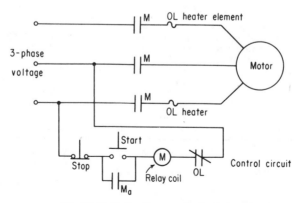

Fig. 15-13. Full-voltage magnetic starter.

starter for a three-phase induction motor. The operation is simple. When the start button is pressed, the relay coil M is energized. This moves the relay armature to its closed position, thereby closing the main contactors M, which in turn apply full voltage to the motor. When the relay armature moves to its energized position, it also closes an auxiliary contactor M_a which serves as an electrical interlock, allowing the operator to release the start button without de-energizing the main relay. Of course contactors M are much larger in size than M_a. The former set must have a current rating that enables it to handle the starting motor current. The latter needs to accommodate just the exciting current of the relay coil. Figure 15-13 also shows that the motor line current flows through two overload heater elements. If the temperature rise becomes excessive, the heater element causes the overload contacts in the control circuit to open. In controller diagrams it is important to remember that all contactors are shown in their de-energized state. Thus the symbol ⊣⊢ means that the contactors M are open when the coil is not energized. Similarly the symbol ⇥⊬ means that these contactors are closed in the de-energized state.

Undervoltage protection is inherent in the magnetic starter of Fig. 15-13. This comes about as a result of designing the coil M so that if the coil voltage drops below a specified minimum, the relay armature can no longer be held in the closed position.

A full-voltage magnetic starter equipped with the control circuitry to permit reversing is illustrated in schematic form in Fig. 15-14. A three-phase induction

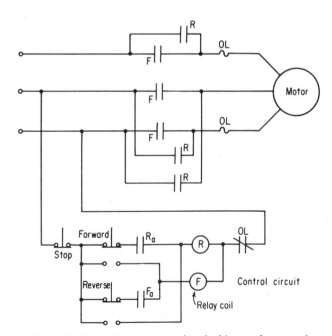

Fig. 15-14. Full-voltage starter equipped with reversing control.

motor is reversed by crossing two of the three line leads going to the motor terminals. In this connection note the criss-cross of two of the R contactors. Pressing
the forward (*FWD*) button energizes coil F which in turn closes the main contactors
F as well as the interlock F_a. This allows the motor to reach its forward operating
speed. To reverse the motor the *REV* button is pushed. This does two things.
One, it de-energizes coil F, thus opening the F contactors. Two, it energizes the R
relay coil, thus closing the R contactors which apply a reversed-phase sequence
to the motor causing it to attain full speed in the reverse direction. Putting the
REV switch in the F_a interlock circuit is a safety measure which prevents having
both the R and F contactors closed at the same time.

An illustration of a reduced-voltage magnetic controller using limiting resistors in the line circuit appears in Fig. 15-15. This unit is frequently referred
to as a three-step acceleration starter because the line resistors are removed in
three steps. Pushing the start button energizes coil M and closes contactors M,
thus applying a three-phase voltage to the motor through the full resistors. In
addition to contactors M and M_a coil M is also equipped with a time-delayed
contactor T_M. This contactor is so designed that it does not close until a preset
time *after* the armature of coil M is closed. The delay is usually obtained through

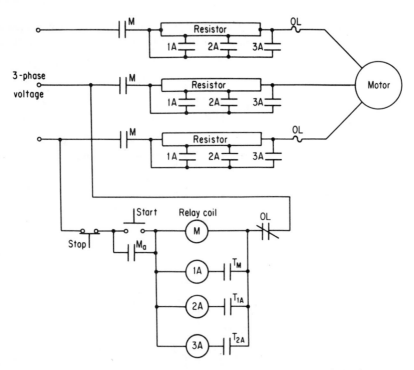

Fig. 15-15. A three-step reduced-voltage starter for a three-phase induction
motor.

a mechanical escapement of some sort which is actuated by the relay armature. Of course the time delay is needed to permit the motor to accelerate to a speed corresponding to the reduced applied voltage. After the elapse of the preset time, the T_M contacts close, energizing coil 1A which in turn closes contactors 1A, short-circuiting the first part of the series resistor. Coil 1A is also equipped with a time-delay contactor T_{1A}, which is designed to allow the motor to accelerate to a higher speed before it closes. When contactor 1A does close, coil 2A is energized. This immediately closes contactors 2A, shorting out the second section of the line resistor. Then, after still another time delay, contactor T_{2A} closes, applying an excitation voltage to coil 3A. With the closing of contactors 3A the full line voltage is applied to the motor. In this manner the motor is brought up to speed in a "soft," smooth fashion without drawing excessive starting current or developing large starting torques.

PROBLEMS

15-1 A three-phase armature winding is shown in Fig. P15-1. Sinusoidal currents having an amplitude of 100 amps flow in the three phases. Each coil consists of 3 turns. Carefully sketch to scale the actual mmf distribution along the air gap for a span of two poles for the indicated time instants t_1 and t_2.

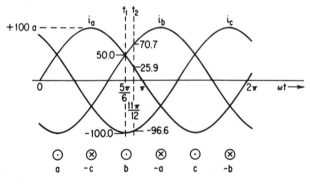

Figure P15-1

15-2 State the general conditions under which it is possible for m alternating fields (with axes fixed in space) to yield a revolving field constant in amplitude and traveling at constant speed.

15-3 Determine the magnitude and direction of the resultant flux field in the machine configuration of Fig. 15-2 corresponding to time instants t_2 and t_4 in Fig. 15-1.

15-4 In the time variation of the phase currents depicted in Fig. 15-1 assume that the amplitude of phase b current is one-half that of phases a and c but that each is displaced by 120° from the other. Find the magnitude and direction of the resultant flux for the time instants t_1, t_3, and t_5, and compare with the results shown in Fig. 15-2.

15-5   An unbalanced three-phase mmf flows through three coils that are space displaced by 120° as shown in Fig. 15-2. The nature of the unbalance is such that phase *b* lags behind phase *a* by 90° instead of 120°. However, each phase has the same amplitude.

Investigate the character of the resultant flux field corresponding to various instants of time.

15-6   A balanced, three-phase, 60-Hz voltage is applied to a three-phase, 4-pole induction motor. When the motor delivers rated output horsepower, the slip is found to be 0.05. Determine the following:

(a) The speed of the revolving field relative to the stator structure, which accommodates the exciting winding.

(b) The frequency of the rotor currents.

(c) The speed of the rotor mmf relative to the rotor structure.

(d) The speed of the rotor mmf relative to the stator structure.

(e) The speed of the rotor mmf relative to the stator field distribution.

(f) Are the conditions right for the development of a net unidirectional torque? Explain.

15-7   Repeat Prob. 15-6 for the case where the rotor structure is blocked, thus preventing rotation in spite of the application of a balanced three-phase voltage to the stator.

15-8   Determine the no-load speed of a 6-pole, wound-rotor, three-phase induction motor, the stator of which is connected to a 60-Hz line and the rotor of which is connected to a 25-Hz line, when:

(a) The stator field and the rotor field revolve in the same direction.

(b) The stator field and the rotor field revolve in opposite directions.

15-9   A 60-Hz polyphase induction motor runs at a speed of 873 rpm at full-load. What is the synchronous speed? Find the frequency of the rotor currents.

15-10   Answer each part briefly:

(a) What is the effect of doubling the air gap of an induction motor on the magnitude of the magnetizing current and on the maximum value of the flux per pole? Neglect the effect of the leakage impedance.

(b) Describe the effect of a reduced leakage reactance on the maximum developed torque, the power factor at full-load, and the starting current of a three-phase induction motor.

15-11   The shaft output of a three-phase, 60-Hz induction motor is 75 KW. The friction and windage losses are 900 watts, the stator core loss is 4200 watts, and the stator copper loss is 2700 watts. The rotor current referred to the stator (primary) is 100 amps. If the slip is 3.75 per cent, what is the per cent efficiency at this output?

15-12   A 15-hp, 220-volt, three-phase, 60-Hz, 6-pole, Y-connected induction motor has the following parameters per phase: $r_1 = 0.128$ ohm, $r'_2 = 0.0935$ ohm, $x_1 + x'_2 = 0.496$ ohm, $r_c = 183$ ohms, $x_\varphi = 8$ ohms. The rotational losses are equal to the stator hysteresis and eddy-current losses. For a slip of 3 per cent find: (a) the line current and power factor; (b) the horsepower output; (c) the starting torque.

15-13   A three-phase induction motor has a Y-connected rotor winding. At standstill the rotor induced emf per phase is 100 volts rms. The resistance per phase is 0.3 ohm, and the leakage reactance is 1.0 ohm per phase.

(a) With the rotor blocked what is the rms value of the rotor current? What is the power factor of the rotor circuit?

(b) When the motor is running at a slip of 0.06, what is the rms value of the rotor current? What is the power factor of the rotor circuit?

(c) Compute the value of the developed power in part (b).

15-14 A three-phase, 12-pole, 60-Hz, 2200-volt induction motor runs at no-load with rated voltage and frequency impressed and draws a line current of 20 amps and an input power of 14 KW. The stator is Y-connected and its resistance per phase is 0.4 ohm. The rotor resistance r_2' is 0.2 ohm per phase. Also, $x_1 + x_2' = 2.0$ ohms per phase. The motor runs at a slip of 2 per cent when it is delivering power to a load. For this condition compute (a) the developed torque and (b) the input line current and power factor.

15-15 A three-phase, 440-volt, 60-Hz, Y-connected, 8-pole, 100-hp induction motor has the following parameters expressed per phase:

$$r_1 = 0.06 \text{ ohm}, \qquad x_1 = x_2' = 0.26 \text{ ohm}$$

$$r_2' = 0.048 \text{ ohm}, \qquad r_c = 107.5 \text{ ohms}$$

$$x_\phi = 8.47 \text{ ohms}$$

The rotational losses are 1600 watts. Using the approximate equivalent circuit determine:

(a) The input line current and power factor.

(b) The efficiency.

15-16 A three-phase, 335-hp, 2000-volt, 6-pole, 60-Hz, Y-connected squirrel-cage induction motor has the following parameters per phase that are applicable at   normal slips:

$$r_1 = 0.2 \text{ ohm}, \qquad x_1 = x_2' = 0.707 \text{ ohm}$$

$$r_2' = 0.203 \text{ ohm}, \qquad r_c = 450 \text{ ohms}$$

$$x_\phi = 77 \text{ ohms}$$

The rotational losses are 4100 watts. Using the approximate equivalent circuit, compute for a slip of 1.5 per cent:

(a) Line power factor and current.

(b) Developed torque.

(c) Efficiency.

15-17 A 6-pole, three-phase, 40-hp, 60-Hz, induction motor has an input when loaded of 35 KW, 51 amps, 440-volts and a speed of 1152 rpm. When uncoupled from the load, the readings are found to be: 440 volts, 21.3 amps, 2.3 KW and 1199 rpm. The resistance measured between terminals for the stator winding is 0.25 ohm for a Y-connection. The stator core losses and the rotational losses are known to be equal. Determine:

(a) The power factor of the motor when loaded.

(b) The motor efficiency when loaded.

(c) The horsepower rating of the load.

15-18 A three-phase, Y-connected, 440-volt, 200-hp induction motor has the following blocked-rotor data: $P_b = 10$ KW, $I_b = 250$ amps, $V_b = 65$ volts, and $r_1 = 0.02$ ohm. Find the value of the rotor resistance referred to the stator.

15-19 A 6-pole, 60 Hz, three-phase, Y-connected wound-rotor (three phases also) induction motor has a standstill induced rotor voltage of 130 volts per phase. At short circuit with the rotor blocked this voltage produces a current of 80 amps at a power factor of 0.3 lagging. At full load the motor runs at a slip of 9 per cent. Find the full-load developed torque.

15-20   A 500-hp, three-phase, 2200-volt, 25-Hz, 12-pole, Y-connected wound-rotor induction motor has the following parameters: $r_1 = 0.225$ ohm, $r'_2 = 0.235$ ohm, $x_1 + x'_2 = 1.43$ ohms, $x_\varphi = 31.8$ ohms, $r_c = 780$ ohms. A no-load and a blocked-rotor test are performed on this machine.

(a) With rated voltage applied in the no-load test, compute the readings of the line ammeters as well as the total wattmeter reading.

(b) In the blocked-rotor test the applied voltage is adjusted so that 228 amps line current is made to flow in each phase. Calculate the reading of the line voltmeter and the total wattmeter reading.

15-21   A three-phase, 2000-volt, Y-connected wound-rotor induction motor has the following no-load and blocked-rotor test data:

No-load: 2000 v,   15.3 amps,   10.1 KW
Blocked-rotor: 440 v,   170.0 amps,   36.4 KW

The resistance of the stator winding is 0.22 ohm per phase. The rotational losses are equal to 2 KW. Calculate all the necessary data for the approximate equivalent circuit at a slip of 2 per cent and draw the circuit showing all parameter values.

(a) The slip at which maximum torque occurs.

(b) The input line current and power factor at the condition of maximum torque.

(c) The value of the maximum torque.

15-22   Refer to the motor of Prob. 15-20 and find the value of resistance that must be externally connected per phase to the rotor winding in order that the maximum torque be developed at starting. What is the value of this torque?

15-23   From the no-load test on a 10-hp, 4-pole, 230-volt, 60-Hz, three-phase, Y-connected induction motor with rated voltage applied, the no-load current was found to be 9.2 amps and the corresponding input power 670 watts. Also, with 57 volts applied in the blocked-rotor test, it was found that the motor took 30 amps and 950 watts from the line. The stator winding resistance was measured to be 0.15 ohm per phase. When this motor is coupled to its mechanical load, the input to the motor is found to be 9150 watts at 28 amps and a power factor of 0.82. The stator core losses are equal to the rotational losses.

(a) Compute the rotor current referred to the stator.

(b) Find the developed torque.

(c) What is the value of the slip?

(d) At what efficiency is the motor operating?

16

three-phase synchronous machines

The synchronous generator is universally used by the electric power industry for supplying three-phase as well as single-phase power to its customers. The single-phase power that is brought to homes, shops, and offices originates from one phase of the three-phase system. Moreover, the assignment of commercial load circuits to each phase is made in an effort to keep the phases balanced.

The basic construction features of these machines are described in Sec. 14-3 and illustrated in Fig. 14-8. It is worth noting here that synchronous generators are classified into two types. The first is the *low-speed* (engine- or water-driven) type, which is characterized physically by having salient poles, a large diameter, and small axial length. The second is the *turbogenerator*, which uses the steam turbine as the prime mover. The usual salient-pole rotor construction is abandoned in these high-speed generators in favor of the cylindrical (or smooth) rotor because the protruding-pole construction gives rise to dangerously high mechanical stresses. For 60-Hz three-phase power the two-pole generator must operate at 3600 rpm. Moreover, since turbogenerators are invariably designed for two poles, it follows that these machines are further characterized by a small diameter and long axial length.

16-1   Generation of a Three-phase Voltage System

How is a three-phase voltage generated? This can be readily understood from a study of Fig. 16-1(a). Depicted here is a two-pole rotor, the field winding of which is assumed energized from a d-c source to create the pole flux. It is further assumed that the pole pieces are shaped to produce a sinusoidal flux field. Appearing in the stator is a balanced three-phase winding with the axis of each phase displaced by 120 degrees. The complete winding of each phase is represented in

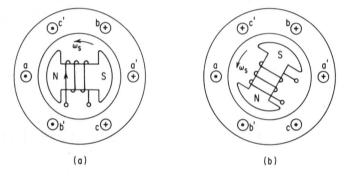

(a)   (b)

Fig. 16-1. Induced-voltage distribution in the stator winding of a synchronous generator:
    (a) phase a at a positive maximum—time t_1 in Fig. 16-2;
    (b) phase b at a negative maximum—time t_2.

Fig. 16-1 by a single coil. Now consider that the rotor is driven by the prime mover in a counterclockwise direction at synchronous speed. Applying the $\bar{v} \times \bar{B}$ rule for the field direction shown reveals that the instantaneous voltages induced in coil sides a, b', c' are directed out of the paper while in coil sides a', b, c they are directed into the paper. Furthermore, since coil sides a and a' are located beneath the maximum value of the flux-density wave, the induced emf in phase a is at its maximum value. The corresponding induced emf in phase b, as identified by coil side b, is seen to be of opposite sign and of smaller magnitude. In fact, since coil sides b and b' are both displaced from the maximum-flux-density position by 60 degrees, it follows that this instantaneous voltage is at one-half (cos 60°) its maximum value. A similar reasoning yields the same result for phase c.

Next consider that the rotor has advanced by 60 degrees in the counterclockwise direction. This puts the amplitude of the north-pole flux directly beneath b', indicating that the induced emf is now a negative maximum in phase b. Also, coil c-c' now finds itself under the influence of flux of reversed polarity as indicated in Fig. 16-1(b). Hence the direction of the emf in c is now out of the paper and into the paper for c'. The instantaneous values of the three phase voltages for the first and second time instants are depicted in Fig. 16-2 and are identified as t_1 and t_2, respectively.

By repeating the foregoing procedure many times and plotting the results

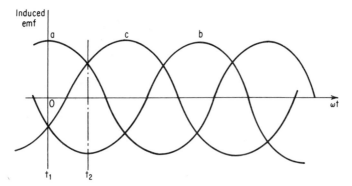

Fig. 16-2. Time history of three-phase induced emf's for machine of Fig. 16-1.

for each instant of time we obtain the complete curves of Fig. 16-2. Note that the induced-emf variation for each phase is identical to the other two except for time displacements of 120° and 240°, respectively. As a matter of fact this variation is a direct consequence of having the beginning of each phase space-displaced by 120 electrical degrees. Physical connection of the ends of each phase (points a', b', c') gives a Y-connected stator winding.

16-2 Synchronous Generator Phasor Diagram and Equivalent Circuit

The principles underlying the development of the phasor diagram are similar to those of the transformer. In fact, more generally one may say that the principles are the same as for any magnetic device subjected to a varying flux. Thus, when the synchronous generator is operating at no-load (i.e. no current flowing through the stator winding, which is the armature winding in this case) there exists a pole flux Φ_f produced by the field winding, which in turn induces an rms voltage per phase which we denote by $\bar{E}_f$. Of course this emf lags the flux that induces it by 90 degrees as depicted in Fig. 16-3. $\bar{E}_f$ is frequently called the *excitation voltage* for obvious reasons. It is always the voltage appearing at the armature terminals at no-load with the field winding excited.

Fig. 16-3. No-load phasor diagram of a synchronous generator.

Now consider that a balanced three-phase unity-power-factor load is placed across the armature terminals so that in the steady state the load draws an rms current I_a at a terminal voltage per phase of $\bar{V}_t$. This situation is depicted on a per phase basis in Fig. 16-4. The resulting phasor diagram is shown in Fig. 16-5. The presence of a three-phase current in the stator winding accounts for the differences

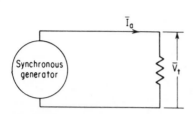

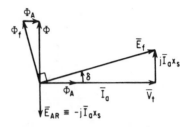

Fig. 16-4. Synchronous generator connected to a unity-pf load.

Fig. 16-5. Synchronous generator phasor diagram for unity-pf load.

in Figs. 16-3 and 16-5. The development of the phasor diagram starts by placing I_a in phase with $\bar{V}_t$ because the load is resistive. However, a three-phase current flowing through the three-phase stator winding gives rise to a rotating field as described in Sec. 15-1. This armature flux field Φ_A combines with the flux field Φ_f, which is produced by the field winding, to yield the resultant flux per pole Φ. If for simplicity the stator-winding resistance and leakage reactance are neglected, then this resultant flux Φ must lead V_t by 90°. Note that Φ_A is shown coincident with I_a, which is consistent with the fact that they are in phase—Φ_A being produced by I_a.

In analyzing the effect of Φ_A we can resort to superposition and consider that Φ_A is a synchronously rotating field which cuts the stator armature winding, thereby inducing an "armature-reaction" voltage per phase denoted by $\bar{E}_{AR}$. This voltage lags the flux that creates it by 90°. Now the excitation voltage, by Kirchhoff's voltage law, must contain not only a component equal to $\bar{V}_t$ but also one equal and opposite to $\bar{E}_{AR}$. To treat the effect of the armature flux as a voltage drop, we replace the armature reaction voltage by an equivalent reactance drop as was done with leakage fluxes in the case of transformers. Thus

$$\bar{E}_{AR} \equiv -jI_a x_s \tag{16-1}$$

The minus sign is used in the definition so that the equal and opposite quantity can be handled with a plus sign. That is, the plus sign is used for the drop and the minus sign for the rise. The parameter x_s is called the *synchronous reactance* per phase of the stator winding. It is a fictitious quantity which replaces the effect of the armature winding mmf on the field flux. By Kirchhoff's voltage law the excitation (or source) voltage must be equal to the sum of the voltage drops. Thus

$$\boxed{E_f\underline{/\delta} = V_t\underline{/0°} + jI_a\underline{/0°}x_s} \tag{16-2}$$

Figure 16-5 shows this addition graphically. Note, too, that once $\bar{E}_f$ is found in this manner, Φ_f can also be shown in the phasor diagram by putting it 90° leading $\bar{E}_f$ and of such a magnitude that when it is added to Φ_A there is yielded the resultant flux Φ.

Appearing in Fig. 16-6 is the phasor diagram of a synchronous generator as

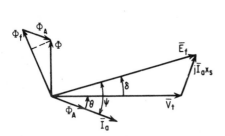

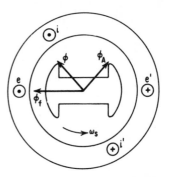

Fig. 16-6. Synchronous generator phasor diagram for lagging-pf load.

Fig. 16-7. Space picture showing the effect of armature flux at lagging pf.

it applies to a lagging-power-factor load. Note that the armature flux field Φ_A causes two effects on the original flux field Φ_f. It directly demagnetizes as well as cross-magnetizes (or distorts) the pole-flux field. This is readily apparent if we consider Φ_A in terms of its two components—one along the line of direction of Φ_f and the other in quadrature. A physical picture of this situation is depicted in Fig. 16-7. The complete three-phase armature winding is again, for convenience, represented by a single coil marked $e\text{-}e'$. In fact this coil is located at the center of the emf distribution of the entire winding. (This corresponds to using coil $a\text{-}a'$ in Fig. 16-1(a) to represent the emf distribution of the complete winding.) A similar ampere-conductor distribution can be identified for the armature winding and represented by a single coil $i\text{-}i'$. The phasor diagram of Fig. 16-6 shows that this current coil must be located in the lag direction (with respect to direction of rotation) by ψ electrical degrees. It is interesting to note that this angle ψ is the same one that appears in the basic torque expression of Eq. (14-39). Applying the right-hand rule to this coil gives an armature flux direction which with respect to the flux field yields the effects already cited.

Equivalent Circuit. The equivalent circuit follows directly from a circuit interpretation of Eq. (16-2). $\bar{E}_f$ is considered the source voltage and x_s is treated as an internal source impedance since it is associated with the armature winding. Figure 16-8 shows the resulting equivalent circuit.

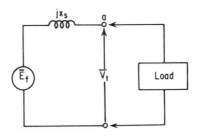

Fig. 16-8. The equivalent circuit of the synchronous generator appears to the left of terminals ab.

16-3 The Synchronous Motor

The significant and distinguishing feature of synchronous motors in contrast to induction motors is that they are doubly excited. Electrical energy is supplied both to the field and the armature windings. When this is done torque can be developed at only one† speed—the *synchronous speed*. At any other speed the average torque is zero. The synchronous speed refers to that rotor speed at which the rotor flux field and the armature ampere-conductor distribution (or the armature flux field) are stationary with respect to each other. To illustrate this point consider the case of the conventional synchronous motor equipped with two poles. It has a d-c voltage applied to the rotor winding, and a three-phase a-c voltage applied to the stator winding which is usually 60 Hz. The three-phase voltage produces in the stator an ampere-conductor distribution which revolves at a rate of 60 Hz. The d-c (or zero-frequency) currents in the rotor set up a two-pole flux field which is stationary so long as the rotor is not turning. Accordingly, we have a situation in which there exists a pair of revolving armature poles and a pair of stationary rotor poles. A little thought should make it clear that for a period of half a cycle (or 1/120 second) a positive torque is developed by the cooperation of the field flux and the revolving ampere-conductor distribution. However, in the next half-cycle this torque is reversed so that the average torque is zero. This is the reason why a synchronous motor per se has no starting torque.

In order to develop a continuous torque, then, it is necessary not only to have a flux field and an appropriately displaced ampere-conductor distribution, *but the two quantities must be stationary with respect to each other.* In singly-fed machines such as the induction motor this comes about automatically. The relative speed between the stator field and the rotor winding induces rotor currents at those slip frequencies which yield a revolving rotor mmf. Superposing this upon the rotor speed gives a revolving rotor mmf, which in turn is stationary relative to the stator field. In the synchronous motor this condition cannot take place automatically because of the separate excitation used for the field and armature windings. (In light of these comments it is reasonable to expect that if an induction motor is doubly energized, it too will behave as a synchronous motor.‡)

On the basis of the foregoing comments it follows that to produce a continuous nonzero torque it is first necessary to bring the d-c excited rotor to synchronous speed by means of an auxiliary device. Sometimes the auxiliary device takes the form of a small d-c motor mounted on the rotor shaft. Most often, however, use is made of a squirrel-cage winding similar to that used in induction motors and imbedded in the pole faces. By means of this winding (also called the *amortisseur winding*) the synchronous motor is brought up to almost synchronous speed. Then if the field winding is energized at the right moment a positive torque

---

† There is a second speed, too, but it is only of academic interest because of the excessive currents that flow.

‡ One such device is the Schrage motor, which is commercially available from the General Electric Company.

will be developed for a sufficiently long period to allow the armature poles to pull the rotor poles into synchronism.

A general expression may be written to identify that frequency of rotation required of the rotor so that the stator and rotor fields are stationary with respect to each other. Let

f_1 denote the frequency of the currents through the stator winding
f_2 denote the frequency of the currents through the rotor winding
f_r denote the frequency of rotation of the rotor structure

Then the defining equation is

$$f_1 = f_2 + f_r \tag{16-3}$$

In the conventional synchronous motor the frequency of the rotor current f_2 is zero, since it is d-c. Hence for an f_1 of 60 Hz the rotor must revolve at 60 Hz in order for nonzero torque to be developed. On the other hand note that if the rotor winding were energized with 20-Hz current in place of d-c, then for an f_1 of 60 Hz the synchronous speed would be 40 Hz.

16-4 Synchronous Motor Phasor Diagram and Equivalent Circuit

The synchronous motor phasor diagram differs from that of the synchronous generator in two respects. The first involves the voltage equation for the stator circuit. In the generator case $\bar{E}_f$, the excitation voltage, played the role of a source voltage and the terminal voltage $\bar{V}_t$ was dependent upon the valve of $\bar{E}_f$ and the synchronous impedance drop. In the motor case the roles are interchanged. Now $\bar{V}_t$ is the source voltage applied to the synchronous motor armature winding and $\bar{E}_f$ is a reaction or counter emf which is internally generated. It is assumed that the terminal voltage originates from an infinite bus system† and so remains invariant. Applying Kirchhoff's voltage law to the synchronous motor then leads us to

$$\boxed{V_t\underline{/0^\circ} = \bar{E}_f + j\bar{I}_a x_s} \tag{16-4}$$

Clearly, this equation states that the applied stator voltage is equal to the sum of the drops. The excitation voltage is treated as a reaction-voltage drop in much the same way as E_1 was treated in the case of the transformer.

The circuit interpretation of Eq. (16-4) leads to the equivalent circuit as it applies to the synchronous motor. This appears in Fig. 16-9.

The second point of view in which the

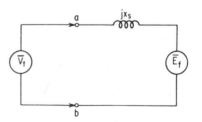

Fig. 16-9. The equivalent circuit of the synchronous motor appears to the right of terminals *ab*.

† A power system of tremendous capacity compared to the rating of the synchronous motor.

phasor diagram of the motor differs from that of the generator involves the angle δ, which is called the power angle for reasons described soon. Physically, for the generator, the phase $\bar{E}_f$ is ahead of $\bar{V}_t$ in time because of the driving action of the prime mover. Keep in mind that $\bar{E}_f$ may be considered associated with the rotor field axis as was done in connection with Fig. 16-7. A similar line of reasoning pertains for the motor. At no-load δ is zero and so the axis of a field pole and the resultant flux associated with the terminal voltage are in time phase. As shaft load is applied, however, the rotor falls slightly behind its no-load position and in this way causes δ to increase but in a sense opposite to that for the generator.

On the basis of these modifications the phasor diagram of the synchronous motor becomes that shown in Fig. 16-10. Note that $\bar{E}_f$ now lags $\bar{V}_t$ and that it is the phasor sum of $\bar{E}_f$ and $j\bar{I}_a x_s$ which equals $\bar{V}_t$ as called for by Eq. (16-4). Another interesting point about this diagram is revealed upon comparison with Fig. 16-6.

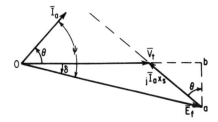

Fig. 16-10. Phasor diagram of the synchronous motor.

In both cases the magnitude of the excitation voltage $\bar{E}_f$ exceeds that of the terminal voltage $\bar{V}_t$. This is referred to as a condition of *overexcitation* (i.e. $|E_f| > |\bar{V}_t|$). Note, however, that overexcitation in the synchronous generator is accompanied by a current of *lagging* power factor whereas in the synchronous motor overexcitation is accompanied by a current of *leading* power factor. When the magnitude of $\bar{E}_f$ is equal to $\bar{V}_t$ the condition is referred to as *100 per cent* excitation. A situation where $|\bar{E}_f| < |\bar{V}_t|$ is described as *underexcitation*.

16-5 Computation of Synchronous Motor Performance

The analysis of the performance of a synchronous motor is readily established in terms of the phasor diagram, the power-flow diagram, and an appropriate expression for the mechanical power developed which is derivable from the phasor diagram. In connection with the phasor diagram one additional item is needed which has not yet been discussed. It is the saturation curve for the magnetic circuit of the synchronous motor. This curve, of course, is a plot of induced armature-winding voltage as a function of field-winding current. Thus corresponding to any computed value of excitation voltage E_f the field current needed to produce it can be specified. Conversely, the excitation voltage is known once the field current is named.

The worth of a power-flow diagram has already been demonstrated in con-

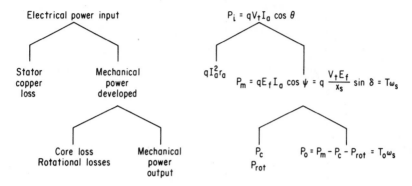

Fig. 16-11. Power-flow diagram of the synchronous motor.

nection with the induction motor. Hence it is presented in Fig. 16-11 without very much additional comment. There is one step less in this flow graph because the a-c source does not have to supply rotor-winding copper loss since the rotor is traveling at synchronous speed. This is not to say that rotor-winding copper losses do not exist. They do, of course, but the power is supplied from the sepa-rated d-c source. Figure 16-11 represents the flow of power from the a-c source to the shaft of the synchronous motor.

A useful expression for the mechanical power developed can be derived from the phasor diagram of Fig. 16-10, which is drawn for negligible armature-winding resistance. Because r_a is assumed zero, the power-flow diagram indicates that the mechanical power developed is equal to the power input. Thus

$$P_m = qE_f V_t \cos \psi = qV_t I_a \cos \theta \qquad (16\text{-}5)$$

This statement is also apparent from Fig. 16-10 when we recognize that the pro-jection of E_f on the I_a phasor is identical to the projection of V_t on I_a. Further-more, from the geometry of the phasor diagram of Fig. 16-10 we note that the quantity ab may be expressed in one of two ways. Thus

$$ab = E_f \sin \delta = I_a x_s \cos \theta \qquad (16\text{-}6)$$

or

$$I_a \cos \theta = \frac{E_f}{x_s} \sin \delta \qquad (16\text{-}7)$$

Insertion of Eq. (16-7) into Eq. (16-5) yields the result being sought. Hence the mechanical power developed in a synchronous motor that has negligible armature resistance may be expressed as

$$\boxed{P_m = q\frac{V_t E_f}{x_s} \sin \delta} \qquad (16\text{-}8)$$

Because of the dependence of the mechanical power developed upon the sine of the angle δ, this angle is called the *power angle*. Equation (16-8) states that if the power angle is zero the synchronous motor cannot develop torque. It further

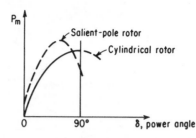

Fig. 16-12. Power angle curves for synchronous motors.

points out that the mechanical power developed is a sinusoidal function of the angle δ as depicted in Fig. 16-12. Note, too, that the maximum developed power for a given excitation occurs at δ equal to 90 degrees.

The application of Eq. (16-8) for computation purposes has a further restriction in addition to that of negligible armature-winding resistance. It assumes that the machine has a uniform air gap. That is, the rotor is assumed to be of the nonsalient-pole type. The use of the phasor diagram of Fig. 16-10 presupposes this, because the diagram applies only to such a machine. For a salient-pole machine the corresponding phasor diagram is more complicated and lies beyond the scope of this book. However, even though synchronous motors are invariably of salient-pole construction, we shall continue to use the phasor diagram of Fig. 16-10 as well as Eq. (16-8) because the general results are the same. Moreover, it is easier to gain insight into the basic operating principles and performance behavior when the secondary effects are mitigated.

The analogous expression for the mechanical power developed by a salient-pole synchronous motor includes a second term besides that appearing in Eq. (16-8). In fact this second term is completely independent of the excitation of the machine; it depends solely upon the terminal voltage and machine parameters, which reflect the nonuniform character of the air gap. In most motors this extra term has a value sometimes as much as 25 per cent of the P_m given by Eq. (16-8). For obvious reasons this term is referred to as *reluctance power*. It causes a change in the power-versus-power-angle plot as indicated by the broken-line curve in Fig. 16-12.

What is the mechanism by which the synchronous motor becomes aware of the presence of a shaft load, and how is the electric energy source made aware of this so that it proceeds to provide energy balance? To answer this question we start first with a study of the conditions prevailing at no-load. As the power-flow diagram indicates, the only mechanical power needed is that which is required to supply the rotational and core losses. This calls for a very small value of δ, as depicted in Fig. 16-13. The machine is assumed to be overexcited since $\bar{E}_f > \bar{V}_t$. Note that δ_0 is just large enough so that the in-phase component of I_a is sufficient to supply the losses. Consider next that a large mechanical load is suddenly applied to the motor shaft. The first reaction is to cause a momentary drop in speed. In turn this appreciably increases the power angle and thereby causes a phasor voltage difference to exist between $\bar{V}_t$ and the excitation voltage $\bar{E}_f$. The result is the flow of an increased armature current at a very much improved power factor compared to no-load. As a matter of fact the speed changes momentarily by a

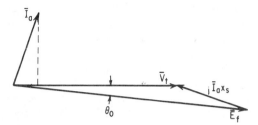

Fig. 16-13. Phasor diagram of synchronous motor at no-load and over-excited.

sufficient amount to allow the power angle to assume that value which enables the armature current and input power factor to take on those values which permit the input power to balance the power required at the load plus the losses. Figure 16-10 is typical of what the final phasor diagram looks like. Note that the power component of the current is considerably increased over the no-load case. The increase in power angle from no-load to the load case can be measured by means of any stroboscopic instrument. The stroboscope furnishes a convincing demonstration of the physical character of the power angle.

EXAMPLE 16-1 A 2300-volt, three-phase, 60-Hz, Y-connected cylindrical-rotor synchronous motor has a synchronous reactance of 11 ohms per phase. When it delivers 200 hp the efficiency is found to be 90 per cent exclusive of field loss, and the power angle is 15 electrical degrees as measured by a stroboscope. Neglect ohmic resistance and determine (a) the induced excitation voltage per phase, E_f, (b) the line current, I_a, (c) the power factor.

solution: (a) The power input and the mechanical power developed are the same. Hence

$$3 \frac{V_t E_f}{x_s} \sin \delta = \frac{200(746)}{0.9}$$

Inserting $V_t = 2300/\sqrt{3}$, $\delta = 15°$,   and $x_s = 11$ yields

$$E_f = 1790 \text{ volts/phase}$$

(b) The armature current follows directly from Eq. (16-4). Thus

$$I_a = \frac{1330\underline{/0°} - 1790\underline{/-15°}}{11\underline{/90°}} = 55.4\underline{/41.4°}$$

Thus since the angle of I_a is positive, the power factor is leading.

(c) The line power factor is merely

$$\text{pf} = \cos 41.4° = 0.755 \text{ leading}$$

16-6 Power-factor Control

For a fixed mechanical power developed (or load) it is possible to adjust the reactive component of the current drawn from the line by varying the d-c field current. This feature is achievable in the synchronous motor precisely because it is a doubly-excited machine. Thus, although operation at a constant applied voltage demands a fixed resultant flux, both the d-c source and the a-c source may cooperate in establishing this resultant flux. If the field-winding current is made excessively large, then clearly the resultant air-gap voltage in the motor tends to be larger than that demanded by the applied voltage. Accordingly, a reaction occurs which causes the armature current to assume such a power-factor angle that the armature mmf exerts that amount of demagnetizing effect which is needed to restore the required resultant flux. Similarly, if the field is underexcited, then the resultant gap flux tends to be too small. This also creates a phasor voltage difference between the line voltage and the motor excitation voltage which acts to cause the armature current to flow at that power-factor angle which enables it to magnetize the air gap to the extent needed to provide the necessary resultant flux.

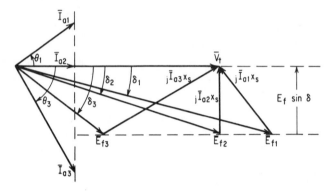

Fig. 16-14. Showing the effect of varying excitation on power factor.

Whether or not the reactive component of current must be leading or lagging readily follows from an investigation of the phasor diagrams depicted in Fig. 16-14 for various values of excitation voltage. The diagram applies for an assumed constant mechanical power developed. Hence as E_f is changed, the $\sin \delta$ must change correspondingly to keep $E_f \sin \delta$ invariant. In Fig. 16-14 this means that as the excitation is varied, the locus of the tip of the $\bar{E}_f$ phasor is the broken line. Furthermore, the in-phase component of the current in each case must be the same. In the first case, where the excitation voltage is $\bar{E}_{f1}$, the field current is producing too much flux. This creates a reaction between the motor and the source which calls for a leading current of such a magnitude that it provides that amount and direction of synchronous reactance drop which when added to $\bar{E}_{f1}$ yields the fixed terminal voltage $\bar{V}_t$. Physically, what is happening is that a leading reactive current is made to flow which acts to demagnetize the flux field to the extent needed. When the excitation is reduced to $\bar{E}_{f2}$, there is no excess flux produced by the field winding. Consequently, the a-c line current contains no reactive component. It merely has the value of in-phase component needed to supply power to the load. At the excitation corresponding to $\bar{E}_{f3}$ the machine is greatly underexcited. To compensate for this a reaction occurs which allows the line to deliver a large lagging reactive current, which helps to establish the value of air-gap flux demanded by the terminal voltage. Note, too, that as the excitation is decreased the power angle must increase for a fixed mechanical power developed.

If a plot is made of the armature current as the excitation is varied for fixed mechanical power developed, the current is observed to be large for underexcitation and overexcitation and passing through a minimum at some intermediate point. The plot in fact resembles a V-shape as demonstrated in Fig. 16-15.

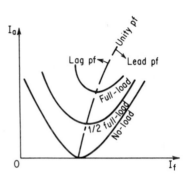

Fig. 16-15. Synchronous motor *V*-curves.

The ability of the synchronous motor to draw leading current when over-excited can be used to improve the power factor at the input lines to an industrial establishment that makes heavy use of induction motors and other equipment drawing power at a lagging power factor. Many electric power companies charge increased power rates when power is bought at poor lagging power factor. Over the years such increased power rates can result in an appreciable expenditure of money. In such instances the installation of a synchronous motor operated over-excited can more than pay for itself by improving the overall input power factor to the point where the penalty clause no longer applies.

16-7 Synchronous Motor Applications

Synchronous motors are rarely used below 50 hp in the medium-speed range because of their much higher initial cost compared to induction motors. In addition they require a d-c excitation source, and the starting and control devices are usually more expensive—especially where automatic operation is required. However, synchronous motors do offer some very definite advantages. These include constant-speed operation, power-factor control, and high operating efficiency. Furthermore, there is a horsepower and speed range where the disadvantage of higher initial cost vanishes, even to the point of putting the synchronous motor to advantage. This is demonstrated in Fig. 16-16. Where low speeds and high horsepower are involved the induction motor is no longer cheaper because it must use large amounts of iron in order not to exceed a gap flux density of 0.7 tesla. In the synchronous machine, on the other hand, a value twice this figure is permissible because of the separate excitation.

Appearing in Table 16-1 are some of the more important characteristics of the synchronous motor along with some typical applications. In addition to the area of application listed there they are also quite prominently found in the following applications, which are characterized by operation at low speeds and high horsepower: large, low-head pumps; flour-mill line shafts; rubber mills and mixers; crushers; chippers; pulp grinders and jordans and refiners used in the papermaking industry.

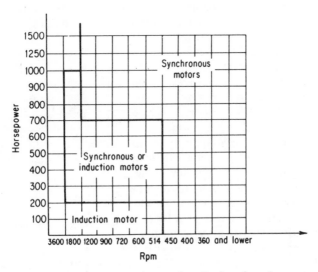

Fig. 16-16. Indicating the general areas of application of synchronous and induction motors.

Table 16-1. Synchronous Motor Characteristics and Applications†

Type designation	Synchronous, high speed about 500 rpm	Synchronous, low speed, below 500 rpm
HP range	25 to several thousand	Usually above 25 to several thousand
Starting torque (% of normal)	Up to 120	Low 40
Pull-out torque (%)	Up to 200	Up to 180
Starting current (%)	500–700	200–350
Slip	Zero	Zero
Power factor	High, but varies with load and with excitation	High, but varies with excitation
Efficiency (%)	Highest of all motors, 92–96	Highest of all motors, 92–96
Typical applications	Fans, blowers, d-c generators, line shafts, centrifugal pumps and compressors, reciprocating pumps and compressors. Useful for power-factor correction. Constant speed. Frequency changes.	Lower-speed direct-connected loads such as reciprocating compressors when started unloaded, d-c generators, rolling mills, band mills, ball mills, pumps. Useful for power-factor control. Constant speed. Flywheel used for pulsating loads.

† By permission from M. Liwschitz-Garik and C. C. Whipple, *Electric Machinery*, vol. II (Princeton, N. J.: D. Van Nostrand Co., Inc., 1946).

PROBLEMS

16-1 A three-phase, 1732-volt (line to line), Y-connected, cylindrical-pole synchronous motor has $r_a = 0$ and $x_s = 10$ ohms per phase. The friction and windage plus the core losses amount of 9 KW. The motor delivers an output of 390 hp. The greatest excitation voltage that may be obtained is 2500 volts per phase.

(a) Calculate the magnitude and power factor of the armature current for maximum excitation at the specified load.

(b) Compute the smallest excitation for which the motor will remain in synchronism for the given power output.

16-2 A 2300-volt, three-phase, 60-Hz, Y-connected, cylindrical-rotor synchronous motor has a synchronous reactance of 11 ohms per phase. When it delivers 200 hp, the efficiency is 90 per cent and the power angle is 15 electrical degrees. Neglect r_a and determine: (a) the excitation emf per phase; (b) the line current; (c) the power factor.

16-3 A synchronous motor is operated at half-load. An increase in its field current causes a decrease in armature current. Does the armature current lead or lag the terminal voltage? Explain.

16-4 A three-phase, Y-connected synchronous motor is operating at 80 per cent leading power factor. The synchronous reactance is 2.9 ohms per phase and the armature winding resistance is negligible. The armature current is 20 amps/phase. The applied line voltage is 440 volts.

(a) Find the excitation voltage and the power angle.

(b) It is desired to increase the armature current to 40 amps/phase and maintain the power factor at 80 per cent leading. Show clearly in a phasor diagram how the field excitation must be changed. Can this be accomplished if the shaft load remains fixed? Explain.

16-5 A Y-connected, three-phase, 60-Hz, 13,500-volt synchronous motor has an armature resistance of 1.52 ohms/phase and a synchronous reactance of 37.4 ohms/phase. When the motor delivers 2000 hp, the efficiency is 96 per cent and the field current is so adjusted that the motor takes a leading current of 85 amps.

(a) At what power factor is the motor operating?

(b) Calculate the excitation emf.

(c) Find the mechanical power developed.

(d) If the load is removed, describe (do not calculate) how the magnitude and power factor of the resulting armature current compare with the original.

16-6 A 1250-hp, three-phase, cylindrical-pole synchronous motor receives constant power of 800 KW at 11,000 volts. Armature winding resistance is negligible and the synchronous reactance per phase is 50 ohms. The armature is Y-connected. The motor has a rated full-load current of 52 amps. If the armature current shall not exceed 135 per cent of this value, determine the range over which the excitation emf can be varied through adjustment of the field current.

16-7 As a synchronous motor is loaded from zero to full-load, can the pf ever become leading if the field excitation is maintained at 75 per cent? Explain.

16-8 A synchronous motor delivers rated power to a load. Is there a lower limit on the excitation current beyond which it may not be reduced? Explain.

16-9 A 440-volt, three-phase, Y-connected synchronous motor has a synchronous reactance of 6.06 ohms per phase. The armature resistance is negligible, and the induced excitation emf per phase is 200 volts. Moreover, the power angle between $\bar{V}_t$ and $\bar{E}_f$ is 36.4 electrical degrees.
  (a) Calculate the line current and pf.
  (b) What values of excitation emf and power angle are necessary to make the pf unity for the same input?

16-10 A synchronous motor at no-load is connected to an infinite bus system. The field circuit is accidentally opened. Explain what happens to the armature current.

16-11 An 8-pole synchronous motor draws 45 kw from the 208-volt, 60 Hz, three-phase power system at a pf of 0.8 lagging. The motor is Y-connected and has a synchronous reactance of 0.6 ohms per phase. Armature resistance is negligible.
  Without any further manipulations on the motor, what is the highest possible value of its steady-state torque?

16-12 Answer whether the following statements are true or false. When the statement is false, describe why.
  (a) Increasing the air-gap of a machine reduces the synchronous reactance and raises the steady-state power limit of the machine.
  (b) For a synchronous motor, the developed electromagnetic torque is in a direction opposite to the direction of rotation.
  (c) In a synchronous machine, the space-phase relation between the field mmf wave and armature-reaction-mmf wave is determined by the pf of the load.
  (d) The sum of the armature mmfs in the three phases of the stator winding is zero at any given instant.
  (e) The sum of the main field flux Φ_f and the armature flux Φ_A always add to give the rsultant flux Φ.
  (f) A synchronous motor operating at leading pf is underexcited.
  (g) The high currents flowing during a short circuit cause a synchronous machine to operate under saturated conditions.
  (h) The load on a synchronous motor is increased. This causes the main field mmf axis to drop further beind the air-gap mmf axis, so as to increase the torque angle.
  (i) An overexcited synchronous motor is a generator of lagging kvars. This means that it supplies lagging kvars to the system to which it is connected.
  (j) For large machines, the armature resistance and leakage reactance are about 10 per cent on the machine rating and the synchronous reactance is about 100 per cent.

17

d-c machines

The basic voltage equation, the developed-torque equation, the construction details, and the need for a commutator have all been described in Chap. 14. In this chapter we analyze the performance of these machines and discuss typical applications.

17-1 D-C Generator Analysis

The d-c machine functions as a generator when mechanical energy is supplied to the rotor and an electrical load is connected across the armature terminals. In order to supply electrical energy to the load, however, a magnetic field must first be established in the air gap. The field is necessary because it serves as the coupling device permitting the transfer of energy from the mechanical to the electrical system. There are two ways in which the field winding may be energized to produce the magnetic field. One method is to excite the field separately from an auxiliary source as depicted in the schematic diagram of Fig. 17-1. But clearly this scheme is disadvantageous because of the need of another d-c source. After all, the purpose of the d-c generator is to make available such a source. Therefore, invariably d-c generators are excited by the second method which involves a pro-

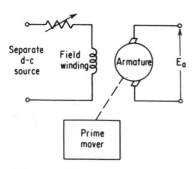

Fig. 17-1. Schematic diagram of a separately excited d-c generator.

cess of self-excitation. The wiring diagram appears in Fig. 17-2, and the arrangement is called the *self-excited shunt generator.* The word shunt is used because the field winding appears in parallel with the armature winding. That is, the two windings form a shunt connection.

To understand how the self-excitation process takes place we must start with the *magnetization curve* of the machine. Sometimes this is called the *saturation curve.* Strictly speaking, the magnetization curve represents a plot of air-gap flux versus field-winding magnetomotive force. However, in the d-c generator where the winding constant K_E is known and the speed n is fixed, the magnetization curve has come to represent a plot of the open-circuit induced armature voltage as a function of

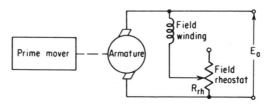

Fig. 17-2. Schematic diagram of a self-excited shunt generator.

the field-winding current. With K_E and n fixed, Eq. (14-47) shows that Φ and E_a differ only by a constant factor. Figure 17-3 depicts a typical magnetization curve, valid for a constant speed of rotation of the armature. It is especially important to note in this plot that even with zero field current an emf is induced in the armature of value Oa. This voltage is due entirely to residual magnetism,

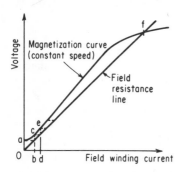

Fig. 17-3. Illustrating the build-up procedure of a self-excited shunt generator.

which is present because of the previous excitation history of the magnetic-circuit iron. The linear curve appearing on the same set of axes is the *field-resistance line.* It is a plot of the current caused by the voltage applied to the series combination of the field winding and the active portion of the field rheostat. Clearly, then, the slope of the linear curve is equal to the sum of the field-winding resistance R_f and the active rheostat resistance R_{rh}. The voltage Oa due to residual magnetism appears across the field circuit and causes a field current Ob to flow. But in accordance with the magnetization curve this field current aids the residual flux and thereby produces a larger induced emf of value *bc*. In turn this increased emf causes an even larger field current, which creates more flux for a larger emf, and so forth. This process of voltage build-up continues until the induced emf produces just enough field current to sustain it. This corresponds to point *f* in Fig. 17-3. Note that in order for the build-up process to take place three conditions must be satisfied: (1) There must be a residual flux. (2) The field winding mmf must act to aid this residual flux. (3) The total field-circuit resistance must be less than the critical value. The *critical field resistance* is that value which makes the resistance line coincide with the linear portion of the saturation curve.

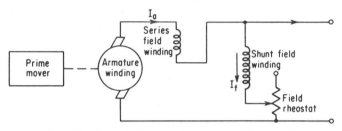

Fig. 17-4. Schematic diagram of a compound d-c generator.

The performance of the d-c generator can be analyzed in terms of an equivalent circuit, a set of appropriate performance equations, the power-flow diagram, and the magnetization curve. To generalize the equivalent circuit consider first the schematic diagram of the compound generator in Fig. 17-4. A *compound generator* is a shunt generator equipped with a series winding. The *series winding* is a coil of comparatively few turns wound on the same magnetic axis as the field winding and connected in series with the armature winding. Since the series field must be capable of carrying the full armature current, its cross-sectional area is much greater than that used in the shunt field winding. The purpose of the series field is to provide additional air-gap flux as increased armature current flows, in order to neutralize the armature-winding resistance drop as well as the voltage drops occurring in the feeder wires leading to the load. In such cases the generator is usually referred to as a *cumulatively* compounded generator, because the series-field flux *aids* the shunt-field flux.

By imposing the appropriate constraint on the connection diagram shown in Fig. 17-4, we can identify any one of the three modes of operation of the d-c generator. Thus, besides the armature winding we have the following:

compound generator: includes shunt- and series-field windings
shunt generator: shunt-field winding
series generator: series-field winding

Since the series generator is rarely used except for special applications, all further treatment of generators is confined to the shunt and compound modes.

Appearing in Fig. 17-5 is the equivalent circuit of the compound generator. The armature winding is replaced by a source voltage having the induced emf

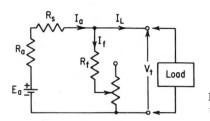

Fig. 17-5. Equivalent circuit of the compound d-c generator.

E_a and a resistance R_a, which represents the armature circuit resistance.† The series field is replaced by its resistance R_s; likewise for the shunt field.

The governing equations for determining the performance are the following:

$$E_a = \left(\frac{pZ}{60a}\right)\Phi n = K_E \Phi n \tag{17-1}$$

$$T = \left(\frac{pZ}{2\pi a}\right)\Phi I_a = K_T \Phi I_a \tag{17-2}$$

$$V_t = E_a - I_a(R_a + R_s) \tag{17-3}$$

$$I_a = I_L + I_f \tag{17-4}$$

These include the two basic relationships for induced voltage and electromagnetic torque as developed in Sec. 17-4. Equations (17-3) and (17-4) are merely statements of Kirchhoff's voltage and current laws as they apply to the equivalent circuit of Fig. 17-5. For the compound generator the expression for the air-gap flux must include the effect of the series field as well as the shunt field. Thus

$$\Phi = \Phi_{sh} + \Phi_s \tag{17-5}$$

where Φ_{sh} denotes the flux produced by the shunt-field winding and Φ_s denotes the flux caused by the series-field winding. For the shunt generator, of course, Φ_s is zero in Eq. (17-5), and so too is R_s in Eq. (17-3).

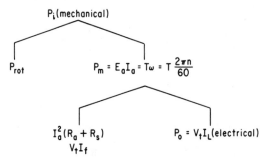

Fig. 17-6. Power-flow diagram of the d-c generator.

† The armature-circuit resistance includes the armature-winding resistance plus the effect of the voltage drop in the carbon brushes.

The power-flow diagram for the d-c generator is depicted in Fig. 17-6. Note the similarity it bears to that of the synchronous generator. The field-winding loss is included in the power flow directly because it is assumed that the generator is self-excited. Of course, if the field winding is separately excited, the field losses are not supplied from the prime mover and so must be handled separately. Moreover, note that the field losses are represented in terms of the product of the field terminal voltage and the field current. This assures that the field-winding rheostat losses are included as well.

EXAMPLE 17-1   The magnetization curve of a 10-KW, 250-volt, d-c self-excited shunt generator driven at 1000 rpm is shown in Fig. 17-7. Each vertical division represents 20 volts and each horizontal unit represents 0.2 amp. The armature-circuit resistance is 0.15 ohm and the field current is 1.64 amps when the terminal voltage is 250 volts. Also, the rotational losses are known to be equal to 540 watts. Find at rated load (a) the armature induced emf, (b) the developed torque, (c) the efficiency. Assume constant-speed operation.

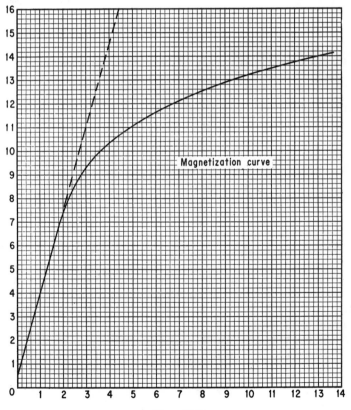

Figure 17-7

solution: (a) With the generator delivering rated load current it follows that

$$I_L = \frac{10,000}{250} \text{ watts} = 40 \text{ amps}$$

Hence from Eq. (17-4) the armature current is

$$I_a = I_L + I_f = 40 + 1.64 = 41.64 \text{ amps}$$

The induced armature voltage then follows from Eq. (17-3):

$$E_a = V_t + I_a R_a = 250 + 41.64(0.15) = 256.26 \text{ volts} \qquad (17\text{-}6)$$

(b) To determine the developed torque it is first necessary to compute the electromagnetic power $E_a I_a$. Thus

$$E_a I_a = 256.25(41.64) = 10,700 \text{ watts}$$

The power-flow diagram reveals that this power may also be found from

$$
\begin{aligned}
E_a I_a &= P_o + I_a^2 R_a + V_t I_f = 10 \text{ KW} + (41.64)^2(0.15) + 250(1.64) \\
&= 10 \text{ kw} + 261 + 410 = 10,671 \text{ watts}
\end{aligned}
\qquad (17\text{-}7)
$$

which checks closely with the preceding calculation. Hence the developed torque is

$$T = \frac{E_a I_a}{\dfrac{2\pi n}{60}} = \frac{10,700}{\dfrac{2\pi(1000)}{60}} = 101.5 \text{ n-m} \qquad (17\text{-}8)$$

(c) The efficiency is found from Eq. (15-20), which for the d-c generator takes the form

$$\eta = 1 - \frac{\sum \text{losses}}{E_a I_a + P_{rot}} \qquad (17\text{-}9)$$

In this case

$$
\begin{aligned}
\sum \text{losses} &= P_{rot} + I_a^2 R_a + V_t I_f \\
&= 540 + 261 + 410 = 1211 \text{ watts}
\end{aligned}
$$

Hence

$$\eta = 1 - \frac{1211}{10,700 + 540} = 1 - \frac{1211}{11,240} = 1 - 0.108 = 0.892$$

The efficiency is therefore 89.2 per cent.

17-2 D-C Motor Analysis

A d-c motor is a d-c generator with the power flow reversed. In the d-c motor electrical energy is converted to mechanical form. Also, as is the case for the generator, there are three types of d-c motors: the *shunt* motor, the *cumulatively compounded* motor, and the *series motor*. The compound motor is prefixed with the word cumulative in order to stress that the connections to the series-field winding are such as to assure that the series-field flux *aids* the shunt-field flux. The series motor, unlike the series generator, finds wide application, especially for traction-type loads. Hence due attention is given to this machine in the treatment that follows.

The performance of the d-c motor operating in any one of its three modes can conveniently be described in terms of an equivalent circuit, a set of performance equations, a power-flow diagram, and the magnetization curve. The equivalent circuit is depicted in Fig. 17-8. It is worthwhile to note that now the armature induced voltage is treated as a reaction or counter emf. By imposing constraints similar to those applied to the d-c generator, we obtain the correct equivalent circuit for the desired mode of operation. For example, for a series motor the appropriate equivalent circuit results upon removing R_f from the circuitry of Fig. 17-8.

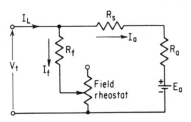

Fig. 17-8. Equivalent circuit of the d-c motor.

The set of equations needed to compute the performance is listed below.

$$E_a = K_E \Phi n \tag{17-1}$$

$$T = K_T \Phi I_a \tag{17-2}$$

$$V_t = E_a + I_a(R_a + R_s) \tag{17-10}$$

$$I_L = I_f + I_a \tag{17-11}$$

The first two equations are identical to those used in generator analysis. Note, however, that the next two equations are modified to account for the fact that for the motor V_t is the applied or source voltage and as such must be equal to the sum of the voltage drops. Similarly the line current is equal to the *sum* rather than the *difference* of the armature current and field currents.

The power-flow diagram, depicting the reversed flow from that which occurs in the generator, is illustrated in Fig. 17-9. The electrical power input $V_t I_L$ originating from the line supplies the field power needed to establish the flux field as well as the armature-circuit copper loss needed to maintain the flow of I_a. This current flowing through the armature conductors imbedded in the flux field causes torque to be developed. The law of conservation of energy then demands that the electromagnetic power, $E_a I_a$, be equal to $T \omega_m$, where ω_m is the steady-state operating speed. Removal of the rotational losses from the developed mechanical power yields the mechanical output power. The use of the foregoing tools in computing motor behavior is illustrated by the following examples.

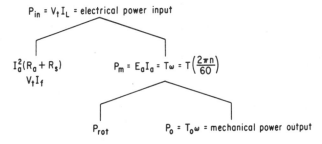

Fig. 17-9. Power-flow diagram of the d-c motor.

EXAMPLE 17-2   A 20-hp, 230-volt, 1150-rpm shunt motor has four poles, four parallel armature paths, and 882 armature conductors. The armature-circuit resistance is 0.188 ohm. At rated speed and rated output the armature current is 73 amps and the field current is 1.6 amps. Calculate: (a) the electromagnetic torque, (b) the flux per pole, (c) the rotational losses, (d) the efficiency, (e) the shaft load.

solution: (a) The torque can be computed from Eq. (17-8), but we first need E_a. Hence

$$E_a = V_t - I_a R_a = 230 - 73(0.188) = 230 - 13.7 = 216.3 \text{ volts}$$

Also

$$\omega_m = \frac{2\pi n}{60} = \frac{2\pi(1150)}{60} = 120 \text{ rad/sec}$$

Therefore

$$T = \frac{E_a I_a}{\omega_m} = \frac{216.3(73)}{120} = 132 \text{ n-m}$$

(b)
$$E_a = K_E \Phi n = \frac{p}{60}\left(\frac{Z}{a}\right)\Phi n = \frac{4}{60}\left(\frac{882}{4}\right)\Phi(1150)$$

$$\therefore \ \Phi = \frac{216.3(60)}{882(1150)} = 0.0128 \text{ weber}$$

(c) From the power-flow diagram

$$P_{rot} = P_m - P_o = E_a I_a - 20(746) = 15,800 - 14,920 = 980 \text{ watts}$$

(d) First find the sum of the losses. Thus

$$\sum \text{losses} = P_{rot} + I_a^2 R_a + V_t I_f = 980 + (73)^2(0.188) + 230(1.6)$$
$$= 980 + 1000 + 368 = 2348 \text{ watts}$$

Hence

$$\eta = 1 - \frac{\sum \text{losses}}{E_a I_a + I_a^2 R_a + V_t I_f} = 1 - \frac{2348}{17,168} = 0.864$$

The efficiency is 86.4 per cent.

(e)
$$T_o = \frac{1}{\omega_m} P_o = \frac{1}{120}(20 \times 716) = 124 \text{ n-m}$$

The difference between this torque and the electromagnetic torque is the amount needed to overcome the rotational losses.

EXAMPLE 17-3   The shaft load on the motor of Example 17-1 remains fixed, but the field flux is reduced to 80 per cent of its value by means of the field rheostat. Determine the new operating speed.

solution: Information about the speed is available from Eq. (17-1). However, in turn, knowledge of flux and armature induced emf E_a is needed. By the statement of the problem the new flux Φ' is related to the original flux by

$$\Phi' = 0.8\Phi$$

To obtain information about E_a, we must determine the change in I_a, if any. By the constant-torque condition we have

$$K_T \Phi I_a = K_T \Phi' I_a'$$

Hence

$$I_a' = \frac{\Phi}{\Phi'} I_a = \frac{1}{0.8}(73) = 91.3 \text{ amps}$$

Consequently

$$E'_a = V_t - I'_a R_a = 230 - 91.3(0.188) = 212.8 \text{ volts}$$

Returning to Eq. (17-1) we can now formulate the ratio

$$\frac{E'_a}{E_a} = \frac{K_E \Phi' n'}{K_E \Phi n}$$

from which the expression for the new operating speed becomes

$$n' = \frac{E'_a}{E_a} \left(\frac{\Phi}{\Phi'} \right) n = \frac{212.8}{216.3} \left(\frac{1}{0.8} \right) 1150 = 1415 \text{ rpm}$$

17-3 Motor Speed-Torque Characteristics. Speed Control

How does the d-c motor react to the application of a shaft load? What is the mechanism by which the d-c motor adapts itself to supply to the load the power it demands? The answers to these questions can be obtained by reasoning in terms of the performance equations that appear on p. 567. Initially our remarks are confined to the shunt motor, but a similar line of reasoning applies for the others. For our purposes the two pertinent equations are those for torque and current. Thus

and

$$\boxed{T = K_T \Phi I_a} \qquad (17\text{-}2)$$

$$\boxed{I_a = \frac{V_t - K_E \Phi n}{R_a}} \qquad (17\text{-}12)$$

Note that the last expression results from replacing E_a by Eq. (17-1) in Eq. (17-10). With no shaft load applied, the only torque needed is that which overcomes the rotational losses. Since the shunt motor operates at essentially constant flux, Eq. (17-2) indicates that only a small armature current is required compared to its rated value to furnish these losses. Equation (17-12) reveals the manner in which the armature current is made to assume just the right value. In this expression V_t, R_a, K_E, and Φ are fixed in value. Therefore the speed is the critical variable. If, for the moment, it is assumed that the speed has too low a value, then the numerator of Eq. (17-12) takes on an excessive value and in turn makes I_a larger than required. At this point the motor reacts to correct the situation. The excessive armature current produces a developed torque which exceeds the opposing torques of friction and windage. In fact this excess serves as an accelerating torque, which then proceeds to increase the speed to that level which corresponds to the equilibrium value of armature current. In other words, the acceleration torque becomes zero only when the speed is at that value which by Eq. (17-12) yields just the right I_a needed to overcome the rotational losses.

Consider next that a load demanding rated torque is suddenly applied to the motor shaft. Clearly, because the developed torque at this instant is only sufficient to overcome friction and windage and not the load torque, the first reaction is for the motor to lose speed. In this way, as Eq. (17-12) reveals, the armature current can be increased so that in turn the electromagnetic torque can increase. As a matter of fact the applied load torque causes the motor to assume that value of speed which yields a current sufficient to produce a developed torque to overcome the applied shaft torque and the frictional torque. Power balance is thereby achieved, because an equilibrium condition is reached where the electromagnetic power, $E_a I_a$, is equal to the mechanical power developed, $T\omega_m$.

A comparison of the d-c motor with the three-phase induction motor indicates that both are *speed-sensitive* devices in response to applied shaft loads. An essential difference, however, is that for the three-phase induction motor developed torque is adversely influenced by the power-factor angle of the armature current. Of course no analogous situation prevails in the case of the d-c motor.

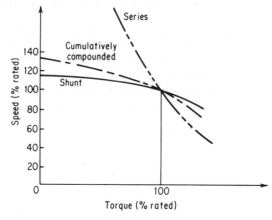

Fig. 17-10. Typical speed-torque curves of d-c motors.

On the basis of the foregoing discussion it should be apparent that the speed-torque curve of d-c motors is an important characteristic. Appearing in Fig. 17-10 are the general shapes of the speed-torque characteristics as they apply for the shunt, cumulatively compounded, and series motors. For the sake of comparison the curves are drawn through a common point of rated torque and speed. An understanding of why the curves take the shapes and relative positions depicted in Fig. 17-10 readily follows from an examination of Eq. (17-1), which involves the speed. For the shunt motor the speed equation can be written as

$$ n = \frac{E_a}{K_E \Phi_{sh}} = \frac{V_t - I_a R_a}{K_E \Phi_{sh}} \tag{17-13} $$

The only variables involved are the speed n and the armature current I_a. At rated output torque the armature current is at its rated value and so, too, is the speed. As the load torque is removed, the armature current becomes correspondingly smaller, making the numerator term of Eq. (17-13) larger. This results in higher speeds. The extent to which the speed increases depends upon how large the armature-circuit resistance drop is in comparison to the terminal voltage. It is usually around 5 to 10 per cent. Accordingly, we can expect the per cent change in speed of the shunt motor to be about the same magnitude. This change in speed is identified by a figure of merit called the *speed regulation*. It is defined as follows:

$$\text{per cent speed regulation} = \frac{(\text{no load speed}) - (\text{full-load speed})}{\text{full-load speed}} \, 100 \qquad (17\text{-}14)$$

The speed equation as it applies to the cumulatively compounded motor takes the form

$$n = \frac{V_t - I_a(R_a + R_s)}{K_E(\Phi_{sh} + \Phi_s)} \qquad (17\text{-}15)$$

A comparison with the analogous expression for the shunt motor bears out two differences. One, the numerator term also includes the voltage drop in the series-field winding besides that in the armature winding. Two, the denominator term is increased to account for the effect of the series-field flux Φ_s. Starting at rated torque and speed, Eq. (17-14) makes it clear that as load torque is decreased to zero there is an increase in the numerator term which is necessarily greater than it is for the shunt motor. At the same time, moreover, the denominator term decreases because Φ_s reduces to zero as the torque goes to zero. Both effects act to bring about an increase in speed. Therefore the speed regulation of the cumulatively compounded motor is greater than for the shunt motor. Figure 17-10 presents this information graphically.

The situation regarding the speed-torque characteristic of the series motor is significantly different because of the absence of a shunt-field winding. Keep in mind that the establishment of a flux field in the series motor comes about solely as a result of the flow of armature current through the series-field winding. In this connection, then, the speed equation for the series motor becomes

$$n = \frac{E_a}{K_E \Phi_s} = \frac{V_t - I_a(R_a + R_s)}{K'_E I_a} \qquad (17\text{-}16)$$

where K'_E denotes a new proportionality factor which permits Φ_s to be replaced by the armature current I_a. When rated torque is being developed the current is at its rated value. The flux field is therefore abundant. However, as load torque is removed less armature current flows. Now since I_a appears in the denominator of the speed equation, it is easy to see that the speed will increase greatly. In fact, if the load were to be disconnected from the motor shaft, dangerously high speeds would result because of the small armature current that flows. The centrifugal forces at these high speeds can easily damage the armature winding. For this reason a series motor should never have its load uncoupled.

Because the armature current is directly related to the air-gap flux in the series motor, Eq. (17-1) for the developed torque may be modified to read as

$$T = K_T \Phi I_a = K'_T I_a^2 \qquad (17\text{-}17)$$

Thus the developed torque for the series motor is a function of the square of the armature current. This stands in contrast to the linear relationship of torque to armature current in the shunt motor. Of course in the compound motor an intermediate relationship is achieved. It is interesting to note, too, that as the series motor reacts to develop greater torques, the speed drops correspondingly. It is this capability which suits the series motor so well to traction-type loads.

Speed Control. One of the attractive features the d-c motor offers over all other types is the relative ease with which speed control can be achieved. The various schemes available for speed control can be deduced from Eq. (17-13), which is repeated here with one modification:

$$n = \frac{V_t - I_a(R_a + R_e)}{K_E \Phi} \tag{17-18}$$

The modification involves the inclusion of an external armature-circuit resistance R_e. Inspection of Eq. (17-18) reveals that the speed can be controlled by adjusting any one of the three factors appearing on the right side of the equation: V_t, R_e, or Φ. The simplest to adjust is Φ. A field rheostat such as that shown in Fig. 17-8 is used. If the field-rheostat resistance is increased, the air-gap flux is diminished, yielding higher operating speeds. General-purpose shunt motors are designed to provide a 200 per cent increase in rated speed by this method of speed control. However, because of the weakened flux field the permissible torque that can be delivered at the higher speed is correspondingly reduced, in order to prevent excessive armature current.

A second method of speed adjustment involves the use of an external resistor R_e connected in the armature circuit as illustrated in Fig. 17-11. The size and cost of this resistor are considerably greater than those of the field rheostat because R_e must be capable of handling the full armature current. Equation (17-18) indicates that the larger R_e is made, the greater will be the speed change. Frequently the external resistor is selected to furnish as much as a 50 per cent drop in speed from the rated value. The chief disadvantage of this method of control is the poor efficiency of operation. For example, a 50 per cent drop in speed is achieved by having approximately half of the terminal voltage V_t appear across R_e. Accordingly, almost 50 per cent of the line input power is dissipated in the form of heat in the resistor R_e. Nonetheless, armature-circuit resistance control is often used—especially for series motors.

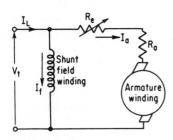

Fig. 17-11. Speed adjustment of a shunt motor by an external armature-circuit resistance.

A third and final method of speed control involves adjustment of the applied terminal voltage. This scheme is the most desirable from the viewpoint of flexibility and high operating efficiency. But it is also the most expensive because it requires its own d-c supply. It means purchasing a motor-generator set with a capacity at least equal to that of the motor to be controlled. Such expense is not generally justified except in situations where the superior performance achievable with this scheme is indispensable, as is the case in steel mill applications. Armature terminal voltage control is referred to as the *Ward-Leonard* system.

17-4   Applications of D-C Motors

The d-c motor is often called upon to do the really tough jobs in industry because of its high degrees of flexibility and ease of control. These features cannot easily be matched by other electromechanical energy-conversion devices. The d-c motor offers a wide range of control of speed and torque as well as excellent acceleration and deceleration. For example, by the insertion of an appropriate armature-circuit resistance, rated torque can be obtained at starting with no more than rated current flowing. Also, by special design of the shunt-field winding speed adjustments over a range of 4:1 are readily obtainable. If this is then combined with armature-voltage control, the range of speed adjustment spreads to 6:1. In some electronic control devices that are used to provide the d-c energy to the field and armature circuits, a speed range of 40:1 is possible. The size of the motor being controlled, however, is limited.

Table 17-1 lists some of the salient characteristics and typical applications of the three types of d-c motors. It is interesting to note that the maximum torque in the case of the d-c motor is limited by commutation and not, as with all other motor types, by heating. Commutation refers to the passage of current from the brushes to the commutator and thence to the armature winding itself. The passage from the brushes to the commutator is an arc discharge. Moreover, as a coil leaves a brush the current is interrupted, causing sparking. If the armature current is allowed to become excessive, the sparking can become so severe as to cause flash-over between brushes. This renders the motor useless.

Another point of interest in the table is the considerably higher starting torque of the compound motor by comparison with the shunt motor. This feature is attributable to the contribution of the series-field winding. The same comment is valid as regards the maximum running torque. In each case, of course, the limit for the armature current is the same.

17-5   Starters and Controllers for D-C Motors

The limitations imposed by commutation as well as voltage-dip restrictions on the source as set forth by the electric utility company make it necessary to use a starter or controller on all d-c machines whose ratings exceed two horsepower. A glance at Eq. (17-12) discloses that at starting ($n = 0$) the armature current is limited solely by the armature-circuit resistance. Hence if full terminal voltage is applied, excessive armature currents will flow. This is especially so where large machines are involved because the armature resistance gets smaller as the rating increases. In addition to limiting the armature current, controllers fulfill other useful functions as described in Sec. 15-8.

Depicted in Fig. 17-12 is a simple line starter which is used for small d-c motors. The operation is straightforward. Pushing the start button energizes the main coil which then closes the main contactors M and the interlock M_a. Note that when the main contactors close, voltage is applied to the armature winding

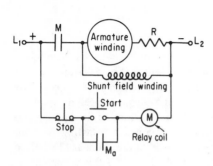

Fig. 17-12. Line starter for low-hp d-c motor.

through the starting resistor R and simultaneously to the field winding. This arrangement prevents "shock" starting because some time is needed before the field flux is fully established. Note, too, that this starter has no provisions for removing the starting resistor once the motor has attained its operating speed. In small motors this is of little consequence and it makes for an inexpensive starter. However, the starter resistor in this case does serve another purpose. When the motor is stopped, the field winding is disconnected from the line. The energy stored in the magnetic field then discharges through the starting resistor, preventing possible damage to the field winding.

Table 17-1. Characteristics and

Type	Starting torque (%)	Max. running torque, momentary (%)
Shunt, constant speed	Medium—usually limited to less than 250 by a starting resistor but may be increased	Uusally limited to about 200 by commutation
Shunt, adjustable speed	Same as above	Same as above
Compound	High—up to 450, depending upon degree of compounding	Higher than shunt—up to 350
Series	Very high—up to 500	Up to 400

† By permission from M. Liwschitz-Garik and C. C. Whipple, *Electric Machinery*, vol. I

Three types of magnetic controllers are in use today for controlling the starting current, the starting torque, and the acceleration characteristics of d-c motors. One type is the *current-limit* controller, which works on the principle of keeping the current during the starting period between specified minimum and maximum limits. It is not too commonly used because its success depends upon a tricky electrical interlock arrangement which must be kept in excellent operating condition at all times. No further consideration is given to this type. The second type is the *counter-emf* type which is illustrated in Fig. 17-13. Two kinds of relays are used in this controller. One is a light, fast-acting unit called an accelerating relay (AR). The other is the strong, heavy-duty type previously discussed. The accelerating relays appearing across the armature circuit in Fig. 17-13 are voltage-sensitive devices. They are designed to close when the voltage across the coil exceeds a preset value. For the controller under discussion the accelerating relay 1AR is usually adjusted to "pick up" at 50 per cent of line voltage and 2AR is adjusted to close at 80 per cent of line voltage.

Pressing the start button energizes coil M which then closes interlock M_a

Applications of D-C Motors†

Speed-regulation or characteristic (%)	Speed control (%)	Typical application and general remarks
5–10	Increase up to 200 by field control; decrease by armature-voltage control	Essentially for constant-speed applications requiring medium starting torque. May be used for adjustable speed not greater than 2:1 range. For centrifugal pumps, fans, blowers, conveyors, woodworking machines, machine tools, printing presses.
10–15	6:1 range by field control, lowered below base speed by armature voltage control	Same as above, for applications requiring adjustable speed control, either constant torque or constant output.
Varying, depending upon degree of compounding—up to 25-30	Not usually used but may be up to 125 by field control	For drives requiring high starting torque and only fairly constant speed; pulsating loads with flywheel action. For plunger pumps, shears, conveyors, crushers, bending rolls, punch presses, hoists.
Widely variable, high at at no-load	By series rheostat	For drives requiring very high starting torque and where adjustable, varying speed is satisfactory. This motor is sometimes called the traction motor. Loads must be positively connected, not belted. For hoists, cranes, bridges, car dumpers. To prevent overspeed, lightest load should not be much less than 15 to 20 per cent of full-load torque.

(Princeton, N. J.: D. Van Nostrand Co., Inc., 1946).

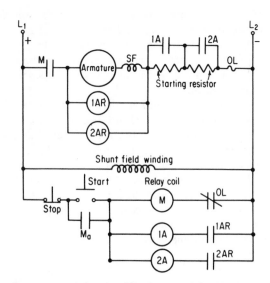

Fig. 17-13. Counter emf magnetic controller for a d-c motor.

and the main contactors M. Since the armature is initially stationary, the accelerating relays are de-energized so that both steps of the starting resistor are in the circuit. As the armature gains speed and develops an induced emf exceeding 50 per cent of line voltage, coil 1AR snaps closed, closing contactors 1AR. In turn coil 1A is energized, closing contactors 1A, which short out the first section of the starting resistor. This then applies increased voltage to the armature, which furnishes further acceleration. When the armature induced emf exceeds 80 per cent of line voltage, accelerating relay 2AR closes. This excites coil 2A, which shorts out the second section of the starting resistor. The motor then accelerates to its full-voltage operating speed.

The counter-emf controller has the advantage of providing a contactor closing sequence which adjusts itself automatically to varying load conditions. Furthermore, this is accomplished in a manner that maintains uniform accelerating current and torque peaks. There can be no question about the desirability of such starting performance; however, there is one disadvantage. The contactor closing sequence is based on the assumption that the motor will start on the first step. If it fails to do so, all subsequent operations cannot take place. Furthermore, the starting resistor is in danger of burning up. To avoid such occurrences general-purpose d-c motors are most often equipped with *definite time-limit controllers*. Figure 17-14 shows the schematic diagram of such a controller. After a relay coil is energized, the corresponding contactors do not close until the elapse of a preset time delay. The time delay is achieved either by means of magnetic flux decay, a pneumatic device, or a mechanical escapement.

Pressing the start button energizes coil M and closes interlock M_a and the main contactors M, thereby applying voltage to the armature winding through

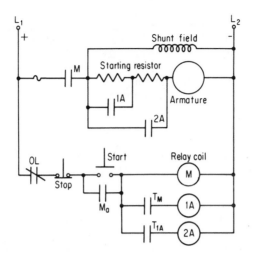

Fig. 17-14. Definite time-limit controller for d-c motor.

the starting resistor. A definite time after the armature of relay coil M closes, contactors T_M in the control circuit close, regardless of whether the rotor is turning. This energizes coil 1A which closes contactors 1A, thus shorting out the first section of the starting resistor. At a preset time delay after coil 1A is energized, contactors T_{1A} close. Coil 2A becomes energized, removing the entire starting resistor. The motor then assumes its normal operating speed.

In the definite time-limit controller the preset time intervals between the closing of contactors are adjusted to obtain smooth acceleration and uniform current peaks for average load conditions. If a heavy starting condition occurs and the motor fails to start on the first step, the first accelerating (time-delay) contactors close anyway. This allows an increased starting torque to be developed. Thus the motor is made to "work harder" if it does not start on the first step. Accordingly, whenever a controller must be selected for a general-purpose motor, it is wiser to prescribe the definite time-limit type. The reason is that as a rule for general-purpose applications the starting conditions are not well known.

PROBLEMS

17-1 Explain your answer to each part:

(a) Can a separately excited d-c generator operate below the knee of its magnetization curve?

(b) Can a d-c shunt generator operate below the knee of its magnetization curve?

17-2 The magnetization curve for a d-c shunt generator driven at a constant speed of 1000 rpm is shown in Fig. 17-7. Each vertical division represents 20 volts; each horizontal division represents 0.2 amp.

(a) Compute the critical field resistance.

(b) What voltage is induced by the residual flux of this machine?

(c) What must be the resistance of the field circuit in order that the no-load terminal voltage be 240 volts at a speed of 1000 rpm?

(d) Determine the field current produced by the residual flux voltage when the field circuit resistance has the value found in part (c).

(e) At what speed must the generator be driven in order that it would fail to build up when operating with the field circuit resistance of part (c)?

17-3 A d-c shunt generator has a magnetization curve given by Fig. 17-7, where each ordinate unit is made equal to 20 volts and each abscissa unit is set equal to 0.2 amp and the speed of rotation is 1000 rpm. The field circuit resistance is 156 ohms.

Determine the voltage induced between brushes when the generator is operated at a reduced speed of 800 rpm.

17-4 It is desired to reverse the terminal voltage polarity of a generator that has been operating properly as a cumulative compound generator. The machine is stopped. Residual magnetism is reversed by temporarily disconnecting the shunt field and separately exciting it with reversed current. The connections are then restored exactly as they were before.

(a) Does the terminal voltage build up? Explain.

(b) If answer to (a) is yes, will the compounding be cumulative or differential? Explain.

17-5 The no-load characteristic of a 10-KW, 250-volt, d-c self-excited shunt generator driven at 1000 rpm is shown in Fig. 17-7. Each vertical division denotes 20 volts and each horizontal division represents 0.2 amp. The armature circuit resistance is 0.3 ohm and the field circuit (shunt field winding plus field rheostat) resistance is set at 110 ohms. Determine: (a) the critical resistance of the shunt field circuit; (b) the voltage regulation; (c) the shunt field current under rated load conditions; (d) the approximate no-load speed at which this machine would run when connected to a 220-volt line as a motor.

17-6 A 10-hp, 230-volt shunt motor has an armature circuit resistance of 0.5 ohm and a field resistance of 115 ohms. At no-load and rated voltage the speed is 1200 rpm and the armature current is 2 amps. If load is applied, the speed drops to 1100 rpm. Determine: (a) the armature current and the line current; (b) the developed torque; (c) the horsepower output assuming the rotational losses are 500 watts.

17-7 A 230-volt, 50-hp, d-c shunt motor delivers power to a load drawing an armature current of 200 amps and running at a speed of 1100 rpm. The magnetization curve is given by Fig. 17-7, where each vertical unit represents 20 volts and each horizontal units represents 2 amps.

(a) Find the value of the armature induced emf at this load condition.

(b) Compute the motor field current.

(c) Compute the value of the load torque. The rotational losses are 600 watts.

(d) At what efficiency is the motor operating?

(e) At what percentage of rated power is it operating?

17-8 Refer to Prob. 17-7 and assume the load is reduced so that an armature current of 75 amps flows.

(a) Find the new value of speed.

(b) What is the new horsepower being delivered to the load?

17-9 A 20-hp, 230-volt, 1150-rpm, 4-pole, d-c shunt motor has a total of 620 conductors arranged in two parallel paths and yielding an armature circuit resistance of 0.2 ohm. When it delivers rated power at rated speed, the motor draws a line current of 74.8 amps and a field current of 3 amps. Compute: (a) the flux per pole; (b) the developed torque; (c) the rotational losses; (d) the total losses expressed as a percentage of the rated power.

17-10 A 230-volt, 10-hp, d-c series motor draws a line current of 36 amps when delivering rated power at its rated speed of 1200 rpm. The armature circuit resistance is 0.2 ohm and the series field winding resistance is 0.1 ohm. The magnetization curve may be considered linear.

(a) Find the speed of this motor when it draws a line current of 20 amps.

(b) What is the developed torque at the new conditions?

(c) How does this torque compare with the original value? Why?

17-11 A 250-volt, 50-hp, 1000-rpm, d-c shunt motor drives a load that requires a constant torque regardless of the speed of operation. The armature circuit resistance is 0.04 ohm. When this motor delivers rated power, the armature current is 160 amps.

(a) If the flux is reduced to 70 per cent of its original value, find the new value of armature current.

(b) What is the new speed?

17-12 When a 250-volt, 50-hp, 1000-rpm, d-c shunt motor is used to supply rated output power to a constant-torque load, it draws an armature current of 160 amps. The armature circuit has a resistance of 0.04 ohm and the rotational losses are equal to 2 KW. An external resistance of 0.5 ohm is inserted in series with the armature winding. For this condition compute: (a) the speed; (b) the developed power; (c) the efficiency assuming the field loss is 1.6 KW.

17-13 A d-c shunt motor has the magnetization curve shown in Fig. 17-7, where one vertical unit represents 20 volts and one horizontal unit represents 2 field amps. The armature circuit resistance is 0.05 ohm. At a specified load and a speed of 1000 rpm the field current is found to be 12 amps when a terminal voltage of 245 volts is applied to the motor. The rotational losses are 2.5 KW.

(a) Find the armature current.

(b) Compute the developed torque.

(c) What is the efficiency?

17-14 Depicted in Fig. P17-14 is the reversing controller circuitry for a d-c shunt motor. Identify the unmarked armature contacts and explain the reversing operation. Explain, too, why the auxiliary interlock contacts (F_a and R_a) are placed in series with the *REV* and *FWD* switches.

17-15 A current-limit type controller for a d-c shunt motor is shown in Fig. P17-15. The accelerating relays (AR) respond to armature current. These relays have a lightweight armature so that they easily pick up and open their contacts whenever the armature current exceeds the rated value. By means of a step-by-step procedure describe the operation of this controller. Be careful to identify the proper closing and opening sequence wherever two or more contactors appear in series.

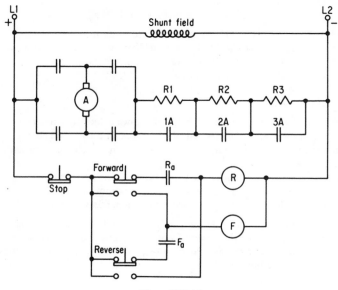

Figure P17-14

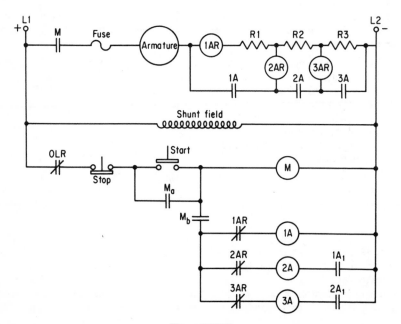

Figure P17-15

18

single-phase induction motors

By far the vast majority of single-phase induction motors are built in the fractional-horsepower range. Single-phase motors are found in countless applications doing all sorts of jobs in homes, shops, offices, and on the farm. An inventory of the appliances in the average home in which single-phase motors are used would probably number close to a dozen. An indication of the volume of such motors can be had from the fact that the sum total of all fractional-horsepower motors in use today far exceeds the total of integral horsepower motors of all types.

The treatment of the single-phase motor as developed in the following pages is concerned chiefly with their method of operation, the classification and characteristics of the various types, and their typical applications. The analysis of the performance of such motors is beyond the scope of this book.

18-1 How the Rotating Field Is Obtained

In its pure and simple form the single-phase motor usually consists of a distributed stator winding (not unlike one phase of a three-phase motor) and a squirrel-cage rotor. The a-c supply voltage is applied to the stator winding, which in turn creates a field distribution. Since there is a single coil carrying an alternating current, a little thought reveals that the air-gap flux is characterized by being fixed in space and pulsating in magnitude. If hysteresis is neglected, the flux is a maximum when the current is instantaneously a maximum and it is zero when the current is zero. Such an arrangement gives the single-phase motor no starting torque. To understand this in terms of the concepts discussed in Sec. 14-1 refer to Fig. 18-1. As a matter of convenience the distributed single-phase winding is represented by a coil wrapped around protruding pole pieces; some motors do in fact use this configuration. Assume instantaneously that the flux-density wave is increasing in the upward direction as shown. Then by transformer action a voltage is induced in the rotor having that distribution which enables the corresponding rotor mmf to oppose the changing flux. To accomplish this, current flows out of the right-side conductors and into the left-side conductors as illustrated in Fig. 18-1. Note that this resulting ampere-conductor distribution corresponds to a space phase angle of $\psi = 90°$. By Eq. (14-39) the net torque is therefore zero. Of course what this means is that beneath each pole piece there are as many conductors producing clockwise torque as there are producing counterclockwise torque. This condition, however, prevails only at standstill. If by some means the rotor is started in either direction, it will develop a nonzero net torque in that direction and thereby cause the motor to achieve normal speed. The problem therefore is to modify the configuration of Fig. 18-1 in such a way that it imparts to the rotor a nonzero starting torque.

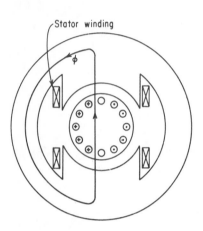

Fig. 18-1. Simplified diagram of the single-phase motor.

The answer to this problem lies in so modifying the motor that it closely approaches the conditions prevailing in the *two-phase* induction motor. In accordance with the general results developed in Sec. 15-1 we know that to obtain a revolving field of constant amplitude and constant linear velocity in a two-phase induction motor two conditions must be satisfied. One, there must exist two coils (or windings) whose axes are space-displaced by 90 electrical degrees. Two, the currents flowing through these coils must be time-displaced by 90 electrical degrees and they must have such magnitudes that the mmf's are equal. If the currents are less than 90° apart in time but greater than 0°, a rotating field will still be developed but the locus of the resultant flux vector will be an ellipse rather than a circle. Hence in such a case the linear velocity of the field varies from one point in time to another. Also, if the currents are 90° apart but the mmf's of the two coils are unequal, an elliptical locus for the rotating field again results. Finally, if the currents are neither 90° time-displaced nor of a magnitude to furnish equal mmf's, a rotating magnetic field will continue to be developed but now the locus will be more elliptical than in the previous cases. However, the important aspect of all this is that a revolving field can be so obtained even if its amplitude is not constant during its time history, and satisfactory performance can be achieved with such a revolving field. Of course such performance items as power factor and efficiency will be poorer than for the ideal case, but this is not too serious because the motors are of relatively small power.

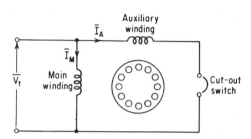

Fig. 18-2. Schematic diagram of the resistance split-phase motor.

Appearing in Fig. 18-2 is the schematic diagram which shows the modifications needed to give the single-phase motor a starting torque. A second winding called the *auxiliary* winding is placed in the stator with its axis in quadrature with that of the main winding. Usually the main winding is made to occupy two-thirds of the stator slots and the auxiliary winding is placed in the remaining one-third. In this way the space-displacement condition is met exactly. The time displacement of the currents through the two windings is obtained at least partially by designing the auxiliary winding for high resistance and low leakage reactance. This is in contrast to the main winding, which has low resistance and higher leakage reactance. Figure 18-3 depicts the time displacement existing between the auxiliary winding current I_A and the main winding current I_M at standstill. Frequently in motors of this design the I_A and I_M phasors are displaced by about 45° in time. Thus with the arrangement of Fig. 18-2 a revolving field results and so the motor achieves normal speed. Because of the high-resistance character of the auxiliary winding, this motor is called the *resistance-start split-phase* induction motor. Also, the auxiliary winding used in these motors has a short time power rating and therefore must be removed from the line once the operating speed is reached. To do this a cut-out switch is placed in the auxiliary winding circuit which, by centrifugal action, removes the auxiliary winding from the line when the motor speed exceeds 75 per cent of synchronous speed.

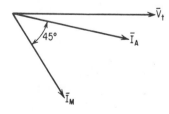

Fig. 18-3. Phasor diagram showing the time-phase displacement between the auxiliary and main winding currents at standstill.

18-2 The Different Types of Single-phase Motors

Many different types of single-phase motors have been developed primarily for two reasons. One, the torque requirements of the appliances and applications in which they are used vary widely. Two, it is desirable to use the lowest-priced motor that will drive a given load satisfactorily. For example, a high-torque version of the split-phase induction motor just discussed is designed almost exclusively for washing-machine applications. This motor is available in a single speed rating and in just two horsepower ratings. This, coupled with the large volume of sales, enables it to be the least expensive motor in its category. See line two of Table 18-1 for more details.

The Capacitor-start Induction-run Motor. The chief difference between the various types of single-phase motors lies in the method used to start them. In the

Table 18-1. Characteristics and Applications

Type designation	Starting torque (% of normal)	Approx. Comparative price	Breakdown torque (% of normal)	Starting current at 115v (amps)
General-purpose split-phase motor	90–200 Medium	85%	185–250 Medium	23 1/4 hp
High-torque split-phase motor	200–275 High	65%	Up to 350	32 High 1/4 hp
Permanent-split capacitor motor	60–75 Low	155%	Up to 225	Medium
Permanent-split capacitor motor	Up to 200 Normal	155%	260	
Capacitor-start general-purpose motor	Up to 435 Very high	100%	Up to 400	
Capacitor-start capacitor-run motor	380 High	190%	Up to 260	
Shaded-pole motor	50	—	150	

† By permission from M. Liwschitz-Garik and C. C. Whipple, *Electric Machinery*, vol.

case of this motor a capacitor is placed in the auxiliary winding circuit so selected that it brings about a 90° time displacement between I_A and I_M. See Fig. 18-4. The result is a much larger starting torque than is achievable with resistance split-phase starting. This motor is widely used for general-purpose applications.

The Capacitor-start Capacitor-run Motor. As Fig. 18-5 shows, two capacitors are used in the auxiliary circuit of this motor. By keeping one capacitor in during normal operation, improved performance is obtained because the motor then behaves more like the balanced two-phase motor. The improved performance is manifested in terms of less noise and higher efficiency and power factor. A second capacitor is needed at starting because the reactive component of the input impedance of the auxiliary circuit is considerably different at standstill than at full speed.

The Permanent-split Capacitor Motor. In this motor a single capacitor is used both for starting and for running. To take advantage of improved running

of Single-Phase A-C Motors[†]

Power factor (%)	Efficiency (%)	Hp range	Application and general remarks
56–65	62–67	1/20 to 3/4	Fans, blowers, office appliances, food-preparation machines. Low- or medium-starting torque, low-inertia loads. Continuous-operation loads. May be reversed.
50–62	46–61	1/6 to 1/3	Washing machines, sump pumps, home work-shops, oil burners. Medium- to high-starting torque loads. May be reversed.
80–95	55–65	1/20 to 3/4	Direct-connected fans, blowers, centrifugal pumps. Low-starting-torque loads. Not for belt drives. May be reversed.
80–95	55–65	1/6 to 3/4	Belt-driven or direct-drive fans, blowers, centrifugal pumps, oil burners. Moderate-starting-torque loads. May be reversed.
80–95	55–65	1/8 to 3/4	Dual voltage. Compressors, stokers, conveyors, pumps. Belt-driven loads with high static friction. May be reversed.
80–95	55–65	1/8 to 3/4	Compressors, stokers, conveyors, pumps. High-torque loads. High power factor. Speed may be regulated.
30–40	30–40	1/300 to 1/20	Fans, toys, hair dryers, unit heaters. Desk fans. Low-starting-torque loads.

II (Princeton, N. J.: D. Van Nostrand Co., Inc., 1946).

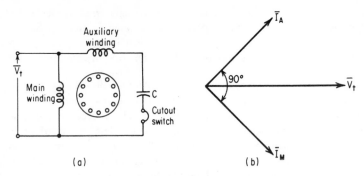

Fig. 18-4. Capacitor-start induction-run motor:
(a) schematic diagram;
(b) phasor diagram at starting.

performance, the value of the capacitor used is that needed at full speed. Consequently, starting torque must be sacrificed. The schematic wiring diagram appears in Fig. 18-6.

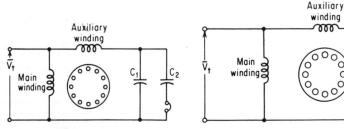

Fig. 18-5. Capacitor-start capacitor-run motor.

Fig. 18-6. Permanent-split capacitor motor.

The Shaded-pole Motor. The shaded-pole motor is very extensively used in applications that require 1/20 hp or less. The construction is extremely rugged, as Fig. 18-7 reveals. There is little that can go wrong with this motor aside from overheating. Note that it contains no cut-out switch, which could be a source of trouble, nor does it have an auxiliary winding, which could burn up—especially if the cut-out switch became faulty. The shaded-pole motor consists essentially of copper wound on iron and not much more.

The manner of obtaining a moving-flux field, however, is different in this motor than it is in those considered previously. Here use is made of a copper ring imbedded in each salient-pole piece. As the gap flux is changing in response to the alternating coil current, transformer action causes currents to be induced in the copper rings, always so directed as to oppose the changing flux. The net effect of this action is that for one portion of the cycle the ring currents cause the flux to concentrate in that part of the pole piece which is free of the ring. At a subsequent portion of the cycle the ring currents act to crowd the gap flux through

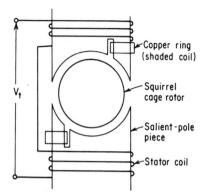

Fig. 18-7. Construction features of the shaded-pole motor.

that part of the pole piece around which the ring is wrapped. In this fashion the air-gap flux undergoes a sweeping motion across the pole face. It appears to be moving from the unshaded to the shaded (or ring) portion of the pole. The sweeping action of the flux occurs periodically, thereby producing a starting as well as a running torque. The starting torque is normally 50 per cent of rated value. The breakdown torque is also relatively low.

18-3. Characteristics and Typical Applications

The principal performance features as well as typical applications of single-phase a-c motors appear in Table 18-1. Note that for applications involving less than 1/20 hp the shaded-pole motor is invariably used. On the other hand, for applications involving 1/20 hp to 3/4 hp the choice of the motor depends upon such factors as the starting and breakdown torque and even quietness of operation. Where a minimum of noise is desirable and low starting torque is adequate, such as for driving fans and blowers, the motor to choose is the permanent-split type. If quietness is to be combined with high starting torque, as might be the case where a compressor drive is needed, then the suitable choice would be the capacitor-start capacitor-run motor. Of course, in circumstances where the compressor is located in a noisy environment, clearly the choice should be the capacitor-start induction-run motor because it is less expensive.

PROBLEMS

18-1 Two field coils are space displaced by 45 electrical degrees. Moreover, sinusoidal currents that are time displaced by 45 electrical degrees flow through these windings. Will a revolving flux result? Explain.

18-2 Two coils are placed in a low reluctance magnetic circuit with their axes in quadrature—the first along the vertical and the second along the horizontal. Sinusoidal currents that are time displaced by 90 electrical degrees and of the same frequency ω are made to flow through the coils. Moreover, the mmf of the first coil is twice that of the second.

(a) Assuming that at time ωt_0 the mmf of the first coil is at its positive peak value, determine the relative magnitude and direction of the resultant mmf.

(b) Repeat part (a) for a time ωt_1 that is 45° later than ωt_0.

(c) Repeat part (a) for a time ωt_2 that is 90° later than ωt_0.

(d) What conclusion can you draw from parts (a), (b), and (c)?

18-3 The two coils of Prob 18-2 are now energized with sinusoidal currents that are 45 electrical degrees out of phase but yield equal mmfs.

(a) Assuming that at time ωt_0 the mmf of the first coil is at its positive peak value, find the relative magnitude and direction of the resultant mmf.

(b) Repeat part (a) for a time ωt_1 that is 45° later than ωt_0.

(c) Repeat part (a) for a time ωt_2 that is 90° later than ωt_0.

(d) What conclusion can you draw from parts (a), (b), and (c)?

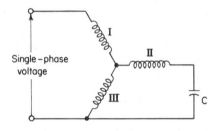

Figure P18-4

18-4 In Fig. P18-4, I, II, and III denote the three coils of a balanced three-phase winding which are spaced displaced by 120° from one another. If a single-phase voltage is applied to the three-phase, squirrel-cage motor in the manner shown, will the motor operate? Justify your answer.

18-5 At standstill, the currents in the main and auxiliary windings of a single-phase induction motor are given by $\tilde{I}_M = 10\underline{/0°}$ and $\tilde{I}_A = 5\underline{/60°}$. There are five turns per pole on the auxiliary winding for every four turns on the main winding. Moreover, these windings are in space quadrature.

What should be the magnitude and phase of the current in the auxiliary winding to produce an mmf wave of constant amplitude and velocity?

19

direct energy-conversion methods

All the electromechanical devices discussed in the preceding chapters are assumed to obtain their energy from an electrical source, the prime mover of which is a steam turbine, a water wheel, or a gasoline engine. However, in recent years a considerable amount of research and development has been done in the area of direct energy conversion. This effort has been motivated principally by the stringent requirements of the outer space program of the United States Government, where low weight and high reliability are indispensable. Accordingly, in this chapter a brief description is given of the principles underlying those methods of direct energy conversion which look most promising for the future, along with their salient characteristics.

19-1 Photovoltaic Energy Conversion: The Solar Cell

Energy conversion through the photovoltaic process is one of two methods of energy conversion that does not require an intermediate stage, i.e., conversion to thermal energy, before obtaining electrical energy. The solar cell converts radiated energy from the sun directly into electrical energy. For this reason it is the simplest conversion device known to man; it requires no auxiliary equipment or source of heat. It is the most frequently used direct energy converter today, and is chiefly responsible for making it possible to operate electrical equipment for a long time.

The photoconversion device that has the highest efficiency is the *p-n* junction. To understand the photovoltaic conversion process, it is necessary first to know the *p-n* junction device and semiconductor behavior. In this connection, therefore, the reader is urged to review Sec. 7-1 up to the material relating to Fig. 7-12.

Each photon of light contains an amount of energy that is proportional to the frequency; or, more specifically, each photon carries an amount of energy given by

$$W = h\nu \tag{19-1}$$

where ν is the frequency and h denotes Planck's constant. If this quantum of energy enters the p material in the neighborhood of the junction and if $h\nu$ is greater than the gap energy eE_g, then this energy is absorbed by the germanium and so creates an electron-hole pair. The influence of the electric field caused by the uncovered immobile charges then accelerates the electron to the right and the hole to the left. As more solar energy is absorbed, more and more electrons are caused to accumulate in the n region and a corresponding number of holes are caused to gather in the p region. The net effect is the emergence of a potential difference between the terminals of the *p-n* junction device, so that upon the application of a load a current can be made to flow. This current can be sustained as long as solar energy is allowed to impinge upon the *p-n* junction.

It is important to note in this process that there is a limit to the number of electrons that can accumulate at the n end and the number of holes that can gather at the p end. The reason is simple. As these electrons and holes gather at opposite ends of the *p-n* junction device, they create an electric field of their own, which is opposed to that caused by the uncovered mobile charges. Gradually, as the two fields become equal, there is no longer an internal field that serves to separate additionally created electron-hole pairs. In fact, it is this condition that determines the open-circuit voltage of the solar cell. Therefore, it follows that a larger open-circuit voltage can be obtained only through the use of a material with a larger band gap. Appearing in Fig. 19-1 is a plot of the open-circuit voltage of a silicon solar cell versus illumination. When the radiation energy is high enough, the value of the open-circuit voltage flattens out to about 0.6 volt. Of course, if it is desirable to have a higher voltage source, it is necessary to employ a series arrangement of many *p-n* junction cells.

Silicon cells have proved to be the most successful of the solar cells exhibiting conversion efficiencies from 12 to 14 per cent. Although this conversion efficiency is not nearly as good as those that are obtainable by other conversion techniques, this is not a serious shortcoming in view of the vast amount of solar energy that floods the earth constantly. The chief drawback of the solar cell as a prime source of energy is its high cost and the need for a large surface area. The major factor in this high cost is the necessity for a large amount of expensive single-crystal silicon. It is hoped that with the anticipated perfection of thin-film techniques the cost per kilowatt of solar cells will be appreciably diminished in the future.

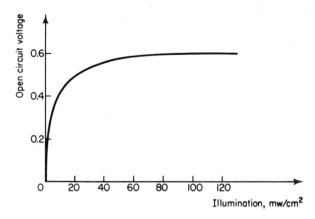

Fig. 19-1. Open circuit voltage vs. illumination for a silicon solar cell.

19-2 Thermoelectric Energy Conversion

We are all familiar with the *thermocouple* as an instrument for measuring temperature. It will be recalled that the thermocouple is formed by joining two dissimilar metals at one end and leaving the other end open. When the fused end is placed in an environment whose temperature exceeds that of the open end, a potential difference is produced across the open end which is proportional to the temperature difference. This behavior is called the *Seebeck effect* and was discovered in the early part of the nineteenth century. For the common metals the order of magnitude of this voltage is measured in millivolts. However, today, with the use of semiconductor materials that exhibit much higher thermoconversion efficiencies, voltages in the vicinity of a volt can be realized for temperature differences of about 1200°F.

The principle underlying the operation of the thermoelectric generator can best be illustrated by referring to the series of diagrams depicted in Fig. 19-2. Figure 19-2(a) shows heat energy applied to the center portion of an *n*-doped material. Recalling that there is an excess of free electrons in the *n* material by virtue of the pentavalent impurity, these electrons are found to diffuse from the center to either end of the material because of the temperature gradient. If the heat source should now be applied to one end of the material, as illustrated in Fig. 19-2(b), then clearly the action of the temperature gradient would be to cause the electrons to move from the left end to the right end. Next, consider that heat is applied to a closed ring composed of *n* type material as shown in Fig. 19-2(c). If the upper end of the ring is placed in ice, the application of heat at the bottom end causes the free electrons to move from bottom to top along paths both to the right and left side of the ring. This means that conventional current flows in the directions indicated by the arrows in the ring. An equilibrium condition is eventually reached when a sufficient number of electrons accumulate at the top end, which causes an electric field to exist in the body of the ring and it acts to prevent

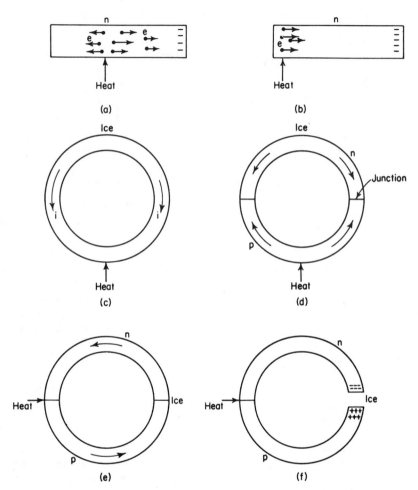

Fig. 19-2. Illustrating the principle of the thermoelectric generator: (a) and (b) scattering of electrons owing to temperature gradient; (c) and (d) configurations where Seeback effect is zero; (e) configuration which yields a net current flow; (f) the basic thermoelectric cell.

the further accumulation of electrons at the cool end. It is worthwhile to note that even when the ring consists of an *n* and a *p* type material, as shown in Fig. 19-2(d), the result is the same. A temperature gradient continues to exist as long as heat is applied to the bottom end while the top end is immersed in ice. Since the charge carriers are holes in the *p* material, these are diffused away from the heat source and across the junction to the cool end. In a similar fashion in the *n* region the temperature gradient causes the charge carriers (which are electrons here) to move from the junction areas toward the ice. Hence the conventional current flow is from the ice toward the junction areas. The total net effect is for zero current to flow in the ring. In other words, the resultant potential difference within the ring is zero.

The situation, however, is markedly different if the heat is applied at one junction while the second junction is immersed in ice. Refer to Fig. 19-2(e). Again it is important to keep in mind that the heat energy causes a diffusion of charge carriers away from the source. Accordingly, in the *n* material, electrons are made to flow to the cold junction via the upper half of the ring. The conventional current flow is therefore counterclockwise in this section. Moreover, since the charge carriers in the *p* section are holes, these move from the hot to the cold junction via the lower half of the ring. Therefore the conventional current-flow direction in this lower section is also counter-clockwise. Consequently, *whenever heat is applied at a junction, there is net current flow different from zero because now the effects of the movement of charge carriers in the n and p materials reinforce rather than neutralize each other.* In short, a net potential difference exists to maintain a current flow in the ring.

An interesting aspect of this description relates to the amount of heat applied to the left junction. If this heat is just sufficient to raise the energy level of the electrons to sustain current flow without raising the temperature of the material itself, it is called the *Peltier heat.*

Appearing in Fig. 19-2(f) is the representation of the basic thermoelectric cell. It shows the ring open at the cold end. Application of heat energy in this configuration causes an accumulation of electrons at the open end of the *n* material and an accumulation of holes at the corresponding end of the *p* material, which results in a measurable open-circuit voltage—the so-called "Seebeck effect." For present-day materials this potential difference is of the order of 1 volt. The limitation is brought about by the electric field in the body of the material caused by the accumulation of these charges. As the field strength increases, fewer charge carriers can overcome the opposing force of this internal field.

A modified form of the thermoelectric cell is depicted in Fig. 19-3. Because all good thermoelectric materials are poor conductors of heat, use is made of an intermediate conductor (such as copper) in the manner illustrated. Copper is an excellent heat conductor and so can conveniently serve to keep one end of the

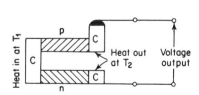

Fig. 19-3. Modified form of the thermoelectric cell. *C* denotes copper; temperature $T_1 > T_2$.

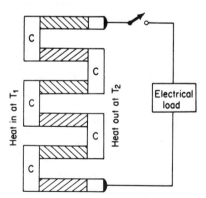

Fig. 19-4. Illustrating a series connection of thermoelectric cells. Temperature $T_1 > T_2$.

p and *n* sections at the same high temperature. Figure 19-4 shows a series connection of the basic thermoelectric cell to yield a higher load voltage. The materials that are used in present-day thermoelectric cells are lead telluride and lead selenide. These are preferred because they yield conversion efficiencies of about 10 per cent.

The chief attraction of the thermoelectric generator is that it is a maintenance-free source of d-c power. Moreover, it is quiet and portable. In addition it has been found that in the power range between 1 watt and 1 KW, the gas-fired thermoelectric generator is more economical than batteries, solar cells, and internal-combustion-engine generators. A 580-watt unit has been successfully tested on a satellite in orbit for more than six weeks.

19-3   Electrochemical Energy Conversion: The Fuel Cell

The direct conversion of energy by electrochemical means is the second of the two methods which does not involve the conversion of the primary energy source into thermal form before producing electrical energy. This electrochemical conversion is frequently identified as the *fuel cell*. In it chemicals are continuously added and caused to produce reactions at one electrode for the purpose of liberating electrons. These electrons are then allowed to flow to a second electrode (through a suitable electrical load) where they are then combined in a second reaction to complete the cycle. The process is illustrated in Fig. 19-5 for the hydrogen-oxygen fuel cell. The cell is composed of a fuel chamber, air-oxidizer chamber, a suitable electrolyte, and two electrodes. The electrolyte in the H_2-O_2 cell is an ion exchange membrane having a thickness of 1 mm. Its function is to allow passage of the hydrogen ions H^+ but not the oxygen. Moreover, the membrane is coated with a material that serves as a catalyst in order to facilitate the chemical reactions.

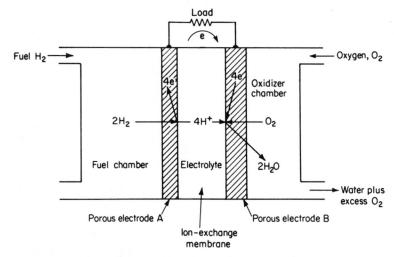

Fig. 19-5. Composition of the hydrogen-oxygen fuel cell.

These membranes also serve as the electrodes and they are of a porous nature so that they can more readily be penetrated by the hydrogen gas.

The chemical reaction that takes place at the electrode on the fuel-chamber side (i.e., electrode A) is

$$2H_2 \longrightarrow 4H^+ + 4e$$

The liberated electrons then find a path through the electrical load and into electrode B. The hydrogen ions pass through the electrolyte to electrode B where they combine with the electrons and the oxygen to produce water. The chemical reaction at the second porous electrode, B, is described as follows:

$$4e + 4H^+ + O_2 \longrightarrow 2H_2O$$

The overall reaction of the fuel cell is

$$2H_2 + O_2 \longrightarrow 2H_2O + 4eE + Q - 3RT$$

where $4eE$ denotes the released electrical energy, Q denotes the released heat, and $3RT$ represents the mechanical energy absorbed in the process. Keep in mind that the hydrogen and oxygen sources are doing work in forcing these quantities into the cell. Of course, R is the universal gas constant and T is the temperature.

The value of the generated emf E of the fuel cell is limited by thermodynamic considerations. It can be shown to be related to the free energy† of the reactants and products. Since its value lies at about 1 volt, it clearly requires a series connection of many cells to obtain a suitable operating voltage level. This indeed was the scheme employed in the fuel cell of the Gemini project.

Another version of the fuel cell is depicted in Fig. 19-6. The hydrogen fuel is replaced by carbon. The carbon also serves as one of the electrodes. In this cell, too, the electrolyte must not allow passage of the oxygen while at the same time

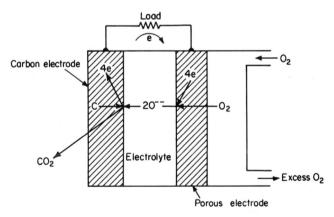

Fig. 19-6. The carbon-oxygen fuel cell. The by-product here is carbon dioxide.

† See S. S. L. Chang, *Energy Conversion* (Prentice-Hall, Inc., Englewood Cliffs, N.J., 1963), p. 199.

it must display good conductivity toward oxide ions. Because carbon is relatively unreactive at low temperature, the carbon-oxygen fuel cell must be operated at high temperature. The chemical reaction that takes place at the carbon electrode is

$$C + 2O^{--} \longrightarrow CO_2 + 4e$$

while at the porous electrode the reaction is

$$4e + O_2 \longrightarrow 2O^{--}$$

It is interesting to note that in addition to the disadvantage of operating at high temperature, the carbon-oxygen fuel cell produces a carbon dioxide product compared to water for the hydrogen-oxygen fuel cell. In space satellite applications the hydrogen-oxygen fuel cell is obviously preferred because water is useful for meeting other needs. Moreover, when hydrogen is used as the fuel, there is no need to operate at high temperature because hydrogen is a highly reactive element.

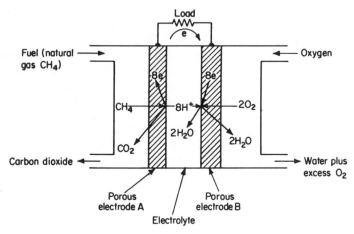

Fig. 19-7. Showing the reactions in a fuel cell employing natural gas.

Figure 19-7 depicts a fuel cell which employs natural gas CH_4 as the fuel. The chemical reaction at the porous electrode on the fuel-chamber side is given by

$$CH_4 + 2H_2O \longrightarrow CO_2 + 8e + 8H^+$$

Again, the liberated electrons flow through the load while the hydrogen ions pass through the electrolyte. At the second electrode these electrons and ions combine with oxygen to produce water. The reaction is represented by

$$8H^+ + 8e + 2O_2 \longrightarrow 4H_2O$$

Note that carbon dioxide is withdrawn as an end product from the fuel chamber, while water and excess oxygen are the quantities that exit from the oxidizer chamber.

The use of natural gas in fuel cells is desirable for its availability and economy. In fact, if this type of fuel cell could be perfected, it would bring about a major revolution. It is a well-known fact that the cost of delivered natural gas is only a fraction of the cost of delivered electricity. Hence, by equipping homes and factories with their own fuel cells, considerable savings in the cost of electrical energy could be achieved. There is yet another notable advantage to the fuel cell as a source of electrical energy: *it offers the possibility of more effective use of fossil fuel than the steam turbine-generator method of producing electrical energy.* Principally the reason is that the fuel cell is not bound by the limitations that surround heat engines. Carnót showed that the conversion of energy into heat is limited to a specific fraction of the high-temperature heat input, which for steam turbines is approximately 60 per cent. This usually leads to an overall energy-conversion efficiency in steam turbogenerators that is in the vicinity of 40 per cent because the conversion process involves several other steps (fuel to heat to mechanical energy to electrical energy), each of which consumes a portion of the total energy. The theoretical conversion-efficiency limit of the fuel cell is about 80 to 90 per cent for most fuels of interest. These figures are determined by the thermodynamic quantities involved in the fuel oxidation process. However, the current state of the art in building fuel cells is such that practical values of 60 per cent are now being achieved. But even this lower figure exceeds by 50 per cent that which is achievable with the steam turbogenerator. A third outstanding advantage is that the weight, efficiency, and cost per kilowatt is independent of the power rating. These advantages help to explain why research interests in fuel cells continue unabated.

Fuel cells are not yet widely accepted for many reasons among which are the following: (1) High investment cost. (2) The need for highly improved electrolytes. If neutral molecules pass into the electrolyte, a short circuit or even an explosion can occur. Moreover, the by-products of the chemical reactions can be a source of contaminants for the electrolyte. (3) The necessity of improved and more effective electrodes. The rate of ionization and oxidation is limited by the fact that the reactions take place along the lines of triple contact involving the reactant (fuel), the electrode, and the electrolyte. (4) The need to eliminate all corrosion and side effects. (5) The provision for appropriate disposal of waste products.

The external characteristic of a typical fuel cell is represented by a plot of cell voltage versus current density as depicted in Fig. 19-8. The quasilinear sloping part is due to the ohmic resistance of the electrolyte, while the sharp tail-off at high current densities results from a space charge effect associated with a concentration gradient of ions in the electrolyte.

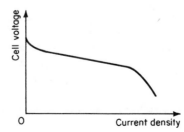

Fig. 19-8. External characteristic of a fuel cell.

19-4   Magnetohydrodynamic Energy Conversion

An immediate distinguishing feature of the magnetohydrodynamic (MHD) generator is that it is a *volume* device compared with the preceding methods treated in this chapter, which are essentially *area* devices. Consequently, much higher power densities are achievable with MHD generators. Moreover, voltages of the order of 10,000 volts can be generated with ordinary magnetic field strengths. In light of these facts, it is easy to understand why this method of energy conversion presents the only real alternative to power generation by the steam turbogenerator. The largest MHD generator being built has a rating of 40 megawatts of direct current. Unfortunately, up to now the generation of a-c power by MHD techniques has proved disappointing because of the very serious low power factor problems. Accordingly attention in this section is confined to the d-c MHD generator.

Principle of Operation. The essential construction features of the MHD generator are depicted in Fig. 19-9. The MHD generator consists of a pair of electrodes (plates P_1 and P_2), field coils to produce a magnetic field B (directed into the paper),

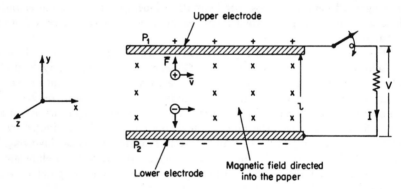

Fig. 19-9. Cross-sectional view of the generator duct of the MHD generator. Plasma flow is along the *x*-axis from left to right.

and a moving *plasma* (directed perpendicular to the magnetic field). The term plasma is used to denote a gas that is sufficiently ionized to be considered a good conductor of electricity. This often implies that the gas has a small *mean free path* (i.e., distance between collisions), which makes operation at high pressure and high temperature mandatory. The application of thermal energy causes gas molecules to be dissociated into positive ions and negative ions.† As the volume of gas plasma moves from left to right between the electrodes under the influence of the magnetic field, the charge carriers experience a force described by

$$\bar{F} = q\bar{v} \times \bar{B} \tag{19-2}$$

† The negative ion results when an electron becomes attached to an atom or molecule.

where q is the charge, v denotes the velocity of the charged particle, and $\bar{B}$ denotes the magnetic field. As indicated in Fig. 19-9, when q refers to a positive ion, the force associated with Eq. (19-2) is directed upward, or in the positive y direction. When q refers to an electron, the force is reversed and so is directed downward. Hence there occurs an average drift of positive ions to the upper plate and electrons to the lower plate in spite of collisions with surrounding uncharged gas molecules. The net effect is that a voltage appears across the electrodes, which can then be applied to a suitable electrical load causing the current I to flow. In this fashion an amount of power equal to VI is delivered to the load, where V is the voltage appearing across the load terminals. Of course the source of this electrical energy originates from the velocity of the gas. Thus it can be said that the MHD generator is a device that allows energy to be extracted from a moving gas and converted into electrical energy. By conservation of energy considerations, it should be expected that the conversion process would result in a decrease in gas velocity. A more detailed examination makes this apparent.

When the voltage V is made to appear across the load as a result of the drift of electrons to the lower plate and of positive ions to the upper plate, an electric field is produced between the electrodes by these charges. If the distance between the parallel plates is called l, the expression for this electric field $\mathscr{E}_y$ is

$$\mathscr{E}_y = -\frac{V}{l} \quad \text{volts/meters} \tag{19-3}$$

where the minus sign indicates that the field lies in the negative y direction. The resultant electric field between electrodes can then be written as

$$\mathscr{E}'_y = vB - \frac{V}{l} \tag{19-4a}$$

The first component on the right side of the last equation is the electric field associated with movement of the charged plasma in the magnetic field and is directed in the positive y direction. Rewriting Eq. (19-4a) in the form

$$\mathscr{E}'_y = \frac{1}{l}(Blv - V) \tag{19-4b}$$

permits the resultant electric field between the electrodes to be identified in terms of an equivalent internal voltage drop described by the quantity in the parentheses. Equation (19-4b) may also be expressed as

$$\mathscr{E}'_y = \frac{E_g - V}{l} \tag{19-5}$$

where

$$E_g \equiv \text{generated emf} = Blv \tag{19-6}$$

In a sense the quantity $E_g - V$ may be interpreted as the equivalent internal resistance drop needed to maintain the proper value of resultant electric field between electrodes consistent with the law of conservation of energy. It is also worthwhile to note the similarity of Eq. (19-6) to Eq. (3-6).

In dealing with plasmas in MHD generators it is not uncommon to consider the electrical conductivity to be attributable solely to the electrons in the plasma. This is justified because of the small mass and high degree of mobility of the electrons compared with the positive ions. Accordingly, when a current I is assumed to flow from plate P_2 to plate P_1, we can for all practical purposes consider that electrons are drifting out of the plasma to plate P_2. As the current I flows from P_2 to P_1, it interacts with the magnetic field to produce a force *opposed* to the plasma flow. This force acts along the negative x direction as is evident from the force equation; its expression is

$$\bar{F}_x = l\bar{I} \times \bar{B} \qquad (19\text{-}7)$$

where for the configuration of Fig. 19-9 the quantities I, $\bar{B}$, and $\bar{F}_x$ are all in quadrature with one another. Because $\bar{F}_x$ acts directly to oppose the plasma flow, the amount of power removed from the gas flow can then be written as

$$P = F_x v \qquad (19\text{-}8)$$

But from Eq. (19-7)

$$F_x = BlI \qquad (19\text{-}9)$$

Hence

$$P = F_x v = BlvI = E_g I \qquad (19\text{-}10)$$

The last equation is simply a statement of the conservation of energy, indicating that the mechanical energy that is removed from the gas flow is converted to electrical energy through the generation of an emf E_g and the associated current I.

Although the foregoing analysis is approximate, it does serve to illustrate the mechanism by which the energy conversion process takes place and to identify the factors that determine the operating voltage. In a practical situation it is necessary to take due account of such matters as the variation in temperature and pressure of the plasma flow and of such other effects as the Hall currents.

The Hall Effect. The Hall effect refers to a situation that can arise in MHD generators and can appreciably reduce the generated power density. To appreciate the problem, two factors must be kept in mind. One, it is desirable to operate at high flux densities in order to achieve high operating voltages. This means operating with a small mean radius of trajectory, r, which is known to be inversely proportional to the flux density. Two, it is usually desirable to keep the plasma at relatively low pressures in order to increase the density of charge carriers. Lower pressures mean larger mean free paths l, which in turn allow electrons more time between collisions. In this way electrons can acquire sufficient energy to dislodge other electrons from neutral molecules whenever a collision does occur. Therefore, in the absence of a superior consideration, it appears desirable to operate the MHD generator with $l < r$. A more detailed investigation, however, reveals that operation under such a condition results in a reduced electrical conductivity of the plasma as viewed from the electrodes.

To understand why, refer to Fig. 19-10(a). Recall from Eq. (19-4) that a net electric field $\mathscr{E}_y'$ exists in the positive y direction. It is also necessary to recall from

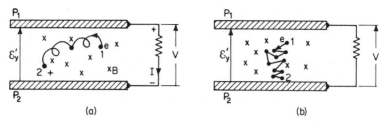

Fig. 19-10. Describing the path of a plasma electron: (a) $l > r$ where l is the mean free path and r denotes the mean radius of trajectory; (b) $l < r$.

electrostatics that the movement of a charged particle in a strong magnetic field describes a circular path. Accordingly, if we were to track the motional behavior of an electron in the plasma of the geometry of Fig. 19-10(a), it would look very much like the path drawn in this illustration between points 1 and 2. Because the length between collisions is large compared with the mean radius of trajectory, a clearly definable cycloidal motion is generated before collision occurs (see heavy dots). The chief effect is a displacement of the electron in a predominantly axial direction. The electron flow is from 1 to 2; hence the current flow is from 2 to 1. Since this current is displaced with respect to the electric field direction, the current can no longer be treated as a scalar quantity but rather as a tensor. It is particularly important to note that in the illustration of Fig. 19-10(a) the component of current along the flow axis (x direction) exceeds that along the y direction (from electrode P_2 to P_1). The reduction in conductivity is directly related to the rotation of the current vector from a plate-to-plate orientation to one that is almost axial. The axial current component is called the *Hall current.*

The degree of rotation of the current vector from the desired direction is dependent upon the ratio of the mean free path to the mean radius of trajectory. When conditions are such that the mean free path is much less than the mean radius of trajectory, the Hall current becomes negligibly small. This situation is depicted in Fig. 19-10(b). The path between collisions is now essentially a straight line. Moreover, the frequency of collisions (which is now greater because of the smaller value of l) causes on the average relatively little displacement of the electron in the axial direction. However, by comparison, the electron drift towards the bottom electrode is quite appreciable and this in turn means good electrical conductivity.

If the same value of flux density is needed in the configurations of Fig. 19-10 (a and b), a reduction in mean free path can be achieved by employing higher operating pressures. This is one of the costs of good conductivity. Another factor that favorably influences the conductivity is operation at high temperatures. The higher the temperature, the greater is the density of ionized particles. Because of the thermal limitations of the present-day materials that are used for electrodes and other enclosure parts, the operating temperature of the gas is limited to about 2500°K for sustained operation. Unfortunately, this is not high enough to provide sufficient ionization for most gases. Hence, *seeding* of the gas is often necessary. The use of 1 per cent of potassium by volume is common in this connection.

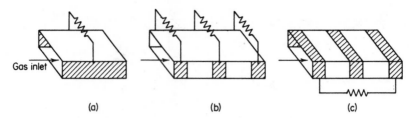

Fig. 19-11. Illustrating various types of linear MHD generators: (a) continuous electrode; (b) segmented electrode; (c) Hall generator.

Linear MHD Generator Types. In Fig. 19-11 are shown three types of linear MHD generators having rectangular cross sections. In the first type the electrodes are continuous and run the full length of the generator section. When a situation arises in which the gas pressure is as high as practical and the flux density has such a large value that the ratio of l to r is unfavorably large, the effect of Hall currents can be minimized by resorting to such a segmented electrode construction as is depicted in Fig. 19-11(b). The disadvantage of such an arrangement is the need for multiple loads at different potentials. Suitable interconnections are possible, however, to permit the loads to operate at the same potential. In the configuration of Fig. 19-11(c) operation occurs with $l < r$, i.e., the Hall effect is emphasized. Such an arrangement results in a potential difference occurring in the direction of gas flow. Electrodes are then placed in segments completely surrounding the rectangular tube, with the load placed as indicated. For obvious reasons this is called the *Hall generator*, and it can be shown that maximum efficiency for such a geometry occurs near the condition of short circuit.

MHD Systems. Figure 19-12 depicts an open-cycle MHD generator system which is designed to recapture most of the energy in the hot plasma as it leaves the generator duct. Fossil fuel (often coal) is mixed with preheated compressed air (the oxidizer) in the combustion chamber, where the products of combustion are in turn mixed with a suitable seed material to yield a productive plasma. As the plasma passes through the generator duct, energy is extracted from it and converted to d-c power. An appropriate d-c to a-c converter then changes the d-c power to three-phase alternating current so that it can be efficiently transmitted to the region of consumption. The high-temperature plasma which exits from the generator duct is fed to an air preheater where it raises the temperature of the compressed air. Additional energy is recovered from the exhaust gas by passing it to a boiler where the generated steam is used to drive a steam turbine. In turn the steam turbine drives a compressor to prepare the air for the combustion chamber under pressure and also to drive a conventional rotating-coil three-phase generator to furnish auxiliary power.

How does the generation of electrical power by an MHD generator system such as the one illustrated in Fig. 19-12 compare with that by the steam turbo-generator method currently in use? It can be shown that the overall efficiency of the MHD generator is approximately 56 per cent, which is attractive in light of

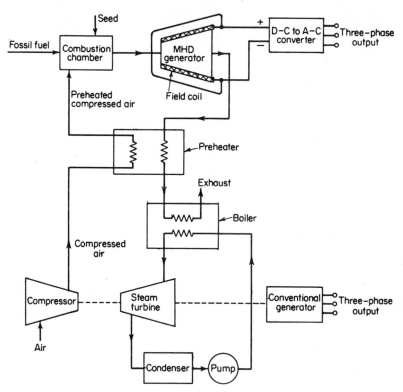

Fig. 19-12. An open-cycle MHD generator with provisions for recovery of energy in the hot gas leaving the MHD duct.

the 40 per cent figure that applies to the steam turbogenerator. However, this advantage is considerably diminished when it is weighed against the investment required in field magnets, generator ducts, d-c to a-c converters, and air pre-heaters. Furthermore, in MHD systems that are designed to recover the expensive seed material from the exhaust gases, there is the additional investment in scrubbers and precipitators. Much research and development yet remain to be done before MHD systems can become truly competitive with steam turbogenerators.

IV

CONTROL SYSTEMS
AND COMPUTERS

20

principles of automatic control[†]

The essential feature of many automatic control systems is feedback. Feedback is that property of the system which permits the output quantity to be compared with the input command so that upon the existence of a difference an actuating signal arises which acts to bring the two into correspondence. This principle of feedback is really not new to us; it surrounds every phase of everyday living. It underlies the coordinated motions executed by the human body in walking, reaching for objects, and driving an automobile. It plays an equally important role in the countless applications of control system engineering in the fields of manufacturing, process industries, control of watercraft and aircraft, special-purpose computers for many types of military equipment, and in many other fields including the home and the office.

In this chapter the mechanism through which the feedback action exhibits itself is studied in control systems ranging from elementary ones to those possessing a high degree of complexity. In each, the same functional behaviors will be observed, and the general nature of the method of analysis will thus be emphasized.

† Sections 20-1 to 20-4 adapted by permission from V. Del Toro and S. R. Parker, *Principles of Control Systems Engineering* (New York: McGraw-Hill Book Company, 1960).

The distinction between an elementary system and a complex one lies primarily in the difficulty of the task to be performed. The more difficult the task, the more complex the system. In fact with many present-day systems this complexity has reached such proportions that system design has virtually become a science. The functional behavior of each system will be treated in terms of a block-diagram notation and its associated terminology.

20-1 Distinction between Open-loop and Closed-loop (Feedback) Control

It is important in the beginning that the distinction between closed-loop and open-loop operation be clearly understood. Both terms will often be used throughout this part of the book. The following simple definitions point out the difference:†

An *open-loop system* is one in which the control action is independent of the output (or desired result).

A *closed-loop system* is one in which the control action is dependent upon the output.

The key term in these definitions is *control action*. Basically, it refers to the actuating signal of the system, which in turn represents the quantity responsible for activating the system to produce a desired output. In the case of the open-loop system the input command is the sole factor responsible for providing the control action, whereas for a closed-loop system the control action is provided by the difference between the input command and the corresponding output.

To illustrate the distinction, consider the control of automobile traffic by means of traffic lights placed at an intersection where traffic flows along north-south and east-west directions. If the traffic-light mechanism is such that the green and red lights are on for predetermined, fixed intervals of time, then operation is open-loop. This conclusion immediately follows from our foregoing definition upon realizing that the desired output, which here is control of the volume of traffic, in no way influences the time interval during which the light shines green or red. The input command originates from a calibrated timing mechanism, and this alone establishes how long the light stays green or red. The control action is thereby provided directly by the input command. Accordingly, the timing mechanism has no way of knowing whether or not the volume of traffic is especially heavy along the north-south direction and therefore in need of longer green-light intervals. A closed-loop system of control would provide precisely this information. Thus, if a scheme is introduced which measures the volume of traffic along both directions, compares the two, and then allows the difference to control the green and red time periods, feedback control results because now the actuating signal is a function of the desired output.

To complete the comparison of closed-loop versus open-loop operation, it is instructive next to list briefly some of the performance characteristics of each.

† More precise definitions are given in the next section.

Generally, the open-loop system of control has two outstanding features. First, its ability to perform accurately is determined by its calibration. As the calibration deteriorates, so too does its performance. Second, the open-loop system is usually easier to build since it is not generally troubled with problems of instability. One of the noteworthy features of closed-loop operation is its ability faithfully to reproduce the input owing to feedback. This in a large measure is responsible for the high accuracy obtainable from such systems. Since the actuating signal is a function of the deviation of the output from the input, the control action persists in generating sufficient additional output to bring the two into correspondence. Unfortunately, this very factor (feedback) is also responsible for one of the biggest sources of difficulty in closed-loop systems, namely, the tendency to oscillate. A second important feature of closed-loop operation is that, in direct contrast to the open-loop system, it usually performs accurately even in the presence of nonlinearities.

20-2   Block Diagram of Feedback Control Systems. Terminology

Every feedback control system consists of components that perform specific functions. A convenient and useful method of representing this functional characteristic of the system is the block diagram. Basically this is a means of representing the operations performed in the system and the manner in which signal information flows throughout the system. The block diagram is concerned not with the physical characteristics of any specific system but only with the functional relationship among various parts in the system. In general, the output quantity of any linear component of the system is related to the input by a gain factor and combinations of derivatives or integrals with respect to time. Accordingly, it is possible for two entirely different and unrelated physical systems to be represented by the same block diagram, provided that the respective components are described by the same differential equations.

The general form of the block diagram of a feedback control system is shown in Fig. 20-1.† Since the differential equation or the gain factor of each block is not specifically identified, lower-case letters are used to represent the input and output variables. When this information is known, capital letters are used. This notation is followed throughout this part of the book. The symbols in the block diagram were selected to avoid those which imply the mechanics of the system such as θ for angle, p for pressure, etc. This helps to preserve the generic nature of the block diagram.

Before we consider the significance and advantages of the block-diagram notation, it is very important that the meanings of the terms used in Fig. 20-1 be clearly understood and remembered. These terms and others are described below:

A *feedback control system* is a control system that tends to maintain a pre-

† A.I.E.E. Committee Report, "Proposed Symbols and Terms for Feedback Control Systems," *Elec. Eng.*, **70** (1951), pt. 2, pp. 905–909.

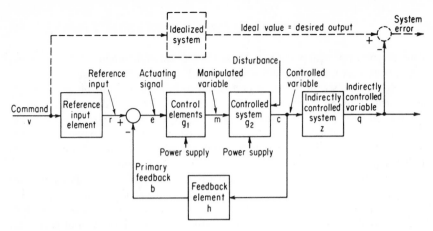

Fig. 20-1. Block diagram of a feedback control system, showing terminology.

scribed relationship of one system variable to another by comparing functions of these variables and using the difference as a means of control.

The *controlled variable* is that quantity or condition of the controlled system which is directly measured and controlled.

The *indirectly controlled variable* is that quantity or condition which is controlled by virtue of its relation to the controlled variable and which is not directly measured for control.

The *command* is the input which is established or varied by some means external to and independent of the feedback control system under consideration.

The *reference input* is a signal established as a standard of comparison for a feedback control system by virtue of its relation to the command.

The *primary feedback* is a signal that is a function of the controlled variable and that is compared with the reference input to obtain the actuating signal. (This designation is intended to avoid ambiguity in multiloop systems.)

The *actuating signal* is the reference input minus the primary feedback.

The *manipulated variable* is that quantity or condition which the controller (g_1) applies to the controlled system.

A *disturbance* is a signal (other than the reference input) that tends to affect the value of the controlled variable.

A *parametric variation* is a change in system properties that may affect the performance or operation of the feedback control system.

The *controlled system* is the body, process, or machine, a particular quantity or condition of which is to be controlled.

The *indirectly controlled system* is the body, process, or machine that determines the relationship between the indirectly controlled variable and the controlled variable.

The *feedback controller* is a mechanism that measures the value of the controlled variable, accepts the value of the command, and, as a result of comparison, manipulates the controlled system in order to maintain an established relationship between the controlled variable and the command.

The *control elements* comprise the portion of the feedback control system that is required to produce the manipulated variable from the actuating signal.

The *reference-input elements* comprise the portion of the feedback control system that establishes the relationship between the reference input and the command.

The *feedback elements* comprise the portion of the feedback control system that establishes the relationship between the primary feedback and the controlled variable.

The *summing point* is a descriptive symbol used to denote the algebraic summation of two or more signals.

The reference-input element usually consists of a device that converts control signals from one form into another. Such devices are known as *transducers*. Most transducers have an output signal in the form of electrical energy. Thus a potentiometer can be used to convert a mechanical position to an electrical voltage. A tachometer generator converts a velocity into a d-c or an a-c voltage. A pressure transducer changes a pressure drop or rise into a corresponding drop or rise in electric potential. The transducer provides the appropriate translation of the command into a form usable by the system.

The feedback element also frequently consists of a transducer of the same kind as the reference-input element. When the two signals of the input and feedback transducers are compared, the result is the actuating signal. This portion of the closed-loop system is very important, and it is usually more descriptively identified as the *error detector* (see Fig. 20-2). Error detectors may take on a variety of forms depending upon the nature of the feedback control system. Some of the more frequently used error detectors include the following: two bridge-connected potentiometers, a control-transmitter control-transformer synchro combination, two linear differential transformers, two tachometer generators, differential gear, E-type transformers, bellows, gyroscopes.

The primary function of the control element g_1 of the block diagram is to provide amplification of the actuating signal since it is generally available at very low power levels.† Many types of amplifying devices are used either singly or in combination. Some of the more common types are the following: electronic, solid-state (transistor), electromechanical, hydraulic, pneumatic, magnetic. Note that the amplifier must have its own source of power. This is indicated in Fig.

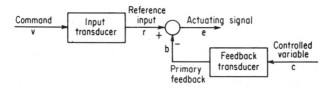

Fig. 20-2. The error-detector portion of a feedback control system.

† The control element also frequently serves the function of changing the basic time character of the signal to control the transient system response.

20-1 by the use of an external arrow leading to the g_1 block. It should be emphasized that the *closed-loop block diagram* refers to the flow of the *control power* and not to the main source of energy for the system. In essence, then, the block diagram represents the manner in which the control signal manipulates the main source of power in order to adjust the controlled variable in accordance with the command.

The block identified as the controlled system g_1 in Fig. 20-1 refers to that portion of the system which generates the controlled variable in accordance with the dictates of the manipulated variable. The controlled system obviously can take on many forms. It may represent an aircraft frame, the body of a ship, a gun mount, a chemical process, or a temperature or pressure system. The milling machine and the lathe are examples where the controlled system represents a machine as distinguished from a process or body.

The operation of any feedback control system can be described in terms of the block diagram of Fig. 20-1. The application of a specific command causes a corresponding signal at the reference input through the action of the input transducer. Since the controlled variable cannot change instantaneously because of the inertia of the system, the output of the feedback element no longer is equal and opposite to the reference input. Accordingly, an actuating signal exists, which in turn is received by the control element and amplified, thereby generating a new value of controlled variable. This means that the primary feedback signal changes in such a direction as to reduce the magnitude of the actuating signal. It should be clear that the controlled system will continue to generate a new level of output as long as the actuating signal is different from zero. Therefore only when the controlled variable is brought to a level equal to the value commanded will the actuating signal be zero.† The system then will be at the new desired steady-state value.

To have feedback control systems operate in a stable fashion, they must be provided with *negative* feedback, which simply means that the feedback signal is opposite in sign to the reference input. This helps to assure that, as the controlled variable approaches the commanded value, the actuating signal approaches zero. A moment's reflection reveals that, if the feedback is positive and the controlled variable increases, the actuating signal will also increase. This causes an increased value of controlled variable, which results in a further increase in actuating signal, and so on. The result is an unbounded‡ increase in controlled variable and loss of control by the command source.

Summarizing, we see that the block diagram of a feedback control system is, first of all, a representation of its functional characteristics which permits a description of the manner in which the control-signal energy flows through the system. Second, it is a means of emphasizing that the closed-loop system is composed of three principal parts: the error detector, the control element (amplifier and output device), and the controlled system.

† There are situations in which the steady-state level of the actuating signal is finite.

‡ The nonlinearities inherent in the system would ultimately limit the magnitude of the output.

20-3   Position Feedback Control System. Servomechanism

A common industrial application of a feedback system is to control position. These systems occur so frequently in practice and are so important that they are given a special name—*servomechanisms*. A servomechanism is a power-amplifying feedback control system in which the controlled variable is mechanical position. The term also applies to those systems which control time derivatives of position, as, for example, velocity and acceleration. The schematic diagram of a servomechanism is shown in Fig. 20-3.

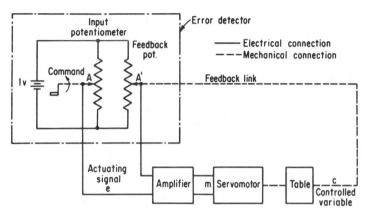

Fig. 20-3. Example of a servomechanism.

Consider that the system of Fig. 20-3 is to be used to control the angular position of a table in accordance with a command signal originating from a remotely located station. The first step in obtaining the block diagram and understanding the operation of the system is to identify the error-detector portion of the system. As pointed out in Fig. 20-2, this requires merely finding the input and feedback transducers and the summing point of their output signals. For the system of Fig. 20-3 the error detector consists of the section blocked off with the broken line. The second principal part of the system—the control elements—consists of the amplifier unit and the output motor. The latter is more commonly referred to as a servomotor, since it requires special design considerations. The table, the position of which is being controlled, represents the controlled system. In a specific application it may be a gun mount or the support table for a lathe or milling machine. The block diagram for the system is shown in Fig. 20-4.

To understand the operation of the system, assume that initially the slider arms of the input and feedback potentiometers are both set at $+50$-mv. The voltage of the input potentiometer is the reference input. For this condition the actuating signal is zero, and so the motor has zero output torque. Next consider that the command calls for a new position of the table, namely, that corresponding to a potentiometer voltage of $+60$-mv. When arm A is placed at the $+60$-mv position, arm A' remains instantaneously at the $+50$-mv position because of the

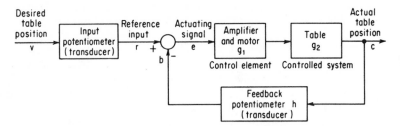

Fig. 20-4. Block diagram of the servomechanism of Fig. 20-3.

table inertia. This situation creates a $+10$-mv actuating signal, which is really a measure of the lack of correspondence between the actual table position and the commanded position. A $+10$-mv input to the amplifier applies an input to the servomotor, which in turn generates an output torque, which repositions the table. With negative feedback present the table moves in a direction which causes the potential of A' to increase beyond $+50$-mv. As this takes place, the actuating signal gets smaller and finally reaches zero, at which time A' has the same potential as A (this is referred to as the *null* position). The actual table position is therefore equal to the commanded value. Note that, if the table is not exactly at the commanded position, the actuating signal will be different from zero, and hence a motor output torque persists, forcing the table to take the commanded position.†
On the basis of the foregoing explanation, we can begin to appreciate that a system of this type possesses an integrating property; i.e., the motor output angle is proportional to the integral of the actuating signal and thus stops changing only when the actuating signal itself is zero. This follows from the fact that for any given actuating signal there results a corresponding motor velocity. However, the feedback potentiometer is sensitive not to velocity but rather to motor position, which is the integral of velocity and therefore also of the actuating signal.

One distinguishing feature that the amplifier of this type of control system must have is *sign sensitivity*. It must function properly whether the command places arm A at a higher or a lower voltage than the original value. In terms of position this means that the command should be capable of moving the table in a clockwise or counterclockwise direction. Thus, if arm A had been placed at $+40$ rather than $+60$-mv by the command, the actuating signal would have been -10-mv. The amplifier must be capable of interpreting the intelligence implied in the minus sign by applying a reversed signal to the servomotor, thereby reversing the direction of rotation.

The manner in which a feedback control system reacts to an externally applied disturbance may be illustrated by considering further the example of the servomechanism. Assume the disturbance to be in the form of an external torque applied to the table. The effect of this torque will be to offset the feedback po-

---

† This is true provided that the components are all assumed to be perfect. Actually, imperfections such as motor dead zone cause the commanded and actual position to be out of correspondence by a small amount.

tentiometer from its null position. Upon doing this, the actuating signal no longer is zero but takes on such a value that when multiplied by the torque constant (units of torque per volt of actuating signal) the developed torque of the servomotor is equal and opposite to the applied disturbance. It should be clear that, as long as the external torque is maintained, the actuating signal cannot be zero. How close it will be to zero depends upon the torque constant of the servomotor and the magnitude of the external force. Of course, the larger the torque constant, the smaller the actuating signal required. Since the actuating signal can no longer be zero, the potential of arm A' is no longer equal to the commanded value. Hence, for a system of the type described in Fig. 20-3, a steady-state position error exists in the presence of a constantly maintained external disturbance.† This characteristic of the system of readjusting itself to counteract the disturbance is referred to as the self-correcting, or automatic control, feature of closed-loop operation.

On the basis of the foregoing description of the servomechanism, a few general characteristics of this type of control system can be stated. A feedback control system is classified as a servomechanism if it satisfies all three of the following conditions: (1) It is error-actuated (i.e., closed-loop operation). (2) It contains power amplification (i.e., operates from low signal levels). (3) It has a mechanical output (position, velocity, or acceleration). Essentially, the servomechanism is a special case of a feedback control system. In regard to performance features we can say at this point that among its salient characteristics are automatic control, remote operation (of the command station), high accuracy, and fast response. The aspect of remote operation deserves further comment since it is one of the chief motivations for the development of these systems. It should be apparent from Fig. 20-3 that the input potentiometer (often called the command station) can be far removed from the output device. Thus, if the controlled system is the rudder of a ship, its heading can be easily and accurately manipulated from the ship's bridge.

20-4 Typical Feedback Control Applications

To assure a more thorough understanding of the principles of feedback control, we shall next apply these principles to several widely different situations. For each system the description will revolve about the block-diagram representation, thus again emphasizing the functional nature of the approach. Also, a thorough comprehension of the procedure should develop within us a facility that will make further thinking in terms of block diagrams a routine matter. It is important to establish such a facility before dealing with the general problem of the dynamic behavior of feedback control systems.

Pressure Control System for Supersonic Wind Tunnel. Of considerable importance in the aircraft and outerspace industry nowadays is obtaining information about aircraft stability derivatives and aerodynamic parameters prevailing at

---

† This position error may be eliminated by introducing additional control elements.

Mach† numbers ranging from 0.5 to 5 and higher: Several of the leading aircraft companies have built wind tunnels to accomplish tests at these speeds. The wind tunnel is often of the blowdown type, which means that compressors pump air to a specified pressure in a huge storage tank and, when a test is to be performed, the stored air is bled through a valve into a settling chamber, where air flow at high Mach numbers is realized. In such tests, keeping the Mach number constant is of the utmost importance. A problem exists since, as the air is bled from the storage tanks, its pressure drops and so too will the Mach number in the test section unless the valve is opened more. The need for a control system therefore presents itself. The nature of the control system depends upon the means available for monitoring the quantity it is desired to control, which in this case is the Mach number. Since the Mach number can be identified in terms of pressure, it turns out that to keep the Mach number fixed requires keeping the settling-chamber pressure constant. Hence the use of a pressure-activated control system suggests itself. It is interesting to note that the Mach number is now the indirectly controlled variable while the pressure is the directly controlled variable.

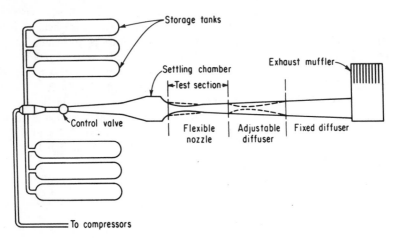

Fig. 20-5. A high-speed wind tunnel.

A schematic diagram of the wind tunnel‡ without the control system is shown in Fig. 20-5. The block diagram of the complete system including the components required for control is shown in Fig. 20-6. As indicated in Fig. 20-5, the air to operate the tunnel is stored in six tanks at a pressure of 600 psia at 100°F. The tanks discharge into a 30-in.-diameter manifold, which in turn leads to a 24-in.-diameter rotating-plug control rotovalve. The discharge side of this valve is con-

---

† *Mach number* refers to the ratio of the speed of the aircraft at a given altitude to the speed of sound at the same altitude.

‡ The wind tunnel discussed here is installed at Convair, a division of General Dynamics Corporation at San Diego, Calif.

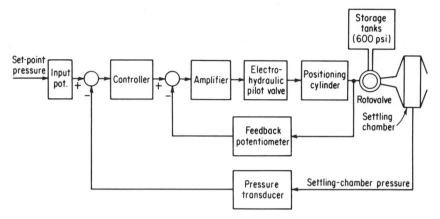

Fig. 20-6. Block diagram of a supersonic wind-tunnel pressure control system.

nected to the settling chamber by means of a duct. In Fig. 20-6 the pressure transducer is a device that converts pressure to a corresponding voltage, which is then compared with the reference-input voltage, and the difference constitutes the actuating signal. The controller refers to an amplifying stage plus electrical networks that integrate and differentiate the input signal. (The reason for integration and differentiation will be discussed in the next chapter.)

The electrohydraulic value is an electrically operated two-stage hydraulic amplifier. It consists of a coil-magnet motor (which receives the electrical output of the preceding amplifier), a low-pressure pilot valve, and a high-pressure pilot valve. The output motion of the latter drives a positioning cylinder, which in turn positions the rotovalve. For improved performance a position feedback is put round the electrohydraulic valve. The controller, the electrohydraulic valve and its feedback path, and the rotovalve represent that part of the feedback control system identified as the *control elements* g_1. The *error detector* consists of the input potentiometer plus the feedback transducer, while the *controlled system* is the process of pressure build-up in the settling chamber.

Assume that a wind-tunnel test is to be performed at Mach 5, to which the settling-chamber pressure of 250 psia corresponds. Operation is started by putting the input potentiometer at a set-point pressure of 250 psia. An actuating signal immediately appears at the controller, which in turn causes the electrohydraulic pilot valve and its positioning cylinder to open the rotovalve and thereby build up pressure in the chamber. As the rotovalve is opened, the input to the pilot-valve amplifier decreases because of the position feedback voltage. Moreover, as pressure builds up, the actuating signal to the controller decreases. When the settling-chamber pressure has reached the commanded 250 psia, the actuating signal will be zero and no further movement of the rotovalve plug takes place. The time needed to accomplish this is relatively small (about 5 sec), and the attendant decrease in tank pressure is also very small. Consequently, an air flow of Mach 5 is established in the settling chamber. However, as time passes and more and more air is bled from the storage tanks, the storage pressure will decrease and, unless a further opening of the rotovalve is introduced, the settling-chamber pressure will drop also. Of course additional rotovalve opening will occur because of the pressure feedback. Specifically, as the settling-chamber pressure decreases, the corresponding feedback voltage drops and, since the reference input voltage is fixed, an actuating signal appears at the controller, which causes the pilot valve to reposition the rotovalve. This in turn maintains the desired chamber pressure. It should be clear that this corrective action will prevail as long as there is any tendency for the chamber pressure to drop and provided that the rotovalve is not at its limit position, i.e., not fully open.

Automatic Machine-tool Control. Application of the principles of feedback control techniques to machine tools, together with the ability to feed the machine tool programmed instructions, has led to completely automatic operation with increased accuracy as well. Fundamentally, three requirements need to be satisfied to obtain this kind of control. First, the machine tool must receive instructions regarding the final size and shape of the workpiece. Second, the workpiece must be positioned in accordance with these instructions. Third, a measurement of the desired result must be made in order to check that the instructions have been carried out. This, of course, is accomplished through feedback.

To understand better the procedure involved, consider the system depicted in Fig. 20-7. It represents the programmed carriage drive for a milling machine. It could just as well represent the vertical and lateral drive of the cutting tool. In practice, for the cutting of a three-dimensional object, three such systems are provided, and so whatever is said about one holds for all three. The information

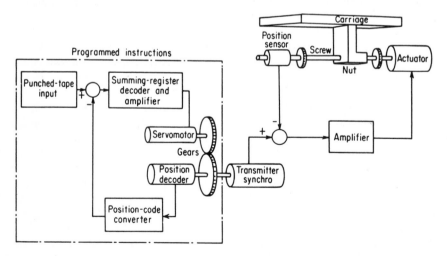

Fig. 20-7. A feedback control system for the carriage drive of a milling machine.

regarding the size and shape of the workpiece usually orginates from engineering drawings, equations, or models and is subsequently converted to more usable forms such as punched cards, magnetic tapes, and cams. In Fig. 20-7 this is the box identified as the punched taped input. Frequently this part of the system is referred to as the input memory. The information contained in this memory is then applied to a computer, whose function is to make available a command signal indicating a desired position and velocity of the carriage. The computer in the case illustrated is a data-interpreting system and decoding servomechanism. Together with the input memory, this constitutes the total programmed instructions, or command input, to the carriage feedback control system. The transmitter synchro is the transducer, which converts the programmed command to a voltage which is the reference input to the control system. This voltage appears at the amplifier, which generates an output to drive the actuator. The actuator may be of the hydraulic, pneumatic, or electric type or perhaps a clutch. The lead screw is then driven and the carriage positioned in accordance with the instruction. Assurance that the instruction has been followed is provided by the feedback position sensor. Should the instruction fail to be fully carried out, the position sensor will allow an actuating signal to appear at the input terminals of the amplifier, thereby causing additional motion of the carriage. When the commanded position is reached, the actuating signal is nulled to zero until the next instruction comes along.

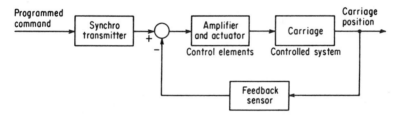

Fig. 20-8. Block diagram of Fig. 20-7.

The block diagram for the system of Fig. 20-7 is shown in Fig. 20-8. The programmed command refers to the broken-line enclosure and is identified as the programmed instructions in Fig. 20-7. The control elements consist of the amplifier and actuator. The controlled system is the carriage. The error detector is made up of the input and feedback transducers as usual. It is interesting to note, too, that the data-interpreting system is itself a servomechanism.

Automobile Power-steering Servomechanism. One of the most common servomechanisms is the power-steering unit found in the automobile. A simplified schematic diagram of the system appears in Fig. 20-9. The corresponding block diagram is given in Fig. 20-10. The purpose of the system is to position the wheels in accordance with commands applied to the steering wheel by the driver. The inclusion of the hydraulic amplifier means that relatively small torques at the

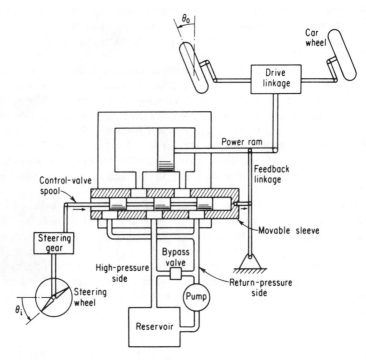

Fig. 20-9. Example illustrating the principle of an automobile power-steering servomechanism.

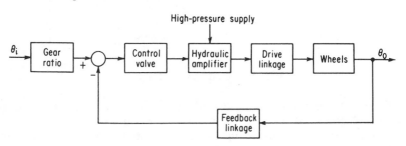

Fig. 20-10. Block diagram of Fig. 20-9.

steering wheel will be reflected as much larger torques at the car wheels, thereby providing ease of steering.

The operation is simple and can be explained by applying the same approach that has been applied to the foregoing systems. Initially, with the steering wheel at its zero position (i.e., the crossbar horizontal), the wheels are directed parallel to the longitudinal axis of the car. In this position the control-valve spool is centered so that no pressure differential appears across the faces of the power ram. When the steering wheel is turned to the left by an amount θ_i, the control-valve spool is made to move toward the right side. This opens the left side of the power cylinder to the high-pressure side of the hydraulic system and the right side to the return, or low-pressure, side. Accordingly, an unbalanced force appears on the power ram, causing motion toward the right. Through proper drive linkage a torque is applied to the wheels, causing the desired displacement θ_0. As the desired wheel position is reached, the control valve should be returned to the centered position in order that the torque from the hydraulic unit will be returned to zero. This is assured through the action of the feedback linkage mechanism. The linkage is so arranged that, as the power ram moves toward the right, the movable sleeve is displaced toward the right also, thereby sealing the high-pressure side. Such action signifies that the system has negative feedback.

The control valve and the power cylinder are part of the same housing, and these, together with the mechanical advantage of the linkage ratio, constitute the control element of the system. The controlled system in this case refers to the wheels. The error detector consists of the net effect of the feedback output position of the movable sleeve and the displacement introduced by the steering wheel at the control-valve spool. A centered valve position means no lack of correspondence between the command and the output.

Figure 20-10 is a representation of the position feedback loop only. Actually there are several more loops involved, such as the velocity loop and the load loop, which account for such things as car dynamics and tire characteristics. These are omitted for simplicity.

20-5 Feedback and Its Effects

Besides the feature of automatic control described in the preceding sections, the feedback principle has many other advantages, effects, and uses. However, before we turn to these matters, it is helpful first to obtain mathematically the relationship between input and output quantities both for the open- and closed-loop systems. Depicted in Fig. 20-11(a) is the block-diagram notation of an open-loop system which has a *direct transmission path function* denoted by G. The symbol G represents the *direct transfer function* of the system. It is that quantity which when multiplied by the input E yields the output C. Thus

$$C = GE \qquad\qquad (20\text{-}1)$$

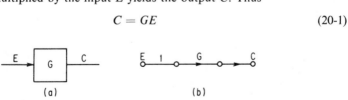

(a)   (b)

Fig. 20-11. An open-loop system:
    (a) the block-diagram representation in terms of the input and output variables E and C and the direct transmission function G;
    (b) the corresponding signal-flow graph.

This transfer function, which is defined as the ratio of output C to input E, is clearly given by

$$\frac{C}{E} = G \tag{20-2}$$

G is characterized entirely by the components that make up the system. In general, in addition to a gain factor, it also contains operations of differentiation and integration. In such instances the transfer function G can be expressed algebraically in terms of the Laplace operators.

Appearing in Fig. 20-11(b) is an alternative way of representing the block diagram. Here use is made of nodes (the small circles) and directed branches marked by the appropriate corresponding transfer function which relates the output and input quantities of any two successive nodes. Appearing at the nodes are the variables of the system. The arrows emphasize the unilateral nature of the signal flow. Transmission in block G can occur from E to C and not vice versa. This representation is called a *signal-flow graph*.

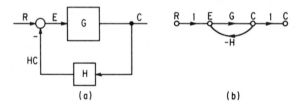

Fig. 20-12. The closed-loop system:
    (a) block-diagram representation;
    (b) signal-flow graph. The feedback function is denoted by H.

The general block diagram of a closed-loop system is shown in Fig. 20-12(a) and the corresponding signal-flow graph in Fig. 20-12(b). The feedback function is denoted by H. In general H, too, can be expressed as an algebraic function of the Laplace transform variable s. From the point of view of the direct function G, the output C still depends upon the input signal E. However, the significant difference in the closed-loop case is that the actuating signal E is dependent upon a portion of the output signal, HC, as well as the reference input, R. That is,

$$E = R - HC \tag{20-3}$$

The feedback signal is opposite in sign from the reference input because, for reasons previously cited, *negative* feedback is desired. Therefore the expression for the output quantity with feedback becomes

$$C = G(R - HC) \tag{20-4}$$

or

$$C = \frac{G}{1 + HG} R \tag{20-5}$$

Moreover, it follows from Eq. (20-5) that the transfer function of the closed-loop system is

$$T \equiv \frac{C}{R} = \frac{G}{1 + HG} \qquad (20\text{-}6)$$

where T denotes the closed-loop transfer function. The quantity HG is called the *loop transfer function* since it is obtained by taking the product of the transfer functions appearing in a traversal of the closed loop. A little thought indicates that a more general way of expressing Eq. (20-6) is to write

$$\text{closed-loop transfer function} = \frac{\text{direct transfer function}}{1 + \text{loop transfer function}} \qquad (20\text{-}7)$$

The manner in which these transfer functions are found, once the system components are known, is treated in Chapter 22.

Effect of Feedback on Sensitivity to Parameter Changes. A particularly noteworthy feature of feedback lies in its ability to reduce significantly the sensitivity of a system's performance to parameter changes. To understand how this is achieved, we define the sensitivity S of the closed loop system T to parameter change p in the direct transmission function G as

$$\text{sensitivity} = \frac{\text{per unit change in closed-loop transmission}}{\text{per unit change in open-loop transmission}} \qquad (20\text{-}8)$$

or, expressed mathematically,

$$S_p^T = \frac{\partial T/T}{\partial G_p/G_p} \qquad (20\text{-}9)$$

where G_p is used to denote that the derivative is to be taken with respect to the parameter p of the function G.

The partial derivative of the closed-loop transfer function with respect to the p parameter of the direct transfer function follows directly from Eq. (20-6)

$$\partial T = \frac{-HG_p \partial G_p}{(1 + HG_p)^2} + \frac{\partial G_p}{1 + HG_p} \qquad (20\text{-}10)$$

or

$$\partial T = \frac{\partial G_p}{(1 + HG_p)^2} \qquad (20\text{-}11)$$

Inserting Eqs. (20-11) and (20-6) into Eq. (20-9) yields

$$S_p^T = \frac{\dfrac{\partial T}{T}}{\dfrac{\partial G_p}{G_p}} = \frac{\dfrac{\partial G_p}{(1 + HG_p)^2} \dfrac{(1 + HG_p)}{G_p}}{\dfrac{\partial G_p}{G_p}} = \frac{1}{1 + HG_p}$$

or

$$\frac{\partial T}{T} = \left(\frac{1}{1 + HG_p}\right)\frac{\partial G_p}{G_p} \qquad (20\text{-}12)$$

A study of the last expression reveals that any per unit (or percentage) change in the direct transmission function brought about by a change in the p parameter carries over as a change in the closed-loop system which is only $1/(1 + HG_p)$ as large. Clearly, by making HG_p a large number, it is possible to reduce changes in G to insignificant proportions in T.

Let us illustrate these ideas by applying the procedure to the grounded-source amplifier shown in Fig. 20-13. The amplifier is enclosed by a broken-line box

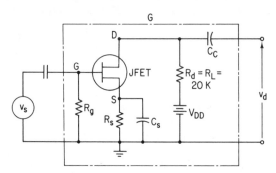

Fig. 20-13. The grounded-source JFET amplifier.

called G to denote that this is the direct transmission function. The input signal in this instance is v_s and the varying component of the corresponding output is v_d. From Eq. (9-174), the output is related to the input by

$$v_d = A_V v_s = \frac{-\mu}{1 + r_d/R_L} v_s \tag{20-13}$$

A comparison with Eq. (20-1) indicates that

$$G = \frac{-\mu}{1 + r_d/R_L} \tag{20-14}$$

Appearing in Fig. 20-14 is the same JFET amplifier of Fig. 20-13, but modified to include a portion of the output in the gate-to-source voltage. The feedback connection is achieved by applying the output quantity across the resistor R_f and then taking a portion, H, of this amount and placing it in series with the signal voltage. The net input to the G portion of the feedback amplifier of Fig. 20-14 thus becomes $v_s + v_f$. The varying component of the output voltage, v_d, is related to the net input by G. Expressed mathematically, we have

$$(v_s + v_f)G = v_d \tag{20-15}$$

It is important to note here that v_f is not preceded with a minus sign (which is often used to denote the negative feedback feature) because G already includes the minus sign owing to the phase-reversal feature of the electronic amplifier. This is evident from Eq. (20-14), which upon insertion into Eq. (20-15) yields

$$(v_s + v_f)\left[\frac{-\mu}{1 + r_d/R_L}\right] = v_d \tag{20-16}$$

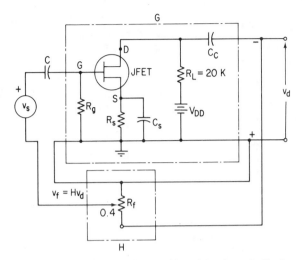

Fig. 20-14. A JFET amplifier provided with voltage feedback.

In the feedback circuit we have

$$v_f = H v_d \qquad (20\text{-}17)$$

Introducing the last expression into Eq. (20-16) and collecting terms leads to

$$v_s\left[\frac{-\mu}{1 + r_d/R_L}\right] = \left[1 + \frac{\mu H}{1 + r_d/R_L}\right]v_d \qquad (20\text{-}18)$$

Finally, upon formulating the ratio of output to input, there results the transfer function of the complete feedback amplifier.

$$T = \frac{v_d}{v_s} = \frac{\dfrac{-\mu}{1 + r_d/R_L}}{1 + \dfrac{\mu H}{1 + r_d/R_L}} \qquad (20\text{-}19)$$

The feedback quantity H is simply a numeric in this case. Equation (20-19) discloses that the feedback amplifier also possesses the phase-reversal feature, i.e., a *positive* change in input signal yields an amplified *negative* change in output signal. Furthermore, it is instructive to observe that the presence of the $+$ sign in the denominator corroborates the use of *negative* feedback.

For ease of illustration, assume that the JFET employed in Fig. 20-14 is the 2N4222 and that for the specified load resistor of 20 kilohms the operating point has the following associated parameters: $\mu = 40$ and $r_d = 20$ kilohms. Then when H is taken to be 0.4, the expression for the gain of the feedback amplifier becomes

$$T = \frac{v_d}{v_s} = \frac{-0.5\mu}{1 + 0.2\mu} \qquad (20\text{-}20)$$

This result can also be obtained directly from the block-diagram representation of Fig. 20-14 (which is depicted in Fig. 20-15) and the use of Eq. (20-6).

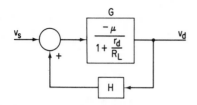

Fig. 20-15. Block diagram representation of the feedback amplifier of Fig. 20-14. Note the use of the plus sign at the output of H. No physical reversal at H is necessary because the phase reversal inherent in the transistor amplifier furnishes the required negative feedback.

The amplification factor μ is left explicit in order that we might now proceed to investigate the effect of changes in μ on the output quantity. The amplification factor can readily change with age as well as vary from transistor to transistor. Variations of the order of 20 per cent or more are not uncommon. When the JFET amplifier is used without feedback, the direct transmission function expressed in terms of μ is simply

$$G = -0.5\mu \qquad (20\text{-}21)$$

Accordingly, if there occurs a 20 per cent drop in μ, the output quantity drops correspondingly by 20 per cent. Thus, the per unit change in the direct transmission of the open-loop arrangement becomes

$$\frac{\partial G}{G} = \frac{-0.5\partial\mu}{-0.5\mu} = \frac{\partial\mu}{\mu} = -0.2 \qquad \text{for a drop} \qquad (20\text{-}22)$$

On the other hand, with feedback the situation is significantly different. The transmission function is now described by Eq. (20-20). Hence, the per unit decrease in the closed-loop case now becomes, by Eq. (20-12)

$$\frac{\partial T}{T} = \frac{1}{1 + 0.5\mu H}\left(\frac{\partial G}{G}\right) = \frac{1}{1 + 0.2\mu}\left(\frac{\partial\mu}{\mu}\right)$$

$$= \frac{1}{1 + 0.2(40)}(-0.2) = -0.022 \qquad (20\text{-}23)$$

Thus, as a feedback amplifier, a 20 per cent change in the μ parameter of the direct transfer function reflects only as a 2.2 per cent decrease in the closed-loop transmission.

The insensitivity to parameter changes associated with negative feedback is not obtained without a price. There always occurs a reduction in gain for a fixed G. This conclusion is immediately apparent upon comparing Eqs. (20-1) and (20-6). For the same G the gain as a closed-loop system is $1/(1 + HG)$ of the value as an open-loop system. For the example at hand the open-loop gain has a value of $G = -0.5\mu = -20$. But as a closed-loop system the over-all gain by Eq. (20-20) is

$$T = \frac{-0.5\mu}{1 + 0.2\mu} = \frac{-20}{+18} = -2.2$$

Clearly, if it is desirable to maintain the same nominal gain of 20 for the closed-loop system between the output and input terminals, it becomes necessary to increase the direct transfer function accordingly. This matter is illustrated by the example that follows.

EXAMPLE 20-1   A voltage amplifier without feedback has a nominal gain of 400. Deviations in the amplifier parameters, however, cause the gain to very in the range from

380 to 420, thus yielding a total per unit change in open-loop transmission of 0.1. It is desirable to introduce feedback to reduce the per unit change to 0.02 while maintaining the original gain of 400. Determine the new direct transfer gain G and the feedback factor H needed to achieve this result.

solution: Let G' be the gain without feedback. Then

$$\frac{\partial G'}{G'} = \frac{420 - 380}{400} = 0.1$$

From Eq. (20-12)

$$\frac{\partial T}{T} = \frac{\partial G}{G}\left(\frac{1}{1 + HG}\right) = 0.02 \tag{20-24}$$

where G denotes the new direct transmission gain with feedback. Also, from Eq. (20-6)

$$T = \frac{G}{1 + HG} = 400 \tag{20-25}$$

From Eq. (20-24)

$$1 + HG = \frac{0.1}{0.02} = 5 \tag{20-26}$$

Inserting this into Eq. (20-25) yields

$$G = 400(1 + HG) = 400(5) = 2000 \tag{20-27}$$

Finally from Eq. (20-26)

$$H = \frac{5 - 1}{2000} = 0.002 \tag{20-28}$$

Effect of Feedback on Dynamic Response and Bandwidth. The effect of feedback on the transient response of a circuit or system can be conveniently described by showing how it modifies the behavior of a system containing one energy-storing element. In this connection assume the direct transfer function† of the system to be given by

$$\frac{C}{E} = G = \frac{K}{1 + s\tau} = \frac{K/\tau}{s + 1/\tau} \tag{20-29}$$

Figure 20-16(a) depicts the open-loop system. The presence of the s term indicates that there exists a time derivative relating the input and output variables. Since the denominator of Eq. (20-29), when equated to zero, is the characteristic equation of the system, it follows from out knowledge of the inverse Laplace transform that the transient solution of the system described by G must be of the form

$$c_t = A\epsilon^{-t/\tau} \tag{20-30}$$

The transient in this system thus decays in accordance with a time constant of τ seconds.

Consider next the case where a feedback path having the transfer function H is wrapped around the function G in the manner illustrated in Fig. 20-16(b).

---

† This transfer function is readily obtained from the differential equation describing the relationship between the input (E) and output (C) variables of the system. It merely requires Laplace transforming the differential equation and then formulating the ratio of output to input variables. See. Chap. 22.

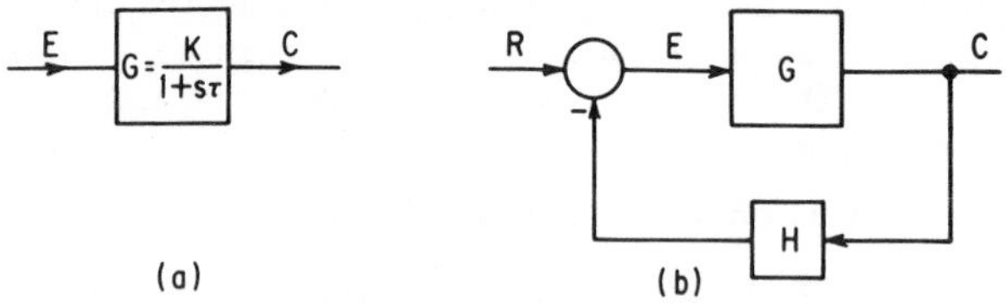

Fig. 20-16. Studying the effect of feedback on dynamic response and bandwidth of the open-loop system shown in (a) and the closed-loop system of (b).

Now the transmission between input and output is represented by the closed-loop transfer function. Hence

$$T = \frac{C}{R} = \frac{\dfrac{K/\tau}{s + 1/\tau}}{1 + \dfrac{HK/\tau}{s + 1/\tau}} = \frac{K/\tau}{s + \dfrac{1 + HK}{\tau}} \tag{20-31}$$

It follows therefore that the corresponding transient solution of the closed-loop system has the form

$$c_{tf} = A_f \epsilon^{-[(1+HK)/\tau]t} \tag{20-32}$$

where c_{tf} denotes the transient response of the output variable with feedback.

A comparison of the last equation with the corresponding equation of the open-loop case, Eq. (20-30), immediately reveals that the time constant with feedback is smaller by the factor $1/(1 + HK)$. Hence the transient decays faster.

The effect of feedback on the bandwidth of the system of Fig. 20-16(a) can also be determined from the foregoing equations. To speak of bandwidth is to speak of a frequency description of the system transfer function. This is readily accomplished by looking upon s as the sinusoidal frequency variable $j\omega$. Proceeding on this basis, then, for the direct transfer function of Eq. (20-29), it follows that the bandwidth spreads over a range from zero to a frequency of $1/\tau$ radians per second. Note that when $s = j\omega = j(1/\tau)$ is inserted in Eq. (20-29) the value of the gain is $0.707K$. On the other hand, for the system employing feedback Eq. (20-31) reveals that the bandwidth spreads from zero to $(1 + HK)/\tau$ radians per second. Thus the bandwidth has been augmented by increasing the upper frequency limit by $(1 + KH)$.

Effect of Feedback on Stability. By the judicious choice of the feedback transfer function H, it is possible to convert an inherently unstable system to a stable one. Appearing in Fig. 20-17(a) is the block-diagram representation of a system that is unstable. The transfer function is

$$G = \frac{C}{E} = \frac{K}{1 - s\tau} = \frac{-K/\tau}{s - 1/\tau} \tag{20-33}$$

To understand why this system is unstable, assume that the input quantity E is

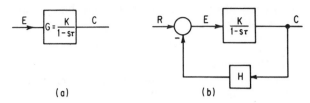

Fig. 20-17. (a) Direct transfer function of an unstable first-order system.
(b) The same system with a feedback path wrapped around it.

a unit-step function. Then the expression for the output becomes

$$C(s) = \frac{-K/\tau}{s - 1/\tau} E(s) = \frac{-K/\tau}{s(s - 1/r)} \qquad (20\text{-}34)$$

By employing the procedure described in Chap. 4, we find the corresponding time solution to be

$$c(t) = K(1 - \epsilon^{t/\tau}) \qquad (20\text{-}35)$$

The second term on the right is the transient component of the solution and, because the exponent has a positive power, it increases with time. Since the response increases without limit as time passes, a constant steady-state output consistent with the steady-state input is never reached. For this reason the system is said to be unstable. A stable system must have negative powers of the exponential terms associated with the transient components.

By placing a feedback path H around the direct transfer function in the manner depicted in Fig. 20-17(b) and applying the feedback formula of Eq. (20-6) we obtain

$$T = \frac{C}{R} = \frac{\dfrac{K}{1 - s\tau}}{1 + \dfrac{KH}{1 - s\tau}} = \frac{K}{1 - s\tau + HK} \qquad (20\text{-}36)$$

The plus sign for HK is used in the denominator because negative feedback is assumed. A comparison of Eqs. (20-33) and (20-35) reveals that the increasing time response is attributable to the presence of the minus sign in the denominator of the direct transfer function [Eq. (20-33)]. We may describe this more elegantly by saying that the direct transfer function has a pole located in the right half-plane at $s = 1/\tau$. Examination of Eq. (20-36), however, indicates that by choosing

$$H = as \qquad (20\text{-}37)$$

the closed-loop transfer function is modified to

$$T = \frac{K}{1 - s\tau + saK} \qquad (20\text{-}38)$$

Then, if we impose the condition that

$$aK > \tau \qquad (20\text{-}39)$$

it follows that the pole of the closed-loop transfer function is located in the left-half s plane. Accordingly, the time response of the closed-loop system, when subjected to a unit-step becomes

$$c_f(t) = K(1 - \epsilon^{-t/(aK-\tau)}) \qquad (20\text{-}40)$$

where $c_f(t)$ denotes the response with feedback. With the constraint of Eq. (20-39) imposed, the exponential term of Eq. (20-40) decays to zero so that the response reaches steady state. The insertion of feedback has thus caused the unstable direct transmission system represented by G to become stable. This technique is frequently used to stabilize missiles, and space rockets, which are inherently unstable because of their large length-to-diameter ratios.

PROBLEMS

20-1 A heavy mass rests on a horizontal slab of steel. The slab is hinge-supported on one side and rests on an adjustable support on the other side. The hinged support is subjected to random vertical displacements. The mass is to be kept level by changing the height of the adjustable support. Devise a feedback control system to achieve this task.

20-2 Devise a control system to regulate the thickness of sheet metal as it passes through a continuous rolling mill (see Fig. P20-2). The control system should not only keep the thickness within suitable tolerances but permit the adjustment of the thickness as well. Show a clearly labeled diagram of the system.

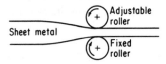

Figure P20-2

20-3 The control system for remotely positioning the rudder of a ship is shown in Fig. P20-3. The resistor coil of the potentiometer is fastened to the frame of the ship. The desired heading is determined by the gyroscope setting (which is independent of the actual heading). Explain how the system operates in following a command for a northbound direction.

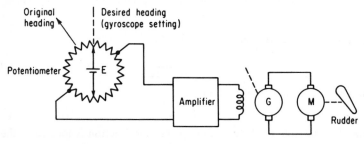

Figure P20-3

20-4   Devise an elementary temperature-control system for a furnace fed by a valve-controlled fuel line. Use a Wheatstone-bridge error detector and a temperature-sensitive resistor such as the thermistor. Assume that the disturbance takes the form of changes in ambient temperature about the furnace. Describe how control is obtained. Also indicate how the desired reference temperature is established.

20-5   The speed of a gasoline engine is to be controlled in accordance with a command that is in the form of a voltage. Devise a control system that is able to provide this kind of control. Describe the operation of your system.

20-6   A medical clinic calls upon you to design a servo anesthetizer for the purpose of regulating automatically the depth of anesthesia in a patient in accordnace with the level of energy output of the brain-wave activity as measured by an electroencephalograph (EEG). The anesthesia is to be administered by means of a hypodermic syringe which is activated by a stepping relay fed from an appropriate amplifying source. Show a diagram of the control system, and state clearly what constitutes the error detector, the feedback, and the corrector. Explain the way the system operates.

20-7   Negotiating a turn in the process of driving an automobile is an example of feedback control. In this instant the eyes, the brain, the arms, and the automobile come into play. Draw a block diagram depicting the interrelationship of these elements and describe the operation.

20-8   Describe in block-diagram form a control system that allows trucks to be weighed as they pass over a platform. The readings are to be recorded at a station remotely located from the platform.

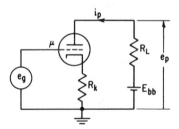

Figure P20-9

20-9   The vacuum-tube triode amplifier circuit depicted in Fig. P20-9 can be analyzed in terms of feedback principles. The control action takes place between the grid and cathode elements where the signal voltage (e_g) and the variable part of the voltage ($i_p R_k$) developed across the cathode resistor are compared.
   (a) Draw a completely labeled block diagram of this amplifier circuit as it applies to varying inputs. The vacuum tube has an amplification factor μ and a plate resistance r_p.
   (b) Compare the gain of this circuit with feedback with the gain without feedback (i.e., $R_k = 0$).
   (c) Find the expression for the sensitivity of this amplifier circuit.

20-10   A vacuum-tube triode is used in a standard grounded-cathode amplifier circuit. The amplification factor is 100 and the dynamic plate resistance is 10 K. The load resistance has a value of 50 K.

(a) What is the a-c voltage gain?

(b) Owing to uncontrollable factors, the plate resistance increases to 20 K. Find the per-unit change in the a-c voltage gain.

(c) The foregoing circuit is modified by tapping off 50 per cent of the output voltage from R_L and placing it into the input circuit in series with the signal voltage with reversed polarity. Compute the voltage gain of the modified circuit.

(d) Determine the per-unit change in the gain of the closed-loop amplifier when r_p changes as specified in part (b).

20-11 Refer to the circuitry of Fig. 20-14.

(a) Draw the block diagram for this feedback amplifier.

(b) Derive the general expression for the voltage gain using the feedback formula of Eq. (20-6). Express the result in terms of the feedback factor H where $0 \leq H \leq 1$.

20-12 Because of aging, the parameters of the FET in a grounded-source amplifier undergo a change that causes a net per-unit decrease in gain of 25 per cent. Several identically constructed amplifier circuits (called stages) having a nominal gain of 80 are available. It is desirable, however, that despite aging an amplifier be designed that yields a gain between the input-output terminals of 80 with the change at no time exceeding 0.1 per cent.

(a) Determine the minimum number of stages required to meet the specifications.

(b) Compute the corresponding feedback factor H.

20-13 It is desirable to have an amplifier with an over-all gain of 1000 ± 20. The gain of any one stage is known to drift from 10 to 20. Find the required number of stages and the feedback function needed to meet these specifications. [*Hint:* Treat the closed-loop system as one that has an initial value of 980 and a final value of 1020. Similarly, consider each stage as having an initial value of 10 and final value of 20.]

20-14 A system has a direct transmission function expressed as

$$G(s) = \frac{10}{s^2 + s - 2}$$

(a) Explain clearly why the system as it stands is unstable.

(b) A feedback path having the transfer function H is put around $G(s)$. Is it possible to make the feedback system stable? Explain. Be specific.

20-15 Repeat Prob. 20-14 for the case where

$$G(s) = \frac{10}{s(s^2 + s - 2)}$$

21

dynamic behavior of
control systems

In the course of the development of the subject matter of this part of the book the dynamic behavior of feedback control systems is analyzed by several methods, each of which has a place and purpose in the over-all picture. A necessary starting point in each is to obtain the differential equation describing the system. The first approach, which constitutes the chief subject matter of this chapter, involves a direct solution of the differential equation by using the procedure outlined in Chap. 4.

Usually the complete solution of the differential equation provides maximum information about the system's dynamic performance. Consequently, whenever it is convenient, an attempt is made to establish this solution first. Unfortunately, however, this is not easily accomplished for high-order systems, and, what is more, the design procedure that permits reasonably direct modification of the transient solution is virtually impossible to achieve. In such cases we are forced to seek out other, easier, more direct methods, such as the frequency-response method of analysis, which is treated at length in the next chapter.

The existence of transients is characteristic of systems that possess energy-storing elements and that are subjected to disturbances. Usually the disturbance is applied at the input or output end or both. Often oscillations are associated with the transients, and, depending upon the magnitude of the system parameters, such oscillations may even be sustained.

21-1 Dynamic Response of the Second-order Servomechanism

The system to be analyzed is shown in Fig. 21-1. It contains two energy-storing elements in the form of inertia and shaft stiffness, making it a second-order system. Here stiffness refers to the torsional elasticity of the servoamplifier-servomotor combination which causes the system to act as though an elastic shaft connected the input to the inertial load. The application of a command in the form of a displacement of the slider arm of the input potentiometer causes an actuating signal e to exist. The amplified version of e is then applied to the servomotor, which in turn develops a torque and causes movement of the load.

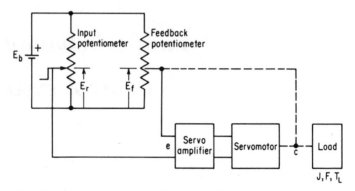

Fig. 21-1. Second-order servomechanism. A direct connection is assumed between the servomotor and the feedback potentiometer as well as the load.

The motor developed torque, T_d, may thus be represented by

$$T_d = eK_aK_m \tag{21-1}$$

where K_a denotes the amplifier gain factor in volts/volt (v/v)

K_m denotes the motor developed torque constant in lb-ft/volt

Moreover, the actuating signal e may be expressed in terms of the potentiometer displacements as

$$e = K_p(r - c) \tag{21-2}$$

where K_p is the potentiometer transducer constant expressed in volts/radian, r is the input command in radians, and c is the output displacement in radians. For simplicity it has been assumed that there in no gear reduction between the output shaft and the load or between the output shaft and the feedback potentiometer. Introducing Eq. (21-2) into Eq. (21-1) allows the motor developed torque to be expressed as

$$T_d = (r - c)K_p K_a K_m = (r - c)K \tag{21-3}$$

where

$$K \equiv K_p K_a K_m \; \frac{\text{lb–ft}}{\text{rad}} \tag{21-4}$$

If, as indicated in Fig. 21-1, it is assumed that there appears at the motor shaft an effective inertia J, an equivalent viscous friction F, and a resultant load torque T_L, it then follows that the developed motor torque must be equal to the sum of the opposing torques represented by these quantities. Thus

$$T_d = \sum \text{opposing torques} \tag{21-5}$$

or

$$(r - c)K = J\frac{d^2c}{dt^2} + F\frac{dc}{dt} + T_L \tag{21-6}$$

The J parameter accounts for the inertia of the load itself as well as the inertia of the rotating member of the servomotor. The viscous friction parameter F is very often determined to a large extent by the drooping slope of the torque-speed characteristic of the servomotor.

Rearranging Eq. (21-6) leads to the preferred form of expressing the defining differential equation for obtaining the dynamic behavior of the system. Thus

$$\boxed{J\frac{d^2c}{dt^2} + F\frac{dc}{dt} + Kc + T_L = Kr} \tag{21-7}$$

Step-position Input. It is assumed that the command r applied to the system described by Eq. (21-7) is a step of magnitude r_0 radians. Furthermore, it is assumed that the system is resting at null prior to the application of the command. Hence the displacement and velocity immediately after the input is applied must be zero, because no movement of the output member can occur in infinitesimal time in the presence of a finite forcing function. The two required initial conditions for the solution of Eq. (21-7) may thus be written as

$$c(0^+) = 0 \quad \text{and} \quad \frac{dc}{dt}(0^+) = \dot{c}(0^+) = 0 \tag{21-8}$$

By Laplace transforming Eq. (21-7), it becomes

$$J[s^2 C(s) - sc(0^-) + c(0^-)] + F[sC(s) - c(0^-)]$$
$$+ KC(s) + T_L(s) = KR(s) \tag{21-9}$$

where $C(s) = \mathscr{L}c(t)$ and $c(t)$ denotes the desired time solution of the controlled variable. Moreover, $T_L(s)$ and $R(s)$ represent the Laplace transforms of the disturbances applied at the output and input ends of the servomechanism. When a step change in load of magnitude T_L and a step command of magnitude r_0 are assumed applied simultaneously for the initial conditions of Eq. (21-8), the preceding equation becomes

$$C(s)[s^2 J + sF + K] = \frac{Kr_0}{s} - \frac{T_L}{s} \tag{21-10}$$

where $R(s) = r_0/s$ and $T_L(s) = T_L/s$. Thus the solution for the controlled variable in the s domain is

$$C(s) = \frac{(K/J)r_0}{s\left(s^2 + s\dfrac{F}{J} + \dfrac{K}{J}\right)} - \frac{T_L/J}{s\left(s^2 + s\dfrac{F}{J} + \dfrac{K}{J}\right)} \tag{21-11}$$

It is interesting to note in passing that, with the Laplace transform method of solving the system differential equation, the solution for two disturbances is found with little more effort than that required for one.

Even at this early point in the solution some useful information may be obtained from an inspection of Eq. (21-11). For example, if T_L is zero, then at steady state $c_{ss} = r_0$ so that exact correspondence exists between the command and the controlled variable. On the other hand, in the presence of a fixed load torque the steady-state value of the controlled variable becomes

$$c_{ss} = r_0 - \frac{T_L}{K} \tag{21-12}$$

Therefore an error in position exists which is directly proportional to the load torque and inversely proportional to the gain K. It will be recalled that these are the same conclusions arrived at in Sec. 20-3 through a process of physical reasoning. Another notable observation is that the denominators of both terms on the right side of Eq. (21-11) involve the same quadratic expression. In fact this quadratic is precisely the characteristic equation of the system which follows from Eq. (21-7) when the external disturbances are set equal to zero. Consequently the source-free modes are of the same character whether the disturbance with which they are associated originates at the input or at the output end.

To determine the time-domain solution corresponding to the s-plane solution of Eq. (21-11), we must find the roots of the characteristic equation. Because of the frequent occurrence of such quadratic terms in control systems work, particular attention is directed to this case with a view toward generalizing the results as was done for the second-order R-L-C circuit in Sec. 5-3. Starting with the characteristic equation as it appears in Eq. (21-11), we have

$$s^2 + \frac{F}{J}s + \frac{K}{J} = 0 \tag{21-13}$$

The roots are

$$s_{1,2} = -\frac{F}{2J} \pm \sqrt{\left(\frac{F}{2J}\right)^2 - \frac{K}{J}} \tag{21-14}$$

Depending on the radical term, the response can be any one of the following: (1) overdamped if $(F/2J)^2 > K/J$; (2) underdamped if $(F/2J)^2 < K/J$; (3) critically damped if $(F/2J)^2 = K/J$. Often a desirable feature of a servomechanism is that it be fast-acting. Unless it is otherwise specified, therefore, we shall confine our attention to the underdamped case because it responds by moving quickly to its new commanded state in spite of the fact that it may then oscillate about the new level before settling down.

Critical damping occurs whenever the damping term (in this case F) is related to the gain and inertia parameters by

$$F_c = 2\sqrt{KJ} \qquad (21\text{-}15)$$

For the underdamped situation the roots may be written as

$$s_{1,2} = -\frac{F}{2J} \pm j\sqrt{\frac{K}{J} - \left(\frac{F}{2J}\right)^2} \qquad (21\text{-}16)$$

To put these roots in a form that will give greater significance in terms of the resulting time response, we introduce the following definition for the damping ratio:

$$\zeta \equiv \frac{\text{total damping}}{\text{critical damping}} = \frac{F}{2\sqrt{KJ}} \qquad (21\text{-}17)$$

and the following definition for the system natural frequency:

$$\omega_n \equiv \sqrt{\frac{K}{J}} \qquad (21\text{-}18)$$

Accordingly

$$\frac{F}{2J} = \frac{\zeta F_c}{2J} = \frac{2\zeta\sqrt{KJ}}{2J} = \zeta\omega_n \qquad (21\text{-}19)$$

and

$$\sqrt{\frac{K}{J} - \left(\frac{F}{2J}\right)^2} = \sqrt{\omega_n^2 - \zeta^2\omega_n^2} = \omega_n\sqrt{1 - \zeta^2} \qquad (21\text{-}20)$$

Therefore Eq. (21-16) may be written as

$$s_{1,2} = -\zeta\omega_n \pm j\omega_n\sqrt{1 - \zeta^2} = -\zeta\omega_n \pm j\omega_d \qquad (21\text{-}21)$$

where

$$\omega_d \equiv \omega_n\sqrt{1 - \zeta^2} = \text{damped frequency of oscillation} \qquad (21\text{-}22)$$

A distinguishing feature of Eq. (21-21) is that both the real and j parts have the unit of inverse seconds, which is frequency. This in fact gives rise to the term *complex frequency*. The practical interpretation is that the real part of the complex frequency represents the damping factor associated with the decaying transients, while the imaginary (or j part) refers to the actual frequency of oscillation at which the underdamped response decays.

In terms of the quantities ζ and ω_n Eq. (21-11) can be rewritten

$$C(s) = \frac{\omega_n^2 r_0}{s(s^2 + 2\zeta\omega_n s + \omega_n^2)} - \frac{T_L/J}{s(s^2 + 2\zeta\omega_n s + \omega_n^2)} \qquad (21\text{-}23)$$

To determine the time solution corresponding to Eq. (21-23) we need a partial fraction expansion. First, however, we shall set $T_L = 0$ in order to simplify the algebraic manipulations. Nothing is thereby lost because the forms of the two terms on the right side of the last equation are similar. Thus the equation to be solved is reduced to

$$C(s) = \frac{\omega_n^2 r_0}{s(s^2 + 2\zeta\omega_n s + \omega_n^2)} \tag{21-24}$$

As a consequence of the simplicity of this expression, a formal partial fraction expansion is unnecessary. A little thought discloses that the right side of Eq. (21-24) may be rearranged as

$$C(s) = \frac{r_0}{s} - \frac{(s + 2\zeta\omega_n)r_0}{s^2 + 2\zeta\omega_n s + \omega_n^2} \tag{21-25}$$

Moreover, in the interest of putting Eq. (21-25) in a form each term of which is immediately identifiable in Table 4-1, it can be written as

$$C(s) = \frac{r_0}{s} - r_0 \left[\frac{(s + \zeta\omega_n)}{(s + \zeta\omega_n)^2 + \omega_d^2} + \zeta \frac{\omega_n}{\omega_d} \frac{\omega_d}{(s + \zeta\omega_n)^2 + \omega_d^2} \right] \tag{21-26}$$

The first term in brackets is readily seen to be a damped cosine function and the second term is the damped sine function. Accordingly, the time solution corresponding to Eq. (21-26) is

$$c(t) = r_0 - r_0 \epsilon^{-\zeta\omega_n t} \left(\cos \omega_d t + \frac{\zeta}{\sqrt{1 - \zeta^2}} \sin \omega_d t \right) \tag{21-27}$$

This last equation can be further simplified as outlined in Sec. 5-4 to

$$c(t) = r_0 \left[1 - \frac{\epsilon^{-\zeta\omega_n t}}{\sqrt{1 - \zeta^2}} \sin (\omega_d t + \theta) \right] \tag{21-28}$$

where

$$\theta = \tan^{-1} \frac{\sqrt{1 - \zeta^2}}{\zeta} \tag{21-29}$$

and is valid for a step input, $0 \leq \zeta \leq 1$ and $T_L = 0$.

Undamped Solution. One extreme condition of Eq. (21-28) occurs when $\zeta = 0$. This makes $\theta = 90°$ so that

$$c(t) = r_0[1 - \sin (\omega_d t + 90°)] = r_0(1 - \cos \omega_n t) \tag{21-30}$$

The solution for the controlled variable is therefore a sustained oscillation. This result is entirely expected because the system has two energy-storing elements and zero damping (see Fig. 21-2). By Eq. (21-22) the frequency of oscillation is

$$\omega_d = \omega_n = \sqrt{\frac{K}{J}} \tag{21-31}$$

which is the natural frequency of the system. It depends upon the two energy-storing elements K (stiffness) and J (inertia).

Universal Curves for Second-order Systems. It is useful to depict the response of second-order systems graphically for specific values of damping ratios independently of the magnitude of the input step command or the system parameters.

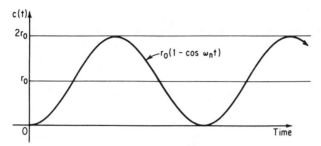

Fig. 21-2. Solution for the controlled variable in the servomechanism of Fig. 21-1 when the damping is zero and the command is a step of magnitude r_0.

The response of Eq. (21-28) can readily be made independent of the magnitude of the step input by plotting $c(t)/r_0$ rather than $c(t)$. Thus the solution is normalized with respect to the value of the step input. This means that for any given system with a fixed damping ratio a single curve may be used to represent the response irrespective of the magnitude of the input signal r_0.[†]

A germane question at this point, however, is: Can this same curve be used for a different second-order system having the same ζ but another value of ω_n? A little thought makes it clear that this cannot be done because of the role played by $\omega_n t$ in the exponential and in the argument of the sine term in Eq. (21-28). In order to use a single response curve so that it applies for all linear second-order systems, it is necessary to plot time in a nondimensional form which masks the effect of the natural frequency parameter. Thus, if the magnitude-normalized response is plotted not as a function of time t directly but rather as a function of $\omega_n t$, then the resulting plot is truly universal, since it applies to all linear second-order systems. These plots are shown in Fig. 21-3 for various values of ζ. They describe the response to step input commands.

Figures of Merit Identifying the Transient Response. As was pointed out in the analysis of the *R-L-C* circuit in Sec. 5-3, the dynamic behavior of any linear second-order system is readily described in terms of two figures of merit: ζ and ω_n. To understand what each quantity conveys about the transient response, let us determine the time at which the maximum overshoot occurs. Differentiating Eq. (21-28) with respect to time yields upon simplification

$$\frac{d}{dt} c(t) = \frac{\omega_n}{\sqrt{1 - \zeta^2}} \, \epsilon^{-\zeta \omega_n t} \sin \omega_d t \tag{21-32}$$

Clearly, then, the first and, therefore, the peak overshoot occurs when

$$\omega_d t = \pi \tag{21-33}$$

or

$$\omega_n t = \frac{\pi}{\sqrt{1 - \zeta^2}} \tag{21-34}$$

[†] It is assumed of course that saturation effects do not exist. In any practical system this is a real restriction.

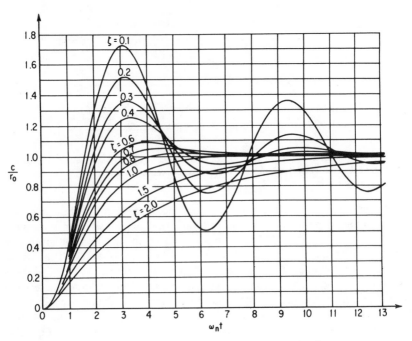

Fig. 21-3. Universal transient response curves for second-order systems subjected to a step input.

If this result is put into Eq. (21-28), the maximum instantaneous value of the controlled variable is found to be

$$\frac{c_{\max}}{r_0} = 1 + \epsilon^{-\zeta\pi/\sqrt{1-\zeta^2}} \tag{21-35}$$

Hence the maximum per cent overshoot becomes

$$\boxed{\frac{c_{\max} - c_{ss}}{c_{ss}} = \frac{c_{\max} - r_0}{r_0} = 100\epsilon^{-\zeta\pi/\sqrt{1-\zeta^2}} \text{ per cent}} \tag{21-36}$$

where c_{ss} is the steady-state value of the output. Figure 21-4 shows the plot of Eq. (21-36). It is important to note that the maximum overshoot depends solely upon the value of ζ. Therefore, the damping ratio is looked upon as a figure of merit which provides information about the maximum overshoot in the system when it is excited by a step forcing function. This is a common and very useful way of obtaining information about the dynamic behavior of systems.

The information conveyed by the value of the system's natural frequency ω_n can be deduced from an examination of the output response to a step input plotted nondimensionally for a given ζ. Such a plot is shown in Fig. 21-5. Note the inclusion of a ± 5 per cent tolerance band. A system can often be considered as having reached steady state once the response stays confined within the limits

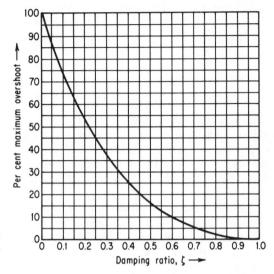

Fig. 21-4. Per cent maximum overshoot versus damping ratio for linear second-order system.

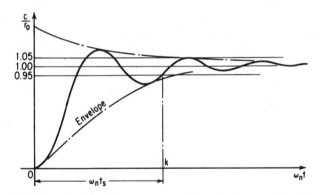

Fig. 21-5. Setting time of a step response identified in terms of a ± 5 per cent tolerance band.

of this band. The time it takes do this is called the *settling time*. As a practical matter the response time of all systems is necessarily gauged in terms of a suitable tolerance band, however large or small the particular specifications may demand. For the situation illustrated in Fig. 21-5 it follows that the settling time is $t_s = k/\omega_n$. The important fact to recognize here is that the settling time is inversely proportional to the natural frequency. Thus, if the response curve shown in Fig. 21-5 applies to two different systems having the same ζ, then clearly the one with the larger natural frequency will have the smaller settling time in responding to input commands or load disturbances. Consequently, the natural frequency of a system may be interpreted as a figure of merit which provides a measure of the settling time.

Another way to arrive at this conclusion is to refer to Eq. (21-28) and to note that the magnitude of the transient term is controlled by the value of the exponential $\epsilon^{-\zeta\omega_n t}$. It is possible to speak of a time constant for the second-order system:

$$T = \frac{1}{\zeta\omega_n} \tag{21-37}$$

Equation (21-37) reveals too that for a fixed ζ, the time constant is inversely proportional to ω_n. Expressed in terms of the time constant, it can then be said that the response to a step input will reach 95 per cent of its steady-state value within three time constants. That is, the settling time is no greater than $t_s = 3/\zeta\omega_n$. For a 1 per cent tolerance band the settling time will be no greater than $t_s = 5/\zeta\omega_n$.

Transfer Function Analysis. For the sake of completeness and in the interest of applying the results described in Sec. 20-4, we shall next analyze the closed-loop response of the system of Fig. 21-1 by developing the appropriate block diagram and then applying Eq. (20-6). The general procedure in finding the transfer function of a component is to identify first the equation that relates the output variable to the input variable. Then in those instances where time derivatives and time integrals are involved, the next step is to Laplace transform the defining integrodifferential equation assuming all initial conditions are zero. This assumption is made because in determining the transfer function we are looking to identify the characteristic equation, which depends solely upon the parameters and which establishes the nature of the dynamic response. Recall that the initial conditions affect only the magnitudes of the transient terms but not their basic decaying characteristics. Finally, the transfer function is obtained by formulating the ratio of the output to input variable.

Let us apply this procedure to each of the components that make up the system of Fig. 21-1. The input command takes the form of a displacement of r radians applied to the slider arm of the input potentiometer. In turn this makes available at the electrical wire attached to the slider arm of the potentiometer a new level of voltage. For a given r displacement the amount of voltage change depends, of course, upon how much excitation voltage E_b is used. It should be apparent, then, that the potentiometer is a transducer that converts a mechanical input variable r expressed in radians to a corresponding output voltage expressed in volts. Accordingly we can say that the potentiometer has a transfer function

$$\frac{E_r}{R} = K_p \qquad \text{volts/rad} \tag{21-38}$$

where E_r is the output voltage at the slider arm corresponding to displacement R. Capital letters are used to stress that we are dealing with the Laplace transform versions of the variables. The potentiometer in the feedback path has the same transfer constant except that the output and input quantities are called E_f and C, respectively. Figure 21-6 shows both potentiometer transfer constants at the proper places in the block diagram.

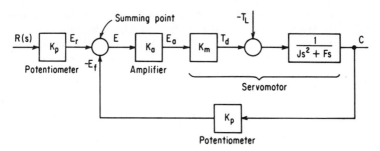

Fig. 21-6. Block diagram of the servomechanism of Fig. 21-1, showing the transfer function of each component.

Continuing with the development of the block diagram, note in Fig. 21-1 that the electrical connections to the servoamplifier are such that its input quantity E is obtained as the difference between E_r and E_f. This is represented in the block diagram of Fig. 21-6 by the circle, which is the symbol for a summing point. The transfer function of the amplifier in this case is merely a gain factor which raises the level of the input voltage E to E_a. Thus

$$\frac{E_a}{E} = K_a \qquad \text{volts/volt} \tag{21-39}$$

If the application of a voltage E_a to the input terminals of the servomotor is assumed to develop an output which is the developed torque T_d, the gain factor associated with the motor is clearly

$$\frac{T_d}{E_a} = K_m \qquad \frac{\text{lb–ft}}{\text{volt}} \tag{21-40}$$

Finally the manner in which this developed torque manifests itself as a displacement angle c is described by the differential equation relating these two variables, namely

$$T_d = J\frac{d^2c}{dt^2} + F\frac{dc}{dt} + T_L \tag{21-41}$$

The presence of a load disturbance may be conveniently treated by placing it on the left side in Eq. (21-41).

$$(T_d - T_L) = J\frac{d^2c}{dt^2} + F\frac{dc}{dt} \tag{21-42}$$

Again a small circle representing the summing operation called for on the left side of the last equation is used in the block-diagram notation. Laplace transforming Eq. (21-42) for zero initial conditions leads to

$$T_d(s) - T_L(s) = Js^2C(s) + FsC(s) \tag{21-43}$$

Formulating the ratio of output to input yields the transfer function, which describes how torque is carried over to output displacement. Thus,

$$\frac{C(s)}{T_d(s) - T_L(s)} = \frac{1}{Js^2 + Fs} \tag{21-44}$$

By interconnecting the output of one component to the input of the following component in the system, we arrive at the complete block diagram of Fig. 21-6. This block diagram can be simplified by observing that, since the actuating signal is given by $E = K_p(R - C)$, the potentiometer transfer factor may be removed from the input and feedback paths and placed instead in the direct transmission path as shown in Fig. 21-7. Also, for simplicity, T_L is assumed to be zero.

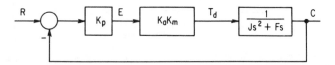

Fig. 21-7. Simplified form of Fig. 21-6 with the load torque assumed to be zero.

The direct transmission function for the system of Fig. 21-7 can clearly be identified as

$$G(s) = \frac{K_p K_a K_m}{Js^2 + Fs} = \frac{K}{Js^2 + Fs} \qquad (21\text{-}45)$$

By Eq. (20-2) the corresponding closed-loop transfer function is then

$$T(s) = \frac{C(s)}{R(s)} = \frac{G(s)}{1 + HG(s)} = \frac{\dfrac{K}{Js^2 + Fs}}{1 + \dfrac{K}{Js^2 + Fs}} = \frac{K}{Js^2 + Fs + K}$$

or

$$\frac{C(s)}{R(s)} = \frac{\dfrac{K}{J}}{s^2 + \dfrac{F}{J}s + \dfrac{K}{J}} \qquad (21\text{-}46)$$

It is important to note that the output variable as well as the input command are represented in terms of their Laplace transforms. To stress the point capital letters are employed along with the "of s" notation; i.e., $C(s)$ and $R(s)$. This procedure is entirely consistent with the fact that the transfer function often involves operations of time, which in the Laplace transform language means that the transfer function is an algebraic function of s. A glance at the right side of Eq. (21-46) corroborates this for the case at hand. Another notable observation in Eq. (21-46) is that the denominator expression of the closed-loop transfer function, when equated to zero, yields the characteristic equation of the closed-loop system. Thus it is that the transfer function is the means through which information about the dynamic behavior of systems can be readily and conveniently accounted for in the process of analysis. It should also be apparent that the use of the Laplace transformation not only lends great facility to the algebraic manipulations involved in arriving at solutions but also makes available a convenient procedure of converting from the s domain to the time domain and vice versa.

Once the transfer function of a system is determined and put in a form similar to that of Eq. (21-46), the transformed solution can be written down forthwith.

For the system described by Eq. (21-46) this becomes

$$C(s) = R(s)\frac{\dfrac{K}{J}}{s^2 + \dfrac{F}{J}s + \dfrac{K}{J}} = R(s)\frac{\omega_n^2}{s^2 + 2\zeta\omega_n s + \omega_n^2} \qquad (21\text{-}47)$$

The particular form of the forced solution depends upon the type of forcing function used. If a step input is applied of magnitude r_0, then $R(s) = r_0/s$ so that the complete transformed solution becomes

$$C(s) = \left(\frac{r_0}{s}\right)\frac{\omega_n^2}{s^2 + 2\zeta\omega_n s + \omega_n^2} \qquad (21\text{-}48)$$

This is the same result shown in Eq. (21-24), which was found directly from the differential equation describing the closed-loop operation. Inspection of Eq. (21-48) reveals that the transient terms resulting from a partial fraction expansion are associated with the poles of the denominator of the closed-loop transfer function. On the other hand the steady-state solution is generated by the pole associated with $R(s)$.

EXAMPLE 21-1   A feedback control system has the configuration shown in Fig. 21-8. The relationship between the output variable c and the input to the direct transmission path e is

$$\frac{d^2c}{dt^2} + 8\frac{dc}{dt} + 12c = 68e \qquad (21\text{-}49)$$

(a) When the system is operated open-loop (i.e., the feedback path removed), find the complete solution for the output variable for $e = u(t)$. Assume the system is initially at rest.

(b) Describe the nature of the dynamic response

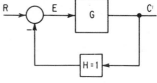

Fig. 21-8. Block diagram for Example 21-1.

of the controlled variable when the feedback loop is connected and the forcing function is again a unit step, i.e., $R(s) = 1/s$. Find the solution by working with the differential equation for the closed-loop system.

(c) Identify the closed-loop behavior by working solely in terms of transfer functions.

solution: (a) The complete solution of the open-loop response is the solution to Eq. (21-49). If we Laplace transform this equation, we obtain for zero initial conditions

$$s^2C(s) + 8sC(s) + 12C(s) = \frac{68}{s} \qquad (21\text{-}50)$$

Hence the transformed solution is

$$C(s) = \frac{68}{s(s^2 + 8s + 12)} = \frac{68}{s(s + 2)(s + 6)} = \frac{K_0}{s} + \frac{K_1}{s + 2} + \frac{K_2}{s + 6} \qquad (21\text{-}51)$$

Evaluating the coefficients of the partial fraction expansion in the usual fashion yields

$$C(s) = \frac{17}{3}\left(\frac{1}{s}\right) - \frac{17}{2}\left(\frac{1}{s + 2}\right) + \frac{17}{6}\left(\frac{1}{s + 6}\right) \qquad (21\text{-}52)$$

The corresponding time solution is then

$$c(t) = \frac{17}{3} - \frac{17}{2}\epsilon^{-2t} + \frac{17}{6}\epsilon^{-6t} \qquad (21\text{-}53)$$

Equation (21-53) indicates that the transient response of this open-loop system takes the form of two exponentially decaying terms—one having a time constant of $\frac{1}{2}$ second and the other a time constant of $\frac{1}{6}$ second. This general information could be deduced directly from Eq. (21-51) once the root factors of the denominator are known.

(b) The essential difference between the differential equation describing open-loop operation and that describing closed-loop operation is that in the latter the variable e is replaced by $(r - c)$. Thus

$$\frac{d^2c}{dt^2} + 8\frac{dc}{dt} + 12c = 68(r - c) \tag{21-54}$$

Now the differential equation describing the relationship between the output and input variables of the closed-loop system takes the form

$$\frac{d^2c}{dt^2} + 8\frac{dc}{dt} + 80c = 68r \tag{21-55}$$

To describe the nature of the dynamic response of the closed-loop system as represented by Eq. (21-55) it is no longer necessary to obtain a formal solution for $c(t)$. Instead, use is made of the information conveyed by the two figures of merit ζ and ω_n. Both these quantities follow almost immediately from the characteristic equation of the closed-loop system, which for this case is

$$s^2 + 8s + 80 = 0 \tag{21-56}$$

Of course Eq. (21-56) comes from Eq. (21-55) when the input is set equal to zero and the left side is Laplace transformed with zero initial conditions. The general form of this equation, as it applies to all linear second-order systems, is

$$s^2 + 2\zeta\omega_n s + \omega_n^2 = 0 \tag{21-57}$$

Comparing coefficients leads to

$$\omega_n = \sqrt{80} = 8.95 \text{ rad/sec} \tag{21-58}$$

and

$$\zeta = \frac{8}{2\omega_n} = \frac{4}{8.95} = 0.448 \tag{21-59}$$

From Fig. 21-4 we find the maximum overshoot to be 17 per cent. Also, since $\zeta\omega_n = 4$, it follows that the controlled variable reaches within 1 per cent of its steady-state value after the elapse of

$$t_s = \frac{5}{\zeta\omega_n} = \frac{5}{4} = 1.25 \text{ seconds} \tag{21-60}$$

(c) The transfer function of the direct transmission path follows from Eq. (21-49). Thus

$$\frac{C(s)}{E(s)} = G(s) = \frac{68}{s^2 + 8s + 12} \tag{21-61}$$

The closed-loop transfer function then becomes

$$T(s) = \frac{C(s)}{R(s)} = \frac{G(s)}{1 + HG(s)} = \frac{\dfrac{68}{s^2 + 8s + 12}}{1 + \dfrac{68}{s^2 + 8s + 12}} \tag{21-62}$$

or

$$\frac{C(s)}{R(s)} = \frac{68}{s^2 + 8s + 80} \tag{21-63}$$

Because the denominator expression of Eq. (21-63) is identical to that of Eq. (21-56), the damping ratio and the natural frequency have the same values as found in part (b). Hence the dynamic behavior can again be described as oscillatory, having a maximum overshoot of 17 per cent, and requiring 1.25 seconds to reach within 1 per cent of steady state.

21-2 Error-rate Control

To meet the requirements for the steady-state performance as well as the dynamic performance of the feedback control system of Fig. 21-1, it is necessary to provide independent control of both performances. In the system of Fig. 21-1 the viscous friction coefficient F is associated with the speed-torque characteristic of the output actuator and is generally not adjustable. Furthermore, the inertia term is kept as small as is physically possible in the interest of achieving a large natural frequency. Therefore the sole adjustable parameter as the system now stands is the loop gain K. This is not sufficient to yield satisfactory steady-state and dynamic performance. The reason is apparent from an examination of Eqs. (21-12) and (21-17). If a command r_0 is applied to the control system in the presence of an external torque, Eq. (21-12) indicates that in the steady state the output displacement will lag behind the command by the amount T_L/K. Clearly, to diminish this position lag error to tolerable limits, it becomes necessary to use high values of the loop gain K. In such circumstances Eq. (21-17) shows that the dynamic behavior may significantly deteriorate because of the inverse relationship between ζ and K. Therefore, if K is to be adjusted to meet accuracy requirements in the steady state, another parameter—the damping coefficient—must be made adjustable to furnish acceptable dynamic behavior at the same time.

The need to control the accuracy performance in steady state is often present even in cases where there are no external disturbances to speak of, but where very high accuracy is to be achieved. Under such circumstances the imperfections of the system components loom large—such as coulomb friction, bearing friction, and dead-zone effects in potentiometers, servomotors, and the like. The use of high loop gains can greatly minimize these effects.

Although there are several methods that can be used to furnish the control being sought, only two are truly practical and have thereby gained widespread acceptance. The first of these makes use of error-rate damping to control the dynamic response, while the second employs output-rate damping and is discussed in the next section.

A system possesses error-rate damping when the generation of the output in some way depends upon the rate of change of the actuating signal. For the system of Fig. 21-1 a convenient way of introducing error-rate damping is to design the amplifier so that it provides an output signal containing a term proportional to the derivative of the input as well as one proportional to the input itself. Thus, if the input voltage to the amplifier is $e = K_p(r - c)$, the output voltage of the amplifier e_a will then be

$$e_a = \left(K_a + K_e \frac{d}{dt}\right)e = K_a e + K_e \frac{de}{dt} \tag{21-64}$$

where K_e denotes the error-rate gain factor of the amplifier. The developed torque of the servomotor is found by multiplying Eq. (21-64) by K_m, the motor torque constant. Thus the differential equation for the system becomes

$$K_a K_m e + K_e K_m \frac{de}{dt} = J\frac{d^2c}{dt^2} + F\frac{dc}{dt} + T_L \tag{21-65}$$

If we next insert $e = K_p(r - c)$ and rearrange terms, Eq. (21-65) may be expressed as

$$\boxed{J\frac{d^2c}{dt^2} + (F + Q_e)\frac{dc}{dt} + Kc = Kr + Q_e\frac{dr}{dt} - T_L} \tag{21-66}$$

where

$$K = \text{loop proportional gain factor} = K_p K_a K_m \tag{21-67}$$

$$Q_e = \text{loop error-rate gain factor} = K_p K_e K_m \tag{21-68}$$

Equation (21-66) is the governing differential equation for the system when error-rate damping is included. A comparison with Eq. (21-7), which applies to the linear second-order system without error-rate damping, reveals that the significant difference lies in the makeup of the coefficient of dc/dt. For the system with error-rate control this coefficient is augmented by the amount Q_e and is separately adjustable by K_e. Accordingly the characteristic equation is modified from that of Eq. (21-13) to the following:

$$s^2 + \frac{F + Q_e}{J}s + \frac{K}{J} = 0 \tag{21-69}$$

yielding the roots

$$s_{1,2} = -\frac{F + Q_e}{2J} \pm j\sqrt{\frac{K}{J} - \left(\frac{F + Q_e}{2J}\right)^2} \tag{21-70}$$

for the case of underdamping. From the last expression the critical viscous friction is found to be

$$(F + Q_e)_c = 2\sqrt{KJ} \tag{21-71}$$

which shows that, whether the system is just viscously damped or viscously damped plus error-rate damped, the critical value of the damping coefficient continues to be dependent upon the loop proportional gain K and the system inertia J. Moreover, since the damping ratio of a second-order system is always expressed as the ratio of the total damping coefficient to the critical damping coefficient, the expression for the damping ratio in the viscous plus error-rate case is

$$\boxed{\zeta = \frac{F + Q_e}{2\sqrt{KJ}}} \tag{21-72}$$

A comparison with Eq. (21-17) shows that now the damping ratio can be independently adjusted through Q_e. Of course, the expression for the natural frequency remains as before: $\omega_n = \sqrt{K/J}$. This is obvious from Eq. (21-69).

The steady-state solution for a step input r_0 in the presence of a load torque T_L is seen from Eq. (21-66) to be the same whether or not error-rate damping is present. Hence the use of high loop gains can effect a closer correspondence between the actual and commanded values of the controlled variable. However, the advantage offered by the error-rate system is that this can be achieved without deteriorating the transient response because of the influence of the Q_e term in establishing the damping ratio [see Eq. (21-72)]. Herein lies the strength of error-rate damping—*it allows higher gains to be used without adversely affecting the damping ratio and in this manner makes it possible to satisfy the specifications for the damping ratio as well as for the steady-state performance.* It offers the additional advantage of increasing the system's natural frequency, which in turn means smaller settling times.

The viscous plus error-rate damped system can also be analyzed through use of the block diagram and the appropriate transfer functions of the system components. For the system that includes error-rate, the only modification required concerns the expression for the transfer function of the servoamplifier. Instead of just the gain K_a, Eq. (21-64) indicates that now the transfer function is

$$\frac{E_a(s)}{E(s)} = K_a + sK_e \qquad (21\text{-}73)$$

The complete block diagram is depicted in Fig. 21-9.

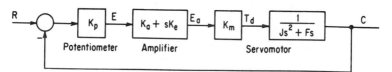

Fig. 21-9. Block diagram of a viscous plus error-rate damped system.

The direct transmission function in this case takes the form

$$G(s) = \frac{K_p K_m (K_a + sK_e)}{Js^2 + Fs} \qquad (21\text{-}74)$$

This yields the closed-loop transfer function

$$\frac{C(s)}{R(s)} = T(s) = \frac{G(s)}{1 + HG(s)} = \frac{K + sQ_e}{Js^2 + (F + Q_e)s + K} \qquad (21\text{-}75\dagger)$$

where K and Q_e are defined by Eqs. (21-67) and (21-68). Note again the recurrence of the proper form of the characteristic equation in the denominator term of the closed-loop transfer function.

† Equation (21-75) also shows an additional numerator term when compared with Eq. (21-46), which applies for a system with only viscous damping. Although at first glance it seems that the additional terms appearing in the solution for the controlled variable may result in a poor dynamic response, this is in fact not the case. It is shown in the next chapter that the entire numerator term serves to increase stability by furnishing greater phase margin.

21-3   Output-rate Control

A system is said to have output-rate damping when the generation of the output quantity in some way is made to depend upon the rate at which the controlled variable is changing. Its introduction often involves the creation of an auxiliary loop, making the system multiloop. For the servomechanism group of feedback control systems a common way of obtaining output-rate damping is by means of a tachometer generator, driven from the servomotor shaft. In order to illustrate the general results and to demonstrate the method of analysis, this study is developed with a particular system in mind, namely, that of a second-order servomechanism employing tachometric feedback damping. Although most of the derived results apply specifically to this system, it should be understood that the procedures and conclusions apply to feedback control systems in general, irrespective of the particular system composition. Moreover, the system performance will again be analyzed by the differential equation method and the transfer function approach. Attention is first directed to the former method.

Let the output-rate signal be denoted by $K_0(dc/dt)$ where K_0 is the output-rate gain factor and is expressed in volts per radian per second. This signal with reversed polarity is combined with the signal e from the error detector to make up the input signal to the servoamplifier. Thus

$$\text{servoamplifier input} = e - K_0\frac{dc}{dt} = K_p(r - c) - K_0\frac{dc}{dt} \tag{21-76}$$

It is important to subtract these signals, otherwise the output-rate term will deteriorate rather than improve the dynamic response as will be shown presently. If the servoamplifier input signal is then multiplied by the amplifier gain K_a and the motor torque constant K_m, the expression for the developed torque is obtained. Upon equating this to the opposing torques, we obtain

$$\left(e - K_0\frac{dc}{dt}\right)K_aK_m = J\frac{d^2c}{dt^2} + F\frac{dc}{dt} + T_L \tag{21-77}$$

or

$$(r - c)K_pK_aK_m = J\frac{d^2c}{dt^2} + (F + K_0K_aK_m)\frac{dc}{dt} + T_L \tag{21-78}$$

Rearranging terms leads to the final form of the governing differential equation of the viscous plus output-rate system, namely

$$\boxed{J\frac{d^2c}{dt^2} + (F + Q_o)\frac{dc}{dt} + Kc = Kr - T_L} \tag{21-79}$$

where

$$Q_o = \text{loop output-rate gain factor} = K_0K_aK_m \tag{21-80}$$

$$K = \text{loop proportional gain factor} = K_pK_aK_m \tag{21-81}$$

The characteristic equation is therefore

$$s^2 + \frac{F + Q_o}{J}s + \frac{K}{J} = 0 \tag{21-82}$$

which yields as the expression for the damping ratio

$$\zeta = \frac{F + Q_o}{2\sqrt{KJ}} \qquad (21\text{-}83)$$

As was the case with error-rate damping, output-rate damping adds a term to the numerator of the expression for the damping ratio, thus providing control over ζ when K is used to meet accuracy requirements.

The manner in which output-rate damping asserts itself in controlling the transient response is most easily demonstrated by assuming a step input applied to the system. If for reasons of accuracy the gain is assumed to be very high, then immediately upon applying the input a large developed torque is produced due to the proportional gain. Explained in terms of Eq. (21-77), at $t = 0^+$ the term dc/dt is zero, and so torque is produced only by the proportional term. This torque acts to bring the controlled variable quickly to the commanded steady-state level. However, as it does so with the elapse of time, the system begins to generate a large rate of change of output. In turn this means that the output-rate signal, $K_o(dc/dt)$, in Eq. (21-77) appears in opposition to the proportional signal, thus removing the tendency for an excessively oscillatory response. This discussion should also make clear why the sign of the output-rate signal must be connected in opposition to the proportional signal. Certainly, greater overshoots would result if the output-rate signal were made to aid the proportional signal. Improper polarity means a reversal in the sign of Q_o in Eq. (21-83), from which it then becomes obvious that the transient behavior deteriorates rather than improves.

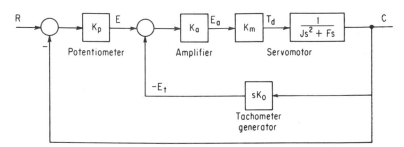

Fig. 21-10. Block diagram of a viscous plus output-rate damped system.

The complete block diagram of the servomechanism employing output-rate damping is depicted in Fig. 21-10. The output voltage of the tachometer generator that furnishes the output-rate signal is given by

$$e_t = K_o \frac{dc}{dt} \qquad \text{volts} \qquad (21\text{-}84)$$

Hence the corresponding transfer function is

$$\frac{E_t(s)}{C(s)} = sK_o \qquad (21\text{-}85)$$

In Fig. 21-10 the output of the tachometer generator is negatively summed with the signal originating from the error detector as called for by Eq. (21-76). The feedback relationship of Eq. (20-6) may be applied to this minor loop to obtain the functional relationship between C and E. Thus

$$\frac{C(s)}{E(s)} = \frac{\dfrac{K_a K_m}{Js^2 + sF}}{1 + \dfrac{sK_o K_a K_m}{Js^2 + Fs}} = \frac{K_a K_m}{Js^2 + (F + Q_o)s} \tag{21-86}$$

Accordingly the block diagram of Fig. 21-10 reduces to that shown in Fig. 21-11.

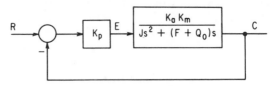

Fig. 21-11. Reduced form of the block diagram of Fig. 21-10.

The closed-loop transfer function for the complete system results when the feedback relationship is applied to Fig. 21-11. Thus

$$\frac{C(s)}{R(s)} = T(s) = \frac{\dfrac{K}{Js^2 + (F + Q_o)s}}{1 + \dfrac{K}{Js^2 + (F + Q_o)s}} = \frac{K}{Js^2 + (F + Q_o)s + K} \tag{21-87}$$

Once again note the equivalence of the denominator expression with the characteristic equation. See Eq. (21-82).

EXAMPLE 21-3 A feedback control system, having the configuration depicted in Fig. 21-1, has the following parameter values:

$$K_p = 0.5 \text{ v/rad} \qquad\qquad F = 1.5 \times 10^{-4} \text{ lb-ft/rad/sec}$$
$$K_a = 100 \text{ v/v} \qquad\qquad J = 10^{-5} \text{ slug-ft}^2$$
$$K_m = 2 \times 10^{-4} \text{ lb-ft/v}$$

(a) Describe the dynamic response of this system when a step input command is applied. Assume the system is initially at null.

(b) A load disturbance of 10^{-3} lb-ft is present on the system when a step command is applied. Find the position lag error in radians.

(c) To what value must the amplifier gain be changed in order that the position lag error of part (b) be no greater than 0.025 rad?

(d) Compute the damping ratio for the gain of part (c). What is the maximum per cent overshoot?

(e) Find the value of the output-rate gain factor which for the gain of part (c) makes the maximum overshoot 25 per cent.

solution: (a) By Eq. (21-4) the loop proportional gain is

$$K = K_p K_a K_m = \tfrac{1}{2}(100)(2 \times 10^{-4}) = 10^{-2} \text{ lb-ft/rad}$$

$$\therefore \ \zeta = \frac{F}{2\sqrt{KJ}} = \frac{1.5 \times 10^{-4}}{2\sqrt{10^{-2} \cdot 10^{-5}}} = 0.236$$

Hence from Fig. 21-4 the maximum overshoot is found to be 47 per cent. Also

$$\omega_n = \sqrt{\frac{K}{J}} = \sqrt{\frac{10^{-2}}{10^{-5}}} = 31.8 \text{ rad/sec}$$

Thus the commanded value of the controlled variable reaches within 1 per cent of its final value in

$$t_s = \frac{5}{\zeta\omega_n} = \frac{5}{(0.236)(31.8)} = 0.67 \text{ sec}$$

The damped oscillations occur at a frequency of

$$\omega_d = \omega_n\sqrt{1 - \zeta^2} = 30.8 \text{ rad/sec}$$

(b) From Eq. (21-12) the position lag error is seen to be

$$\frac{T_L}{K} = \frac{10^{-3}}{10^{-2}} = 0.1 \text{ rad}$$

(c) The loop gain must be increased by a factor of 4. Hence the new value of amplifier gain is

$$K_a' = 4K_a = 4(100) = 400 \text{ v/v}$$

(d) Equation (21-17) shows that when the gain is quadrupled the damping ratio is halved. Hence

$$\zeta' = \frac{\zeta}{2} = 0.118$$

This yields a maximum overshoot of 70 per cent.

(e) For a 25 per cent overshoot Fig. 21-4 reveals that $\zeta = 0.4$. Then by Eq. (21-83) we have

$$0.4 = \frac{F + Q_o}{2\sqrt{KJ}} = \frac{1.5 \times 10^{-4} + Q_o}{12.72 \times 10^{-4}}$$

$$\therefore \ Q_o = 0.4(12.72)10^{-4} - 1.5 \times 10^{-4} = 3.6 \times 10^{-4}$$

Hence

$$K_o = \frac{Q_o}{K_a'K_m} = \frac{3.6 \times 10^{-4}}{4 \times 10^2 \times 10^{-2}} = 0.9 \times 10^{-4} \text{ v/rad/sec}$$

21-4 Integral-error (or Reset) Control

A control system is said to contain integral-error control when the generation of the output in some way depends upon the integral of the actuating signal. Integral control is readily achieved by designing the servoamplifier so that it makes available an output voltage that contains an integral term as well as a proportional term. Expressed mathematically we have

$$e_a = eK_a + K_i \int_0^t e \, dt \qquad (21-88)$$

where e is the input to the servoamplifier

e_a is the output of the servoamplifier, which is also the motor input voltage

K_i is the proportionality factor of the integral-error component

Since the motor developed torque is $T_d = K_m e_a$, the equation of motion for the system can be written as

$$eK_aK_m + K_iK_m \int_0^t e\, dt = J\frac{d^2c}{dt^2} + F\frac{dc}{dt} + T_L \tag{21-89}$$

Differentiating both sides to remove the integral leads to

$$K_aK_m\frac{de}{dt} + K_iK_me = J\frac{d^3c}{dt^3} + F\frac{d^2c}{dt^2} + \frac{dT_L}{dt} \tag{21-90}$$

Inserting $e = K_p(r - c)$ yields

$$K_aK_mK_p\left(\frac{dr}{dt} - \frac{dc}{dt}\right) + K_iK_pK_m(r - c) = J\frac{d^3c}{dt^3} + F\frac{d^2c}{dt^2} + \frac{dT_L}{dt} \tag{21-91}$$

Rearranging gives

$$J\frac{d^3c}{dt^3} + F\frac{d^2c}{dt^2} + K\frac{dc}{dt} + Q_ic + \frac{dT_L}{dt} = Q_ir + K\frac{dr}{dt} \tag{21-92}$$

where

$$Q_i = K_iK_pK_m \tag{21-93}$$

and K is as defined before.

A glance at Eq. (21-89) reveals the interesting fact that the integral term does not combine with the viscous-friction term as was the case when error-rate and output-rate signals were introduced. Instead the integral term stands alone in the differential equation, thus stressing that its influence on system performance differs basically from the previous compensation schemes. Equation (21-93) shows that integral-error compensation has changed the order of the system from second to third. The inclusion of the integral term means therefore that a third independent energy-storing element is present. This stands in sharp contrast to the influence that error rate or output rate exerts on the same basic proportional system. For rate control the effect is confined to altering the coefficient of the first derivative term in the governing differential equation, and for this reason does not appear as an additional independent energy-storing element.

It follows from Eq. (21-92) that the characteristic equation now takes the form of a cubic, namely

$$s^3 + \frac{F}{J}s^2 + \frac{K}{J}s + \frac{Q_i}{J} = 0 \tag{21-94}$$

Finding the solution of this equation is more involved than it is for the second-order system. Moreover, if standard time-domain procedures are used, it becomes considerably more difficult to design *directly* for a specific dynamic behavior as is done in Example 21-2. In third- and higher-order systems it is laborious and difficult to see what the effect on the dynamic behavior is when the value of a particular parameter is modified. However, if the analysis is done in the frequency domain, the solution of the problem is easier and more direct. This matter is treated in Chapter 22.

In view of the fact that integral compensation increases the order of the system and thus makes satisfactory dynamic behavior more difficult to achieve, what advantage is to be gained by its inclusion? The answer readily follows from an examination of Eq. (21-92). Note that in the steady state, even in the presence of a fixed external load, exact correspondence exists between the actual and commanded values of the output. That is, the position lag error, which exists with error-rate and output-rate control, disappears with integral-error control. In fact it is characteristic of integral control that it greatly improves the steady-state performance (system accuracy). At the same time, however, it makes the dynamic behavior more difficult to cope with successfully.

The analysis of integral-error control by means of the transfer function approach leads, of course, to the same results. Since the servoamplifier is the only component in the block diagram that is modified in the integral-error controlled system, the new transfer function is readily found from Eq. (21-88). Thus

$$\frac{E_a(s)}{E(s)} = K_a + \frac{K_i}{s} \tag{21-95}$$

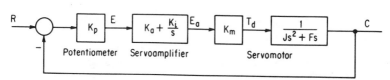

Fig. 21-12. Block diagram of a proportional plus integral-error controlled system.

The complete block diagram then takes the form depicted in Fig. 21-12, from which it immediately follows that the direct transmission function is

$$G(s) = \frac{K_p K_m (K_i + sK_a)}{Js^3 + Fs^2} \tag{21-96}$$

The corresponding closed-loop transfer function then becomes

$$\frac{C(s)}{R(s)} = \frac{Q_i + sK}{Js^3 + Fs^2 + Ks + Q_i} \tag{21-97}$$

This expression is equivalent to Eq. (21-92) for the case where T_L equals zero. The characteristic equation is the same.

PROBLEMS

21-1 Measurements conducted on a servomechanism show the output response to be

$$\frac{c(t)}{r_0} = 1 - 1.66\epsilon^{-8t} \sin (6t + 37°)$$

when the input is a step displacement of magnitude r_0.
   (a) Determine the natural frequency of the system in radians per second.
   (b) What is the damped frequency of oscillation?
   (c) Find the value of the damping ratio.

(d) The inertia of the output member is known to be 0.01 lb-ft-sec^2, and the viscous coefficient is 0.16 lb-ft/rad/sec. What is the loop gain?

(e) If the damping ratio is not to be less than 0.4, how much can the loop gain be increased?

21-2 A feedback control system is represented by the following differential equation:

$$\frac{d^2c}{dt^2} + 6.4\frac{dc}{dt} = 160e$$

where $e = r - 0.4c$ and c denotes the output variable.

(a) Find the value of the damping ratio.

(b) What information does this convey about the tranisent behavior of the system? Be specific.

21-3 A feedback control system has the configuration shown in Fig. P21-3.

(a) Obtain the expression for the closed-loop transfer function.

(b) Find the complete expression for the output response as a function of time when the input is a unit step.

(c) Plot each component of the output response as a function of time. Also sketch the resultant response and find graphically the time required to reach 95 per cent of the final value.

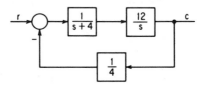

Figure P21-3

21-4 A control system is composed of components whose transfer functions are those specified in the block diagram of Fig. P21-4.

(a) What is the expression for the closed-loop transfer function?

(b) At what frequency does the output variable oscillate in responding to a step command before reaching steady state?

(c) What is the maximum per cent overshoot in part (b)?

(d) How many seconds are required for the output to reach within 99 per cent of steady state in part (b)?

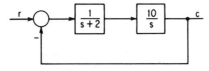

Figure P21-4

21-5 The step response of a unity feedback control system is given by

$$\frac{c(t)}{r_0} = 1 - 1.66\epsilon^{-8t} \sin(6t + 37°)$$

where r_0 is the magnitude of the input quantity.

(a) Find the closed-loop transfer function.

(b) What is the corresponding open-loop transfer function?

(c) Determine the complete output response for a unit-step input when the system is operated open-loop.

21-6 The governing differential equation of a feedback speed control system is given by

$$\left(\frac{J}{F}\right)\frac{d\omega_e}{dt} + (1 + K)\omega_e = \left(\frac{J}{F}\right)\frac{d\omega_r}{dt} + \omega_r$$

where ω_e is the error in speed between the reference speed ω_r and the output speed ω_o, i.e., $\omega_e = \omega_r - \omega_o$. Moreover, J is the system inertia, F is the system viscous-friction coefficient, and K is the system loop gain.

(a) What is the expresion for the steady-state speed error?

(b) Determine the complete expression for the speed error when the reference speed is applied to the system in the form of a step command.

(c) Can any setting of the gain K cause sustained oscillations in the response of the system? Explain.

21-7 The dynamic behavior of a deivce is described by the equation

$$\frac{dc}{dt} + 10c = 40e$$

(a) Find the expression for the complete response of this sytem when $e = u(t)$.

(b) What is the expression for the direct transmission function?

(c) A unity-feedback path is placed around the device. Find the response of the feedback system to a unit-step command.

(d) Compare the solutions to parts (a) and (c) and draw appropriate conclusions.

21-8 Solve Prob. 21-7 for the case where a unit-ramp function, i.e., $e(t) = tu(t)$, is used as the driving force in parts (a) and (c).

21-9 A servomechanism has the following parameters:

$K = $ open-loop gain $= 18 \times 10^{-4}$ lb-ft/rad

$J = $ system inertia $= 10^{-5}$ slug-ft^2

$F = $ system viscous-friction coefficient $= 161 \times 10^{-6}$ lb-ft/rad/sec

(a) Find the value of the damping ratio for the constants specified.

(b) To meet the accuracy requirements during steady-state operation, it is found that the loop gain must be increased to 185.5×10^{-4} lb-ft/rad. The damping ratio is to remain the same as in part (a). Determine the error-rate damping coefficient needed to achieve this result.

21-10 A feedback system employing proportional plus error-rate control is depicted in Fig. P21-10. The output actuator is a servomotor that develops a torque in accordance with

$$T_d = -D\frac{dc}{dt} + Be_a$$

where $D = 10^{-5}$ lb-ft/rad/sec

$B = 0.375 \times 10^{-5}$ lb-ft/volt

$e_a = $ amplifier output $= $ servomotor input voltage

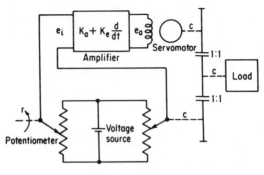

Figure P21-10

The output voltage of the amplifier is given by $e_a = K_a e_i + K_e(de_i/dt)$. The other constants of the system are

K_p = potentiometer transducer constant = 1.615 volt/rad

J = system inertia = 1.56×10^{-8} slug-ft^2

F_L = load viscous-friction coefficient = 0.625×10^{-5} lb-ft/rad/sec

(a) Determine the complete differential equation of this system relating the output variable c to input r.
(b) Obtain the transfer function of the output actuator relating the output variable $C(s)$ to the input variable $E_a(s)$.
(c) The amplifier gain is set at a value $K_a = 40$ to meet accuracy requirements. Find the value of K_e that allows the dynamic behavior to be critically damped.

21-11 A second-order feedback control system has the configuration shown in Fig. 21-10 and the following parameters:

K_p = transducer constant = 0.5 volt/rad

K_m = motor torque constant = 2×10^{-4} lb-ft/volt

F = viscous-friction coefficient = 1.5×10^{-4} lb-ft/rad/sec

J = system inertia = 10^{-5} slug-ft^2

A load torque of 10^{-3} lb-ft appears constantly on the output shaft. It is required that in the steady state the effect of this load torque is not to cause a position-lag error between the command and the response exceeding 0.01 rad. Moreover, the dynamic behavior of the system is to be characterized by a maximum overshoot that does not exceed 25 per cent when subjected to a step input. Specify the amplifier gain and the output-rate factor that will enable the system to meet these specifications.

21-12   Depicted in Fig. P21-12 is the system shown in Fig. P21-4 modified to include error-rate damping. Determine the value of the error-rate factor K_e so that the damping ratio of the modified characteristic equation is 0.6.

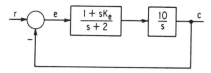

Figure P21-12

21-13   For the system depicted in Fig. P21-13 find the value of the output-rate factor that yields a response to a step command having a maximum overshoot of 10 per cent.

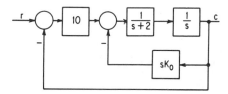

Figure P21-13

21-14   Repeat Prob. 21-13 for the system whose block diagram appears in Fig. P21-14.

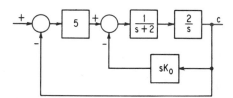

Figure P21-14

22

dynamic behavior by the
frequency-response method

The frequency-response method is a steady-state analysis in which the sinusoidal response of the system is found as the frequency is varied from very low to very high values. As is pointed out in Chap. 6, the response of any linear circuit or system to a sinusoidal forcing function can be identified in terms of a change in *magnitude* and *phase angle*. Both these quantities are influenced by the parameters of the system and the frequency of the forcing function. The big attraction the frequency response offers as a method of analysis lies in its ability to convey information about the dynamic response in the time domain in terms of the way in which the amplitude and phase angle of the response vary as the frequency is changed. It is interesting to note here that there does exist an exact mathematical relationship enabling us to write the time response precisely in terms of the summation of the steady-state sinusoidal frequency response taken over the entire frequency spectrum.† But this relationship is not a simple one and is best avoided. Instead, use is made of appropriate frequency-domain figures of merit, such as phase margin and gain margin, which convey essentially the same information. The cost of using this easier approach is that we do not obtain an exact expression

† The reference here is to the inverse Fourier transform.

for the time solution. The time response is identified, instead, in terms of some equivalent damping ratio and natural frequency—and very often this is all that is needed.

The frequency-response method of analysis has several notable advantages, which account for its great popularity among engineers. First, the level of mathematics it requires does not go beyond a knowledge of complex numbers. The transient behavior of a system in the time domain carries over as variations in amplitude and phase in the frequency domain. Second, the effect on the overall time solution of changing any single parameter of the system is simply and directly observable in the frequency domain irrespective of the order of the system. As is pointed out in Sec. 21-4, this is certainly not possible when a time-domain analysis is made. Third, the frequency-response method lends itself conveniently to measurements in the laboratory. In view of the abundance of sine-wave generators and the steady-state nature of the analysis, measurements can be carried out quite easily and by persons who need not be extraordinarily skilled and trained.

The subject matter of this chapter is developed in a way that stresses the role of the frequency response in establishing the nature of the dynamic time response. In this connection the derivation and analysis of the transfer functions of typical system components is a necessary first step. The transfer functions of the components are then combined to yield the overall system transfer functions. These, in turn, are used to classify control systems as well as to determine their stability and performance characteristics. The chapter ends with a short description of the root locus method as an alternative and/or supplement to the frequency-response method when the situation warrants.

22-1 Transfer Functions of System Components

The typical system components whose transfer functions are developed here represent a necessary selection. Because of the nature of this book they are primarily chosen from the electrical field; a few are from the mechanical area. Examples from the pneumatic, hydraulic, and process field are omitted, not only because of space considerations but also because insofar as the procedure is concerned nothing new is gained. No matter what type of component we consider, the general procedure for deriving the transfer function is always the same. It involves the following three steps: (1) Determine the governing equation for the component expressed in terms of the output and input variables. (2) Laplace transform the governing equation, assuming all initial conditions are zero. (3) Rearrange the equation to formulate the ratio of the output to input variable.

The Lag Network (or Integrating Circuit). It often is necessary to reshape the frequency-response characteristic of a system's transfer function in order to meet the performance specifications. Integrating and differentiating circuits are very commonly used to achieve these results. Effectively they provide the system with integral- and error-rate control in a simple and economical fashion.

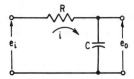

Fig. 22-1. An integrating circuit or lag network.

The circuit shown in Fig. 22-1 is called an integrating circuit because under certain conditions the output signal e_o is effectively the integral of the input signal e_i. To understand this, recall that the output voltage across the capacitor terminals is

$$e_o = \frac{1}{C} \int i \, dt \qquad (22\text{-}1)$$

If the frequency of e_i is assumed to be such that the capacitive reactance is negligible compared to the resistance, it follows that the capacitor current is given approximately by

$$i \approx \frac{e_i}{R} \qquad (22\text{-}2)$$

Inserting Eq. (22-2) into Eq. (22-1) yields

$$e_0 \approx \frac{1}{RC} \int e_i \, dt \qquad (22\text{-}3)$$

which shows that the output signal is the integral of the input signal. However, this circuit ceases to be an integrating circuit at very low frequencies because of the correspondingly high capacitive reactance.

An exact description of the properties of the circuit of Fig. 22-1 is readily had from the transfer function. The governing equations involving the input and output variables are:

$$e_i = iR + \frac{1}{C} \int i \, dt = R\frac{dq}{dt} + \frac{q}{C} \qquad (22\text{-}4)$$

$$e_o = \frac{1}{C} \int i \, dt = \frac{q}{C} \qquad (22\text{-}5)$$

The corresponding Laplace transforms are:

$$E_i(s) = RsQ(s) + \frac{Q(s)}{C} \qquad (22\text{-}6)$$

$$E_o(s) = \frac{Q(s)}{C} \qquad (22\text{-}7)$$

where $Q(s) = \mathscr{L}q(t)$. Formulating the ratio of output to input then yields

$$\frac{E_o(s)}{E_i(s)} = \frac{\frac{1}{sC}}{R + \frac{1}{sC}} = \frac{1}{1 + sRC} \qquad (22\text{-}8)$$

Note that $Q(s)$ cancels out in the formulation. In fact the middle expression of Eq. (22-8) shows that, where networks are involved, the transfer function can be found directly by using the voltage-divider theorem and the operational form of the impedance of the circuit elements.

For sinusoidal input functions the frequency s is replaced by $j\omega$† so that the

† See Chaps. 4 and 6.

final expression for the transfer function becomes

$$\boxed{\frac{E_o(j\omega)}{E_i(j\omega)} = \frac{1}{1 + j\omega RC} = \frac{1}{1 + j\omega\tau}}$$
(22-9)

where $\tau = RC$. It is important to understand that the frequency ω appearing in the transfer function always refers to the frequency of variation of the input signal. When this frequency has a very low value (i.e. $\omega \to 0$), the capacitor behaves as an open circuit and so a direct transmission of the signal occurs. Thus the circuit does not integrate. On the other hand, as the frequency increases (i.e. $\omega\tau \gg 1$), the transfer function of Eq. (22-9) becomes approximately $1/j\omega\tau$. The presence of the factor $1/j\omega$ in the transfer function means that the sinusoidal input function appears across the capacitor terminals as an integrated quantity. Recall that integration is algebraically denoted by $1/s$ or $1/j\omega$. Finally, it should also be noted that, when this circuit performs an as integrator, the input signal is appreciably attenuated because $\omega\tau \gg 1$. In such circuits the integration property goes hand in hand with attenuation of the signal level.

Attention is next focused on the graphical representation of the transfer function of Eq. (22-9) as a function of the signal frequency ω. Two methods are generally preferred—the polar plot and the Bode plot. The polar plot is merely a plot of the amplitude and phase of Eq. (22-9) at each frequency in the complex plane. The result is shown in Fig. 22-2(a). Note that the locus of the output-to-input phasor is a semicircle in the fourth quadrant. The location in this quadrant stresses the fact that the phase angle is always a *lagging* one. In fact, integration in the time domain is characteristically described by phase lag in the frequency domain. When the lag angle in sinusoidal steady-state analysis of the output of a linear component is 90°, the corresponding behavior in the time domain is one of pure integration. When the lag angle lies between 0 and 90°, the result in the time domain involves a proportional term as well as an integral term.

In the Bode-diagram representation of the transfer function the amplitude and phase characteristics are individually plotted as a function of frequency on a logarithmic scale. Moreover, it is customary to plot the amplitude expressed in units of *decibels*, which is defined as twenty times the logarithm of the voltage ratio in this case and is frequently called the log-modulus characteristic. The log-modulus characteristic as a function of frequency can be expressed mathematically by

$$A = 20 \log \frac{1}{|1 + j\omega\tau|} = 20 \log 1 - 20 \log |1 + j\omega\tau|$$
$$= -20 \log |1 + j\omega\tau|$$
(22-10)

The absolute signs are included to emphasize that we are dealing only with the magnitude of the complex number for each value of ω. The phase characteristic is given by

$$\phi = -\tan^{-1}\omega\tau$$
(22-11)

Usually the attenuation characteristic can be quickly plotted by drawing

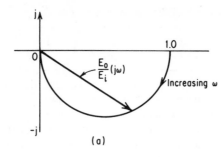

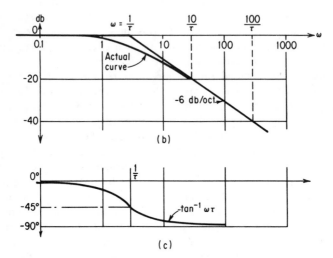

Fig. 22-2. Graphical representation of the lag network transfer function:

    (a) polar plot;
    (b) attenuation in db versus frequency;
    (c) phase versus frequency.

two straight lines originating from the point $\omega\tau = 1$ or $\omega = 1/\tau$. To understand this, consider the behavior of Eq. (22-10) at low and high frequencies. In this instance a low frequency is one that is small compared to $1/\tau$. Examination of Eq. (22-10) discloses that when $\omega\tau \ll 1$, the log-modulus attenuation characteristic is just a straight horizontal line at the zero-db level. That is,

$$A = -20 \log 1 = 0 \text{ db} \qquad \text{for } \omega\tau \ll 1 \qquad (22\text{-}12)$$

At high frequencies, Eq. (22-9) becomes

$$A \approx -20 \log |j\omega\tau| \qquad \text{for } \omega\tau \gg 1 \qquad (22\text{-}13)$$

Accordingly, at

$$\omega_1\tau = 10 \qquad A_1 = -20 \log |j10| = -20 \text{ db} \qquad (22\text{-}14)$$

$$\omega_2\tau = 20, \qquad A_2 = -20 \log |j20| = -26 \text{ db} \qquad (22\text{-}15)$$

$$\omega_3\tau = 100, \qquad A_3 = -20 \log |j100| = -40 \text{ db} \qquad (22\text{-}16)$$

A plot of these high-frequency values of the amplitude identifies a straight line having a slope of -6 db/octave or -20 db/decade. Since $\omega_2 = 2\omega_1$, ω_2 is said to be an *octave* larger than ω_1. Similarly, ω_3 is said to be a *decade* higher than ω_1 because it is ten times greater. The extension of this sloping line intersects the frequency axis at the value $1/\tau$. This frequency is a key quantity in plotting the log-modulus characteristic because it marks the frequency value at which the asymptotic representation of the transfer function changes slope. Hence, to distinguish it from other frequencies, it is called the *breakpoint* or *corner* frequency of the transfer function. It is readily found by setting the j part of the transfer function equal to unity and solving for ω. The log-modulus characteristic is depicted in Fig. 22-2(b). Note that the actual curve and the asymptotic approximations are in very close agreement everywhere in the frequency spectrum except for the small region about the breakpoint frequency. A study of Eq. (22-10) shows that the maximum deviation between these plots occurs at the breakpoint frequency and has a value of

$$20 \log |1 + j1| = 20 \log |\sqrt{2}| = 3 \text{ db}$$

Since this deviation is a determinable quantity, the procedure in plotting the log-modulus characteristic of a transfer function is to use the straight line asymptotic approximations and then to introduce corrections only when the situation warrants it.

The phase characteristic of Eq. (22-11) plots simply as the arc tangent curve. This is shown in Fig. 22-2(c). Note that at the breakpoint frequency the phase shift is $-45°$.

Lag Network with Fixed Maximum Attenuation. At high frequencies the capacitor in the circuit of Fig. 22-1 appears almost as a short circuit so that the output signal level is practically zero. In the interest of limiting the high-frequency attenuation this circuit is modified to the practical form shown in Fig. 22-3. Taking advantage of the validity of using the voltage-divider theorem in finding network transfer functions, we get for this circuit

$$\frac{E_o(s)}{E_i(s)} = \frac{R_2 + \dfrac{1}{sC}}{R_1 + R_2 + \dfrac{1}{sC}} = \frac{1 + sR_2C}{1 + s(R_1 + R_2)C} \qquad (22\text{-}17)$$

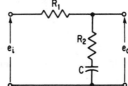

Fig. 22-3. Practical form of integrating circuit to limit attenuation at high frequencies.

Calling

$$\tau \equiv R_2 C \tag{22-18}$$

$$\alpha \equiv \frac{R_1 + R_2}{R_2} \tag{22-19}$$

Equation (22-17) may be rewritten as

$$\frac{E_o(s)}{E_i(s)} = \frac{1 + s\tau}{1 + s\alpha\tau} \tag{22-20}$$

For sinusoidal inputs this becomes

$$\boxed{\frac{E_o(j\omega)}{E_i(j\omega)} = \frac{1 + j\omega\tau}{1 + j\omega\alpha\tau}} \tag{22-21}$$

The corresponding polar plot is shown in Fig. 22-4(a). Note that at $\omega = 0$ the transfer function ratio is unity. At $\omega \to \infty$ the value is $1/\alpha$. Also, since the radius of this semicircle is

$$\left(1 - \frac{1}{\alpha}\right)\frac{1}{2} = \frac{\alpha - 1}{2\alpha}$$

and the distance from the origin to this center is

$$\left(\frac{1}{\alpha} + \frac{\alpha - 1}{2\alpha}\right) = \frac{\alpha + 1}{2\alpha}$$

it follows that the maximum phase lag angle possible with this circuit is

$$\phi_m = \sin^{-1}\frac{\alpha - 1}{\alpha + 1} \tag{22-22}$$

The quantity ϕ_m is solely dependent upon α.

The Bode diagrams of attenuation and phase are depicted in Figs. 22-4(b) and (c). The asymptotic attenuation characteristic is found by superposing the effects of the numerator and denominator terms in Eq. (22-21). The denominator is handled as before. Its breakpoint frequency is $\omega_{b_1} = 1/\alpha\tau$. Hence we have a zero-db line originating from zero frequency and going up to ω_{b_1}. At this point we have a negatively sloping line of -20 db/decade which continues over the remainder of the frequency range. The numerator term is treated in a similar fashion. A zero-db line proceeds from zero frequency to the breakpoint frequency $\omega_{b_2} = 1/\tau$. This frequency occurs higher in the scale than ω_{b_1}. Beyond ω_{b_2} the numerator term provides an increase in decibels at the rate of $+20$ db/decade. Since we are using logarithms to express the amplitude variations, the resultant attenuation characteristic is found by simply adding the contributions of the numerator and denominator terms. This leads to the resultant asymptotic curve shown in Fig. 22-4(b). It is worth noting that below ω_{b_2} the resultant curve is the same as that of the denominator term. Above ω_{b_2} the $+20$ db/dec slope of the numerator term cancels the effect of the -20 db/dec slope of the denominator term, thus yielding a horizontal line placed at a value corresponding to the fixed high-frequency attenuation of $-20 \log \alpha$.

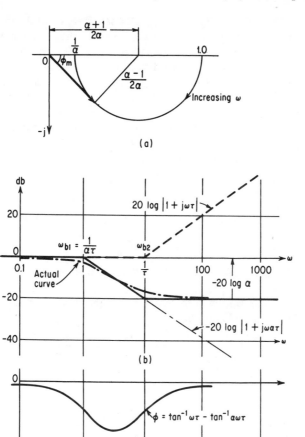

Fig. 22-4. Graphical representation of the transfer function of the network of Fig. 22-3:

    (a) polar plot;

    (b and c) Bode plots of attenuation and phase versus frequency.

The phase characteristic is illustrated in Fig. 22-4(c). It is a plot of the equation

$$\phi = \tan^{-1} \omega\tau - \tan^{-1} \alpha\omega\tau \qquad (22\text{-}23)$$

The first term is the angle of the numerator and the second is that of the denominator. Since α is defined always greater than unity, the resultant phase angle at any finite frequency is negative, indicating that the output lags the input. Another notable observation regarding the Bode diagram is that the phase angle follows the changes in attenuation. Thus when the slope of the attenuation curve is changing rapidly so too does the phase curve. In fact, mathematical relationships are available that allow one characteristic to be derived from the other.

The Lead Network (or Differentiating Circuit). Rate control can be obtained by using the differentiating circuit of Fig. 22-5. Like the integrating circuit, the differentiating circuit provides the derivative of the input only in a limited portion of the frequency spectrum. For example, if e_i varies at relatively high frequencies where the capacitive reactance is negligibly small, the circuit of Fig. 22-5 shows that the input signal appears directly at the output terminals without undergoing a time operation. On the other hand, at low frequencies, where the capacitive reactance predominates, the current can be written approximately as

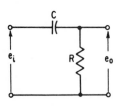

Fig. 22-5. A differentiating circuit or lead network.

$$i \approx C\frac{de_i}{dt} \qquad (22\text{-}24)$$

Hence the output voltage as obtained across the resistor becomes

$$e_o = iR \approx RC\frac{de_i}{dt} \qquad (22\text{-}25)$$

Clearly, then, this circuit behaves as a differentiating circuit in the restricted frequency range.

A more exact description of the properties of the circuit can be obtained from the transfer function, which in this case is

$$\frac{E_o(s)}{E_i(s)} = \frac{R}{R + \dfrac{1}{sC}} = \frac{sRC}{1 + sRC} = \frac{s\tau}{1 + s\tau} \qquad (22\text{-}26)$$

For sinusoidal inputs Eq. (22-26) becomes

$$\frac{E_o(j\omega)}{E_i(j\omega)} = \frac{j\omega\tau}{1 + j\omega\tau} \qquad (22\text{-}27)$$

This transfer function makes it clear that when $\omega\tau \ll 1$, the ratio of output to input signals is approximately $j\omega\tau$. Since the term $j\omega$ is the operational representation of differentiation where sinusoids are involved, the circuit behaves as a differentiator for the restriction specified. Again note that the output signal level is attenuated when the circuit provides differentiation. Finally, in terms of a frequency-domain description, a differentiating circuit can be described as a lead network because at any finite frequency the output phasor *leads* the input phasor by the resultant phase angle of the transfer function. Equation (22-27) shows this phase lead angle to be

$$\phi = +90° - \tan^{-1}\omega\tau \qquad (22\text{-}28)$$

A polar plot of Eq. (22-27) appears in Fig. 22-6(a). The locus in this case is a semicircle located in the first quadrant, which is consistent with the fact that the phase angle is a positive or lead angle. The log-modulus attenuation characteristic in decibels is

$$A = 20\log\left|\frac{E_o}{E_i}\right| = 20\log|j\omega\tau| - 20\log|1 + j\omega\tau| \qquad (22\text{-}29)$$

A little thought discloses that the first term plots as a straight line passing through

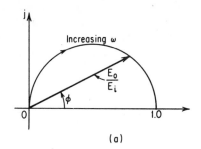

(a)

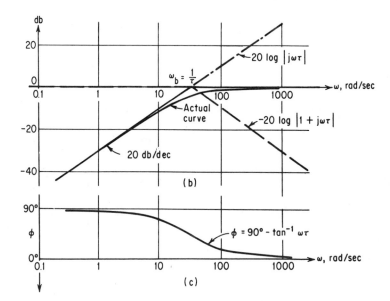

Fig. 22-6. Graphical representation of the transfer function of the circuit of Fig. 22-5:

    (a) polar plot;

    (b) attenuation versus frequency plot (note the high-pass character of the circuit above ω_b);

    (c) phase versus frequency plot.

$\omega_b = 1/\tau$ with a slope of 20 db/decade. The second term is plotted as two straight lines—a horizontal line at zero db spreading from zero frequency to $\omega_b = 1/\tau$, and another straight line with a slope of -20 db/decade spreading over the frequency range from ω_b to infinity. The resultant asymptotic amplitude curve is the solid curve in Fig. 22-6(b). The phase characteristic, given by Eq. (22-28), is depicted in Fig. 22-6(c).

Lead Network with Fixed Low-frequency Attenuation. A serious limitation of the circuit of Fig. 22-5 is that it permits no signal transfer when the system reaches steady state (i.e., $\omega = 0$). Accordingly, if a system is so composed that it demands a finite signal level at steady state to generate the output quantity, the use of this circuit forbids it. To avoid the difficulty the practical form of the lead network shown in Fig. 22-7 is commonly used. Its transfer function can be derived as follows:

$$\frac{E_o(s)}{E_i(s)} = \frac{R_2}{R_2 + \dfrac{R_1}{1 + sR_1C}} = \frac{R_2}{R_1 + R_2}\left[\frac{1 + sR_1C}{1 + s\left(\dfrac{R_1R_2}{R_1 + R_2}\right)C}\right] \quad (22\text{-}30)$$

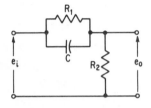

Calling

$$\alpha \equiv \frac{R_1 + R_2}{R_2} \quad (22\text{-}31)$$

$$\tau \equiv \frac{R_1R_2}{R_1 + R_2}C = \frac{R_1C}{\alpha} \quad (22\text{-}32)$$

Fig. 22-7. Practical form of a rate network.

and inserting into Eq. (22-30) yields

$$\frac{E_o(s)}{E_i(s)} = \frac{1}{\alpha}\left(\frac{1 + s\alpha\tau}{1 + s\tau}\right) \quad (22\text{-}33)$$

The expression for the transfer function for sinusoidal inputs then becomes

$$\boxed{\frac{E_0(j\omega)}{E_i(j\omega)} = \frac{1}{\alpha}\left(\frac{1 + j\omega\alpha\tau}{1 + j\omega\tau}\right)} \quad (22\text{-}34)$$

The polar plot of Eq. (22-34) is depicted in Fig. 22-8(a). It is identical to Fig. 22-4(a) with the exception that the semicircle is now located in the first quadrant because lead angles are involved. The expressions for the log-modulus and phase characteristics are given by

$$A = 20 \log \frac{1}{\alpha} + 20 \log |1 + j\omega\alpha\tau| - 20 \log |1 + j\omega\tau| \quad (22\text{-}35)$$

and

$$\phi = \tan^{-1} \omega\alpha\tau - \tan^{-1} \omega\tau \quad (22\text{-}36)$$

By plotting each term of Eq. (22-35) and summing, we obtain the resultant curve of Fig. 22-8(b). Note that with this circuit the attenuation at low frequencies (i.e., $\omega \ll 1/\alpha\tau$) is fixed at $1/\alpha$ where α is defined by Eq. (22-31). A plot of Eq. (22-36) yields the phase characteristic shown in Fig. 22-8(c).

The Underdamped Quadratic Factor. It is not uncommon for a system component or a subsystem of a large system to be described by the second-order linear differential equation

$$\frac{d^2y}{dt^2} + 2\zeta\omega_n\frac{dy}{dt} + \omega_n^2 y = \omega_n^2 x \quad (22\text{-}37)$$

where y denotes the output and x denotes the input variable. If we Laplace transform this expression for zero initial conditions and formulate the ratio of output to input, the following transfer function results:

$$\frac{Y(s)}{X(s)} = \frac{\omega_n^2}{s^2 + 2\zeta\omega_n s + \omega_n^2} = \frac{1}{\left(\dfrac{s}{\omega_n}\right)^2 + \dfrac{2\zeta}{\omega_n}s + 1} \tag{22-38}$$

For sinusoidal inputs the expression is

$$\boxed{\frac{Y(j\omega)}{X(j\omega)} = \frac{1}{1 - \left(\dfrac{\omega}{\omega_n}\right)^2 + j2\zeta\left(\dfrac{\omega}{\omega_n}\right)}} \tag{22-39}$$

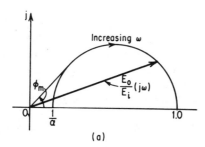

(a)

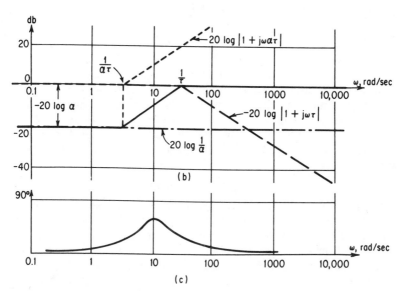

Fig. 22-8. Graphical representation of the transfer function of the circuit of Fig. 22-7:

    (a) polar plot;
    (b) attenuation versus frequency;
    (c) phase versus frequency.

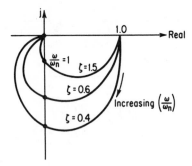

Fig. 22-9. Polar plot of Eq. (22-39) for a linear second-order system component.

An inspection of the polar plot for this function, which is drawn in Fig. 22-9 for several values of ζ, discloses that the output always lags the input at finite steady-state frequencies. At very high frequencies the phase lag approaches $-180°$. Note, too, that for low values of ζ the amplitude ratio at first rises and then diminishes as the frequency increases.

The log-modulus characteristic of the quadratic factor is given by

$$A = -20 \log \left| 1 - \left(\frac{\omega}{\omega_n}\right)^2 + j2\zeta\left(\frac{\omega}{\omega_n}\right) \right| \tag{22-40}$$

When $\omega/\omega_n \ll 1$, the amplitude ratio has a magnitude of zero db, which means that the output and input quantities are equal. Moreover, at the high end of the frequency scale, $\omega/\omega_n \gg 1$, Eq. (22-40) becomes

$$A]_{\omega/\omega_n \gg 1} \approx -20 \log \left| -\left(\frac{\omega}{\omega_n}\right)^2 \right| = -40 \log \left| \frac{\omega}{\omega_n} \right| \tag{22-41}$$

This expression plots as a straight line having a slope of -40 db/decade and spreading from ω_n to infinity. The behavior of the magnitude ratio curve in the vicinity of the natural frequency, however, very critically depends upon ζ and can deviate appreciably from the straight-line asymptotic approximations. Figure 22-10(a) shows the attenuation characteristic for several values of ζ. Note that at $\omega = \omega_n$ for $\zeta = 0.05$ the error between the actual value and that of the straight-line approximation is 20 db. This immediately follows from Eq. (22-40) when it is written for $\omega = \omega_n$. Thus

$$A]_{\omega=\omega_n} = -20 \log |j2\zeta| \tag{22-42}$$

For $\zeta = 0.05$ Eq. (22-42) assumes a value of 20 db. On the other hand, note that when $\zeta = 0.5$ no error occurs at $\omega = \omega_n$ between the actual and approximate curves. Clearly, the conclusion to be drawn from this discussion is that when dealing with underdamped quadratic factors we must give due attention to the value of the damping ratio. If its value is in the vicinity of 0.5, it is proper to use the straight-line approximations to plot the attenuation characteristic. If its value is quite low, then the curve called for in Fig. 22-10(a) must be used.

The phase characteristic, which is given by

$$\phi = -\tan^{-1} \frac{2\zeta\omega/\omega_n}{1 - (\omega/\omega_n)^2} \tag{22-43}$$

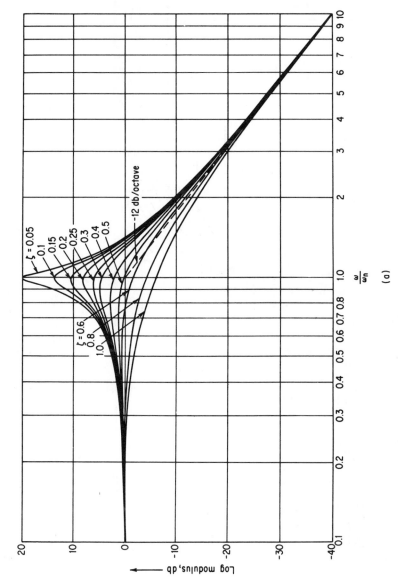

Fig. 22-10. (a) Attenuation versus frequency plot of Eq. (22-39) for various values of ζ.

673

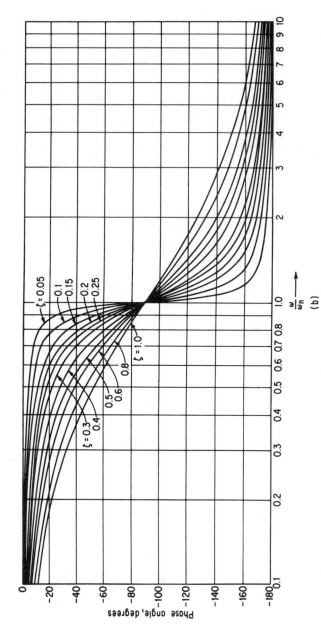

Fig. 22-10. (b) Phase versus frequency plot of Eq. (22-39) for various values of ζ.

plots as shown in Fig. 22-10(b). Note that at $\omega = \omega_n$ the phase lag is the same irrespective of ζ. However, for other values of frequency the rate of change of phase depends on ζ rather critically.

The Linear Accelerometer. The linear accelerometer has gained wide-spread use in high-speed aircraft and space vehicle applications. In these applications it can be employed either as a measuring device, which provides valuable data on the extent to which the spacecraft or jet aircraft is subjected to vibrations and shock, or else as an important, inherent part of the spacecraft or aircraft guidance system.

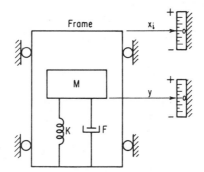

Fig. 22-11. Construction features of a linear accelerometer constrained for a single degree of freedom.

Our attention is directed here to a linear accelerometer having a single degree of freedom as shown in Fig. 22-11. The frame of the accelerometer is assumed attached to the missile frame. Let

$$y = \text{motion of } M \text{ relative to inertial space†}$$

$$M = \text{accelerometer mass}$$

$$x_i = \text{motion of frame relative to inertial space}$$

and

$$x_0 = y - x_i = \text{motion of } M \text{ relative to frame} \qquad (22\text{-}44)$$

The quantity x_0 is considered the output of the accelerometer. The input quantity may be taken to be either the displacement x_i or the input acceleration $s^2 x_i$. The equation of motion for the system is found by equating to zero the sum of the forces associated with an assumed displacement of the mass and the frame. Thus

$$M\frac{d^2 y}{dt^2} + F\left(\frac{dy}{dt} - \frac{dx_i}{dt}\right) + K(y - x_i) = 0 \qquad (22\text{-}45)$$

Three transfer functions can be identified for the accelerometer. One of these follows directly from Eq. (22-45). Thus, rearranging and taking the Laplace transform yields

$$s^2 Y + \frac{F}{M}sY + \frac{K}{M}Y = \frac{F}{M}sX_i + \frac{K}{M}X_i \qquad (22\text{-}46)$$

† Inertial space is the reference frame in which a force-free body is unaccelerated; it is celestial space determined by the "fixed stars."

Introducing

$$\frac{F}{M} = 2\zeta\omega_n \quad \text{and} \quad \omega_n = \sqrt{\frac{K}{M}}$$

and formulating the ratio of output to input results in the expression

$$\frac{Y(s)}{X_i(s)} = \frac{1 + (2\zeta/\omega_n)s}{(s/\omega_n)^2 + (2\zeta/\omega_n)s + 1} \tag{22-47}$$

The transfer function of Eq. (22-47) applies whenever the output is measured relative to inertial space.

A more useful form of the output of the accelerometer, however, occurs when the displacement of the mass M is measured relative to the frame and not to inertial space. To provide this transfer function, Eq. (22-44) is inserted in Eq. (22-45) to yield

$$\frac{d^2x_0}{dt^2} + 2\zeta\omega_n\frac{dx_0}{dt} + \omega_n^2x_o = -\frac{d^2x_i}{dt^2} \tag{22-48}$$

Upon taking the Laplace transform and formulating the ratio of output to input, we obtain the desired result. Thus

$$\frac{X_o(s)}{X_i(s)} = -\frac{s^2}{s^2 + 2\zeta\omega_n s + \omega_n^2} = -\frac{1}{\omega_n^2}\frac{s^2}{(s/\omega_n)^2 + (2\zeta/\omega_n)s + 1} \tag{22-49}$$

The attenuation and phase characteristics of this transfer function are shown in Fig. 22-12. It is interesting to note that these curves are mirror images of the characteristics plotted in Fig. 22-10.

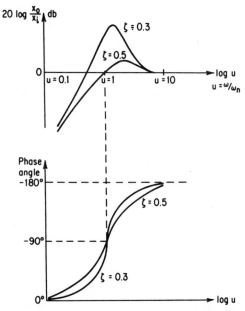

Fig. 22-12. Attenuation and phase of X_o/X_i versus frequency for a linear accelerometer.

Examination of Eq. (22-49) or of its corresponding Bode plots reveals some useful information about the application limitations of linear accelerometers. For example, the question may arise regarding the advisability of using the accelerometer as a measuring device for detecting low-frequency oscillations in the spacecraft air frame. Equation (22-49) shows that the transfer function under such circumstances is approximately

$$\frac{X_0}{X_i} \approx -\frac{1}{\omega_n^2} s^2 \tag{22-50}$$

However, the accelerometer's natural frequency is designed to be very high by virtue of a small mass and a very stiff spring. Accordingly, there is very little observable output displacement if it is used for such a purpose. In fact the output signals may very well fall within the noise level of the system, thereby making such measurements unreliable. By the same token, however, it follows that to detect and measure an oscillation having a frequency comparable with that of the natural frequency of the accelerometer, the use of the accelerometer is satisfactory.

Of most importance is the transfer function that relates the output displacement to input acceleration. The right side of Eq. (22-48) represents the input acceleration, which we shall now call a_x. Consequently the desired transfer function becomes

$$\frac{X_o}{a_x} = -\frac{1}{s^2 + 2\zeta\omega_n + \omega_n^2} = -\frac{1}{\omega_n^2} \frac{1}{(s/\omega_n)^2 + (2\zeta/\omega_n)s + 1} \tag{22-51}$$

which is readily recognized as the standard quadratic form. Since the input quantity is acceleration, it follows that the output displacement X_o is proportional to input acceleration. Recalling the nature of the second-order response curves, we can easily understand why the damping ratio in these units is kept between 0.4 and 0.7. In this way large overshoot and therefore false indications of peak accelerations are avoided. Furthermore, the large natural frequency is designed into the system to assure the ability of the accelerometer to measure acceleration pulses with a high degree of accuracy.

Since the magnitude of X_o provides the means of measuring spacecraft acceleration, the accelerometer serves an important function in the spacecraft autopilot. Through the inclusion of an accelerometer in a feedback path of the spacecraft air frame, it follows that, whenever the spacecraft output acceleration deviates from the command value, an actuating signal arises which acts to remove this lack of correspondence. Thus, high performance is assured.

The Gyroscope. The gyroscope has achieved wide use in the aircraft industry as an indicating instrument and as a component in fire control, automatic flight control, and navigational systems. The principle of the gyroscope is well known and is based upon the law of conservation of angular momentum.

When this instrument is used as a practical component, the motion of the spinning wheel of the gyroscope is usually restrained by mounting it in a set of bearings called *gimbals*. It is customary to classify gyroscopes in terms of the number of degrees of freedom of angular motion of the spinning wheel. Figure 22-13(a) shows a system with one degree of freedom and Fig. 22-13(b) a system

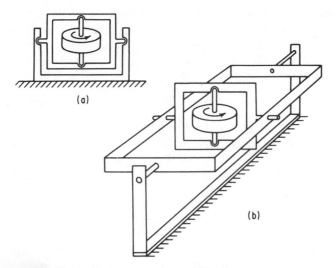

Fig. 22-13. Gyroscope degrees of freedom:
(a) one-degree-of-freedom gyroscope;
(b) two degrees-of-freedom gyroscope.

with two degrees of freedom. There are many types of gyroscopes in practical use, including the rate gyro, the integrating rate gyro, and the directional gyro of the one-degree-of-freedom class, and the displacement gyro, the vertical gyro, and the gyrocompass of the two-degrees-of-freedom class. Because of the general complexity of the subject and the limited treatment possible here, only the transfer functions of some simple one-degree-of-freedom components are derived.

Figure 22-14 illustrates a general one-degree-of-freedom system with an

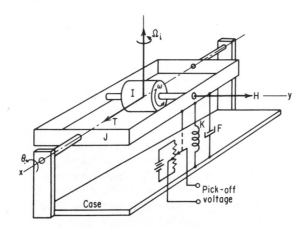

Fig. 22-14. A generalized one-degree-of-freedom gyroscope.

elastic restraint (spring K) and a viscous damper (dashpot F) about the free, or output, axis x. If the gyro case is rotated about the z axis with an angular velocity Ω_i, then a torque $T = H\Omega_i$ is developed as shown, which tends to rotate the gyro through an angle θ about the gimbal axis x. The unit tends to rotate so as to align its angular-momentum vector H with the input angular-velocity vector Ω_i. This torque is opposed by the inertia of the system about the output x axis, the spring restraint K, and the viscous damping F. If the moment of inertia about the x axis is denoted by J and the angular displacement by θ, then the equation for the dynamic balance of the system is given by

$$H\Omega_i = J\frac{d^2\theta}{dt^2} + F\frac{d\theta}{dt} + K\theta \qquad (22\text{-}52)$$

The transfer function between the output angle θ and the input angular velocity is therefore given by

$$\frac{\theta}{\Omega_i} = \frac{H}{Js^2 + Fs + K} \qquad (22\text{-}53)$$

Mathematically this equation is similar in form to Eq. (22-37), and the sinusoidal steady-state characteristics are as shown in Fig. 22-10. Usually the output angle is transformed to a voltage by means of a "pick-off" transducer, which is pictured as a potentiometer in Fig. 22-14 although differential transformers and other types of pick-offs are also used.

This type of gyro is known as a *rate gyro* and is commonly used for measuring aircraft motion and for computing rates of turn for use in autopilots and gunfire control systems. The application of this device as a rate-of-turn indicator may be illustrated by considering the steady-state output deflection for a constant angular-velocity Ω_{i_o} input. The steady-state deflection about the output axis is therefore given by (letting s go to zero)

$$\theta_{ss} = \frac{H}{K}\Omega_{io} \qquad (22\text{-}54)$$

Thus the output angular displacement is proportional to the constant input angular velocity. Some damping is always necessary in this gyroscope to prevent an oscillatory response.

It is important to note that a constant deflection θ_{ss} introduces an error into the system because it causes the H vector to move away from the y axis, and the input vector Ω_i is therefore no longer in line with the z axis. The resulting error in angular-rate measurement is said to be due to *geometric cross coupling*. In order to keep its effect to a minimum, the angular rotation about the output axis must be restrained to small values by increasing the spring constant K. This requires that the pick-off transducer be extremely sensitive, which in turn gives rise to problems of drift and noise. The effects of frictional torques in the output bearings and pick-off transducer are also a major source of error in the system. Thus designing a precision gyroscope becomes a difficult engineering problem.

If the elastic restraint of the rate gyro is removed, then the instrument becomes an *integrating rate gyro*. This unit is also known as an integrating gyro or

HIG (hermetically sealed integrating gyro). In this case the equation for dynamic balance about the output axis is given by

$$H\Omega_i = J\frac{d^2\theta}{dt^2} + F\frac{d\theta}{dt}$$ (22-55)

The transfer function corresponding to Eq. (22-55) depends on how the input and output variables are defined, which, of course, depends upon the particular application. If the output pick-off transducer is sensitive to angular deflection about the output axis, then θ becomes the output variable. Usually the input to the instrument is taken as the angular displacement about the input z axis (denoted by the letter θ_i) rather than the input angular velocity Ω_i (note that $\Omega_i = d\theta_i/dt$). For these variables the transfer function becomes

$$\frac{\theta}{\theta_i} = \frac{H}{Js + F} \approx \frac{H}{F} \quad \text{for negligible } J$$ (22-56)

If the gyroscope input consists of a constant angle θ_{io}, then the steady-state deflection about the output axis is given by

$$\theta_{ss} = \frac{H}{F}\theta_{io}$$ (22-57)

Thus the steady-state output angular displacement is proportional to the input angular displacement. The problem of geometric cross coupling also exists in the integrating rate gyro. Typical values for H/F are of the order of 1 to 10 so that the problem cannot be neglected. Because of this, integrating rate gyros are usually restricted to applications such as stabilized platforms, where the output angle is returned close to the zero position by feedback.

Output Actuators. Very often control systems contain components that generate the controlled variable or the manipulated variable in a manner that exhibits a pure integration property. In such cases the transfer function of the component assumes the general form given by Eq. (22-58):

$$\frac{Y(s)}{X(s)} = \frac{K}{s(1 + s\tau)}$$ (22-58)

where $Y(s)$ is the Laplace transform of the output variable and $X(s)$ is the Laplace transform of the input variable. Our knowledge of the Laplace transform of a time integral tells us that the presence of s as a root factor in the denominator means that the output variable is a function of the integral of the input variable. The pure integration property is a very useful one in control systems because it eliminates the need for errors in the steady state. This matter was briefly treated in Sec. 20-4.

The servomotor in the servomechanism of Fig. 20-3 has a transfer function the form of which is described by Eq. (22-58), provided the output quantity is considered to be the motor displacement angle c and the input is the voltage to the servomotor control winding m. Thus

$$\frac{C(s)}{M(s)} = \frac{K}{s(1 + s\tau_m)}$$ (22-59)

where K is an appropriate motor constant expressed in inverse volt-seconds and τ_m is the mechanical time constant of the motor expressed as the ratio of its inertia to viscous friction coefficient. The same is true for the output actuator of Fig. 20-7.

Another output device commonly found in control systems is the hydraulic actuator. It is used, for example, in the systems of Figs. 20-9 and 20-6. For the system of Fig. 20-6 both the electrohydraulic pilot valve and the positioning cylinder have transfer functions that are expressed by Eq. (22-58). In these cases, however, $Y(s)$ denotes the output power-ram piston position and $X(s)$ denotes the input-valve-spool displacement.

22-2 Transfer Functions of Systems

The overall system transfer function is composed of the transfer functions of the individual components of the system. The total direct transfer function, for example, is obtained by taking the product of the transfer function of each component encountered in traversing the path from the input point to the output point. In deter-mining the transfer function of the components, we assume that any loading effect that a succeed-ing stage has on a preceding one is taken into account. The transfer function G in Fig. 22-15 can be considered to be made up in this fashion. The feedback transfer functions H_1 and H_2 are also assumed to be so composed. The overall system transfer function is then found in terms of G, H_1, and H_2 by applying the feedback rela-tionship of Eq. (20-6).

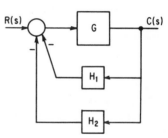

Fig. 22-15. A two-loop feedback system.

The expression for the actuating signal in the configuration of Fig. 22-15 for negative feedback is the input signal R minus the signals emerging from both feedback paths. Thus

$$E(s) = R(s) - H_1 C(s) - H_2 C(s)$$

Multiplying $E(s)$ by the direct transmission function yields the output

$$CE(s) = C(s) \tag{22-60}$$

or

$$G[R(s) - H_1 C(s) - H_2 C(s)] = C(s) \tag{22-61}$$

Collecting terms and rearranging leads to

$$\frac{C(s)}{R(s)} = \frac{G}{1 + H_1 G + H_2 G} \tag{22-62}$$

It is interesting to note that this result could have been written directly by applying Eq. (20-7) to the system of Fig. 22-15. The direct path from $R(s)$ to $C(s)$ is G; hence this is placed in the numerator. Appearing in the denominator must be: $1 + \sum$ loop transfer functions. In this instance there are two loop transfer functions: $H_1 G$ and $H_2 G$. Keep in mind that the signal flow associated with transfer functions is unilateral, i.e. flow is permissible only from the arrow side. For this reason H_1H_2 does not constitute a valid loop transfer function. On the other hand, H_1G and H_2G do, because for each a loop can be closed by progressing in the direction of the arrows.

EXAMPLE 22-1   Find the closed-loop transfer function of the system depicted in Fig. 22-16.

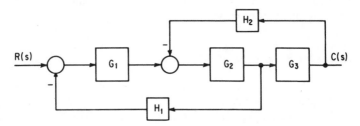

Fig. 22-16. System configuration for Example 22-1.

solution: By employing Eq. (20-7) we may write in this case

$$\frac{C(s)}{R(s)} = \frac{\text{direct transfer function}}{1 + \text{loop transfer functions}} \tag{22-63}$$

The direct path connecting input to output is described by $G_1G_2G_3$. Moreover, two loop transfer functions can be identified: $H_1G_1G_2$ and $H_2G_2G_3$. Inserting this information into Eq. (22-63) yields the desired result:

$$\frac{C(s)}{R(s)} = \frac{G_1G_2G_3}{1 + G_1H_1G_2 + H_2G_2G_3} \tag{22-64}$$

Sometimes feedback systems are subjected to more than one input simultaneously. Figure 22-17 illustrates such a system. What is the output under these circumstances? Since we are dealing with linear systems, the principle of superposition is applicable. The output to each input may thus be found separately and then summed. For the input $R_1(s)$ the direct transfer function is clearly $G_1G_2G_3$. Hence the closed-loop transfer function for this input is

$$\frac{C_1(s)}{R_1(s)} = \frac{G_1G_2G_3}{1 + HG_1G_2G_3} \tag{22-65}$$

where $C_1(s)$ is the component of the output due to $R_1(s)$. The direct transfer function

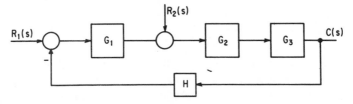

Fig. 22-17. Single-loop feedback system with multiple inputs.

for input $R_2(s)$ is G_2G_3 and the feedback factor in this case is HG_1. Accordingly, the closed-loop transfer function for this input is

$$\frac{C_2(s)}{R_2(s)} = \frac{G_2G_3}{1 + HG_1G_2G_3} \tag{22-66}$$

The total output may then be expressed as

$$C(s) = C_1(s) + C_2(s) = \frac{G_1G_2G_3R_1(s) + G_2G_3R_2(s)}{1 + HG_1G_2G_3} \tag{22-67}$$

A comparison of Eqs. (22-65) and (22-66) shows that the denominator expressions are the same. A little thought discloses that this result is entirely expected, because this denominator represents the characteristic equation of the system and the nature of the dynamic response is the same no matter where an input or a disturbance is applied anywhere in the loop.

22-3 System Classification

    Feedback control systems are classified in accordance with their steady-state performance, which in turn depends upon the number of pure integrations that appear explicitly in the expression for the loop transfer function. One convenient way of identifying the steady-state performance of a control system is to determine the final value of the actuating signal for stand-

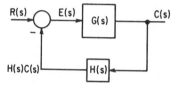

Fig. 22-18. Feedback control system block diagram.

ard input functions such as the step, the ramp, and the parabola. The expression that relates the actuating signal $E(s)$ to the input command $R(s)$ follows readily from Fig. 22-18. Thus

$$E(s) = R(s) - H(s)C(s) \tag{22-68}$$

or

$$R(s) = E(s) + H(s)C(s) \tag{22-69}$$

Dividing through by $E(s)$ yields

$$\frac{R(s)}{E(s)} = 1 + H(s)\frac{C(s)}{E(s)} \tag{22-70}$$

But

$$\frac{C(s)}{E(s)} = G(s) \tag{22-71}$$

Hence

$$\frac{R(s)}{E(s)} = 1 + H(s)G(s) \tag{22-72}$$

or

$$\boxed{E(s) = \frac{R(s)}{1 + H(s)G(s)}} \tag{22-73}$$

which is the desired result. The quantity $H(s)G(s)$ is the loop transfer function. Sometimes it is called the *open-loop* transfer function because, if the loop is opened at the point where the feedback is combined with the input signal, the transmission function between the input and that point is $H(s)G(s)$. This expression, in fact, plays the key role in establishing the classification of control systems.

The general expression of the loop transfer function is of the form

$$H(s)G(s) = \frac{K(1 + s\tau_a)(1 + s\tau_b) \cdots \left[\left(\dfrac{s}{\omega_{na}}\right)^2 + \dfrac{2\zeta_a}{\omega_{na}}s + 1\right]}{s^N(1 + s\tau_1)(1 + s\tau_2) \cdots \left[\left(\dfrac{s}{\omega_{n1}}\right)^2 + \dfrac{2\zeta_1}{\omega_{n1}}s + 1\right]} \qquad (22\text{-}74)$$

Often it is directly available in the factored form indicated because it is obtained by multiplying together the transfer functions of the individual components encountered as the input signal passes through the direct and feedback paths. For reference purposes systems are classified in accordance with the particular value of N in Eq. (22-74). Thus we have

for $N = 0$,   a Type 0 system

for $N = 1$,   a Type 1 system

for $N = 2$,   a Type 2 system

for $N = 3$,   a Type 3 system

Clearly, such a classification scheme places emphasis on the integration properties of the loop transfer function.

Equation (22-73) is expressed entirely in terms of the Laplace transform variable s. This is useful because it allows the final-value theorem of the Laplace transform to be used directly in identifying the steady-state performance for various input functions. Thus the steady-state value of the actuating signal e_{ss} may be expressed as

$$e_{ss} = \lim_{t \to \infty} e(t) = \lim_{s \to 0} sE(s) \qquad (22\text{-}75)$$

Step Input. Let us now investigate the steady-state behavior of each system type to a step input command. If the step command has a value r_0, the corresponding Laplace transform is $R(s) = r_0/s$. Hence by Eq. (22-73) the transformed expression for the actuating signal becomes

$$E(s) = \frac{1}{1 + H(s)G(s)}\left(\frac{r_0}{s}\right) \qquad (22\text{-}76)$$

When this is inserted into Eq. (22-75), we obtain

$$e_{ss} = \lim_{s \to 0} s\frac{1}{1 + H(s)G(s)}\left(\frac{r_0}{s}\right) = \frac{r_0}{1 + \lim_{s \to 0} H(s)G(s)} \qquad (22\text{-}77)$$

For convenience we define

$$K_p \equiv \lim_{s \to 0} H(s)G(s) \qquad (22\text{-}78)$$

Equation (22-77) can then be written as

$$e_{ss} = \frac{r_0}{1 + K_p} \qquad (22\text{-}79)$$

where K_p may be found for each system type by means of Eq. (22-78). Accordingly, we find the following results:

Type 0:

$$K_p = \lim_{s \to 0} \frac{K(1 + s\tau_a) \cdots}{(1 + s\tau_1)(1 + s\tau_2) \cdots} = K; \qquad \therefore e_{ss} = \frac{r_0}{1 + K}$$

Type 1:

$$K_p = \lim_{s \to 0} \frac{K(1 + s\tau_a) \cdots}{s(1 + s\tau_1)(1 + s\tau_2) \cdots} = \infty; \qquad \therefore e_{ss} = \frac{r_0}{1 + \infty} = 0$$

Types 2 and 3:

$$K_p = \infty; \qquad\qquad\qquad\qquad \therefore e_{ss} = 0$$

These results clearly point out that for any system that contains pure integration there is no need for a steady-state actuating signal. This means that there is exact correspondence between the command and the controlled variable in the steady state. Only the Type 0 system calls for a lack of correspondence. In situations where such errors can be tolerated, it is advisable to choose the Type 0 system because it is usually cheaper to build and easier to stabilize.

On the basis of the manner in which K_p is used, it is easy to understand why it is called the figure of merit that identifies the steady-state performance of systems. In particular it is called the *position* constant, because it was first used in connection with servomechanisms where the controlled variable was position. A more general name would be the *step* const nt, since the command is calling for a step change in controlled variable—be it position, temperature, pressure, or whatever.

Ramp Input. Consider next the application of a ramp input command to each system type. A ramp command means that we are calling for a change in the controlled variable that changes linearly with time in the final state. If the ramp input is $r(t) = r_1 t u(t)$, the corresponding Laplace transform is $R(s) = r_1/s^2$. Hence the expression for the steady-state value of the actuating signal becomes

$$e_{ss} = \lim_{s \to 0} s \frac{1}{1 + H(s)G(s)} \left(\frac{r_1}{s^2}\right) = \lim_{s \to 0} \frac{r_1}{s + sH(s)G(s)}$$

$$= \frac{r_1}{\lim_{s \to 0} sH(s)G(s)} \qquad (22\text{-}80)$$

or

$$e_{ss} = \frac{r_1}{K_v} \qquad (22\text{-}81)$$

where

$$K_v \equiv \lim_{s \to 0} sH(s)G(s) \qquad (22\text{-}82)$$

By means of Eqs. (22-81) and (22-82) the final-state performance of each system type is readily found. Thus

Type 0:

$$K_v = \lim_{s \to 0} s \frac{K(1 + s\tau_a)\cdots}{(1 + s\tau_1)\cdots} = 0; \qquad \therefore e_{ss} = \frac{r_1}{0} \longrightarrow \infty$$

Type 1:

$$K_v = \lim_{s \to 0} s \frac{K(1 + s\tau_a)\cdots}{s(1 + s\tau_1)\cdots} = K; \qquad \therefore e_{ss} = \frac{r_1}{K}$$

Types 2 and 3:

$$K_v = \lim_{s \to 0} s \frac{K(1 + s\tau_a)\cdots}{s^2(1 + s\tau_1)\cdots} = \infty; \qquad \therefore e_{ss} = \frac{r_1}{\infty} = 0$$

The first result above shows that a Type 0 system requires an infinite actuating signal to follow a ramp command in the steady state. As the actuating signal increases with time, saturation of one or more of the system components sets in and makes the Type 0 system incapable of following the ramp command. In the case of the Type 1 system the ramp command can be followed but with a fixed lag error in the controlled variable in the final state. Of course the magnitude of this error can be reduced through the use of higher loop gains. Type 2 systems and higher are capable, in the steady state, of following the ramp command exactly—that is, not only at the same rate of change as the controlled variable but also at the commanded level of the controlled variable.

The constant K_v is the figure of merit that identifies the steady-state performance of a system to a ramp command. It is frequently called the *velocity* constant because it was first introduced in connection with the servomechanism where a ramp command of position meant a velocity command. A more appropriate name is the *ramp* constant.

Parabolic Input. A parabolic command calls for the controlled variable to change in accordance with the square of time. For convenience assume $r(t) = \frac{1}{2}r_2 t^2 u(t)$. Then the corresponding Laplace transform is $R(s) = r_2/s^3$. This leads to

$$e_{ss} = \lim_{s \to 0} s \frac{1}{1 + H(s)G(s)} \left(\frac{r_2}{s^2}\right) = \frac{r_2}{\lim\limits_{s \to 0} s^2 H(s)G(s)} \tag{22-83}$$

or

$$\boxed{e_{ss} = \frac{r_2}{K_a}} \tag{22-84}$$

where

$$\boxed{K_a \equiv \lim_{s \to \infty} s^2 H(s)G(s)} \tag{22-85}$$

Upon evaluating the K_a value for each system type and inserting into Eq. (22-84), we obtain the steady-state performance. Thus

Type 0:

$$K_a = \lim_{s \to 0} s^2 \frac{K(1 + s\tau_a) \cdots}{(1 + s\tau_1) \cdots} = 0; \qquad \therefore e_{ss} = \frac{r_2}{K_a} \longrightarrow \infty$$

Type 1:

$$K_a = \lim_{s \to 0} s^2 \frac{K_a(1 + s\tau_a) \cdots}{s(1 + s\tau_1) \cdots} = 0; \qquad \therefore e_{ss} \longrightarrow \infty$$

Type 2:

$$K_a = \lim_{s \to 0} s^2 \frac{K(1 + s\tau_a) \cdots}{s^2(1 + s\tau_1) \cdots} = K; \qquad \therefore e_{ss} = \frac{r_2}{K}$$

Type 3:

$$K_a = \lim_{s \to 0} s^2 \frac{K(1 + s\tau_a) \cdots}{s^3(1 + s\tau_1) \cdots} = \infty; \qquad \therefore e_{ss} = \frac{r_2}{\infty} = 0$$

These results make clear that neither the Type 0 nor the Type 1 system can follow a parabolic command. Hence in such situations one must employ at least a Type 2 system, which contains two pure integrators. This system will then be able to generate a parabolic output but always with a lag error of r_2/K with respect to the commanded value. The use of a Type 3 system is required if the lag error is to be eliminated entirely. The Type 3 system not only is capable of generating the parabolic output but does so at exactly the commanded level of the controlled variable.

The constant K_a is the figure of merit that identifies the steady-state performance of a system to a parabolic command of the controlled variable. As was the case with the poition and velocity constants, it came to be called an *acceleration* constant because it was first used in systems where the controlled variable was acceleration. A more appropriate name is the *parabolic* constant.

For ready reference the foregoing results are tabulated in Table 22-1. Note that the useful portion of the table for various system and input types is marked off in heavy lines.

22-4 System Stability

The nature of the dynamic response of a linear feedback control system is determined by the roots of its characteristic equation. If these roots are negative real, or in the case of complex roots have negative real parts, the transient terms all decay with time. Such a system is called *absolutely stable*. On the other hand, if any one of the real roots is positive, or if any of the complex roots has a positive real part, the transient terms associated with these roots increase with time. In a linear system the increase in output would be without limit. In practical situations, however, the large outputs, when fed back, cause one or several components to saturate, with the end result that the system goes into a state of sustained oscillation. Such a system is called *absolutely unstable*. It follows that the simple criterion for establishing the stability of a system is to determine whether the solution to the system characteristic equation contains roots that lie entirely within the left-half s plane. Although this rule is simple, the amount of labor involved

Table 22-1. Tabulation of the Steady-state Actuating Signal e_{ss} for Various System Classifications and Inputs[†]

$$K_p = \lim_{s\to 0} HG(s) \qquad K_v = \lim_{s\to 0} sHG(s) \qquad K_a = \lim_{s\to 0} s^2 HG(s)$$

System type	Input type		
	A step $r(t) = r_0 u(t)$. $R(s) = \dfrac{r_0}{s}$	A ramp $r(t) = r_1 tu(t)$. $R(s) = \dfrac{r_1}{s^2}$	A parabola $r(t) = \dfrac{1}{2} r_2 t^2 u(t)$. $R(s) = \dfrac{r_2}{s^3}$
$N = 0$ $HG(s) = \dfrac{K(1+s\tau_a)\cdots}{(1+s\tau_1)\cdots}$	$e_{ss} = \dfrac{r_0}{1+K_p} = \dfrac{r_0}{1+K}$ Step output with constant actuating signal	Actuating signal increases with time	Actuating signal increases with time
$N = 1$ $HG(s) = \dfrac{K(1+s\tau_a)\cdots}{s(1+s\tau_1)\cdots}$	Step output with zero actuating signal	$e_{ss} = \dfrac{r_1}{K_v} = \dfrac{r_1}{K}$ Ramp output with constant actuating signal	Actuating signal increases with time
$N = 2$ $HG(s) = \dfrac{K(1+s\tau_a)\cdots}{s^2(1+s\tau_1)\cdots}$	Step output with zero actuating signal	Ramp output with zero actuating signal	$e_{ss} = \dfrac{r_2}{K_a} = \dfrac{r_2}{K}$ Parabolic output with constant actuating signal
$N = 3$ $HG(s) = \dfrac{K(1+s\tau_a)\cdots}{s^3(1+s\tau_1)\cdots}$	Step output with zero actuating signal	Ramp output with zero actuating signal	Parabolic output with zero actuating signal

† From V. Del Toro and S. R. Parker, *Principles of Control Systems Engineering* (New York: McGraw-Hill Book Company, 1960), by permission.

in finding the solution for systems beyond third order can be considerable. Fortunately, procedures are available that enable us to determine the presence of a positive root in the solution to the characteristic equation without formally finding the solution. Two such schemes are described in this section without proof. These are the *Routh criterion* and the *Nyquist criterion*. The latter, we shall see, is the more useful technique.

As we consider the Routh and Nyquist methods of determining stability, it will be useful to keep in mind the general form of the characteristic equation of a feedback control system:

$$1 + H(s)G(s) = 0 \tag{22-86}$$

This result follows directly from Eq. (20-6) upon recalling that the denominator of the closed-loop transfer function, when set equal to zero, is the characteristic equation of the closed-loop system. This particular representation of the characteristic equation is especially useful because it is expressed in terms of the open-loop transfer function. It is also interesting to note that this same function serves as the basis for system classification.

Routh Criterion.† The open-loop transfer function is readily found by taking the product of the individual system components as one progresses round the loop with the loop assumed open at the point of primary feedback. In general, $H(s)G(s)$ may be expressed as the ratio of a numerator and a denominator polynomial in s. Thus

$$H(s)G(s) = \frac{N(s)}{D(s)} \tag{22-87}$$

If Eq. (22-87) is inserted into Eq. (22-86), we find that the characteristic equation may be expressed as

$$N(s) + D(s) = 0 \tag{22-88}$$

Assume that these steps have been carried out for an nth-order system and that the resulting form of the characteristic equation is found to be

$$B_n s^n + B_{n-1} s^{n-1} + B_{n-2} s^{n-2} + \cdots + B_1 s + B_0 = 0 \tag{22-89}$$

To apply the Routh method it is necessary to form an array of numbers. The first two rows of the array are taken directly from the coefficients of the characteristic equation. The first row consists of the coefficient of the s^n term in Eq. (22-89) plus all other coefficients of those powers of s that differ from n by two or multiples of two. It does not matter whether n is odd or even. The second row is formed by taking the coefficient of the s^{n-1} term and all other coefficients of those powers of s that differ from $n - 1$ by two or multiples of two. Thus, for Eq. (22-89) the first two rows look like

$$\begin{array}{cccc} B_n & B_{n-2} & B_{n-4} & \cdots \\ B_{n-1} & B_{n-3} & B_{n-5} & \cdots \end{array}$$

† E. J. Routh, *Dynamics of a System of Rigid Bodies*, 3rd ed. (London: Macmillan and Co., Ltd., 1877).

All succeeding rows are then derived in accordance with the following formulations. The terms in the third row are computed from

$$a_1 = \frac{B_{n-1}B_{n-2} - B_n B_{n-3}}{B_{n-1}}; \quad a_2 = \frac{B_{n-1}B_{n-4} - B_n B_{n-5}}{B_{n-1}}; \quad \cdots \quad (22\text{-}90)$$

The array then takes the form

$$
\begin{array}{cccc}
B_n & B_{n-2} & B_{n-4} & \cdots \\
B_{n-1} & B_{n-3} & B_{n-5} & \cdots \\
a_1 & a_2 & a_3 & \cdots
\end{array}
$$

The terms in the fourth row are computed in a similar fashion. Thus

$$b_1 = \frac{a_1 B_{n-3} - B_{n-1}a_2}{a_1}; \quad b_2 = \frac{a_1 B_{n-5} - B_{n-1}a_3}{a_1}; \quad \cdots \quad (22\text{-}91)$$

This procedure is continued until the last computable row is reached. The final array would then assume a form similar to the following:

$$
\begin{array}{cccc}
B_n & B_{n-2} & B_{n-4} & \cdots \\
B_{n-1} & B_{n-3} & B_{n-5} & \cdots \\
a_1 & a_2 & a_3 \\
b_1 & b_2 \\
c_1 & c_2 \\
d_1 \\
e_1
\end{array}
$$

After the array is completed, the following criterion is applied: *the number of changes in sign for the terms in the first column equals the number of roots of the characteristic equation with positive real parts.* Hence, by the Routh criterion, for a system to be stable the array resulting from its characteristic equation must have a first column with terms of the same sign.

To illustrate the procedure in terms of a numerical example, consider that the characteristic equation of a feedback control system is

$$s^5 + 6s^4 + 3s^3 + 2s^2 + s^1 + 1 = 0 \qquad (22\text{-}92)$$

This leads to the following array of terms:

$$
\begin{array}{ccl}
1 & 3 & 1 \\
6 & 2 & 1 \\
16 & 5 & \text{after multiplication by 6} \\
2 & 16 & \text{after multiplication by 16} \\
-123 & 0 \\
16
\end{array}
$$

The terms in any row may be multiplied by a constant without changing the results. It is convenient to use this procedure to avoid fractions in the array.

Examination of the first column of the array reveals that there are two changes in sign—from $+2$ to -123 and from -123 to $+16$. Therefore the characteristic equation has two roots with positive real parts; hence the system is unstable.

A little thought about formation of the array discloses that, if the characteristic equation of (22-89) has any order of s missing or coefficients that are not of the same sign, the system is unstable.

For a system of any given order it is possible to use the Routh criterion to determine the conditions that must exist between the coefficients of the characteristic equation for the system to be absolutely stable. Consider, for example, the characteristic equation of a third-order system where the coefficients are positive numbers. Thus

$$B_3 s^3 + B_2{}^2 + B_1 s + B_0 = 0 \qquad (22\text{-}93)$$

The corresponding array of terms is

$$
\begin{array}{cc}
B_3 & B_1 \\
B_2 & B_0 \\
\dfrac{B_2 B_1 - B_3 B_0}{B_2} & \\
B_0 &
\end{array}
$$

Inspection of the first column leads to the conclusion that a necessary and sufficient condition for absolute stability in a linear third-order system is the fulfillment of the following inequality:

$$B_2 B_1 > B_3 B_0 \qquad (22\text{-}94)$$

Nyquist Criterion. The Routh criterion has several serious shortcomings. First, it assumes that the characteristic equation is available in polynomial form. This may not always be so, especially in such areas as the process and aircraft industries where information about the dynamics of some system components often is available only in terms of steady-state sinusoidal frequency-response data. Second, although the Routh criterion gives information about absolute stability, it conveys little or no information about how close the system may be to becoming unstable. The Routh array may show no change in sign in the first column but the ensuing dynamic response may be characterized by overshoots so excessive as to render the system useless for control purposes. Thus the system may be *relatively unstable* in spite of the fact that it is absolutely stable. Finally, the Routh method does not provide the facility for selecting in a simple and direct fashion the parameters of a system component to stabilize the system when it is found to be absolutely unstable.

These disadvantages are effectively overcome by the Nyquist stability criterion. The Nyquist stability test is basically a graphical method that can be applied to experimental data of the sinusoidal open-loop transfer function or to computed results. Information about relative stability is readily obtained through the use of appropriate figures of merit that are easily found from the Nyquist plot. Moreover, the effect on the relative or absolute stability of changing any system parameter is directly observable in terms of the manner in which the figures of merit are changed through the reshaping of the Nyquist plot.

Information about the existence of roots in the right-half s plane is provided by the Nyquist method through the rotational properties of the open-loop transfer function as it is plotted in the complex plane for all values of ω. We shall now describe in a brief and nonrigorous fashion how this is accomplished. If we insert Eq. (22-87) into Eq. (22-86), the characteristic equation can be expressed as

$$F(s) \equiv \frac{N(s) + D(s)}{D(s)} = 0 \qquad (22\text{-}95)$$

where $N(s)$ and $D(s)$ are the numerator and denominator polynomials in s of the open-loop transfer function. For the present we shall focus attention on the rotational properties of the function $F(s)$. In this connection assume that all the root factors of $F(s)$ are known. Equation (22-95) can then be written as

$$F(s) = \frac{(s - s_a)(s - s_b) \cdots (s - s_p)}{(s - s_1)(s - s_2) \cdots (s - s_q)} = 0 \qquad (22\text{-}96)$$

where $s_a, s_b, \cdots, s_p$ are the assumed known zeros of $F(s)$ and $s_1, s_2, \cdots, s_q$ are the assumed known poles of $F(s)$. These poles and zeros may be complex quantities.

Before proceeding further, we should take note that we are dealing with the problem in reverse order. We have assumed that the roots of the characteristic equation $(s_a, s_b, \cdots, s_p)$ are already known. This assumption is made so that we can better see how the choice of values of s in Eq. (22-96) influences the rotational properties of $F(s)$.

In general, for each value of s that is used in Eq. (22-96) there results a value of $F(s)$ that is a complex number. If a series of values is selected for s, a series of corresponding values is found for $F(s)$. Upon joining the selected s points as well as the $F(s)$ points, we can identify definite curves. For example, if values of s are chosen on the curve C_1 of Fig. 22-19(a) and inserted into Eq. (22-96), the resulting curve for $F(s)$ may take the form depicted in Fig. 22-19(b). It is important to note that, if a closed curve is selected in the s plane, the corresponding plot in the $F(s)$ plane is also a closed curve. The way in which the F_1 function closes in the $F(s)$ plane is very critically dependent upon whether any of the poles or

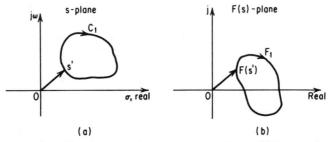

Fig. 22-19. Illustrating how values of the complex variable s selected on curve C_1 of (a) plot into curve F_1 in (b) through the use of Eq. (22-96).

zeros of $F(s)$ is included inside the closed curve C_1. If neither a zero nor a pole of $F(s)$ is included, then the rotational characteristic of F_1 will be such that the radius vector does not encircle the origin of the $F(s)$ plane as it moves along the F_1 curve.

Consider next the situation depicted in Fig. 22-20. Here it is assumed that a zero of the characteristic equation, s_a, is located in the right half-plane and that curve C_2 is chosen to enclose it. Now, if values of s are chosen on the curve C_2 in a clockwise fashion and inserted into Eq. (22-96), the resulting plot in the $F(s)$ plane must be such that the radius vector undergoes a net clockwise encirclement of the origin as the curve F_2 is traversed. This net clockwise encirclement is called for by the fact that as values of s are selected on C_2 in clockwise fashion, the root factor $(s - s_a)$, which appears in the numerator of Eq. (22-96), undergoes a clockwise rotation thus reflecting in the $F(s)$ function.

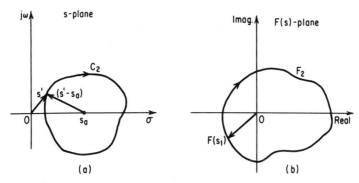

Fig. 22-20. (a) Encirclement of a root of the characteristic equation by C_2.
(b) Corresponding plot of $F(s)$ function. Note encirclement of origin.

Consider next the situation where the s-plane path encloses two zeros and a pole of the $F(s)$ function as depicted in Fig. 22-21(a). Examination of Eq. (22-96) now shows that there are two clockwise encirclements of the origin in the $F(s)$

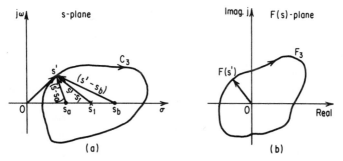

Fig. 22-21. (a) An s-plane path enclosing two zeros and one pole of the $F(s)$ function;
(b) the corresponding $F(s)$ plot.

plane due to the two zeros and one counterclockwise encirclement due to one pole. Hence the $F(s)$ function F_3 must show a net clockwise encirclement. In general it can be concluded that the number of net encirclements N of the origin in the $F(s)$ plane is

$$N = Z_r - P_r \qquad (22\text{-}97)$$

where Z_r is the number of zeros encircled in the right half-plane by the s-plane path and P_r is the number of poles of $F(s)$ so enclosed.

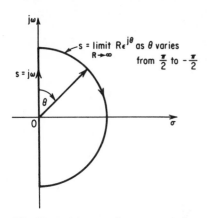

Fig. 22-22. The Nyquist path enclosing the entire right-half plane.

Up to this point we have assumed that the root factors of the $F(s)$ function were available and we have seen how, by taking paths encircling zeros and poles of $F(s)$ in the right-half plane, encirclement of the origin in the $F(s)$ plane resulted. The complex function theory shows that reasoning conversely is also valid. Hence, if a path in the right half-plane is chosen so as to assure the enclosure of any zero or pole of $F(s) = 1 + H(s)G(s)$, information about the existence of such roots may be obtained by investigating the rotational characteristics of $F(s)$ as s is allowed to vary along this path. In this way we can ascertain the presence of roots with positive real parts without solving for the roots of the characteristic equation. Depicted in Fig. 22-22 is the s-plane path which is used to assure the enclosure of poles and/or zeros of $F(s)$ lying in the right half-plane. It is called the *Nyquist path*.

When we deal with the rotational properties of the $F(s)$ function our concern is with encirclement of the origin because we are interested in the characteristic equation

$$F(s) = 1 + H(s)G(s) = 0 \qquad (22\text{-}98)$$

However, a glance at this equation discloses that

$$\boxed{H(s)G(s) = -1} \qquad (22\text{-}99)$$

is equivalent to $F(s) = 0$. Therefore we may either plot $F(s)$ and investigate enclosure of the origin or else plot $H(s)G(s)$ and investigate enclosure of the $(-1, 0)$ point. Because the open-loop transfer function is more readily available and is easier to work with, we obtain information about the existence of roots in the right half-plane by plotting $H(s)G(s)$ for values of s on the Nyquist path. Moreover, considerable simplicity results when we deal with $H(s)G(s)$ because in almost all practical systems the order of s in $D(s)$ is greater than in $N(s)$. This means that all the points on the infinite semicircle of the Nyquist path plot into an infinitesimal semicircle about the origin in the complex plane where $H(s)G(s)$ is plotted.

Furthermore, as a result of the complex conjugate property of complex numbers, the plot of the open-loop transfer function for negative real frequencies (i.e., $s = -j\omega$) is the mirror image of that obtained for positive real frequencies ($s = j\omega$). Through these simplifications the rotational character of $H(s)G(s)$ is entirely revealed by plotting the open-loop transfer function for real frequencies from some suitably low value to a suitably high value. Keep in mind that this is the kind of information that can readily be obtained experimentally or conveniently computed.

The Nyquist method of determining the absolute stability of a system can be summarized, then, as follows: (1) Plot the open-loop transfer function $H(j\omega)G(j\omega)$. The data may either be computed or obtained experimentally. (2) Draw the mirror image of $H(j\omega)G(j\omega)$. (3) With a radius vector originating at the $-1 + j0$ point, follow the plot of $H(j\omega)G(j\omega)$ from $-j\infty$ to $+j\infty$ and note the number of encirclements, N, if any. Clockwise encirclements carry the plus sign. (4) Determine the value of P_r in Eq. (22-97). This is found from inspection of $D(s)$, which is usually available in factored form. (5) Insert the values of N and P_r into Eq. (22-97) and solve for Z_r, which is the number of roots of the characteristic equation located in the right half-plane. The system is absolutely stable only if Z_r is zero.

EXAMPLE 22-2 A Type 0 feedback control system is described by the open-loop transfer function

$$HG = \frac{K}{(1 + s\tau_1)(1 + s\tau_2)(1 + s\tau_3)} = \frac{K}{(1 + j\omega\tau_1)(1 + j\omega\tau_2)(1 + j\omega\tau_3)} \quad (22\text{-}100)$$

The polar plot, which is obtained by substituting values of ω from zero to infinity, is depicted in Fig. 22-23. Determine whether this system will be absolutely stable when the feedback loop is closed.

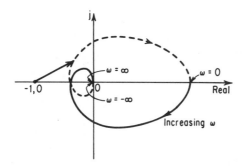

Fig. 22-23. The Nyquist plot of a third-order Type 0 system. For the gain specified, this system is absolutely stable.

solution: The curve drawn with the solid line is for positive values of ω. The mirror image is sketched with a broken line and is valid for negative values of ω. As the radial line originating from $-1 + j0$ is made to traverse the Nyquist plot from $-j\infty$ to $+j\infty$, it is noted that no net encirclement of the $-1 + j0$ point occurs. Hence $N = 0$. Furthermore, inspection of the denominator of Eq. (22-100) discloses that $P_r = 0$, i.e., the poles of $D(s)$ are all located in the left half s plane at $s = -1/\tau_1, -1/\tau_2$, and $-1/\tau_3$. Therefore it follows from Eq. (22-97) that $Z_r = 0$. Hence this system is absolutely stable.

EXAMPLE 22-3   The system of Example 22-2 has its gain increased to the point where the Nyquist plot appears as shown in Fig. 21–24. Is this system absolutely stable?

solution: The shape of the plot in this case is similar to that of Fig. 22-23. At any given frequency the angle of the transfer function is the same. The larger gain in no way influences this quantity; it merely increases the magnitude of the complex number.

By following the radial line from the $-1 + j0$ point along the Nyquist plot for values of ω from $-\infty$ to $+\infty$, we find that the radial line makes two net clockwise encirclements. Since $P_r = 0$, it follows that

$$Z_r = N + P_r = 2 + 0 = 2$$

Therefore, there exist two roots of the characteristic equation that lie in the right half s plane for the gain specified, and so the system is absolutely unstable. It is interesting to note that information about the stability of the *closed-loop* system is being obtained from a steady-state sinusoidal frequency analysis of the *open-loop* transfer function.

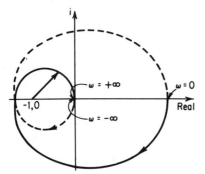
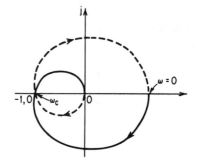

Fig. 22-24. The Nyquist diagram of the system of Fig. 22-23 for increased gain. The system is absolutely unstable.

Fig. 22-25. Illustrating the boundary condition for instability for the system of Fig. 22-23. A stable system results when the gain is reduced.

A comparison of the polar plots of Figs. 22-23 and 22-24 makes it apparent that the boundary between stability and instability occurs when the gain is such that the Nyquist plot passes through the $-1 + j0$ point as depicted in Fig. 22-25. Under these conditions the system goes into a state of sustained oscillations at the frequency ω_c. At this frequency the open-loop transfer function has the value

$$\frac{B}{E}(j\omega_c) = HG(j\omega_c) = -1 = 1\underline{/180°} \tag{22-101}$$

where $B(j\omega_c)$ is the signal in the primary feedback path (see Fig. 20-1). If it is recalled that the negative feedback connection introduces an additional $-180°$ phase shift, it follows that the output signal appearing at the feedback point is identical in magnitude and phase to whatever input signal appears at the input side of the direct transfer function. Hence when the loop is closed for such a system and a disturbance occurs anywhere in the system, conditions are right for oscillations to be sustained at the frequency ω_c. Equation (22-101) points out that two conditions must be satisfied for sustained oscillations. One, the signal must experience a total phase lag of 180° as it goes round the loop. Two, the magnitude of the signal fed back must be equal to the input. In any stable system the degree of deviation from either of these conditions may be used as a measure of the system's relative stability.

   Relative Stability. Consider the system depicted in Fig. 22-26. On the basis of the Nyquist criterion we know that this system is absolutely stable. However, we can make the Nyquist plot yield information about how close the given system is to instability by noting how close the plot comes to the $-1 + j0$ point. This proximity can be measured by two quantities: phase margin and gain margin.

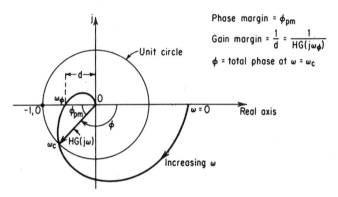

Fig. 22-26. Illustrating the definitions of phase margin and gain margin.

A study of Fig. 22-26 discloses how each is defined. Thus if it is assumed that the magnitude condition for sustained oscillations is satisfied at the frequency ω_c, one can then compute the amount of additional phase lag needed in the system for the phase angle of the open-loop transfer function to be $-180°$. This quantity is called the *phase margin*, and it is represented mathematically by

$$\phi_{pm} = 180° + \phi \qquad (22\text{-}102)$$

where ϕ is the total phase lag of the open-loop transfer function at ω_c. For the system represented by Eq. (12-100) this phase lag is computed from

$$\phi = -\tan^{-1}\omega_c\tau_1 - \tan^{-1}\omega_c\tau_2 - \tan^{-1}\omega_c\tau_3 \qquad (22\text{-}103)$$

   Consider next that point on the Nyquist plot at which the phase lag caused by the system components is exactly $-180°$. The frequency at which this occurs is identified in Fig. 22-26 as ω_ϕ. Note that at this frequency the open-loop transfer function has a value less than unity. In other words, there is a gain margin between what actually exists and what is needed for sustained oscillations. In terms of an equation, gain margin can be defined as

$$\text{gain margin} = \frac{1}{d} = \left| \frac{1}{H(j\omega_\phi)G(j\omega_\phi)} \right| \qquad (22\text{-}104)$$

Accordingly, if in a given instant the value of d is 0.8, the gain margin is then 1.25. This means that the gain at the frequency ω_ϕ may be increased by 25 per cent before sustained oscillations occur.

EXAMPLE 22-4   A third-order, Type 1 system is described by the open-loop transfer function

$$HG(s) = H(j\omega) = \frac{K}{s(1 + s\tau_1)(1 + s\tau_2)} = \frac{K}{j\omega(1 + j\omega\tau_1)(1 + j\omega\tau_2)} \qquad (22\text{-}105)$$

The Nyquist path in this case is modified somewhat at the origin to that shown in Fig. 22-27(a). The path along the $j\omega$ axis detours on an infinitesimal semicircle ABC in order to avoid passing through the pole at the origin that appears in Eq. (22-105). The corresponding polar plot for positive real frequencies is depicted by the solid curve in Fig. 22-27(b). Determine whether or not the closed-loop system is stable.

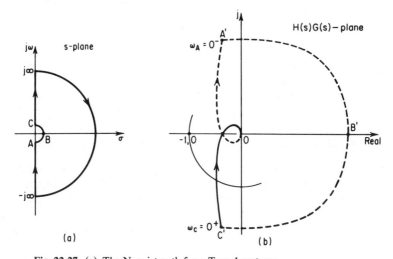

Fig. 22-27. (a) The Nyquist path for a Type 1 system;
(b) the corresponding polar plot of the open-loop transfer function.

solution: Before determining the number of encirclements of the $-1 + j0$ point, it is necessary to show how points on the infinitesimal semicircle in the s plane plot into the $H(s)G(s)$ plane. Here it is useful to quote a plotting theorem that applies to complex functions. It states that by the property of conformal mapping any angle in the s plane is preserved in the $H(s)G(s)$ plane in magnitude and direction. Thus as one moves along the Nyquist path of Fig. 22-27(a) from $-j\infty$ to the infinitesimal semicircle there occurs at point A, where the frequency is $\omega_A = 0^-$, a 90° turn to the right. Accordingly, in the $H(s)G(s)$ plane the plot must also turn 90° to the right. Furthermore, all points on the infinitesimal semicircle ABC carry over as points on the semicircle $A'B'C'$, the radius of which approaches infinity. These large values of HG follow as a consequence of the sole s in Eq. (22-105). If the Nyquist criterion is now applied to the complete plot, it is found that no encirclement of the $-1 + j0$ point occurs. Moreover, since P_r is zero, Z_r must be zero. Hence the system is stable.

A simplified version of the Nyquist criterion is possible for systems that are Type 3 or lower and for which P_r is zero. It is as follows: Plot the sinusoidal open-loop transfer function for increasing frequency from $\omega = 0$ to $\omega = \infty$ (the solid curves in Figs. 22-23 to 22-27). If the $-1 + j0$ point lies to the right of the curve as it is traversed in the direction of increasing frequency, the system is unstable; if the $-1 + j0$ point lies to the left of the curve, the system is stable. By applying this simplified criterion to the systems discussed in the foregoing, we reach the same conclusions regarding system stability.

The absolute stability of a feedback control system may also be determined by computing the phase margin. If the phase margin is positive, the system is absolutely stable; if the phase margin is computed to be a negative quantity, the system is unstable. This procedure is particularly useful when the Bode diagram of the open-loop transfer function is plotted, because the frequency ω_c at unity gain is immediately placed in evidence. This matter is treated in the next section.

Correlation between Phase Margin and Damping Ratio. Recall that the damping ratio is a figure of merit associated with the time-domain analysis of a control system. Phase margin, on the other hand, is a figure of merit defined on the basis of the frequency-response characteristic of the open-loop transfer function. It can be shown† that for a Type 1 second-order system the phase margin is solely dependent upon the damping ratio of that second-order system. Figure 22-28 depicts this relationship graphically. For each value of phase margin there is

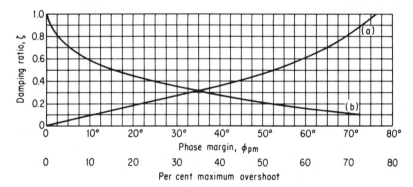

Fig. 22-28. (a) Damping ratio versus phase margin, second-order system;
(b) damping ratio versus per cent maximum overshoot, second-order system step response.

† See V. Del Toro and S. R. Parker, *Principles of Control Systems Engineering* (New York: McGraw-Hill Book Company, 1960), p. 286.

a corresponding value of damping ratio. Since, in turn, damping ratio is a measure of the maximum overshoot of the dynamic time response, clearly phase margin may be used as a frequency-domain figure of merit conveying the same information.

Although the data in Fig. 22-28 apply strictly to the second-order Type 1 system, they are also useful for higher-order systems. This is based on the assumption that the transient response of stable systems is determined usually in terms of a predominant pair of complex conjugate roots of the characteristic equation for the closed-loop system. These roots may then be considered as belonging to an equivalent second-order system as far as the transient response is concerned. The limitations of such an approximation are obvious, but nonetheless it is a useful and practical correlation which serves to tie together the frequency-response and the time-response analyses.

22-5 Predicting System Performance from the Open-loop Transfer Function

This section will show how it is possible to describe the performance of a feedback control system in an easy and direct fashion, even for high-order systems, solely in terms of the system's frequency response. The analysis is based on the Bode-diagram representation of the system's open-loop transfer function. The Bode diagram is preferred over the polar plot for several reasons. One, the open-loop transfer function can be quickly plotted by means of straight-line approximations. Two, the use of a logarithmic scale for frequency allows equal emphasis to be placed on each decade of that portion of the frequency spectrum that lies in the vicinity of the important $-1 + j0$ point. The distribution of points with frequency on the polar plot is highly nonuniform with considerable bunching of points at the higher frequency values. Finally, it is very easy to see on the Bode diagram the effect on performance of changing a system parameter.

As a first illustration of the procedure involved, let us find the performance of a feedback control system whose open-loop transfer function is given by

$$H(j\omega)G(j\omega) = \frac{180}{j\omega\left(1 + j\frac{\omega}{2}\right)\left(1 + j\frac{\omega}{6}\right)} \tag{22-106}$$

The attenuation characteristic is expressed by

$$A = 20 \log 180 - 20 \log |j\omega| - 20 \log \left|1 + j\frac{\omega}{2}\right| - 20 \log \left|1 + j\frac{\omega}{6}\right| \tag{22-107}$$

If we follow the procedure outlined in Sec. 22-1, the resultant magnitude characteristic plots as curve (a) in Fig. 22-29. Note the presence of the two breakpoint frequencies at 2 and 6 radians/second. As pointed out on p. 699, we may directly ascertain the absolute stability of this system by finding the phase margin. It should be clear from Eq. (22-106) that the expression for the phase angle is

$$\phi = -90° - \tan^{-1}\frac{\omega}{2} - \tan^{-1}\frac{\omega}{6} \tag{22-108}$$

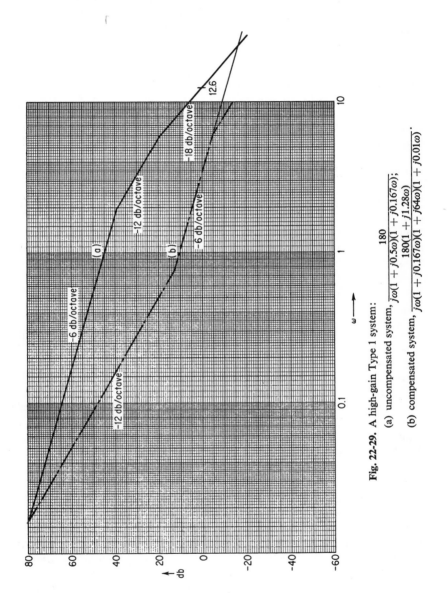

Fig. 22-29. A high-gain Type 1 system:

(a) uncompensated system, $\dfrac{180}{j\omega(1+j0.5\omega)(1+j0.167\omega)}$;

(b) compensated system, $\dfrac{180(1+j1.28\omega)}{j\omega(1+j0.167\omega)(1+j64\omega)(1+j0.01\omega)}$.

Of course the $-90°$ results from the pure integration term. The frequency at which the magnitude ratio is unity is found from Fig. 22-29 as the point where the attenuation characteristic [curve (a)] intersects the 0-db axis and is called the *crossover frequency*. The value here is 12.6 rad/sec. Thus at $\omega = \omega_c = 12.6$ rad/sec, Eq. (22-108) yields

$$\phi = -90° - \tan^{-1}\frac{12.6}{2} - \tan^{-1}\frac{12.6}{6} = -235.7° \qquad (22\text{-}109)$$

Accordingly by Eq. (22-102) the phase margin is found to be

$$\phi_{pm} = 180° + \phi = 180° - 235.7° = -55.7° \qquad (22\text{-}110)$$

Therefore the system is unstable and we need not proceed further with a description of the performance. Before leaving this example, it is worthwhile to note that the complete phase characteristic was not necessary and so was not plotted. All the information in the phase-angle characteristic is actually implied in the attenuation characteristic for linear systems.

By the insertion of appropriate rate and integral circuits the attenuation characteristic of the unstable system can be reshaped so that not only will the compensated system be absolutely stable but satisfactory relative stability can be achieved as well. This may be illustrated by examination of the open-loop transfer function for the compensated system, which is represented by

$$H(j\omega)G(j\omega) = \frac{180(1 + j1.28\omega)}{j\omega\left(1 + j\dfrac{\omega}{6}\right)(1 + j64\omega)\left(1 + j\dfrac{\omega}{100}\right)} \qquad (22\text{-}111)$$

In this instance the attenuation characteristic exhibits four breakpoint frequencies as shown by curve (b) of Fig. 22-29. The first breakpoint frequency occurs at $\frac{1}{64}$ rad/sec and serves to change the slope from -6 db/oct for the uncompensated system to -12 db/oct for the compensated system. Then, at $\omega = 1/1.28$ rad/sec, the second breakpoint frequency comes into consideration and performs the important function of changing the slope of the resultant attenuation characteristic from -12 db/oct to -6 db/oct as it crosses the zero-db axis. It very often happens that the asymptotic attentuation characteristic of a stable system crosses the zero-db axis at a slope of -6 db/oct. Moreover, it can be frequently concluded that such a system behaves essentially as an equivalent second-order system irrespective of the actual order of the system. The effects of the compensation devices help to assure this result. The third break-point frequency occurs at 6 rad/sec, beyond which point the slope again becomes -12 db/oct. After the fourth breakpoint frequency of 100 rad/sec, the slope changes to -18 db/oct and remains at that value for all higher frequencies. This large negative slope is not harmful from the viewpoint of absolute stability so long as it occurs sufficiently beyond the unity gain condition (i.e. below the zero-db axis). Reference to curve (a) of the uncompensated system shows that the magnitude characteristic crosses the zero-db axis at a slope of -18 db/oct; the result is an unstable system.

A glance at curve (b) of Fig. 22-29 for the compensated system shows that the new crossover frequency at unity gain is $\omega_c = 3.5$ rad/sec. Also, the expression for the phase angle from Eq. (22-111) is seen to be

$$\phi = -90° - \tan^{-1}\frac{\omega}{6} - \tan^{-1}64\omega - \tan^{-1}\frac{\omega}{100} + \tan^{-1}1.28\,\omega \qquad (22\text{-}112)$$

Insertion of $\omega = \omega_c = 3.5$ rad/sec yields

$$\phi = -90° - 30.2° - 90° + 2° + 77.4° = -134.8° \qquad (22\text{-}113)$$

Accordingly the phase margin is found to be a positive quantity given by

$$\phi_{pm} = 180° + \phi = 180° - 134.8° = 45.2° \qquad (22\text{-}114)$$

Hence the compensated system is absolutely stable.

Information about relative stability and a description of the dynamic performance in the time domain may also be obtained in an approximate fashion. Proceeding on the assumption that the compensated system is for all practical purposes an equivalent second-order system, we can estimate by means of Fig. 22-28 and Eq. (22-114) that the compensated system undergoes a maximum overshoot of 24 per cent when subjected to a step input. An estimate of the settling time is also possible. It can be shown that the crossover frequency ω_c is a fair approximation of the natural frequency of the equivalent second-order system. Furthermore, Fig. 22-28 shows that for a phase margin of 45.2° the damping ratio is 0.41. If we now use Eq. (21-37), which is the expression for the time constant of the second-order system, and replace ω_n by ω_c, we get

$$T = \frac{1}{\zeta\omega_n} \approx \frac{1}{\zeta\omega_c} = \frac{1}{0.41(3.5)} = 0.7\text{ sec} \qquad (22\text{-}115)$$

Then, on the assumption that steady state is achieved within five time constants, the approximate time it takes for the system to settle is 3.5 seconds.

22-6 The Root-locus Method

The determination of system performance by the sinusoidal frequency-response method is a straightforward procedure. It requires mathematics no more sophisticated than complex numbers. By means of Bode-diagram representations the attenuation characteristic of the open-loop transfer function is easily and quickly drawn so that such figures of merit as phase margin and crossover frequency can be readily found. Moreover, the effect on system behavior of changing the value of a system parameter can be directly identified by computing the change it brings about in phase margin and crossover frequency. The one notable drawback of this procedure is that an exact expression for the time-domain solution is not obtained. This is a natural consequence of not dealing directly with the roots of the characteristic equation. The root-locus method, however, removes this shortcoming. It is the purpose of this section to indicate how this is accomplished by applying the method to a third-order Type 1 system. Space considerations prevent a formal and detailed exposition of the root-locus method here. Readers who are interested in learning more about the method are referred to the bibliography at the end of the book.

The root-locus method is a graphical procedure for finding the locus of points in the complex-frequency plane that are solutions of the characteristic equation for values of open-loop gain varying from zero to infinity. Before we proceed with a description of this method, it is helpful to describe how the sinusoidal frequency response can also be determined graphically. For the sake of illustration we direct attention to the system whose open-loop transfer function is given by Eq. (22-116), repeated here with s replacing $j\omega$ for greater generality. Thus

$$H(s)G(s) = \frac{180}{s\left(1 + \dfrac{s}{2}\right)\left(1 + \dfrac{s}{6}\right)} = \frac{2160}{s(s + 2)(s + 6)} \qquad (22\text{-}116)$$

The second form of $H(s)G(s)$ in Eq. (22-116) is preferred because in a graphical procedure based on the s plane the variable s should stand alone in the root factors.

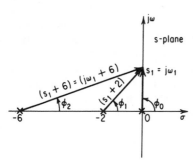

Inspection of Eq. (22-116) indicates that the open-loop transfer function has three poles located at $s = 0$, -2, and -6. These are plotted in the s plane and are identified by the cross marks in Fig. 22-30. The sinusoidal frequency response can now be found by confining s to the $j\omega$ axis. Thus, if the magnitude and phase of the open-loop transfer function are desired at, say, the frequency $s_1 = j\omega_1$, this frequency point is located on the $j\omega$ axis along with each factor appearing in the expression for the open-loop transfer function. In Fig. 22-30 note that we can readily locate the factor $(s_1 + 2)$ by drawing an arrow from the pole (or zero, when such is present) to the frequency point in question. When drawn to scale, both the magnitude and phase angle of $(s_1 + 2)$ can be measured. Assuming that these values are respectively r_{p_1} and ϕ_{p_1}, we can then write

Fig. 22-30. Illustrating the graphical procedure for finding the frequency-response of the open-loop transfer function.

$$(s_1 + 2) = (j\omega_1 + 2) = r_{p1}\underline{/\phi_{p1}} \qquad (22\text{-}117)$$

Similarly for the pole at the origin we have

$$s_1 = j\omega_1 = r_{p0}\underline{/\phi_{p0}} = \omega_1\underline{/90°} \qquad (22\text{-}118)$$

Finally for the pole at -6

$$(s_1 + 6) = (j\omega_1 + 6) = r_{p2}\underline{/\phi_{p2}} \qquad (22\text{-}119)$$

Keep in mind that r_{p0}, r_{p1}, r_{p2}, ϕ_{p0}, ϕ_{p1}, and ϕ_{p2} can be easily measured by a suitable instrument.† Inserting Eqs. (22-117), (22-118), and (22-119) into Eq. (22-116) yields

$$H(s_1)G(s_1) = H(j\omega_1)G(j\omega_1) = \frac{2160}{r_{p0}r_{p1}r_{p2}\underline{/\phi_0 + \phi_1 + \phi_2}}$$

$$= \frac{2160}{r_{p0}r_{p1}r_{p2}}\underline{/-(\phi_0 + \phi_1 + \phi_2)} \qquad (22\text{-}120)$$

† The spirule; see V. Del Toro and S. R. Parker, *Principles of Control Systems Engineering* (New York: Mc Graw-Hill Book Co., 1960), Chap. 10.

In this way both the magnitude

$$2160/r_{p0}r_{p1}r_{p2}$$

and the phase angle

$$-(\phi_0 + \phi_1 + \phi_2)$$

of the open-loop transfer function at the frequency $s_1 = j\omega_1$ are obtained. By repeating this procedure for a sufficient number of points on the $j\omega$ axis we can plot the magnitude and phase characteristics of the open-loop transfer function.

It should, of course, be obvious that it is much easier to obtain the frequency response by using the straight-line approximations called for in the Bode-diagram representation of the magnitude characteristic of the open-loop transfer function. The alternative procedure is outlined here in order to stress the restrictive nature of the frequency-response method. It is important to understand that with the frequency-response method s is allowed to vary only along a single line, namely the $j\omega$ axis, in the entire complex plane. In a sense this restriction lies behind our inability to obtain *in a simple way* an exact description of the time-domain solution. The root-locus method recognizes this limitation and removes it. In fact, in the root-locus method s is allowed to take on any value in the entire complex plane that satisfies the chracteristic equation of the closed-loop system.

Since the characteristic equation is expressible in the form

$$\boxed{H(s)G(s) = 1\underline{/180°}(2k + 1), \qquad k = 0, \pm1, \pm2, \cdots} \qquad (22\text{-}121)$$

it can be said that the root-locus method is in essence a search for those points in the s plane which satisfy this relation. When any value of the complex-frequency variable s is inserted into $H(s)G(s)$, the result is in general a complex number having a magnitude R and a total phase angle ϕ. Thus

$$H(s)G(s) = R\underline{/\phi} \qquad (22\text{-}122)$$

Now, when the value of s is a root of the characteristic equation, then

$$R = 1 \quad \text{and} \quad \phi = (2k + 1)\pi \quad \text{where } k = 0, \pm1, \pm2, \cdots \qquad (22\text{-}123)$$

which means that Eq. (22-121) is satisfied. The procedure for finding those s-plane points which qualify as roots of the characteristic equation first requires that the phase condition of Eq. (22-123) be fulfilled. That is, the sum of the angles of the phasors drawn from the zeros to the point s in question minus the sum of the angles of the phasors drawn from the poles to s must add up to a resultant of 180° or an odd multiple thereof. Once this condition is satisfied, the value of open-loop gain that makes the magnitude of the open-loop transfer function unity is computed. It can then be said that this value of s is a root of the characteristic equation for the computed gain.

It should not be concluded that the search for those points in the s plane which satisfy Eq. (22-121) is an aimless one. In fact, by the application of a few simple rules, the general shape of the branches of the root locus is clearly brought into evidence—even with only a cursory examination of the open-loop transfer function. This statement is true even for high-order systems where $H(s)G(s)$ contains many zeros and poles. The procedure, together with statements of the rules, will be illustrated by finding the root locus of the system represented by

$$H(s)G(s) = \frac{K'}{s(s + 2)(s + 6)} \qquad (22\text{-}124)$$

This system is the same as the one described by Eq. (22-116) except that now the gain is treated as a variable ranging from zero to infinity. It is important to keep in mind that K' in Eq. (22-124) is not the same as K, the open-loop proportional gain. It differs from it by the inclusion of the system time constants and other factors of the open-loop transfer function which appear whenever $H(s)G(s)$ is put in the form that isolates s.

   Rule 1. Locate the poles and zeros of $H(s)G(s)$ in the complex-frequency plane. Be sure to use the same scale for the abscissa and ordinate axes, otherwise the resulting root-locus plot will be meaningless. The location of each pole of Eq. (22-124) is shown in Fig. 22-31 by the cross marks.

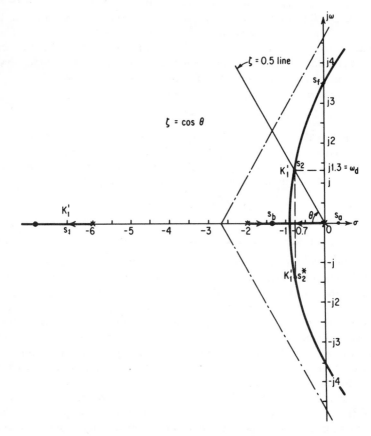

Fig. 22-31. Root-locus plot of $HG(s) = \dfrac{K'}{s(s + 2)(s + 6)}$.

Rule 2. Mark off those sections of the real axis which lie to the left of an odd number of poles and/or zeros. The validity of this statement follows from an examination of Eq. (22-124) in connection with fulfilling the phase criterion for points along the real axis of the *s* plane. Consider, for example, a value of *s* such as s_a. When s_a is inserted into Eq. (22-124), will it be found to yield a phase lag of 180°? Recalling that the factors s_a, $(s_a + 2)$, and $(s_a + 6)$ are located by drawing phasors from the poles of the open-loop transfer function to the value of *s* in question, it should be apparent that, as long as these poles (and/or zeros) lie to the left, the angle of each factor is zero. Hence, in the case of s_a, the resultant angle of the open-loop transfer function is zero and so the point cannot qualify as a root of the characteristic equation. However, for a point such as s_b the factor of the open-loop transfer function originating from the origin contributes an angle of $-180°$; whereas those originating from -2 and -6 contribute zero, yielding a resultant of $-180°$. Thus any point lying on the negative real axis between zero and -2 qualifies as a root of the characteristic equation. By similar reasoning it is found that no point lying between -2 and -6 can qualify, whereas all points from -6 to $-\infty$ do qualify.

Rule 3. The branches of the root locus originate at the poles of the open-loop transfer function corresponding to $K' \rightarrow 0$ and terminate at the zeros corresponding to $K' \rightarrow \infty$. The number of branches of the root locus is equal to the order of *s* in the denominator of the expression for $H(s)G(s)$. For the example at hand, for each value of K' there must exist three roots of the characteristic equation. In turn this means that we must be able to identify three branches in the root locus of the system. Each branch yields a root for the specified K'. The section of the negative real axis spreading from -6 to $-\infty$ is one such branch. For any value of K' there exists a point on this branch that is a solution to the characteristic equation. As already pointed out, this result follows forthwith from the phase criterion.

To show that the branches of the root locus originate at the poles of the open-loop transfer function, let us find the value of K' associated with that root of the characteristic equation having the value $s = -1.999$. We know this qualifies as a root because it fulfills the phase condition. If we insert this value of *s* into Eq. (22-124), we get

$$H(-1.999)G(-1.999) = \frac{K'}{-1.999(0.001)(4.001)} \approx \frac{K'}{0.008} \underline{/+180°} \qquad (22\text{-}125)$$

Now, to satisfy the magnitude condition called for by Eq. (22-121), the value of K' must be such that the open-loop transfer function has a unity gain. Thus

$$K' = 0.008 \qquad (22\text{-}126)$$

A little thought discloses that the closer *s* is to -2, the smaller must be the value of K'. A similar conclusion prevails at each of the other poles. Accordingly, as K' is increased in Eq. (22-124), the roots of the closed-loop characteristic equation migrate from the pole at -6 toward the left, from the pole at -2 toward the right and from the pole at the origin toward the left. The direction of the movement of the roots as K' increases is depicted in Fig. 22-31 by the arrowheads.

Equation (22-124) reveals that there are no explicit zeros in $H(s)G(s)$. However, all branches of the root locus that originate at the poles of $H(s)G(s)$ must terminate at its zeros. Where explicit zeroes are missing in the expression for $H(s)G(s)$, the branches terminate at the implicit zeros that necessarily occur at infinity. For Eq. (22-124) there are three zeros at infinity. A study of Eq. (22-124) should make it clear that when s is allowed to take on values approaching infinity, but still subject to fulfilling the phase condition, K' in turn must assume increasingly larger values also approaching infinity to assure satisfaction of the unity-magnitude condition.

Rule 4. Compute the angles of the asymptotes for those branches of the root locus which terminate at the implicit zeros of $H(s)G(s)$ by using

$$\phi_{\text{asymp}} = \frac{(2k + 1)\pi}{P - Z}, \qquad k = 0, \pm 1, \pm 2, \cdots \qquad (22\text{-}127)$$

where P denotes the number of poles of $H(s)G(s)$ and Z denotes the number of explicit zeros of $H(s)G(s)$.

Furthermore, compute the intersection point of the asymptotes from

$$\sigma_c = \frac{\sum \text{poles} - \sum \text{zeros}}{P - Z} \qquad (22\text{-}128)$$

Equation (22-127) is a means of assuring that, as s is allowed to take on large values, only those values will be considered which satisfy the phase condition. Thus, for the system of Eq. (22-124) we have $P = 3$ and $Z = 0$, so that from Eq. (22-127)

$$\phi_{\text{asymp}} = \frac{\pi}{3} = +60° \qquad \text{for } k = 0$$

$$\phi_{\text{asymp}} = \frac{-\pi}{3} = -60° \qquad \text{for } k = -1 \qquad (22\text{-}129)$$

$$\phi_{\text{asymp}} = \frac{3\pi}{3} = 180° \qquad \text{for } k = +1$$

There are as many asymptotes as the value of $P - Z$. For very large values of s the expression for the open-loop transfer function can be represented by

$$\lim_{s \to \infty} H(s)G(s) = \frac{K'}{s^{P-Z}} = \frac{K'}{s^3} \qquad (22\text{-}130)$$

which follows directly from Eq. (22-124). This last equation makes it clear that, when s is far removed from the origin, only those values will be roots of the characteristic equation which lie on the asymptotes of $\pm 60°$ and $180°$. Of course the insertion of any one of these angles into Eq. (22-130) yields a resultant angle of $180°$ or an odd multiple thereof.

The asymptotes of Eq. (22-129) cannot be drawn anywhere in the complex plane. Instead their particular placement is uniquely determined by the specific distribution of the poles and zero of $H(s)G(s)$.† For our example the asymptotes

---

† For a derivation of Eq. (22-128) see reference 19 in the bibliography.

appear to originate from

$$\sigma_c = \frac{\sum \text{poles} - \sum \text{zeros}}{P - Z} = \frac{(0 - 2 - 6) - (0)}{3} = -2.67 \qquad (22\text{-}131)$$

The asymptotes and their point of origin are shown in Fig. 22-31.

Rule 5. Locate the breakaway points where appropriate. A glance at Fig. 22-30 shows that as K' increases, the roots on the branch from the origin and those on the branch from the pole at -2 move towards each other. Accordingly, at some point between 0 and -2 the roots will coalesce, thus yielding a double root for the K' that applies at that point. A further increase in K' causes the branches to move away from the real axis into the s plane and towards the asymptotes. In this case one branch proceeds towards the $+60°$ asymptote and the other towards

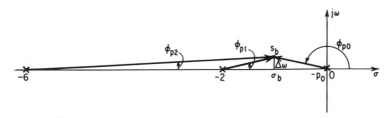

Fig. 22-32. Finding the breakaway point in the root-locus plot.

the $-60°$ asymptote. To illustrate how this breakaway point can be found, consider the trial point s_b shown in Fig. 22-32 which has a very small j part, namely, $\Delta\omega$. The three phasors of the open-loop transfer function are also shown. The phase angles are

$$\phi_{p0} = \pi - \tan^{-1}\left|\frac{\Delta\omega}{\sigma_b}\right|$$

$$\phi_{p1} = \tan^{-1}\left|\frac{\Delta\omega}{\sigma_b + 2}\right| \qquad (22\text{-}132)$$

$$\phi_{p2} = \tan^{-1}\left|\frac{\Delta\omega}{\sigma_b + 6}\right|$$

To satisfy the phase criterion the sum of these angles must equal π degrees. Thus

$$\pi - \tan^{-1}\left|\frac{\Delta\omega}{\sigma_b}\right| + \tan^{-1}\left|\frac{\Delta\omega}{\sigma_b + 2}\right| + \tan^{-1}\left|\frac{\Delta\omega}{\sigma_b + 6}\right| = \pi \qquad (22\text{-}133)$$

Because $\Delta\omega$ is very small, Eq. (22-133) may be written as

$$\left|\frac{\Delta\omega}{\sigma_b + 2}\right| + \left|\frac{\Delta\omega}{\sigma_b + 6}\right| = \left|\frac{\Delta\omega}{\sigma_b}\right| \qquad (22\text{-}134)$$

which simplifies to

$$\frac{1}{|\sigma_b + 2|} + \frac{1}{|\sigma_b + 6|} = \frac{1}{|\sigma_b|} \qquad (22\text{-}135)$$

Upon selecting σ_b so that Eq. (22-135) is satisfied, the value of the breakaway point is determined. After several trials, this quantity is found to be $\sigma_b = -0.9$.

It is also possible to find σ_b by using a spirule. For an explanation of how to use the spirule see reference 19 of the bibliography.

Rule 6. When the form of $H(s)G(s)$ is such that $P - Z \geq 2$, as some branches of the root locus move toward the left for increasing values of K' other branches must move towards the right. This statement follows from the law of algebra that states that if the characteristic equation is in normalized form, i.e.

$$s^P + bs^{P-1} + \cdots + K' = 0$$

then the negative sum of the roots is a constant equal to the coefficient of the s^{P-1} term, namely, b. This sum is independent of K' provided that $P - Z \geq 2$, which is often the case. For the example depicted in Fig. 22-31 note that as the roots migrate towards the left on the branch from -6 to $-\infty$ there occurs a simultaneous compensating migration of the roots towards the right on the other two branches of the root locus. This statement is particularly obvious for large values of s.

Once the sections of the real axis that are part of the locus are marked off, and the asymptotes are drawn, and the breakaway point is found, the remainder of the root locus can be completed in either of two ways. One, with the asymptotes and rule 6 as guides, the locus can be approximately sketched in. A phase-angle check can be made on several points for control purposes. This procedure is frequently acceptable where high accuracy is not important. Two, the spirule or other similar instrument is used to complete the root locus. The spirule is essentially a device that permits the net phase angle of the phasors drawn from the poles and zeros to any s-plane point to be determined in a matter of seconds. Moreover, the spirule is equipped with a logarithmic spiral curve that enables the value of K' for any root point to be found very quickly. The complete root-locus plot depicted in Fig. 22-31 was determined with this instrument.

It is important to understand that the root locus is a plot of the roots of the characteristic equation of the closed-loop system. The analysis, however, is based on Eq. (22-121), which involves only the open-loop transfer function. Also important to keep in mind is that the variable quantity on the plot is the open-loop gain K'. To show how the root locus makes available the time-domain response, let it be desired to find the system response for a gain K_1'. If we refer to Fig. 22-31, we find on each branch of the root locus one root of the characteristic equation for the specified value of gain. These are identified as s_1, s_2, and s_2^*. Note that the gain is such that a pair of complex conjugate roots results. The root s_1 is associated with a transient term which decays rapidly because the time constant is smaller than $\frac{1}{6}$ second. On the other hand the time constant associated with the complex roots is $1/0.7$ or 1.43 seconds. This certainly makes it clear that the dynamic response is determined by the complex roots. Further examination of Fig. 22-31 shows that damped frequency of oscillation of the transient is 1.3 rad/sec and that the damping ratio is 0.5. As indicated in Fig. 22-31 the damping ratio is computed as $\cos \theta$ where θ is defined by the construction shown. In view of the fact that we are dealing directly with the roots of the characteristic equation there are no approximations involved in this description of the dynamic response. The only errors involved are those associated with the graphical construction and the assumption of linearity in the components of the system.

PROBLEMS

22-1 Find the transfer function for the circuit shown in Fig. P22-1. What is the purpose of the unity-gain amplifier?

22-2 Derive the transfer function for the network shown in Fig. P22-2. Manipulate the result in the form of either a phase lag or a phase lead (whichever applies), and define an appropriate α and τ of the element values of the circuit.

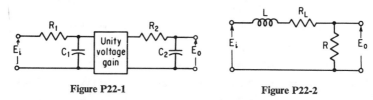

Figure P22-1   Figure P22-2

22-3 Derive the transfer function for the circuit of Fig. P22-3. Express the result in terms of an appropriate α and τ.

22-4 Find the transfer function for the circuit of Fig. P22-4, and express the result in terms of an appropriate α and τ.

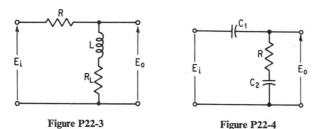

Figure P22-3   Figure P22-4

22-5 For the network depicted in Fig. P22-5, show that

$$\frac{E_o(s)}{E_i(s)} = \frac{1}{\tau_1 \tau_2 s^2 + (\tau_1 + \tau_2 + \tau_{12})s + 1}$$

where $\tau_1 = R_1 C_1$, $\tau_2 = R_2 C_2$, and $\tau_{12} = R_1 C_2$.

22-6 Determine the transfer function for the circuit shown in Fig. P22-6.

22-7 Derive the transfer function $X_o(s)/X_i(s)$ for the mechanical system shown in Fig. P22-7. Put the transfer function in the form of a phase lead or phase lag by using the α and τ notation.

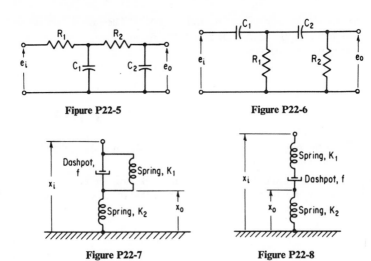

Fipure P22-5   Figure P22-6

Figure P22-7   Figure P22-8

22-8 Find the transfer function of the mechanical system depicted in Fig. P22-8, and express the result in terms of an appropriate α and τ.

22-9 Determine the transfer function $X_o(s)/F(s)$ for the mechanical system of Fig. P22-9. To which standard form does this transfer function belong?

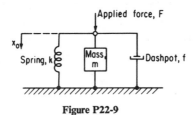

Figure P22-9

22-10 Sketch the attenuation and phase characteristics of the system components whose transfer functions are given by

(a) $T(s) = \dfrac{10(1 + j\omega\tau)}{(j\omega\tau)^2}$,   where $\tau = 1$ sec.

(b) $T(s) = \dfrac{1}{3 + j2\omega}$.

(c) $T(s) = \dfrac{1}{(1 + j\omega)(1 + j10\omega)}$.

22-11 A system component is described by the differential equation

$$c(t) = r(t) + \tau \frac{dr(t)}{dt}$$

where $c(t)$ is the output quantity and $r(t)$ is the input variable. This device is said to provide proportional plus derivative control. Sketch the log-modulus and phase-angle characteristics as well as the polar plot.

22-12   A proportional-plus-integral controller is described by the equation

$$c(t) = r(t) + \frac{1}{\tau} \int c(t)\, dt$$

where $c(t)$ and $r(t)$ are the output and input variables, respectively. Sketch the attenuation and phase-angle characteristics as well as the polar plot.

22-13   The asymptotic log-modulus for a system component is shown in Fig. P22-13.
(a) Find the corresponding transfer function.
(b) Sketch the polar plot.

22-14   A control system has the log-modulus attenuation characteristic depicted in Fig. P22-14.
(a) Determine the transfer function.
(b) Sketch the corresponding polar plot.

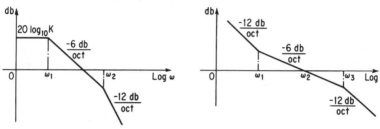

| Figure P22-13 | Figure P22-14 |

22-15   Sketch the log-modulus and phase-angle characteristics described by the transfer function $T(j\omega) = (j\omega)^2 + j2\zeta\omega + 1$ for $\zeta = 0.1$, 0.3, and 1.0. Compare these curves with those of Fig. 22-10.

22-16   (a) Derive the transfer function of the circuit shown in Fig. P22-16.
(b) Sketch the log-modulus attenuation characteristic for the case where $C_1 = 0.1\ \mu f$, $C_2 = 4.86\ \mu f$, $R_1 = 12.8$ megohms, and $R_2 = 0.103$ megohms.
(c) Comment on the nature of the phase-angle characteristics.

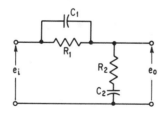

Figure P22-16

22-17   Determine the closed-loop transfer function for the system depicted in Fig. P22-17.

22-18   Find the closed-loop transfer function for the system shown in Fig. P22-18.

22-19   In the system of Fig. P22-19 determine the expression for the transformed output $C(s)$ in terms of the inputs $R_1(s)$, $R_2(s)$, and $R_3(s)$ and the specified system components. What is the form of the characteristic equation for closed-loop operation?

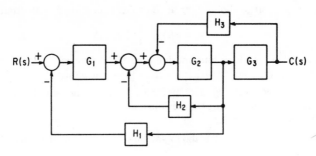

Figure P22-17

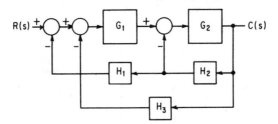

Figure P22-18

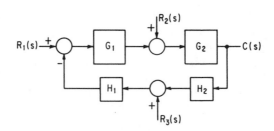

Figure P22-19

22-20   A feedback control system has a polynomial input signal given by

$$r(t) = r_0 + r_1 t + \tfrac{1}{2} r_2 t^2$$

Describe the steady-state performance of the control system when (a) it is a Type 0 system, (b) it is a Type 1 system, and (c) it is a Type 2 system. Is there any advantage to the use of a Type 3 system in this case? Explain.

22-21   The open-loop transfer function of a system is

$$H(s)G(s) = \frac{10}{s(s + 2)(s + 5)}$$

(a) What type system is it?
(b) What is the value of the ramp constant?
(c) What is the value of the steady-state error when the input signal is a step of magnitude 2?

(d) What is the value of the steady-state error when the input signal is a ramp of magnitude 2?

22-22 A Type 2 system has an open-loop gain of 100 sec^{-2} and two time constants having values of 0.5 second and 4 seconds.
(a) Write the expression for the open-loop transfer function.
(b) Find the closed-loop transfer function.
(c) Is this system stable? Explain.

22-23 A servomechanism has the open-loop transfer function

$$H(s)G(s) = \frac{50(s + 0.8)}{s^2(1 + 0.01s)(1 + 0.002s)}$$

(a) What is its classification?
(b) Find the value of the acceleration constant.
(c) A parabolic input $r(t) = 5t^2$ is applied to the system. Does the controlled quantity follow a parabolic variation in the steady state? Explain. Assume $H = 1$.
(d) In part (c) is there any deviation in the value of controlled variable and the command?

22-24 Apply the Routh criterion to determine whether or not the systems having the following characteristic equations are stable.
(a) $5s^3 + 2s^2 + 12s + 6 = 0$.
(b) $s^5 + 6s^4 + 3s^3 + 2s^2 + s + 1 = 0$.
(c) $4s^3 + 3s^2 + 2s + 5 = 0$.

22-25 Apply the Routh cirterion to determine whether or not the system of Prob. 22-21 is absolutely stable.

22-26 A feedback control system has the open-loop transfer function

$$HG(j\omega) = \frac{200}{j\omega(5 + j\omega)}$$

(a) Apply the Routh criterion to determine stability.
(b) Draw the Nyquist diagram and determine whether or not zeros of the characteristic equation exist in the right-half s plane.
(c) Draw the asymptotic log-modulus attenuation characteristic and find the phase margin. Is the conclusion about stability consistent with those of parts (a) and (b)?

22-27 Repeat Prob. 22-26 for a system described by the transfer function

$$HG(j\omega) = \frac{200}{j\omega(5 + j\omega)(10 + j\omega)}$$

22-28 Repeat Prob. 22-26 for a feedback control system whose transfer function is

$$HG(j\omega) = \frac{200}{(j\omega)^2(5 + j\omega)}$$

22-29 Repeat Prob. 22-26 for the case where $HG(j\omega) = \frac{10(1 + j\omega)}{(j\omega)^2}$.

22-30 A feedback control system has an open-loop transfer function given by

$$HG(j\omega) = \frac{40}{j\omega[1 + j(\omega/5)]}$$

(a) Draw the asymptotic log-modulus characteristic.

(b) Find the phase margin.

(c) Compute the approximate maximum per cent overshoot as this system responds to step commands.

(d) Compute the approximate settling time in part (c). Assume a tolerance band of 5 per cent.

22-31 Repeat Prob. 22-30 for the case where

$$HG(j\omega) = \frac{40}{j\omega[1 + j(\omega/5)][1 + j(\omega/10)]}$$

22-32 Repeat Prob. 22-30 for the case where

$$HG(j\omega) = \frac{10(1 + j\omega)}{(j\omega)^2}$$

22-33 The direct transfer function of a control system is given by

$$G(s) = \frac{1}{s^2(s + 1)}$$

A unity-feedback path is used for this system.

(a) Draw the Nyquist stability diagram and determine whether or not the system is stable.

(b) The feedback transfer function is changed from unity to $H(s) = 1 + 4s$. Draw the new Nyquist diagram and again determine whether or not the system is stable.

(c) Describe the approximate performance of the system of part (b) to a step command.

22-34 The pole-zero configuration of a control system is given by

$$\text{zero at } -5$$

$$\text{poles at } 0, -2$$

(a) Write the expession for the open-loop transfer function.

(b) Locate those portions of the real axis that are part of the locus.

(c) Are there any asymptotes?

(d) Sketch the complete root locus for varying values of the open-loop gain.

22-35 The open-loop transfer function of a control system is

$$HG(s) = \frac{K'}{s(s + 5)}$$

(a) Locate those portions of the real axis of the s plane that are part of the root locus.

(b) How many asymptotes are there? What are the angles of the asymptotes? At what point do they intersect?

(c) Sketch the root locus for values of K' between zero and infinity.

(d) Find the value of loop gain that makes the value of the damping ratio equal to 0.5.

(e) Is there any value of K' for which the system becomes unstable? Explain.

22-36 Repeat Prob. 22-35 for the system whose open-loop transfer function is given by

$$HG(s) = \frac{K'}{s(s + 5)(s + 10)}$$

22-37 A feedback control system is described by the open-loop transfer function

$$HG(s) = \frac{K'(s + 2)}{s(s + 3)(s^2 + 2s + 2)}$$

(a) Identify those parts of the real axis that are part of the locus.

(b) How many roots of the characteristic equation are there for each value of loop gain?

(c) Can the roots of the characteristic equation *ever* be all complex conjugates? Explain.

(d) Find the angles and point of intersection of the asymptotes.

(e) Identify the general direction in which the branches of the root locus from the complex poles must move as K' increases. Justify your answer.

23

electronic analog-computer techniques[†]

The material of the foregoing three chapters is confined entirely to linear systems because it provides the foundation for all subsequent efforts. Now that this objective is accomplished, attention is turned to the analysis of systems containing inherent nonlinearities such as saturation, coulomb friction, backlash, and so forth. Basically there are two methods of approaching this problem. One requires that the control engineer learn about the techniques available for treating nonlinear systems such as the describing-function method or the phase-plane method.[‡] The second approach makes use of the analog computer as an engineering tool of analysis and design. Effectively the linear as well as the nonlinear elements of the system are simulated, and the solutions are subsequently found by machine methods. The latter approach is more attractive for several reasons. One is that it usually enables the engineer to make a far more comprehensive analysis with considerably less time and effort. Thus, after the problem has been simulated on the analog computer, the solution for the controlled quantity, corresponding to

[†] Adapted by permission from V. Del Toro and S. R. Parker, *Principles of Control Systems Engineering* (New York: McGraw-Hill Book Company, 1960).

[‡] J. G. Truxal, *Automatic Feedback Control System Synthesis* (New York: McGraw-Hill Book Company, 1955), chaps. 10 and 11.

various types of command inputs as well as various combinations of control and system parameters, is obtained with great ease and speed. Moreover, the results are easily interpreted because of the graphical form in which the computer output equipment presents the solutions. This makes it possible then to optimize the values of a nonlinear system's control parameters easily. The corresponding amount of analytical work needed to accomplish the same results is enormous by comparison. A second reason is this: The evaluation of many feedback control systems depends upon the behavior of such systems in the presence of random noise, which exists either as an unwanted quantity or as a system input of a statistical nature. The electronic analog computer equipped with a suitable random-noise generator can be used to study problems of this kind. Such a procedure is especially valuable if the system is nonlinear, because there is no known general analytical method for treating such cases. For these reasons (as well as others described below) this chapter is devoted to an explanation and description of electronic analog-computer techniques and their application to control systems engineering. However, it should not be construed that this choice in any way belittles the importance of acquiring a solid background in nonlinear analytical techniques. The latter is always useful, especially in applying the analog computer more intelligently.

23-1 Classification of Analog Computers

An analog (or a model) of a problem can always be determined if the same equations or set of equations can be used to describe the behavior in the problem as well as in the model. Thus, if the differential equation describing the behavior of a mechanical system containing mass M, spring constant K, and viscous-friction coefficient F is

$$M\frac{d^2x}{dt^2} + F\frac{dx}{dt} + Kx = f \text{ (force)} \qquad (23\text{-}1)$$

and if the differential equation describing the flow of charge in an R-L-C electric circuit is

$$L\frac{d^2q}{dt^2} + R\frac{dq}{dt} + \frac{1}{C}q = E \text{ (voltage)} \qquad (23\text{-}2)$$

then, since the two equations are of the same form, it is possible to build an electrical analog (or model) of the mechanical system. For the analog to be a correct one, it is required that the electrical parameters and variables play precisely the roles called for in the original problem. To accomplish this, it is necessary that self-inductance L be made to take on the role of mass M, R be analogous to F, and $1/C$ be analogous to K. Moreover, the dependent variable x in the original problem is represented by charge q in the analog. Also, the applied force in the mechanical system is represented by a voltage in the electrical model. These results are apparent from a comparison of Eqs. (23-1) and (23-2). Analog computers which are constructed on this principle are called *general-purpose analog computers* of the *direct type*.

A second approach can be employed in solving for the displacement variable of the mechanical system of Eq. (23-1). A study of this equation reveals that the solution can be obtained by interconnecting appropriate equipment capable of performing the mathematical operations called for in Eq. (23-1). Thus, by making available two computing devices capable of performing integration with respect to time as well as a unit capable of summing, the solution to Eq. (23-1) can be found. An analog computer built on this principle is called an *indirect-type general-purpose computer*. It is this type of computer with which this chapter is concerned. Here all variables are represented as voltages.

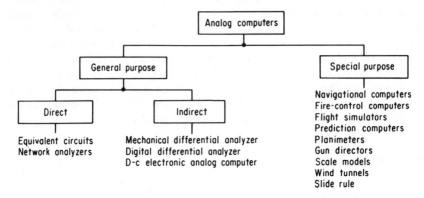

Fig. 23-1. Classification of analog computers. (Reprinted by permission from *Analog Computer Techniques*, C. L. Johnson, Copyright 1956, Mc-Graw-Hill Book Company.)

Appearing in Fig. 23-1 is a classification of all types of analog computers. The special-purpose analog computers employ the same basic principles as those used in the general-purpose units. The difference is that they are designed to solve a specific set of equations corresponding to specific types of inputs. In essence, then, they lack flexibility. They are designed for a special purpose and no other. The mechanical differential analyzer was the first of the indirect types of general-purpose analog computers built. It was constructed at MIT by Dr. V. Bush, employing mechanical ball-and-disk integrators. Although this computer is capable of a fair degree of accuracy, such units are not being built any more because on the one hand they are not as accurate as digital computers and on the other hand they cannot match the ease and speed of operation offered by the d-c electronic analog computer. With regard to the direct-type general-purpose analog computer it is interesting to note that in 1958 the world's largest computer of this kind was delivered to the Convair Division of General Dynamics Corporation at Fort Worth, Texas. It measures 29 by 46 by 91 ft and is used to solve complex design problems of high-speed, high-altitude aircraft.

Since our attention from this point on is exclusively on the d-c electronic analog computer, it is worthwhile to mention a few additional characteristic features. Because of the method used to obtain integration, time is usually the independent variable. Therefore in this respect the electronic analog computer is different from the digital computer as well as the mechanical differential analyzer because they do not share this restriction. For this reason it should be apparent that the solution of partial differential equations is difficult to achieve on the electronic analog computer.

The accuracy of solution obtained with the electronic analog computer is usually two, three, or four significant figures. Essentially this restriction is imposed by the nature of the analog components with which the computer is built. For example, the integrating capacitors at best have an accuracy of 0.1 to 0.01 per cent so that solution accuracies exceeding those of the components cannot be expected. Moreover, in addition to the limitations of the components, there are errors introduced by the integration process (which is subject to drift, etc.), the summation process, and so forth.

It is pointed out at the beginning of the chapter that one of the big areas of application of the electronic analog computer is the analysis and design of non-linear control systems subjected to random-noise inputs. A second and perhaps equally important area is the analysis of such complex problems as are encountered in heat-flow problems, process control problems, and six-degree-of-freedom aircraft control problems. In areas of this kind, when the behavior is of a dynamical nature, the electronic computer shows to considerable advantage. Moreover, it frequently offers economy. For example, in the case of the design of an autopilot for a high-performance aircraft it is possible to simulate the flight equations for various altitudes and Mach numbers on the electronic analog computer and then feed the appropriate aircraft output variables to a cockpit installed alongside the computer in the laboratory. In this way, with the pilot in the cockpit, the design can be optimized without incurring the high cost per hour associated with actual flight testing.

The aircraft industry is one of the biggest users of electronic analog computers. One reason for this preference over digital computers, for example, is that frequently the accuracy of the aerodynamic parameters used in the equations of motions is not known to better than 5 or 10 per cent. Accordingly, the limited accuracy of the analog computer is not a drawback, and in fact the analog computer has much to offer in terms of simplicity and ease of operation. Of course in situations where a high degree of precision is needed, as, for example, in computing the trajectory of a long-range spacecraft, the digital computer must be preferred.

Another worthwhile feature of the electronic analog computer is illustrated by the following case reported by the RAND Corporation. Several reports were written on the theoretical aspects of applying the calculus of variation to optimizing aircraft flight paths. However, no conclusions were reached because of the complexity of the equations. At this point an electronic analog computer was used, and not only were practical answers obtained, but by means of the computer suggested revisions of the theory were verified.

23-2 Basic Computations in D-C Electronic Analog Computers

The electronic analog computer is made up of building blocks capable of performing specific mathematical operations. In this section attention is directed to the manner in which multiplication, addition, integration, and differentiation are accomplished.

Multiplication by Constant Coefficients. Very often in analog simulation it is necessary to multiply a machine variable by a constant coefficient. In those instances where the constant is a number less than unity the multiplication is accomplished by means of a potentiometer, as illustrated in Fig. 23-2. If it is assumed that there is no loading on the output arm of the potentiometer (that is, $R_L \rightarrow \infty$), the output quantity is related to the input by

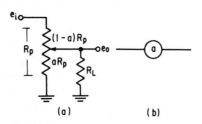

$$e_o = ae_i \qquad (23-3)$$

where a is the slider-arm setting of the potentiometer. In most of the commercially available computers these potentiometers are designed to make 10 full revolutions and can be set to three places in the range $0.000 \leq a \leq 1.000$. The total resistance R_p of the potentiometer varies from 10,000 to 100,000 ohms, depending upon the manufacturer. A common value is 30,000 ohms. In the normal application of these potentiometers the resistor R_L is usually in the range from 0.1 to 1.0 megohm so that the loading effect is not always negligible. A more exact analysis of the ratio of output to input shows the actual ratio to be

Fig. 23-2. Coefficient setting potentiometer:
(a) schematic diagram;
(b) symbolic diagram.

$$\frac{e_o}{e_i} = \frac{a}{1 + a(1 - a)(R_p/R_L)} \qquad (23-4)$$

where a is the ideal ratio value. Clearly the ratio error can be kept small by making the value of R_p/R_L small. The potentiometer resistance R_p cannot be made too small; otherwise it will draw a current in excess of the capacity of the amplifier which supplies it with the input-voltage variable.

In order to multiply a machine variable by a number greater than unity, use is made of the operational amplifier. The operational amplifier is a direct-coupled feedback amplifier having an open-loop gain often in excess of 10^8. It

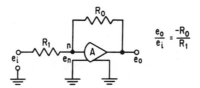

$$\frac{e_o}{e_i} = \frac{-R_0}{R_1}$$

Fig. 23-3. Illustrating multiplication by a constant greater than unity. $R_0 > R_1$.

is the basic building block used in most of the computing components of the d-c electronic analog computer. To understand how multiplication by a constant coefficient greater than unity is accomplished, refer to Fig. 23-3. The operational amplifier having a gain A is represented by the symbol $\text{---}\!\!\langle A \rangle\!\!\text{---}$. Applying Kirchhoff's current law to the nodal point n and assuming negligible input current permits writing

$$\frac{e_i - e_n}{R_1} + \frac{e_o - e_n}{R_0} = 0 \tag{23-5}$$

But

$$e_0 = Ae_n \tag{23-6}$$

Inserting Eq. (23-6) into Eq. (23-5) yields

$$\frac{e_o}{e_i} = \frac{AR_0}{R_1[(R_0/R_1) + (1 - A)]} \tag{23-7}$$

Since in most commercial units $A = -10^8$ or more, it follows that Eq. (23-7) may be written as

$$\frac{e_o}{e_i} = -\frac{R_0}{R_1} \tag{23-8}$$

Thus by selecting $R_0 > R_1$ the desired result is accomplished. For $R_0 = 1$ megohm and $R_1 = 0.1$ megohm the output voltage is ten times greater than the input. Furthermore, this multiplication is performed with a precision that is determined by the input and feedback resistors alone. Hence, if 0.1 per cent resistors are used in the arrangement of Fig. 23-3, then multiplication is accomplished with the same degree of precision. A glance at Eq. (23-5) reveals that the result of Eq. (23-8) could be obtained directly by considering the node to be virtually at ground potential. The use of a feedback circuit round the very high open-loop gain assures this. This same result is obtained by considering that the output voltage is usually limited to 100 volts for linear operation. Therefore, for a gain of $A = -10^8$ it follows that e_n must be merely 1 μv above ground to generate the 100-volt output. The node is accordingly at virtual ground and shall be so considered in all subsequent analysis.

When a machine variable is to be multiplied by a noninteger constant coefficient, as, for example, obtaining $e_0 = 3.69e_1$, the arrangement appearing in Fig. 23-4(a) is used. By setting the potentiometer for a reading of 0.369 and then

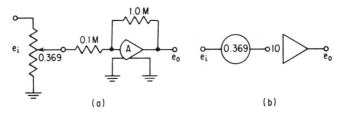

(a)   (b)

Fig. 23-4. Illustrating multiplication by a noninteger coefficient:
(a) schematic diagram;
(b) symbolic representation.

multiplying by 10, the desired result is obtained (assuming no potentiometer loading). Figure 23-4(b) shows the symbolic diagram for this mathematical operation. Note that the high-gain d-c amplifier, together with the input and feedback resistors, is represented by a triangle and by the multiplying factor being specifically indicated.

Adding of Several Variables. The addition of n variables may be performed by the arrangement shown in Fig. 23-5(a) and (b). Applying Kirchhoff's current

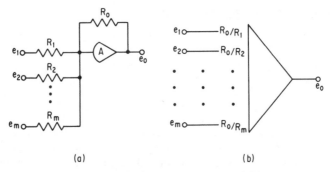

(a)   (b)

Fig. 23-5. Illustrating the addition of n variables:
(a) schematic diagram;
(b) symbolic diagram.

law in nodal form at the node point and assuming zero node voltage gives

$$\frac{e_1}{R_1} + \frac{e_2}{R_2} + \cdots + \frac{e_n}{R_n} = -\frac{e_o}{R_o} \tag{23-9}$$

Therefore

$$e_o = -\left(\frac{R_0}{R_1}e_1 + \frac{R_0}{R_2}e_2 + \cdots + \frac{R_0}{R_n}e_n\right) \tag{23-10}$$

This expression demonstrates that not only is the output the sum of several variables but each variable may be multiplied by an appropriate scale factor if desirable. In the standard electronic analog-computing equipment manufactured by the Reeves Instrument Company the configuration used for the summing amplifiers is as illustrated in Fig. 23-6. Note that there are seven input terminals, two of which provide multiplication by 10, two by 4, and three by unity. With this arrangement, and by the use of parallel connections of the input variable, the latter can be multiplied by any integer from 1 to 31. Other manufacturers use other arrangements.

Integration. The method for obtaining integration with respect to time in the electronic analog computer is depicted in Fig. 23-7. Writing Kirchhoff's law at the node point yields

$$\frac{e_1}{R_1} + \frac{e_o}{1/sC} = 0$$

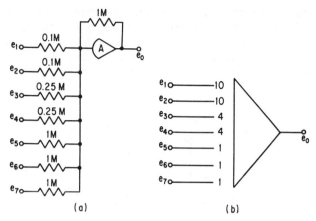

Fig. 23-6. Configuration used in the standard summing amplifier of the REAC (Reeves electronic analog computer):
(a) schematic diagram;
(b) symbolic diagram.

Fig. 23-7. Circuit arrangement for integration:
(a) schematic diagram;
(b) symbolic diagram.

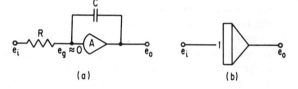

Therefore

$$\frac{e_o}{e_i} = -\frac{1}{sRC} \tag{23-11}$$

A glance at this result indicates that the transfer function is that of a pure integrator. Accordingly, when Eq. (23-11) is interpreted in the time domain, it states that the output of the circuitry shown in Fig. 23-7 is the integral of the input. Moreover, the output may also be modified by the scale factor $1/RC$. It is customary in commercial computers for the feedback capacitor to be fixed at the value of 1 μf and the scaling accomplished by the appropriate selection of R. The standard electronic integrator manufactured by Electronic Associates, Inc., has the configuration shown in Fig. 23-8. Thus multiplication by 1 or 10 may be accomplished if no input potentiometers are used.

Whenever an analog simulation requires integrators, it means that the

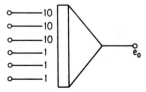

Fig. 23-8. Standard electronic integrator configuration as manufactured by Electronic Associates, Inc.

corresponding physical problem contains energy-storing elements. Consequently, in order for the simulation to be correct, it is necessary at the start of the computer solution of the problem to establish at the output of each integrator that value of voltage which properly represents the initial condition in the original problem. To accomplish this, circuitry similar to that shown in Fig. 23-9 is used.

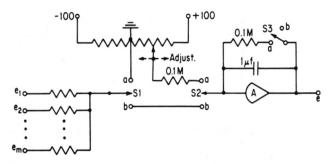

Fig. 23-9. Illustrating how initial conditions are applied to integrators.

When the master control knob of the analog computer is placed at the so called *reset* position, switches $S1$, $S2$, and $S3$ are placed at position a. Switch $S1$ puts the input terminals to ground, thereby making the simulation inoperative. The switches $S2$ and $S3$ establish a circuit permitting the proper initial-condition voltage to appear across the capacitor. This voltage is applied to the capacitor in accordance with a time constant of 0.1 sec. When the master control knob is placed at the *compute* position, switches $S1$, $S2$, and $S3$ go to position b in Fig. 23-9. The original simulation is re-established, and the 0.1-megohm resistor is removed from the capacitor so that it cannot upset the initial-condition voltage.

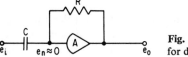

Fig. 23-10. Circuit arrangement for differentiation.

Differentiation. The method for obtaining differentiation in the electronic analog computer is illustrated in Fig. 23-10. Writing Kirchhoff's law at the node point leads to

$$\frac{e_1}{1/sC} + \frac{e_o}{R} = 0$$

Hence

$$\frac{e_o}{e_i} = -sRC \tag{23-12}$$

Inspection of this equation reveals that this transfer function corresponds to that of a pure differentiator in the time domain with a multiplying factor of RC. Thus, if the input signal e_1 is a ramp, then ideally the output signal e_o, as required by Eq. (23-12), should be a step (see Fig. 23-11). The height of the output step clearly

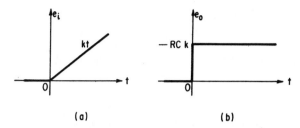

Fig. 23-11. Input and output signals of the differentiator of Fig. 23-10:
(a) input signal;
(b) output signal.

depends upon the slope of the input ramp. It is significant to note that, when the input signal to a differentiator contains high-frequency noise (i.e., is not a pure ramp in the example considered), the noise components are amplified to a greater extent than the signal component because of the greater slopes of the noise signals. Consequently, the output signal of the differentiator has a much poorer signal-to-noise ratio. Moreover, it is quite likely that the high amplification of the noise will cause the amplifier to overload. This behavior of the differentiator stands in sharp contrast to the way in which integrators influence noise components. Since noise is generally a high-frequency phenomena, Eq. (23-11) reveals that the higher the frequency components in the noise the greater the attenuation caused by the integrator. This is the frequency-domain viewpoint. In terms of the time domain it is apparent that, since integration is an averaging process, the integrator exerts a smoothing effect on the input signal, thus furnishing an output signal with an improved signal-to-noise ratio. Chiefly for these reasons, then, the electronic analog computer provides solutions to problems by performing successive operations of integration rather than of differentiation.

"Impure" Integration and "Impure" Differentiation. Often in the computer study of physical problems, especially in control systems engineering, it is necessary to simulate such important items as proportional plus integral operations, proportional plus derivative operations, simple time lags, quadratic transfer functions, and so forth. The manner of accomplishing this is straightforward. It requires merely a recognition of the fact that in general the output is related to the input in the configuration of Fig. 23-12 by

$$\frac{e_o}{e_i} = -\frac{Z_0}{Z_1} \qquad (23\text{-}13)$$

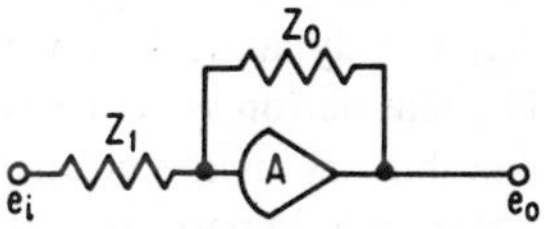

Fig. 23-12. Operational amplifier with general input and feedback impedances.

where Z_0 and Z_1 are, respectively, the feedback and input impedance functions, which need not be restricted to resistors or capacitors but may be varied combinations of these. Consider, for example, the situation where Z_0 is a parallel combination of a resistor and a capacitor. Then the operational impedance of the parallel combination is

$$Z_0 = \frac{R_0}{1 + sR_0C_0} \tag{23-14}$$

If the input impedance is also made a parallel combination of R_1 and C_1, it follows that

$$Z_1 = \frac{R_1}{1 + sR_1C_1} \tag{23-15}$$

Inserting the last two equations into Eq. (23-13) yields

$$\frac{e_o}{e_i}(s) = -\frac{R_0}{R_1}\frac{1 + sR_1C_1}{1 + sR_0C_0} \tag{23-16}$$

Examination of this result makes it apparent that the standard lead and lag networks discussed in Chapter 22 can readily be simulated by such a configuration. Table 23-1 shows some of the more commonly used forms. Note that for case 1 the simulation of a proportional plus derivative operation can be had by removing the feedback capacitor C_0. On the other hand the removal of the input capacitor C_1 instead furnishes a means of simulating a simple time lag.

23-3 Simulation Procedures. Differential-equation Approach. Magnitude Scale Factors

How are the basic mathematical building blocks that are described in the preceding section combined to solve problems? The answer to this question is illustrated by applying the required techniques to the solution of a linear second-order differential equation. In the process of developing the simulation diagram, it becomes necessary to discuss the very important topics of magnitude scaling and time scaling. Although the treatment of these topics is in connection with a linear system, the techniques are applicable in general.

Assume that the behavior of a control system is described by the differential equation

$$\ddot{y} + 5\dot{y} + 100y = f \tag{23-17}$$

where f is the forcing function and y the controlled variable. Let it be desired that the dynamic behavior is to be studied on an electronic analog computer. The simulation is to be such that it allows a study to be made of the complete time history of the controlled variable as well as its velocity and acceleration as it responds to various types and magnitudes of the forcing function. Also, assume for the purpose of illustration that the maximum value of f is 200 units of forcing function and that all initial conditions are zero.

The first step in simulating Eq. (23-17) is to isolate the highest derivative term on one side of the equation. Thus

$$\ddot{y} = f - 5\dot{y} - 100y \tag{23-18}$$

This is required in view of the fact that it is desirable to use only integrators to obtain the solution because of the reasons previously cited. Hence, by summing and then performing two successive integrations of Eq. (23-18), the solution for the dependent variable is obtained. For the moment let us disregard the need for magnitude scale factors and proceed directly to a simulation of the last equation. If attention is confined solely to the operations called for in Eq. (23-18), it should be clear that to obtain the second derivative quantity $\ddot{y}$ it is necessary to sum f, $-5\dot{y}$, and $-100y$. This can be done by means of a summing amplifier, provided that $\dot{y}$ and y are available. Actually, the latter quantities are made available through feedback connections as they are generated in the analog simulation. The forcing function f is available, since it is the known input quantity. The result of the summation of these quantities yields $-\ddot{y}$, as depicted in Fig. 23-13(a). The negative sign is due to the phase reversal of the operational amplifier. Integration of $-\ddot{y}$ yields $\dot{y}$, as illustrated in Fig. 23-13(b). A second integration yields the control variable $-y$, as shown in Fig. 23-13(c). Combining these operations in the manner shown in Fig. 23-13(d) furnishes the complete simulation diagram for solving Eq. (23-17). Note that the output of amplifier $A2$ needs to be reversed before it can be applied as an input at amplifier $A1$. This is accomplished by means of the inverter amplifier $A4$.

Although Fig. 23-13(d) is a correct mathematical simulation of Eq. (23-17), it does not necessarily follow that the correct solution can be measured with this

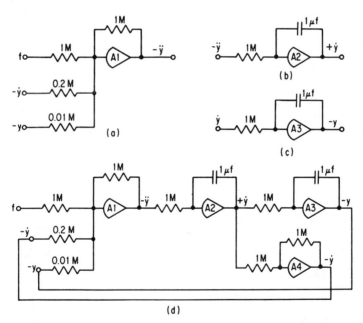

Fig. 23-13. Simulation procedure for a linear second-order system:
  (a) summation to obtain acceleration quantity;
  (b) integration of acceleration to obtain velocity;
  (c) integration of velocity to obtain the controlled variable;
  (d) complete simulation diagram.

Table 23-1. Simulation of Complex Functions

Case 1:

$$Z_0 = \frac{R_0}{1 + sR_0C_0}$$

$$Z_1 = \frac{R_1}{1 + sR_1C_1}$$

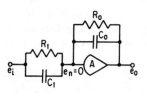

Therefore $\dfrac{e_o}{e_i} = -\dfrac{Z_0}{Z_1} = -\dfrac{R_0}{R_1}\dfrac{1 + sR_1C_1}{1 + sR_0C_0}$

Subcase 1a—C_0 removed:

Therefore $\dfrac{e_o}{e_i} = -\dfrac{R_0}{R_1}(1 + sR_1C_1)$

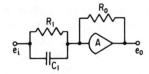

Subcase 1b—C_1 removed:

Therefore $\dfrac{e_o}{e_i} = -\dfrac{R_0}{R_1}\dfrac{1}{1 + sR_0C_0}$

Case 2:

$$Z_0 = R_0 + \frac{1}{sC_0} = \frac{1 + sR_0C_0}{sC_0}$$

$$Z_1 = \frac{R_1}{1 + sR_1C_1}$$

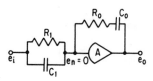

Therefore

$$\frac{e_o}{e_i} = -\frac{Z_0}{Z_1} = -\frac{(1 + sR_1C_1)(1 + sR_0C_0)}{sR_1C_0}$$

Subcase 2a—R_0 removed:

$$\frac{e_o}{e_i} = -\frac{1 + sR_1C_1}{sR_1C_0}$$

Subcase 2b—C_1 removed:

$$\frac{e_o}{e_i} = -\frac{1 + sR_0C_0}{sR_1C_0}$$

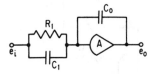

Table 23-1. Simulation of Complex Functions (*Continued*)

Case 3:

$$Z_0 = R_0 + \frac{1}{sC_0} = \frac{1 + sR_0C_0}{sC_0}$$

$$Z_1 = R_1 + \frac{1}{sC_1} = \frac{1 + sR_1C_1}{sC_1}$$

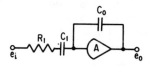

Therefore $\dfrac{e_o}{e_i} = -\dfrac{Z_0}{Z_1} = -\dfrac{C_1}{C_0}\dfrac{1 + sR_0C_0}{1 + sR_1C_1}$

Subcase 3a—R_0 removed:

Therefore $\dfrac{e_o}{e_i} = -\dfrac{C_1}{C_0}\dfrac{1}{1 + sR_1C_1}$

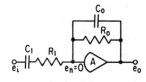

Subcase 3b—R_1 removed:

$$\frac{e_o}{e_i} = -\frac{C_1}{C_0}(1 + sR_0C_0)$$

Case 4:

$$Z_0 = \frac{R_0}{1 + sR_0C_0}$$

$$Z_1 = \frac{1 + sR_1C_1}{sC_1}$$

Therefore

$$\frac{e_o}{e_i} = -\frac{Z_0}{Z_1} = -\frac{sR_0C_1}{(1 + sR_0C_0)(1 + sR_1C_1)}$$

Subcase 4a—C_0 removed:

$$\frac{e_o}{e_i} = -\frac{sR_0C_1}{1 + sR_1C_1}$$

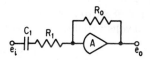

Subcase 4b—R_1 removed:

$$\frac{e_o}{e_i} = -\frac{sR_0C_1}{1 + sR_0C_0}$$

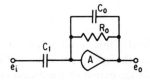

731

simulation. An incorrect solution results whenever improper magnitude scaling is used in the problem. To illustrate this, let it be assumed that the useful operating range of the d-c amplifiers is ± 100 volts, which is typical of most commercial electronic analog-computing equipment. Clearly, then, if f has a maximum value of 200 units, this quantity at best can be represented by 100 volts. Thus, there is a scale factor for this simulation of $\frac{1}{2}$ volt/unit forcing function. Once this scheme is used to represent f as a voltage and no modifications are made anywhere in the circuitry of Fig. 23-13(d), it follows that each of the remaining three variables, viz., $\ddot{y}$, $\dot{y}$ and y, are to be interpreted in accordance with this same scale factor. Thus, in looking at the output of amplifier $A3$ in units of volts, the corresponding value of the physical variable is found by dividing the measured output voltage of $A3$ by $\frac{1}{2}$ volt/unit y. Here then lies the difficulty with Fig. 23-13(d). When f has its maximum value of 200 units (assume f to be a step input), Eq. (23-17) shows that the steady-state value of the controlled variable is

$$y_{max} = \frac{f_{max}}{100} = \frac{200}{100} = 2$$

Now, because the scale factor for the simulation diagram was chosen to be $\frac{1}{2}$ volt per unit variable, it follows that the maximum steady-state voltage level appearing at amplifier $A3$ is 1 volt. Yet the full dynamic voltage range of the amplifier is 100 volts. This is indeed a case of poor magnitude scaling which is very likely to lead to serious errors in the measurement of the quantity y. In fact, when f is not at its maximum value, the output of $A3$ will be less than 1 volt, which practically puts the reading of $A3$ in the amplifier noise region, thereby affording very poor accuracy.

This line of reasoning is likely to prompt one to suggest using as the scale factor for the circuit a value which will allow $A3$ to operate at a maximum level of 100 volts. Proceeding in this manner leads to another serious difficulty. Although the conditions at the output of $A3$ are apparently improved, this is not the case at the remaining amplifiers in the simulation. Operating with a scale factor of 50 volts/variable, which is selected to make full use of the output voltage range of $A3$, calls for a highly excessive voltage at $A1$ if the maximum acceleration of 200 units is to be realized. Equation (23-17) reveals that, if f is given a maximum value of 200 units, the maximum acceleration is also 200 units. However, for the output of $A1$ to represent correctly this physical quantity in terms of a scale factor of 50 volts/variable, it is necessary for the output voltage to reach a level of

$$\frac{e_{A1} \text{ (volts)}}{50 \text{ (volts/unit acceleration)}} = 200 \text{ units of acceleration}$$

or

$$e_{A1} = 10{,}000 \text{ volts}$$

Clearly, this result is unrealistic. The output-voltage level is 100 times greater than the maximum value of the amplifier output voltage. It is instructive to note that the extent of the excess corresponds to the coefficient of y in Eq. (23-17).

On the basis of the foregoing discussion it is apparent that some modification is needed in order to make Fig. 23-13(d) a workable simulation diagram. To appreciate better the nature of this modification, a closer examination of the source of the difficulty is in order. It has already been pointed out that the maximum steady-state value of y is 2 units. What then are the maximum values of $\dot{y}$ and $\ddot{y}$? The answers to this question lie in the coefficients of the differential equation of concern. A little thought should make it apparent that, if the coefficients were each unity, the difficulty described in the foregoing would not occur, because the value of scale factor would be the same regardless of how it was arrived at. However, when the coefficients differ by as much as 100 : 1, as in the case of Eq. (23-17), then the maximum values of y, $\dot{y}$, and $\ddot{y}$ can differ considerably. Accordingly, if the diagram is not appropriately modified to reflect this situation, the simulation is not a very useful one.

An estimate of the maximum value of $\dot{y}$ can be obtained by reasoning in terms of the differences in values of y, $\dot{y}$, and $\ddot{y}$ when the system of Eq. (23-17) responds to a command. Clearly, this information is available from the solution to the characteristic equation. Upon using the s notation for derivatives, it follows that for Eq. (23-17) the characteristic equation is

$$s^2 + 5s + 100 = 0 \qquad (23\text{-}19)$$

Since the general form of the characteristic equation of second-order linear systems is

$$s^2 + 2\zeta\omega_n s + \omega_n^2 = 0 \qquad (23\text{-}20)$$

it is apparent that for the case at hand

$$\omega_n = 100 \text{ radians/sec}$$

Furthermore, the complementary solution is given by

$$y_c = k\epsilon^{-\zeta\omega_n t} \sin(\omega_d t + \phi) \qquad (23\text{-}21)$$

so that the complete solution can then be expressed as

$$y = y_{ss} + y_c = y_{ss} + k\epsilon^{-\zeta\omega_n t} \sin(\omega_d t + \phi) \qquad (23\text{-}22)$$

When dealing with magnitude scale factors, it is advisable to direct attention to the worst case that is likely to arise. Hence assume that $\zeta = 0$. This allows Eq. (23-22) to be rewritten as

$$
\begin{aligned}
y &= y_{ss} + k \sin(\omega_n t + 90°) \\
  &= y_{ss} + k \cos \omega_n t
\end{aligned}
\qquad (23\text{-}23)
$$

It is not uncommon in the last equation for k to be equal to the steady-state solution or to be fairly close to it. For example, if at $t = 0^+$, y is known to be zero, it follows that k is equal to $-y_{ss}$, so the complete solution becomes

$$y = y_{ss} - y_{ss} \cos \omega_n t \qquad (23\text{-}24)$$

Examination of Eq. (23-24) discloses that in the worst case the first derivative quantity, $\dot{y}$, can have a maximum value that is ω_n times as large as the steady-state value. Thus differentiating Eq. (23-24) yields

$$\dot{y} = \omega_n y_{ss} \sin \omega_n t \tag{23-25}$$

Note that the coefficient of the sinusoid is, in fact, the amplitude (or maximum value) of the velocity quantity. A further differentiation of Eq. (23-25) reveals information about the peak value of acceleration. Thus

$$\ddot{y} = \omega_n^2 y_{ss} \cos \omega_n t \tag{23-26}$$

Note that the second derivative quantity can have a value in the worst case that is ω_n^2 times as large as y_{ss}. A moment's reflection makes it obvious from the foregoing analysis that when $\omega_n = 1$ rad/sec, the maximum values of all three variables are the same and, consequently, no scaling difficulty arises. However, when ω_n is 10 rad/sec, as in the case here, then a definite problem exists, because the second derivative variable has a maximum value that is 100 times greater than that of the controlled variable. If the same computer reference voltage is to represent the maximum value of each variable, then the magnitude scale factors cannot be chosen equal. Rather, the scale factors should be chosen to reflect this difference.

One convenient way to accomplish this is to introduce modified machine variables that bear out the difference in scale factors. Thus, if it is known (or estimated) that the maximum steady-state value of y is 2 units, the modified machine variable for y becomes

$$\frac{y}{y_{max}} 50 \text{ (volts)} = y\left(\frac{50 \text{ volts}}{2 \text{ units of } y}\right) = [25y] \text{ volts} \tag{23-27}$$

This approach assumes that k in Eq. (23-22) is comparable to the steady-state value of y. Note that 50 volts rather than the maximum available 100 volts is used. The reason is to allow enough margin so that even if y undergoes a 100 per cent overshoot during the transient state the operational amplifiers will not exceed the 100-volt limit. Note, too, that the modified machine variable $25y$ carries a number that is in reality the scale factor for the physical variable y. The brackets enclosing the quantity $25y$ are introduced to serve as a reminder that this quantity is to be treated as an *entity*. It is this total quantity (bearing units of volts) that represents in the computer the actual physical variable y.

Proceeding in like manner, the modified machine variables for velocity and acceleration become

$$\dot{y} \text{ (units of velocity)} \frac{50}{\dot{y}_{max}}\left(\frac{\text{volts}}{\text{units of velocity}}\right) = \dot{y}\frac{50}{10(2)} = [2.5\dot{y}] \text{ volts} \tag{23-28}$$

$$\ddot{y} \text{ (units of acceleration)} \frac{50}{\ddot{y}_{max}}\left(\frac{\text{volts}}{\text{units of acceleration}}\right)$$

$$= \ddot{y}\frac{50}{100(2)} = [\tfrac{1}{4}\ddot{y}] \text{ volts} \tag{23-29}$$

In order to simulate the physical problem [described by Eq. (23-17)] in a manner consistent with the modified variables, the procedure is as follows: First, manipulate Eq. (23-17) in such a way that it brings into evidence the modified machine variable $[\tfrac{1}{4}\ddot{y}]$. Clearly, this requires multiplying each term by $\tfrac{1}{4}$. Thus

$$\left[\frac{1}{4}\ddot{y}\right] + \frac{5}{4}\dot{y} + [25y] = \left[\frac{f}{4}\right] \tag{23-30}$$

It is desirable, too, at this point to manipulate the coefficients of the other variables so that the modified form of each of the variables appears. Hence Eq. (23-30) becomes

$$\left[\frac{1}{4}\ddot{y}\right] + \frac{1}{2}[2.5\dot{y}] + [25y] = \left[\frac{f}{4}\right] \tag{23-31}$$

The simulation of the summation indicated in this equation to yield acceleration is depicted in Fig. 23-14(a). Note that the modified machine variable representing velocity is multiplied by $\frac{1}{2}$, as called for by Eq. (23-31). When f has its maximum value of 200 units, this is represented in the simulation by a voltage of 50 volts.

The integration of the modified form of the machine variable (output of $A1$) representing acceleration must be done in such a way that the modified form of the velocity appears. Clearly, this requires that the integrator also introduce a scale factor change of 10 : 1—as illustrated in Fig. 23-14(b). Of course the significance of such a procedure is that both amplifiers $A1$ and $A2$ be capable of employ-

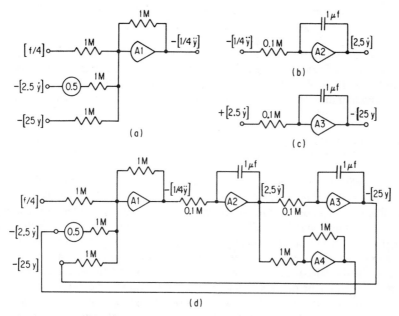

Fig. 23-14. Simulation procedure for a linear second-order system using proper scale factors:

    (a) summation to obtain acceleration;

    (b) integration of acceleration voltage with scale-factor adjustment to obtain velocity;

    (c) integration of velocity voltage with scale-factor adjustment to obtain the controlled variable;

    (d) complete simulation diagram.

ing the full dynamic range from -100 to $+100$ volts in providing the solution of the variable involved. A similar procedure is followed to get y. This is depicted in Fig. 23-14(c). The complete simulation diagram appears in Fig. 23-14(d). By the selection of appropriate scale factors a simulation diagram can be developed which not only simulates the given physical equation but does so in a fashion which allows each amplifier to operate over its full dynamic range. Figure 23-14(d) is a useful simulation diagram because the computing amplifiers are not subjected to overloads or to low voltage levels. Figure 23-15 is the symbolic representation of Fig. 23-14(d).

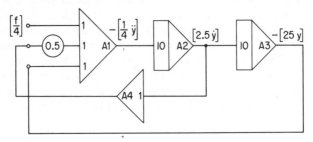

Fig. 23-15. Symbolic representation of Fig. 23-14(d).

The presence of the scale factors in the simulation diagram also serves as a convenience in interpreting the meaning of the output voltages of the individual amplifiers. Thus, if at a particular time the output of $A2$ is found to be 5 volts, the corresponding value of the physical quantity is readily obtained by noting that

$$[2.5\dot{y}] = 5 \text{ volts}$$

from which it follows that the physical quantity, namely, velocity, has the value $\dot{y} = 5/2.5 = 2$ units of velocity.

Simplified Magnitude Scaling Procedure. The foregoing method of magnitude scaling a problem for solution on the analog computer leads to highly satisfactory results, but it has one serious shortcoming. It requires knowledge of the approximate maximum values of all the variables. Although this is a straightforward matter for second-order problems, it is certainly not so for higher-order cases. For example, if we were concerned with finding the computer solution of a third-order differential equation, we would need first to find the roots of the characteristic equation before we could conveniently estimate the maximum values of the variables. But this would make a computer solution anticlimactic, because if the roots of the characteristic equation are known, certainly ample information about the dynamical behavior of the solution is deducible from these roots. We are almost forced to conclude, therefore, that if the computer is to be truly useful as a tool for finding solutions to problems, there must exist a more direct approach to scaling.

A careful analysis of the simulation diagrams appearing in Figs. 23-13 and 23-14 provides important clues for developing a direct scaling approach. It is instructive to note in these diagrams that the gain in the y loops in each case is

100, which is consistent with the coefficient of y in Eq. (23-17). However, the manner in which this gain is distributed among amplifiers 1, 2, and 3 is significant and, in fact, makes the difference between a simulation diagram that provides a solution and one that fails to do so. In the diagram of Fig. 23-13, the gain of 100 is placed at the summing amplifier, $A1$, at the y input terminal, whereas in Fig. 23-14 this same gain is distributed as a gain of 10 at the first integrator amplifier ($A2$), and a further gain of 10 at the second integrator amplifier ($A3$). In view of the fact that the peak values of the first and second derivative quantities are influenced by the natural frequency of the system, this distribution of the gain ($\omega_n^2 = 100$) at the integrators is a result that is perfectly consistent with the results appearing in Eqs. (23-25) and (23-26). For ω_n greater than unity, the maximum value of the first derivative ($\dot{y}$) will be greater than that of the dependent variable (y) by approximately ω_n. It should be entirely expected, therefore, that the value of the natural frequency be involved in some way as a gain factor when passing through an integrator in the computer diagram to convert from a voltage representing velocity to one representing displacement. A similar argument holds for the integrator that converts acceleration to velocity. Under no circumstances should we expect large gain factors to appear at summing amplifiers. The function of the summing amplifier is to do just that: to sum. The use of high gain factors at summing or even inverter amplifiers will lead to overload difficulties, which means improper magnitude scaling. *One cardinal rule of good magnitude scaling is to distribute the gain factors at the integrators, while staying as close to unity as is practical at the summing amplifiers.* With this in mind, let us now develop the simulation diagram for Eq. (23-17) by a direct approach. It is important to remember that all the information about maximum values is contained in the coefficients of this equation.

Equation (23-18) can serve as the starting point, but it is now rewritten with the following emphasis:

$$[\ddot{y}] = [f] - [5\dot{y}] - [100y] \qquad (23\text{-}32)$$

The brackets are used to stress the fact that this equation is now expressed in volts. Thus, a quantity such as $[\ddot{y}]$ means that the physical quantity acceleration ($\ddot{y}$) is assumed multiplied by 1 volt/unit of acceleration to yield a voltage variable $[1 \times \ddot{y}]$ that represents acceleration in the computer. Similarly, the quantity $[5\dot{y}]$ is the voltage variable in the computer that represents the actual problem variable velocity in accordance with a scale factor of 5 volts per unit of velocity. The displacement term is treated in a like fashion, except that here the scale factor is 100 volts/unit of displacement. Recall that the bracketed quantities are to be treated as entities. They, in fact, represent the variables of the problem. Thus, if it is desirable to generate the computer variable for acceleration, $[\ddot{y}]$, it is necessary to sum the three bracketed quantities on the right side of Eq. (23-32). It is significant to observe here that each bracketed quantity on the right has a multiplying factor of unity. As a matter of fact, the brackets were placed in the manner shown in Eq. (23-32) precisely to achieve this result. Consequently, the computer variable for acceleration can be obtained by using only unity gain factors, as illustrated in Fig. 23-16.

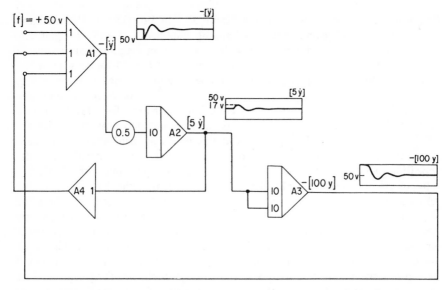

Fig. 23-16. Simulation diagram as derived from Eq. (23-32). Responses are for a step input of $+50$ v.

To obtain the computer variable for velocity, it is necessary to multiply the computer variable for acceleration by 5 as well as to integrate it. This then yields the computer variable for velocity, $[5\dot{y}]$. The computer variable for displacement is obtained in a similar fashion. Since the original manipulation of Eq. (23-32) brought into evidence the scale factor 100, it is necessary to multiply the computer variable for velocity by 20, while integrating it to obtain displacement, as depicted in Fig. 23-16. Inspection of the distribution of the gain factors in this diagram discloses that there is a gain factor of 5 at the first integrator amplifier ($A2$) and a gain factor of 20 at the second integrator amplifier. Also, depicted in Fig. 23-16 are the actual time histories of each of the variables in response to a step command of 50 volts, i.e., $[f] = 50$ volts. Examination reveals that the dynamic voltage range of amplifiers $A1$ and $A3$ are being well used in this simulation. This is indicative of the selection of appropriate scale factors for acceleration and displacement. The choice of the scale factor for the velocity quantity, however, could have been better selected.

A better choice of the scale factor for velocity usually results if we relax the restriction of unity gain factor for each bracketed term in Eq. (23-32). The use of unity multiplying factors was introduced initially for simplicity. Actually, even in this simple second-order case there are a couple of good reasons for deviating from the unity factors. One, the differentiation process that gives rise to a difference in the maximum values of displacement and velocity is the very same process that gives rise to the difference in the maximum values of velocity and acceleration. Hence we should expect the gain factor at each integrator to be the same. Two, since the standard multiplying factor with which computers are equipped is either 1 or 10, it is convenient to have the scale factors differ by a ratio of 10, provided that such a practice is not inconsistent with the original equation. Keeping these points in mind, Eq. (23-18) can be manipulated more appropriately into the following form:

$$[\ddot{y}] = [f] - \tfrac{1}{2}[10\dot{y}] - [100y] \tag{23-33}$$

Observe again that this equation is identical to the original equation, but the use of brackets to identify computer variables makes this a computer equation expressed in volts. When manipulating the original equation, care must be taken not to change the equation. The computer version must remain identical to the original equation. The coefficients of the bracketed terms in Eq. (23-33) now range from

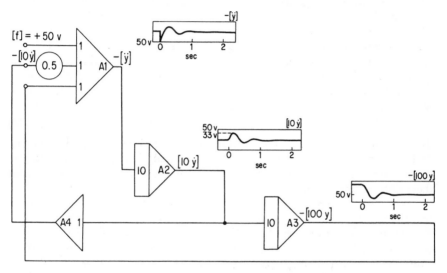

Fig. 23-17. Simulation diagram using the simplified magnitude scaling procedure but with equal gain factors at the integrators.

$\tfrac{1}{2}$ to 1. Figure 23-17 depicts the complete simulation diagram as it proceeds from Eq. (23-33). The computer variable that represents velocity is now $[10\dot{y}]$. Since the physical quantity $\dot{y}$ is not at all altered by the choice of scale factor, it follows that the dynamic voltage range in this case of the velocity amplifier, $A2$, should be twice that appearing in Fig. 23-16, because the scale factor is twice as large. A comparison of the computer readouts on the oscillographic recordings at amplifier $A2$ for the same step input of $[f] = 50$ volts furnishes verification. Note, too, that the computer readouts for $A1$ and $A3$ are identical. This is entirely consistent with the fact that in either case the computer variables carry the same scale factor.

Compare the simulation diagram of Fig. 23-17 with that of Fig. 23-15. Clearly, the *circuitry* is identical in each case. The only differences one can see in this comparison are the scale factors. Why are they not the same? They are not the same because Fig. 23-17 was developed without any knowledge whatsoever of the magnitude of the forcing function in the physical problem. Knowledge of the forcing function is not a prerequisite to establish proper magnitude scale factors, notwithstanding the fact that this is the way the subject matter was introduced at the start of this section. Knowledge of the forcing function was used merely to furnish the background needed to understand the direct approach to magnitude scaling. The direct approach depends solely upon the coefficients in the original equation. It requires no *a priori* knowledge of the forcing function nor of the maximum values of the variables. This was illustrated in arriving at the simulation appearing in Fig. 23-17. The scale factors appearing in this diagram provide information about the ratios that must prevail between the scale factors in any specific problem, provided only that the characteristic equation remains intact. This leads then to the significant conclusion that if *any one* scale factor is known for a specific forcing function, all the others are also known. For example, if in Eq. (23-18) the physical quantity f is assumed to be 200 units of force, it follows from this equation that y_{ss} will be 2 units. The scale factor for y thus becomes $\frac{50}{2}$ or 25 volts/unit of displacement. Hence, the computer variable for displacement becomes $[25y]$. A study of Eq. (23-33) then shows that to bring into evidence a computer variable for y that is $[25y]$, it is necessary to divide each term by 4, which leads to

$$[\tfrac{1}{4}\ddot{y}] = [\tfrac{1}{4}f] - \tfrac{1}{2}[2.5\dot{y}] - [25y] \tag{23-34}$$

The simulation of this equation leads directly to the circuitry appearing in Fig. 23-15. When specific values are assigned to the forcing function, the scale factors of the computer variables will take on values consistent with it. But the ratios of these scale factors remain intact.

Although the simulation procedure that leads to Fig. 23-17 is preferable to that leading to Fig. 23-16, it should not be inferred that the results obtained from Fig. 23-16 are unsatisfactory. Let it be desired to find the maximum velocity in the physical problem for the step forcing function used in Figs. 23-16 and 23-17. From the computer oscillographic readout of $A2$ in Fig. 23-17 we have

$$[10\dot{y}] = 33 \text{ volts}$$

$$\therefore \ \dot{y} = \tfrac{33}{10} = 3.3 \text{ units of velocity}$$

Similarly, from the computer oscillographic readout of $A2$ in Fig. 23-16 we have

$$[5\dot{y}] = 17 \text{ volts}$$

$$\therefore \ \dot{y} = \tfrac{17}{5} = 3.4 \text{ units of velocity}$$

The discrepancy here lies well within the accuracy limits of the graphical recording devices.

EXAMPLE 23-1 Obtain the properly magnitude-scaled simulation diagram for studying the solution to the following linear differential equation:

$$\dddot{y} + 40\ddot{y} + 250\dot{y} + 4500y = f \tag{23-35}$$

The source function f is not specified and no information is available concerning the maximum values of the variables. All variables up to the third derivative are to be available for recording.

solution: The original differential equation will be manipulated in such a manner that appropriate computer variables expressed in volts will be identified. However, it is important that this operation be performed consistent with the physical capability of the computing equipment available.

It is first necessary to solve for the highest-order derivative. Thus

$$\dddot{y} = f - 40\ddot{y} - 250\dot{y} - 4500y \tag{23-36}$$

Next, put the equation in an appropriate manipulated form. One such form might be

$$[\dddot{y}] = [f] - 1.33[30\ddot{y}] - \tfrac{5}{6}[300\dot{y}] - 1.5[3000y] \tag{23-37}$$

The computer variable for the second derivative is chosen to be $[30\ddot{y}]$ rather than $[40\ddot{y}]$ for the following practical reason. After the quantities on the right side of Eq. (23-37) are summed to yield $-[\dddot{y}]$, the computer variable for the second derivative is found by integrating and multiplying by the appropriate coefficient. If the coefficient is chosen to be 40, a problem arises, because the standard operational amplifier makes available a gain of 30 but not 40. The gain of 30 is achieved by placing the three 10 terminals in parallel. Once $[30\ddot{y}]$ is chosen as the computer variable for the second derivative, the computer variables for the first derivative and the dependent variable are then chosen as convenient multiples. In this instance, appropriate values are $[300\dot{y}]$ and $[3000y]$, respectively. Note that the gain factors that have to appear at the summing amplifier are modest, ranging from 0.833 to only 1.5. Most of the gain will automatically occur at the integrator amplifiers because of the way in which the computer variables have been chosen. Equation (23-37) applies at the summing amplifier, which is $A1$ in Fig. 23-18. Then the machine variable $[30\ddot{y}]$ is generated from the output of $A1$ by passing the latter through an integrator and multiplying by a factor of 30 in the process. Similarly, at amplifier $A3$ we have

$$10\int_0^t [30\ddot{y}]\,dt = [300\dot{y}]$$

and at $A4$

$$10\int_0^t [300\dot{y}]\,dt = [3000y]$$

When the complete simulation diagram is drawn, as in Fig. 22-18, several useful checks are worthwhile making. To guard against inadvertently omitting an inverter amplifier, count the number of operational amplifiers appearing in each loop. If the original differential equation contains all plus (or all minus) signs, the number of amplifiers

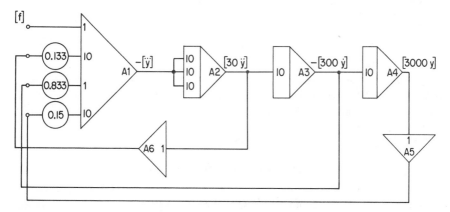

Fig. 23-18. Simulation diagram for Example 23-1.

in the loop must be *odd*. Next, check that the net gain factors present in each loop correspond, respectively, to the coefficients in the original differential equation. Thus, Eq. (22-36) states that the gain factor should be 40 for the second derivative loop of Fig. 22-18. The second derivative loop consists of amplifiers $A1$, $A2$, and $A6$. The gain of $A1$ for the $\ddot{y}$ terminal is 1.33. The gain for $A2$ is 30, and the gain of $A6$ is unity. The product thus yields 40, which checks. In a similar fashion, the loop gain for the first derivative is found to be 250, and for y it is 4500.

23-4   Time Scaling

The dynamic behavior of a system containing energy-storing elements is determined by the roots of the characteristic equation. Whether or not velocity or acceleration terms will have excessive values depends upon the magnitude of the complex conjugate roots. Thus, if the natural frequency of these roots is large, there is need for magnitude scaling. On the other hand, if the natural frequency of the predominant complex conjugate roots of a system has a natural frequency of 1 radian/sec (or thereabouts), magnitude scaling is unnecessary because the scale factor for the controlled variable as well as its first- and higher-order derivatives is the same. In situations where large natural frequencies are known to exist, therefore, it is reasonable to ask whether or not it is possible to introduce an appropriate time-scale change so that the system behaves as if the natural frequency were 1 radian/sec. To pursue this thought, let us return to Eq. (23-17), which is repeated here for convenience as

$$\frac{d^2y}{dt^2} + 5\frac{dy}{dt} + 100y = f \tag{23-38}$$

The rate of change symbol d/dt refers to the rate at which changes take place in the physical problem with respect to time. However, in the simulation of this problem there is no reason why the time required for the controlled variable to respond to commands and disturbances cannot be slowed down or speeded up as long as the functional relationship called for in Eq. (23-38) is satisfied. Thus, it should be possible to take 10 sec of machine time to accomplish what takes 1 sec in the physical problem. Mathematically expressed, we have

$$\frac{d}{dt} = \alpha_t \frac{d}{d\tau} \tag{23-39}$$

where α_t represents the desired time-scale change and $d/d\tau$ denotes the rate of change with respect to machine time as contrasted to the rate of change with respect to problem time d/dt. By choosing $\alpha_t = 10$ the rates of change in the analog-computer machine must be one-tenth the rates which occur in the physical problem; i.e., it computes on a slowed-down basis. Inserting the equality expressed in the last equation into Eq. (23-38) yields

$$\alpha_t^2 \frac{d^2y}{d\tau^2} + 5\alpha_t \frac{dy}{d\tau} + 100y = f = 200$$

or

$$\frac{d^2y}{d\tau^2} + \frac{5}{\alpha_t}\frac{dy}{d\tau} + \frac{100}{\alpha_t^2}y = \frac{200}{\alpha_t^2} \tag{23-40}$$

A glance at Eq. (23-40) makes it apparent that the selection of α_t equal to 10 makes the computer solution behave in terms of a natural frequency of 1 radian/sec. Hence the magnitude-scale factors for the controlled variable and its first and second derivatives may be the same. By using the dot notation for derivatives as well as the capital letter for the controlled variable and for derivatives with respect to machine time τ in order to emphasize the associated time-scale change, Eq. (23-40) may be written as

$$\ddot{Y} + \tfrac{1}{2}\dot{Y} + Y = 2 \qquad (23\text{-}41)$$

Since the maximum value of Y is not influenced by the time-scale change, the scale factor for the simulation of Eq. (23-41) is chosen to be that of the controlled variable. As given by Eq. (23-27), this has the value of 25 volts/variable. Before proceeding with a simulation of Eq. (23-41), it is desirable to convert it to a computer equation (expressed in volts) by multiplying each term by the common-scale factor 25 volts/variable. Thus Eq. (23-41) becomes

$$[25\ddot{Y}] + \tfrac{1}{2}[25\dot{Y}] + [25\,Y] = 50 \text{ volts} \qquad (23\text{-}42)$$

The simulation of this equation in symbolic notation appears in Fig. 23-19.

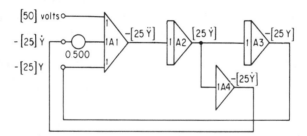

Fig. 23-19. Simulation diagram for the solution of Eq. (23-38).

A comparison of Figs. 23-13(d), 23-14(d), and 23-19 reveals an interesting result. In Fig. 23-13(d) note that the loop-gain factor in the y loop is 100 and that in the $\dot{y}$ loop is 5. These results are consistent with the values of the coefficients of y and $\dot{y}$, as called for in Eq. (23-17). An examination of Fig. 23-14(d) indicates that the use of proper magnitude-scale factors in no way alters the total value of these loop gains. What has happened is that the total gain no longer is put at a single amplifier but rather is distributed throughout the loop in a fashion to assure full use of the amplifiers' dynamic voltage range. However, the situation is decidely different in the case of Fig. 23-19. Here the loop gain for the controlled variable is now 1 as compared with 100 for Figs. 23-13(d) and 23-14(d). Also, the loop gain for the first derivative quantity is 0.5 compared with 5. Thus, it is apparent that the introduction of *time scaling changes the loop gains.* This is clearly borne out by Eq. (23-41), which reveals that the coefficient of the Y and $\dot{Y}$ terms are, respectively, 1 and $\tfrac{1}{2}$. It is important to keep in mind, too, that Eq. (23-41) is still a physical equation whose proportional gain (or spring constant) has been reduced by a factor of 100. The application of the appropriate scale factor to this equation then yields the computer version of the equation expressed in volts. The solutions which are obtained from the electronic analog computer are of course solutions to Eq. (23-41). To make these solutions applicable to the original problem, one merely needs to introduce the appropriate modification in the time axis. Thus, in the case where the solution is slowed down by a factor of 10, a recorded computer solution which yields $\dot{Y}$ equal to 1 in./sec machine time actually corresponds to $\dot{y} = \alpha_t \dot{Y} = 10(1)$ in./sec of problem time.

Another interesting conclusion which comes from a comparison of Figs. 23-14(d) and 23-19 is that the time-scale change appearing in Fig. 23-19 is effectively accomplished by increasing the time constant of both integrators of Fig. 23-14(d) by the time-scale factor α_t. In fact this is a rule which can be applied in general: *To change the time scale of a given simulation merely requires that all time constants be changed by the same factor.* If α_t is chosen greater than unity, the computer solution is slower than in the physical problem. Conversely, if α_t is chosen less than unity, the computer takes less time to generate the solution than is required in the physical problem.

Although one of the most important factors responsible for introducing a time-scale change is the need to avoid high loop gains, there are other good reasons for doing this. One concerns the limited bandwidth of the galvanometer-type oscillographic recorders which are used to measure the magnitudes of the problem variables. The majority of these units have a bandwidth of 100 Hz; some have a bandwidth of 300 Hz. Accordingly, if a situation arises which calls for frequencies in excess of 300 Hz, time scaling is a necessity if accuracy is to be preserved. Servo-driven variplotters and the use of servomultipliers in the total simulation impose similar restrictions. Another factor which influences the need for time scaling is the limit on the available maximum recorder speed. If the problem solution occurs at a speed far in excess of the maximum recorder speed, the time scale will be so compressed as to make it impossible to read the galvanometer oscillographic recordings with any degree of suitable accuracy.

23-5   Simulation Procedures. Transfer-function Approach

Often the mathematical description of a physical problem is given in terms of the transfer functions of the components rather than a set of differential equations. This occurs frequently in the area of control systems engineering. In fact sometimes the output-input relationship cannot be expressed by an appropriate differential equation of a component. Instead, on the basis of measured attenuation and phase vs. frequency characteristics, it becomes necessary for an equivalent transfer function to be written for the component. Now, although it is true that once the transfer function is available the corresponding differential equation can be written, it is usually unnecessary to do this because the simulation diagram can be arrived at directly. For the sake of illustration, assume that the open-loop transfer function of a feedback control system is given by

$$\frac{C(s)}{E(s)} = G(s) = \frac{K}{s(1 + s)(1 + s\tau_1)} \tag{23-43}$$

and that it is desirable to investigate† with the computer the effect on the dynamic performance of varying the loop gain K and the time constant τ_1. Since the transfer function is given in factored form, the simulation is readily determined by making use of the configurations discussed in Sec. 23-1 for performing the operations called for in Eq. (23-43). Note that building blocks are needed which can provide a pure integration and two simple lags. The simulation diagram of Eq. (23-43) appears in Fig. 23-20(a). Amplifier $A1$ furnishes the required 1-sec time constant. The circuitry associated with $A2$ makes available the variation in the time constant τ_1. The arrangement shown allows τ_1 to be varied between zero and 1 sec. If greater values of τ_1 are needed, the feedback resistor must be changed to a larger value. The circuitry associated with $A3$ furnishes integration as well as the variable gain feature called for.

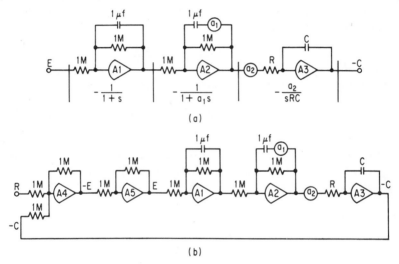

(a)

(b)

Fig. 23-20. Simulation of a closed-loop system starting from the open-loop transfer function:

        (a) simulation of the open-loop transfer function of Eq. (23-43);

        (b) simulation of the closed-loop system of (a).

Since the purpose of this computer study is to establish information concerning the dynamic behavior of the closed-loop system, it is necessary to provide the configuration of Fig. 23-20(a) with a feedback connection. The result is shown in Fig. 23-20(b). Note that the output of $A3$, which is $-C$, is summed with the command signal at $A4$ to yield the negative of the actuating signal. However, since $A1$ calls for $+E$, it follows that an additional inverter amplifier is needed. This is shown as $A5$ in Fig. 23-20(b).

---

† Again it is pointed out that a computer is really not needed for this purpose, because the available analytical techniques are adequate. However, if there were a nonlinearity such as backlash in the feedback path, then indeed the computer would show to advantage.

When the simulation diagram of a transfer function is developed in the fore-going manner, it is important to understand that all initial conditions are assumed equal to zero and that the magnitude-scale factors for the controlled variable and its derivatives are assumed equal. Of course, depending upon the natural frequency of the predominant complex roots of the characteristic equation, this procedure could lead to amplifier overloading. If this should happen, one remedy is to change all time constants by an appropriate time-scale factor. The matter of initial conditions is not important here if the investigation is concerned chiefly with a study of dynamic behavior.

The simulation of a transfer function when it is expressed in factored form is easily accomplished because it usually requires a cascading of conventional building blocks. The procedure needs to be modified, however, when the transfer function is in unfactored form. To illustrate this, consider the simulation of the transfer function given by

$$\frac{e_o}{e_i} = \frac{As^2 + Bs + C}{s^3 + Ds^2 + Es + F} \tag{23-44}$$

Dividing the numerator and denominator by s^3 permits writing

$$\frac{e_o}{e_i} = \frac{A/s + B/s^2 + F/s^3}{1 + D/s + E/s^2 + F/s^3} \tag{23-45}$$

Examination of the denominator discloses that it is in the form of $(1 + \sum$ loop transfer functions). The numerator consists of the output of integrators appropriately multiplied and summed. A graphical representation of Eq. (23-45) in terms of the signal-flow graph appears in Fig. 23-21(a). Recall that the plus signs in the denominator of the last equations mean negative feedback in the signal-flow graph. The analog-computer simulation diagram that corresponds to the signal-

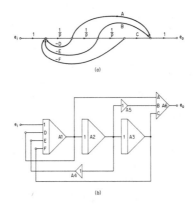

Fig. 23-21. Simulation procedure for transfer functions in unfactored form:
(a) signal flow graph representation of Eq. (22-44);
(b) analog computer simulation.

flow graph is shown in Fig. 23-21(b). Note that the first amplifier, $A1$, serves both as a summing and integrating amplifier. Since the output of $A1$ is negative for a positive e_i, it is brought directly to that input terminal where multiplication by D occurs. The output of $A2$ is positive. Hence, $A4$ is needed to reverse its sign before bringing it back to $A1$. Moreover, $A5$ is needed so that all the quantities appearing at the input terminals of $A6$ will have the same sign as called for by Eq. (23-44). The output of $A6$ will be e_o with the proper sign.

23-6 Multiplication and Division of Variables

The electronic analog computer shows to considerable advantage in solving nonlinear problems as well as linear problems having time-varying coefficients. It possesses this capability by virtue of equipment which enables the multiplication and the division of variables to be performed conveniently and accurately. It also possesses this capability because it contains components which allow it to generate many types of nonlinearities. The latter is the subject matter of the next section. This section is confined to a brief treatment of how multiplication and division of variables is accomplished.

One of the most commonly used methods for obtaining the product of two variables is the servomultiplier. The circuit arrangement is shown in Fig. 23-22. Usually it is equipped with six potentiometers. Five of these, A through E, are used for multiplication; the sixth one is used as a feedback potentiometer. To understand the operation of this circuit, consider that two variables A and X are applied to the servomultiplier as shown in Fig. 23-22. Corresponding to a specific

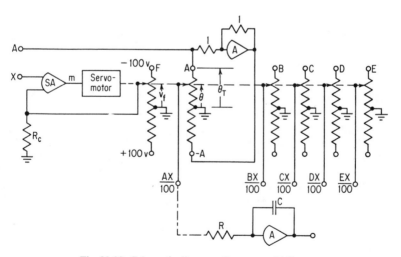

Fig. 23-22. Schematic diagram of a servomultiplier.

magnitude of X, there are fed back and summed at the servoamplifier (SA) both X and V_f, where the latter quantity denotes the voltage picked off the feedback potentiometer F. The servoamplifier output can be expressed as

$$(X - V_f)\mu = m \tag{23-46}$$

Here μ is the servoamplifier gain, and the negative sign denotes negative feedback. Since μ is very large, it follows that

$$X = V_f \tag{23-47}$$

Moreover, for a linear feedback potentiometer the magnitude of V_f may be expressed as

$$V_f = \frac{\theta}{\theta_T}(100) \tag{23-48}$$

where θ is the displacement of the slider arm of the potentiometer on which appears the feedback voltage and θ_T is the total travel from the center tap to either end of the potentiometer. The reference voltage applied to the feedback potentiometer is -100 volts at one end and $+100$ volts at the other end. Rearranging Eq. (23-48) and inserting Eq. (23-47) then yields

$$\frac{\theta}{\theta_T} = \frac{V_f}{100} = \frac{X}{100} \tag{23-49}$$

Furthermore, for a given displacement θ of the feedback potentiometer, the multiplying potentiometer undergoes the same displacement because each is ganged to the common servomotor shaft. Accordingly, the voltage appearing on the slider arm of the first multiplying potentiometer is

$$e_{OA} = \frac{\theta}{\theta_T} A \tag{23-50}$$

Substituting Eq. (23-49) into the last equation furnishes the desired result, namely

$$e_{OA} = \frac{AX}{100} \tag{23-51}$$

The form of Eq. (23-51) is such as to avoid any amplifier-overloading difficulty. Thus, if the variables A and X are supplied to the servomultiplier with proper magnitude scaling, that is, $A_{max} = 100$ volts and $X_{max} = 100$ volts, then, clearly, the output voltage will not exceed 100 volts even though both variables may be at their maximum values at the same time. The equation for the output of each of the remaining multiplying potentiometers is, of course, the same as Eq. (23-51) with the one modification that the variable A is replaced by B, or C, or D, or E. It is apparent, then, that with the arrangement of Fig. 23-22 it is possible to obtain computer voltage signals which provide the product of one variable (X) by each of five other variables (A, B, C, D, E).

Before leaving Fig. 23-22, two additional items are worth noting. The first concerns the use of an operational amplifier placed between the ends of the first multiplying potentiometer A. It is included in order to allow four quadrant multiplication. Thus, Eq. (23-51) gives a correct description of the output quantity irrespective of the combination of signs used for A and X. The second item refers to the presence of a resistor R_c placed between ground and the slider arm of the feedback potentiometer. This is done in order to preserve the equivalence of the potential distributions existing along the feedback and the multiplying potentiometers under conditions where the slider arm of the multiplying potentiometer is fed to a succeeding stage through an input resistor R as depicted in Fig. 23-22. Error in the computation can be avoided by selecting $R_c = R$.

The servomultiplier is used because of its simplicity, its relative economy, and its capability of providing four-quadrant multiplication with a good degree of accuracy. However, in certain applications it does have one serious disadvantage. Because it is essentially an electromechanical device containing appreciable inertia, it has a limited frequency response which varies from 5 to 50 Hz. Accordingly, if the machine solution to a physical problem requires the use of a multiplier and it is known that oscillatory responses having frequencies in excess of 50 Hz will occur, then clearly the servomultiplier cannot be used. However, it can be used if the machine solution is time-scaled. As a matter of fact this is a good reason for introducing time scaling in the simulation of the problem. In those instances where time scaling is not permitted because of other considerations, then it is necessary to employ an all-electronic scheme to provide the product of variables. One method is to use the time-division multiplier; a second method is to use the quarter-squares multiplier. Both are capable of furnishing fast and accurate multiplication of variables. For information about these methods consult the bibliography.

The division of two variables can be effected by the use of the servomultiplier and a summing amplifier arranged as illustrated in Fig. 23-23. The division feature of this configuration may be demonstrated as follows: Writing Kirchhoff's law at the node of the operational amplifier yields

$$\frac{Y}{R} + \frac{ae_o}{R} = 0 \tag{23-52}$$

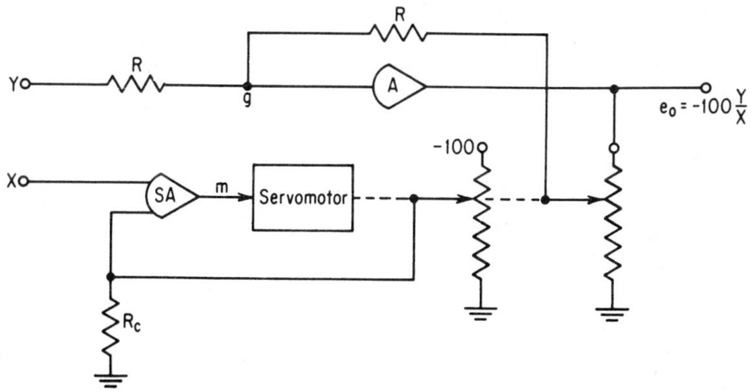

Fig. 23-23. Circuit arrangement for the divison of two variables.

where a is the slider-arm setting corresponding to the particular value of X. Solving for e_o gives

$$e_o = -\frac{Y}{a} \tag{23-53}$$

But the quantity a is given by $a = X/100$, so that Eq. (23-53) becomes

$$e_o = -\frac{Y}{X}(100) \tag{23-54}$$

Thus the variable Y is divided by the variable X in the manner called for by Eq. (23-54). Inspection of this result makes it apparent that, if the output is not to exceed 100 volts, then it is necessary that at all times $X \geq Y$. This restriction is in addition, of course, to that of limited bandwidth.

23-7 Simulation of Common Nonlinear Functions

Physical systems always contain nonlinearities to a greater or lesser extent. Quite often the nonlinearities have their origin in the imperfections of the components of which the system is composed. The most commonly found nonlinearties of this kind are limiting (or saturation), dead zone, backlash, and coulomb friction. Attention is now directed to the various ways of simulating these nonlinearities with the electronic analog computer.

Limiting. The character of this nonlinearity is depicted in Fig. 23-24. It can occur in many ways, as, for example, saturation in electronic and magnetic amplifiers, hitting the stop of a control surface, or reaching velocity and acceleration limits of output actuators. The principle utilized by the electronic analog computer to generate such a characteristic is illustrated by the circuitry appearing in Fig. 23-25. As the input signal increases from zero in the positive direction, the output voltage increases in direct proportion until it reaches the value V_{B1}. The voltage V_{B1} is the positive potential at which the cathode of diode $D1$ is held. As e_i attempts to increase beyond this point, diode $D1$ conducts, thereby holding the potential of the output lead at V_{B1}. Of course, as a result of the finite resistance of the potentiometer as well as the diode resistance, the cutoff will not be abrupt.

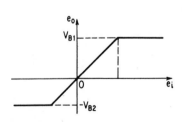

Fig. 23-24. Illustrating the nonlinear character of limiting.

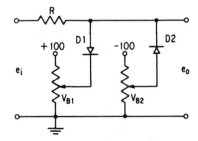

Fig. 23-25. Limiter circuit to illustrate the principle involved.

Output shunt limiting can also be used to generate a limiter characteristic. See Fig. 23-26. The output voltage of the amplifier is applied to one end of the potentiometer while the reference voltage of 100 volts is applied to the other end. As long as the amplifier output voltage is positive, the upper diode cannot conduct. However, when this output voltage becomes negative, corresponding to a positive input signal e_i, a point will be reached for increasing values of e_i which will cause this diode to conduct. The exact value of amplifier output voltage for which conduction occurs depends upon the potentiometer setting. The relationship between this potentiometer setting a, the 100-volt reference, and the saturation (or limiting) voltage V_B is

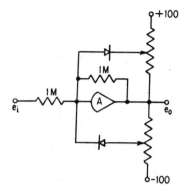

$$a = \frac{V_B/100}{1 + V_B/100}$$

When e_i exceeds V_B in magnitude, the amplifier output voltage remains clamped at the bias value V_B so that no further change in amplifier output voltage occurs for increasing input voltages.

Fig. 23-26. Generation of limiter characteristic by using output shunt limiting.

Dead Zone. This term is used to denote the nonlinear variation shown in Fig. 23-27. It may be expressed mathematically as

$$e_o = 0, \qquad -V_B < e_i < V_B$$
$$e_o = e_i - V_B, \qquad e_i > V_B$$
$$e_o = e_i + V_B, \qquad e_i < -V_B$$

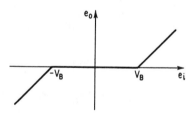

Fig. 23-27. Illustrating dead zone.

It is encountered in mechanical and electromechanical transducers such as pressure gauges, gyroscopes, accelerometers, and so forth. Effectively it refers to a region in which there is no sensitivity to the input signal. Appearing in Fig. 23-28 is a practical circuit used to generate this characteristic.

Backlash. This is a type of hysteresis which is frequently associated with mechanical systems and in particular with coupling devices such as spur gears, rack and pinion gears, and so on. To illustrate the character of the output-input relationship, refer to Fig. 23-29(a) which shows a rod A which is being used to

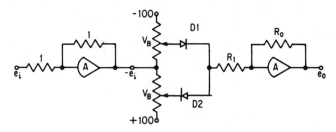

Fig. 23-28. Practical circuit used to generate dead zone.

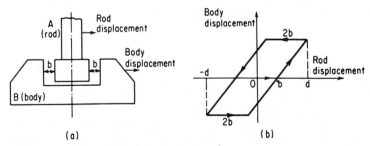

Fig. 23-29. Illustrating backlash:
(a) torque-transmitting device;
(b) backlash.

transmit motion to a body *B* (as in the case of gears). When the rod is centered, there is a space of *b* units on either side as depicted in the figure. Now, as the rod is displaced, no movement of the body occurs until a displacement of *b* units is reached. Beyond this point there is a one-to-one correspondence between the rod displacement and that of the body. If, after a total positive displacement of *d* units, the direction of the rod is reversed, then, clearly, the body will not move again until the rod has gone through a displacement of *2b* units. At this point body

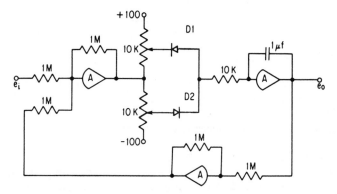

Fig. 23-30. Backlash simulator of the closed-loop type.

motion is again restored. It should be apparent that, if the rod is displaced between the limits $\pm d$ in a cyclic fashion, the output-input displacement curve appears as illustrated in Fig. 23-29(b).

In the electronic analog computer the input and output displacements are conveniently represented by voltages. A practical circuit for simulating backlash is shown in Fig. 23-30. Each diode is set at a value equal to one-half the total backlash.

Coulomb Friction. The term coulomb friction is used in reference to the frictional force that exists between two bodies when one moves with respect to the other. The magnitude of the force is equal to the product of the coefficient of friction and the normal force. It is independent of the velocity, and it always opposes motion. A graphical description of this relationship appears in Fig. 23-31. Coulomb friction is frequently found in output actuators such as hydraulic motors, electric motors, etc., and in transducers such as potentiometers. Figure 23-32 shows an analog-computer simulation of this characteristic. When the velocity of

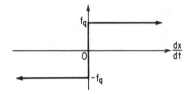

Fig. 23-31. Coulomb friction.

the controlled variable dx/dt is the slightest bit positive, the output of the high-gain amplifier (without feedback) becomes greatly negative, which causes diode $D2$ to conduct almost instantly, thus applying the coulomb-friction force to the second amplifier. If dx/dt becomes the slightest bit negative, diode $D1$ immediately conducts, thus changing the sign of the output quantity.

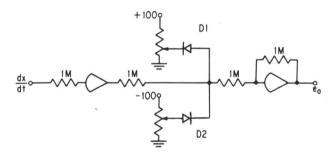

Fig. 23-32. Coulomb friction simulator.

23-8 Example and Concluding Remarks

In the interest of illustrating the procedure involved in simulating a nonlinear differential equation, which requires a servomultiplier as well as unequal magnitude-scale factors, consider a physical system that is described by the equation

$$\ddot{y} + 5\dot{y}y + 100y = f = 200u(t) \tag{23-55}$$

This is a nonlinear equation, and, accordingly, it is desirable to study the solution on the electronic analog computer. As is usual, before proceeding with the simulation, an estimate is needed of the maximum values of y, $\dot{y}$, $\ddot{y}$. It is sufficient here to use the same maximum values employed in the simulation of the linear version of this equation. The corresponding scale factors are given by Eqs. (23-27), (23-28a), and (23-29).

The first step in the procedure requires multiplying Eq. (23-55) by $\frac{1}{4}$ and then solving for the highest derivative term. Thus

$$[\tfrac{1}{4}\ddot{y}] = 50 - \tfrac{1}{2}(2.5\dot{y})y - 25y \qquad \text{volts} \qquad (23\text{-}56)$$

A question immediately arises at this point concerning the manner of handling the product $y\dot{y}$. Since a servomultiplier is to be used to obtain this product and because the input voltages to the multiplier will be the properly scaled versions of the physical quantities y and $\dot{y}$, it follows that the multiplier output voltage under these circumstances is necessarily

$$\frac{[2.5\dot{y}][25y]}{100} = [0.625\dot{y}y] \qquad \text{volts} \qquad (23\text{-}57)$$

A comparison with the correspondng term of Eq. (23-56) reveals that for correctness the output signal of the servomultiplier must be multiplied by a factor of 2 before it is summed with the other terms on the right side of Eq. (23-56) to yield the properly scaled acceleration. The remainder of the simulation procedure is then identical to that of the linear case. The final simulation diagram appears in Fig. 23-33.

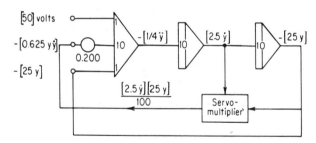

Fig. 23-33. Simulation diagram for the nonlinear equation $\ddot{y} + 5y\dot{y} + 100y = 200$.

It is significant to realize that this simulation diagram is not much more complicated than it is for the linear case. However, the simulation of the nonlinear system enables the engineer to study the effects upon the controlled variable and its derivatives of using various types of forcing functions, even to the extent where noise is included. The attractive part, too, is that very little time is needed to obtain the results. The solution is simple, direct, and complete.

Essentially it has been the purpose of this chapter to provide an introduction to electronic analog computers. Accordingly, a description has been given of the most commonly used methods of computation employed in the analysis of linear as well as nonlinear systems. Also, a careful treatment of magnitude and time scaling has been offered and illustrated because useful computer results would be difficult to obtain without such an understanding. There is, of course, a good deal more to the subject matter than appears here. For additional information, therefore, the bibliography should be consulted. However, before concluding it is worthwhile to mention that, even with the limited background provided in this chapter, it is possible for the control system engineer to see the vast possibilities that the electronic analog computer offers as a tool of analysis and design. Thus, when confronted with a complex system design, the engineer can begin with a simple paper analysis to guide his thinking, then resort to the computer for accuracy and flexibility, then return to the analysis to derive to full benefit from the computer results, and finally modify the computer results consistent with the latest analytical study. Doubtlessly, the best way to appreciate the merits of such an approach is for the engineer to try it in the solution of his problems.

PROBLEMS

23-1   Derive Eq. (23-4).

23-2   Obtain the potentiometer calibration curves by plotting Eq. (23-4) versus Eq. (23-3) for $R_p/R_L = 0.3$, 0.15, and 0.03.

23-3   For the given input waveshapes and circuitry shown in Fig. P23-3 sketch the output waveshapes.

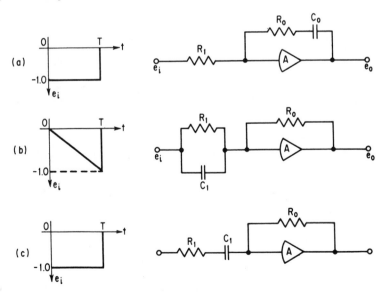

Figure P23-3

23-4 Derive an expression for the per cent error caused by the leakage resistance of the capacitor (R_l) for a step input. Put the result in terms of the computing time t, the per cent error ϵ, and the time constant represented by the leakage resistance $T_l = R_l C$. Assume $e_g = 0$. (See Fig. P23-4.)

For a value of $R_l = 10^{14}$ ohms compute the permissible computing time within which the error due to this source is 0.1 per cent or less. $C = 1 \ \mu f$.

23-5 Derive the transfer function e_o/e_i, for the computer circuitry illustrated in Fig. P23-5. Assume infinite d-c gain, that is, $e_g = 0$.

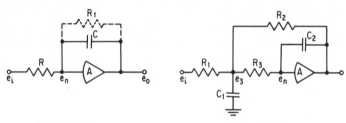

Figure P23-4   **Figure P23-5**

23-6 Determine the transfer function which can be simulated with the circuitry of Fig. P23-6. Point out the basic difference in the computational functions performed by the circuits of Figs. P23-5 and P23-6.

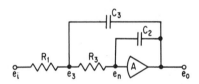

Figure P23-6

23-7 Develop the simulation diagram to solve Eq. (23-17) subject to the restriction that only three operational amplifiers are to be used. The assumption is that acceleration is not to be recorded. Use appropriate scale factors.

23-8 Develop the simulation diagram to solve the equation $4\ddot{x} + \dot{x} + 4x = f$.

23-9 Develop the simulation diagram to solve the equation $\dddot{y} + 29\ddot{y} + 520\dot{y} + 2000y = f$.

23-10 Repeat Prob. 23-9, introducing a time-scale change that allows the magnitude scale factors for each of the variables to be the same.

23-11 The differential equation that describes the behavior of a servomechanism is

$$\frac{d^2c}{dt^2} + 2\zeta\omega_n\frac{dc}{dt} + \omega_n^2 c = \omega_n^2 r$$

Simulate this equation for a time-scale change of $\omega_n/1$. The solution must be such that the value of ζ may be set between 0.000 and 1.000 by means of a single potentiometer adjustment.

23-12 Describe the operation of the circuitry of Fig. 23-25 in generating the dead-zone nonlinearity.

23-13 Describe the operation of the circuitry of Fig. 23-27 in generating the backlash nonlinearity.

23-14 A system is defined by the following differential equation:

$$\dddot{y} + 60\ddot{y} + 500\dot{y} + 20,000y = 20,000f$$

Develop an appropriate simulation diagram for this equation and indicate whether magnitude scaling alone, or time scaling alone, or both are needed. Clearly mark all machine variables on the diagram.

23-15 A thermal process is described by

$$10^4\dddot{y} + 10^3\ddot{y} + 40\dot{y} + 4y = 4f$$

(a) Develop a proper magnitude-scaled simulation diagram.
(b) On the basis of the coefficients appearing in the describing differential equation, would you say that the dynamic response is relatively fast or slow? Why?
(c) Modify the simulation diagram of part (a) to include a time-scale factor that makes magnitude scaling unnecessary. Be sure to specify the time-scale factor.

23-16 A thermal process is described by the following differential equation:

$$\dddot{y} + 0.2\ddot{y} + 0.02\dot{y} + 0.001y = 0.001f$$

(a) Develop a proper magnitude-scaled analog simulation diagram.
(b) Modify the simulation of part (a) to include a time-scale factor that makes magnitude scaling unnecessary.
(c) List two conditions under which time scaling becomes a necessity rather than a choice.

23-17 A system has a direct transmission function given by $G(s) = 25/s^2$.
(a) Simulate this function by means of standard computer components.
(b) In the interest of providing satisfactory dynamic behavior, a lead compensator having the transfer function $(s + 4)/(s + 20)$ is to be used in tandem with $G(s)$. Show how you would simulate the lead compensator using only standard computer components.
(c) This lead compensator is then placed in tandem with $G(s)$ and a unity feedback loop is placed around the combination. Show the complete analog simulation diagram of the closed-loop system.

23-18 (a) Show how to simulate the absolute value function.
(b) Simulate $e_o = e_i\epsilon^{-\alpha t}$, where e_i is an available input voltage.
(c) Simulate $y = x\epsilon^{-\alpha t}$, where x and y are both time variables.

23-19 Describe the variation of the signal appearing at the output terminal of circuits (a) and (b) in Fig. P23-19 as the input signal is allowed to vary arbitrarily through positive and negative values.

23-20 Find the homogeneous differential equation that is being simulated by the diagram appearing in Fig. P23-20.

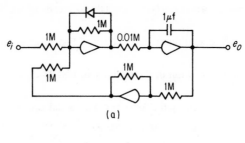

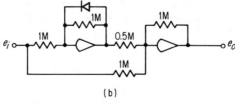

(b)

Figure P23-19

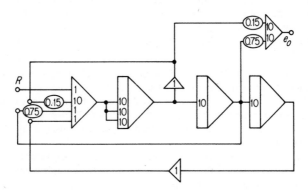

Figure P23-20

APPENDICES

appendix a

Table A-1. Units

Quantity	Symbol	Name of unit	Dimension
Fundamental			
Length	l, L	meter	L
Mass	m, M	kilogram	M
Time	t	second	T
Current	i, I	ampere	I
Mechanical			
Force	F	newton	MLT^{-2}
Torque	T	newton-meter	ML^2T^{-2}
Angular displacement	θ	radian	—
Velocity	v	meter/second	LT^{-1}
Angular velocity	ω	radian/second	T^{-1}
Acceleration	a	meter/second^2	LT^{-2}
Angular acceleration	α	radian/second^2	T^{-2}
Spring constant (translation)	K	newton/meter	MT^{-2}
Spring constant (rotational)	K	newton-meter	ML^2T^{-2}
Damping coefficient (translational)	D, F	newton-second/meter	MT^{-1}
Damping coefficient (rotational)	D	newton-meter-second	ML^2T^{-1}
Moment of inertia	J	kilogram-meter^2	ML^2
Energy	W, w	joule (watt-second)	ML^2T^{-2}
Power	P, p	watt	ML^2T^{-3}
Electrical			
Charge	q, Q	coulomb	TI
Electric potential	v, V, E	volt	$ML^2T^{-3}I^{-1}$
Electric field intensity	$\mathscr{E}$	volt/meter (or newton/coulomb)	$MLT^{-3}I^{-1}$
Electric flux density	D	coulomb/meter^2	$L^{-2}TI$
Electric flux	ψ, Q	coulomb	TI
Resistance	R	ohm	$ML^2T^{-3}I^{-2}$
Resistivity	ρ	ohm-meter	$ML^3T^{-3}I^{-2}$
Capacitance	C	farad	$M^{-1}L^{-2}T^4I^2$
Permittivity	ϵ	farad/meter	$M^{-1}L^{-3}T^4I^2$
Magnetic			
Magnetomotive force	$\mathscr{F}$	ampere(-turn)	I
Magnetic field intensity	H	ampere(-turn)/meter	$L^{-1}I$
Magnetic flux	ϕ	weber	$ML^2T^{-2}I^{-1}$
Magnetic flux density	B	tesla	$MT^{-2}I^{-1}$
Magnetic flux linkages	λ	weber-turn	$ML^2T^{-2}I^{-1}$
Inductance	L	henry	$ML^2T^{-2}I^{-2}$
Permeability	μ	henry/meter	$MLT^{-2}I^{-2}$
Reluctance	$\mathscr{R}$	ampere/weber	$M^{-1}L^{-2}T^2I^2$

Table A-2. Conversion Factors

Quantity	Multiply number of	by	to obtain
Length	meters	100	centimeters
	meters	39.37	inches
	meters	3.281	feet
	inches	0.0254	meters
	inches	2.54	centimeters
	feet	0.3048	meters
Force	newtons	0.2248	pounds
	newtons	10^5	dynes
	pounds	4.45	newtons
	pounds	4.45×10^5	dynes
	dynes	10^{-5}	newtons
	dynes	2.248×10^{-6}	pounds
Torque	newton-meters	0.7376	pound-feet
	newton-meters	10^7	dyne-centimeters
	pound-feet	1.356	newton-meters
	dyne-centimeters	10^{-7}	newton-meters
Energy	joules (watt-seconds)	0.7376	foot-pounds
	joules	2.778×10^{-7}	kilowatt-hours
	joules	10^7	ergs
	joules	9.480×10^{-4}	British thermal units
	foot-pounds	1.356	joules
	ev	1.6×10^{-19}	joules
Power	watts	0.7376	foot-pounds/second
	watts	1.341×10^{-3}	horsepower
	horsepower	745.7	watts
	horsepower	0.7457	kilowatts
	foot-pounds/second	1.356	watts

appendix b: periodic table of the elements

	I	II	III	IV	V	VI	VII	VIII
1	H 1 1.0081							He 2 4.002
2	Li 3 6.940	Be 4 9.02	B 5 10.82	C 6 12.01	N 7 14.008	O 8 16.000	F 9 19.00	Ne 10 20.183
3	Na 11 22.997	Mg 12 24.32	Al 13 26.97	Si 14 28.06	P 15 31.02	S 16 32.06	Cl 17 35.457	A 18 39.944
4	K 19 39.096	Ca 20 40.08	Sc 21 45.10	Ti 22 47.90	V 23 50.95	Cr 24 52.01	Mn 25 54.93	Fe 26 55.84 Co 27 58.94 Ni 28 58.69
	Cu 29 63.57	Zn 30 65.38	Ga 31 69.72	Ge 32 72.6	As 33 74.91	Se 34 78.96	Br 35 79.916	Kr 36 83.7
5	Rb 37 85.48	Sr 38 87.63	Y 39 88.92	Zr 40 91.22	Cb 41 92.91	Mo 42 96.0	Te 43 ...	Ru 44 101.7 Rh 45 102.91 Pd 46 106.7
	Ag 47 107.880	Cd 48 112.41	In 49 114.76	Sn 50 118.70	Sb 51 121.76	Te 52 127.61	I 53 126.92	Xe 54 131.3
6	Cs 55 132.91	Ba 56 137.36	La 57 138.92	Hf 72 178.6	Ta 73 180.88	W 74 184.0	Re 75 186.31	Os 76 191.5 Ir 77 193.1 Pt 78 195.23
	Au 79 197.2	Hg 80 200.61	Tl 81 204.39	Pb 82 207.21	Bi 83 209.00	Po 84 ...	At 85 ...	Rn 86 222
7	Fr 87 ...	Ra 88 226.05	Ac 89 ...	Th 90 232.12	Pa 91 231	U 92 238.70		

The number to the right of the symbol for the element gives the atomic number. The number below the symbol for the element gives the atomic weight. This table does not include the Rare Earths and the synthetically produced elements above 92.

appendix c

Table C-1. Resistivity and Resistance Temperature Coefficients

Material	Resistivity (ρ)		Resistance temp. coefficient at 20°C (α)
	(microhm-cm at 20°C)	*(ohm-cir. mils per foot at 20°C)*	
Aluminum	2.828	—	0.0039
Brass	—	40	0.0017
Copper (std. annealed)	1.724	10.37	0.00393
Nichrome	100.	—	0.0004
Silver	1.63	—	0.0038
Tungsten	—	33.2	0.0045

Table C-2. Round Copper-Wire Data

AWG number	Area (cir. mils)	Resistance (ohms/1000 ft)	Weight (lb/1000 ft)	Allowable current † (amps)
0000	212,000	0.0490	640	358
000	168,000	0.0618	508	310
00	133,000	0.0779	402	267
0	106,000	0.0983	319	230
1	83,700	0.1240	253	196
2	66,400	0.156	201	170
3	52,600	0.197	159	146
4	41,700	0.248	126	125
5	33,100	0.313	100	110
6	26,300	0.395	79.5	94
8	16,500	0.628	50	69
10	10,400	0.999	31.4	50
12	6,530	1.59	19.8	37
14	4,110	2.52	12.4	29

† For type RH insulation—National Electric Code.

appendix d: derivation
of the torque equations

Developed Torque for Sinusoidal B and Sinusoidal Ni (A-C Machines). The armature winding ampere-conductor distribution refers to a point-by-point plot of the current flowing through the individual conductors of the armature as depicted in Figs. D-1(a) and (b). Note that the instantaneous value of the current existing in the conductors is assumed to be sinusoidally distributed. This situation is readily achieved in practical machines. For the twelve conductors shown in Fig. D-1(a) six carry current into the paper and the other six carry current out of the paper, thus giving the appearance of two poles. Actually the number of poles associated with the ampere-conductor distribution of the armature winding is always equal to the number of field poles produced by the field winding.

Note too that the greater the number of conductors the more nearly sinusoidal will be the ampere-conductor distribution. In fact, with a truly large number of conductors a current sheet of sinusoidal density as shown in Fig. D-1(c) can be used to represent the ampere-conductor distribution. Of course this means that the latter now takes on a purely sinusoidal variation. Moreover, the armature winding can be considered as consisting of a single sheet of one turn having the units of ampere-turns/radian. A per unit radian notation is used because in Fig. D-1(c) the distribution is no longer discrete.

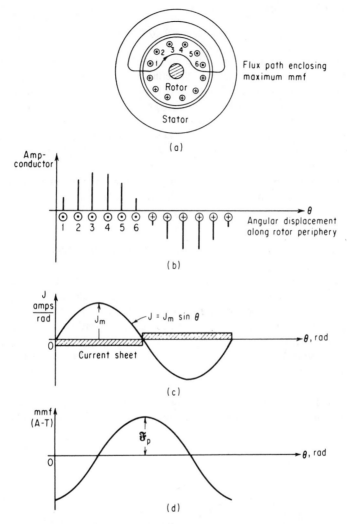

Fig. D-1. Development of current sheet of sinusoidal density:
(a and b) finite ampere-conductor distribution;
(c) current-sheet representation;
(d) corresponding mmf wave.

A study of Fig. D-1(a) reveals that for the ampere-conductor distribution shown the armature winding, taken by itself, looks like a solenoid, the mmf of which is directed towards the right. For any specified armature-winding distribution the total value of this mmf is found by applying Ampere's circuital law. As a convenience the procedure is applied to the current-sheet distribution of Fig. D-1(c). Accordingly, for the flux path depicted in Fig. D-1(a) the maximum mmf associated with the current sheet is

$$\mathscr{F} = \int_0^\pi J_m \sin \theta \, d\theta = 2J_m \tag{D-1}$$

where J_m denotes the peak value of the current-sheet density and $\mathscr{F}$ denotes the mmf per pole pair. It is assumed that the reluctance of the iron paths is negligible compared to the two air gaps involved in the specified flux path. By taking a flux path that spans from zero to π electrical degrees (i.e. one pole pitch) the maximum contribution to the integral is obtained. A little thought should make it apparent that a path spanning less than or more than π electrical degrees encloses a net mmf smaller than that given by Eq. (D-1). A graphical representation of Ampere's circuital law appears in Fig. D-1(d).

Because the flux path used in obtaining the result of Eq. (D-1) involves two air gaps, it follows that the mmf per pole (or per air-gap crossing) is

$$\mathscr{F}_p = \frac{\mathscr{F}}{2} = J_m \tag{D-2}$$

This result is useful in converting the basic electromagnetic torque equation to its practical forms for a-c machines.

We are now in a position to develop a basic torque formula as it applies to a device that has a sinusoidally distributed flux density along the periphery of the air gap and an ampere conductor distribution which for convenience is represented by a current-sheet density. Also, for generality, it is assumed that the current-sheet density is out of phase with the B-distribution by ψ degrees as depicted in Fig. D-2. The procedure is really not much different from that which leads to Eq. (14-9), but several modifications are needed. We can again start with the fundamental force expression Eq. (14-7) and replace B by its sinusoidal variation $B_m \sin \theta$. Furthermore, in place of the conductor and the current flowing through it, we now deal with a current element which is a differential strip of width $d\theta$ of

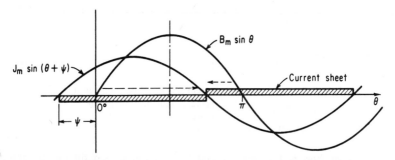

Fig. D-2. Configuration for development of the torque equation showing the sinusoidally distributed flux density produced by the field winding and the current sheet which represents the armature winding. The broken-line arrows denote relative magnitude and direction of the electromagnetic torque developed per pole. The magnetic field and the current sheet are stationary relative to each other.

the current sheet. Accordingly, for i in Eq. (14-7) we may use

$$i = J \, d\theta = I_m \sin (\theta + \psi) \, d\theta \qquad \text{(D-3)}$$

The force appearing on the elemental current strip is then expressed as

$$F_e = Bli = (B_m \sin \theta)l[J_m \sin (\theta + \psi) \, d\theta] \qquad \text{(D-4)}$$

where the subscript e denotes elemental current strip. Of course the corresponding torque is obtained by multiplying by the rotor radius r. Hence

$$T_e = B_m J_m lr \sin \theta \sin (\theta + \psi) \, d\theta \qquad \text{(D-5)}$$

To determine the total torque developed on the current sheet beneath one pole of the field distribution, we must take the sum of the elemental torques produced by each element of the current sheet lying beneath the pole. In other words, it is necessary to integrate Eq. (D-5) over π electrical degrees, which is the span of one pole. Hence

$$T_p = \sum \text{torques on current elements beneath one pole}$$
$$= B_m J_m lr \int_0^\pi \sin \theta \sin (\theta + \psi) \, d\theta \qquad \text{(D-6)}$$

where T_p denotes the torque developed per pole. Clearly for a field distribution which has p poles the total resultant electromagnetic torque is

$$T = p B_m J_m lr \int_0^\pi \sin \theta \sin (\theta + \psi) \, d\theta \qquad \text{(D-7)}$$

By introducing the trigonometric identity

$$\sin \theta \sin (\theta + \psi) = \tfrac{1}{2} \cos \psi - \tfrac{1}{2} \cos (2\theta + \psi) \qquad \text{(D-8)}$$

into Eq. (D-7) we obtain

$$T = p B_m J_m lr \left[\int_0^\pi \frac{1}{2} \cos \psi \, d\theta - \int_0^\pi \frac{1}{2} \cos (2\theta + \psi) \, d\theta \right] \qquad \text{(D-9)}$$

The second term in brackets has a value of zero, so the final expression for the developed electromagnetic torque is

$$T = \pi \frac{p}{2} J_m B_m lr \cos \psi \qquad \text{(D-10)}$$

Keep in mind that this equation is valid for sinusoidal distributions of the flux density and ampere-conductors (or current sheet).

An inspection of Eq. (D-10) reveals that the torque is dependent upon the phase displacement angle ψ. Thus when $\psi = 0°$ the best *field pattern* for developing torque prevails because the entire current sheet beneath a pole has a unidirectional torque. When $\psi = 45°$, as depicted in Fig. D-2, note that a portion of the current sheet beneath a pole will produce an opposing torque, thereby resulting in a reduced torque if all other things are equal. It should not be inferred, however, that because $\psi \neq 0°$ the developed torque cannot be a maximum. In some a-c machines operation at $\psi \neq 0°$ causes a manifold increase in J_m, thereby giving a considerable increase in developed torque in spite of a nonoptimum field pattern.

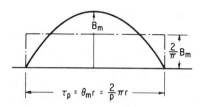

Fig. D-3. Relationships for computing
flux per pole.

Often it is convenient to express Eq.
(D-10) in terms of the total flux per pole,
which is the flux associated with any one
pole of the flux-density curve. One such pole
is shown in Fig. D-3. But before proceeding
further it is necessary to distinguish between
mechanical and electrical degrees. It has
already been pointed out that regardless of
the number of poles in the flux-density
curve, each pole spans π electrical degrees. Now since there are always 2π mechanical degrees in a configuration such as appears in Fig. D-1(a), it follows that
when the flux-density curve has two poles the mechanical and electrical degrees
are the same. However, when the flux-density curve has four poles, then clearly
the number of mechanical degrees is one-half the number of electrical degrees.
For a p-pole flux-density curve the relationship is

$$\theta_m = \frac{2}{p}\theta \qquad (D-11)$$

where θ_m denotes mechanical degrees and θ denotes electrical degrees. Therefore,
in Fig. D-3 the pole pitch expressed in meters is

$$\tau_p \equiv \text{pole pitch} = \theta_m r = \frac{2}{p}\pi r \qquad (D-12)$$

where r is the radius of the rotor in meters. Moreover, for a rotor axial length of
l the area associated with a pole of the flux-density curve is

$$A_p = l\tau_p = \frac{2}{p}\pi l r \qquad (D-13)$$

A further simplication in the computation leading to the expression for the
flux per pole results upon recalling that the average value of the positive (or negative) half of a sine wave is

$$B_{av} = \frac{2}{\pi}B_m \qquad (D-14)$$

Thus the sine wave of Fig. D-3 may be replaced by a rectangular wave having
a height of B_{av} and spanning over the entire pole pitch as indicated by the broken
line in Fig. D-3. The expression for the flux per pole then follows from

$$\Phi = B_{av}A_p \qquad (D-15)$$

Inserting Eqs. (D-13) and (D-14) into Eq. (D-15) yields

$$\Phi = \frac{4}{p}B_m l r \qquad (D-16)$$

Again keep in mind that this result is valid for sinusoidal variations of flux density.

By means of Eq. (D-16) the equation for the developed electromagnetic
torque may be expressed in terms of the flux per pole. Insertion of Eq. (D-16)

into Eq. (D-10) yields

$$T = \frac{\pi}{8}p^2\Phi J_m \cos\psi \quad \text{n–m} \tag{D-17}$$

A study of Eq. (D-17) points out more clearly the three conditions that must be satisfied for the development of torque in conventional electromechanical energy-conversion devices. There must be a field distribution represented by Φ. There must be an ampere-conductor distribution represented by J_m. Finally, there must exist a favorable space displacement angle between the two distributions. Note that it is possible to have both a field and an ampere-conductor distribution and still the torque can be zero if the field pattern is such that equal and opposite torques are developed by the conductors i.e. $\psi = 90°$.

It is also worthwhile to note that no really new concepts have been used to derive the torque formula of Eq. (D-17). It is merely an extension of Ampere's law.

Torque for Nonsinusoidal B and Uniform Ampere-conductor Distribution. The typical distributions are depicted in Fig. D-4. Again for generality the current

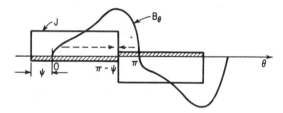

Fig. D-4. Current-sheet and flux-density curves characteristic of d-c machines.

sheet that is representative of the ampere-conductor distribution is shown displaced in phase by the angle ψ. These distributions are commonly found in d-c machines. The starting point of the derivation is Ampere's law, again as expressed in Eq. (14-7). Of course modifications are needed consistent with the assumed distributions. Because the armature winding is represented by a current sheet we shall deal with an elemental strip of the current sheet rather than with a particular conductor and the current it carries. Thus in Eq. (14-7) we introduce

$$i = J\, d\theta \tag{D-18}$$

and

$$B = B_\theta \tag{D-19}$$

where B_θ denotes the nonsinusoidal variation of flux density along the air gap. An exact analytical expression for B_θ is difficult to obtain, but this is not disturbing because one deals rather with the flux per pole, which is readily available. Hence the force developed by an elemental strip of the current sheet is

$$F_e = B_\theta l J\, d\theta \tag{D-20}$$

The corresponding torque for a p-pole machine is then

$$T_e = pJB_\theta lr \, d\theta \quad \text{n-m} \tag{D-21}$$

The total resultant torque is obtained upon integrating Eq. (D-21) over one pole pitch. Thus

$$T = pJ\left[\int_0^{\pi-\psi} B_\theta lr \, d\theta - \int_{\pi-\psi}^{\pi} B_\theta lr \, d\theta\right] \tag{D-22}$$

The second term in brackets carries the minus sign in order to account for the fact that J is negative beneath the positive pole of the B_θ curve in the region from $\pi - \psi$ to π. Clearly, without an analytical expression for B_θ, a closed-form solution for the torque is not readily available. However, in electromechanical energy-conversion devices where these distributions apply, the designer by intention makes $\psi = 0$. Unlike the preceding case this always provides maximum torque. Introducing $\psi = 0$ simplifies the expression for the electromagnetic torque to

$$T = pJ\int_0^{\pi} B_\theta lr \, d\theta \quad \text{n-m} \tag{D-23}$$

Now in the interest of avoiding B_θ we can express the torque in terms of the flux per pole—a quantity which is available from other considerations. In terms of B_θ the flux per pole Φ is

$$\Phi = \int_0^{\pi} B_\theta dA_p \tag{D-24}$$

But the area per pole can be expressed as

$$A_p = lr\theta_m = lr\frac{2}{p}\theta \tag{D-25}$$

Hence Eq. (D-24) becomes

$$\Phi = \int_0^{\pi} B_\theta lr \frac{2}{p} d\theta = \frac{2}{p}\int_0^{\pi} B_\theta lr \, d\theta \tag{D-26}$$

Therefore

$$\int_0^{\pi} B_\theta lr \, d\theta = \frac{p}{2}\Phi \tag{D-27}$$

Substituting Eq. (D-27) into Eq. (D-23) yields the final form for the electromagnetic torque developed when the distributions are as depicted in Fig. D-4 with $\psi = 0$. Thus

$$\boxed{T = \frac{p^2}{2}J\phi} \tag{D-28}$$

It is worthwhile to note again that electromagnetic torque is dependent upon a field distribution represented by Φ, an ampere-conductor distribution represented by J and the angular displacement between the two distributions. Of course ψ does not appear in Eq. (D-28) because for simplicity the current sheet was assumed in phase with the flux-density curve, thereby assuring that all parts of the current sheet experience a unidirectional torque. A glance at Fig. D-4 should make it apparent that for $\psi = 90°$ the resultant electromagnetic torque is zero.

appendix e: selected transistor and vacuum-tube characteristics

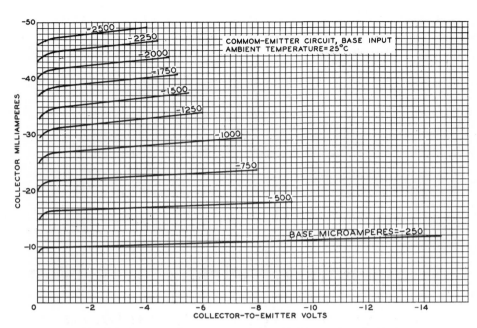

Fig. E-1. Average collector characteristics of RCA transistor 2N104 in common-emitter mode.

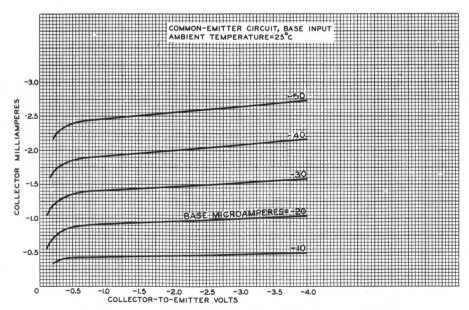

Fig. E-2. Average collector characteristics of RCA transistor 2N104 in common-emitter mode, high current.

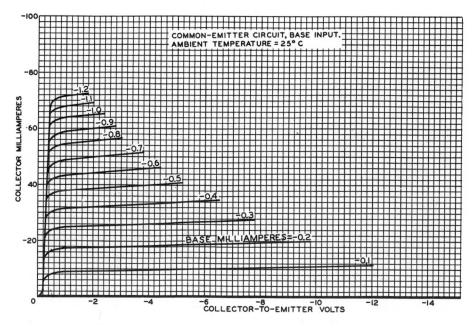

Fig. E-3. Average collector characteristics of RCA transistor 2N109 in common-emitter mode.

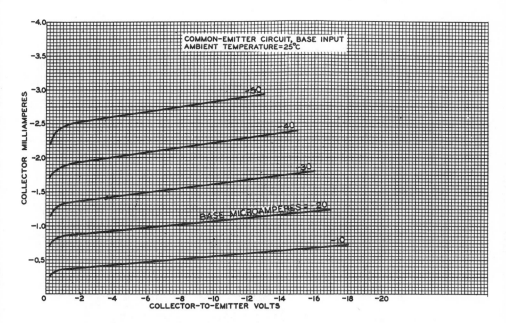

Fig. E-4. Average collector characteristics, common-emitter connection, RCA transistor 2N139.

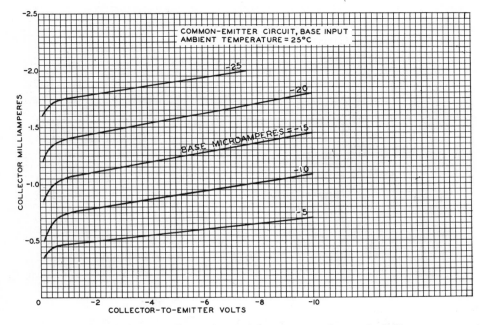

Fig. E-5. Average collector characteristics, common-emitter mode, RCA transistor 2N175.

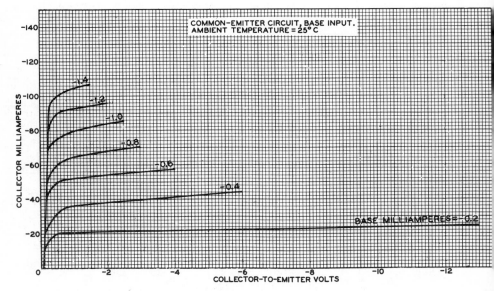

Fig. E-6. Average collector characteristics, common-emitter mode, RCA transistor 2N270.

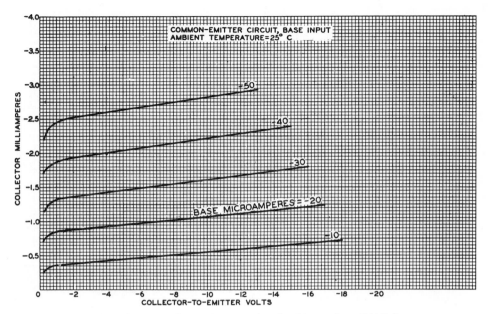

Fig. E-7. Average collector characteristics of RCA transistor 2N140 in common-emitter mode.

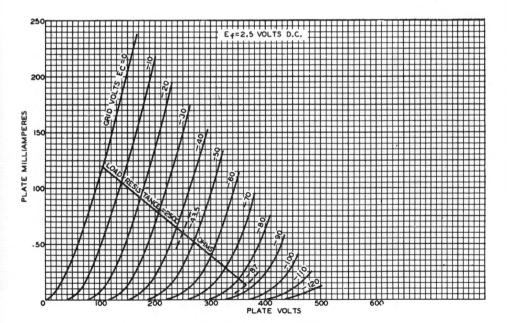

Fig. E-8. Average plate characteristics of 2A3 power triode.

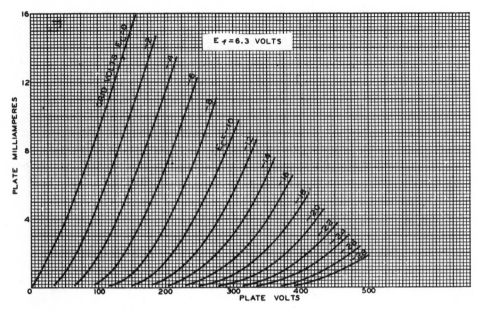

Fig. E-9. Average plate characteristics of 6C5 triode.

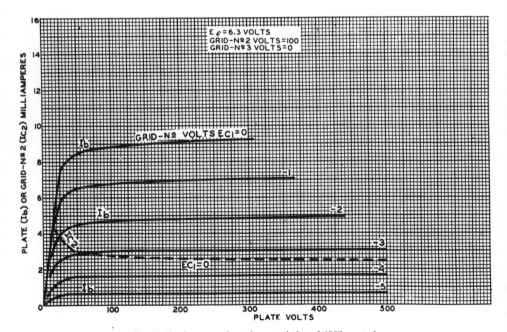

Fig. E-10. Average plate characteristics of 6SJ7 pentode.

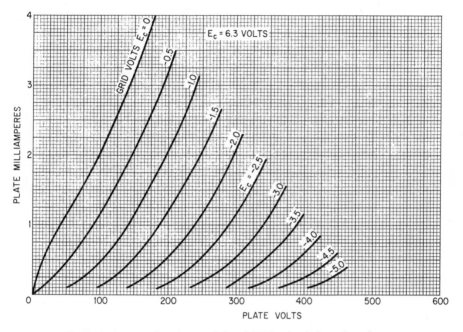

Fig. E-11. Average plate characteristics of 6EU7 twin triode, each section.

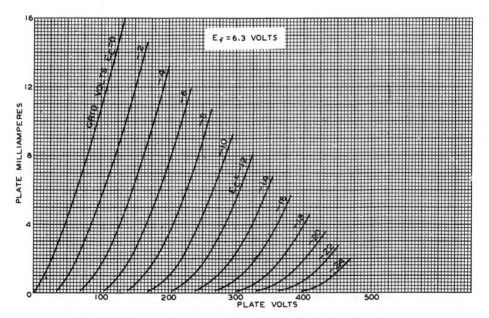

Fig. E-12. Average plate characteristics of 6SN7 twin triode, each section.

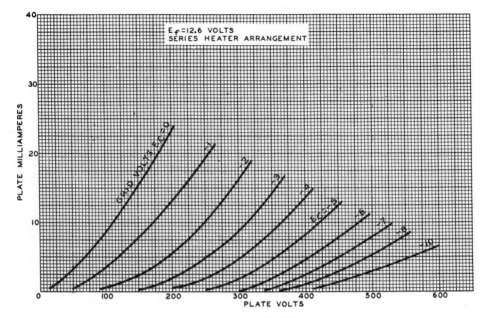

Fig. E-13. Average plate characteristics of 12AT7 twin triode, each section.

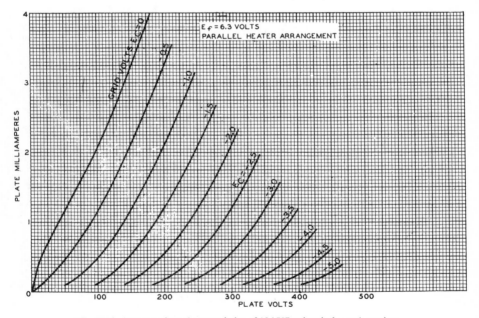

Fig. E-14. Average plate characteristics of 12AX7 twin triode, each section.

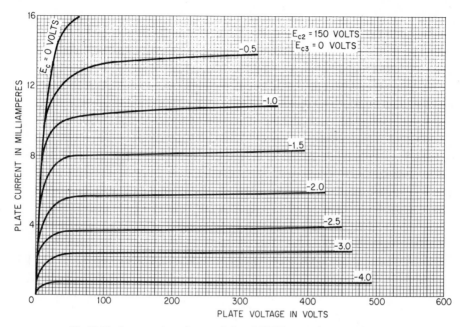

Fig. E-15. Average plate characteristics of 6AU6, pentode connection.

appendix f: bibliography

The following is a selected list of books that the reader will find useful as a supplementary source of information in the study of this text.

I Electric Circuit Theory

1. W. H. Hayt, Jr. and J. E. Kemmerly, *Engineering Circuit Analysis*. New York: McGraw-Hill Book Co., 1962.

2. R. P. Ward, *Introduction to Electrical Engineering* (3rd ed.). Englewood Cliffs, N. J.: Prentice-Hall, Inc., 1960.

3. Members of the Department of Electrical Engineering, M.I.T., *Electric Circuits*. New York: John Wiley & Sons, Inc., 1940.

4. E. Brenner and M. Javid, *Analysis of Electric Circuits*. New York: McGraw-Hill Book Co., 1959.

5. H. A. Foecke, *Introduction to Electrical Engineering Science*. Englewood Cliffs, N. J.: Prentice-Hall, Inc., 1961.

6. H. H. Skilling, *Electrical Engineering Circuits*. New York: John Wiley & Sons, Inc., 1957.

II Electronics

7. J. Millman, *Vacuum-tube and Semiconductor Electronics.* New York: Mc-Graw-Hill Book Co., 1958.

8. H. A. Romanowitz, *Fundamentals of Semiconductor and Tube Electronics.* New York: John Wiley & Sons, Inc., 1962.

9. J. D. Ryder, *Electronic Fundamentals and Applications* (3rd ed.). Englewood Cliffs, N. J.: Prentice-Hall, Inc., 1964.

10. K. R. Spangenberg, *Fundamentals of Electron Devices.* New York: McGraw-Hill Book Co., 1957.

11. R. B. Hurley, *Junction Transistor Electronics.* New York: John Wiley & Sons, Inc., 1958.

12. D. Le Croissette, *Transistors.* Englewood Cliffs, N. J.: Prentice-Hall, Inc., 1963.

13. J. F. Pierce, *Transistor Circuit Theory and Design.* Columbus, Ohio: Charles E. Merrill Books, Inc., 1963.

III Electromechanical Energy Conversion

14. Members of the Department of Electrical Engineering, M.I.T., *Magnetic Circuits and Transformers.* New York: John Wiley & Sons, Inc., 1943.

15. W. T. Hunt, Jr. and R. Stein, *Static Electromagnetic Devices.* Boston: Allyn and Bacon, Inc., 1963.

16. M. Liwschitz-Garik, assisted by C. C. Whipple, *Electric Machinery*, Vols. I and II. Princeton, N. J.: D. Van Nostrand Co., Inc., 1946.

17. A. E. Fitzgerald and C. Kingsley, Jr., *Electric Machinery* (2nd ed.). New York: McGraw-Hill Book Co., Inc., 1961.

18. G. W. Heumann, *Magnetic Control of Industrial Motors* (2nd ed.). New York: John Wiley & Sons, Inc., 1954.

IV Control Systems and Computers

19. V. Del Toro and S. R. Parker, *Principles of Control Systems Engineering.* New York: McGraw-Hill Book Co., Inc., 1960.

20. E. O. Doebelin, *Dynamic Analysis and Feedback Control.* New York: Mc-Graw-Hill Book Co., Inc., 1962.

21. R. N. Clark, *Introduction to Automatic Control Systems.* New York: John Wiley & Sons, Inc., 1962.

22. F. H. Raven, *Automatic Control Engineering.* New York: McGraw-Hill Book Co., Inc., 1961.

23. J. N. Warfield, *Introduction to Electronic Analog Computers*. Englewood Cliffs, N. J.: Prentice-Hall, Inc., 1959.

24. W. A. Lynch and J. G. Truxal, *Introductory Systems Analysis*, Vol. I. New York: McGraw-Hill Book Co., Inc., 1961.

25. J. R. Ashley, *Introduction to Analog Computation*. New York: John Wiley & Sons, Inc., 1963.

answers to selected problems

Chapter 1

1-1. (a) $\varepsilon = 0.018$ v/m; (b) $W = 0.0576 \times 10^{-19}$ joule; (c) $F = 115.2 \times 10^{-22}$ newton.
1-5. $V_A = 12$ v, $V_B = 4$ v, $V_c = 6$ v, $V_D = 14$ v. **1-6.** $e = 20$ v. **1-9.** $R_3 = 5$ ohms. **1-11.** $R = 17.5$ ohms.

Chapter 2

2-1. 62.5 ohms. **2-2.** (a) 21¢; (b) 129,600 coulombs; (c) 8.08×10^{23}; (d) 1380 w.
2-4. (a) 556 ohms; (b) 556 ohms. **2-6.** 55 °C. **2-9.** 79.4 feet. **2-11.** 10 h.
2-13. 0.89 h. **2-15.** (a) 1.03 μf; (b) 0.00462 joule; (c) 1 ma; (d) 0. **2-17.** Capacitance of 1 μf. **2-20.** (b) $i(1) = 20a$; $i(2) = 32.72$ amp; $i(4) = 6.7$ amp.

Chapter 3

3-4. (a) 70 μf; (b) 14×10^{-3} coulomb; (c) $Q_1 = 2 \times 10^{-3}$; $Q_2 = 4 \times 10^{-3}$, $Q_3 = 8 \times 10^{-3}$ coulomb. **3-5.** (a) 40/7 μf; (b) 2×10^{-3}; (c) $V_1 = 200$ v, $V_2 = 100$ v, $V_3 = 50$ v. **3-6.** (a) $Q_1 = 240$ μcoul, $Q_2 = 180$ μcoul, $Q_3 = 60$ μcoul; (c) $W_1 = 4.8 \times 10^{-3}$ joule, $W_2 = 5.4 \times 10^{-3}$ joule, $W_3 = 1.8 \times 10^{-3}$ joule. **3-10.** (a) 4/75 h; (b) 50 v. **3-11.** 4.5 and 14.5 h. **3-15.** 165 v. **3-16.** 4.78 ohms. **3-18.** 18.75 ohms. **3-19.** (a) 150 v; (b) 40 v. **3-20.** 15.44 ohms. **3-23.** 7 ohms. **3-24.** 1 amp. **3-26.** $I_{3-ohm} = I_{8-ohm} = I_{6-ohm} = 0$, $I_{12-ohm} = 0.835$ amp. **3-27.** (a) 12, 9 resistors and 3 voltage sources; (b) 10; (c) 3; (d) 5; (e) 3. **3-30.** 1.67 amp from bottom to top. **3-32.** 0.448 amp. **3-35.** $V_{oc} = 130$ v, $R_i = 22$ ohms. **3-39.** 1.89 watts. **3-40.** 300 watts. **3-42.** $V_{oc} = 12$ v, $R_i = 48$ ohms. **3-44.** 2.86 amp.

Chapter 4

4-1. (c) $(s^2 + 3s + 2)C(s) = \dfrac{1}{s^2} - s - 1$.   **4-2.** (a) $\mathcal{L}f(t) = \dfrac{1}{s^2(s + \alpha)}$,

(c) $\mathcal{L}\epsilon^{-\alpha t} \sin(\omega t + \theta) = \dfrac{\omega \cos \theta}{(s + \alpha)^2 + \omega^2} + \dfrac{(s + \alpha) \sin \theta}{(s + \alpha)^2 + \omega^2}$.

4-3. $\mathcal{L}f(t) = \dfrac{A}{s}(\epsilon^{-sT_1} - 2\epsilon^{-sT_2} + \epsilon^{-sT_3})$.   **4-4.** $\mathcal{L}f(t) = \dfrac{(1 - \epsilon^{-s})^2}{s^2}$.

4-5. $\mathcal{L}f(t) = \dfrac{1}{s^2 T_1} - \dfrac{\epsilon^{-sT_1}}{s} - \dfrac{\epsilon^{-sT_1}}{s^2 T_1}$.

4-7. $x(t) = \frac{1}{8} + 1.79\epsilon^{-4t} - 0.92\epsilon^{-10t}$.   **4-9.** $f(s^t) = 2$.

4-10. $f(t) = 2\epsilon^{-t} - t\epsilon^{-2t} - 2\epsilon^{-2t}$.   **4-12.** $f(t) = 5\epsilon^{-2t} \sin 2t$.

4-14. $f(t) = 3\epsilon^{-3t} \cos 4t - \frac{1}{4}\epsilon^{-3t} \sin 4t$.

Chapter 5

5-1. $i(t) = 2 + 1 - \epsilon^{-25t}$ amp.   **5-3.** (a) $i(t) = 2(1 - \epsilon^{-25/3t})$ amp; (b) 2 amp; (c) 9/25 sec.   **5-4.** (a) $v(t) = \frac{1}{2}(1 - \epsilon^{-4t})$ m/sec; (b) $\frac{1}{2}$ m/sec; (c) 1.25 secs.   **5-6.** $i(t) = 5(1 - \epsilon^{-200t})$ amp.   **5-7.** (a) $2\epsilon^{-10t}$; (b) 6 joules; (c) 6 joules.   **5-9.** 1.15 secs, 2.77 secs.   **5-10.** $4 \times 10^{-3}\epsilon^{-t}$ counterclockwise.   **5-12.** (a) $s^2 + 1.25s + 0.25 = 0$; (b) 1 sec, 4 secs.   **5-19.** (a) A/RT; (b) $i(t) = \dfrac{A}{RT}\epsilon^{-R/Lt}$.   **5-21.** (a) 20 v; (b) 20v; (c) $v(t) = 7.5 + 12.5\epsilon^{-4/15t}v$; (d) 3.75 secs.   **5-22.** (a) 0; (b) $2 \times 10^{-3}\epsilon^{-t}$ amp; (c) 5 secs; (d) $i(t) = q_0\omega \sin \omega t$.   **5-25.** $10 - 6\epsilon^{-10t}$ amp.   **5-26.** (a) $i(t) = 15 - 3\epsilon^{-5.4t}$ amp; (b) 0.185 sec.   **5-28.** $i(t) = \frac{3}{2}(1 - \epsilon^{-32t})$ amp.   **5-31.** (a) ϵ^{-30t}; (b) 1000 v; (c) tenfold increase.   **5-37.** (a) 0.4; (b) 25%; (c) 7.32 rad/sec; (d) 0.936 sec.   **5-38.** (a) $2 \cos(20t - 53°)$; (b) 1.2 amp; (c) $-1.2\epsilon^{-15t}$.   **5-41.** $i(t) = 1.6 - 0.4\epsilon^{-20/3t}$ amp.   **5-42.** (a) 24.8 ohms, 12.4h; (b) positive; (c) peak value.

Chapter 6

6-1. (a) 0, I_m; (b) $0.5I_m$, $0.707\ I_m$.   **6-2.** (a) 0, $0.577\ V_m$; (b) $0.25\ V_m$, $V_m/\sqrt{6}$.   **6-4.** (a) $0.5\ I_p$, $0.777\ I_p$; (b) 0, $0.577\ I_p$; (c) $0.25\ I_p$, $I_p/\sqrt{6}$.   **6-6.** 862 w.   **6-7.** (a) $8.63\underline{/-5.1°}$; (b) $12.2 \sin(377t - 5.1°)$; (c) $863\ v - a$; (d) 0.966.   **6-11.** (a) $15\underline{/53.2°}$; (b) $300\sqrt{2} \sin(377t + 53.2°)$; (c) 0.0317 h.   **6-13.** 80 v.   **6-16.** (a) $39.4 \sin(377t + 15.1°)$.   **6-22.** $\overline{V}_{ab} = 84\underline{/28°}$, $\overline{V}_{bc} = 47.3\underline{/-56.9°}$.   **6-23.** 33.3 ohms. **6-24.** $X_1 = 4$, $X_2 = 3$ and $\overline{I} = 39.6\underline{/-45°}$.   **6-25.** (a) $6\underline{/36.8°}$; (b) $6\sqrt{2} \sin(\omega t + 36.8)$; (c) $\overline{V}_{ab} = 24\underline{/-53°}$, $\overline{V}_{bc} = 107.4\underline{/10.3°}$.   **6-26.** $11.8\underline{/-80.2°}$.   **6-29.** (a) $11.72 + j47.1$; (b) $-j47.1$; (c) 53.2 μf; (d) $6.8\underline{/0°}$.   **6-30.** (a) $5\underline{/-36.8}$; (b) $L = 1.5$ h; (c) $12.5\underline{/-90°}$ v.   **6-38.** (a) 1000 rad/sec; (b) 100; (c) 10 rad/sec; (d) 995 and 1005 rad/sec; (e) 100 v.   **6-40.** (a) $0.233\underline{/90°}$ ma; (b) $0.567\underline{/-90°}$ ma.   **6-44.** (a) $17.32\underline{/-60°}$; (b) 5710 w.   **6-45.** (a) $15.21\underline{/-83.14°}$, $15.21\underline{/-203.14°}$, $15.21\underline{/36.7°}$; (b) 1160 w; (c) 0.   **6-47.** $47.8\underline{/-83.14°}$.   **6-49.** (a) $i(\omega t) = \dfrac{4}{\pi}\ I_m\ (\sin \omega t + \frac{1}{3} \sin 3\omega t + \frac{1}{5} \sin 5\omega t + \cdots +)$;   (b) $i(\omega t) = \dfrac{I_m}{2} + \dfrac{2}{\pi}\ I_m\ (\sin \omega t + \frac{1}{3} \sin 3\omega t + \frac{1}{5} \sin 5\omega t + \cdots +)$.

6-50. $v(\theta) = \dfrac{8}{\pi^2}\left(\sin\theta - \dfrac{1}{3^2}\sin 3\theta + \dfrac{1}{5^2}\sin 5\theta - \dfrac{1}{7^2}\sin 7\theta + \cdots +\right)\cdot$ *6-52.* $i(\theta) =$

$\dfrac{Ip}{2} + \dfrac{1}{\pi} I_p (\sin\theta + \tfrac{1}{2}\sin 2\theta + \tfrac{1}{3}\sin 3\theta + \cdots +).$

Chapter 7

7-1. (a) False; (b) False; (c) False; (d) False.   *7-2.* (a) 4.4×10^{14} holes/cm³;
(b) 6.6×10^{12} electrons/cm³; (c) See Figs. 7-13 and 7-16 of text.   *7-4.* (a) 418 ohms;
(b) 108 ohms.   *7-5.* 150 ma.   *7-6.* 472 ma and 0.235 ma.   *7-7.* 7.6 ohms.
7-10. (a) -8.12 v; (b) -0.388 v; (c) 149.   *7-13.* 2 ma/v and 240.   *7-14.* (a)
4 ma; (b) $16.6K$; (c) 1.55 ma/v; (d) 25.7.

Chapter 8

8-1. Oxide-coated.   *8-2.* (a) 4 ma; (b) 38.4 ma.   *8-3.* 16 ma, voltage across
first diode is 100 v and 150 v across the second diode.   *8-6.* (a) 65 ma; (b) $V_{5BC_3} = 20$ v,
$V_{5Y_3} = 40$ v; (c) 308 and 616 ohms.   *8-7.* 100 v.   *8-11.* (a) 2.6 ma; (b) -41 v.
8-12. -9.2 v.   *8-14.* (a) 20; (b) 8.7 ma; (c) 7.92 kilohms; (d) 2.52 millimhos.

Chapter 9

9-1. (a) 6.3 v; (b) 80; (c) 24.5; (d) 1960.   *9-2.* (a) Yes; (b) 0.2 v; (c) 0.74 v.
9-4. (a) 5 mv; (b) 400; (c) 50; (d) 400.   *9-5.* (a) 2550 ohms.   *9-7.* (a) 0.46 ma;
(b) 30 K.   *9-11.* 25,180 ohms.   *9-12.* $R_E = 500$ ohms, $R_B = 44$ K.   *9-16.* (b)
1.07 μa; (c) 68.4 μa; (d) 34.2.   *9-17.* (a) 34.1; (b) 17.3; (c) 590; (d) 1.92 K; (e)
2.46 K.   *9-19.* (b) 241.5 mv; (c) 46.7, 80.5 and 2185.   *9-20.* (a) 356 μa; (b) 291 μa;
(c) 4.1 v and 20.5 ma; (d) 2420; (e) 205; (f) 496,000.   *9-24.* (a) 36.6; (b) 1.26 Hz;
(c) 89,300 Hz; (d) -109.8 μa.

Chapter 10

10-1. (a) 6.3 ma, 270 v; (b) 55; (c) 2 ma; (d) 1.5 ma.   *10-2.* (b) 2.66 ma; (c) 52.2 v;
(d) 139 milliwatts; (e) Yes.   *10-4.* (a) 32.5; (b) 0.144 w; (c) 270 v.   *10-6.* (a)
300 K; (b) 116.5 K; (c) 36 v; (d) 0.707 v; (e) 0.37 ma.   *10-7.* (a) 205 v and 8.2 ma;
(c) 8.6 ma; (d) 3.1 watts.   *10-8.* $g_m = 1.5 \times 10^{-3}$ mho, $r_p = 66.7$ K, $\mu = 100$.   *10-9.*
7 ma, 220 v and -7.6 v.   *10-11.* (a) 3.6 ma, 210 v and -8.5 v; (b) 10.7 v rms; (c) 34.4
v rms; (d) 0.475 w.   *10-12.* (b) 51.2 Hz to 29,100 Hz; (c) 97.6 K.   *10-14.*
$10.2\underline{/-197.3°}$ v.

Chapter 11

11-1. (b) 5 v; (c) 7.07 v.   *11-2.* (b) 1.1 amp.   *11-3.* (b) 0.022 amp; (c) full-
wave rectification.   *11-5.* $3E_m$.   *11-6.* $4E_m$.   *11-8.* $2E_m$.   *11-9.* (a) $\dfrac{V_o}{V_b} =$
$h_{fe}\dfrac{R_L}{h_{ie} + (1 + h_{fe})R_E}$; (c) Place a capacitor in parallel with R_E which yields $\dfrac{V_o}{V_b} = h_{fe}\dfrac{R_L}{h_{ie}}\cdot$
11-10. (a) 660 K; (b) 56 ohms; (d) 5 v.   *11-14.* carrier frequency, 10^8 rad/sec; first
lower-side frequency $= \omega_C - \omega_{m1} = 10^8 - 10^3$; first upper-side frequency $= \omega_C + \omega_{m1} =$
$10^8 + 10^3$; second lower-side frequency $= \omega_C - \omega_{m2} = 10^8 - 1.8 \times 10^3$; second upper-

side frequency $= \omega_C + \omega_{m2} = 10^8 + 1.8 \times 10^3$; (b) For ω_{m1}, $m_1 = 0.4$; for ω_{m2}, $m_2 = 0.2$.
11-22. $ABC = D$.   **11-23.** $A + B + C = \overline{D}$.

Chapter 12

12-1. (a) $4/\pi$; (b) 32×10^{-7} n/m; (c) 0.32 n/m.   **12-5.** 20 amp.   **12-7.** (a)
98 AT; (b) 1.225 amp.   **12-8.** 4.96 amp.   **12-9.** 183.6 AT.   **12-11.** 0.0045 weber.
12-12. 0.003125 weber.   **12-14.** 361 AT applied in direction to cause flux in right leg
to flow CCW.   **12-17.** (a) 1500 AT; (b) 2300 AT/m; (c) 2.9 w/m²; (d) 1.25×10^6;
(e) 1.26×10^{-3}; (f) 1010.   **12-19.** 20.1 watts.   **12-20.** At 60 Hz: $P_h = 1400$ w,
$P_e = 400$ w; at 90 Hz: $P_h = 2100$ w and $P_e = 900$ w.   **12-22.** $F = \dfrac{B^2}{\mu} bg$.   **12-23.**
(a) 5 joules; (b) 2 w-secs; (c) 20 h; (d) 3.35 w-secs; (e) No, it comes from electrical
source.

Chapter 13

13-1. (a) 0.0108 weber; (b) 1.44 in².   **13-4.** A 50% reduction.   **13-5.** (a) 114
watts; (b) 360 watts; (c) 97.7%; (d) 0.055%.   **13-6.** (a) 2340.8 v; (b) 98%.   **13-8.**
4.82%.   **13-10.** 103 amps on high side, 206 amps on low side.   **13-11.** $150\underline{/-30°}$.
13-12. $906\underline{/-44°}$.   **13-14.** 87.8 v.   **13-16.** A lead pf angle of 28.6° with respect
to V_1.

Chapter 14

14-1. (a) 0.0024 weber; (b) 0.144 weber-turns; (c) $\lambda = 0.144 \cos 5\pi t$; (d) 2.26 v;
(e) 1.96 v.   **14-2.** (a) $e = \pi \sin 10\pi t$; (b) πv.   **14-3.** (a) $\Phi = Blr (\pi - 2\theta_0)$; (b)
$e = 2NBlv = 4.8$ v; (c) 7.54 v; (d) 4.8 v.   **14-6.** 531 AT/pole.   **14-7.** (a) 0.0925
weber; (b) 34.7°.   **14-9.** (a) 8640 AT; (b) 53.1 n − m; (c) 104 v; (d) 0.00122 weber.

Chapter 15

15-3. $1.5\Phi_m\underline{/30°}, 1.5\Phi_m\underline{/-30°}$.   **15-6.** (a) 1800 rpm; (b) 3 Hz; (c) 90 rpm; (d) 1800
rpm; (e) 0; (f) Yes, in direction of revolving field.   **15-8.** (a) 700 rpm; (b) 1700 rpm.
15-9. 900 rpm, 1.8 Hz.   **15-11.** 87.45%.   **15-12.** (a) 44.7 amp at 0.87 pf lag;
(b) 17.85 hp; (c) 125 n − m.   **15-14.** (a) 6770 n − m; (b) 127 amps at 0.942 pf lag.
15-16. (a) 88.3 at a pf of 0.958 lag; (b) 2260 n − m; (c) 94%.   **15-18.** 0.0333 ohm/
phase.   **15-19.** 69.5 n − m.   **15-22.** (a) 0.162; (b) 90.3 amps at 0.505 pf lag;
(c) 560 n − m.   **15-24.** (a) $22.4\underline{/-18°}$; (b) 45 n − m; (c) 3.56%; (d) 86.1%.

Chapter 16

16-1 (a) $163.5\underline{/52.8°}$; (b) 1000 v/phase.   **16-2.** (a) 1790 v/phase; (b) $55.4\underline{/41°}$;
(c) 0.755 lead.   **16-4.** (a) $293\underline{/-9.1°}$; (b) No.   **16-6.** 4120 v to 9408 v.   **16-9.**
(a) 24.85 amps at 0.79 lagging pf; (b) $295\underline{/-30.7°}$ v.

Chapter 17

17-2. (a) 366.7 ohms; (b) 12 v; (c) 176.5 ohms; (d) 0.068 amp; (e) 617 rpm.   **17-3.**
201 v.   **17-5.** (a) 375 ohms; (b) 5.08%; (c) 2.27 amps; (d) 880 rpm.   **17-7.** (a)
226 v; (b) 207.8 amps; (c) 388 n − m; (d) 93.33%.   **17-9.** (a) 0.00905 weber;

(b) 129 n $-$ m; (c) 580 watts; (d) 15.2%.   *17-11.* (a) 229 amps; (b) 1414 rpm.
17-13. (a) 240 amps; (b) 535 n $-$ m; (c) 86.5%.

Chapter 18

18-1. Yes.   *18-4.* Yes.   *18-5.* $8\underline{/90°}$.

Chapter 20

20-9. (b) $\dfrac{e_p}{e_g} = \dfrac{-\mu R_L}{r_p + (\mu + 1)R_k + R_L}$; (c) $S = \dfrac{r_p + R_L + R_k}{r_p + (\mu + 1)R_k + R_L}$.   *20-10.* (a)
-83.3; (b) -0.167; (c) -1.95; (d) 0.00391.   *20-12.* (a) 3 stages; (b) 0.487×10^{-3}.
20-13. 5 stages.

Chapter 21

21-1. (a) 10 rad/sec; (b) 6 rad/sec; (c) 0.8; (d) unity; (e) 4 times.   *21-2.* (a) 0.4;
(b) 25% maximum overshoot to step command.   *21-4.* (a) $\dfrac{10}{s^2 + 2s + 10}$; (b) 3 rad/sec;

(c) 35%; (d) 5 secs.   *21-6.* (a) $\dfrac{\omega_r}{1 + K}$; (b) $\omega_e(t) = \dfrac{1}{1 + K} + \dfrac{K}{1 + K}\epsilon^{-((1+K)/(J/F))t}$; (c)

No.   *21-8.* (a) $c(t) = 4t - 0.4(1 - \epsilon^{-10t})$; (b) $\dfrac{40}{s + 10}$; (c) $c(t) = \frac{4}{5}t - \frac{4}{250}(1 - \epsilon^{-50t})$.
21-9. (a) 0.6; (b) 355×10^{-6}.   *21-11.* $K = 1000$ v/v and $K_o = 3.25 \times 10^{-3}$.   *21-12.*
$K_e = 0.179$.   *21-14.* $K_o = 0.9$.

Chapter 22

22-1. $1/(1 + s\tau_1)(1 + s\tau_2)$.   *22-3.* $\dfrac{e_o}{e_i} = \dfrac{1}{\alpha}\left(\dfrac{1 + s\alpha\tau}{1 + s\tau}\right)$ where $\alpha = \dfrac{R + R_L}{R}$ and

$\tau = \dfrac{L}{R + R_L}$.   *22-7.* $\dfrac{x_o}{x_1} = \dfrac{1}{\alpha}\left(\dfrac{1 + s\alpha\tau}{1 + s\tau}\right)$ where $\alpha = \dfrac{k_1 + k_2}{k_1}$ and $\tau = \dfrac{f}{k_1 + k_2}$.

22-9. $\dfrac{x_o}{F} = \dfrac{1/M}{s^2 + s\dfrac{f}{M} + \dfrac{k}{M}}$ where $\omega_n = \sqrt{\dfrac{k}{M}}$, $\zeta = \dfrac{f}{2\sqrt{kM}}$.

22-13. $T(j\omega) = \dfrac{K}{\left(1 + j\dfrac{\omega}{\omega_1}\right)\left(1 + j\dfrac{\omega}{\omega_2}\right)}$.   *22-14.* $T(j\omega) = \dfrac{\left(1 + j\dfrac{\omega}{\omega_1}\right)}{\dfrac{(j\omega)^2}{\omega_1\omega_2}\left(1 + j\dfrac{\omega}{\omega_3}\right)}$.

22-17. $\dfrac{c}{r} = \dfrac{G_1G_2G_3}{1 + H_2G_2 + H_1G_1G_2 + H_3G_2G_3}$.

22-18. $\dfrac{c}{r} = \dfrac{G_1G_2}{1 + H_3G_1G_2 + H_2G_2 + H_1H_2G_1G_2}$.

22-21. (a) Type 1; (b) $K_v = 1$; (c) 0; (d) 2.   *22-25.* System is stable.   *27-27.* (a)
System is stable; (b) Polar plot intersects negative real axis at $-\frac{4}{15}$ for a frequency of 7.07
rad/sec. Hence no enclosure of point $(-1, 0)$ occurs; (c) $\phi_{PM} = 29.6°$.   *22-30.* (b)
21.8°; (c) 52.5% approx.; (d) $t_s \cong 1.2$ secs.   *22-33.* (a) absolutely unstable; (b) absolutely stable; (c) 8% maximum overshoot and 2.5 secs settling time.   *22-35.* (a)

Part between 0 and -5; (b) Two, $90°$, -2.5; (d) 5; (e) No.   **22-36.** (a) Portions between 0 and -5; also between -10 and $-\infty$; (b) 3, $\pm 60°$, and -180, -5; (d) 2.52; (e) Yes, any value of gain exceeding 2.52.

Chapter 23

23-1. $\dfrac{e_o}{e_i} = \dfrac{a}{1 + a(a-1)R_p/R_L}$.   **23-4.** 2×10^5 secs.

23-5. $\dfrac{e_o}{e_i} = -\dfrac{R_2}{R_1}\left(\dfrac{1}{s^2(R_2 R_3 C_1 C_2) + sC_2\left(\dfrac{R_2 R_3}{R_1} + R_2 + R_3\right) + 1}\right)$.

23-6. $\dfrac{e_o}{e_i} = -\dfrac{1}{s(sR_1 R_3 C_2 C_3 + R_3 C_2 + R_1 C_2 + R_1 C_3)}$

index